Handbook of Climate Change Resilience

Walter Leal Filho
Editor

Handbook of Climate Change Resilience

Volume 1

With 563 Figures and 494 Tables

Springer

Editor
Walter Leal Filho
International Climate Change Information
and Research Programme (ICCIRP)
Faculty of Life Sciences
Hamburg University of Applied Sciences
Hamburg, Germany

Research and Transfer Centre Sustainable Development and
Climate Change Management
Hamburg University of Applied Sciences
Hamburg, Germany

ISBN 978-3-319-93335-1 ISBN 978-3-319-93336-8 (eBook)
ISBN 978-3-319-93337-5 (print and electronic bundle)
https://doi.org/10.1007/978-3-319-93336-8

This Springer imprint is published by the registered company Springer Nature Switzerland AG.
The registered company address is: Gewerbestrasse 11, 6330 Cham, Switzerland

Preface

As the Fifth Assessment Report (AR5) produced by the Intergovernmental Panel on Climate Change (IPCC) has shown, climate change is expected to have widespread impacts on societies.

The fact that the impacts of climate change are already being felt, not only in terms of increases in temperature but also in respect of agriculture (with lower crop yields) and the availability of water resources, among others, suggests that action is urgently needed, in order to increase the resilience of both the physical environment and communities.

In particular, there is a perceived need to identify, test, and deploy appropriate processes, methods, and tools that may help societies and nations to better adapt to the many impacts of a changing climate and become more resilient. There is also a perceived need to showcase successful examples of how this can be achieved and to illustrate how the quest to pursue climate resilience may face the many social, economic, and political problems associated with it.

It is against this background that the *Handbook of Climate Change Resilience* has been prepared by numerous scientists from across all geographical regions, under the auspices of the International Climate Change Information and Research Programme (ICCRIP). This handbook is a truly interdisciplinary publication, mobilizing scholars from over 30 countries undertaking research and/or executing climate change projects around the world.

This book focuses on "fostering resilience and capacity to adapt," meaning that it serves the purpose of showcasing experiences from research, field projects, and best practice in climate change adaptation across many countries, which may be useful and suitable for replication of perhaps even upscaling elsewhere.

Consistent with the need for more cross-sectoral interactions among the various stakeholders working in the field of climate change adaptation, this book aims to

(i) Provide research institutions, universities, NGOs, and enterprises with an opportunity to display and present their works in the field of climate change adaptation and resilience
(ii) Disseminate information, ideas, and experiences acquired in the execution of climate change adaptation projects, especially successful initiatives and good practice

(iii) Introduce methodological approaches and experiences deriving from case studies and projects, which aim to show how climate change adaptation may be implemented in practice

Last but not least, a further aim of this book is to document and disseminate the wealth of experiences on climate change resilience available today.

Enjoy your reading!

August 2019
Germany

Walter Leal Filho

Acknowledgments

The editor would like to thank the valuable support of the Deputy Editor, Professor Desalegn Ayal, as well as the many members of the Editorial Board. Thanks to your hard work, expertise, and constructive criticisms, we were able to produce a publication with the highest standards.

I would also thank the authors for their willingness to share their knowledge, know-how, and experiences, as well as the many peer reviewers, who have helped us to ensure the quality of the manuscripts.

Finally, thanks are due to the International Climate Change Information and Research Programme (ICCIRP) at the Hamburg University of Applied Sciences, Germany, for the coordination of the project and mobilization of a wide range of stakeholders and authors around the world.

Contents

Volume 1

Volume 4

About the Editor

Prof. Walter Leal Filho (B.Sc., Ph.D., D.Sc., D.Phil., D.L., D.Litt., D.Ed.) is Professor and Head of the Research and Transfer Centre "Sustainable Development and Climate Change Management" at the Hamburg University of Applied Sciences in Germany and holds the Chair of Environment and Technology at Manchester Metropolitan University, UK. He is a Lead Author at AR6's Working Group II (Climate Change Adaptation) at the Intergovernmental Panel on Climate Change (IPCC), is Founding Editor of the *International Journal of Climate Change Strategies and Management*, and heads the International Climate Change Information and Research Programme (ICCIRP). He is also Editor-in-Chief of the Climate Change Management series with Springer. Professor Leal serves on the editorial board of various journals. He has in excess of 400 publications to his credit, among which are ground-breaking books such as *Universities and Climate Change*; *The Economic, Social and Political Aspects of Climate Change*; the *Handbook of Climate Change Adaptation*; and the *Handbook of Climate Change Communication*. He has nearly 30 years of field experience on project management and has a particular interest in the connections between sustainability, climate change adaptation, and human behavior.

About the Deputy Editor

Desalegn Yayeh Ayal is Assistant Professor of Geography and Environmental Studies at the Center for Food Security Studies, College of Development Studies, Addis Ababa University. Desalegn holds a Ph.D. in Geography. Desalegn serves as the Deputy Editor of *International Journal of Climate Change Strategies and Management*. He has published more than 30 publications including books, book chapters, and refereed journal articles and is East Africa Vice President for Interconnections for Making Africa Great Empowered and Sustainable Initiative. Desalegn has also presented papers on climate adaptation and related issues at many international and national conferences. His principal areas of research interest include climate change and adaptation, climate resilience, climate change mitigation and related issues, indigenous weather forecasting, integrated natural resources rehabilitation and management, livelihoods, and food security nexus, among others. He thoroughly understands the link between natural and human-induced hazards with sustainable development and works hard to familiarize with current tools of climate change impact assessment on livelihood and the wider environment. Desalegn has been actively involved in climate resilience and integrated natural resources rehabilitation and management research, and development interventions to improve food security.

Editorial Board

Editorial Advisory Board

Contributors

Ghulam Abbas Department of Environmental Science, COMSATS University Islamabad, Vehari Campus, Vehari, Pakistan

Omer A. Abdallah Department of Agricultural Engineering, Faculty of Agriculture, University of Khartoum, Khartoum, Sudan

M. O. Abioja Department of Animal Physiology, College of Animal Science and Livestock Production, Federal University of Agriculture, Abeokuta, Ogun, Nigeria

Praise Abraham Department of Social Sustainability, Centre for Sustainable Development, University of Ibadan, Ibadan, Oyo, Nigeria

Bashir Yusuf Abubakar Department of Botany, Ahmadu Bello University, Zaria, Nigeria

Sarah Achola International Centre of Insect Physiology and Ecology, Nairobi, Kenya

Mariola Acosta International Institute of Tropical Agriculture (IITA), Kampala, Uganda

Myriam Adam Centre de Coopération Internationale en Recherche Agronomique pour le Développement (CIRAD), UMR AGAP/PAM, Montpellier, France

Institut de l'Environnement et de la Recherche Agricole (INERA), Bobo Dioulasso, Burkina Faso

International Crops Research Institute for Semi-Arid Tropics (ICRISAT), Bamako, Mali

Bamidele Adanikin Department of Plant Science and Biotechnology, Ekiti State University, Ado Ekiti, Ekiti State, Nigeria

Oluwabunmi Opeyemi Adejumo Institute for Entrepreneurship and Development Studies, Obafemi Awolowo University, Ile-Ife, Nigeria

M. O. Adekunle Department of Animal Physiology, College of Animal Science and Livestock Production, Federal University of Agriculture, Abeokuta, Ogun, Nigeria

Ibidun O. Adelekan Department of Geography, Faculty of Social Science, University of Ibadan, Ibadan, Oyo State, Nigeria

Mosunmola Lydia Adeleke Department of Fisheries and Aquaculture Technology, FUTA, Akure, Nigeria

Romola Adeola Centre for Human Rights, Faculty of Law, University of Pretoria, Pretoria, South Africa

Peter Adegbenga Adeonipekun Department of Botany, University of Lagos, Akoka-Yaba, Lagos, Nigeria

O. M. Adesope Department of Agricultural Economics and Extension, University of Port Harcourt, Choba/Port Harcourt, Nigeria

William Adzawla Faculty of Agribusiness and Communication Sciences, University for Development Studies, Tamale, Ghana

Oludare Agboola Department of Botany, University of Lagos, Akoka-Yaba, Lagos, Nigeria

Savita Aggarwal Department of Development Communication and Extension, Institute of Home Economics, University of Delhi, New Delhi, India

Yaw Boakye Agyeman Department of Ecotourism, Recreation and Hospitality, University of Energy and Natural Resources, UENR, Sunyani, Brong Ahafo Region, Ghana

Abeer Ahmed Qaed Ahmed Pharmacy Department, Al-Saeed University, Osaifrah, Taiz, Yemen

A. Ajao Amsol Bio Company, Lagos, Nigeria

Oluyede Clifford Ajayi Technical Centre for Agriculture and Rural Co-operation, Wageningen, The Netherlands

I. A. Ajibefun Department of Agriculture and Resource Economics, FUTA, Akure, Nigeria

The Federal University of Technology, (FUTA), Akure, Nigeria

Javaid Akhtar Institute of Soil and Environmental Sciences, University of Agriculture, Faisalabad, Pakistan

J. G. Akinbomi Department of Chemical and Polymer Engineering, Faculty of Engineering, Lagos State University, Lagos, Nigeria

Akinlabi Akintuyi Department of Geography, University of Lagos, Akoka-Yaba, Lagos, Nigeria

Bosede Olufunmilayo Akinwalere Department of Agricultural Extension and Communication, FUTA, The Federal University of Technology, Akure, Nigeria

Tunrayo Alabi Geographic Information Systems Unit, International Institute of Tropical Agriculture, Ibadan, Nigeria

Erwin A. Alampay National College of Public Administration and Governance, University of the Philippines, Diliman, Quezon City, Philippines

Mohammed Al-Azzawi Department of Civil, Geo and Environmental Engineering, Technical University of Munich, Garching, Germany

Eike Albrecht Brandenburg University of Technology Cottbus – Senftenberg, Senftenberg, Germany

Nigatu Alemayehu Consultant, Addis Ababa, Ethiopia

Mohammad Al-Saidi Center for Sustainable Development, College of Arts and Sciences, Qatar University, Doha, Qatar

Zakou Amadou Faculty of Agricultural Sciences, Department of Rural Economics and Sociology, Tahoua University, Tahoua, Niger

T. T. Amos Department of Agriculture and Resource Economics, FUTA, Akure, Nigeria

Peter Adewale Amosun Department of Teacher Education, Faculty of Education, University of Ibadan, Ibadan, Oyo State, Nigeria

Benjamin Anabaraonye Benjy Poetry and Music Global Concepts, Awka, Anambra State, Nigeria

Isaac Gershon Kodwo Ansah Faculty of Agribusiness and Communication Sciences, University for Development Studies, Tamale, Ghana

Toyib Aremu Department of Environmental Sustainability, Centre for Sustainable Development, University of Ibadan, Ibadan, Oyo, Nigeria

Kinfe Asayehegn School of environment, Gender and development Studies, Hawassa University, Hawassa, Ethiopia

Center for International Cooperation in Agricultural Research and Development (CIRAD), Montpellier, France

Department of Agricultural Economics, Universidad Politécnica de Madrid, Madrid, Spain

S. O. Asifat Lopek Engineering and Construction Ltd, Lagos, Nigeria

Joanes Atela African Centre for Technology Studies, Nairobi, Kenya

Musiliyu K. Atoyebi National Centre for Technology Management, Federal Ministry of Science and Technology, Obafemi Awolowo University, Ile-Ife, Nigeria

Amina Nihad Awartani College of Arts and Sciences, Qatar University, Doha, Qatar

David Olufemi Awolala Department of Agricultural and Resource Economics, Federal University of Technology Akure, Akure, Nigeria

Research Fellow, Adaptation Finance Fellowship Programme of the Thailand Research Development Institute Bangkok, Thailand and Frankfurt School of Finance and Management, Frankfurt, Germany

Abiodun Emmanuel Awoyemi Department of Agricultural Economics, Faculty of Agriculture and Forestry, University of Ibadan, Ibadan, Oyo State, Nigeria

Desalegn Yayeh Ayal Center for Food Security Studies, College of Development Studies, Addis Ababa University, Addis Ababa, Ethiopia

Ayansina Ayanlade Department of Geography, Obafemi Awolowo University, Ile-Ife, Nigeria

Neda Azam Department of Water Engineering, College of Agriculture, Isfahan University of Technology, Isfahan, Iran

Olubukola Oluranti Babalola Food Security and Safety Niche Area, Faculty of Natural and Agricultural Sciences, North-West University, Mmabatho, Mafikeng, South Africa

Oladapo Opeyemi Babalola Afe Babalola University, Ado-Ekiti, Nigeria

Mansur Matazu Bako Nigerian Meteorological Agency (NiMet), Abuja, Nigeria

Mohammed Lawal Balarabe Department of Biology, Ahmadu Bello University, Zaria, Nigeria

Mulubrhan Balehegn Department of Animal, Rangeland and Wildlife Sciences, Mekelle University, Mekelle, Ethiopia

Paul Basudde Sector Planning and Policy Analysis Department, Ministry of Energy and Mineral Development (MEMD), Kampala, Uganda

Lin Bautze Research Institute of Organic Agriculture (FiBL), Frankfurt am Main, Germany

Jules Bayala World Agroforestry Centre, West and Central Africa Regional Office – Sahel Node, Bamako, Mali

Paulin Bazié Institut de l'Environnement et de Recherche Agricole (INERA), Ouagadougou, Burkina Faso

Tamrat Bekele Department of Plant Biology and Biodiversity Management, College of Natural and Computational Sciences, Addis Ababa University, Addis Ababa, Ethiopia

Kassa Belay Pan African University Africa Union, Addis Ababa, Ethiopia

Adedoyin Bello Department of Urban and Regional Planning, Crescent University, Abeokuta, Ogun State, Nigeria

Ramkumar Bendapudi Watershed Organisation Trust (WOTR), Pune, India

Mohan Kumar Bera Institute of Economic Growth, New Delhi, India

Hillary K. Bett Department of Agricultural Economics and Agribusiness Management, Egerton University, Njoro, Kenya

Martha Bohm School of Architecture and Planning, State University of New York at Buffalo, Buffalo, NY, USA

Adjoua Nadège Boko-Koiadia Department of Sociology, University of Félix Houphouët-Boigny, Abidjan, Côte d'Ivoire

P. Bosbœuf Lab'Urba – Paris-Est University, Marne-la-Vallée, France

G. D. Bothun Department of Physics, University of Oregon, Eugene, OR, USA

Bennett Brooks Consensus Building Institute, New York, NY, USA

Kathryn Brown Marine Research Centre, School of Environment, Science and Engineering, Southern Cross University, Lismore, NSW, Australia

Claudia Capitani Environment Department, University of York, York, UK

Tales Carvalho Resende University of Strathclyde, Glasgow, UK

Winifred Chepkoech Center for Rural Development (SLE), Humboldt Universität zu Berlin, Berlin, Germany

Dumisani Chirambo Department of Civil and Public Law with references to Law of Europe and the Environment, Brandenburg University of Technology Cottbus-Senftenberg, Cottbus, Germany

Lieven Claessens International Institute of Tropical Agriculture (IITA), Arusha, Tanzania, Kenya

Helene Jacot Des Combes PaCE-SD, Lower Campus, Laucala, USP, Suva, Fiji

Florence Crick Grantham Research Institute on Climate Change and the Environment, London School of Economics and Political Science, London, UK

Tijana Crnčević Institute of Architecture and Urban and Spatial Planning of Serbia, Belgrade, Serbia

Marcella D'Souza The WOTR-Centre for Resilience Studies (W-CReS), Pune, India

Taiwo M. Dada Department of Aquaculture and Fisheries Management, University of Ibadan, Ibadan, Nigeria

Petr Daněk Department of Geography, Masaryk University, Brno, Czech Republic

Kyle Frankel Davis The Earth Institute, Columbia University, New York, NY, USA

The Nature Conservancy, New York, NY, USA

Data Science Institute, Columbia University, New York, NY, USA

Dennis dela Torre Center for Local and Regional Governance, NCPAG, University of the Philippines, Diliman, Quezon City, Philippines

Solomon Desta MARIL (Managing Risk for Improved Livelihood) Consulting Firm, Addis Ababa, Ethiopia

Aliou Diouf Climate Change, Natural Resource Management, Rural and Urban Development Enda Energie-Environnement-Développement, Dakar, Senegal

Samuel A. Donkoh Faculty of Agribusiness and Communication Sciences, University for Development Studies, Tamale, Ghana

Felix Kwabena Donkor College of Agriculture and Environmental Sciences, University of South Africa (UNISA), UNISA Science Campus. Corner of Christiaan de Wet Road and Pioneer Avenue, Florida, South Africa

School of Animal, Plant and Environmental Sciences, University of the Witwatersrand WITS, Johannesburg, South Africa

Jörg E. Drewes Department of Civil, Geo and Environmental Engineering, Technical University of Munich, Garching, Germany

Jack Alban Dyer Australia Maritime College, University of Tasmania, Launceston, TAS, Australia

Abule Ebro International Livestock Research Institute, Addis Ababa, Ethiopia

Glory Ikponmwosa Edwards Climate Change, Development and Policy, School of Global Studies, University of Sussex, Falmer, UK

Jackson Efitre Department of Zoology, Entomology and Fisheries Sciences, Makerere University, Kampala, Uganda

Emmanuel Ejim-Eze National Centre for Technology Management, Obafemi Awolowo University, Ile-Ife, Ile-Ife, Nigeria

M. U. Ekong Department of Forestry and Wildlife Resources Management, University of Calabar, Calabar, Cross River State, Nigeria

Ali W. Elamin Department of Agricultural Engineering, Faculty of Agriculture, University of Khartoum, Khartoum, Sudan

Haitham R. Elramlawi Center of Dryland Farming Research and Studies (CDFRS), Faculty of Agricultural and Environmental Sciences, University of Gadarif, Gadarif, Sudan

David Yisrael Epstein HaLevi Department of Educational Policy and Leadership, University at Albany, Albany, NY, USA

M. O. Esilaba Department of Environmental Science, Egerton University, Egerton, Kenya

Saeid Eslamian Department of Water Engineering, College of Agriculture, Isfahan University of Technology, Isfahan, Iran

Nsikak-Abasi A. Etim Department of Agricultural Economics and Extension, University of Uyo, Uyo, Akwa Ibom State, Nigeria

NseAbasi N. Etim Department of Animal Science, Akwa Ibom State University, Obio Akpa Campus, Abak, Akwa Ibom State, Nigeria

K. N. N. Ezike Department of Agricultural Economics, Management and Extension, Ebonyi State University, Abakaliki, Ebonyi State, Nigeria

O. A. Fagbenro Department of Fisheries and Aquaculture Technology, FUTA, Akure, Nigeria

Augustine E. Falaye Department of Aquaculture and Fisheries Management, University of Ibadan, Ibadan, Nigeria

Belarmain A. Fandohan Université nationale d'agriculture École des sciences et techniques de conservation et de transformation des produits agricoles BP 114, Sakété, Bénin

T. Fane Faculty of Economic Sciences and Management, University of Bamako, Bamako, Mali

M. Farrely Tanzania Organic Movement for Agriculture, Dar es Salaam, Tanzania

Mayowa Fasona Department of Geography, University of Lagos, Akoka-Yaba, Lagos, Nigeria

Aryo Feldman Crops for the Future (CFF), Jalan Broga, Selangor Darul Ehsan, Malaysia

Sabastine Dekaa Francis National Weather Forecasting and Climate Research Centre, Nigerian Meteorological Agency, Abuja, Nigeria

Chao Fu Key Laboratory of Ecosystem Network Observation and Modeling, Institute of Geographic Sciences and Natural Resources Research, Chinese Academy of Sciences, Beijing, China

Hubert Fudjumdjum Research and Transfer Centre Sustainable Development and Climate Change Management, Hamburg University of Applied Sciences, Hamburg, Germany

Charles K. K. Gachene Department of Land Resource Management and Agricultural Technology (LARMAT), University of Nairobi, Nairobi, Kenya

Kenya Agriculture and Livestock Research Organisation, Nairobi, Kenya

Eva N. Gacheru Kenya Agriculture and Livestock Research Organisation, Nairobi, Kenya

Prithviraj Gaikwad Watershed Organisation Trust (WOTR), Pune, India

WOTR Centre for Resilience Studies (W-CReS), Pune, India

Thomas Gaiser Institute of Crop Science and Resource Conservation, University of Bonn, Bonn, Germany

Kate Elizabeth Gannon Grantham Research Institute on Climate Change and the Environment, London School of Economics and Political Science, London, UK

Komla Kyky Ganyo Centre d'Etude Régional pour l'Amélioration de l'Adaptation à la Sécheresse (CERAAS), BP, Thiès, Sénégal

Institut Togolais de Recherche Agronomique (ITRA), Lomé, Togo

Weyessa Garedew Jimma University College of Agriculture and Veterinary Medicine, Jimma, Ethiopia

Andreas Gattinger Faculty 9 – Agricultural Sciences, Nutritional Sciences, and Environmental Management, Justus Liebig University (JLU), Institute of Agronomy and Plant Breeding II, Gießen, Germany

Tosin K. W. Gbadegesin Centre for Sustainable Development (CESDEV), University of Ibadan, Ibadan, Nigeria

Yohannes GebreMichael Department of Geography and Environmental Studies, Addis Ababa University, Addis Ababa, Ethiopia

Getachew Gebru International Livestock Research Institute, Addis Ababa, Ethiopia

Patrick T. Gicheru Kenya Agriculture and Livestock Research Organisation, Embu, Kenya

Victoria Gioto Institute of Climate Change and Adaptation (ICCA), University of Nairobi, Nairobi, Kenya

Laurent Gnonlonfin Doctoral School of Agronomic and Water Sciences/Department of Sustainable Management of Natural Resources, Laboratory of Ecology and Forest Research (LERF), University of Parakou, Parakou, Borgou, Republic of Benin

Daphne Gondhalekar Department of Civil, Geo and Environmental Engineering, Technical University of Munich, Garching, Germany

Degye Goshu Department of Economics, Faculty of Business and Economics, Kotebe Metropolitan University (Asso.), Addis Ababa, Ethiopia

Gérard Gouwakinnou Laboratory of Ecology, Botanic and Plant Biology (LEB), University of Parakou, Parakou, Borgou, Republic of Benin

Aliou Guissé Département de Biologie Végétale, Université Cheikh Anta Diop de Dakar (UCAD), Dakar, Sénégal

Marwa R. Hafez Institute of Graduate Studies and Research (IGSR), Alexandria University, Alexandria, Egypt

Ghrmawit Haile Ministry of Environment Forest and Climate Change and Addis Ababa University, Addis Ababa, Ethiopia

Mitiku Haile Department of Land Resource Management and Environmental Protection, Mekelle University, Mekelle, Ethiopia

Razlin Azman Halimi Crops for the Future (CFF), Jalan Broga, Selangor Darul Ehsan, Malaysia

Southern Cross Plant Science, Southern Cross University, East Lismore, NSW, Australia

Jarso Harou Ministry of Water, Energy and Climate Change, Isiolo County Government, Isiolo, Kenya

Ali Hasantabar-Amiri Department of Civil Engineering, Lenjan Branch, Islamic Azad University, Lenjan, Iran

Peni Hausia Havea PaCE-SD, Lower Campus, Laucala, USP, Suva, Fiji

Sarah L. Hemstock Bishop Grosseteste University, Lincoln, UK

A. Henri-Ukoha Department of Agricultural Economics and Extension, University of Port Harcourt, Choba/Port Harcourt, Nigeria

J. S. (Pat) Heslop-Harrison Department of Genetics and Genome Biology, University of Leicester, Leicester, UK

James Hope Ahmadu Bello University, Zaria, Kaduna State, Nigeria

Md. Shahadat Hossain Institute of Water Modeling, Dhaka, Bangladesh

Towanou Olivier Houetchegnon Doctoral School of Agronomic and Water Sciences/Department of Sustainable Management of Natural Resources, Laboratory of Ecology and Forest Research (LERF), University of Parakou, Parakou, Borgou, Republic of Benin

Mark Howells Unit of Energy Systems Analysis, KTH – Royal Institute of Technology, Stockholm, Sweden

Imad Antoine Ibrahim Institute of Law, Politics, and Development (Dirpolis), Sant'Anna School of Advanced Studies, Pisa, Italy and gLAWcal – Global Law Initiative for Sustainable Development, Essex, UK

Mustapha Adeojo Ibrahim Department of Biology, Ahmadu Bello University, Zaria, Nigeria

Funmilayo C. Idowu National Biotechnology Development Agency, Ogbomoso, Nigeria

N. M. Ifebueme Department of Forestry and Wildlife Resources Management, University of Calabar, Calabar, Cross River State, Nigeria

Samuel Aderemi Igbatayo Department of Economics and Management Studies, Afe Babalola University, Ado-Ekiti, Nigeria

Simeon Igbe Association of Entrepreneurs and Technology Managers of Nigeria (AETMAN), Eco-system Based Adaptation For Food Security Assembly (EBAFOSA), Makurdi, Nigeria

Ana Iglesias Department of Agricultural Economics, Universidad Politécnica de Madrid, Madrid, Spain

Oluwatosin E. Ilevbare National Centre for Technology Management, Federal Ministry of Science and Technology, Obafemi Awolowo University, Ile-Ife, Nigeria

Christopher O. Ilori Crop Protection and Environmental Biology, University of Ibadan, Ibadan, Nigeria

Masengo Francois Ilunga Department of Civil Engineering, University of South Africa, Pretoria, South Africa

Antonina Ivanova Universidad Autonoma de Baja California Sur, La Paz, Mexico

Abdulai Jalloh West and Central African Council for Agricultural Research and Development (CORAF), Dakar, Senegal

Adamu James National Weather Forecasting and Climate Research Centre, Nigerian Meteorological Agency, Abuja, Nigeria

S. Jari Department of Crop Production and Protection, Federal University Dutsin-Ma Katsina State, Katsina, Nigeria

Safieh Javadinejad Water Resource Management, University of Birmingham, Edgbaston, UK

A. Jebiwott Department of Environmental Science, Egerton University, Egerton, Kenya

Obert Jiri Agricultural Practice, Faculty of Agriculture, University of Zimbabwe, Harare, Zimbabwe

School of Agricultural, Earth and Environmental Sciences, University of KwaZulu-Natal, Pietermaritzburg, South Africa

Lara Johannsdottir Environment and Natural Resources, School of Business, University of Iceland, Reykjavík, Iceland

Tino Johansson Department of Geosciences and Geography, University of Helsinki, Helsinki, Finland

Eshwer Kale The WOTR-Centre for Resilience Studies (W-CReS), Pune, India

Antoine Kalinganiré World Agroforestry Centre, West and Central Africa Regional Office – Sahel Node, Bamako, Mali

A. M. Kane Egerton University, Njoro, Kenya

Institut d'Economie Rurale (IER), Bamako, Mali

Rebekah Karimi Enonkishu Conservancy, Narok County, Kenya

Joshua Kayode Department of Plant Science and Biotechnology, Ekiti State University, Ado Ekiti, Ekiti State, Nigeria

Hugh Kelly Permezone, Santa Barbara, CA, USA

Jagriti Kher Department of Development Communication and Extension, Institute of Home Economics, University of Delhi, New Delhi, India

Method Kilasara Department of Soil and Geological Sciences, Sokoine University of Agriculture, Morogoro, Tanzania

Stephen K. Kimani KALRO- Muguga South, Nairobi, Kenya

Caroline King-Okumu The Borders Institute (TBI), Nairobi, Kenya

GeoData Institute, Southampton, UK

Geography and Environment, University of Southampton, Southampton, UK

Isaac V. Kinhonhi Electricity Regulatory Authority (ERA), Kampala, Uganda

Monicah Kinuthia Ministry of Devolution and Planning, Nairobi, Kenya

Rukia A. Kitula Institute of Marine Sciences, University of Dar es Salaam, Zanzibar, Tanzania

L. Kleerekoper Amsterdam University of Applied Sciences, Amsterdam, Netherlands

J. Kluck Amsterdam University of Applied Sciences, Amsterdam, Netherlands

Dominic Kniveton Climate Science and Society (Geography), University of Sussex, Falmer, UK

Aristea Kounani Research Centre of Environmental Education and Communication, Department of Environment, University of the Aegean, Mytilene, Greece

Nitin Kumbhar Watershed Organisation Trust (WOTR), Pune, India

WOTR Centre for Resilience Studies (W-CReS), Pune, India

Catherine Ky-Dembele World Agroforestry Centre, West and Central Africa Regional Office – Sahel Node, Bamako, Mali

Stéphane La Branche PACTE – Grenoble Institute of Political Studies, Grenoble, France

Peter Läderach International Centre for Tropical Agriculture (CIAT), International Center for Tropical Agriculture (CIAT) – Asia Regional Office c/o Agricultural Genetics Institute, Hanoi, Vietnam

J. K. Lagat Egerton University, Njoro, Kenya

Makarius C. S. Lalika Department of Geography and Environmental Studies, Solomon Mahlangu College of Science and Education, Sokoine University of Agriculture, Morogoro, Tanzania

J. K. Langat Egerton University, Njoro, Kenya

Walter Leal Filho International Climate Change Information and Research Programme (ICCIRP), Faculty of Life Sciences, Hamburg University of Applied Sciences, Hamburg, Germany

Research and Transfer Centre Sustainable Development and Climate Change Management, Hamburg University of Applied Sciences, Hamburg, Germany

David K. Lelgut KALRO- Muguga South, Nairobi, Kenya

Chunlan Li School of Geographic Sciences, East China Normal University, Shanghai, China

Wu Liang Research Center for World Geography and Resources, Institute of Geographic Sciences and Natural Resources Research, Chinese Academy of Sciences, Beijing, China

Tran Thi Kim Lien Deutsche Gesellschaft für Internationale Zusammenarbeit (GIZ) GmbH, Ha Noi, Vietnam

Ondari Lilian University of Toledo, Toledo, OH, USA

Clayco Construction, Inc., Chicago, IL, USA

J. F. Lindahl International Livestock Research Institute, Uppsala University and Swedish University of Agricultural Sciences, Nairobi, Kenya

Crispino Lobo Watershed Organisation Trust (WOTR), Pune, India

R. Loeve Amsterdam University of Applied Sciences, Amsterdam, Netherlands

Hermann Lotze-Campen Potsdam Institute for Climate Impact Research (PIK), Potsdam, Germany

Department of Sustainable Land Use and Climate Change, Humboldt-Universität zu Berlin, Berlin, Germany

Kytt MacManus Center for International Earth Science Information Network (CIESIN), The Earth Institute, Columbia University, New York, NY, USA

Malgosia Madajewicz Center for Climate Systems Research, The Earth Institute, Columbia University, New York, NY, USA

Paramu L. Mafongoya School of Agricultural, Earth and Environmental Sciences, University of KwaZulu-Natal, Pietermaritzburg, South Africa

A. Q. M. Mahbub Department of Geography and Environment, Faculty of Earth and Environmental Sciences, University of Dhaka, Dhaka, Bangladesh

Athuman Juma Mahinda Department of Agricultural Engineering, University of Dar es Salaam, Dar es Salaam, Tanzania

Department of Land Resource Management and Agricultural Technology (LARMAT), University of Nairobi, Nairobi, Kenya

Environmental Science and Technology, Kyoto University, Kyoto, Japan

Tanzania Agricultural Research Institute (TARI)-Makutupora Centre, Dodoma, Tanzania

Riffat Mahmood Department of Geography and Environment, Faculty of Life and Earth Sciences, Jagannath University, Dhaka, Bangladesh

Daniel Mailumo Department of Government and Public Administration, College of Advanced and Professional Studies, Makurdi, Nigeria

Association of Entrepreneurs and Technology Managers of Nigeria (AETMAN), Eco-system Based Adaptation For Food Security Assembly (EBAFOSA), Makurdi, Nigeria

Pius Mailumo Association of Entrepreneurs and Technology Managers of Nigeria (AETMAN), Eco-system Based Adaptation For Food Security Assembly (EBAFOSA), Makurdi, Nigeria

University of Calabar, College of Education, Katsina-Ala, Nigeria

Immaculate Maina KALRO Headquarters, Nairobi, Kenya

S. M. Makindi Department of Environmental Science, Egerton University, Egerton, Kenya

Erick Maklago Research and Business Development, National Construction Authority, Nairobi, Kenya

Michael L. Mann Department of Geography, George Washington University, Washington, DC, USA

Kristin Marcell NYS Department of Environmental Conservation and Cornell University, New Paltz, NY, USA

Wilfred Mariki Selian Agriculture Research Institute, Arusha, Tanzania

Mary Masafu University of South Africa, Pretoria, South Africa

Sani Abubakar Mashi Nigerian Meteorological Agency (NiMet), Abuja, Nigeria

Sean Mayes Crops for the Future (CFF), Jalan Broga, Selangor Darul Ehsan, Malaysia

Kevin Mearns College of Agriculture and Environmental Sciences, University of South Africa (UNISA), UNISA Science Campus, Florida, South Africa

Matthias Meier Department of Socio-Economic Sciences, Research Institute of Organic Agriculture (FiBL), Frick, Switzerland

Neil R. Miller Canadian Foodgrains Bank, Arusha, Tanzania

Jane Mills Center for International Earth Science Information Network (CIESIN), The Earth Institute, Columbia University, New York, NY, USA

Greg William Misiaszek Institute of Education Theories, Beijing Normal University, Beijing, People's Republic of China

Sara Abdelhakim Mohammad College of Engineering, Qatar University, Doha, Qatar

Hassan I. Mohammed Department of Agricultural Engineering, College of Agricultural Studies, Sudan University of Science and Technology, Khartoum, Sudan

Mohammad Mousavi Department of Civil Engineering, Najafabad Branch, Islamic Azad University, Najafabad, Iran

Monicah Mucheru-Muna Kenyatta University, Nairobi, Kenya

Paschal Arsein Mugabe College of Engineering and Technology (COET), University of Dar es Salaam, Dar es Salaam, Tanzania

Charles Mugarura Broadfield Enterprises Uganda – Permaculture Group (Resilient African Network- MUK), Kampala, Uganda

Jayne Mugwe Kenyatta University, Nairobi, Kenya

Kingstone Mujeyi International Maize and Wheat Improvement Center (CIMMYT), Southern Africa Regional Office (SARO), Harare, Zimbabwe

Angeline Mujeyi College of Agricultural Engineering and Science, Discipline of Agricultural Economics, University of KwaZulu-Natal, Pietermaritzburg Campus, Scottville, South Africa

Adrian Muller Department of Socio-Economic Sciences, Research Institute of Organic Agriculture (FiBL), Frick, Switzerland

Bertrand Muller Centre d'Etude Régional pour l'Amélioration de l'Adaptation à la Sécheresse (CERAAS), BP, Thiès, Sénégal

Centre de Coopération Internationale en Recherche Agronomique pour le Développement (CIRAD), UMR AGAP/PAM, Montpellier, France

M. Mulumpwa Sengabay Fisheries Research Centre, Salima, Malawi

Nancy W. Mungai Department of Crops, Horticulture and Soils, Njoro, Kenya

M. Muntaka Department of Agricultural Extension and Rural Sociology, Federal University Dutsin-Ma, Katsina State, Nigeria

Fumiaki Murakami Rural and Agriculture Department, Nippon Koei Co., Ltd, Tokyo, Japan

Albanus Mutiso Enonkishu Conservancy, Narok County, Kenya

Fiona Mwaniki Climate Change Communication, Kilimo Media International, Nairobi, Kenya

Majaliwa Mwanjalolo Department of Geography, Geo Informatics and Climatic Sciences, College of Agriculture and Environmental Science, School of Forestry, Environmental and Geographical Science, Makerere University (Asso.), Kampala, Uganda

Caroline Mwongera International Centre for Tropical Agriculture (CIAT), Africa Regional Office, Nairobi, Kenya

Chris M. Mwungu International Centre for Tropical Agriculture (CIAT), Africa Regional Office, Nairobi, Kenya

R. T. Nabinta Department of Agricultural Extension and Rural Sociology, Federal University Dutsin-Ma, Katsina State, Nigeria

Herve Alain Napi Wouapi Department of Agricultural Extension and Rural Sociology, Dschang School of Agronomy and Environmental Sciences, University of Dschang (UDs), Dschang, Cameroon

Sarah Ndonye International Centre of Insect Physiology and Ecology, Nairobi, Kenya

Alice Neht Institute of Urban and Transport Planning, RWTH Aachen University, Aachen, Germany

Mohsen Nekooei Department of Water Engineering, College of Agriculture, Isfahan University of Technology, Isfahan, Iran

Sileshi Nemomissa Department of Plant Biology and Biodiversity Management, College of Natural and Computational Sciences, Addis Ababa University, Addis Ababa, Ethiopia

Humphrey M. Ngibuini Forestry Development Trust, Iringa, Tanzania

Rosemary N. Ngotho-Esilaba Veterinary Sciences Research Institute, Veterinary Research Centre, Kenya Agricultural and Livestock Research Organisation, Nairobi, Kenya

Gian L. Nicolay Department of International Cooperation, Research Institute of Organic Agriculture (FiBL), Frick, Switzerland

Pinghouinde Michel Nikiema Agence Nationale de la Météorologie du Burkina Faso, Ouagadougou, Burkina Faso

F. Njau Institute of Rural Development Planning (IRDP), Dodoma, Tanzania

Peterson N. M. Njeru KALRO- Muguga South, Nairobi, Kenya

Mwangi Peter Njiiri Kenya Wildlife Service, Nairobi, Kenya

Julie Noble City of Kingston, Kingston, NY, USA

Ijaz Rasool Noorka Department of Plant Breeding and Genetics College of Agriculture, University of Sargodha, Sargodha, Pakistan

Alison Nord Department of Plant, Soil and Microbial Sciences, Michigan State University, East Lansing, MI, USA

Chinedum Nwajiuba Alex Ekwueme Federal University Ndufu-Alike Ikwo (AE-FUNAI), Ikwo, Nigeria

Nkem J. Nwosu Department of Agronomy, University of Ibadan, Ibadan, Oyo State, Nigeria

George Nyarko Faculty of Agriculture, University for Development Studies, Tamale, Northern Region, Ghana

Martine Nyeko Faculty of Agriculture and Environment, Gulu University, Gulu, Uganda

Emmanuel C. Nzegbule Nigerian Environmental Study Action Team (NEST), Ibadan, Nigeria

Michael Okpara University of Agriculture Umudike, Umudike, Nigeria

Patrick J. O'Reilly Crops for the Future (CFF), Jalan Broga, Selangor Darul Ehsan, Malaysia

Nelson Ifeanyi Obine Centre for Sustainable Development, University of Ibadan, Ibadan, Nigeria

Kehinde Abraham Odelade Food Security and Safety Niche Area, Faculty of Natural and Agricultural Sciences, North-West University, Mmabatho, Mafikeng, South Africa

Idowu Kolawole Odubote School of Agricultural Sciences, Zambian Open University, Lusaka, Zambia

Samson Samuel Ogallah Solidaridad/Pan African Climate Justice Alliance (PACJA), Nairobi, Kenya

Chika Ogbonna Brandenburg University of Technology Cottbus – Senftenberg, Senftenberg, Germany

G. M. Ogendi Department of Environmental Science, Egerton University, Egerton, Kenya

Dryland Research Training and Ecotourism Centre, Chemeron, Egerton, Kenya

A. U. Ogogo Department of Forestry and Wildlife Resources Management, University of Calabar, Calabar, Cross River State, Nigeria

Olusegun Michael Ogundele Research Group in Health and Environment- Africa, (RGHE-AFRICA), Ibadan, Nigeria

International University Network on Cultural and Biological Diversity, Nigeria International Students' Coordination Unit, (IUNCBD N-ISCU), University of Ibadan, Ibadan, Nigeria

Oluwatoyin Ogundipe Department of Botany, University of Lagos, Akoka-Yaba, Lagos, Nigeria

Damilola Grace Ogunrotimi Department of Plant Science and Biotechnology, Faculty of Science, Ekiti State University, Ado Ekiti, Nigeria

Stephen M. Ojebisi Department of Geography, Obafemi Awolowo University, Ile-Ife, Nigeria

Olumuyiwa Idowu Ojo Department of Civil Engineering, University of South Africa, Pretoria, South Africa

Ladoke Akintola University of Technology, Ogbomoso, Nigeria

Joachim Chukwuma Okafor Department of Political Science, University of Nigeria, Nsukka, Enugu State, Nigeria

David Okali Nigerian Environmental Study Action Team (NEST), Ibadan, Nigeria

Paul B. Okon Department of Soil Science, University of Calabar, Calabar, Nigeria

Yasmein Okour Department of City Planning and Design, Jordan University of Science and Technology, Irbid, Jordan

O. Oladeji Amsol Bio Company, Lagos, Nigeria

O. Adeola Olajide Department of Agricultural Economics, Faculty of Agriculture and Forestry, University of Ibadan, Ibadan, Oyo State, Nigeria

Kehinde Oluwatosin Olajubu Department of Fisheries and Aquaculture Technology, FUTA, Akure, Nigeria

Samuel Adelowo Olakojo Institute of Agricultural Research and Training, Obafemi Awolowo University, Ibadan, Nigeria

Abdulafeez Olalekan Olaniyan Ladoke Akintola University of Technology, Ogbomoso, Nigeria

Olumide Olaniyan National Weather Forecasting and Climate Research Centre, Nigerian Meteorological Agency, Abuja, Nigeria

Olawale Emmanuel Olayide Centre for Sustainable Development, University of Ibadan, Ibadan, Nigeria

Sunday J. Olotu National Centre for Technology Management, Federal Ministry of Science and Technology, Obafemi Awolowo University, Ile-Ife, Nigeria

Christopher Oludhe Department of Meteorology and Institute of Climate Change and Adaptation (ICCA), University of Nairobi, Nairobi, Kenya

O. J. Oluwatosin Department of Fisheries and Aquaculture Technology, FUTA, Akure, Nigeria

Anne Nyatichi Omambia National Environment Management Authority, Nairobi, Kenya

J. N. Ombui Department of Public Health, Pharmacology and Toxicology, College of Agriculture and Veterinary Sciences, University of Nairobi, Nairobi, Kenya

Ademola Omojola Department of Geography, University of Lagos, Akoka-Yaba, Lagos, Nigeria

Gbenga Emmanuel Onibi Department of Animal Production and Health, FUTA, The Federal University of Technology, Akure, Nigeria

J. O. Onono Department of Public Health, Pharmacology and Toxicology, College of Agriculture and Veterinary Sciences, University of Nairobi, Nairobi, Kenya

Robert Onyeneke Alex Ekwueme Federal University Ndufu-Alike Ikwo (AE-FUNAI), Agriculture (Agricultural Economics and Extension Programme), Ikwo, Nigeria

Victor A. Orindi ADA, National Drought Management Authority (NDMA), Nairobi, Kenya

Isimemen Osemwegie International University of Grand Bassam, Grand-Bassam, Côte d'Ivoire

Kaveh Ostad-Ali-Askari Department of Civil Engineering, Isfahan (Khorasgan) Branch, Islamic Azad University, Isfahan, Iran

Christine N. A. Ouinsavi Doctoral School of Agronomic and Water Sciences/ Department of Sustainable Management of Natural Resources, Laboratory of Ecology and Forest Research (LERF), University of Parakou, Parakou, Borgou, Republic of Benin

Emmah Owidi International Centre of Insect Physiology and Ecology, Nairobi, Kenya

Oyediran O. Oyebola Department of Aquaculture and Fisheries Management, University of Ibadan, Ibadan, Nigeria

Obasanjo Joseph Oyedele Department of Mass Communication, Faculty of Social and Management Sciences, Bowen University Iwo, Osun State, Nigeria

Olawale Oyemade Oyekanmi Department of Teacher Education, University of Ibadan, Ibadan, Oyo State, Nigeria

Maureen Papas Faculty of Law, The University of Western Australia, Perth, WA, Australia

Nyeko Pen-Mogi Gulu University, Gulu, Uganda

Anuradha Phadtare The WOTR-Centre for Resilience Studies (W-CReS), Pune, India

Surendra Poonia ICAR-Central Arid Zone Research Institute, Jodhpur, Rajasthan, India

Adithya Pradyumna Watershed Organisation Trust (WOTR), Pune, India

Society for Community Health Awareness Research and Action (SOCHARA), Bangalore, India

Kathrin Prenger-Berninghoff Institute of Urban and Transport Planning, RWTH Aachen University, Aachen, Germany

Geeta Punhani Department of Development Communication and Extension, Institute of Home Economics, University of Delhi, New Delhi, India

Muhammad Abdur Rahaman Climate Change Adaptation, Mitigation, Experiment and Training (CAMET) Park, Noakhali, Bangladesh

Mohammad Mahbubur Rahman Network on Climate Change, Bangladesh (NCC,B) Trust, Dhaka, Bangladesh

Nicholas B. Rajkovich School of Architecture and Planning, State University of New York at Buffalo, Buffalo, NY, USA

Laxmi Ramasubramanian Department of Urban Policy and Planning and The Institute for Sustainable Cities, Hunter College, CUNY, New York, NY, USA

Eunice Pereira Ramos Unit of Energy Systems Analysis, KTH – Royal Institute of Technology, Stockholm, Sweden

A. S. Rao ICAR-Central Research Institute for Dryland Agriculture, Hyderabad, India

Lance Robinson International Livestock Research Institute, Nairobi, Kenya

Sajal Roy Institute for Culture and Society, Western Sydney University, Sydney, NSW, Australia

Department of Women and Gender Studies, Begum Rokeya University, Rangpur (BRUR), Bangladesh

Onkangi N. Ruth Research and Business Development, National Construction Authority, Nairobi, Kenya

Ruchi Sachan School of International Studies, Centre for African Studies, Jawaharlal Nehru University, New Delhi, India

Sakshi Saini Department of Development Communication and Extension, Institute of Home Economics, University of Delhi, New Delhi, India

A. Saleem Khan Center for International Earth Science Information Network (CIESIN), The Earth Institute, Columbia University, New York, NY, USA

Zakari Saley Bana Disaster Risks Reduction Program Officer (DRR/O), Catholic Agency For Overseas Development (CAFOD), Niamey, Niger

David Olabanjo Sanni Department of Geography, Obafemi Awolowo University, Ile Ife, Nigeria

Maruf Sanni National Centre for Technology Management, Federal Ministry of Science and Technology, Obafemi Awolowo University, Ile-Ife, Nigeria

Josias Sanou Institut de l'Environnement et de Recherche Agricole (INERA), Ouagadougou, Burkina Faso

Muhammad Saqib Institute of Soil and Environmental Sciences, University of Agriculture, Faisalabad, Pakistan

Sheena Shah Permaculture Research Institute Kenya, Nairobi, Kenya

Kelvin M. Shikuku International Centre for Tropical Agriculture (CIAT), Africa Regional Office, Nairobi, Kenya

Samson Shimelse Department of Plant Biology and Biodiversity Management, College of Natural and Computational Sciences, Addis Ababa University, Addis Ababa, Ethiopia

College of Dryland Agriculture and Natural Resources, Mekelle University, Mekelle, Ethiopia

Yogesh Shinde The WOTR-Centre for Resilience Studies (W-CReS), Pune, India

Arega Shumetie Department of Economics, College of Business and Economics, Haramaya University, Haramaya, Ethiopia

Constantina Skanavis Research Centre of Environmental Education and Communication, Department of Environment, University of the Aegean, Mytılene, Greece

Sef Slootweg Natural Resources Management Project – Ngorongoro District Council, Deutsche Gesellschaft für Internationale Zusammenarbeit (GIZ) GmbH, Loliondo, Tanzania

Sieglinde Snapp Department of Plant, Soil and Microbial Sciences, Michigan State University, East Lansing, MI, USA

Alabi Soneye Department of Geography, University of Lagos, Akoka-Yaba, Lagos, Nigeria

Vignesh Sridharan Unit of Energy Systems Analysis, KTH – Royal Institute of Technology, Stockholm, Sweden

Silke Stöber Center for Rural Development (SLE), Humboldt Universität zu Berlin, Berlin, Germany

Adeniyi Suleiman Gbadegesin Ladoke Akintola University of Technology, Ogbomoso, Nigeria

E. Y. Swai Agricultural Research Institute (ARI) Hombolo, Dodoma, Tanzania

Nava Tabak Scenic Hudson, Poughkeepsie, NY, USA

Abdel Aziz A. M. Taha Department of Agricultural Engineering, University of Gadarif, Gadarif, Sudan

Hosein Talebmorad Department of Water Engineering, College of Agriculture, Isfahan University of Technology, Isfahan, Iran

Constantinos Taliotis Unit of Energy Systems Analysis, KTH – Royal Institute of Technology, Stockholm, Sweden

Premsagar Tasgaonkar Watershed Organisation Trust (WOTR), Pune, India

Azage Tegegne International Livestock Research Institute, Addis Ababa, Ethiopia

B. Teme Institut d'Economie Rurale (IER), Bamako, Mali

Ludovic Temple Center for International Cooperation in Agricultural Research and Development (CIRAD), Montpellier, France

Molu Tepo Merti Integrated Development Programme (Mid-P), Nairobi, Kenya

Jérôme Ebagnerin Tondoh WASCAL and UFR des Sciences de la Nature, Nandjui Abrogoua University, Abidjan, Côte d'Ivoire

Rosemary Egodi Ubaekwe Research Group in Health and Environment- Africa, (RGHE-AFRICA), Ibadan, Nigeria

International University Network on Cultural and Biological Diversity, Nigeria International Students' Coordination Unit, (IUNCBD N-ISCU), University of Ibadan, Ibadan, Nigeria

Paul Akeh Ugbah National Weather Forecasting and Climate Research Centre, Nigerian Meteorological Agency, Abuja, Nigeria

Collins Ugochukwu Alberta Environmental and Parks, Government of Alberta, Edmonton, AB, Canada

Gloria Ujor Nigerian Environmental Study Action Team (NEST), Ibadan, Nigeria

Aliyu Tambuwal Umar Department of Geography, Usmanu Danfodiyo University, Sokoto, Nigeria

Philippe Vaast Center for International Cooperation in Agricultural Research and Development (CIRAD), Montpellier, France

Swapnil Vyas The WOTR-Centre for Resilience Studies (W-CReS), Pune, India

Hafiz Abdul Wahab Institute of Soil and Environmental Sciences, University of Agriculture, Faisalabad, Pakistan

James R. Wallace Group Communications and Corporate Responsibility, Allianz, Munich, Germany

Liam James Walsh Sydney, NSW, Australia

David Walunya Ong'are National Environment Management Authority, Nairobi, Kenya

Shem Wandiga Department of Chemistry, University of Nairobi, College of Biological and Physical Sciences, Nairobi, Kenya

James M. Warner International Food Policy Research Institute, Addis Ababa, Ethiopia

H. O. Wesonga Veterinary Sciences Research Institute, Veterinary Research Centre, Kenya Agricultural and Livestock Research Organisation, Nairobi, Kenya

Andreas Witte Institute of Urban and Transport Planning, RWTH Aachen University, Aachen, Germany

Muluneh Woldetisadik Addis Ababa University, Addis Ababa, Ethiopia

Lippa Wood Mara Training Centre, Narok County, Kenya

Omonlola Nadine Worou The International Crops Research Institute for the Semi-Arid Tropics (ICRISAT), Bamako, Mali

Dipak Zade The WOTR-Centre for Resilience Studies (W-CReS), Pune, India

Libby Zemaitis NYS Department of Environmental Conservation and Cornell University, New Paltz, NY, USA

Jelena Živanović Miljković Institute of Architecture and Urban and Spatial Planning of Serbia, Belgrade, Serbia

Part I

Climate Change Resilience in Transportation, Energy, Forestry, and Water/Coastal Infrastructure

Sequestrated Carbon Content in Tree Species and Diurnal Temperature Influence for Adaptive Climate Change Resilience in Nigeria

1

Mustapha Adeojo Ibrahim, Bashir Yusuf Abubakar, and Mohammed Lawal Balarabe

Contents

M. A. Ibrahim · M. L. Balarabe
Department of Biology, Ahmadu Bello University, Zaria, Nigeria
e-mail: imustapha2008@gmail.com; mlbalarabe@abu.edu.ng

B. Y. Abubakar (✉)
Department of Botany, Ahmadu Bello University, Zaria, Nigeria
e-mail: byabubakar@abu.edu.ng

W. Leal Filho (ed.), *Handbook of Climate Change Resilience*,
https://doi.org/10.1007/978-3-319-93336-8_4

Abstract

Trunk wood, bark, and leaf carbon concentration was assessed based on Walkley-Black in three (3) tropical tree species (*Eucalyptus camaldulensis* Dehnh., *Khaya senegalensis* (Desr.) A. Juss, and *Tectona grandis* L.f.) as influenced by diurnal temperature. This is to provide the baseline data for adaptive climate change resilience in one of the developing countries, Nigeria. The data demonstrated that carbon content was highly variable within and between the wood and bark of the species. The bark of *E. camaldulensis* and *K. senegalensis* had the highest percentage of total carbon content (C_{TOT}) at 61.92% and 49.5% when compared to the woody total carbon content at 55.04% and 44.77%, respectively, while *T. grandis* had the highest total carbon content (C_{TOT}) in the woody tissue at 50.32% when compared to 41.19% in the bark tissue. Also *E. camaldulensis* had a significant positive relationship between their trunk diameter and carbon content at different levels, while *K. senegalensis* and *T. grandis* had a weak positive and negative relationship in their trunk diameter and carbon content at different levels. A strong negative relationship in leaf carbon content and diurnal temperature was observed in *K. senegalensis* and *T. grandis,* while *E. camaldulensis* had a weak negative and unclear relationship in their leaf carbon content and diurnal temperature. The carbon content decreases as the tree diameter reduces with increase in the tree height; therefore, sampling at tree diameter at breast height (DBH) may provide a good indicator of whole trunk carbon content. The relationship in the tree leaf carbon content with diurnal temperature indicated that trees respond differentially to daily temperature.

Keywords

Carbon sequestration · Tree species · Climate change · Diurnal temperature · Resilience · Savannah forest · Nigeria

Introduction

Climate change is defined by the Intergovernmental Panel on Climate Change (IPCC 2007) as statistically significant variations that persist for an extended period, typically decades or longer. It includes shifts in the frequency and magnitude of sporadic weather events as well as the slow continuous rise in global mean surface temperature, 1.4–5.8 °C (Houghton et al. 2001). This change manifests in a number of ways. They include changes in average climatic conditions, drier or wetter on average in some regions, climate variability, erratic rainfall events in some regions, changes in the frequency and magnitude of extreme weather, and changes in sea levels. Climate change is a process of global warming, in part attributable to the greenhouse gases generated by human activity (Bret 2009). These greenhouse gases are generated from the burning of fossil fuels such as coal, oil, and gas. These fuels contain carbon and carbon dioxide as some of the gases which contribute to global warming. Carbon sequestration by the plant species is a way by which gases like atmospheric CO_2 can be reduced in our ecosystem through the vital function of photosynthesis. The role of CO_2 in photosynthetic process of plants is further influenced by other environmental factors that in turn affect the net productivity of the species. In assessing climate change in the tropics, Thorsten (2014) opined that knowledge on species composition and distribution is essential for discussing the future of tropical biodiversity and ecosystem function. *Tectona grandis*, *Gmelina arborea*, *Dalbergia sissoo*, *Bambusa vulgaris* var. *vulgaris*, *and Swietenia macrophylla* were selected for field study under wasteland condition to assess carbon sequestration in Tamil Nadu, India (Prasath et al. 2016). These authors opined that it is also significant that even though *Dalbergia sissoo* recorded a highest growth and productivity, it showed a lowest intercellular CO_2 concentration than other moderately performing trees. Similarly, Chen et al. (2016) reported using correlation analyses between various environmental variables, and CO_2 sequestration rates indicated that air temperature and soil water content were likely the main factors influencing carbon sequestration of *F. griffithii* at their study site in Southern Taiwan. Their results demonstrate that air temperature, soil water content, and leaf area are the major factors affecting CO_2 sequestration in this species.

Ecological Implications of Climate Change in Nigeria

Hengeveld et al. (2002) reported that increasing temperature (global warming) and decreasing precipitation in most parts of the world are the greatest impacts of climate change. These bring about either negative or positive ecological impacts in different parts of the world. The increasing temperature has led to increased land-based ice instability and its melting. The thawing of the Arctic, cool and cold temperate ice, the increasing rainfall in some parts of the world, and expansion of the oceans as water warms have started impacting on sea level rise, coastal inundation, and erosion.

The current global estimate of sea level rise is 0.2 m, and it is projected to increase to 1 m by the year 2100 (Hengeveld et al. 2005). The implication is that the present 0.2 m sea level rise has inundated 3400 km^2 of the coastal region of Nigeria, and if the sea level rise attains the projected 1 m on or before 2100, then 18,400 km^2 of the coastal region may be inundated (NEST 2003). Coastal settlements like Bonny, Forcados, Lagos, Port Harcourt, Warri, and Calabar, among others, that are less than 10 m above the sea level would be seriously threatened by a meter rise of sea level. The sea incursion due to sea level rise means saltwater intrusion into the freshwater and invasion and destruction of mangrove ecosystems, coastal wetlands, and coastal beaches. The worst impact is population displacement, which may result in communal crisis. The coastal inundation and erosion with their associated population displacement are currently major environmental problems in Nembe, Eket and other coastal settlements in Bayelsa, Delta, Cross River, Rivers, and Lagos states of Nigeria. It is estimated that a meter rise in sea level will displace about 14 million people from the coastal areas of Nigeria (Abu 2007).

The Sahara Desert is observed to be expanding to all directions trying to engulf the Sahelian region of Africa with annual expansion of 1–10 km (Odjugo and Ikhuoria

2003; Yaqub 2007). Odjugo and Ikhuoria (2003) also observe that Nigeria 12°N is under severe threat of desert encroachment, and sand dunes are now common features of desertification in states like Yobe, Borno, Sokoto, Jigawa, and Katsina. The migrating sand dunes have buried large expanse of arable lands, thus reducing viable agricultural lands and crops' production. This has prompted massive emigration and resettlement of people to areas less threatened by desertification. Such emigration gives rise to social effects like loss of dignity and social values. It often results in increasing spate of communal clashes among herdsmen and farmers, and such clashes resulted in the death of 186 people in six northern states of Nigeria between 1998 and 2006 (Yaqub 2007). Akonga (2001) also shows that most of the destitute that emigrated as a result of drought and desertification usually move to nearby urban areas to beg for alms, thereby compounding the already tense urbanization problems.

Climate change will alter all aspects of the hydrological cycle ranging from evaporation through precipitation, runoff, and discharge (Mcquire et al. 2002). The global warming and decreasing rainfall together with the erratic pattern of rainfall produce a minimal recharge of groundwater resources, wells, lakes, and rivers in most parts of the world especially in Africa, thereby creating water crisis. In Nigeria, many rivers have been reported by Yugunda (2002) to have dried up or are becoming more seasonally navigable, while Lake Chad shrunk in area from 22,902 km^2 in 1963 to a mere 1304 km^2 in 2000. This shows that what is left of Lake Chad in the year 2000 is just 5.7% of 1963 (Odjugo 2007). Yaqub (2007) also confirms the fact that Lake Chad has shrunk by 95% since the 1960s, and Aral Sea in Central Asia was the fourth largest lake in the planet in 1960, but by 2007 it had shrunk to 10% of its original size. Lake Chad and so many rivers in Nigeria, especially in Northern Nigeria, are in the danger of disappearing. The water scarcity will create the tendency for concentration of users around the remaining limited sources of water. Under such circumstances, there is increased possibility of additional contamination of the limited sources of water and transmission of waterborne diseases like cholera, typhoid fever, guinea worm infection, and river blindness.

Deweerdt (2007) notes that the increasing temperature will mean northward migration of mosquitoes and malaria fever which will extend from the tropical region to warm temperate region, while the sporogony of the protozoa causing the malaria accelerates from 25 days at 10 °C to 8 days at 32 °C. The excessive heat, increasing water stress, air pollution, and suppressed immune system occasioned by climate change will result in increasing incidence of excessive death due to heat exhaustion, famine, water-related diseases (diarrhea, cholera, and skin diseases), inflammatory and respiratory diseases (cough and asthma), depression, skin cancer, and cataract. One of the greatest impacts of climate change is the worsening condition of extreme weather events like drought, flood, rainstorms, windstorms, thunderstorms, landslides, avalanches, and tsunamis, among others (Odjugo 2001; Changnon 2001). Odjugo (2008) notes that the frequency and magnitude of wind and rainstorms did not only increase; they also killed 199 people and destroyed property worth 85.03 billion (USD 236 million) in Nigeria between 1992 and 2007. Buadi and Ahmed (2006) had similar result when they reported that rainstorms claimed 42 lives in southern Cameroon between 2000 and 2005.

Climate change has started to and will continue to impact negatively on agriculture and food security especially in tropical and subtropical regions because greenhouse gas emissions would increase the risk of hunger by additional 80 million people by 2080 in Africa and southern Asia (Nwafor 2006, 2007; DeWeerdt 2007). Odjugo (2008) shows that climate change has led to a shift in crops cultivated in Northern Nigeria. Ahmed (1978) cited in Odjugo (2008) reveals that as of 1978, the preferred crops the farmers cultivated were guinea corn followed by groundnut and maize, but due to increasing temperature and decreasing rainfall amount and duration occasioned by climate change, the farmers as a means of adaptation in 2007 shifted to the production of millet followed by maize and beans. Another major problem to agriculture in Nigeria due to climate change is the reduction of arable lands. While the sea incursion is reducing the arable land of the coastal plains, the desert encroachment with its associated sand dunes is depriving farmers of their agricultural farmlands and grazing rangelands. Moreover, the frequent droughts and lesser rains have started shortening the growing season, thereby causing crop failure and food shortage. It has been shown that drought, desert encroachment, and coastal inundation have started affecting the country's ecosystem leading to ecological destabilization due to climate change impact in the semiarid region of Northern Nigeria (Odjugo and Ikhuoria 2003; Ayuba et al. 2007).

CO_2 Emissions in Nigeria

In Nigeria, the contribution of locally GHGs to global warming has been assessed by different authors. Edeoja and Edeoja (2015) evaluated the management of carbon emissions in the Nigerian construction industry by measuring the amount of CO_2 being emitted from constructional activities within some selected organizations. The study was pursued via three-case study analysis within the industry; scope 1 and scope 2 emissions according to GHG protocol classification were considered emission sources. It was found that though there was a significant emission from Nigeria construction industry, the provision for emission monitoring and management was lacking, general awareness regarding carbon emissions and environmental issues across the various organizations evaluated was poor, and the construction organization had no singular person directly in charge of carbon or environmental issues. Anomohanran (2011) investigated the amount of CO_2 emission released from consumption of petroleum products in Nigeria between 1999 and 2009. Tier 1 method of GHG determination using the reference approach was adopted; this method involves the estimation of CO_2 emission from the supply of petroleum product to the economy rather than from the actual emission at the combustion plant. The result indicated that Nigeria was responsible for 0.26% of global CO_2 emission within the time frame. The study also reported that the consumption of premium motor spirit, automated gas oil, and household kerosene constituted 85% of the total emission for the previous 20 years and that there was a yearly increase in CO_2 emission by 4.7% as against the global 1.9% within the time frame.

A similar study also conducted by Anomohanran (2012) on the greenhouse gas emission resulted from gas flaring activities in Nigeria between the period of 1999 and 2009. The study made use of the records of natural gas produced and natural gas flared in Nigeria within the time frame, and the reference approach method of greenhouse gas determination was also adopted. The result showed that the total quantity of CO_2 emitted within the time frame was 23.1% when compared with that of the global value, and it represents one of the largest emissions from a single source. Nwaichi and Uzazobona (2011) delve into exploring the level and distribution of CO_2 and other associated potential contaminant in the Niger Delta. In the study, some flare sites were investigated to examine their possible environmental characteristics. Two flow stations of Shell Petroleum Corporation were monitored for a period of 3 months. It was ascertained that the concentration of CO_2 and other contaminants was significant and that the flaring activities have substantially impacted on the environment through emission and as a result require mitigation measures to avoid the inherent biomagnification with time. Iduh et al. (2016) highlight the potential reduction of CO_2 emissions from gas flaring in Nigeria's oil and gas industry through alternative productive use. The study investigated the potentials of converting flared gas from the Nigeria's oil and gas industry to compressed natural gas (CNG) which could be an alternative fuel for the 220 Lagos Bus Rapid Transit (BRT-Lite) while reducing CO_2 emissions. In addition, the study provided an overview of gas flaring in the oil and gas industry and energy utilization in some selected sectors in the country. The study concluded that the use of CNG as an alternative fuel for Lagos BRT-Lite will significantly reduce CO_2 emissions in Nigeria's oil and gas industry, and other utilization options for flared gas from this industry include liquefied natural gas (LNG), liquefied petroleum gas (LPG), and power generation.

Okhimamhe and Okelola (2013) in a study of the assessment of the levels of atmospheric carbon dioxide emission at road junctions in three major cities in the southwest of Niger State in Nigeria, namely, Minna, Bida, and Suleja identify measures that improve traffic operations while reducing the emission levels that could have implications on global warming and hence climate change. The results established that the emission levels in these three cities were approximately eight times more than the internationally accepted safe limits of 350 ppm for atmospheric CO_2, but less than the Occupational Safety and Health Administration permissible exposure limits of 5000 ppm which have adverse health effects and may contribute to climate change, in the long term, if unmitigated. Samson et al. (2013) probe into CO_2 electricity generation prospect in Nigeria. The study exploited the ability of CO_2 exhaust gas from the power plant with the aid of the carbon dioxide data obtained from IEA through the Ministry of Environment in Nigeria and the knowledge of bottoming power generation. Also, qualitative amount of power was estimated from the nation industrial CO_2 potential generation. The result shows that an optimum amount of 564.7 MW of electricity per year could be estimated from this power source; this is equivalent to 10.8% of projected power required for year 2030. Igwenagu (2011) examined the position of Nigeria in relation to emission readiness for emission trading as proposed in Kyoto Protocol as a measure of reducing global warming. It was discovered that Nigeria emits only 0.4% of the total world CO_2 emission which indicate that Nigeria will be possible seller of emission as contained in the protocol.

Climate change impacts are felt on agricultural production, health, biodiversity, and social and economic conditions and affect people and the environment in general (Smit and Skinner 2002). It is predicted to worsen the incidence of drought and desertification, and millions of people will become refugees as a result (Hir 2010). Mohammed (2002) reported that the desert, which now covers about 35% of Nigeria's land mass, is advancing at an estimated 0.6 km per annum, while deforestation is taking place at 3.5% per annum. The desert belt has moved from Kebbi, Kano, and Maiduguri to new Bussa, Kaduna, Jos, and Sheleng, while the savannah now interfaces between the desert and forest along Oyo, Osun, Kogi, and Benue states. Moreover, the Sudano-Sahelian region of Nigeria has suffered a decrease in rainfall in the range of 3–4% per decade since the beginning of the nineteenth century and concluded that impacts of climate change are being felt by both developed and developing countries, and in Nigeria, for example, more than two thirds of the country is thought to be prone to desertification. Another dimension as to the effect of elevated CO_2 in the atmosphere is that the current global risk of protein deficiency was estimated to be 12.2%, which is expected to rise to 15.1% (or a total of 1.4 billion people) by 2050 through demographic changes alone (i.e., not accounting for changing CO_2 and protein levels). Furthermore, if changes in protein levels resulting from CO_2 concentrations of $>$500 ppm are incorporated, the authors estimate this number to increase by a further 1.57% or 148.4 million (Medek et al. 2017).

Building resilience on the consequences of climate change must be put in place to generate data that can be used for mitigation efforts. One of the approaches to reducing carbon and carbon dioxide concentration in the atmosphere is through carbon (C) sequestration, the process of removing C from the atmosphere and depositing it in a reservoir. The critical level of air pollution along with government regulation and subsidies on greenhouse gas sequestration makes it desirable and attractive to capture emitted carbon dioxide (Buesseler 2008). The main methods of CO_2 sequestration are geological, chemical, and biological. Geological sequestration is among the most actively implemented and researched in the industry. One of the modes of geo-sequestration is ocean storage. During ocean storage, CO_2 is being pumped to the bottom of the ocean. When CO_2 is pumped to a depth below 3000 m, the gas becomes denser than water which results in the formation of under ocean CO_2 lakes. Another actively investigated mode of geo-sequestration by Goldberg (1998) is the injection of CO_2 into depleted natural reservoirs of oil and gas.

Bio-sequestration of CO_2 uses the ability of photosynthetic organisms to capture atmospheric CO_2. Biological carbon mitigation (BCM) is the process whereby autotrophic organisms and plants convert this CO_2 into organic carbon through photosynthesis producing large amounts of biomass (Stephan et al. 2001). All biological media contain carbon, and the major stores of carbon can be found in vegetation like forestry, soils, peat, as well as a large portion being sequestered over time naturally in the ocean (Worrall et al. 2010). Through photosynthesis and other metabolic pathways, carbon becomes incorporated into the cells of these organisms

(Petela 2008). Careful management of the carbon cycle would ensure that biomass could be utilized for various commercial applications while ensuring sufficient carbon is stored in biological media, thereby maintaining safe levels of CO_2 in the atmosphere (Harun et al. 2010). These authors have the feeling that carbon sequestering through the use of plant species may the best employed in the developing world. This is based on the consideration that establishment of forest cover through afforestation schemes is more affordable that can cover large acreages by government efforts or even by community or individual effort to serve as firewood lot and indirect biodiversity enhancement. Access to carbon credit that was recently enhanced by developed nations at the last climate change summit of Paris would also have to rely on basic information of carbon sequestration by the plant species. In a recent report by IISD (2017), it was stated that in the Paris Agreement on climate change, countries agreed to make "finance flows consistent with a pathway towards low greenhouse gas (GHG) emissions and climate-resilient development." This report further observed that developing countries will receive financial resources for both mitigation and adaptation actions, while developed countries are expected to continue leading in mobilizing climate finance from a variety of sources, with public funds playing a significant role in reaching the previously agreed US$100 billion annual target by 2020. This chapter is therefore aimed at assessing the bio-sequestered carbon content in *K. senegalensis*, *T. grandis*, and *E. camaldulensis* as influenced by diurnal temperature at Savanna Forestry Research Station, Samaru, Zaria, Nigeria.

Materials and Methods

Study Area

The study was carried out in Savanna Forestry Research Station (SFRS) along Zaria Sokoto Road, Samaru, Zaria Kaduna State, Nigeria (Fig. 1). SFRS is located at latitude $11^{\circ}10^{1}$N and longitude $7^{\circ}37^{1}$E. The area has a tropical wet and dry climate with warm weather year-round. It has also a wet season lasting from April to September and a drier season from October to March. The average day temperature ranges from 27 to 38 °C, while the average annual rainfall is 750.8 mm. Relative humidity is extremely low in this region for major part of the year which ranges between 36% and 69%, while it is highest (85%) during rainy season (IAR, Metrological center ABU, Zaria).

Sampling Procedure

Twenty (20) trees per species *Eucalyptus camaldulensis* (river red gum), *Khaya senegalensis* (African mahogany), and *Tectona grandis* (teak) were selected purposively and tagged for diameter at breast height (DBH) and normal bole form measurements, whereas ten (10) replicates for each species were sampled randomly

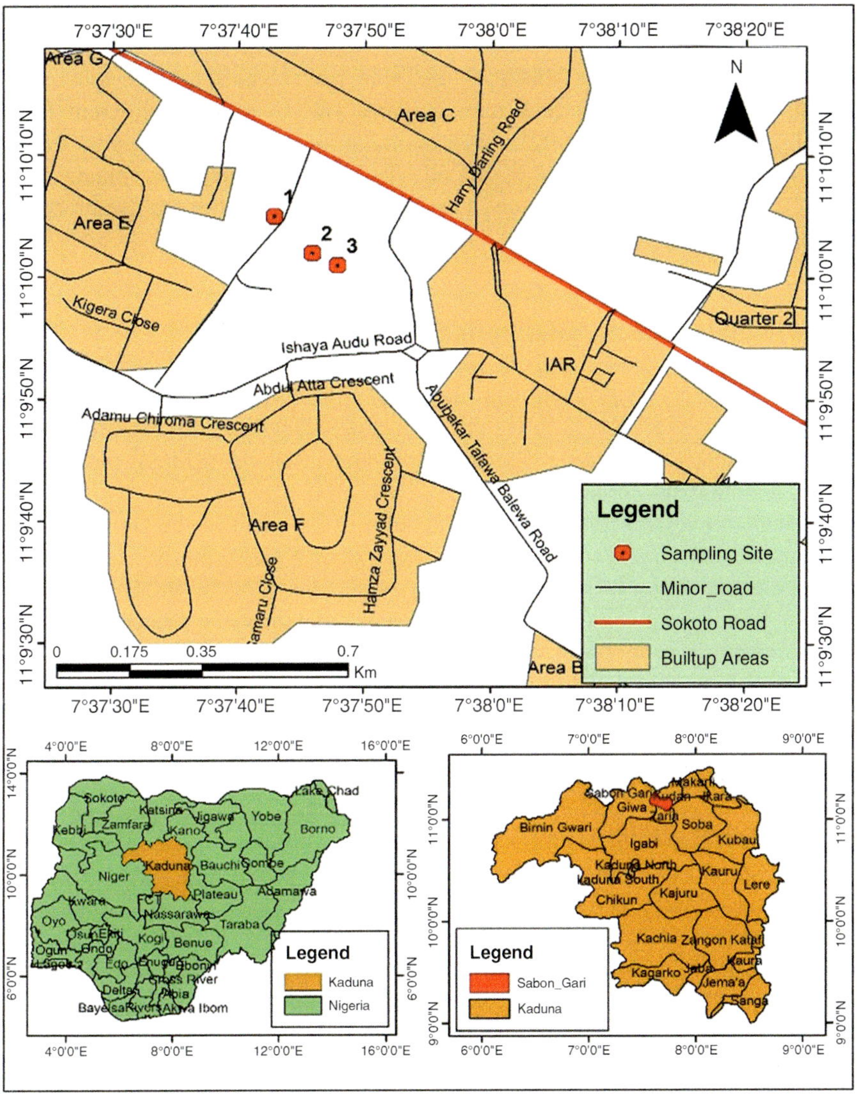

Fig. 1 Map of Savanna Forestry Research Station (SFRS) with the sampling sites (1, 2, and 3)

for wood and bark sampling. Uniformly, 5 m height for each trunk was measured from the ground level and was further partitioned into regions of basal, middle, and top levels at intervals of 1.33 m. Also, three (3) replicates for each species were used for leaf sampling. In line with the recommended practices of the Intergovernmental Panel on Climate Change (IPCC 2007) with slight modification to estimate the total carbon content in the aboveground biomass, the tree diameter at each level height was measured, and thereafter, samples were analyzed for their carbon content.

Biophysical Measurements

The tree height and the diameter are two main biophysical measurements which were measured for each tree sampled. Tree diameter at different levels (basal, middle, and top) was calculated by measuring the tree trunk circumference at 1.33 m intervals from the ground. Tree height (TH) was obtained by direct measurement with surveyors' tape aided by a pole following the procedure given (Chavan and Rasal 2012).

Sample Collection for Biophysical Measurement

A total of ninety (90) wood and bark samples from the tree species were collected at interval of 1.33 m for the three levels (basal, middle, and top) along the tree trunk in Savanna Forestry Research Station (SFRS) plot across three genera and order (*Eucalyptus camaldulensis*, *Khaya senegalensis*, and *Tectona grandis*) following identification of the plant in the Department of Biology Herbarium, Faculty of Life Sciences, A.B..U, Zaria. The species were selected from three different plots which were 20 m away from Zaria Sokoto Road. Zaria Sokoto Road is a major road linking some states in the north western Nigeria; these make the study area suitable for this chapter because of the high vehicular emissions.

Cutlass was used in peeling off the bark of the trees before using a 5 cm diameter increment borer for collecting wood cores at intervals of 1.33 m on the trunk from ground levels for ten (10) species each per tree. To avoid biases due to the presence of compression or tension wood, only individual stems with straight growth forms were sampled. Trees with crooked stems, substantial heartrot, or other forms of stem damage were excluded, and when necessary, cores were taken in directions parallel to slopes. One half of the collected sample (both the bark and wood core) was placed in a freezer within 4 h of extraction to minimize loss of volatiles, while the other half was taken to Savanna Forestry Research Laboratory for oven-drying treatment, making it 90 samples per each drying treatment (freeze- and oven-drying). The sampled (90) wood core and bark were divided into two halves, where one half of the samples was placed in separate polythene bag and taken to National Research Institute for Chemical Technology (NARICT), Zaria, for freeze-drying treatments, and the other half in separate polythene bags also was dried in Savanna Forestry Research Laboratory using oven dryer at 80 °C.

The dried samples were carried to the General Purpose Laboratory in Soil Science Department A.B..U, Zaria, were the samples were pulverized into a homogenous powder. The dried powdered sample was analyzed for their carbon content using wet oxidation method (Walkley-Black).

Determination of the Total Carbon Conversion Factor

Based on IPCC (2006) procedure on total carbon conversion factors (C_{tot}) that correct for mass and carbon loss during sample drying treatment, the total carbon conversion factor (Ctot) was determined for each sample using

$$C_{TOT} = C_{OV} + C_{VOL}$$

where

C_{TOT} is C fraction from oven-dried sample analysis
C_{VOL} is the C fraction in volatiles relative to oven-dried mass, such that

$$C_{VOL} = C_{FZ} \times \left(1_{/(1-VMF)}\right) - C_{OV}$$

C_{FZ} is C content of freeze-dried samples, and **VMF** represents the species' mean mass in volatile compound lost upon heating:

$$VMF = (C_{FZ} - C_{OV})/C_{FZ}$$

(Martin and Thomas 2011).

Data Analysis

The means of the data obtained from the wood and bark were analyzed using descriptive statistics. One-way analysis of variance (ANOVA) was used to detect significant difference in the interspecific % C content variation in the wood and barks of the tree species. Pearson's correlation was used to test the relationship between tree diameter and carbon content. Pearson's correlation was also used to show the relationship between leaf carbon content and diurnal temperature at both $p < 0.01$ and $p < 0.05$.

Results and Discussion

Variation in % Carbon Content Within the Wood and Bark of the Tree Species

The results in Fig. 2 show that the % carbon content variation within the wood and bark of the tree species was highly significant. The total carbon content varied significantly within each of the tree wood and bark, averaging 55.04 ± 2.13 and 61.92 ± 2.18, 44.77 ± 1.86 and 49.50 ± 1.57, and

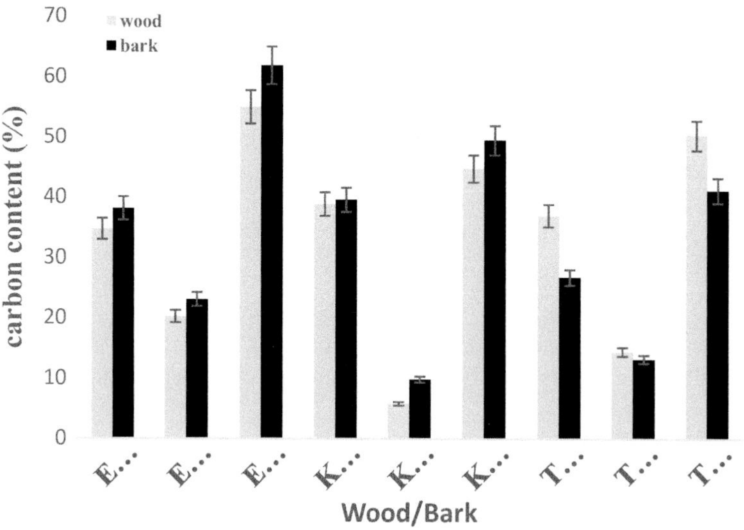

Fig. 2 Variation in % carbon content within the wood and bark of the sampled tree species. Key: $E.C_{ov}$ *E. camaldulensis* oven carbon, $E.C_{vol}$ *E. camaldulensis* volatile carbon, $E.C_{tot}$ *E. camaldulensis* total carbon, $K.C_{ov}$ *K. senegalensis* oven carbon, $K.C_{vol}$ *K. senegalensis* volatile carbon, $K.C_{tot}$ *K. senegalensis* total carbon, $T.C_{ov}$ *T. grandis* oven carbon, $T.C_{vol}$ *T. grandis* volatile carbon, $T.C_{tot}$ *T. grandis* total carbon

50.31 ± 1.19 and 41.89 ± 1.95 in wood and bark of *E. camaldulensis*, *K. senegalensis*, and *T. grandis*, respectively. The total carbon contents within the wood and bark of *E. camaldulensis* and *K. senegalensis* increase from the wood (55.04 ± 2.13 and 44.77 ± 1.86) down to the bark, with the bark having the maximum % carbon contents (61.92 ± 2.18 and 49.50 ± 1.57); while *T. grandis* showed different variations, the wood % carbon content was higher (50.31 ± 1.19) when compared to the bark (41.89 ± 1.95). The percentage oven carbon content (C_{OV}) also varied within the wood and bark of these tree species, averaging 34.78 ± 1.45 and 38.81 ± 1.13, 38.92 ± 1.53 and 39.62 ± 1.16, and 37.02 ± 0.90 and 26.78 ± 1.52 in wood and bark of *E. camaldulensis*, *K. senegalensis*, and *T. grandis*, respectively. Both % C_{OV} for *E. camaldulensis* and *K. senegalensis* also increase from the wood (34.78 ± 1.45 and 38.92 ± 1.53) down to the bark tissue with the bark having the highest (38.81 ± 1.13 and 39.62 ± 1.16) % C_{OV}, whereas the woody (37.02 ± 0.90) % C_{OV} of *T. grandis* was the highest when compared to the bark (26.78 ± 1.52). It was however observed that the volatile percentage carbon content (C_{VOL}) follows the same trends as in oven and total % carbon contents for the three plants, respectively. *E. camaldulensis* and *K. senegalensis* had highest % C_{OV} in their barks (23.10 ± 1.29), while *T. grandis* had the highest volatile in their woody tissues (13.27 ± 0.88).

Table 1 Interspecific % carbon content variation between the woods of the tree species

Tree species	% carbon content		
	C_{OV}	C_{VOL}	C_{TOT}
E. camaldulensis	34.78 ± 1.16^{c}	20.25 ± 0.97^{a}	55.04 ± 2.13^{a}
K. senegalensis	38.92 ± 1.53^{a}	5.88 ± 0.92^{c}	44.77 ± 1.86^{c}
T. grandis	37.02 ± 0.90^{b}	13.27 ± 0.88^{b}	50.32 ± 1.19^{b}

Mean values, standard error, and different superscript letter columns depicting significant difference at 0.05 level using ANOVA test
C_{TOT} total carbon, C_{VOL} volatile carbon, C_{OV} oven-dried carbon

Table 2 Interspecific % carbon content variation between the barks of the sampled tree species

Tree species	% Carbon content		
	C_{OV}	C_{VOL}	C_{TOT}
E. camaldulensis	38.82 ± 1.13^{b}	23.10 ± 1.29^{a}	61.92 ± 2.17^{a}
K. senegalensis	39.62 ± 1.16^{a}	9.89 ± 0.80^{c}	48.05 ± 1.04^{b}
T. grandis	26.78 ± 1.52^{c}	14.46 ± 0.83^{b}	41.18 ± 1.95^{c}

Mean values, standard error, and different superscript letter columns depicting significant difference at 0.05 level using ANOVA test
C_{TOT} total carbon, C_{VOL} volatile carbon, C_{OV} oven-dried carbon

Interspecific Carbon Content Variations in the Wood and Bark of Tree Species

The results in Tables 1 and 2 shows the interspecific % C content variation among the sampled tree wood and bark species, which were highly significant at $p > 0.05$. In Table 1, species shows different wood C concentrations with *E. camaldulensis* having the highest total and volatile % carbon concentration ($C_{TOT} = 55.04 \pm 2.13$ and $C_{VOL} = 20.25 \pm 0.97$), with the lowest oven-dried carbon fraction ($C_{OV} = 34.78 \pm 1.16$). *T. grandis* was observed to be next to *E. camaldulensis* in terms of their total and volatile carbon concentrations ($C_{TOT} = 50.32 \pm 1.19$ and $C_{VOL} = 13.27 \pm 0.88$), and the species also had an oven-dried % carbon content ($C_{OV} = 37.02 \pm 0.90$) which was slightly lower than that of *K. senegalensis* ($C_{OV} = 38.92 \pm 1.53$). *K. senegalensis* had the highest oven-dried % carbon fraction ($C_{OV} = 38.92 \pm 1.53$) with the lowest total and volatile % carbon concentration ($C_{TOT} = 44.77 \pm 1.86$ and $C_{VOL} = 5.88 \pm 0.92$).

Table 2 shows the interspecific % carbon content observed in the barks of the tree species. *E. camaldulensis* had the highest bark total and volatile % carbon content ($C_{TOT} = 61.92 \pm 2.17$ and $C_{VOL} = 23.10 \pm 1.29$), with the second highest in oven-dried % carbon content ($C_{OV} = 38.82 \pm 1.13$). *K. senegalensis* had a total carbon content (C_{TOT}) of 48.05 ± 1.04 which were next to the concentration observed in *E. camaldulensis*, with the lowest ($C_{VOL} = 9.89 \pm 0.80$) and highest in their oven carbon content ($C_{OV} = 39.62 \pm 1.16$). *T. grandis* had the lowest bark oven and total carbon content ($C_{OV} = 26.78 \pm 1.52$ and $C_{TOT} = 41.18 \pm 1.95$).

Relationship Between Tree Diameter and Their Percentage Carbon Content

Table 3 shows the relationship with tree diameter at different levels and their percentage wood carbon content. In *E. camaldulensis*, a strong positive correlation between tree diameter and wood carbon content which were highly significant at $P < 0.01$ and $P < 0.05$ was observed at E_1 and E_2; also, positive correlation was observed at E_3, and the relationship were however not significantly different. The strength of the overall relationship decline as the tree levels increase from E_1 to E_2 and E_3. The relationship between *K. senegalensis* diameter at different levels and their carbon content showed a weak positive correlation at K_1 and K_2 which were not significantly different. A weak negative correlation was observed at K_3 which were not statistically significant. *T. grandis* showed a weak negative relationship in their diameter at different levels with % carbon content across the three levels (T_1, T_2, and T_3).

Variations in the % Carbon Contents Within the Wood and Bark of the Tree Species

The carbon concentration within the wood and barks of *E. camaldulensis*, *K. senegalensis*, and *T. grandis* is highly variable. The total % carbon content varies from 55.04 ± 2.13 and 61.92 ± 2.18, 44.77 ± 1.86 and 49.50 ± 1.57, and 50.31 ± 1.19 and 41.89 ± 1.95 in wood and barks of *E. camaldulensis*, *K. senegalensis*, and *T. grandis*, respectively. In general, the total carbon content (C_{TOT}) increases as both oven (C_{OV}) and volatile (C_{VOL}) % carbon contents increase.

In Fig. 3, the total carbon content increases from the woody tissue down to the bark tissue with the bark tissues having the maximum carbon concentration in *E. camaldulensis* and *K. senegalensis*. Conversely, % carbon content was highest in the woody tissues when compared with the bark tissues for *T. grandis*. Previous studies had shown that carbon concentration could vary within tree tissues. For example, Javad et al. (2016) show that parts of tree species had different capacity in C sequestration. They further reported *Z. spina-christi* having higher C in the leaf when compared to other species, while *P. juliflora* and *E. camaldulensis* had higher carbon sequestration in their wood and litter, respectively. The results obtained by these authors agree with the results obtained for *T. grandis* in the present chapter with higher woody % carbon content when compared with the bark tissues.

The overall variation in % carbon content within the wood and bark of the tree species in this chapter could be accounted for by their chemical composition as the chemical composition of trees varies with tissues type. Romberger et al. (2004) reported a variation in tree chemical composition as a function of tree part (root, bark, and stem or branch). Bark contains a similar range of chemical constituents to wood. Thus, cellulose, hemicelluloses, and lignin plus extractives (including fats, sterols, terpenes, various polyphenols, etc.) are present. Gifford (2002) obtained higher bark carbon content in *Pinus radiata* and concluded that, due to its high levels

Table 3 Relationships between tree diameter and their carbon content at different levels

	E_{TD}	E_1	K_{TD}	K_1	T_{TD}	T_1	E_{TD2}	E_2	K_{TD2}	K_2	T_{TD2}	T_2	E_{TD3}	E_3	K_{TD3}	K_3	T_{TD3}	T_3
E_{TD}	1																	
E_1	**0.951****	1																
K_{TD}	−0.251	−0.266	1															
K_1	−0.240	−0.021	**0.196**	1														
T_{TD}	−0.452	−0.221	−0.406	0.427	1													
T_1	-0.413	−0.439	0.845**	0.054	**−0.342**	1												
E_{TD2}	0.893**	0.879**	−0.467	−0.310	−0.144	−0.595	1											
E_2	0.727*	0.807**	−0.201	0.239	−0.021	−0.328	**0.666***	1										
K_{TD2}	0.074	−0.022	0.662*	0.264	−0.645*	0.595	−0.326	0.027	1									
K_2	−0.239	0.004	0.124	0.763*	0.557	0.027	−0.177	0.171	**0.179**	1								
T_{TD2}	−0.712*	−0.576	−0.295	0.225	0.823**	0.090	−0.488	−0.354	−0.448	0.332	1							
T_2	−0.149	−0.275	0.796**	−0.179	−0.690*	0.781**	−0.470	−0.409	0.740*	−0.285	**−0.314**	1						
E_{TD3}	0.654*	0.661*	−0.449	−0.464	0.034	−0.542	0.897**	0.459	−0.622	−0.143	−0.270	−0.506	1					
E_3	0.720*	0.760*	−0.215	0.309	−0.107	−0.473	0.674*	0.837**	0.113	0.030	−0.530	−0.422	**0.378**	1				
K_{TD3}	−0.099	−0.211	0.728*	0.162	−0.675*	0.491	−0.379	−0.114	0.759*	0.005	−0.463	0.712*	−0.474	−0.069	1			
K_3	−0.314	−111	0.507	0.638*	0.354	0.367	−0.248	−0.091	0.091	0.801**	0.176	0.071	−0.164	−0.086	**−0.078**	1		
T_{TD3}	−0.494	−0.334	−0.270	−0.048	0.789**	0.002	−0.160	−0.317	−0.740*	0.224	0.807**	−0.370	0.187	−0.494	−0.631	0.246	1	
T_3	−0.167	−0.334	0.740*	−0.267	−0.679*	0.676*	−0.330	−0.420	0.575	−0.275	−0.341	0.840**	−0.279	−0.391	0.803**	0.151	**−0.220**	1

E_{TD} *E. camaldulensis* tree diameter, *E E. camaldulensis*, k_{TD} *K. senegalensis* tree diameter, *K K. senegalensis*, T_{TD} *T. grandis* tree diameter, *T T. grandis*

**Correlation is significant at the 0.01 level, *Correlation is significant at the 0.05 level (two-tailed)

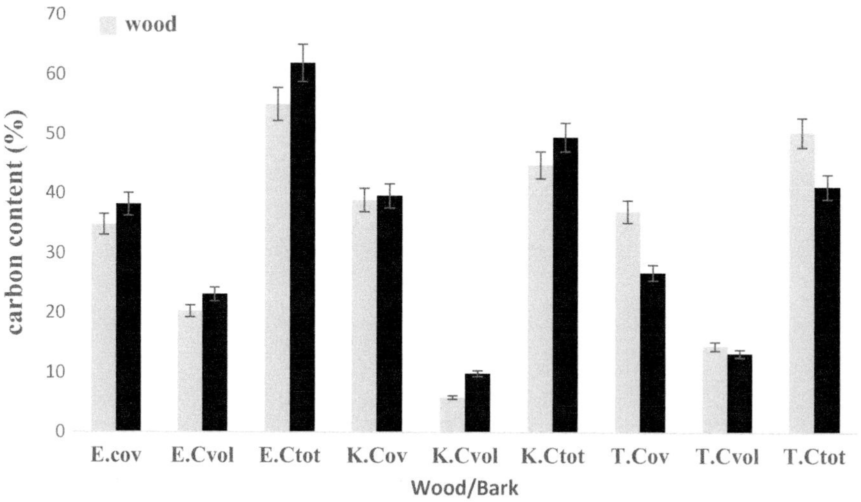

Fig. 3 Variation in % carbon content within the wood and bark of the sampled tree species. $E.C_{ov}$ *E. camaldulensis* oven carbon, $E.C_{vol}$ *E. camaldulensis* volatile carbon, $E.C_{tot}$ *E. camaldulensis* total carbon, $K.C_{ov}$ *K. senegalensis* oven carbon, $K.C_{vol}$ *K. senegalensis* volatile carbon, $K.C_{tot}$ *K. senegalensis* total carbon, $T.C_{ov}$ *T. grandis* oven carbon, $T.C_{vol}$ *T. grandis* volatile carbon, $T.C_{tot}$ *T. grandis* total carbon

of extractives, lignin, and tannins, the bark is that part of the pine with the highest carbon concentration. This is similar to the results obtained for *E. camaldulensis* and *K. senegalensis* in the present chapter and disagrees with the result obtained for *T. grandis*. Bert and Danjon (2005) also reported a higher bark carbon concentration in the different compartments of *Pinus pinaster* studied which are also similar to the results obtained for *E. camaldulensis* and *K. senegalensis* and also disagree with the result of *T. grandis*. However, the development of specialized bark tissues also produces polymeric materials peculiar to bark (Fradinho et al. 2002); this might probably account for the higher bark % carbon content observed in *E. camaldulensis* and *K. senegalensis*. On the other hand, the result for *T. grandis* agrees with the work of Nunes et al. (1996) who reported that some trees might have a reduced or lack specialized bark in trees that are responsible for the production of polymeric materials which may account for the low bark % carbon content in *T. grandis* when compared to the woody tissue.

Interspecific Variation in Carbon Content Among the Sampled Tree Species

The wood and bark C content (expressed as a percentage of wood and bark dry mass) was highly variable among the studied tropical species (Tables 1 and 2) with *E. camaldulensis* flanking the highest carbon concentration in both tables and on average significantly higher than assumed in the scientific literature. The appearance of *E. camaldulensis* as the highest in both wood and bark carbon contents (55.04 ± 2.13 and 61.92 ± 2.17, respectively) could be accounted for by their large tree diameter and tree height and probably due to their closeness to the roadside. This chapter also confirms that the volatile carbon fraction (C_{VOL}) is an important component of total wood bark C content in tropical species, indicating that neglecting this fraction will significantly underestimate total wood and bark C content.

Generally, it was observed that volatile carbon fraction contributed significantly to the among species variation in the wood and bark total carbon content; however, the larger component of the variation might be a reflection in the differences in their solid-phase chemical constituents, the most abundant of which are cellulose and lignin (Rana et al. 2010). Proportions of these compounds are variable in tropical trees, and on a dry mass basis, cellulose (including hemicellulose and cellobiose) constitutes 65–75% of woody tissues, while lignin constitutes 20–50% (Rana et al. 2010; Pastore et al. 2004). The C content of these compounds differs greatly, with cellulose containing 40–44% C and lignin 60–72% (Lamlom and Savidge 2006). Thus cellulose/lignin ratios between 2.5 and 4 likely account for much of the variation in wood C content in tropical hardwoods. Bark contains a similar range of chemical constituents to wood. Thus, cellulose, hemicelluloses, and lignin plus extractives (including fats, sterols, terpenes, various polyphenols, etc.) are also present (Fradinho et al. 2002). This trend is supported by existing data (Pastore et al. 2004). When studying carbon content from 11 hardwoods to 9 lightwoods through the combustion analysis, Ragland et al. (1991) determined that hardwoods have an average of 50% in carbon content and lightwoods possess an average of 53%. When analyzing the causes for such variation, the conclusion was that the differences are due to the lignin content and the extractives in each of the woods. Nogueira et al. (2008), for example, calculated error estimations when the heartwood was not considered in biomass estimations. This variation reflects differences in species' chemical makeup. For instance, lipids, lignin, and proteins have elevated carbon concentrations, while organic acids and minerals contain little and no carbon, respectively (Poorter and Bergkotte 1992). However, the present chapter fails to show the direct relationship between the wood and bark chemical constituent and their carbon content. Analyzing the correlation between cellulose/lignin ratios and carbon content for tropical tree species would have been necessary to confirm the generality of this relationship.

Carbon content values obtained using Walkley-Black method are values that are above 0.50 in *E. camaldulensis* wood and bark, except for *K. senegalensis* which presents values of 0.40, while *T. grandis* present 0.50 and 0.40 in their wood and bark tissues, respectively. These values coincide with values established for living organisms, i.e., 0.50 in carbon from dry weight of wood and bark (Woodcock and Shier 2003). However, other studies such as Elias and Potvin (2003) also report variations in carbon content, from 0.45 to 0.53, a range that is close to values found for the three species studied in Savanna Forestry Research Station, Samaru. A relevant aspect to point out from these values is that carbon content determination procedure is not detailed. On the other hand, Kumar et al. (2010) also found carbon content variations in seven forest species used in India to produce firewood. The latter authors reported a range of 0.378 and 0.466 of carbon content, similar to that in the present chapter, but the determination of carbon content in the India species was done using the Walkley-Black method (1934) as used in the present chapter, which is based on the oxidation of organic carbon shown by the sample in the presence of potassium dichromate in a strongly acid medium. This method was developed to determine carbon in soils and oxidizes only 77% of the total C in that condition; nevertheless, adjustment in the calculation must be done before using it for other materials. Other interesting research was carried out by Elias and Potvin (2003), who measured carbon content of 32 species of tropical trees in Panama using an automated elemental analyzer (model EA 1108, FISONS Instruments, Milan, Italy). They found that carbon content varied between 0.44 and 0.49, with *Ormosia macrocalyx* and *Tectona grandis* flanking the lower and upper limits, respectively.

Relationship Between Tree Diameter and Their Carbon Content at Different Levels

Conventionally, wood core sampling has been collected within tree diameter at breast height (DBH) (Martin and Thomas 2011); the present chapter attempts to show the relationship between tree diameter at different levels on the trunk and their carbon content. The highly significant strong positive relationship observed in *E. camaldulensis* at E_1 and E_2 which declines at E_3 implies that the trunk carbon content increases with increases in tree diameter and decreases with increases in the tree height. And a weak positive relationship observed in *K. senegalensis* (K_1 and K_2) and the weak negative relationship observed in *T. grandis* could be accounted for by their growth form and probably their low varieties of trunk diameter. Also, the relationship observed in *E. camaldulensis* might be a result of their significant reduction in the tree diameter as the tree height increases.

Navarro and Moya (2013) reported that variation in carbon content was higher in *A. tibourbou* (coefficient of variation of 25.78%) than in the other species they studied, which can be explained by the fact that sampled trees present a high variety of diameters. Diameters for this species ranged from 9.00 to 45.30. The results are similar to the relationships observed in the studied tree species in Savanna Forestry Research Station, Samaru. The observed relationship in the present chapter is consistent with the report of Elias and Potvin (2003) on the variation in carbon content with tree height, and they related this to changes in the tree diameters as the tree height increases. The observed relationships at E_1, K_1, and T_1 are in strong agreement with the work of Jibrin (2014) on exploring the key predictors of carbon stock density in savannah woodland area, Niger State, Nigeria, that indicate diameter at breast height (DBH) as one of the key predictors of the carbon stock. In a related work by Chavan and Rasal (2012), they concluded that the total standing biomass of *Mangifera indica* in 2847 ha of Aurangabad City is 104.41tha-1, while the sequestered carbon stalks in aboveground and belowground standing biomass of *M. indica* are 44.73 tha-1 and 11.63 tha-1, respectively, while the total sequestered carbon of *M. indica* in 2847 ha area is 56.36 tha-1. They also showed that the atmospheric CO_2 captured by *M. indica* from the Aurangabad City is 206.84 tCO_2 ha-1, and the allometric regression equations indicated high correlation and accurate relationship between aboveground biomass as a function of both variables DBH and height in the *M. indica* in this chapter.

Leaf Carbon Content as Influenced by Diurnal Temperature

Leaf carbon content as a result of photosynthesis and leaf level photosynthesis resulted from the interaction of environmental factors and the sensitivity of the species to these factors which may include temperature, light, and CO_2 (Sage and Kubien 2007). This is aligned with the observation in the present chapter (Figs. 4, 5, and 6) that the tree's leaf carbon content responds differentially to the effect of diurnal temperature. The results in Figs. 5 and 6 (*K. senegalensis* and *T. grandis*, respectively) correlated negatively well with diurnal temperature; their leaf carbon content increases at a lower temperature and reduces at a higher temperature, while the relationship was however unclear in Fig. 4 as some of the day's leaf carbon content increases irregularly around the diurnal temperature. According to Mcquire et al. (2002), temperature has an effect on three different processes, which might affect the photosynthetic rate of plant. Increase in temperature causes an increase in cellular respiration, an increase in the diffusion coefficient of CO_2 (Fick's law), and a decrease in the solubility of CO_2 in water (Henry's law) and, hence, increase of gaseous CO_2. The interplay of all of these processes determines the relationship between the leaf carbon content and diurnal temperature as leaf carbon content is a result of photosynthesis. In assessing the influence of temperature on within-canopy acclimation and variation in leaf photosynthesis, Bauerle et al. (2007) reported that comparison of predicted versus actual responses indicates that this is not the case, where the influence of temperature acclimation on enzyme activity and the respective influence on photosynthetic parameters differed substantially. Thus, the development of a mechanistic understanding of these responses is still a major challenge to pursue if we hope to quantify accurately the interactions among temperature, physiological genetics, and ecosystem function. In a recent work (Hartfield and Prueger 2015) on how extreme temperatures affect plant growth and development is dependent upon plant species, under an increasing climate change scenario, there is greater likelihood of air temperatures exceeding the optimum range for many species. As a result of this, large impact on grain yield than vegetative growth would be experienced because of the increased minimum temperatures. Biologically, they showed that the effects are evident in an increased rate of senescence which reduces the ability of the crop to efficiently fill the grain of fruit.

The result in Fig. 4 is consistent with the work of Bakshi (1996) in which he reported high diurnal uptake of CO_2 in *Eucalyptus* species at a higher temperature and related this behavior to their high transpiration and stomatal conductance in order to maintain a high productivity. Prasath et al. (2016) also reported that the

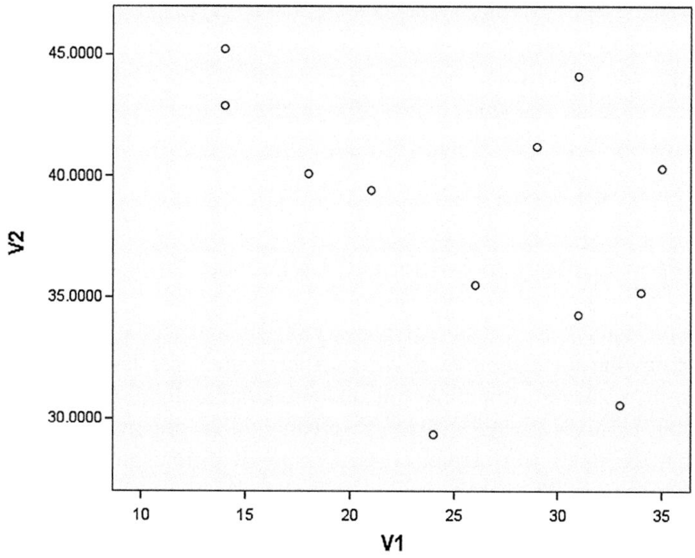

Fig. 4 Estimated % carbon content in the leaves of *E. camaldulensis* in response to diurnal temperature: V1, diurnal temperature, V2, carbon content

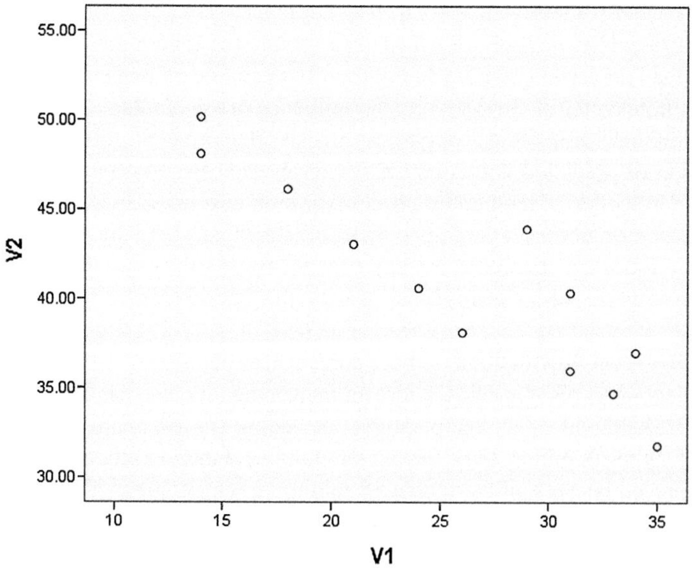

Fig. 5 Estimated % carbon content in the leaves of *K. senegalensis* in response to diurnal temperature: V1, diurnal temperature, V2, carbon content

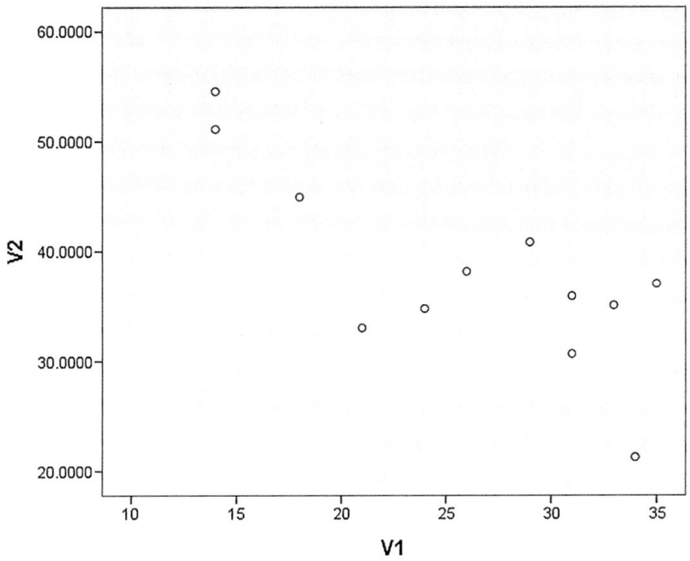

Fig. 6 Estimated % carbon content in the leaves of *T. grandis* in response to diurnal temperature: V1, diurnal temperature, V2, carbon content

carbon sequestration potential of *Dalbergia sissoo* and *Bambusa vulgaris* var. *vulgaris* was high as they maintained high level of photosynthetic rate, transpiration rate, and other ecophysiological activities throughout the day and exhibited their suitability for wasteland afforestation and high carbon sequestration. The poor performing species of *Tectona grandis* and *Gmelina arborea* maintained a very high ecophysiological activity without any prominent midday closure leading to failure in regulating the stomata according to the environmental conditions which resulted in very low productivity and their non-suitability for wasteland afforestation and high carbon sequestration. Saveyn et al. (2007) also observed similar discrepancies in some of the tree species studied as observed in the present chapter (Fig. 4).

They however attributed these behaviors to other factors such as better water status or nutrient availability in addition to diurnal temperature which might have contributed to some of this day's increase in leaf carbon content. The result in Figs. 5 and 6 is substantial with the report of Hari et al. (2016) that stated that trees may undergo midday depression in their leaf carbon content because of their lower CO_2 uptake at higher temperature. The models of Jon and Graham (2008) are also in agreement with the present chapter who reported that reduction in CO_2 uptake at leaf temperature above 30 °C may occur as results lead to reduction of leaf carbon content, which are almost accountable for in terms of reduction in stomatal conductance in response to higher leaf to air vapor pressure deficit. As evaporative demand increases due to higher temperature, stomata tend to close to reduce the rate of water loss through transpiration. Associated with this stomatal closure is a reduction in

CO_2 assimilation rate due to a reduction in the rate of supply of CO_2 to the chloroplast, which as a result leads to the reduction of the leaf carbon content of these tree species. Diurnal temperature (Dt°C) had a strong negative relationship with the leaf carbon content of both *K. senegalensis* and *T. grandis*, while *E. camaldulensis* had a weak negative and sometimes unclear relationship.

Conclusions

It is evident that functional role of tree and other plant species signifies their niche in our ecosystem providing mitigation effect on climate change and by extension ensuring food security to our increasing human population. It therefore becomes imperative that factors influencing this physiological role of plant species as carbon sink need to be elaborately studied using many species to further enrich our understanding of the present environmental changes. Finally, studies like this not only utilize the tree vegetation cover in estimating the biological role of plant species in carbon content estimation as a result of functional photosynthesis but also may provide data to map out resilience strategies for mitigation.

References

Abu B (2007) Sea level rise and the Niger Delta of Nigeria. J Wetl 3(1):44–52

Ahmed HA (1978) The pattern of farming systems and crops production in Borno State, Nigeria. J Agric Sci 2(1):9–16

Akonga AZ (2001) The causes and impacts of drought. ANIS Monogr 3:1–16

Anomohanran O (2011) Estimating greenhouse gas emission from petroleum products in Nigeria. J Appl Sci 11(17):3209–3214. https://doi.org/10.3923/jas.2011.3209.3214

Anomohanran O (2012) Determination of greenhouse gas emission resulting from gas flaring activities in Nigeria. Energy Policy 45:666670. https://doi.org/10.1016/j.enpol.2012.03.018

Ayuba HK, Maryah UM, Gwary DM (2007) Climate change impact on plant species composition in six semi-arid rangelands of Northern Nigeria. Nigerian Geogr J 5(1):35–42

Bakshi M (1996) Diurnal and seasonal patterns of photosynthetic characteristics of leaves of *Eucalyptus* hybrid. Ann For 4(1):78–84

Bauerle WL, Bowden JD, Wang GG (2007) The influence of temperature on within-canopy acclimation and variation in leaf photosynthesis: spatial acclimation to microclimate gradients among climatically divergent *Acer rubrum* L. genotypes. J Exp Bot 58(12):3285–3298

Bert R, Danjon F (2005) Carbon concentration variations in the roots, stem and crown of mature Pinus pinaster (Ait.). Silvae Genet 44:10–19

Bret H (2009) Food and agriculture, features, climate change threat to food security. Available at http://www.peopleandplanet.Net/doc.Php?Id=3482

Buadi PV, Ahmed AN (2006) The impacts of wind related hazards in Southern Cameroon. J Phys Sci 3(2):82–88

Buesseler KO (2008) Ocean iron fertilization-moving forward in a sea of uncertainty. Environ Sci 319:5860

Chavan BL, Rasal GB (2012) Total sequestered carbon stock of *Mangifera indica*. J Environ Earth Sci 2(1):37–48

Chen CI, Wang YN, Lih HW, Yu JC (2016) Three-year study on diurnal and seasonal CO_2 sequestration of a young *Fraxinus griffithii* plantation in Southern Taiwan. Forest 7(230):1–11

DeWeerdt S (2007) Climate change coming home: global warming effect on populations. World Watch 2(3):6–13

Edeoja JA, Edeoja AO (2015) Carbon emission management in the construction industry-case studies of Nigerian construction industry. Am J Eng Res 4(7):112–122

Elias M, Potvin C (2003) Assessing inter- and intra-specific variation in trunk carbon concentration for 32 neotropical tree species. Can J For Res 33:1039–1045

Fradinho DM, Pascoal Neto C, Evtuguin D, Jorge FC, Irle MA, Gil MH, Jesus JP (2002) Chemical characterisation of bark and alkaline bark extracts from maritime pine grown in Portugal. Ind Crop Prod 16:23–32

Gifford RM (2002) Carbon content of above-ground tissues of forest and woodland trees. Australian Greenhouse Office, National Carbon Accounting System, Technical Report 22, 18 pp

Goldberg J (1998) Ethanol for a sustainable energy future. Science (315):808–810. https://doi.org/10.1126/science.1137013

Hari B, Prasath CN, Balasubramanian A, Prasanthrajan M, Radhakrishanan S (2016) Performance evaluation of different tree species for carbon sequestration under wasteland condition. Int J For Crop Improv 7(1):7–13. https://doi.org/10.15740/HAS/IJFCI/7.1/7-13

Hartfield JC, Prueger JH (2015) Temperature extremes: effect on plant growth and development. Weather Clim Extremes 10:4–10

Harun R, Singh M, Forde GM, Danquah MK (2010) Bioprocess engineering of microalgae to produce a variety of consumer products. Renew Sust Energ Rev 14:1037–1047

Hengeveld GH, Bush E, Edwards P (2002) Frequently asked questions about climate change science. Environment Canada, Canada

Hengeveld H, Whitewood B (2005) Understanding climate change 2005: a synthesis of climate change science. Environment Canada, Canada, pp 31–35

Hir J (2010) Sand dunes threaten Northern Nigeria-Climate Change group. http://dailytrust.dailytrust.com/index.php?option=com_content&view=article&id=4517:sand-dunes-threaten-northern-nigeria-climate-changegroup&catid=10:environment&Itemid=11. Retrieved 20 Feb 2011

Houghton JT, Ding Y, Griggs DJ, Noguer M, Linden PJ, Dai X, Maskell K, Johnson CA (eds) (2001) Climate change 2001: the scientific basis. A contribution of working group I to the third assessment report of the intergovernmental panel on climate change (IPCC). Cambridge University Press, Cambridge, UK, p 881

Iduh JJ, Murray P, Enongene KE (2016) The potential reduction of carbon dioxide (CO_2) emissions from gas flaring in Nigeria oil and gas industry through alternative productive use. Proceedings of SBE16 Dubai, 17–19 January 2016, Dubai UAE (paper number: SBE16D135)

Igwenagu CM (2011) Principal component analysis of global warming with respect to carbon(iv) oxide emission in Nigeria: an exploratory study. Asian J Math Stat 4(2):71–80. https://doi.org/10.3923/ajms.2011.71.80

IISD (2017) International institute for sustainable development- SDG hob knowledge. 25th February, 2017. http://sdg.iisd.org/news/february-2017-climate-finance-update-focus-on-new-financing-tools/. Accessed 1 Nov 2017

IPCC (2007) Climate change 2007. Climate change impacts, adaptation and vulnerability. Working Group II. IPCC, Geneva

IPCC (2006) Guidelines for national greenhouse gas inventories. Agriculture, forestry and other land use, vol 4. Institute for Global Environmental Strategies, Hayama

Javad M, Mostafa M, Farzad S (2016) Carbon sequestration in the leaf, litter and soil of *Eucalyptus camaldulensis*, *Prosopsis juliflora*, *Ziziphus spina-christi* species. Ecopersia 4(3):1481–1491

Jibrin A (2014) Exploring the key predictors of carbon stock density in Savanna woodland area, Niger State, Nigeria. Int J Geogr Geol 3(12):145–158

Jon L, Graham DF (2008) Effects of rising temperatures and [CO_2] on the physiology of tropical forest trees. Phil Trans R Soc B 363:1811–1817. https://doi.org/10.1098/rstb.2007.0032

Kumar R, Pandey KK, Chandrashekar S (2010) Effect of tree-age on caloric value and other fuel properties of Eucalyptus hybrid. J For Res 21:514–516

Lamlom SH, Savidge RA (2006) Carbon content variation in boles of mature sugar maple and giant sequoia. Tree Physiol 26:459–468

Martin AR, Thomas SC (2011) A reassessment of carbon content in tropical trees. PLoS One 6 (8):1–9

Mcquire B, Mason I, Kilburn C (2002) Natural hazards and environmental change. Arnold, London, pp 53–63

Medek DE, Schwartz J, Myers SS (2017) Estimated effects of future atmospheric CO_2 concentrations on protein intake and the risk of protein deficiency by country and region. Environ Health Perspect 125(8):087002, 1–8. Accessed 1 Nov 2017

Mohammed A (2002) Socio-economic and cultural impacts of desert encroachment in Nigeria. J Environ Dyn 5(2):19–30

Navarro M, Moya R (2013) Successional variation in carbon content and wood specific gravity of four tropical tree species. BOSQUE 34(1):33–43. https://doi.org/10.4067/S0717-92002013000100005

NEST (Nigerian Environmental Study/Action Team) (2003) Climate change in Nigeria. A communication guide for reporters and educators. NEST, Ibadan, pp 5–16

Nogueira EM, Fearnside PM, Nelson BN, Barbosa R, Hermanus EW (2008) Estimates of forest biomass in the Brazilian Amazon: new allometric equations and adjustments to biomass from wood-volume inventories. For Ecol Manag 256:1853–1867

Nunes E, Quilho T, Pereira H (1996) Anatomy and chemical composition of *Pinus pinaster* bark. IAWA J 17:141–149

Nwafor JC (2006) Environmental impact assessment for sustainable development. The Nigerian perspective. EL 'DEMAK pubs, Enugu, pp 359–394

Nwafor JC (2007) Global climate change, the driver of multiple causes of flood intensity in Sub-Saharan Africa. Paper presented at the International Conference on Climate Change and Economic Sustainability held at Nnamdi Azikiwe University, Enugu, 12–14 June 2007

Nwaichi EO, Uzazobona MA (2011) Estimation of the CO_2 level due to gas flaring in the Niger Delta. Res J Environ Sci 5(6):565–572. https://doi.org/10.3923/rjes.2011.565.572

Odjugo PAO (2007) The impact of climate change on water resources; global and regional analysis. Indones J Geogr 39:23–41

Odjugo PAO (2001) The impacts of global warming on climate change and extreme weather conditions: global and regional evidences. Asia Pac J Environ Dev 7:53–70

Odjugo PAO (2008) Quantifying the cost of climate change impact in Nigeria: emphasis on wind and rainstorms. J Hum Ecol (In Press)

Odjugo PAO, Ikhuoria AI (2003) The impact of climate change and anthropogenic factors on desertification in the semi-arid region of Nigeria. Glob J Environ Sci 2(2):118–126

Okhimamhe AA, Okelola OF (2013) Assessment of carbon dioxide emission at road junctions in southeast of Niger state, Nigeria. Alam Cipta 6(2):59–70

Pastore TCM, Santos KO, Rubim JC (2004) A spectro-colorimetric study on the effect of ultraviolet irradiation of four tropical hardwoods. Bioresour Technol 93:37–42

Petela R (2008) An approach to the exergy analysis of photosynthesis. Sol Energy 82:311–328

Poorter H, Bergkotte M (1992) Chemical composition of 24 wild species differing in relative growth rate. Plant Cell Environ 15:221–229

Prasath CNH, Balasubramanian A, Prasanthrajan M, Radhakrishnan S (2016) Performance evaluation of different tree species for carbon sequestration under wasteland condition. Int J For Crop Improv 7(1):7–13

Ragland KW, Aerst DJ, Baker AJ (1991) Properties of wood for combustion analysis. Bioresour Technol 37:161–168

Rana R, Langenfeld-Heyser R, Finkeldey R, Polle A (2010) FTIR spectroscopy, chemical and histochemical characterisation of wood and lignin of five tropical timber wood species of the family of *Dipterocarpaceae*. Wood Sci Technol 44:225–242

Romberger JA, Hejnowicz Z, Hill JF (2004) Plant structure, function and development. A treatise on anatomy and vegetative development with special reference to Woody plants. The Balckburn Press, Caldwel

Sage B, Kubien R (2007) Biomass, carbon, and nutrient dynamics of secondary forest in a humid tropical region of Mexico. Ecology 80:1892–1907

Samson AA, Oluseyi OA, Folorunso OA (2013) Carbon dioxide electricity generation prospect in Nigeria. J Energy Technol Policy 3(1):32
Saveyn A, Steppe K, Lemeur R (2007) Daytime depression in tree stem CO2 efflux rates: is it caused by low stem turgor pressure? Ann Bot 99:477–485
Smit B, Skinner MW (2002) Adaptation options in agriculture to climate change: a typology. Mitig Adapt Strateg Glob Chang 7:85–114
Stephan DJ, Shockey RE, Moe TA, Dorn R (2001) Carbon dioxide sequestering using microalgal systems. US Department of Energy, Pittsburgh
Thorsten P (2014) Climate change in the tropics, Tropical Forestry Handbook. Springer, Berlin, 1–14 tropical forest trees. Philos Trans Royal Soc Biol Sci 363:1811–1817
Walkley A, Black IA (1934) A critical examination of a rapid method for determining soil organic matter and a proposed modification of the chromic acid titration method. Soil Sci 37:29–37
Woodcock DW, Shier AD (2003) Does canopy position affect wood specific gravity in temperate forest trees? Ann Bot 91:529–5370
Worrall F, Bell MJ, Bhogal A (2010) Assessing the probability of carbon and greenhouse gas benefit from the management of peat soils. Sci Total Environ 408:66–2657
Yaqub CN (2007) Desert encroachment in Africa: extent, causes and impacts. J Arid Environ 4(1):14–20
Yugunda BS (2002) Socio-economic and cultural impacts of desert encroachment in Nigeria. J Environ Dyn 5(2):19–30

Climate Resilience in African Coastal Areas: Scaling Up Institutional Capabilities in the Niger Delta Region

2

Chika Ogbonna, Eike Albrecht, Collins Ugochukwu, Chinedum Nwajiuba, and Robert Onyeneke

Contents

C. Ogbonna (✉) · E. Albrecht
Brandenburg University of Technology Cottbus – Senftenberg,
Senftenberg, Germany
e-mail: ogbonchi@b-tu.de; albrecht@b-tu.de

C. Ugochukwu
Alberta Environmental and Parks, Government of Alberta,
Edmonton, AB, Canada
e-mail: collins.ugochukwu@gov.ab.ca

C. Nwajiuba
Alex Ekwueme Federal University Ndufu-Alike Ikwo (AE-FUNAI),
Ikwo, Nigeria
e-mail: chnwajiuba@yahoo.de

R. Onyeneke
Alex Ekwueme Federal University Ndufu-Alike Ikwo (AE-FUNAI),
Agriculture (Agricultural Economics and Extension Programme), Ikwo, Nigeria
e-mail: robertonyeneke@yahoo.com

W. Leal Filho (ed.), *Handbook of Climate Change Resilience*,
https://doi.org/10.1007/978-3-319-93336-8_9

Abstract

African coastal areas are increasingly prone to coastal challenges. The Niger Delta coastal areas are exposed to physical alterations due to natural and anthropogenic influences. In addition to current and projected extreme events such as flooding, erosion, sea-level rise, and heat waves, other conflicting factors increasing the vulnerability of the coastal Niger Delta range from the rapid shift in demography, urbanization, unsustainable land use, and inadequate implementation of relevant policies to oil spillage and gas flaring. All these issues, in addition to climate variability, increase the vulnerability and threaten the resilience of the human and natural environment. This chapter highlights the effects of climate- and weather-related extremes in the vulnerable riparian Niger Delta, based on existing facts and an empirical study, which gives insight on institutional challenges derived from the views of relevant technocrats, nongovernmental organizations, and stakeholders. Analysis of stakeholder views indicates some weaknesses and potential strengths of relevant institutions in addressing climate change issues through effective governance. Hence, scaling up institutional capabilities would enhance the resilience of communities and improve adaptive capacities. Key strengths involve employing existing institutional frameworks under relevant MDAs to climate-proof future coastal, riverbank, or lakeshores development.

Keywords

African coast · Climate resilience · Institutions · Capabilities · Strengths and weaknesses · Niger Delta region

Introduction

Several studies have anticipated the severe impacts of climate extremes in the low-lying coastal settlements of developing countries (DCs). According to the Fourth Assessment Report of the Intergovernmental Panel on Climate Change (IPCC), African coastal zones are made up of narrow, low-lying coastal belt consisting of continental shelf and coasts of 32 mainland countries. Characterized by variety of ecosystems such as barrier/lagoons, deltas, mountains, wetlands, mangroves, coral reefs, and shelf zones. The width of the ecosystems in African coastal zones differs from few hundred meters in the Red Sea area and to larger ones spanning up to 100 km, for example, the Niger and Nile Deltas (IPCC 2007a). According to the IPCC, the East African coastal zones and wetlands include the near-shore islands off the coast of Tanzania and Mozambique and the oceanic islands of Madagascar, the Seychelles, Comoros, Mauritius, and Reunion. A large proportion of these coastal zones are vulnerable to various physical transformations including overexploitation of natural resources, pollution, biodiversity loss, and climate change. The Fifth Assessment Report of the IPCC reveals that the global coastal systems are sensitive to three major drivers, which are climate change related, including sea-level rise, ocean temperature, and ocean acidity.

Among these three key drivers, impacts of sea-level rise due to global warming are of more concern to the sustainability of coastal systems (IPCC 2014a). The United Nations Framework Convention on Climate Change (UNFCCC) posits that by the end of the twenty-first century, sea-level rise could affect low-lying areas and populations located in vulnerable coastal regions of Africa. The Nigeria's Niger Delta is one of the most important deltas in West Africa, in recent times, very vulnerable to potential effects of extreme weather, climate variability, and climate change (Federal Ministry of Environment 2015). The Nigeria's Nationally Determined Contribution (NDC) reemphasizes on the vulnerability of the Niger Delta to sea-level rise. It states that accelerated sea-level rise (ASLR) of about 0.5 m would lead to the loss of 35% of the highly productive part of the region's coastline, while ASLR of 1.0 m about 75% of the Delta coastline is expected to be lost by the year 2100 (Federal Ministry of Environment 2015).

The rationale behind this research is derived from the review of facts and information from Nigeria's Nationally Determined Contribution (NDC) to the Paris Agreement, which made specific reference to increasing coastal vulnerability in the Niger Delta due to global warming-induced accelerated sea-level rise (Federal Ministry of Environment 2015). Similarly, the findings from the various Fifth Assessment Report (AR5) of Intergovernmental Panel on Climate Change and Nigeria's NDC emphasized on the adverse effects of current and future climate impacts on coastal zones. In particular, the IPCC proclaims that up to 70% of the coastlines worldwide are estimated for a potential sea-level change within ±20% of the global mean (IPCC 2013, 2014b). In this regard, low-lying coastal areas especially those in coastal countries of West Africa are gradually witnessing flooding, submergence, and erosion and sea-level rise due to climate change and intense storms (IPCC 2014a, b). To this end, the significance of this chapter lies in giving perspectives on how existing and potential coastal governance structures could be strengthened to contribute to resilience mechanisms in the Niger Delta coastal zones, wetlands, riverbank, and lakeshores. In addition, it establishes how institutional weaknesses and challenges can be overcome to address risks brought about by current and expected climate-related events in the area. Furthermore, the study also extends to the need to strengthen institutional capabilities essential to apply the strengths to take advantages of potential benefits and how to overcome the weaknesses that could reduce the chances of maximizing essential resilience and adaptation opportunities.

Literature Review

Overview of Vulnerability of African Coastal Zones

African coastal zones cut across the continent extending from West Africa and Central Africa to East Africa and South Africa, Tanzania, Egypt, Mauritania, and Namibia. Climate impacts could "exacerbate existing physical, ecological/

biological, and socioeconomic stressors on the African coastal zones" (IPCC 2007a, 2014a). While the coastal zones in Africa were affected by climate change, the effects are likely to be incongruent as coastal countries of West Africa (West Africa's Atlantic Coast) could experience more negative impacts than those in East Africa (IPCC 2007a, 2014a). The West African coastal zones are vulnerable mainly due to potential relative rise in sea level that could inundate and erode low-lying areas or increase coastal flooding and coastal erosion exacerbated by storm surges and intense rainstorms (IPCC 2007a, 2014a). In West Africa, the coastal zone hosts various economic activities that support the national and regional economic growth.

About 31% (105 million) of West African's population live in coastal areas across sub-Saharan Africa (World Bank 2015, 2016). This population generates 56% of the region's gross domestic product (GDP). The IPCC 2014 report pointed out that large urban centers located on mega-deltas (e.g., Alexandria in Egypt in the Nile Delta and Benin City and Port Harcourt and Aba cities in the Niger Delta region) would experience the impact of climate change. These urban centers are exposed to the challenges of urbanization through migration (IPCC 2014a). The West African coastal zones also accommodate important cities, ports, fisheries landings, and agro-industries. Moreover, estuaries and lagoons account for the huge agricultural production of the region including offshore petroleum exploration and production (World Bank 2015). According to the World Bank Report, the West African coasts account for 56% of West Africa's gross domestic products. The World Bank Report also posits that "More than 1.6 million tons of fish are legally captured in West African waters each year, with an estimated wholesale value of USD 2.5 billion" (World Bank 2015).

In West Africa, a significant number of people who live and sustain their livelihoods in the coastal areas cut across coastal cities of Benin, Cote d'Ivoire, Ghana, Mauritania, Nigeria, Senegal, Sao Tome and Principe, Sierra Leone, and Togo. The changing climatic conditions, recurrent extreme weather, and their potential effects on sea-level rise amplify the overflow of water bodies and the probability of flooding in African coastal areas, as extreme flood events are expected to become more in many coastal areas accommodating vast population (AGRA 2014; Wetzel et al. 2012). According to a World Bank Report, about 500,000 people are affected by coastal floods in West Africa annually (World Bank 2016). A key challenge to resilience in Africa is its rapid urban growth. Urban growth can be seen as both an opportunity and a challenge (UNECA 2017). Urbanization in Africa is estimated to be about 40%. By 2035, Africa's urban population could reach 49% presenting considerable demands on infrastructural services (UNECA 2017). Africa's urban population is expected to triple in 40 years, from 395 million in 2010 to 1.339 billion in 2050, corresponding to 21% of the world's projected urban population (Güneralp et al. 2017; UN 2014).

Projections indicate a whopping loss of 18,000 km^2 of West African coastline from a 1 m rise in sea level (World Bank 2016). Even though West African coastal areas are said to be more vulnerable, an impact assessment on coastal flooding due to sea-level rise in East Africa shows that by 2030, about 10,000–86,000 people would be affected, with an estimated economic cost ranging between USD 7 million and USD 58 million (IPCC 2014b). In this regard, the potential sea-level rise has affected several lagoons and mangrove forests in both East and West Africa and affects urban centers and ports, such as Cape Town, Maputo, and Dar es Salaam (UNFCCC 2007).

The challenges of urbanization may intensify as people migrate from low-lying vulnerable coastal areas to safer areas and from vulnerable rural coastal communities to urban centers to avoid the vulnerability created by extreme weather and coastal climate-related impacts (IPCC 2014a; Ogbonna 2014; Ogbonna and Albrecht 2015). For instance, the current rate of population growth in Nigeria and urbanization is a recipe to increase vulnerability, through inappropriate land use and waste disposal and land fragmentation that has become part of the urban fabric. This situation reduces the resilience to a range of climate risks. Some accounts of recent events are mind-blowing. For instance, the International Displacement Monitoring Centre (IDMC) and Norwegian Refugee Council (NRC) Report, published in 2013, clearly expressed the magnitude of damage brought about by floods which displaced several millions of people in some coastal, lakeshore, and riparian areas of Nigeria and Niger Delta in 2012 (IDMC 2013).

Some urban and rural settlements in Africa are already at risk of climate change-related extreme events from drought, heat waves, floods, inundation, and other hazards that may lead to changes in economic values (UN-Habitat 2011). The Nigeria's NDC posits that low-lying coastal settlements of the Niger Delta region of Nigeria could be vulnerable. In this regard, the IPCC (2014a) asserts that risk on coastal systems is attributable to integrating drivers' associated hazards, exposure, and vulnerability as indicated in Fig. 1.

The adaptation deficit for African countries with regard to coastal flooding is likely to reach over USD 300 billion (Hinkel 2011; IPCC 2014a). Yet, in several African countries, coastal policies have not considered longer-term climate change (Bunce et al. 2010; IPCC 2014a). Apparently, this is a huge threat to climate resilience in the continent. In this regard, the IPCC Fifth Assessment Report suggests that proactive planning by coastal communities in consideration of climate change moderates the risk of harm from such impacts (IPCC 2014a). According to the IPCC, proactive planning minimizes the need for a reactive response to the effects of extreme events. In other words, effective and proactive planning is expected to increase the resilience of the natural and human environment. Moreover, reactive planning could be more expensive and less efficient (IPCC 2014a). However, adaptation strategies can be used in reducing the hazards, exposure, and vulnerability that erupts because of the climatic changes (IPCC 2014a).

According to IPCC (2007a), responses for carrying out coastal adaptation would include protection, accommodation, and retreat (Fig. 2). These responses when conscientiously applied would make coastal zones, lakeshores, wetlands, and riverine areas more resilient. The big challenge is now shouldered on institutional capabilities and intuitive to apply such context-specific measures.

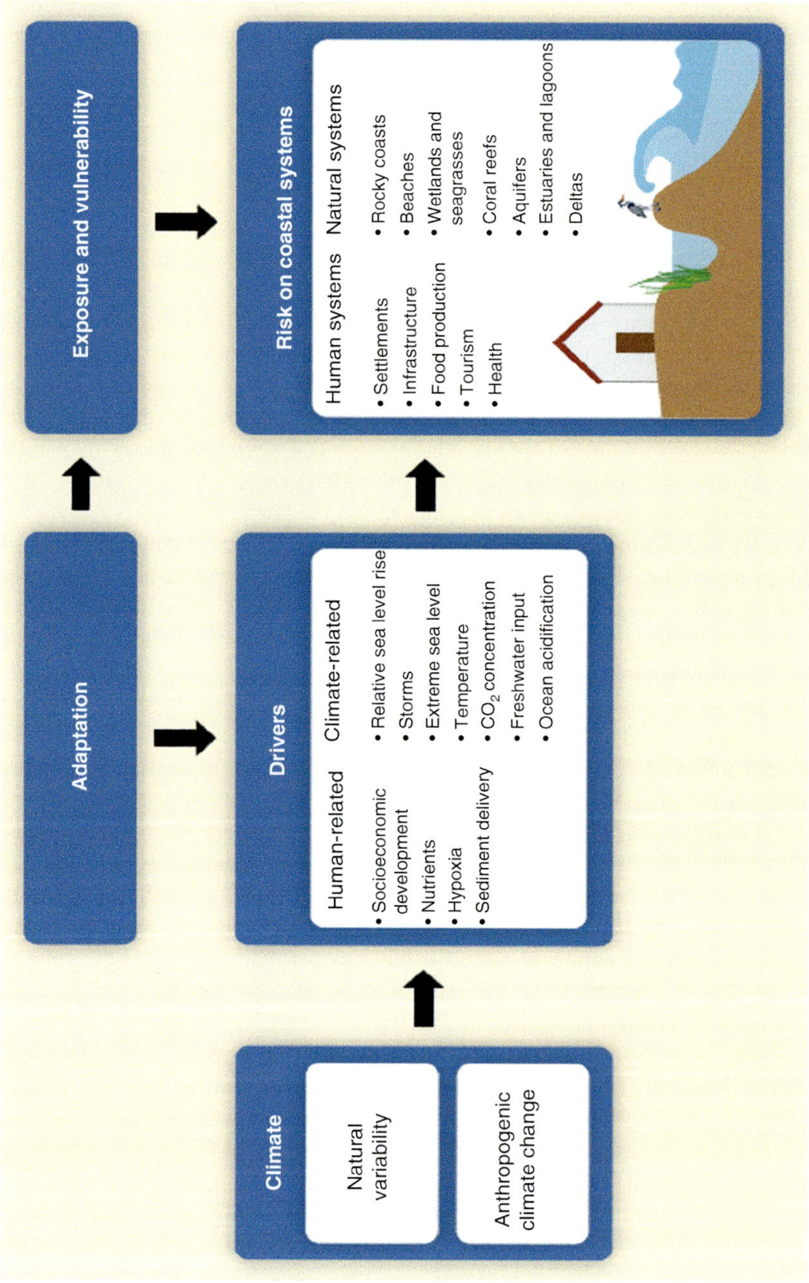

Fig. 1 Climate risk on coastal system. (Source: IPCC 2014a)

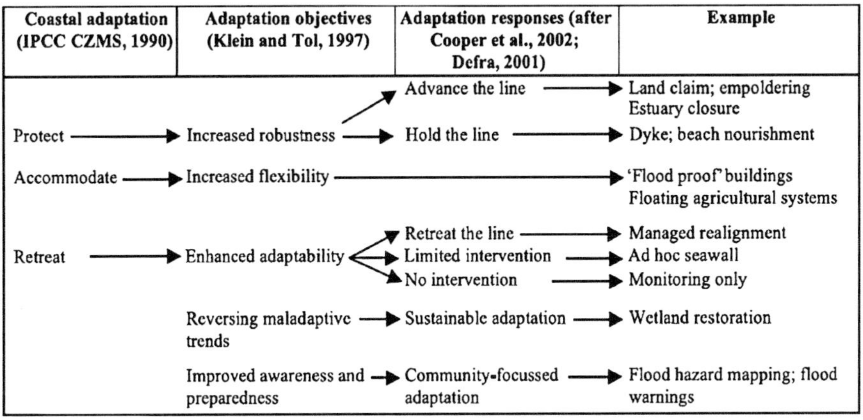

Fig. 2 Planned coastal adaptation actions. (Source: IPCC 2007b)

Climate Risks and Vulnerability in Niger Delta

The coastal wetland covers about 3% of Nigeria's land surface (Delta State Government 2013). The Niger Delta accommodates a major part of Nigeria's coastal and marine environment covering about 70,000 km^2 making it one of the largest wetlands in the world (FRN 2014). Nigeria's Second Communication to the UNFCCC acknowledges the fact that the Niger Delta ranked highest regarding vulnerability when compared with other vulnerable parts of the country (FRN 2014). The vulnerability of the Niger Delta is aggravated by the challenges of climate variability, coastal flooding, and erosion due to rising waters as well as environmental problems and social impacts of oil and natural gas exploration and exploitation. The vulnerability of Niger Delta communities to climate variability and change exacerbates high poverty levels of rural people (Ogbonna et al. 2017). The expected effects and impacts on wetlands include siltation, which leads to the reduced capacity of lakes and rivers that sustain local communities and ecosystems (FGN 2015a).

Flooding has become prevalent in the coastal urban and rural areas of the region due to high precipitation and runoffs from rivers (Agumagu and Todd 2015). In fact, a rise in sea level of 0.462 m (above sea level) was recorded around the 1960s and 1970s (Boateng 2010). Sea-level rise is occurring and has reached several communities located close to shorelines; consequently, many coastal fishing communities have raised the foundation of their houses to keep them above unwanted water (Musa et al. 2016). The problem is worsened by the fact that the Niger Delta is the hub of Nigeria's natural gas production as virtually all the oil and gas resources and mining infrastructures are mostly located on the coastal environment of the Niger Delta (FRN 2014; FGN 2010). The vulnerability of the region is also intensified due to the increasing trend in population growth and urbanization. With a rapid population growth of 2.9% per year in Nigeria with nearly 50% of the population, living in the urban areas portends increases in demand for public infrastructural services (BNRCC 2011; Ogbonna and Albrecht 2015).

Role of Relevant Institutions in Contribution to Climate Resilience

Initiating climate resilience strategies in a region or community entails engaging several institutions; therefore, cooperation and collaboration of various institutions are needed to take actions that focus on reducing the vulnerabilities induced by climate change. Pradhan et al. (2012) pointed out that a strong cooperation between different institutions functioning at multiple planning scales might perhaps result in better resilience and adaptation outcomes. In this regard, McGray and Sokona (2012) assert that a well-set institutional framework in response to climate change is a prerequisite in facilitating collaboration among relevant stakeholders who could offer robust decisions on a specific program across relevant sectors. In the face of climate change, institutions, especially affected ones, may need to adjust organizational rules and norms and adapt substantially to the adverse impacts of climate change.

Coastal Management in Nigeria

The federal government bestows the overall obligation to protect the coastal areas of Nigeria under section 20 of the Nigeria's 1999 Constitution which provides that the "State shall protect and improve the environment and safeguard the water, air and land, forest and wildlife of Nigeria." In this regard, coastal management falls on various agencies and all levels of governments in Nigeria. The Erosion, Flood and Coastal Zone Management of the Federal Ministry of Environment is responsible for controlling flood and erosion along the 853 km of the Nigeria's coastline. The National Environmental Standards and Regulations Enforcement Agency (NESREA) is tasked with the implementation and enforcement of relevant regulations for the coastal zone protection as contained in the National Environmental (Coastal and Marine Area Protection) Regulations 2011. According to Mwalimu (2009), the Nigerian Institute for Oceanography and Marine Research supports protection and monitoring of the coastal zone by providing relevant data that shows widespread erosion and flooding of barrier island as well as in the Niger Delta.

Mmom and Chukwu-Okeah (2011) averred that Nigerian policies on coastal development are not well linked to the nation's policy on the environment, as this does not provide a requisite attention to coastal zone management. In the same vein, the Economic Recovery and Growth Plan (ERGP) listed poor coastal management and weak environmental governance as major problems of environmental sustainability in Nigeria (MBNP 2017). The challenges of coastal management in Nigeria seem to revolve on duplication of duties, unclear responsibilities, and mandates. For instance, one of the challenges of applying the Land Use Decree of No. 6 of 1978 in reducing environmental degradation and coastal protection stems from duplication of powers and control rights, given to different tiers of government in Nigeria, which usually lead to weak enforcement and conflicting interest in control and management of land resources. In retrospect, the litigation concerning the Attorney General of Lagos State versus Attorney General of the Federation in 2003, where Lagos State challenged the physical planning powers of the Nigeria federal government over land in the territory of Lagos State, is such an example of conflicting interests and institutional and administrative challenges. Such issues expressed in litigation could also be detrimental to reducing coastal vulnerabilities due to climate risks; hence, adaptation planning is stronger at the federal level due to financial prowess and institution capabilities, whereas relevant state and local institutions that are supposed to be closer to the grassroots lack appropriate capacities. Implementation of integrated coastal zone management in the Niger Delta has become urgent to improve the livelihoods of coastal communities while building the resilience of coastal communities.

Climate Change Resilience Responses and Governance in Nigeria

Nigeria became a signatory to the United Nations Framework Convention in 1992 as a Non-Annex 1 party and ratified the Kyoto Protocol in 1994 (FGN 2012). Nigeria is committed to fulfilling its national pledge under the Paris Agreement to reduce greenhouse gas emissions and adapt to climate change. The Nigerian government through its Ministry of Environment submitted its initial National Communication on November 17, 2000 and its second National Communication on February 27, 2014. The Federal Ministry of Environment plays a pivotal role in facilitating the provisions of the Convention and the Protocol in the country. In 2006, the Special Climate Change Unit (SCCU) was established under the Ministry of Environment for implementing the Convention and the Protocol. So far, the Ministry of Environment has focused on several climate mitigation options such as Clean Development Mechanisms (CDM) and reforestation strategies (FGN 2012). In the light of implementation purpose, recently the Special Climate Change Unit of the Federal Ministry of Environment was made a full-fledged department. Furthermore, a committee on Environment and Climate Change has also been established in the Nigerian House of Representatives. More so, some states, local government areas, and nongovernmental organizations in Nigeria are making efforts to enhance resilience strategies by introducing awareness campaign, for example, even though challenges remain.

Policy and Programs

In 2011, the National Adaptation Strategy and Plan of Action on Climate Change for Nigeria (NASPA-CCN) was published. The broad objectives of NASPA-CCN are to minimize risks, improve local and national adaptive capacity and resilience, leverage new opportunities, and facilitate collaboration with the global community, all to build resilience and to minimize Nigeria's vulnerability to the adverse impacts of climate change. On the other hand, the National Climate Change Policy Response and Strategy (NCCPRS) 2013 identified climate change as one of the major threats to sustainable economic development and human well-being. The policy documents also listed the coastal areas as the most vulnerable part of the country as it reiterates the prediction that a 0.5–1 m climate change accelerated sea-level rise (ASLR) would worsen the existing environmental conditions of the coastal areas in Nigeria, particularly in the Niger Delta (FME-DCC 2013). This can have a pernicious impact on the socioeconomic sector and the economy of the Nigerian nation as oil mining and natural gas investments in the coastal Niger Delta amount to over USD 13 billion and such investments are already under threat due to the predicted sea-level rise and recurrent flooding (FME-DCC 2013; TACC 2013). The policy document seems to be comprehensive and inclusive therefore augmenting the existing national initiatives to adapt to and mitigate climate change, thereby initiating measures relevant for a resilient society.

Regulatory Response

The Nigeria's Legislative Council initiated the National Climate Change Commission Bill that has been passed by both Houses of National Assembly. The general goal of the National Climate Change Commission Bill is to establish the National Climate Change Commission as a statutory body, charged with the responsibility to regulate and coordinate policies and action plans on climate change and other environment-related matters. The Bill is **still for too long** pending for presidential assent. Nigeria lacks a comprehensive national climate change law to support the National Adaptation Strategy and Plan of Action on Climate Change for Nigeria (NASPA-CCN), but there are a number of environmental legislations and policies, which could be relevant to enhance resilience of coastal and natural environment in the region. However, it is assumed that the new Climate Change and Green House Gas Emission Reduction Bill of 2016 being considered at the National Assembly would support the implementation of the National Adaptation Strategy. The Climate Change Bill of 2016 is an act to provide measures to address climate change with a view to assist in achieving a sustainable future for the country. The Climate Change Bill has passed its second reading at the National Assembly (SFRN 2016). It is expected that such Climate Change Bill will support the achievement of the relevant policies and strategies by facilitating the early development of policies and programs with a focus on addressing climate issues and building resilience of the entire country. The National Assembly adopted the National Policy on Climate Change in 2012; the country also joined other parties in submitting the Nigeria's Nationally Determined Contribution in 2015 in view of achieving the goal of the Paris Agreement. In addition to strengthen its commitment to pursuing climate resilience goals, Nigeria signed the Paris Agreement on September 22, 2016, and ratified it on May 16, 2017, which entered into force on June 15, 2017.

Case Study and Method

Niger Delta pertains to Delta of the Niger River situated on the Gulf of Guinea on the Atlantic Ocean in Nigeria. The Niger Delta is centered right on the Atlantic Coast, where Niger River splits into numerous tributaries.

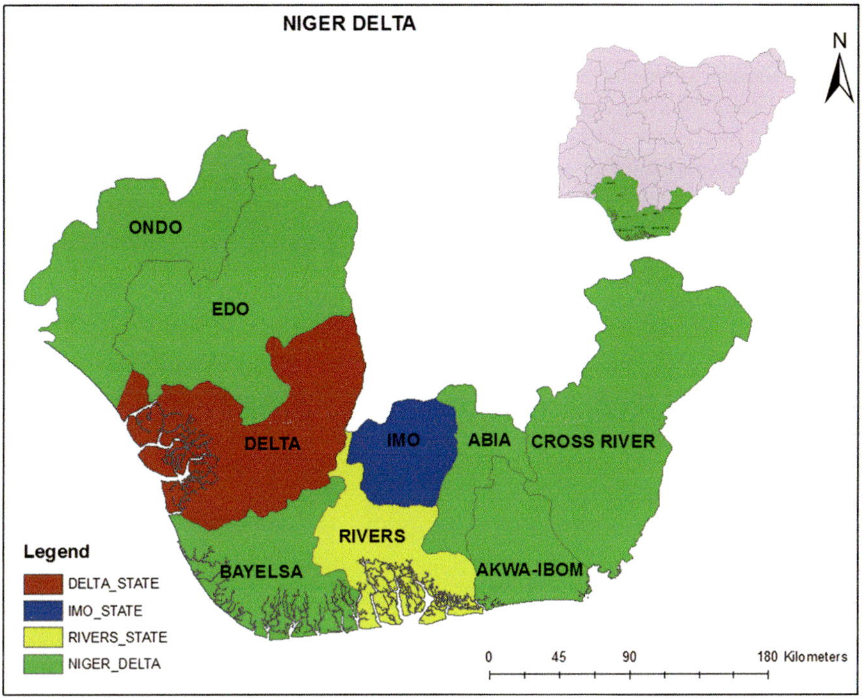

Fig. 3 Map of Nigeria

The area is bordered to the south by the Atlantic Ocean. Figure 1 depicts a map of Nigeria indicating the study area. Nigeria's population is currently the seventh largest in the world; it is estimated to be about 191 million in 2017 (UNECA 2017). Average growth rate of urban population by settlement size from 2010 to 2020 indicates that Port Harcourt is growing at the rate of 6.83% (Bloch et al. 2015). Estimates show that the population of the Niger Delta region could reach about 40 million by 2020 (FRN 2006). The region accommodates about 25% of the country's population and various economic opportunities and resources (FRN 2014). The Niger Delta region is a multicultural region having more than 140 ethnic groups comprising the Annang, Ibibio, Efik, Ijaw, and Igbo people, speaking about 250 dialects (Oyewo 2016). The region is one of the world's largest wetlands with significant biological diversity (UNDP/GEF 2013). The region is made up of nine states (Abia, Akwa Ibom, Bayelsa, Cross River, Delta, Edo, Imo, Ondo, Rivers States); for the purpose of this research, four states where extreme weather and climate impacts have become visible were purposively selected including Bayelsa, Delta, Imo, and Rivers State. Climate resilience approach is therefore imperative to reducing coastal and estuarine decadence in the selected states (Fig. 3).

Description of the Method

The research is based on extensive literature review, legal analyses, and primary data obtained from interviewing relevant stakeholders. Responses from the interviews were coded, transcribed, and grouped into qualitative themes and analyzed descriptively. The chapter stems from the academic research of the first author in the Niger Delta region of Nigeria, which was conducted in 2014 with a follow-up study in 2016. One of the limitations of this chapter is not being able to assemble stakeholders for group discussions. This is because of time constraints on the part of the stakeholders and convenience of the researcher and the interviewees. The interviews targeted stakeholders who are more or less directly involved in environmental management in their various ministries, departments, and agencies (MDAs).

Semi-structured Interviews

This chapter adopted a case study approach. In this regard, semi-structured interviews were crucial data collection method for this study (Barnett et al. 2013). Climate resilience and adaptation policy research is still a relatively new concept in the areas visited; thus, a semi-structured interview was preferred. Interviewees exist as anonymous in the data set for ethical reasons and were represented thus (R#1), meaning the first respondent on the interview list. The study also relied on the analysis of various secondary materials and literature including available climate change policies existing in various states. A variety of sampling techniques were employed to elicit the stakeholders' opinion. For instance, **snowball sampling** was very helpful in connecting different sectors and stakeholders that could provide expertise information relevant to the research in the respective study areas. The interaction with the stakeholders was helpful to elucidate and understand the strengths and weaknesses of institutions in the study areas. The data elicited from relevant institutions were analyzed qualitatively to determine the strengths and weaknesses of relevant institutions in carrying out resilience and adaptation strategies. As climate change resilience and adaptation in the study area involve strategic initiatives, it may be essential to identify institutional strengths and weaknesses.

Characteristics of Institutional Respondents

The stakeholder interviewees included government officials from several government institutions, NGOs, university professors, and environmental consultants. Fifty-two experts who could give relevant information on climate change resilience and adaptation participated in the research. The breakdown of the participants who took part in the interviews is presented in Table 1.

Table 1 Institutional respondents

Institutions	No. of respondents (stakeholders) ($n = 52$)
Dept. water resource, utility, and rural development	2
Environmental nongovernmental organizations	6
Federal and state ministry of environment	12
Federal and state ministry of urban development	8
Ministry of agriculture and forestry	4
Ministry of health	2
National environmental standards and regulation agency	2
NEMA and SEMA	5
Universities and academic institutions	11

NEMA and SEMA means National and States' Emergency Management Agency
All the respondents have more than 10 years of professional experience in environmental protection

Research Findings and Discussions

Institutional Weaknesses and Potential Strengths in Building Climate Resilience in the Niger Delta

This section stems from the SWOT analysis applied to the information gathered during interviews with various staff of institutions that participated in the research. Therefore, this section narrates the weakness and strengths of relevant institutions in building resilience in the selected Niger Delta States. It sought to identify ways to incorporate climate resilience and adaptation into relevant institutional frameworks. Apparently, reduction of climate risks in the selected states will be informed through internal forces; these forces will be better weighed to implement robust adaptation strategies. During the course of this research, the SWOT showcased some elements of strengths and weaknesses of relevant institutions, which are internal factors associated with the institutions' "efforts to reduce potential impacts of climatic changes while building a resilient society" (Yohe and Tol 2002; Hill et al. 2010).

Strengths

One of the key strengths of relevant institutions in the selected states is the existing institutional environmental frameworks under the respective ministries of environment, agencies, and physical planning, which can be adapted to build resilience and carry out adaptation in the vulnerable areas. For instance, Strategic Environmental Assessment (SEA) can be easily applied using existing Environmental Impact Assessment (EIA) institutional structures in climate proofing future implementation of programs and projects. Therefore, the introduction of SEA as an assessment tool for climate change adaptation and other related environmental planning issues could be possible. Some of the respondents who were aware of SEA procedures and good practice were of the opinion that such tool would help in reducing climate risks. In this regard, R#8 stated that:

> Application of SEA produces credible community-based adaptation projects. The goal of community-based adaptation projects is to build up the capacity of communities and increase their resilience to cope with less predictable changes as a result of climate change effects. In other words, it is an approach, which ensures that environmental disharmony does not arise in the course of new program development. On the other hand, according to R#47: "most of the SEA are theoretical, that is, they exist only on paper or are discussed in workshops. They need to be applied and implemented so that they tackle adaptation and poverty strategies in an effort to achieve sustainable development goals."

In addition, strengths of relevant institutions and ministries lie within applying existing knowledge to abate environmental deterrence in the area and build capacities of vulnerable communities. Most technocrats and staff working within relevant ministries such as ministries of environment, urban planning and agriculture, water, and other relevant agencies are eager to broaden their knowledge and offer insights on climate change management. In this case, capacity building and training of administrative staff would be cost-effective. As R#14 pointed out:

> There is need to create more awareness at the community level, organize training programs, share information, organize meetings, show graphic images of what is happening and sensitize people on the exposure to climate change.

This could allow technocrats to apply their skills to reduce vulnerability to impacts of climate change. Availability of climate change desks in Bayelsa and Delta States constitutes institutional strengths within the state ministries of environment, respectively. Additionally, provision of climate change policies to be implemented by the climate change department under the ministries of environment of the two states can be argued to constitute institutional strengths in the face of implementation challenges. Table 2 provides the summary of strengths and weakness factors surrounding resilience and adaptation practices through institutional arrangements in the study areas. There are various resilience strategies that are being put forward in the study area. For instance, R#13 from Delta State highlights such initiative and mandates within the Climate Change Unit of Delta State Ministry of Environment to include:

> Research on vulnerability assessment; community consultation; free, prior and informed consent (FPIC) approach; meeting with vulnerable communities and collaborating with nongovernmental organizations and community-based organizations to see how to address and execute the resilience and adaptation options in their communities.

Weaknesses

Key weaknesses of MDAs in the study area range from the inadequate capacity to reduce climate risks to the lack of reliable information and inventories relating to

Table 2 Summary of strength and weakness factors facing institutional resilience and adaptation

Strengths	Weaknesses
Existing environmental tools (e.g., EIA)	Lack of reliable climate information inventories
Existing institutional knowledge and capacities	Inadequate and accurate scientific baseline data
Climate change policies already in place in Bayelsa and Delta States	Inadequate legal frameworks and political will to execute existing climate change policies in Delta and Bayelsa
Availability of climate change units in Bayelsa and Delta States	Inadequate collaboration between relevant institutions and MDAs
Training of few experts on climate change management in Bayelsa and Delta States	Insufficient institutional knowledge/capacities on climate management in all the states particularly in Imo State

climate information and accurate scientific baseline data, which are prerequisite for coastal resilience and adaptation planning in the area. Furthermore, lack of legal frameworks to execute existing climate change policies amounts to institutional weakness. According to R#25:

> there is no comprehensive law covering the environment, but a law on public sanitation exists. Rather, we apply laws and national regulations made by the federal legislature on environmental issues specific to Bayelsa State.

In this regard, it should be noted that relevant national environmental laws and regulations are mostly silent on addressing and reducing climate risk. In addition, inadequate collaboration among relevant environmental departments and the meteorological department may hamper the development of robust adaptation strategies as well as discourage resilience practices. Notably, collaboration can enhance awareness of stakeholders on the potential impact of the changing climate, thereby leading to an informed decision on the ways to incorporate resilience actions and initiate appropriate adaptation strategies. Another key weakness of relevant institutions relates to ineffective policies, plans, programs, and incentives to promote resilience strategies.

Other Challenges

It has been established that inadequate implementation of relevant policies in vulnerable coastal communities is one of the impending challenges in Nigeria. For instance, due to weak enforcement of Nigeria's Environmental Impact Assessment Act of 1992. To strengthen the process, Ijaiya (2015) suggests the amendment of the EIA Act to provide for public participation at every stage of the EIA process. In this regard, according to Agbonifo (2016), regulatory agencies have failed to protect communities against the impacts of environmental degradation and other consequences of oil and gas exploration activities in the Niger Delta. Apparently, such failures of regulatory agencies reduce resilience and increase the vulnerability of the ecosystems and communities.

Analyses of Regulatory Frameworks

Within the four case study states selected for this research, Delta and Bayelsa States are the two states that have own climate policies. Both states have also established climate change departments under their respective ministries of environment. The Bayelsa State Climate Policy advocates the establishment of legal frameworks, while the Delta State Climate Policy suggests that the policy be revised every 3 years, taking into account emerging issues and trends on climate change at the local, subregional, regional, national, and global levels including the ongoing international climate policy debates (Bayelsa State Government 2011; Delta State Government 2013). However, while Rivers State has some relevant programs that target climate change, Imo State is yet to design such policy. Overall, at the time of the research, the four states do not have any comprehensive legal and regulatory framework on climate change.

Environmental regulations are imperative for future environmental development in the country (Okorodudu-Fubara 2012). At the moment, Nigeria has no efficient legislative oversight framework that can enhance Nigeria's climate resilience even though the impacts of climate change on the environmental sectors and coastal resources are increasingly acknowledged (ICLG and Makinde 2016; This Day 2 August 2016). As stated earlier, the Climate Change Bill of 2016 is expected to provide measures to build a resilient future for the country. Although the relevant NESREA Regulations are silent to climate change, they can be adapted for planning resilience actions and for reducing the impact of climate change and extreme events. Some of such existing environmental regulations are those that target coastal and marine area protection, soil erosion and flood control, wetlands, riverbank, and lakeshores protection.

- **National Environmental (Coastal and Marine Area Protection) Regulations 2011**
  The Coastal and Marine Area Regulations contains the most important aspiration relevant to coastal resilience and adaptation as part of the general objective, precisely Regulation 2(b), is comprised of a regulatory provision for the application of preventive, precautionary, and anticipatory approaches relevant to manage the coastal and marine environment in Nigeria. The challenges of implementing regulations therein stem from the business-as-usual approach of the enforcement agencies. Therefore, relevant institutions including NESREA and State Environmental Agencies and local planning authorities must propel adequate enforcement of the regulations. For instance, local government authorities must be capable and should possess the know-how to deal with coastal challenges and natural hazard risks by applying modern planning approaches. There is a need for a compulsory vulnerability assessment and precautionary principle in making planning applications and development decisions in the coastal and riparian parts of the Niger Delta. For instance, see the case of Gippsland Coastal Board versus South Gippsland Shire Council where the Gippsland Coastal Board made a legal decision using climate change as a reason for refusing a coastal development.

- **National Environmental (Soil Erosion and Flood Control) Regulations 2011**
  In view of the urgent need to resolve potential risks through anticipatory adaptation measures, the National Environmental (Soil Erosion and Flood Control) Regulations of 2011 is of relevance for climate change resilience and adaptation in Nigeria and the Niger Delta. This is so given that the regulation laid down relevant flood prevention activities that can be considered to reduce potential flood risks. The general objective of the flood regulation is to protect human life and the environment and minimize losses due to flood and erosion and their effects on vulnerable areas by controlling accelerated soil erosion, flood, and sedimentation deposition in water bodies and watercourses in order to prevent pollution of water resources in Nigeria (Regulation 2(a–c)). However, the regulation dwells more on approval and requirements of proponents for earth-disturbing activities and infrastructure development but falls short of mentioning how community resilience could be supported and how recovery plans can be executed after storms and extreme flood events and disasters such as the 2012 flood. Often until now, inhabitants of vulnerable communities still apply inadequate makeshift strategies to contain or ward off floodwater from adjacent riverbanks.
- **National Environmental (Wetlands, River Bank and Lakeshores) Regulations 2009**
  The Niger Delta currently has three sites listed as Ramsar Wetlands of International Importance. The sites include the Apoi Creek Forest in Bayelsa State with an area of 29,213 hectares; Oguta Lake in Imo State, which covers about 572 hectares; and the Upper Orashi Forest in Rivers State with a size of 25,165 hectares (FRN 2014). The National Environmental (Wetlands, River Bank and Lakeshores) Regulations of 2009 is a key related regulation that should be enforced to promote resilience of such wetlands to potential climate- and weather-related events. The core objectives of the regulations include to provide for the conservation and wise use of wetlands and their resources, ensure the sustainable use of wetlands for ecological and tourism purposes, ensure that wetlands are protected as habitat for species of fauna and flora, and control pollution. Although the Regulation provides a succinct view on its application, part II of the regulation specifies the need for the government to protect the people, riverbanks, and lakeshores for the benefit of the inhabitants in the surrounding wetland areas. This institutional task is yet to be seen in the Niger Delta, largely because of inadequate enforcement of the regulations. However, interviews with stakeholders pointed to the challenges of over-centralization of powers and inadequate resource availability to accelerate the enforcement of such provisions.

Concluding Remarks

Observations on recent extreme weather and changing climatic conditions and their effects on riparian and coastal areas call for resilience actions to avoid intensification and damages that emanate from river flooding, coastal erosion, as well as sea-level rise so as to reinforce African coastal areas. Considering the outcome of the findings, scaling up institutional capabilities and enforcement and strengthening of planning and regulatory frameworks in the Niger Delta are pivotal for climate resilience. To build climate resilience in and around the coasts, riverbanks, and lakeshores of the Niger Delta, institutional frameworks and capacities must be strengthened to deliver positive environmental outcomes.

To overcome the identified institutional weaknesses and strengthen capacities of staff, adequate mandates are needed to start the initial process of mainstreaming resilience and adaptation into existing coastal zone, wetlands, riverbanks, and lakeshores management policies, programs, and regulations. MDA staff should incorporate resilience strategies in their future development and land-use planning activities. For example, taking soft actions like planning and climate proofing new infrastructure, planting wetland trees/shrubs along the creeks and riverbeds, as well as maintenance of wetlands that serve as storage reservoirs for floodwaters would enhance the resilience of natural and human environment. Nevertheless, removing the hurdles surrounding institutional effectiveness and building stronger institutional strengths would be relevant for enforcement of international conventions and agreements such as the Paris Agreement. Article 6 of the UNFCCC is relevant here and can be achieved through organizing professional seminars and workshops that involve education and training of administrative staff of relevant environmental MDAs, as this would help strengthen institutional capabilities and capacities required for climate resilience.

The relevant planning and regulatory authorities could encourage resilience actions through accelerating the enforcement and compliance of environmental standards especially those that relate to coastal and marine area protection, soil erosion and flood control, and wetlands, riverbanks, and lakeshores regulations. Thus, functional governance systems are imperative to drive climate change management in the study area, especially now that the Nigerian government has ratified the Paris Agreement. Maximizing existing institutional strengths and reducing identified institutional weaknesses would help in addressing climate risks at the same time support efforts in building a resilient future with the aim of supporting sustainable development pathways in the Niger Delta region.

References

Agbonifo PE (2016) Risk management and regulatory failure in the oil and gas industry in Nigeria: reflections on the impact of environmental degradation in the Niger Delta region. J Sustain Dev 9(4):1–10

AGRA (Alliance for a Green Revolution in Africa) (2014) Africa agriculture status report: climate change and smallholder agriculture in sub-Saharan Africa. Nairobi 2:1–216

Agumagu O, Todd M (2015) Modeling the climatic variability in the Niger Delta region: influence of climate change on hydrology. J Earth Sci Clim Change 6(284):1–7

Attorney General of Lagos State vs. Attorney General of the Federation and Others [2003] 12 N. W. L.R. (Part 833)

Barnett J, Waters E, Pendergast S, Puleston A (2013) Barriers to adaptation to sea level rise: the legal, institutional and cultural barriers to adaptation to sea-level rise in Australia. National Climate Change Adaptation Research Facility, Gold Coast

Bayelsa State Government (2011) Bayelsa state climate change policy. Bayelsa State Government Ministry of Environment, Yenegoa

Bloch R, Fox S, Monroy J, Ojo A (2015) Urbanisation and urban expansion in Nigeria. Urbanisation Research Nigeria (URN). Research report. ICF International. Creative Commons Attribution-Non-Commercial-ShareAlike CC By-NC-SA, London

BNRCC (Building Nigeria's Response to Climate Change) (2011) National adaptation strategy and plan of action on climate change for Nigeria (NASPA-CNN). Federal Ministry of Environment – Special Climate Change Unit. Building Nigeria's Response to Climate Change Project, Ibadan

Boateng I (2010) Spatial planning in coastal regions: facing the impact of climate change. In: The International Federation of Surveyors Report. Oriveden, Kirjapaino, pp 13–14

Bunce M, Brown K, Rosendo S (2010) Policy misfits, climate change and cross-scale vulnerability in coastal Africa: how development projects undermine resilience. Environ Sci Policy 13(6):485–497

Constitution of the Federal Republic of Nigeria 1999. The Constitution of the Federal Republic of Nigeria (Promulgation) 1999, No. 24, Section 20

Delta State Government (2013) Delta state climate change policy. Delta State Ministry of Environment (Climate Change Department), Delta State, Asaba

Federal Ministry of Environment (2015) Nigeria's intended nationally determined contribution. Conference of parties to the United Nations Framework Convention on Climate Change (COP-UNFCCC) in preparation for the adoption of climate change agreement at the Paris conference on climate change. Federal Ministry of Environment, Abuja

FGN (Federal Government of Nigeria) (2010) Nigeria vision 20:2020. Economic transformation blueprint for Nigeria, Abuja. http://a4ai.org/wp-content/uploads/2014/03/Nigeria-Vision-2020_0.pdf. Accessed 28 June 2018

FGN (Federal Government of Nigeria) (2012) Climate change and sustainable development in Nigeria. From Durban through Rio to Doha: National Report for COP 18/CMP 8, Federal Ministry of Environment, Abuja

FGN (Federal Government of Nigeria) (2015a) Nigeria's Intended Nationally Determined Contribution. Conference of Parties to the United Nations Framework Convention on Climate Change (COP-UNFCCC) in Preparation for the Adoption of Climate Change Agreement at the Paris Conference on Climate Change. Federal Ministry of Environment, Abuja, 23 pp

FME-DCC (Federal Ministry of Environment – Department of Climate Change) (2013) National Policy on climate change. Federal Ministry of Environment, Abuja

FRN (Federal Republic of Nigeria) (2006) Niger delta regional development master plan. Facilitated by the Niger Delta Development Commission (NDDC), in partnership with the state governments, LGAs, oil companies, civil society organizations and communities of the Niger Delta, Nigeria: Niger Delta Development Commission, Physical Description, Port Harcourt

FRN (Federal Republic of Nigeria) (2014) Nigeria's second communication under the United Nations Framework Convention on Climate Change. http://unfccc.int/resource/docs/natc/nganc2.pdf. Accessed 28 June 2018

Gippsland Coastal Board vs. South Gippsland Shire Council [2008] VCAT 1545 [Victorian Civil Administrative Tribunal [2008] VCAT 1545 (Australia)]

Güneralp B, Lwasa S, Masundire H, Parnell S, Seto KC (2017) Urbanization in Africa: challenges and opportunities for conservation. Environ Res Lett 13(1):1–8

Hill M, Wallner A, Furtado J (2010) Reducing vulnerability to climate change in the Swiss Alps: a study of adaptive planning. Clim Pol 10(1):70–86

Hinkel J (2011) Indicators of vulnerability and adaptive capacity. Towards a clarification of the science policy Interface. Glob Environ Chang 21:198–208

ICLG (International Comparative Legal Guide) and Makinde O (2016) Nigeria. The International Comparative Legal Guide to: A practical Cross-border Insight into Environment and Climate Change Law 13th edition Environment and Climate Change Law 2016. Global Legal Group Ltd, London, 20, pp.150–156

IDMC (International Displacement Monitoring Centre) (2013) Global estimates 2012: people displaced by disasters. Norwegian refugee council (NRC)/International Displacement Monitoring Centre. Châtelaine, Geneva

Ijaiya H (2015) Public participation in environmental impact assessment in Nigeria: prospects and problems. Niger Judicial Rev 13:83–102

IPCC (Intergovernmental Panel on Climate Change) (2007a) Climate change 2007: the regional impacts of climate change contribution of working group I to the third assessment report of the intergovernmental panel on climate change. Cambridge University Press, Cambridge

IPCC (Intergovernmental Panel on Climate Change) (2007b) Climate change 2007: impacts, adaptation and vulnerability. The contribution of working group II to the fourth assessment report of the intergovernmental panel on climate change (eds: Parry ML, Canziani OF, Palutikof JP, van der Linden PJ, Hanson CE). Cambridge University Press, Cambridge

IPCC (Intergovernmental Panel on Climate Change) (2013) Climate change 2013: the physical science basis. The contribution of working group I to the fifth assessment report of the intergovernmental panel on climate change (eds: Stocker TF, Qin D, Plattner G-K, Tignor M, Allen SK, Boschung J, Nauels A, Xia Y, Bex V, Midgley PM). Cambridge University Press, Cambridge

IPCC (Intergovernmental Panel on Climate Change) (2014a) Climate change 2014: impacts, adaptation, and vulnerability. Part A: global and sectoral aspects. The contribution of working group II to the fifth assessment report of the intergovernmental panel on climate change (eds: Field CB, Barros VR, Dokken DJ, Mach KJ, Mastrandrea MD, Bilir TE, Chatterjee M, Ebi KL, Estrada YO, Genova RC, Girma B, Kissel ES, Levy AN, MacCracken S, Mastrandrea PR, White LL). Cambridge University Press, Cambridge

IPCC (Intergovernmental Panel on Climate Change) (2014b) Climate change 2014: synthesis report contribution of working groups I, II and III to the fifth assessment report of the intergovernmental panel on climate change (eds: Core Writing Team, Pachauri RK, Meyer L). IPCC, Geneva

Klein RJT, Tol RSJ (1997) Adaptation to climate change: options and technologies: an overview paper. Technical paper FCCC/TP/1997/3. UNFCC secretariat, Bonn

MBNP (Ministry of Budget & National Planning) (2017) Economic recovery & growth plan 2017–2020. Ministry of Budget and National Planning, Abuja

McGray, H, Sokona Y (2012) Why institutions matter for climate change adaptation in developing countries. http://insights.wri.org/news/2012/05/why-institutions-matter-climate-change-adaptation-developing-countries. Accessed 28 June 2018

Mmom PC, Chukwu-Okeah GO (2011) Factors and processes of coastal zone development in Nigeria: a review. Research Journal of Environmental and Earth Sciences 3(6):625–632

Musa ZN, Popescu I, Mynett A (2016) Assessing the sustainability of local resilience practices against sea level rise impacts on the lower Niger Delta. Ocean Coast Manag 130:221–228

Mwalimu C (2009) The Nigerian Legal System: Public Law, Volume 1, Peter Lang Publishing Inc., New York, 1047 pp.

Ogbonna CU (2014) Adaptation to climate change in developing countries: a need in the Niger Delta region of Nigeria. In: Albrecht E, Schmidt M, Missler-Behr M, Spyra SPN (eds) Implementing adaptation strategies by legal, economic and planning instruments on climate change. Environmental protection in the European Union, vol 4. Springer, Berlin, pp 165–187

Ogbonna CU, Albrecht E (2015) Strategic environmental assessment as a tool to integrate climate change adaptation: a perspective for Nigeria. In: Leal Filho W (ed) Handbook of climate change adaptation. Springer, Berlin, pp 1239–1260

Ogbonna CU, Albrecht E, Schönfelder R (2017) Adaptation opportunities to climate variability and potential effects on sustainable development. The case of Nigeria's Niger Delta region. In: Leal

Filho W, Kalangu J, Kumbirai M, Munishi P, Simane B, Wuta M (eds) Climate change adaptation in an African context: fostering resilience and capacity to adapt. Springer, Berlin, pp 711–726
Okorodudu-Fubara M (2012) Country report. Nigeria IUCN Academy of Environmental Law 1:170–179
Oyewo HT (2016) Nigeria: the challenges of reintegrating Niger Delta militants conflict. Studies Quarterly Issue 17:58–76
Pradhan NS, Khadgi VR, Schipper L, Kaur N, Geoghegan T (2012) Role of policy and institutions in local adaptation to climate change – case studies on responses to too much and too little water in the Hindu Kush-Himalayas. International Centre for Integrated Mountain Development. ICIMOD, Kathmandu
SFRN (Senate of the Federal Republic of Nigeria) (2016) A Bill for an Act to provide for measures to address climate change. Senate of the Federal Republic of Nigeria Order Paper Tuesday 03 May 2016. http://www.nassnig.org/document/download/7979. Accessed 28 June2018
TACC (Territorial Approach to Climate Change) (2013) Biophysical and socio-economic assessment of the Nexus of environmental degradation and climate change, Delta State, Nigeria
This Day News (2016, August 16) Legislative agenda for climate change. http://www.thisdaylive.com/index.php/2016/08/16/legislative-agenda-for-climate-change/. Accessed 28 June 2018
UN (2014) World urbanization prospects: the 2014 revision. United Nations Department of Economic and Social Affairs/Population Division, New York. http://esa.un.org/unpd/wup/. Last accessed 20 Mar 2018
UNDP/GEF (United Nations Development Program/Global Environment Facility) (2013) Niger Delta biodiversity project. The GEF's strategic program for West Africa (SPWA) – sub-component biodiversity, Federal Ministry of Environment, Abuja
UNECA (United Nations Economic Commission for Africa) (2017) Urbanization and industrialization for Africa's transformation. Economic Report on Africa. United Nations Addis Ababa, Ethiopia
UNFCCC (United Nations Framework Convention on Climate Change) (2007) Climate change: impacts, vulnerabilities and adaptation in developing countries. United Nations Framework Convention on Climate Change Secretariat, Bonn
UN-Habitat (United Nations Human Settlements Program) (2006) Urbanization facts and figures. World Urban Forum II. An International UN-Habitat Event on Urban Sustainability Vancouver, Canada, June 2006. Vancouver. https://unhabitat.org/wuf/. Last accessed 28 Mar 2018
UN-Habitat (United Nations Human Settlements Program) (2011) Cities and Climate Change: Global Report on Human Settlements, 2011, United Nations Human Settlements Program, Nairobi, 300 pp
Wetzel FT, Kissling WD, Beissmann H, Penn DJ (2012) Future climate change Driven Sea level rise: secondary consequences from human displacement for island biodiversity. Glob Chang Biol 18(9):2707–2719
World Bank (2015) World Bank 2015: West Africa coastal areas management program. A partnership for saving West Africa's Coastal Assets. World Bank Group. http://pubdocs.worldbank.org/en/622041448394069174/1606426-WACA-Brochure.pdf. Accessed 28 June 2017
World Bank (2016) West Africa coastal areas management program (WACA). Saving West Africa's coastal assets. http://www.worldbank.org/en/news/infographic/2016/09/01/saving-west-africas-coastal-assets. Accessed 28 June 2018
Yohe G, Tol RS (2002) Indicators for social and economic coping capacity – moving toward a working definition of adaptive capacity. Glob Environ Chang 12(1):25–40

Livelihood Resilience of the Indigenous Munda Community in the Bangladesh Sundarbans Forest

3

Sajal Roy

Contents

Abstract

The Indigenous (Adivasi) Munda community in the village of Kalinchi in Shyamnagar upazilla (subdistrict), Satkhira of Bangladesh has undergone severe threats to livelihood due to the long-term effects of climatic disasters (such as: tropical cyclones, floods, salinity intrusion, famine, and heat waves). Kalinchi is situated adjacent to the riverbank of Dhojikhali near the Sundarbans Forest. This is the largest mangrove forest in the world, and it provides livelihood support to a large number of the coastal populations southwest of Bangladesh. The Adivasi Munda Community at Kalinchi has been traditionally earning a livelihood (such as harvesting honey, catching fishes and crabs in the forest surrounding rivers and channels, cutting trees and timbers) in the Sundarbans Forest. The earning of livelihoods was severely threatened, due to the severe Cyclone Aila on 9 May

S. Roy (✉)
Institute for Culture and Society, Western Sydney University, Sydney, NSW, Australia

Department of Women and Gender Studies, Begum Rokeya University, Rangpur (BRUR), Bangladesh
e-mail: roysajal.wgs@gmail.com; s.roy@westernsydney.edu.au

W. Leal Filho (ed.), *Handbook of Climate Change Resilience*,
https://doi.org/10.1007/978-3-319-93336-8_10

2009. This chapter primarily considers the long-term socioeconomic and ecological impacts of Aila on traditional livelihoods in the Sundarbans. The current study then documents resilience with a particular focus on the human livelihood of the untouchable Adivasi Munda Community dwelling near the Bangladesh Sundarbans Forest. An autoethnographic approach combing both focus groups and face-to-face life story interviews has been utilized in this study.

Keywords
Livelihood resilience · Cyclone Aila · Sundarbans forest · Adivasi Munda community · Intersectionality

Introduction: An Overview of Resilience and Livelihoods Literature

Livelihood resilience is a globally pressing subject in critical disaster and cultural studies (Bahadur et al. 2013; Watts 2015). Resilience is conceptualized as the capacity of a system to absorb disturbance and still retain its rudimentary function and structure (Walter et al. 2006). There are two main purposes for building resilience: firstly, to prevent the system from moving to an undesired, alternative regime in the face of climate disaster. Secondly, to nurture and preserve the components of the system that build resilience and allow the system to renew and reorganize after a disturbance (Walter et al. 2006). Broadly, livelihood resilience explores adaptive changes because it provides a way for analyzing how to maintain stability between livelihood resources and livelihood stresses in the face of transformation (Berkes 2006).

There are three main ways for extending the application of resilience in research studies (Tanner and Allouche 2011). Firstly, resilience can be regarded as an "end" process and a product of that process. It is contingent on social values concerning what is deemed important, and how to allocate resources to foster it. Disaster survivors may be continually locked into resilient but undesirable states of poverty and marginalization (Tanner et al. 2014). That is, disaster shocks at times paralyze the community and shackle them into poverty. The southwest coastal inhabitants of Bangladesh are solely dependent on a monolithic livelihood option in the Sundarbans Forest (Roy 2013).

Secondly, values and ideologies are translated into the activities and institutions that characterize livelihood resilience (Tanner et al. 2014) in the Munda Community in Kalinchi. The Adivasi Munda Forest Community has drawn a low level of priority from the local government to deal with hazards, livelihood insecurity and mitigation. The Munda people, who are locally treated as *Buno* (forest people), have inadequate access to political and economic power for earning a livelihood without the Sundarbans (Paggi 2001; Sharmeen 2013). For example, resilience in ecosystem services for human well-being needs to focus more on whose needs are being met, and on the politics of ecosystem management and distribution of benefits (Beymer-Farris et al. 2012). For livelihoods, the Munda Community have been surviving with unequal power relations, differing access to resources, and issues of inequality for many years (Sharmeen 2013; SAMS 2016). The local institutions (such as local government, development NGO) could not ensure livelihood benefits for the marginalized groups (Roy 2013). Hence, it is of utmost importance to ensure the beneficiaries' interests in the process of resilience.

Thirdly, despite disasters, impacts manifest through local ecosystems and the focus of resilience thinking on "natural" systems may lose sight of the people inhabiting these ecosystems. Both disturbances and responses to the shocks of disasters are determined by levels of on-the-ground social inequality, rights and unequal access to resources, poverty, poor infrastructure, lack of representation, and inadequate systems of social protection, planning, and risk management (Tanner et al. 2014). The unevenness of these factors translates climatic fluctuations into disproportionate concentrations of suffering and loss (Tanner et al. 2014). These issues have been historically experienced by the inhabitants of the Munda Community in Kalinchi (Roy 2018; Guhathathakurta and Banu 2017).

The livelihood of a rural poor community includes the capabilities, assets (stores, resources, claims, and access), and activities required for a means of living (Chambers and Conway 1991 cited in Tanner et al. 2014; Roy 2013). Livelihood resilience as a process depends on a number of issues, such as previous condition, the time between disturbances, and their severity. The social infrastructures and institutions are one aspects of socio-ecological resilience which emphasizes peoples' capacities to absorb and recover from hazards and disturbances (Reid and Vogel 2006). This study takes a Munda people-centered perspective on resilience that emphasizes not only the ability to absorb shocks and recover from cyclone Aila but also livelihood improvement, despite hazards and disturbances.

Bangladesh is acknowledged as one of the most vulnerable countries due to its exposure to frequent and extreme climate events, such as cyclones and associated storm surge. According to Center for Research on the Epistemology of Disaster, globally 606,000 lives have been lost and 4.1 billion people have been injured, left homeless, or in need of emergency assistance as a result of climate change induced disasters (Kaluarachchi 2018). According to Paton and Johnston (2006), the magnitude of physical hazards, poor land-use decisions, and unenforced public policy are the main causes of disaster-related death and casualty. Susceptibility of communities to loss from hazards could be reduced by creating a community that is resilient (Johnston et al. 2006). According to the International Strategy for Disaster Risk Reduction of the United Nations (2014), Bangladesh is ranked as the most disaster-prone countries in terms of the impacts of tropical cyclones. Cyclone-related immediate and long-term death rate was highest in Bangladesh amongst other cyclone prone countries at 32.1 deaths per 100,000 people over 100 years (UNDP 2004).

The southwestern coastal areas of Bangladesh were heavily impacted by Cyclone Aila, which struck on May 25, 2009. Despite the fact that Aila (2009) was a weak cyclone by classification, its monetary cost exceeded the effects of Sidr (2007) for the general population of southwestern Bangladesh (Roy 2013). About 2.3 million people were affected by Aila, and many of them were stranded in flooded villages as they had no alternative to save themselves (Mallick et al. 2011). Most of the

Aila-affected people could not reach a safer place due to the rush of seawater intrusion and inundation of roadways. The effect of subsequent saline water intrusion inside embankments caused the destruction of houses, roads, and culverts. This added more obstacles to the post-disaster activities and also increased the sufferings of those affected. The coastal residents were impoverished and lived in poorly constructed houses, and the cyclone shelters were inadequate to protect them all (Paul and Dutt 2010). In a previous cyclone, Gorky in 1991, the loss of lives was largely attributed to a lack of adequate housing capable of providing shelter (Mathbor 2007). Casualties due to Gorky were directly related to the types of housing and shelter seeking activities; no deaths occurred among individuals living in pucca houses (made of brick and concrete) and the ones who sought shelter in these buildings (Mathbor et al. 1993). During Aila, a large number of people lost their houses and livelihoods along with capital equipment (Mallick et al. 2011). These situations called for community capacity building that encompasses housing conditions, livelihoods, and preparedness regarding the cyclone's consequences (Vogt et al. 2009).

People of the Aila affected area employed a range of strategies for their survival. Some strategies involved diversification of income sources by seeking second jobs, cultivating a variety of crops including saline tolerant rice, vegetables on floating beds, and poultry and livestock rearing. At times, the Aila affected people utilized their social and family relationships for continuing survival in the local town of Shyamnagar (Roy 2013). According to Hussein and Nelson (1998), livelihood diversification is normal for most people in rural areas and nonagricultural activities are critical components of the diversification process. People with better socioeconomic circumstances were more likely to cope with impacts and were better prepared in responding to the aftermath of Aila (Roy 2013). However, effective utilization of social capital, such as, social networks, social cohesion, social interaction, and solidarity, are crucial for the capacity building of the community (Mathbor et al. 1993). In this regard, various community activities, such as Community Risk Assessment and Risk Reduction Action Plan, can act as a catalyst towards building a resilient community in post-disaster situations (Department of Livestock Services n.d.). Community people may take part directly in the development process of disaster preparedness activities to reduce the disaster risk (Miththapala 2009). This approach places community people at the heart of decision-making and implementation of disaster risk management activities (Miththapala 2009).

In this chapter, livelihood resilience is defined as the capacity of the Munda people to sustain and improve their livelihood opportunities and wellbeing despite environmental, economic, social, and political disturbances caused by a disaster event, like Cyclone Aila. The resilience, hence, depends on the Munda people's agency and capacity to combat the disturbances within the dynamic social transformations that occurred in the post-Aila period within and beyond Kalinchi. Thus, the Adivasi Munda Community, as the primary player, is engaged with a wider development process concerning their livelihoods in the Sundarbans (Perucca 2013) and beyond.

Case Study: The Adivasi Munda Community

The researcher considered autoethnography as an approach to "research and writing" that seeks to describe and systematically analyze (graphy) personal experience (auto) in order to understand cultural experience (ethno) (Ellis et al. 2011). This approach was utilized along with two focus groups (one with the Munda females, another with the mixed group participants) and fifteen life story (livelihood stories) interviews. In addition, direct observation was conducted in the Sundarbans Forest and households of the Munda people. A gatekeeper (a Munda male, who is a college student aged around 22) was voluntarily recruited before the fieldwork, to introduce the researcher to the Munda Community. This allowed the researcher to establish a rapport and build trust with the Mundas. A total of 30 participants (15 males and 15 females) aged around 35–80 were recruited through purposive sampling technique. This age group was selected to allow the participants to share their past memories of Cyclone Aila, along with its associated consequences on the livelihood and local infrastructures. The researcher conducted 3 months (April 2018 to June 2018) of fieldwork in Kalinchi. Before starting focus groups and interviews, the researcher clearly explained study aims to participants and recorded their consents. Each focus group lasted for 90 min, whereas a life story interview consumed around 50 min. The researcher conducted the life story interviews and moderated the focus groups. An audio recorder was used to tape conversations in focus groups and life story interviews. Participants in life story interviews shared livelihood-earning stories and described impacts of Aila on livelihood activities in the Sundarbans. The transcribed focus groups and interview data (Bengali texts, direct quotes) were coded. Afterwards, the thematic analysis (Braun and Clarke 2006) was utilized to develop key-themes and associated subthemes of the study.

Discussion of the Results and Findings

An Overview of the Adivasi Munda Community in Kalinchi

According to the Article 23A of the Constitution of Bangladesh, "the state shall take steps to protect and develop the unique local culture and tradition of the tribes, minor races, ethnic sects and communities." This article does not recognize Munda people as an indigenous community; rather it defines Munda as an ethnic community. Despite this debate, research studies (Perucca 2013) recognize the Munda Community as an indigenous community. The Munda Community has retained their own language (Mundari), food (a local drink of *Haria*), cultural performances (such as singing, dancing, worship, caste system, marriage, drama, and clothing), and faith in animism (Sharmeen 2013). Due to globalization, cultural transformation, and in the hope to be integrated with the mainstream Bengali Community, the Munda Community in the greater Shyamnagar region has been practicing rituals of Hinduism in the recent past. Apart from this, inhabitants of the Munda community also prefer to represent themselves as an indigenous community. Since 2003, the SAMS (Sundarban Adibasi Munda Sangstha) has been intervening to build cohesion within the Munda Communities.

There are around 1770 Munda males and female inhabitants living in the greater Shyanmanagar region, where 27 Munda households are found at Kalinchi (SAMS 2011). Unfortunately, the Munda Community does not own the land, on which they have been living for many years. According to the elderly male participants, for establishing landownership, the existing Munda people have been fighting with local land grabbers and some elites of the Muslim community, who are involved in the local politics at Kalinchi. The ancestors of the current Mundas also fought for the ownership of land, in which they had lived in.

Since the war of independence of Bangladesh in 1971, this community has experienced severe threats of climatic disasters, such as cyclones (that happened in 1988, 1991, 2009), flash floods, drought (1974), on-going salinity intrusion in the farming land, and severe scarcity of pure drinking water. The long-term effects of these disasters have positioned them to earn their livelihoods from diverse sources including the Sundarbans Forest. Despite this, the socio-environmental perspectives and identities of the Munda community with the Sundarbans forest in the post-Aila largely shape their economic geographies and earning of livelihoods (Perucca 2013).

Socio-Environmental Identities of the Munda People with the Sundarbans

"The relationship between human beings and forests has been important for the development of society" (Ritter and Dauksta 2013, p. 645). This relationship particularly between a natural resource-dependent community (such as, Munda) and the Sundarbans Forest is constructed through the provision of cultural, spiritual, and symbolic roles of forest. Likewise, Munda forest peoples have been essentially utilizing the Sundarbans Forest since the British colonial era for obtaining socio-ecological, entertaining, cultural, and economic support services. Through obtaining such services, the Munda people of the Eastern and Western Kalinchi have developed an in-depth ecological knowledge (Houde 2007) about harvesting resources in the Sundarbans.

The in-depth socioecological connections, the Munda people have developed through their direct contact with the Sundarbans and its resources are unique. Such connections have been built up through cutting wood, gathering honey, fishing, and receiving ecological protection of the forest during and after climatic disasters, such as river erosion, salinization, and floods. These are supported by long-established traditional ecological knowledge (Berkes et al. 2000), which has been following adaptive processes and passed down through generations by cultural transmission for years within the Munda Community and beyond at Kalinchi. The direct involvement of the Munda people with the forest is, of course, specific to Kalinchi, and it includes their associations with trees, conquering the threats of wild animals (tigers, deer, estuarine crocodiles), honey gathering in the risky hives, fishing in the adjacent

Fig. 1 Relics of a colonial-era salt industry, source: researcher's fieldwork

river channels, canals and large tracts of the forestland in the Sundarbans. In addition, the Munda peoples' traditional ecological knowledge integrates techniques, which are related to both legal and illegal hunting inside the forest, fishing, trapping, and conserving forest for obtaining future livelihoods in the Sundarbans.

There is a deep ecological bond between the Munda Community, Sundarbans, and Kalinchi itself. Historically, the construction of ecological bonding of the Munda people commenced when they had started clearing the Sundarbans Forestland for expanding agriculture during the British colonial era. This was done to promote the economic interest of both colonial administrators and local zamindars (landlord) of the greater southwestern region of Satkhira and Jessore. Utilizing the physical labor of the Mundas, the colonial administrators established an industry inside the forest. While conducting observations inside the forest, a relic (Fig. 1) surrounded the river Dhojakhali and small canals were visible to researcher.

While asking the about history of relics to a Munda boatman (The boatman, aged around 55, has been plying motorboat for the domestic and international tourists since 1993. He is originally a high school graduate, and capable of describing the history of Kalinchi and Sundarbans forest to the visitors. His ancestors used to earn livelihood through boating domestic and international tourists in the Sundarbans.), who accompanied the researcher in the forest, he informed that:

> The relics is originally a part of a colonial-era salt industry. It is assumed that the colonial administrator constructed this salt industry. Saline water was utilised to run the industry. This was first explored around 100 years ago by a tiger hunter, who was a brave Munda male. He observed the relics while checking the footprint of a tiger in the forestla5nd. Returning to

> Kalinchi, he informed the local forest office and local residents about the sault industry. Over the years due to the over exploitation of the forest resources and authoritative interest in the post-colonial forest administration, the salt industry (pictured above) has been damaged.

The Munda boatman also confirmed that:

> The first wooden boat that the Munda people utilised for going to the Sundarbans and plough (wooden) for cultivating forest land were made of sundari wood. During that period the edible fishes gathered from the canals and rivers were burned into the firewood and leaves gathered from the Sundarbans. Interestingly, livelihoods of the Munda people were then largely dependent on the traditional agriculture. Munda people cleared the forest to make space for fields and pastures. This hindered regeneration of the mangrove twigs, plants and the natural expansion of the Sundarbans forest land.

There was a lack of regulations for protecting resources of the Sundarbans and its land during the period of colonial administration. This administration developed the first forest act in 1927 for conserving Sundarbans. The laws were not strictly followed thinking the protection of the broader ecological future of the forest. This lack of legal regulations encouraged both colonial administrators and Munda Community to remove a considerable amount of sundari (*Heritiera fomes*) and gewa (*Excoecaria agallocha)* timbers. These were exported to several East Asian countries. This timber trading gained enormous popularity in the colonized Kalinchi. This would link the then Sundarbans Forest to the East Asian countries opening new opportunities for tourism. In contrast, it resulted in a significant depletion of the forest timbers, which were valuable to Kalinchi and the ecology of the Sundarbans itself. Trading timbers generated a considerable amount of revenue to the forest administrators. Unfortunately, the forest was not managed sustainably by the colonial administrators. The amount of revenue earned through timber trading was not disclosed to the Munda forest people. As a consequence, there was no financial incentive to manage the Sundarbans forest sustainably. In addition, the colonial administration and the zamindars forced bolder Munda young males to hunt a great considerable amount of Royal Bengal Tigers (Bagh) and deer from the forest. The hunted tigers' hairs and skins were given as gifts to the leading colonial administrators and zamindars. These were used for decorating the living rooms of the colonizers and zamindars. The local Muslims and Hindus at Kalinchi called the Munda people as *Buno* (which means primitive people of the forest) because they were involved in hunting tigers, gathering forest resources, and clearing the Sundarbans.

While harvesting forest resources and trading timbers were enhancing the economic interests of the colonial administrators, the groups of Munda people brought from Ranchi informally interacted between them and with groups both inside the forest and at Kalinchi. This was done through the maintenance of the internal networking among Munda people. This allowed them to share experiences of harvesting forest resources, adopting the rules and regulations imposed by the colonial administration and continuation of the clearance of the forest. Following animism as a spiritual faith, some Munda males wedded females, who lived in other villages near Kalinchi. Their livelihoods both before and after wedding were earned from the Sundarbans Forest. Historically, the socio-environmental identities of the Munda people represented a collective group centering on hunting and gathering in the greater Sundarbans Forest region. This is also an outcome of the traditional ecological bonding of the Munda people with the forest and Kalinchi. It reflects their contacts and attitudes towards the forest, which are linked to the socio-environmental identities of the Munda Community in the greater Kalinchi region.

The Forest Act of 1927 confirms Sundarbans as a reserve forest, which was regulated by the state-led Forest Department. There was a lack of execution of this law, which enabled the Munda people to frequently and illegally harvest logs from the forest's enormous resources. It also ensured their access to the forest without obligations allowing them to roam the forest for checking the availability of resources and their possible extraction mechanism. While carrying out resources extraction for many years, the Mundas within the Munda Community developed an in-depth understanding that Sundarbans was a forest under their ownership. Perceiving this indirect ownership of the forest, obtaining its protection from natural disasters (such as: cyclone, flood and river erosion) and utilization of resources for recreational services (tourism) confirm the long-established socio-environmental identity of the Munda forest community. Munda people believed that to have access to the forest and harvesting resources expressed authority, a sign of wisdom, shelter for life, and fertility. Primarily practicing these beliefs through worship of *Bonbibi* and other trees and the sun as they were considered as sources of energy, longevity, and welfare. The Munda forest community gradually organized themselves as a collective group at the greater Kalinchi for the further continuation of livelihoods in the Sundarbans Forest.

Globally forests have been manipulated through hunting, gathering, and shifting cultivation, used primarily for agriculture and commercially intensive consumption for industrial products and processes (Russell 1999 quoted in Ritter and Dauksta 2013). Historically, there is evidence that Kalinchi's Munda Community were involved in small-scale hunting of wild animals and fishes for supporting the colonial administration. There were very poor literacy skills among the Munda people due to the use of *Shardi* as their primary language of communication since their arrival to Kalinchi. Lacking literacy skills inadequately positioned the Munda Community to develop a broader sense of self-determination, which limited their relations with the remaining Hindus and minority of Muslims in Kalinchi. This double burden obliged them to strictly follow the colonial rules of clearing forestland and continuation of hunting and gathering in the Sundarbans. A female participant noted that "the British colonial administration and zamindars forced our husbands to serve as slaves. We were compelled to follow the colonial regulations for clearing the forest. We were collectively thinking to get rid of the severity of lives inside forest."

During the British colonial administration, there was no industry available at the broader Shyamnagar region, where forest products could be commercially utilized and processed. At the end of colonial administration, the Munda Community realized that hunting- and gathering-based livelihoods in the Sundarbans made their lives and survival precarious. This perception led them to look for an alternative source of livelihood by reducing their dependency on the Sundarbans. The Munda people then engaged in rain-fed agriculture and started serving as a domestic labor to the Hindu and Muslim households in Kalinchi. The Munda Community and their utilization of the forest for gaining monetary benefit drew significant attention of the post-colonial administration of the then West Pakistan. The forest department of the post-colonial administration imposed an official order on the Munda community for an inadequate use of the forest resources, which intervened in the traditional earning of livelihoods in the Sundarbans.

Hossain (2011) in the Bengali book entitled *Satkhira Zillar Itihas* (The History of Satkhira District) stated that after abolishing the British colonial administration in 1947, two independent nation states – Indian and Pakistan – emerged. This emergence separated the unity between Muslims and Hindus in the greater Indian subcontinent. According to mixed focus group participants, the majority Hindus, some Muslims and people from other castes and creeds were included in current India, whereas majority Muslims and a little number of Hindus were placed in current Pakistan East Pakistan and West Pakistan. Before the war of independence (1971), the current Bangladesh was part of East Pakistan (East Bengal), where Hindus, Muslims, and Adivasi communities lived with solidarity, peace, and unity. The administrative and political power was vested in West Pakistan. Between 1947 and 1971, the administration of the West Pakistan ruled the Eastern Pakistan (current Bangladesh), where a majority were Bengali people. During the formation of Pakistan as an independent nation state, it was governed by the army-backed administration. This administration only included the majority of the elite class inhabitants of the West Pakistan, except some chosen few from the East Pakistan. As a result, having no representation in the governance and political system, majority inhabitants of East Pakistan experienced marginalization occupying secondary status in citizenship; severe unemployment; sociopolitical, cultural, and linguistic exploitation.

According to an elderly participant, the existence of East Pakistan in terms of its land, sovereignty and the economic contribution of the people to the national income were not recognized in the constitution. This participant also reported that the political exclusion and inadequate constitutional recognition of the West Pakistan administration particularly compelled the Hindus and indigenous communities (who were living in and around the southwest coastal regions of the then East Pakistan) to realize the crises of "identities" and "existences." The involvement of the Munda Community in agriculture, forest and landownership, as well as economic marginalization was not drawn to the attention of the local administration of the West Pakistan. Living within the duality of lacking recognition from the Kalinchi's local authority and continuous struggle for a permanent source of earning, a considerable number of young males of the Munda Community expressed a reduction in earning capacity in the Sundarbans Forest. In contrast, the elderly people were involved in the paddy production, serving as domestic servants, and performing at the land cultivation activities. These small-scale paid jobs were done by the Munda males during this period. Females tended to occupy the domestic sphere cooking at home and caring for livestock including cows, pigs, and goats within their homesteads.

Effects of Aila and Associated Climate Stressors on the Forest-Based Livelihood

There have been three main impacts – immediate, short-term, and long-term – which cyclone Aila created, according to the participants in the focus group. These have affected the livelihoods of the inhabitants of the Munda community since Aila in 2009. As a result of Aila, there was severe damage to cyclone centers, roadway, and mud-built houses. The subsequent floods also washed away saline water ponds used for fishing (Gher). Aila's severe wind velocity imperiled a large amount of woodland of the Sundarbans Forest and affected the Munda peoples' everyday forest-going activities.

In the absence of adequate numbers of cyclone shelters, the Munda people sheltered on the rooftop of the primary schools and top branches of trees. Many women and children, who did not know how to swim, died while moving to a safe place. According to female focus group participants, there was an acute scarcity of pure drinking water, dry foods, and sanitary equipment, which are considered as essential requirements for the Mundas immediately after a cyclone. There were differing immediate livelihood impacts as a result of Aila on the inhabitants of Eastern Kalinchi and Western Kalinchi. The west Kalinchi Munda Community is quite isolated from the upazilla of Shyamnagar. This isolation seemingly delayed and slowed down the provision of the support services from local and national NGOs and public services from the Government of Bangladesh, according to a male elderly participant. This also created inertia in managing the immediate shocks experienced by the Munda Community.

A male focus group participant stated that:

> Cyclone Aila's short-term consequences hindered recovery; limited the overall livelihood restoration process; and turned the previous brief saltiness issue into a lasting one. The short-term effect of Aila included complexities around economic and social inequalities (such as: physical security, constraints in finding out a good source of cash income, homelessness), which further led to livelihood insecurities in the long-term both in the Sundarbans and other alternative sources. Due to the salinity water trapped in the fishing pond, there was a sharp decline in the shrimp cultivation in Gher, where Munda people used to work as day labourers.

Due to Aila (2009), seawater entered and submerged farmlands, fresh water fish-ponds, and vegetables growing areas. Subsequently, numerous farmers in the entire southwest waterfront locale, including the ponder zone, could not proceed with ordinary crops and fish cultivation as a result of non-accessibility to freshwater and water system infrastructure. These calamities constrained numerous individuals and groups to move to saline water shrimp cultivating (Baghda). The traditional "Gher" aquaculture had been practiced in coastal areas to grow shrimps and other white fishes long before the introduction of current shrimp culture practices. During the 1960s, a large number of coastal embankments were constructed to protect agricultural land in the coastal areas from tidal waves and saline water intrusion. Since the 1970s, strong international market demand and high prices for shrimp product have encouraged farmers to recommence shrimp farming in polders within the embanked areas. Contrastingly, to cultivate rice was no longer financially viable because the polders had become waterlogged due to poor drainage.

Saline water shrimp cultivation is financially more profitable than the freshwater shrimp cultivation (Galda). The observation data reveals that the conventional Munda fish farmers adequately took part in shrimp culture in the hope of earning high capital. Due to the long-standing floodwater, there was a high level of salinization in the agricultural land of Kalinchi. This further encouraged the fish farmers to be engaged with Galda cultivation. The Munda female fish farmers reported that saline water shrimp cultivation has significantly replaced the traditional rice cultivation in the farmland since 1983 to date. This was anticipated to cover in excess of 75% of rural land by 2020. Historically, the Munda people had undergone severe financial hardship, which further expanded in post-Aila. This ultimately discouraged many farmers in developing shrimp businesses in the study site. In this circumstance, the Munda people were able to continue earning a livelihood in the Sundarbans despite the stress on earning capacity through the environmental shock caused by Aila.

Since Aila (2009) to date, there has been an increase in bribery of forest officials to gain access to the Sundarbans Forest. The Munda forest dependents have usually paid a high-level bribe particularly for harvesting honey, which annually begins in April and ends in May. In the remaining 10 months, Munda forest people go inside the forest to net crabs, catch juvenile shrimps, and white fish. For this, they also need to pay a bribe to both the local forest officials and the ranger inside the forest. Participants frequently reported that after Aila, there were numerous robbers inside the forest. Asking for a ransom from the fisher folk is a good way for the robbers to earn their livelihoods. A participant opined that he was robbed quite a few times by a troop of armed robbers, but he was freed when he gave money amounting to at least Taka (TK) 15,000 each time. In the post-Aila period, earning a livelihood in the Sundarbans has become increasingly difficult, which in the long run has trickled down to the poor economic situation of the Munda Community in Kalinchi. Figure 2 shows that Mundas have been undergoing three key climatic stressors – environmental, economic, and social – affecting traditional livelihoods in the Sundarbans Forest. A participant opined that over the years due to the climatic stressors, there was a significant reduction of cash income in the Sundarbans Forest. In particular, the large-scale extraction of forest resources (honey, trees) through illegal logging has decelerated the regenerative functions of the mangrove saplings. The Forest Department does not undertake reforestation of mangrove plants.

The male focus group participants reported that before Aila, there was only one cyclone center in the Kalinchi that did not accommodate all of the community during a disaster. During Aila, a large number of people lived in a confined space, which created unhygienic sanitation and caused health hazards. The participants suggested that the government could aid locals to build their own housing capable of resisting cyclones rather than investing in poorly managed cyclone centers. However, in the Munda Community, the male focus group participants revealed that they are now moving from one economic activity to another. This is how the stressors are affecting income of the people Munda Community.

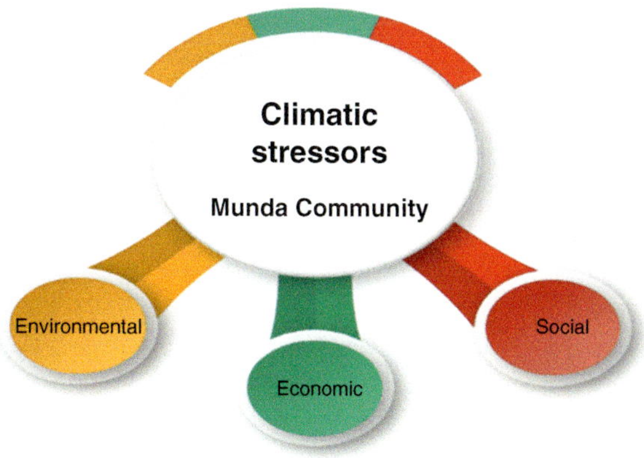

Fig. 2 Climatic stressors that the Munda Community experienced due to Aila

Present Livelihood Context

Cyclone Aila disturbed the ecosystem of the Sundarbans in various ways. The sway of Aila on honey and wax harvesting had significant negative impacts in the Sundarbans. The male focus group of the Munda Community revealed that honey harvesting had plummeted during the Aila period. This caused a severe economic loss to the honey gatherers, who are locally known as *mawali*. During the absence of honey collection, many Munda forest dependents migrated from Kalinchi to the divisional town of Khulna and Barisal in the hope of being employed in the paddy field for harvesting rice. Three years after Aila, the harvesting of honey is back to its normal pace.

Figure 3 shows the varied livelihood approaches that the Mundas have adopted post-Aila. The Munda people have utilized traditional ecological knowledge, which their ancestors used for harvesting honey, chopping wood, trees and timbers in the Sundarbans Forest. The Forest Department did not offer scientific training to the Munda forest people for gathering forest resources in the Sundarbans. Therefore, an excessive extraction of forest resources has caused a significant decrease in honey, timbers, wax, golpata, fish, shrimp, and seashell inside the Sundarbans, as narrated by an adult Munda fish catcher.

Post-Aila, the local Forest Department has become more alert about the extraction of forest resources. The strict enactment of the forest law inadequately enables the people to collect honey and catch crabs and fish in the forest's contiguous channels and small rivers. A focus group participant reported that "pre-Aila, there was availability of deer meat in the forest surrounding coastal villages. The dishonest hunters in collaboration with the officials of the Forest Department killed the deer inside the forest, and sold the meat in Kalinchi. Deer meat was sold at a high price. For example, 1 kg of deer meet cost around TK. 1000. Post-Aila, due to strict monitoring by the forest guards (staff of the Forest Department) in the Sundarbans, the extent of deer hunting has significantly reduced."

Historically, inhabitants of the Munda Community have led their lives under extreme poverty. They have been dependent on the Sundarbans and local agriculture

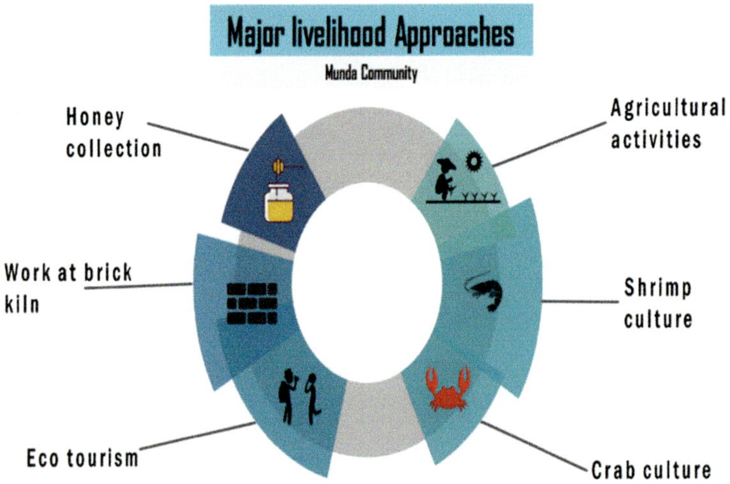

Fig. 3 Livelihood approaches that the Munda Community adopted in Kalinchi

for their subsistence for the last 200 years. The elderly members claimed that they were brought by the zamindars, during the British colonial administration, to clear land to expand agriculture. In the Sundarbans, members of the Munda Community would primarily catch fish and collect wood, which was transported to their homes for their family's consumption such as food, firewood, fencing, house building roofing, and so on. In terms of agriculture, both males and females have served as day laborers for cleaning Gher (saline water fishpond) of weeds. During the 1980s, many males served as domestic servants to the houses of local elites of the Muslims and Hindus. They were recruited for 12 months for the purpose of household activities and working in the farmlands for cultivating rice and vegetables. This employment would ensure that expenses were covered with regard to daily food, accommodation, and health. The Munda males established trust and faith working as domestic servants in the greater Kalinchi region.

Gher cultivation was introduced at Kalinchi in 1983. The Gher landlords are the local Muslim elites who are the principal landowners in the area. Working in the Gher generated a new form of livelihood. The Gher cultivation requires only a minimal amount of labor to initiate saline-water fish culture. Some Munda females were recruited as cheap laborers for ploughing Gher land and subsequently applying fertilizer. There is an income disparity for working in the Gher. The Munda males' 6 h of work at the Gher earns around TK. 200, while for similar work, a female laborer earns only TK. 120. This gendered wage gap reflects patriarchal dominance of the Muslim elites in the study site. Nowadays, the Gher laborers demonstrate a negative perception about Gher cultivation. It was reported that Gher cultivation has increased salinity in the farmland, restricted the movement of cattle, and significantly reduced the production of rice paddies and other crops. Participants considered Gher cultivation as a curse for them, as in the long run it has completely constrained the opportunity for agriculture posing a severe threat to livelihood.

Due to the gradual increase of thieves and fear of the wild animals (such as tigers and estuarine crocodiles) inside Sundarbans in the recent past, there is an uncertainty about the future livelihood of the Munda Community. Cyclone Aila has caused a shift in males' access to an income. A decent portion of Munda young males (aged around 25–35) has employed themselves in the brickfields in Barisal region. This seasonal employment lasting around 6 months (commencing in the Bengali month of Kartick and until the end of Chaitra) involves gathering soil for making bricks, transferring water for converting soil into mud, preparing the kiln, manufacturing bricks, and moving bricks from the kiln to trucks. The value of these 6 months of employment is between TK. 50,000 and 60,000. Treating the human body as a machine for earning livelihood, Munda males (who are the workers in the brickfield) reported that working in the brickfields is an extremely physically labor-intensive job. The working conditions are really poor in the brickfields. Females are usually recruited for cooking in the brickfields though their numbers are few.

After Aila, there is a growing interest in building a crab industry at Kalinchi and at the greater Shyamnagar region. Crabs caught from the Sundarbans Forest surrounding rivers and canals are fattened at home; afterwards, they are sold in the district town, Satkhira at a higher price (a female crab weighting 300 g is sold in TK 600). The large sized crabs have a demand in the international market of China, Hong Kong, and Singapore as reported by a female participant, who is involved in crab fattening in her homestead. This new economic possibility in the post-Aila has generated a new livelihood approach for the Munda people. Interestingly, participants in the male focus group suggested that the existing laws and regulations for crab collection should be revised. Participants also noted that the Forest Department is not intended to revise the existing laws thinking the need of the local residents who have been catching crabs as their sole means for livelihood. It has also been suggested that crab collection should be stopped in the April-May period, which is the reproduction time of the crab. One of the suggestions included "it would be better to revise the current laws governing the crab collection." It also requires keeping in mind about the reproduction phase of the crab. This will allow crabs to grow more so that residents can get more benefits out of it.

Despite the fact that the Munda Community has a historical socio-environmental connection to the Sundarbans forest, SAMS together with Relief International (RI), an NGO, initiated a project entitled "Promotion of Local Culture in the Sundarbans Impact Zone in Bangladesh through Cultural Eco-tourism and Entrepreneurship." This project began in late 2014 with the Munda Community living in and around Kalinchi and the broader Shyamnagar region. The objective of this project was to engage Munda people for expanding the heritage of the Sundarbans forest and culture of the Munda people to domestic and international tourists. Through this project, 10 cottages were built in the Munda Community. The cultural team from the Munda Community received financial support to purchase musical instruments, costumes, and materials required for cultural performances. In addition, members of the community were trained as eco-guides, in hospitality and services, in finance and in security, while their performances have been streamlined to include diversity. Mixed focus group participants revealed that ecological cottages have engaged 18 families of the Eastern Kalinchi Munda Community, allowing them an

Fig. 4 Eco-cottages established by the Relief International in the Munda Community in Kalinchi. (Source: researcher's fieldwork)

opportunity for making money from the domestic and international tourists. There is an influx of tourists in December, January, and February. The tourists visiting Sundarbans are accommodated in the eco-cottages (Fig. 4) built within the homesteads of the Munda Community. The money earned from the tourists are equally distributed among the families involved in eco-tourism. A Munda female reported that:

> I have been cooking for the tourists since 2015 to date. The Relief International has trained me on cooking hygienic food. I have been usually earning around TK. 5000 between December and February. During these months, many tourists come to Kalinchi to visit the Sundarbans forest. The tourists like to take our local food and prefer to entertained through songs and dances performed by the Munda cultural team. Cooking for the tourists has indeed created a new livelihood for me. Nowadays, I do not go to Dhojakhali for fishing.

In the post-disaster period, the diversification of livelihoods is considered as a process in which rural households adopt increasingly wider varieties of livelihood portfolios combining many distinct resources and assets (Niehof 2004). In addition, diversification encompasses a variety of dissimilar income sources of the disaster affected to support their livelihoods in a particular geographical location (Niehof 2004). Livelihood diversities post-Aila can be viewed as adaptive and survival strategies, which create space for continuous basic support by providing extra monetary income and resources to rural households (Ellis 2000). The diversification of livelihoods among the Munda Community in eco-tourism, which was explained earlier, exposed their traditional culture, and ensured an annual fixed income from tourists. Moreover, it has given economic sustainability among the Munda peoples households in the study region. According to the male focus group participants, 70% of Munda males rely on catching several fish varieties (such as: Basa, Vatkee, Passaya, Tangra) and harvesting honey for livelihoods. If the Forest Department enforces the relevant rules and regulations, it obstructs Munda people's livelihoods in the Sundarbans, and it has direct impacts on their family income.

In the post-Aila period, livelihood resilience requires looking beyond a return to the status quo to address the root causes of socioeconomic vulnerabilities of the coastal people and particularly the Munda Community. This needs to "build both resilience to cope with future threats and ability to exploit opportunities" (Pomeroy et al. 2006, p. 787). Cyclone Aila resulted in direct and indirect impacts on livelihood opportunities of the Munda Community in the study site. Aila severely damaged a large number of Ghers and crop fields. Inhabitants in the Munda Community, who could not cultivate crops and expand Gher, were forced to switch their livelihood during the post-Aila period. Munda people utilized distinctive survival mechanisms to enhance income sources by looking for alternative employment, developing an assortment of food production techniques including saline tolerant rice, vegetables on skimming beds, poultry, and raring of cows and goats within homesteads.

In the post-Aila period, building human capital (such as education, developing self-awareness of disaster) is considered as a determinant for livelihood resilience in the Munda Community. SAMS has been running two home-based schools for children aged between 8 and 10 in the community since 2013. A female Munda tutor (who is a college graduate) teaches preliminary subjects, such as the English language alphabet, vocabulary, mathematics, Bengali language, ethics and morality, cultures of Munda Community, concepts of disasters and religion. The school study curriculum helps children to develop a solid understanding on the Munda society, its lifestyle, traditional culture, hygiene, and consequence of a disaster. The tutor also arranges a monthly parents' meeting. In this meeting, either a government officer or a SAMS trainer is invited to briefly discuss sociopolitical rights (e.g., right to vote, having national identity card), disaster preparedness, warning system, and roles to be played in post-disaster. Each meeting consumes around 1 h focusing on a particular subject matter (such as, disaster warning system). This meeting motivates parents to educate their sons and daughters in high schools, colleges, and universities. Both the home-based schooling and a series of parental meetings have radically increased awareness of education and alternative employment (rather than earing livelihoods in the Sundarbans) among the Munda Community. In addition, this created more awareness and resilience against forthcoming disasters. The livelihood damage Munda people experienced in earlier cyclones (during 1988, 1991, and 2009) was higher in comparison with the ongoing climatic disasters (salinization) in Kalinchi.

To mitigate livelihood scarcities in post-Aila, the government of Bangladesh utilized social safety nets such as employment generation for the extreme poor and providing rice to the poor households. The initiative was taken to help people to cope with disasters and anticipated climate impacts so as to protect millions of vulnerable men, women, and children (DMB 2010). Following Aila, the government mobilized resources through different social safety nets to reach out to those who had lost their houses and personal belongings as well as livelihoods. Although safety nets can reduce resilience by increasing reliance on external sources (such as wheat, bottled water, lentil) depending on what form they take and how they are administrated at community level. However, the safety nets given to the Aila affected Mundas included supplying food, providing seasonal employment, and expanding livelihood assets bases for at-risk people. The aforementioned social safety nets created resilience in the Munda community. In addition, the safety nets allowed the Munda people to promote alternative livelihoods. A safety net recipient informed that he received a packet of 15-kg of dry biscuits from the officials of the local government. From these biscuits, he fed his family 25 meals. This assisted to mitigate the initial livelihood crisis created immediately after Aila. Two weeks after receiving the biscuits, he moved to the district town of Satkhira in the hope of looking for an alternative income. He found a rickshaw-pulling job, which is a highly physically laborious task. Another participant reported that there were significant institutional and structural barriers in the local government for distributing safety nets facilities to Munda people. Afterwards, the ongoing financial hardship (before, during, and after Aila) severely deterred Aila-affected Munda people to bounce back to the pursuit of traditional livelihoods in the Sundarbans Forest.

Intersectional Dimensions of Livelihood Resilience of the Munda Community

This section looks at the intersectional dimensions of gender, marital status, and mobility in building livelihood resilience of the Munda community in post-Aila context. Gender plays a significant role in understanding how males and females differently experience climatic disasters (Agarwal 1992; Carr and Thompson 2014). According to Dey and Laila (2017), the Garo Adivasi women's role in their traditional livelihood system has been transformed with the backdrop of the state-community conflicts over right of access to forest and control over forest resources in Bangladesh. However, there is a key gap in the existing literature documenting gendered livelihood resilience of an indigenous community in the southwest of Bangladesh. Thus, adopting a gender lens unpacks complexities around differing sensitivity and adaptive capacity in livelihood resilience of the Munda individuals in Kalinchi.

In the Munda Community, livelihood resilience is largely constructed by socio-economic roles, responsibilities, and entitlements of individuals and social groups. These are historically linked to social positioning and power relations in mitigating challenges for earning livelihood in the Sundarbans and other alternative sources. These are deeply rooted in ethnicity and socioeconomic class (Adger 2006; Reid and Vogel 2006; Segnestam 2009; Djoudi and Brockhaus 2011; Carr and Thompson 2014). In the mixed focus group, six of 10 Munda males completed primary school (capable of signing name and reading a book written in Bengali), while the female participants were illiterate as they did not complete primary school. Despite the seemingly poor literacy level in the study location, each participant was eager to be employed as a full-time worker. According to participants, full-time employment would ensure livelihood security, and reduce the extent of livelihood dependency on the Sundarbans.

The Munda females do not obtain full-time employment either in local NGOs or in Government Offices because they do not hold the required educational qualification. Livelihood earning of females is limited to fishing in the river of Dhojikhali, which is contiguous with the Sundarbans Forest. They are also employed as day laborers for clearing weeds in the Gher. Female unpaid work, which has been performed for years, includes cooking for family members, rearing home-based poultry and goats. The cultivable land is primarily used for shrimp aquaculture and producing paddy, which are considered a male occupation. The Munda males have been traditionally playing the role of the sole livelihood earner of each Munda family. Both before and after Aila, there are no visible changes in the position of the male as a breadwinner and female as a domestic worker and basic care provider in a family. Historically, these divisions of gendered roles and domestic responsibilities have confined Munda female's economic mobility only within in the labor force of Kalinchi.

The Munda females have less mobility than their male counterparts do. As members of an Adivasi community, they are discriminated against by the majority of local Muslims and Hindus in Kalinchi. Historically, females were confined within their homesteads or to visiting a relative living in a nearby village. Between 1980 and 2000, each Munda married female was accompanied by her husband while going to the local market at Vetkhali, for casual shopping and buying groceries. The unmarried adult females were accompanied by either their brother or father while going one place to another. This was done to ensure physical security in the public space. In the post-Aila, due to shyness, stigma, and fear of losing prestige, the Munda females did not take part in the construction of the embankment, which was offered by the local government. There are negative attitudes of Munda males towards females' paid-work, which redefines females as traditional housewives and financially dependent on husbands' income.

There was always severe scarcity of livelihood opportunities for Munda females within and beyond Kalinchi. A few Munda females reported that they worked as cooks in the brickfields situated in Khulna, Barisal, and Gopalgonj. While working in the brickfields, some experienced sexual harassment from other workers and employers. Despite this, they did not complain about sexual harassment to the local police station due to the fear of being fired from their cooking job. Though males and females in the community do not have ownership of the land on which they have been living in for centuries, they are not legally entitled to lease land to develop their own Gher business. There was an inadequate period of time for livelihood restoration after the cyclone Sidr (2007) and Aila (2009). The local roadways and river-based transportation system at Kalinchi are in a fragile condition since cyclone Aila. This geographical isolation has decreased Munda Community's lack of access to asset bases, financial services, and livelihood mobility. This, in the long run, has increased the economic vulnerabilities of both Munda males and females.

The Munda people believe in rituals, which are practiced on birthdays, deaths, engagements before marriages, and during wedding ceremonies. The birth of a boy is considered as an earner and permanent asset for the family, while a girl is seen as a family caretaker. *Dali Takka*, a monetary gift to paternal guardians, is generally paid before the marriage. This is considered as one of the main rituals of life lasting for a weeklong festivity. Marriage as a traditionally male-biased social institution in the

Munda Community encourages a male to be dominant over his wife while making familial decisions and household governance. Participants informed that before marriage an adult male leads an indifferent life financially depending upon his father and senior brother in the family. While after marriage, an adult male is compelled to earn more money for a smooth running of a family. The better income of a husband is treated as a sign of welfare by the wife. The husband is also treated as a primary provider of financial security, source of livelihood, and defense for a wife.

Early marriage (a girl aged between 12 and 14 would marry a male aged 25–34) has been historically practiced in the Munda Community for centuries. This age-related heterosexual marriage inequality shouldered upon adequate familial responsibilities of the newly married husbands and wives. These include everyday activities in the Sundarbans for a viable income, giving birth to children, working for community solidarity, and taking care of wife's parents. These social burdens would affect the earning of livelihoods in the Sundarbans. In contrast, it was expected that a wife would have to be a very good cook and bearer of all domestic duties. If a newly married wife failed to perform the expected duties of her husband's family members, she would have been physically tortured. In addition, a wife had no choice in decisions to have children. There were large amount of unwanted pregnancies, which forced a wife to give birth to a large number of children. This heinous marriage system was abolished after the evolution of SAMS in the late 2000. This NGO ran a series of courtyard meetings to build awareness against early marriage. As a result, the Munda males and females became aware of their future livelihood and economic opportunities. Since Aila to date, there have been fewer cases of early marriage and associated livelihood stresses in the Munda Community in Kalinchi.

Concluding Notes

This paper has documented ways the Adivasi Munda Community in the Bangladesh Sundarbans Forest region responded to and coped with cyclone Aila. It has also tried to reflect how livelihood resilience in the post-Aila has been built. The findings of the study suggest that livelihood resilience of the Adivasi Munda community was inbuilt within Kalinchi, both before and after cyclone Aila. The over extraction of Sundarbans forest resources by the Munda people and residents of the other local elites has been continually causing an imbalance between survival and on-going threats posed by climatic disasters. The loss and damage of physical assets due to Aila and associated climate stressors deterred the livelihood resilience among households in Munda people. Despite the fact that the Munda people were more vulnerable to hazard shocks and severely affected by higher financial, settlement, and physical damage, they showed a comparatively better ability to respond to, cope with, and recover from shock than did the wealthier Muslims and Hindus. This is because the Munda people are extremely hardworking and inclined to make an income to continue their survival in Kalinchi. However, the increasing risks from tropical cyclones are likely to affect livelihoods in the future and the living standards of coastal residents of Munda Community. The intersectional dimensions of gender, marriage, and mobility in livelihood resilience show a clear connection between the Munda people's livelihood-seeking behaviors and their on-going survival in the periods of before and after cyclone Aila.

References

Adger WN (2006) Vulnerability. Glob Environ Chang 2006(16):268–281

Agarwal B (1992) The gender and environment debate: lessons from India. Fem Stud 11:149–164

Bahadur AV, Ibrahim M, Tanner T (2013) Characterising resilience: unpacking the concept for tackling climate change and development. Clim Dev 5:55–65

Berkes F (2006) From community-based resource management to complex systems. Ecol Soc 11(1):45

Berkes F, Colding J, Folke C (2000) Rediscovery of traditional ecological knowledge as adaptive management. Ecol Appl 10(5):1251–1262

Beymer-Farris BA, Bassett TJ, Bryceson I (2012) In: Plieninger T, Bieling C (eds) Resilience in the cultural landscape. Cambridge University Press, Cambridge, pp 283–299

Braun V, Clarke V (2006) Using thematic analysis in psychology. Qual Res Psychol 3(2):77–101

Carr RE, Thompson CM (2014) Gender and climate change adaptation in agrarian settings: current thinking, new directions, and research frontiers. Geo Com 8(3):182–197

Department of Livestock Services (n.d.) Community based resilience in the Southwest of Bangladesh. Government of Bangladesh. Ministry of Fisheries and Livestock. http://old.dls.gov.bd/yearlyreport.php. Accessed 20 Feb 2017

Dey S, Laila R (2017) Reconstruction of Women's Role in *Jhum* Cultivation and Shift in the Gendered Division of Labor among the Garos. In: Meghna G, Ayesha B (eds) Gendered lives, livelihood and transformation. University Press Limited, Dhaka, pp 8–39

Djoudi H, Brockhaus M (2011) Is adaptation to climate change gender neutral?: lessons from communities dependent on livestock and forests in northern Mali. Int For Rev 13(2):123–135

Ellis F (2000) Rural livelihoods and diversity in developing countries. Oxford University Press, Oxford

Ellis C, Adams TE, Bochner AP (2011) Autoethnography: an Overview. Hist Soc Res 36(4): 273–290

Guhathathakurta M, Banu A (eds) (2017) Gendered lives, livelihood and transformation: the Bangladesh context. University Press Ltd., Dhaka

Hossain A (2011) Satkhira Zillar Ithisas. The district administration of Satkhira. Satkhira, Bangladesh

Houde N (2007) The six faces of traditional ecological knowledge: challenges and opportunities for canadian co-management arrangements. Ecol Soc 12(2):34

Hussein K, Nelson J (1998) Sustainable livelihoods and livelihood diversification. IDS 68:1–32

Kaluarachchi Y (2018) Building Community Resilience in the Re-settlement of Displaced Communities. Proc Eng 212:443–450

Mallick B, Rahman KR, Vogt J (2011) Coastal livelihood and physical infrastructure in Bangladesh after cyclone Aila. Mitigt Adapt Strateg Globe Chang 16:629–648

Mathbor GM (2007) Enhancement of community preparedness for natural disasters: the role of social work in building social capital for sustainable disaster relief and management. Intl Soc Work 50:357

Mathbor GM, Bern C, Sniezek J, Siddiqi MS, Ronsmans C, Chowdhury AMR, Choudhury AE, Islam K, Bennish M, Noji E, Glass RI (1993) Risk factors for mortality in the Bangladesh cyclone of 1991. Bull World Health Organ 71(1):73–78

Ministry of Food and Disaster Management (DMB) (2010) National Plan for Disaster Management 2007–2015. Dhaka, Government of Bangladesh

Miththapala S (2009) Incorporating environmental safeguards into disaster risk management. Volume 3: tools, techniques and other resources. Ecosystems and Livelihoods Group, Asia, IUCN, Colombo. viii+142
Niehof A (2004) The significance of diversification for rural livelihood systems. Food Policy 29(4):321–338
Paggi L (2001) The mundas indigenous people in the South-west coastal region of Bangladesh. Unpublished research report, 1–12
Paton D, Johnston D (2006) Disaster resilience an integral approach. Charles C Thomas Publishers Ltd., Springfield
Paul BK, Dutt S (2010) Hazard warnings and responses to evacuation orders: the case of Bangladesh's cyclone sidr. Geogr Rev 100(3):336–355
Perucca C (2013) Social water management among munda people in the Sundarban. University of Liberal Arts. Dhaka, Bangladesh
Pomeroy RS, Ratner BD, Hall SJ, Pimoljinda J, Vivekanandan V (2006) Coping with disaster: rehabilitating coastal livelihoods and communities. Mar Policy 30:786–793
Reid P, Vogel C (2006) Living and responding to multiple stressors in South Africa – glimpses from KwaZulu-Natal. Glob Environ Chang 16(2):195–206
Ritter E, Dauksta D (2013) Human–forest relationships: ancient values in modern perspectives. Env Dev and Sus 15(3):645–662
Roy S (2013) Connectedness with Mangrove forest "Sundarbans": perceptions and interactions of the forested community 'Shora' in Bangladesh. MPhil thesis, University of Bergen
Roy S (2016) Long-term effects of extreme weather events on rural livelihoods and gender relations: a case study of the cyclone Aila's impacts on the Shora (Muslim) and Munda (Hindu) sundarbans forest communities of Bangladesh. Unpublished PhD Research Project Proposal, Institute for Culture and Society, Western Sydney University
Roy S (2018) Book review: gendered lives, livelihood and transformation: the Bangladesh context edited by Meghna Guhathakurata and Ayesha Banu. LSE Review of Books blog
Segnestam L (2009) Division of capitals – what role does it play for gender-differentiated vulnerability to drought in nicaragua? Commun Dev 40(2):154–176
Sharmeen S (2013) Politics of development and articulation of indigenous identity: the formation of munda identity in barind, Bangladesh. IJAPS 9(1):142–160
Sundarban Adibasi Munda Sangstha (SAMS) (2016) About us. http://www.samsbd.org/index.php?page=about. Accessed 18 Mar 2018
Tanner TM, Allouche J (2011) Towards a new political economy of climate change. IDS Bull 43:1–14
Tanner T et al (2014) Livelihood resilience in the face of climate change. Nat Cl Chan 5(2):26
United Nations Development Programme (UNDP) (2004) A global report: disaster risk reduction: a challenge for development. UNDP, New York
United Nations International Strategy for Disaster Risk Reduction (2014) Progress and challenges in disaster risk reduction: a contribution towards the development of policy indicators for the Post2015 framework on disaster risk reduction. Switzerland, Geneva
Vogt J, Mallick B, Mahboob AS (2009) Social supremacy and location of cyclone centre: a determinant of social vulnerability analysis in a coastal union of Bangladesh. In: Dhaka: paper presented in international conference on water and flood management, 15–17 Mar 2009
Walter B, Sault D, Reid W (2006) In resilience thinking: sustaining people and ecosystems in a changing world. Island Press, Washington, DC
Watts M (2015) Adapting to the anthropocene: some reflections on development and climate in the West African Sahel. Geo Res 53(3):288–297

Building Community Capacity in Fragile Environments: Case Study of the Mara Serengeti Ecosystem

4

Rebekah Karimi, Albanus Mutiso, and Lippa Wood

Contents

Abstract

Pastoralists' culture values the livestock, but many lack the capacity to establish Sustainable Rangeland Management (SRM) and optimal livestock management

R. Karimi (✉) · A. Mutiso
Enonkishu Conservancy, Narok County, Kenya
e-mail: info@enonkishu.org

L. Wood
Mara Training Centre, Narok County, Kenya
e-mail: woodlippa@gmail.com

W. Leal Filho (ed.), *Handbook of Climate Change Resilience*,
https://doi.org/10.1007/978-3-319-93336-8_19

to maximize the output for the effort invested. The Grazing for Growth (G4G) curriculum at a training center on the northern edge of the Mara ecosystem aims to enhance disaster resilience of vulnerable communities to end drought emergencies in pastoralist environments by introducing new innovations while building on indigenous knowledge. The difference in overall quality between the experimental and control grazing blocks has improved from a difference of 7.6% in 2014 to 13.5% in 2018, demonstrating overall improvement in grassland managed through SRM. According to observational narrative, the improved quality of the grasslands has improved wildlife biodiversity and populations, although preliminary transect data is as yet inconclusive. Multiple enterprises support the community contributing to its overall resilience. A slaughterhouse eliminates the middleman in cattle trading. Tourism accounts for revenue from a lodge and visitors to a real estate entity capitalizing on the development of a bush-home community. Incorporating multifaceted enterprises is an important aspect of modern conservation of rangelands and promotes coexistence between people and wildlife by rehabilitating rangeland used by all.

Keywords
Biodiversity · Climate change · Disaster resilience · Multiple enterprise model · Natural resources · Sustainable development · Markets · Sustainable rangeland management (SRM) · Livestock

Introduction

Background

Due to the rapidly growing population, accompanied by climatic and political influences, the world's rangelands are facing increasing pressure to avoid future disasters (Fratkin and Mearns 2003; Solomon et al. 2007). Traditional practices of pastoralism are under threat by competition for more lucrative land uses such as mechanized arable farming, which has been shown to result in soil degradation (Van Oudenhoven et al. 2015), increasing fragmentation (Hobbs et al. 2008), thereby decreasing wildlife habitat and overall biodiversity (Homewood et al. 2001). Climate change impacts all agricultural enterprises, pastoralism not excluded. Over the past 4 years, pastoralists have struggled with coping with recurrent drought in increasing frequency manifested by increased competition for resources and losses of livestock from starvation (http://reliefweb.int/disaster/dr-2014-000131-ken). In addition, traditional pastoralism is wrought with issues such as interbreeding, high parasite loads and degenerative overgrazing, and a struggle to reach the market and attain the price their owners are seeking (Fratkin 2001; Serneels and Lambin 2001), which could be avoided through better management practices.

Agriculture is arguably the sector most affected by climate change and people living in arid and semi-arid lands (ASALs) bear the brunt of its negative impacts (Stige et al. 2006). These fragile ecosystems are facing a recurrent combination of natural disasters, food chain crises, conflicts, and protracted crises (Kabubo-Mariara 2009). Within the vulnerable environment of ASALs, it is imperative to anticipate and plan to prevent recurrent crises affecting food security and nutrition (Lankester and Davis 2016). While mitigating the future rise in global temperature has received much attention worldwide, support to help those vulnerable adapt to the inevitable impacts of climate shocks and stresses has been limited (Lybbert et al. 2004). Climate change adaptation in developing countries may cost between US$70 billion and US$100 billion per year for the period between 2010 and 2050 (UNEP 2016). Given that current donor and government funding is neither sufficient nor sustainable, there is an urgent need to engage private capital and enterprise in climate change adaptation (Lybbert et al. 2004; Norton-Griffiths 1996).

Large-scale arable farming is necessary to feed the growing population of the world, but the scope and practices present a challenge to preserving biodiversity, especially in high-risk locations bordering protected areas. Eighty percent of all threatened terrestrial bird and mammal species worldwide are imperiled by agriculturally driven habitat loss (Tillman et al. 2017). The alternative of a mixed use area where wildlife and livestock co-exist is more suitable for buffer zones (Oguta et al. 2016; Tyyrell et al. 2017). However, livestock husbandry also has historically degraded the land (Van Oudenhoven et al. 2015) and livestock farmers are not always willing to cope with the constant struggles presented by human-wildlife conflict (Ogada et al. 2003). Research has shown that if herders are vigilant and mobile bomas are utilized properly at night, the risk of livestock predation is minimal (Ogada et al. 2003; Riginos et al. 2012). On the other hand, predators can encourage the ungulates to vacate areas in the short term, allowing more recovery for the grasslands and eventually more resources for the livestock (Bauer et al. 2015; Bowyer et al. 2005). A complete ecosystem, including apex predators, is ideal in situations where livestock and wildlife coexist, as the natural balance is restored (Bowyer et al. 2005; Western et al. 2015; Ripple et al. 2014). With the innovative approach of Sustainable Rangeland Management (SRM), there is an opportunity to manage the rangelands in such a way that both livestock and wildlife benefit (Porensky et al. 2013; Riginos et al. 2012; Vavra 2005). When livestock are grazed according to a plan, the foot action of the livestock can assist in aerating soil, essentially ploughing their manure into the grassland, increasing the organic content and enabling grass and forbes to take seed (Lankester and Davis 2016; Young et al. 1997). Pastoralism is more resilient against climate change than arable farming because it is adaptable and mobile. In cases of prolonged or serious drought, herd size can be adjusted to cope with climatic events or the livestock can be moved to an area that has received adequate rainfall (Lankester and Davis 2016; Mureithi et al. 2014). SRM demands constant attention and adaptive management, assessing the quality of grassland and planning for the future by allowing land to rest and rejuvenate before commencing intensive grazing (Oba 2012).

Traditionally agro-pastoralists, the Maasai in East Africa, have been faced with challenges including human population growth, land degradation, privatization of rangeland, urban migration, and political conflict (Fratkin 2001). These confounding factors have increased pressure to move into a market-based economy, but traditionally, cattle ownership holds intrinsic value in the Maasai culture (Little et al. 2008). Aside from contributing to their dietary needs by providing milk, meat, and blood, cattle serve as a living savings account, enhancing their social status and providing the family with assets that can be sold in times of financial need (Randolph et al. 2014). The choices made by Maasai landowners are not a simple function of the economic returns potentially accruing from a particular enterprise, but as much or more influenced by who is able to control the different flows of returns from these different types of enterprise (Solomon et al. 2007; Thompson and Homewood 2002). Traditionally, the Maasai view themselves as part of the ecosystem and historically, the ideals they hold embrace sustainability within wildlife dynamics, but the population growth has impeded on their co-existence as much as it has impeded on wildlife habitat (Galvin et al. 2006). The ancestors of the current Maasai population took pride in their coexistence with wildlife and had an understanding that predators were part of the natural balance, as were their livestock and families (Kinga et al. 2018; Western et al. 2015). Therefore, it is imperative for Maasai to understand, implement, and benefit from SRM practices to improve not only their livelihoods, but also for the preservation of rangelands and wildlife around protected areas.

Wildlife holds intrinsic value worldwide, with many people travelling to East Africa to see its wildlife resources. Tourism accounts for 25% of Kenya's GDP, 70% of which stems from wildlife resources (Korir et al. 2013). With the wildlife comes the burden of human-wildlife conflict which falls on communities that are located within the hotspots of biodiversity, often along the borders with protected areas (Lankester and Davis 2016). Therefore, wildlife management needs to be reformed to secure more benefits for the communities that shoulder the burden of conservation policies to ensure those communities benefit from the wildlife (Kabiri 2010). Examples of communities lacking substantial income from supposedly community led tourism ventures are numerous. An investigation conducted in 2012 found that Maasai collected as little as 5% of the tourism revenue with the rest accruing to tourism operators and associated support services. Although initiatives to improve accountability have been initiated, many have fallen short because of poor and corrupt management practices (Bedelian 2012). Conservancies are critical in creating more space for conserving biodiversity and ecological services outside state protected areas and buffering protected areas from growing human impacts pressing on their edges (Oguta et al. 2016; Western et al. 2009). But conservancy members *must* see the benefits of being involved in conservation initiatives. However, the tourism market is fickle, vulnerable to political instability throughout the world, and climate change may exacerbate the vulnerability (Bunce et al. 2010; Loon and Polakow 2001). When Kenya was wrought with postelection violence in 2008, the tourism revenue dropped 56%, a crippling amount in a country where a large portion of the GDP flows from tourism (Bedelian 2012; Bunce et al. 2010; Korir et al. 2013). In addition, high-end tourism and hospitality often calls for specialized training to accommodate guests with certain expectations (Kiss 2004). This disconnect seems to facilitate a lack of ownership within the community and without cohesion with the tourism enterprise, communities are apt to view tourism ventures only for the monetary profit rather than a source from which to share their cultural pride or pride for the wildlife resources they have protected. Seventy percent of Kenya's tourism revenue comes from wildlife tourism (Korir et al. 2013). Ironically, wildlife throughout Kenya has experienced a 70% decline throughout the last three decades (Bedelian 2012). Conservancy members are more likely to buy into the conservation of biodiversity and the wildlife supported by ecosystems on their land if the conservancies become successful and lucrative. Training would be essential to instill the Maasai with pride for the wildlife supported on their land, while directing attention to the evidence of wildlife conservation being directly related to the financial success of the conservancy (Reid et al. 2016).

The Maasai Mara and the annual wildebeest migration was named one of the seven new wonders of the world in 2006. Although many of the Maasai were issued land titles in a land grab in the 1970s and 1980s, they are currently competing with wildlife for space, and pastoral Maasai livestock face the additional challenge of competing with wildlife for food availability (Bedelian 2012; Bowyer et al. 2005). As the generations progress, land parcels are getting smaller and smaller for the descendents of the original Maasai that acquired land titles, which has resulted in a movement of landowners moving to other parts of the country where they own another parcel of land which is more developed enabling them to send their children to better schools. The culture of the Maasai has been significantly watered down, although those who hold on to the pastoral traditions have a lot of pride in the occupation of their ancestors. The battle for preserving the variety of wildlife within the Mara is not a battle that needs to be fought with the Maasai (Reid et al. 2016; Norton-Griffiths 1996). If in fact, a pastoral livelihood is most compatible with wildlife tourism and the livestock are used to enhance the grassland for the wildlife, both could capitalize and the multiple enterprise model of wildlife and livestock coexisting. A lucrative alternative to arable farming could be a resounding success and in the case of Enonkishu Conservancy, the surrounding neighbors could want to be part of that success, which would mean pushing the boundary of the front line of conservation within the Maasai Mara. The success of the model could be replicated in similar situations where conservation areas that are technically protected areas could contribute significantly to the conservation of wildlife species. It would be a benchmark for conservation and one of the rare success stories in the field of conservation biology.

Preserving Biodiversity

In the discussion of building disaster resilience, especially among the disadvantaged rural communities of Kenya, preserving biodiversity is a valid objective. A decline in biodiversity has adverse effects, many of which are not identified until the loss in biodiversity has already occurred. Protected areas are not the only solution to the loss of biodiversity, as the sustainable management of grasslands and rangelands to enhance pastoral livelihoods and the conservation of wildlife habitats is one form of ecosystem-based adaptation that can provide multiple socio-cultural, economic, and biodiversity co-benefits (Osano et al. 2015).

Protected areas could cover 17% of the world's terrestrial area by 2020, but biodiversity is still declining because of the boundary areas and population growth and because most biodiversity resides in human-modified landscapes (Western et al. 2015). Historically, wild areas were prevalent without being categorized as protected areas. However, as the human population has grown and more development has occurred within wildlife habitat, humans are encroaching on many of the world's wildlife habitats, hindering biodiversity (Green et al. 2017). Wildlife conservancies could fill a gap in the designation of protected areas. In Kenya, a wildlife conservancy is defined as a conservation area set aside by an individual landowner, group of owners or a community for purposes of wildlife conservation in accordance with the provisions of the Wildlife Act (Osano et al. 2015). The conservancies surrounding the Maasai Mara National Reserve (MMNR) provide habitat for the majority of the wildlife residing in the Mara Ecosystem, yet receive none of the revenue from the reserve fees paid at the MMNR gates. Therefore, the need for outside funding in order to support the Mara Conservancies and preserve the biodiversity of one of world's hotspots is immense. Each conservancy has developed their own payment scheme to entice landowners to set aside their land for wildlife conservation. Offering an alternative income in addition to tourism revenue has potential to increase the income of the Maasai and satisfy their monetary investment in the land they own.

The preservation of biodiversity and a well-balanced ecosystem relies on the resilience of populations of large carnivores. Large carnivore populations have been declining due to habitat change, conflicts over livestock, utilization of body parts, and depletion of prey (Bauer et al. 2015; Green et al. 2017). Large carnivores are particularly vulnerable to the threat of human population growth around protected areas due to their large home ranges and slow life histories. Large carnivores inside one of the most heavily touristed game reserves in the world may not be adequately buffered from the environmental degradation from arable farming and human settlement occurring along its boundaries, highlighting the importance of solid management within Kenya's conservancies (Green et al. 2017). Anthropogenic disturbance near the edge of a protected area can concurrently influence populations of multiple large carnivore species (Bauer et al. 2015). Carnivores on the edge of protected areas especially with herders and livestock are vulnerable to being attacked by Maasai if the people do not recognize the benefits of having large carnivores around.

Community-based ecotourism has become a popular tool for biodiversity conservation, based on the principle that biodiversity must pay for itself by generating economic benefits, particularly for local people (Kiss 2004). In Kenyan conservancies, the typical model of ecotourism includes revenue sharing with conservancy members. Alone, ecotourism can be incredibly fickle and unsteady, but when supplemented with SRM, livestock husbandry, and appropriate training on the importance of preserving biodiversity, conservancy members only stand to benefit from multiple enterprises (Oguta et al. 2016). Utilizing buffer zones around protected areas by providing management of rangelands will increase the revenue gained when conservancy members can also utilize the pastoral lands responsibly (Oguta et al. 2016), while maximizing the output. Approaching biodiversity conservation with a business mindset has potential to reach people whose primary objective is not conservation alone, but rather maximizing their assets by investing in the conservation of biodiversity. Prioritizing people's development through community conservation is a strategy used by multiple nongovernmental organizations to serve both purposes of uplifting the livelihoods of local people while preserving biodiversity by making it pay for itself. To a certain extent, landowners' livelihoods need to be prioritized over conservation because without the landowners' support, conservation efforts are likely to fail (Pellis et al. 2015). Disseminating science at an understandable level and full engagement in pastoral and wildlife policy development is likely to produce a positive outcome for all involved (Reid et al. 2016). When local people are engaged in such activities, science and local knowledge together insight pride and ownership over conservation initiatives (Reid et al. 2016).

There are several positive pathways to build coexistence in the management of livestock husbandry, wild herbivores, and other wildlife species (Riginos et al. 2012; Young et al. 1997). Nutrient hotspots are created when mobile bomas house livestock overnight and these hotspots can attract specific herbivores and subsequent carnivores, increasing the biodiversity and wildlife populations of rangelands (Riginos et al. 2012). A positive side effect of nutrient hotspots is that the efficiency of livestock production is improved, a major recommendation for increasing the resilience of ecosystems (Schader et al. 2015). As grass-fed livestock engage in multiple uses for rangelands, it reduces the land requirement of supplying feed lots, which are grossly inefficient in the production of food (Schader et al. 2015).

Many wildlife species are threatened with extinction because of human activities. It is imperative that natural lands are protected and serve dual purposes if biodiversity and threatened species are to survive (Tillman et al. 2017). Large carnivores are among such threatened species (Bowyer et al. 2005) and are imperative to the resilience of ecosystems. Preventing the extinction of these species and the loss of their irreplaceable ecological function and importance will require novel, bold, and deliberate action (Ripple et al. 2014). Proactive international efforts to increase crop yields, minimize land clearing and habitat fragmentation, and protect natural lands could increase food security and preserve much of the Earth's remaining biodiversity (Tillman et al. 2017).

Conservancies in Kenya have the opportunity to explore a new mode of conserving biodiversity. Promoting the coexistence of wildlife and livestock pastoralism to the mutual benefit of all may be the silver bullet in ensuring the survival of the Mara Serengeti Ecosystem and the biodiversity therein. With the success of this model, documented with adequate and thorough data, Enonkishu is poised for the expansion of a crucially important ecosystem.

Innovative Solutions

The theory of livestock owners in East Africa participating in a "tragedy of the commons," where pastoralists are selfish and aiming at maximizing short-term offtake with no consistent management of the commons, is not universal (Allsopp et al. 2007; Fratkin and Mearns 2003). Multiple studies show that livestock owners are concerned about the sustainability of quality grasslands and the resulting land degradation of communal areas (Allsopp et al. 2007; Lybbert et al. 2004; Solomon et al. 2007). Furthermore, indigenous local knowledge is essential to building grazing plans that can assist in the recovery of biodiversity (Oba 2012). However, there is a need for a balance between science, social values, economic feasibility, institutional traditions, and political muscle (Azadi et al. 2007). Livestock owners in rural Africa do have the capacity and skills to recognize the need for a long-term plan for sustainable rangeland management (Allsopp et al. 2007). Herders take their cue from the condition of the forage by focusing on key forage plant species and make deductions from livestock production performance to make decisions about changing grazing conditions (Oba 2012). Community management of natural resources does offer promise, but must explicitly consider the linkages between community management, environmental management, and conflict management (Haro et al. 2004). One has to be careful not to over-romanticize local knowledge. The empowerment of previously marginalized groups may have unexpected and potentially negative interactions with existing power structures (Reed et al. 2007), which is why it is essential to build understanding through SRM training.

As outlined, all development ventures have their shortcomings: livestock owners are vulnerable to rainfall patterns; the tourism market is vulnerable to the current political climate; highlighting the importance of diversifying revenue streams by engaging in multiple enterprises (Lybbert et al. 2004; Wairore et al. 2015). Within buffer zones bordering protected areas, it is important to preserve biodiversity by extending rangelands into semi-protected conservancies, which are ideal for mixed land use livestock and wildlife grazing areas, and less than ideal for arable farming (Homewood et al. 2001; Tyrrell et al. 2017).

Methods

Multiple Enterprise Case Study: Enonkishu

Enonkishu conservancy is located on the far northern reaches of the Maasai Mara-Serengeti Ecosystem bordering Ol Chorro conservancy, Lemek Conservancy, and on the northern edge, arable commercial farmland. It is comprised of 4234 acres of land owned by 31 landowners who are paid an annual lease fee for preserving land use within the conservancy boundaries. Restricted land uses include permanent structures, harvesting natural resources (firewood), fences, arable farming, and a regulated number of livestock that must follow the SRM plan implemented by Mara Training Centre. The unique approach is the incorporation of multiple enterprises supporting Enonkishu, diversifying its financial support to curb vulnerability.

A slaughterhouse that specializes in grass-fed conservation beef was established in 2014 within Enonkishu Conservancy. Its premise is that the company provides the market and is ready to buy on a transparent price per kilo basis. The company has solid relationships with supermarkets and other retailers in Nairobi and is able to deliver a consistent supply of refrigerated good quality meat on a daily basis.

Tourism enterprises include Naretoi Holdings, a 1000 acre fenced real estate property comprised of 34 five-acre plots on which high-end homes are constructed. The community contributes conservancy fees for access to game routes within the three participating conservancies (Lemek, OlChorro, and Enonkishu). Within Naretoi is House in the Wild, a small safari lodge in which additional visitors pay rates per bed night for conservancy access.

A training center was built in 2014 with funding from the Africa Enterprise Challenge Fund. It provides trainings to communities centered around social cohesion, improving livestock quality, and preserving biodiversity. Trainings thus far have focused on the implementation of SRM to enhance their ability to deal with drought emergencies. The trainings fall in line with the Sustainable Development Goal 15 (SDG 15: Life on Land) which aims to protect, restore, and promote sustainable use of terrestrial ecosystems by combating desertification, reversing land degradation, and eventually halting biodiversity loss. Enonkishu Conservancy serves as a demonstration site for the SRM taught and implemented by the training center. The beef company assists with the cattle management to inform and enhance landowners' knowledge on methods by which to gain the most benefit, such as improving the genetic quality of Maasai Zebu cattle by introducing Boran bulls to the breeding herds.

The model adapted by Enonkishu Conservancy and its stakeholders has implemented the concept of employing multiple enterprises to support the longevity of the conservancy and build resilience within the community against crises which any single enterprise might have succumbed. Teaching SRM with a tangible demonstration site nearby is an effective tool that could change the way conservation programs are operated across the country.

Research Methods

Grazing for Growth (G4G) Curriculum

The Grazing for Growth Program (G4G) at MTC introduces new techniques while building on indigenous knowledge. Through hands-on learning techniques applied in a group setting, with follow-up extension services over a longer time period, communities learn how to make informed decisions about their livelihood practices and resource use strategies. The purpose of the G4G program is to create a locally lead, locally designed, and locally implemented regenerative grazing plan and to stimulate innovation in livestock enterprises. Alongside this, the program aims to build capacity on the sustainable management of the environment and the importance of protecting the future resource base.

The G4G curriculum first attracts representatives from the communities with an introductory course to engage them by showing evidence of the implementation of SRM. During the 3-day training, subjects covered include social framework, regenerative land management, and sustainable wildlife/livestock systems. After the introductory training, a smaller focus group of "champions" are selected who have influence within their communities and use that influence to motivate change. The champions are invited to attend an 11-day comprehensive course which covers the first three modules more intensely and includes two more modules on managing the livestock/wildlife interface and holistic financial management. Additional specialized trainings are offered to individualize trainings according to the needs of the communities in attendance.

All of the training provides in situ examples using the Holistic Management demonstration site (Enonkishu Conservancy) as well as the cattle farm, where resources are used as case studies. The Enonkishu livestock form the group's study herd, and tools such as illustrations, practical demonstrations, and real-life exhibits are used as learning aids, adapted to meet the needs of some illiterate group members.

Biological Monitoring of Rangeland

Enonkishu has been divided into 13 grazing blocks, demarcated by natural and manmade boundaries such as streams, hills, and roads. Eight of these blocks are utilized in the grazing plan implemented by trained staff at the training center. Fourteen transects have been set up from which five quadrats are examined during each biological monitoring session every quarter. There are four established transects of five quadrats each within control blocks that are not included in Enonkishu's grazing plan ($n = 20$). Ten transects of five quadrats each are located within blocks implementing planned grazing ($n = 50$). Data is collected at an interval by laying a quadrat every 5 m along each transect. Each transect starting point has been marked with a metallic peg which serve as a focal point for photographic record. A GPS coordinate is also taken at the marker peg for easy navigation to the transect sites. Nineteen parameters are examined in each quadrat to describe cover, soil surface description, water movement, litter, and plant species. In addition, insect and animal presence is noted. The corresponding ratings are designed such that a rating of "5" indicates the best possible score, with "0" indicating the worst possible score. As an example, the parameter of Plant Density rates 5 if there is 100% plant cover, with a score of 0 indicating no plant cover. By contrast the parameter of Soil Erosion rates 5 if there is 0% erosion, with a score of 0 indicating 100% erosion.

The first parameter investigated within every quadrat of each transect is plant density, measured by the percentage of the 1×1 m^2 covered in plants. It is the one parameter that has been consistently accounted for during biological monitoring sessions since its inception in November 2014. However, at the onset, transects were monitored once annually, after the grazing season, which varied from 2014 to 2015. In 2016, the methodology was adapted to collecting data at the end of each quarter, with an extra session conducted if necessary to ensure data included the growing season.

Wildlife Transects

Beginning in June 2016, two 1 km transects were established within Enonkishu conservancy. The two transects are driven between 0700 and 0900 twice per month. All species located within 100 m on either side of each transect are counted and recorded. The data collected is intended to be part of a long term database to monitor wildlife as the conservancy progresses in its objectives of enhancing wildlife habitat through the tool of planned livestock grazing. As the dataset grows, it will be interesting to compare wildlife populations and biodiversity to transect data that has been collected elsewhere in the Mara-Serengeti Ecosystem.

Additionally, eight conservancy rangers collect data from daily patrols throughout the conservancy which is recorded in a database. Such patrol data is useful in capturing more elusive or uncommon species such as predators or nocturnal wildlife species.

Continual training and adaptation has been necessary to ensure data reliability, but in addition, a new method of patrol information is being implemented through the Maasai Mara Wildlife Conservancies Association (MMWCA) with their introduction of the Wildlife Information and Landscape Data (WILD) smart phone application. Data includes GPS locations so that wildlife and other incidents involving wildlife are continually mapped. The application allows conservancy visitors to participate in the wildlife monitoring by downloading the application and submitting data to Enonkishu management.

Tracking Cattle Quality

In March 2017, a commercial herd of bulls was purchased by conservancy members in lieu of annual land rent. The intention was to fatten them on the conservancy and then sell them for a net profit. For the duration of the trading herd (4 months), bulls were weighed once per month which provided a valuable research tool to track the weight gain of cattle that are grazed properly according to SRM.

In addition, the conservancy established a breeding herd in September 2017 to contribute to the long-term stability of the conservancy. Funds were raised to purchase 160 cows and pay for the herds' care throughout its establishment of the initial 4 years. The herd is managed by the conservancy management team with the training center administrating the grazing plan, as it does with the cattle belonging to conservancy members. Data is collected monthly on weight gain, calving rate, and body condition. The majority of young heifers purchased from a local market are of the Zebu breed. Heifers were be crossed with a Boran bull to improve their genetics. Boran cattle tend to do very well in Enonkishu's climate and are larger than the local Zebu breed. The initial herd will be sold after calving twice to ensure constant upgrading of the herd's genetics. Monetary value will subsequently be tracked to quantify the value of a SRM program as cattle are fattened and bred in Enonkishu.

Results

Training Centre Effectiveness

Throughout 2017, trainings were funded by the Mau Mara Serengeti Sustainable Water Initiative (MaMaSe) project in partnership with SNV from the Dutch Embassy. In 2017, its inaugural year, the center trained 700 community livestock owners representing 15 conservancies or approximately 70,000 acres of land. Word of mouth has been a very powerful marketing strategy, which exemplifies the need for programs enhancing livestock management techniques and SRM.

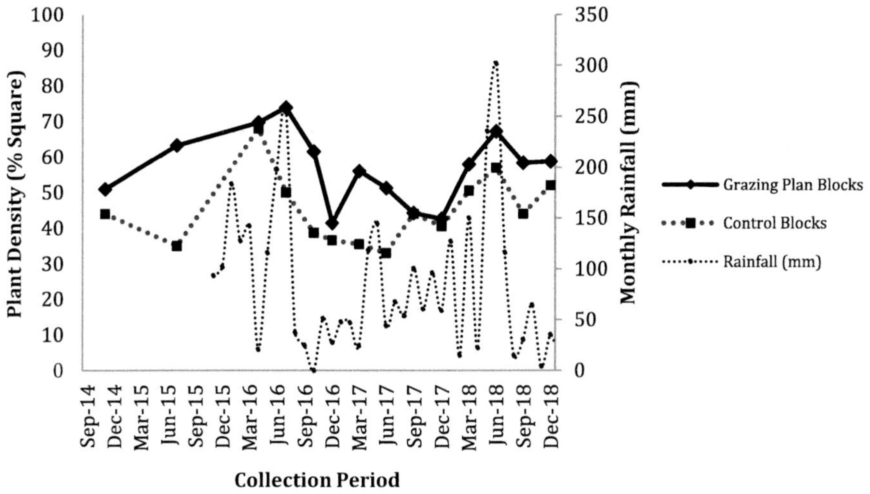

Fig. 1 Plant density measured as percentage of quadrat covered in plants throughout the duration of the biomonitoring effort. Rainfall data is displayed on the secondary axis

Biological Monitoring

An investigation into plant density over the past 3 years shows a notable difference between blocks implemented in the grazing plan and the control blocks (Fig. 1). Data is strongly related to rainfall in the conservancy.

Although the data from 2014 and 2015 is not as comprehensive as the last ten sessions (second quarter of 2016 through fourth quarter of 2018), data has been extrapolated for parameters where it was indicated on the raw data sheets. The resulting average score of all parameters shows a consistent trend of planned grazing blocks having higher scores than the control blocks (Fig. 2). In 2014, overall quality within grazing block quadrats was 7.6% better in the blocks implementing the grazing plan. By contrast, in 2017 (a very dry year) overall quality was 8.1% better in grazing plan blocks than in the control. In the first quarter of 2018, when Enonkishu received some rain and the vegetation began to grow, the overall quality within grazing block quadrats was 13.5% better than quadrats within control blocks that were not part of the grazing plan.

Wildlife Transects

Data from both transects have been collected over 60 days since June 2016 resulting in results for 120 transects. The most common species (>500 individuals) recorded are Thompson's gazelle (*Eudorcas thomsonii)*, white-bearded wildebeest (*Connochaetes taurinus albojubatus*), Burchell's zebra (*Equus quagga burchelli*), and Common impala (*Aepyceros melampus*). The average number of species

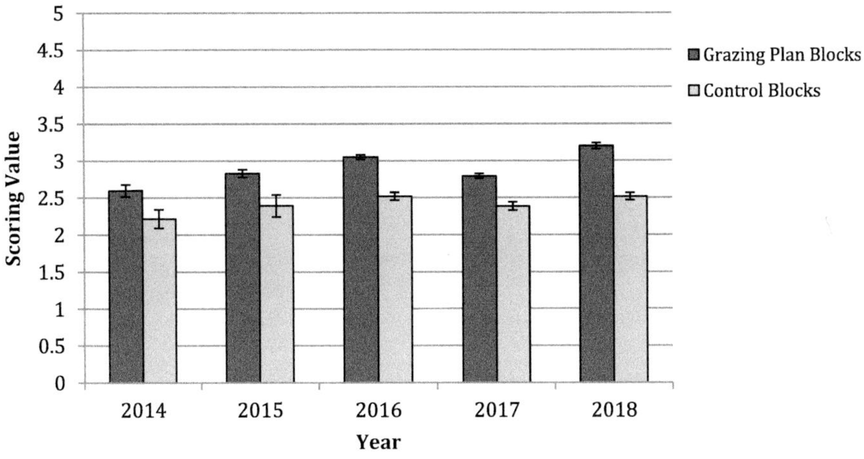

Fig. 2 Average score of up to 19 parameters throughout 14 biomonitoring sessions. The years 2014 and 2015 had one session each, while 2016, 2017, and 2018 had four quarters of data. Error bars indicate standard error of data collected each year

counted is 6.36, ranging between 2 and 12. The average number of animals counted is 119.0, ranging between 16 and 458. Preliminary data do not show an evident trend (Table 1), but the prospect of comparing long term data is promising.

Eighteen wildlife species have been observed along transects within Enonkishu (Table 2). Many more species exist within the conservancy and have been counted on ranger patrols or observed by visitors and employees.

Methodology standardization and training of rangers is ongoing to improve their data collection during daily patrols. Likewise, the data collected through the WILD application is still in its trial stage within Enonkishu, but shows promise for the collection of reliable spatial data in the future which can be compared to data collected elsewhere in addition to the development of more robust monitoring methodology.

Tracking Cattle Quality

Within the trading herd of bulls purchased in March 2017, weight gain averaged 19.28 kg over a period of 3 months. The commercial herd resulted in 21 bulls being sold for slaughter by the conservancy members to Mara Beef at an average price of Ksh.34, 891 compared to the initial buying price which was on average KES 27,435, an increase in value of KES 7456. Although the overall success of the conservancy members' short-lived trading herd was not resounding (five bulls died, impacting the net profit), there is evidence that a fattening-up period following the carefully monitored grazing plan can substantially increase the value of local cattle.

Table 1 Wildlife counted within 100 m of each side along two 1-km transects within Enonkishu Conservancy June 2016–December 2018

	Average # species counted			Average # animals counted		
Month	2016	2017	2018	2016	2017	2018
Jan	–	4.5	6.4	–	91.1	69.6
Feb	–	6.4	7.8	–	115.8	98.5
Mar	–	6.8	8.8	–	170.8	143.4
Apr	–	7.1	7.6	–	134.0	98.1
May	–	5.6	9.1	–	92.8	175.3
Jun	5.4	6.5	7.7	133.9	75.6	109.5
Jul	5.9	7.4	5.7	83.0	191.5	98.5
Aug	6.6	6.9	5,9	112.8	100.0	52.7
Sep	6.1	6.9	7,4	180.5	177.1	96.3
Oct	6.4	5.7	6.6	143.6	131.1	72.2
Nov	7.0	7.5	6.3	92.0	121.6	57.0
Dec	6.9	7.9	5.5	158.0	112.3	27.1

Table 2 Species observed on two 1-km wildlife transects June 2016–December 2018 in Enonkishu Conservancy

Common name	Scientific name
Banded mongoose	*Mungos mungo*
Bat-eared fox	*Otocyon megalotis*
Black-backed jackal	*Canis mesomelas*
Burchell's zebra	*Equus quagga burchelli*
Cape buffalo	*Syncerus caffer*
Coke's hartebeest	*Alcelaphus buselaphus cokii*
Common eland	*Taurotragus oryx*
Common impala	*Aepyceros melampus*
Deffassa waterbuck	*Kobus ellipsiprymnus*
Kirk's dik-dik	*Madoqua kirkii*
Grant's gazelle	*Nanger granti*
Maasai giraffe	*Giraffa camelopardalis tippelskirchi*
Olive baboon	*Papio Anubis*
Ostrich	*Struthio camelus*
Thompson's gazelle	*Eudorcas thomsonii*
Topi	*Damaliscus lunatus jimela*
Warthog	*Phacochoreus africanus*
White-bearded wildebeest	*Connochaetes taurinus albojubatus*

The breeding herd established by donors from the United Kingdom was named Herds for Growth (H4G), as its success is intended to be the driver in growing the conservancy. The initial purchase of 45 heifers took place in September 2017 and has increased in size from month to month. Currently, there are 155 heifers in the herd. The average purchase price was approximately $1.16/kg at the local market. Forty-two have been served by a Boran bull. When livestock first is brought to a new location, there is a transition period in which it is not uncommon for livestock to lose weight from the stress of adapting to a new herd and environment. The 45 members of the initial herd of heifers have gained on average 16.2 kg over the last 6 months.

Discussion

Establishing a multiple enterprise model encompassing SRM is setting a new standard for community conservation that builds on disaster resilience within vulnerable communities. With continued support and adaptive management, the resulting success can be implemented in various rangelands supporting the growth of conservation areas within buffer zones on the edge of protected areas. The multifaceted results offer numerous opportunities for tracking progress as the model progresses.

Training Centre Participants

The training center is in the process of developing a program for monitoring and evaluation that could be implemented across the Mara, in conservancies that will participate in SRM. Obtaining baseline data is essential to tracking the success of the program and will yield exciting results in the near future.

The scaling up of monitoring and evaluation in Mara conservancies has vast potential. The success of the training center in its initial location has been well received by surrounding communities and struggling conservancies throughout the region. Part and parcel of that success has been invitations to introduce a multi-enterprise model to various locations across Kenya (currently three locations are being investigated). The majority of requests have come from severely degraded environments where there are little resources to attract revenue generating ventures. Livestock husbandry is not the only option to build a broader revenue base. Rather than focusing solely on rangeland management, the goal is to develop creative solutions customized to each location. Remaining open to adaptive management is a strategy maintained within the model. Building disaster resilience in such environments will be a challenge, but the success of Enonkishu with improved biodiversity, improved livestock husbandry, and improved livelihoods of the landowners serves as a goal in many of the devastated regions requesting assistance.

Use of Biological Monitoring to Engage Pastoral Communities

With the standardized collection of biomonitoring data, future results are likely to show a larger difference in overall quality of blocks incorporated into the grazing plan compared to the control blocks. Any trend indicating lower quality grassland over time is likely caused by a difference in rainfall from year to year, which logically levels the playing field. An obvious result of low rainfall is decreased productivity within each quadrat, but a less obvious result is increased pressure from herders as they fail to adhere to the grazing plan during periods of low rainfall. Fortunately, the use of a grazing plan is most important during the season where the grass is growing. It is evident from the 2018 first quarter that blocks incorporated into the grazing plan recover faster than blocks that are not included in the grazing plan (Fig. 1). In Enonkishu, there is constant monitoring of the grazing plan and the herders implementing it to ensure adherence. It is highly suggested that any area incorporating a grazing plan closely monitors the daily livestock movements. Without supervision, herders are influenced by their experience to graze where the grass is green, hindering the recovery process within grasslands that have already experienced intense grazing.

The approach used at the training center is very conversational. Herders are given the opportunity to explain why the herds should deviate from the initial plan. There are many variables involved in deciding the best place to graze and administering a grazing plan incorporates adaptive management (Tyrell et al. 2017). As conditions change with above average rainfall or an influx of wildlife, the plan is adjusted accordingly. The most important aspect is that the herds all graze in the same block, bunched together to maximize the ploughing action of their hooves and incorporation of their manure into the soil. While recovery time for each block is also an important aspect, it is logical that each block has its advantages and disadvantages. Enonkishu encompasses Kileleoni Hill, the highest point in the Maasai Mara. It is treacherous to climb and as it is forested, the risk of predation is higher than on the savannah covered grazing blocks. Maasai herders are the employees that know best about the condition of each grazing block, regardless of how often the more scientific method of biomonitoring occurs. Although the grazing plan is implemented by the training center, there is an ongoing conversation with the herders to ensure they are sharing the same objectives of allowing vegetation to recover.

To embrace change in mindsets all around, the inclusion of scientific data collection can encourage strengthening of communities' capacity to manage and promote the health and wellbeing of their rangelands resource. Educating the herders on SRM (and subsequent monitoring) along with the livestock owners further engages the community in conservation. The transformative intervention then hits all levels of management and the benefits of SRM are more universal, therefore uplifting the livelihoods of a wider base within local pastoralist communities. Encouraging implementation of SRM and showcasing the scientific results will engage pastoralists, herders, and livestock owners alike, to design a sustainable future integrated with the conservation of biodiversity.

Wildlife Data and the Way Forward

Collecting information on the wildlife populations within the conservancy is important for several reasons. Wildlife is incorporated into SRM and the carrying capacity must include biomass consumed by wildlife species. Wildlife fluctuations could impact the tourism enterprises incorporated into the model, and populations of ungulates are used to assess the populations of more elusive carnivores. There has been marked improvement in wildlife numbers since the inception of the multifaceted model in 2014, but the data to support observations has been inadequate. However, as rangers are learning how to collect thorough and accurate data, the conservancy is also recording their general impressions of monthly counts of prevalent wildlife species. Conservancy management maintains that a general impression of the number of each species is better than poorly recorded patrol data that tends to be incomplete. With new and varied methodologies in place, it will be exciting to see how SRM affects wildlife populations, which influence the tourism enterprises and can be used as a general assessment of the success of the conservancy.

In addition to the historical wildlife data collected, Enonkishu needs to bolster the availability of wildlife data throughout the conservancy. Therefore, Fixed Route Foot Patrols are now being conducted weekly from each of three field camps. The rangers will also climb Kileleoni Hill (the highest point in the Mara) and observe their surroundings for 2 h, recording any species within their sight. Methodology has been designed in such a way that volunteers can collect data along vehicle transects. Similar data will be collected from a few man-made water points within the conservancy, and in September 2018, a grid transect camera trap survey was initiated to capture and inventory elusive species present within conservancy boundaries.

Enonkishu strives to collect data necessary to track improvements in the ecosystem as well as forming collaborations with species specific researchers who could include Enonkishu in their study area. Particularly interesting is the seasonal sightings of African wild dogs (*Lycaon* pinctus), which were observed on four occasions from June to August 2017, but increased in frequency throughout 2018. The location of Enonkishu has so many implications for research opportunities, because of its location on the northern edge of the Mara-Serengeti Ecosystem and the proximity to human habitation, a model that applies to so many protected area edges at a global scale.

Tracking Livestock Quality and Added Value

Incorporating a breeding herd into the multienterprise model will be advantageous for the members of Enonkishu Conservancy. Rather than livestock owned by a few (roughly 40%), this herd and the resulting revenue will support all members of the conservancy for years to come, generating annual conservancy member rental payments and running costs. However, the initial investment in managing and caring for the herd is substantial. A current advantage, even in this initial phase, is that conservancy members are able to sell their heifers to H4G for some extra income, while also evening out the number of cattle in the conservancy. It is an easy sale with no stressful transition for the heifers purchased. Seventeen of the 155 cows have been purchased from conservancy members, demonstrating their buy-in of a conservancy herd as an ambitious model that can improve their future livelihood. The projected financial model predicts that by 2020, the breeding herd will generate half of the income necessary to lease the land from Maasai conservancy members. With time, the herd could enable the conservancy to increase the rental rate per acre and prove to the conservancy members that livestock and wildlife coexistence can be more lucrative than any likely alternative use of their land.

Collecting continued data on the H4G will be advantageous for MTC to showcase the monetary success of SRM and optimal animal husbandry, which will promote the program further in support of conservation and the multienterprise model proposed by Enonkishu and MTC.

Conclusion

In experimental cases of establishing sustainable rangeland management, the model will take time to develop, but collecting vital data throughout the process will showcase the model so that it can be adapted elsewhere.

Building resilience against drought with the confounding factor of current climate change is of utmost importance. Using a multienterprise model enhanced with training communal pastoralists on livestock husbandry offers a lucrative income to landowners, as well as maintaining biodiversity which is so important within buffer zones on the edge of protected areas. With so many variables playing a role in land degradation, it is prudent to combat the issues with a multi-faceted approach. With detailed data collection to document the benefits of SRM and related trainings, evidence for a multienterprise model will bolster the argument of a sustainable future for rangelands.

Land owning communities are in the best position to take the responsibility of the health of their land – their future resource base. This responsibility is shared with tourism partners and other stakeholders all of whom share a common holistic context, with common values of wanting secure futures, healthy families, and healthy profits, in a healthy environment.

Cross-References

- ▶ Biodiversity, Ecosystem Degradation, and Climate Change Effects on Livelihoods in the Bitumen Area of Nigeria
- ▶ Building Pastoral and Agro-pastoral Community Resilience Against Drought in the Context of the Paris Agreement: The Case of Isiolo County, Kenya
- ▶ Climate Change and Population Growth in Pastoral Communities of Ngorongoro District, Tanzania
- ▶ Climate Change and Variability in the Mixed Crop/Livestock Production Systems of Central Ethiopian Highland
- ▶ Climate Change Vulnerability Among Pastoralists and Farmers in Ethiopia
- ▶ Climate Variability and Change in Guinea Savannah Ecological Zone, Nigeria: Assessment of Cattle Herders' Responses
- ▶ Community-Based Climate Change Adaptation Action Plans to Support Climate-Resilient Development in the Eastern African Highlands

- Ecotourism as an Adaptation Strategy for Mitigating Climate Change Impacts on Local Communities Around Protected Areas in Ghana
- Educating Farmers in Rural Areas on Climate Change Adaptation for Sustainability in Nigeria
- Enhancing Resilience of Livelihoods and Production Systems to Climate Variability and Other Related Risks in Africa
- Enhancing the Capacity of Vulnerable Community to Climate Change: Role of Quality Declared Seed Production Model in Semi-Arid Areas of Central Tanzania
- Herdsmen on the Move: The Burdens of Climate Change and Environmental Migration in Nigeria
- Livestock Market Improvement With Anthropological Approach in Drought Resilience Project in Northern Kenya
- Managing Vulnerability to Drought and Enhancing Smallholder Farmers Resilience to Climate Change Risks in Zimbabwe
- Understanding Kenyan Farmers' Perceptions of and Responses to Climatic Variability to Build their Resilience

References

Allsopp N, Laurent C, Debeaudoin LMC, Samuels MI (2007) Environmental perceptions and practices of livestock keepers on the Namaqualand Commons challenge conventional rangeland management. J Arid Environ 70:740–754

Azadi H, Shahvali MI, van den Berg J, Faghih N (2007) Sustainable rangeland management using a multi-fuzzy model: how to deal with heterogeneous experts' knowledge. J Environ Manag 83:236–249

Bauer H, Chapron G, Nowell K, Henschel P, Funston P, Hunter LTB, Macdonald DW, Packer C (2015) Lion (*Panthera leo*) populations are declining rapidly across Africa, except in intensively managed areas. PNAS 112(48):14894–14899. https://doi.org/10.1073/pnas.1500664112

Bedelian C (2012) Conservation and ecotourism on privatized land in the Mara, Kenya: the case of conservancy land leases. In: Land deal politics initiative working paper. The Land Deal Politics Initiative, ldpi@iss.nl

Bowyer RT, Person DK, Pierce BM (2005) Detecting top-down versus bottom-up regulation of ungulates by large carnivores: implication for conservation of biodiversity. Large carnivores and the conservation of biodiversity. Island Press, Covelo, pp 342–361

Bunce M, Brown K, Rosendo S (2010) Policy misfits, climate change and cross-scale vulnerability in coastal Africa: how development projects undermine resilience. Environ Sci Pol 13(6): 485–497

Fratkin E (2001) East African pastoralism in transition: Maasai, Boran, and Rendille cases. Afr Stud Rev 44(3):1–25

Fratkin E, Mearns R (2003) Sustainability and pastoral livelihoods: lessons from East African Maasai and Mongolia. Hum Organ 62(2):112–122

Galvin KA, Thornton PK, de Pinho JR, Sunderland J, Boone RB (2006) Integrated modeling and its potential for resolving conflicts between conservation and people in the rangelands of East Africa. Hum Ecol 44(5):350–356. https://doi.org/10.1007/s10745-006-9012-6

Green DS, Johnson-Ulrich L, Couraud HE, Holekamp KE (2017) Anthropogenic disturbance induces opposing population trends in spotted hyenas and African lions. Biodivers Conserv 27(4):871–889

Haro GO, Doyo GJ, McPeak JG (2004) Linkages between community, environmental, and conflict management: experiences from northern Kenya. World Dev 33(2):285–299

Hobbs NT, Galvin KA, Stokes CJ, Lackett JM, Ash AJ, Boone RB, Reid RS, Thornton PK (2008) Fragmentation of rangelands: implications for humans, animals, and landscapes. Glob Environ Chang 18(4):776–785. https://doi.org/10.1016/j.gloenvcha.2008.07.011

Homewood K, Lambin EF, Coast E, Kariuki A, Kikula I, Kivella J, Said M, Serneels S, Thompson M (2001) Long-term changes in Serengeti-Mara wildebeest and land cover: pastoralism, population, or policies? PNAS 98(22):12544–12549. https://doi.org/10.1073/pnas.221053998

Kabiri N (2010) The political economy of wildlife conservation and decline in Kenya. J Environ 19(4):424–445

Kabubo-Mariara J (2009) Global warming and livestock husbandry in Kenya: impacts and adaptations. Ecol Econ 68(7):1915–1924

Kinga GW, Mironga J, Odadi WO (2018) Analysis of the spatial relationship between cattle and wild ungulates across different land-use systems in a tropical savanna landscape. Int J Ecol 2018:2072617

Kiss A (2004) Is community-based ecotourism a good use of biodiversity conservation funds? Trends Ecol Evol 19(5):232–237

Korir J, Muchiri J, Kamwea J (2013) Wildlife-based tourism, ecology and sustainability of protected areas in Kenya. J Nat Sci Res 3(3):40–48

Lankester F, Davis A (2016) Pastoralism and wildlife: historical and current perspectives in the East African rangelands of Kenya and Tanzania. Rev Sci Tech Off Int Epizootics 35(2):473–484

Little PD, McPeak JG, Barrett CB, Kristjanson P (2008) Challenging orthodoxies: understanding poverty in pastoral areas of East Africa. Dev Chang 39(4):587–611

Loon RM, Polakow D (2001) Ecotourism ventures: rags or riches? Ann Tour Res 28(4):892–907

Lybbert TJ, Barrett CB, Desta S, Coppock DL (2004) Stochastic wealth dynamics and risk management among a poor population. Econ J 114(498):750–777

Mureithi SM, Verdoodt A, Njoka JT, Gachene CKK, Warinwa F, Van Ranst E (2014) Impact of community conservation management on herbaceous layer and soil nutrients in a Kenyan semi-arid savannah. Land Degrad Dev 27(8):1820–1830

Norton-Griffiths M (1996) Property rights and the marginal wildebeest: an economic analysis of wildlife conservation options in Kenya. Biodivers Conserv 5(10):1557–1577

Oba G (2012) Harnessing pastoralists' indigenous knowledge for rangeland management: three African case studies. Pastoralism: Res Policy Pract 2(1):1

Ogada MO, Woodroffe R, Oguge NO, Frank LG (2003) Limiting depredation by African carnivores: the role of livestock husbandry. Conserv Biol 17(6):1521–1530

Oguta JO, Piepho HP, Said MY, Ojwang GO, Njino LW, Kifugo SC, Wargute PW (2016) Extreme wildlife declines and concurrent increase in livestock numbers in Kenya: what are the causes? PLoS One 11(19):e0163249. https://doi.org/10.1371/journal.pone.0163249

Osano PM, Said MY, de Leeuw J, Moiko SS, Ole Kaelo D, Schomers S, Birner R, Ogutu JO (2015) Pastoralism and ecosystem-based adaptation in Kenyan Masailand. Int J Clim Chang Strateg Manag 5(2):198–214

Porensky LM, Bucher SF, Veblen KE, Treydte AC, Young TP (2013) Megaherbivores and cattle alter edge effects around ecosystem hotspots in an African savanna. J Arid Environ 96:55–63

Randolph TF, Schelling E, Grace D, Nicholson CF, Leroy JL, Cole DC, Demment MW, Omore A, Zinsstag J, Ruel M (2014) Invited review: role of livestock in human nutrition and health for poverty reduction in developing countries. J Anim Sci 85(11):2788–2800

Reed MS, Dougill AJ, Taylor MJ (2007) Integrating local and scientific knowledge for adaptation to land degradation: Kalahari rangeland management options. Land Degrad Dev 18(3):249–268

Reid RS, Nkedianye N, Said MY, Kaelo D, Neselle M, Makui O, Onetu L, Kiruswa S, Ole Kamuaro N, Kristjanson P, Ogutu J, BumSilver SB, Goldman MJ, Boone RB, Galvin KA, Dickson NM, Clark WC (2016) Evolution of models to support community and policy action with science: balancing pastoral livelihoods and wildlife conservation in savannas of East Africa. PNAS 113(17):4579–4584

Reliefweb (2018) Kenya: drought- 2014–2018. United Nations Office for the Coordination of Humanitarian Affairs (OCHA). http://reliefweb.int/disaster/dr-2014-000131-ken. Accessed 15 Jan 2018

Riginos C, Porensky LM, Veblen KE, Odadi WO, Sensenig RL, Kimuyu D, Keesing F, Wilkerson ML, Young TP (2012) Lessons on the relationship between livestock husbandry and biodiversity from the Kenya long-term exclosure experiment (KLEE). Pastoralism: Res Policy Prac 2(1):10

Ripple WJ, Estes JA, Beschta RL, Wilbers CC, Ritchie EG, Hebblewhite M, Berger J, Elmhagen B, Letnic M, Nelson MP, Schmitz OJ, Smith DW, Wallach AD, Wirsing AJ (2014) Status and ecological effects of the world's largest carnivores. Science 343(6167):1241484

Schader C, Muller A, Scialabba NE, Hecht J, Isensee A, Erb KH, Smith P, Makkar HPS, Klocke P, Leiber F, Schwegler P, Stolze M, Niggli U (2015) Impacts of feeding less food-competing feedstuffs to livestock on global food system sustainability. J R Soc Interface 12(113):20150891

Serneels S, Lambin EF (2001) Proximate causes of land-use change in Narok District, Kenya: a spatial statistical model. Agric Ecosyst Environ 85(1–3):65–81

Solomon TB, Snyman HA, Smit GN (2007) Cattle-rangeland management practices and perceptions of pastoralists towards rangeland degradation in the Borana zone of southern Ethiopia. J Environ Manag 82(4):481–494

Stige LC, Stave J, Chan KS, Ciannelli L, Pettorelli N, Galntz M, Herren HR, Stenseth LC (2006) The effect of climate variation on agro-pastoral production in Africa. PNAS 103(9):3049–3053

Thompson M, Homewood K (2002) Entrepreneurs, elites, and exclusion in Maasailand: trends in wildlife conservation and pastoralist development. Hum Ecol 30(1):107–138

Tillman D, Clark M, Williams DR, Kimmel K, Polasky S, Packer C (2017) Future threats to biodiversity and pathways to their prevention. Nature 546(7656):73

Tyrrell P, Russell S, Western D (2017) Seasonal movements of wildlife and livestock in a heterogenous pastoral landscape: implications for coexistence and community based conservation. Glob Ecol Conserv 12:59–72

UNEP (2016) The adaptation finance gap report (2016). United Nations Environment Programme (UNEP), Nairobi

Van Oudenhoven APE, Veerkamp CJ, Alkemade R, Leemans R (2015) Effects of different management regimes on soil erosion and surface runoff in semi-arid to sub-humid rangelands. J Arid Environ 121:100–111

Vavra M (2005) Livestock grazing and wildlife: developing compatibilities. Rangel Ecol Manag 58(2):128–134

Wairore JN, Mureithi SM, Wasonga OV, Nyberg G (2015) Benefits derived from rehabilitating a degraded semi-arid rangeland in private enclosures in West Pokot County, Kenya. Land Degrad Dev 27(8):1853–1862

Western D, Russell S, Cuthill I (2009) The status of wildlife in protected areas compared to non-protected areas of Kenya. PLoS One 4(7):e6140. https://doi.org/10.1371/journal.pone.006140

Western D, Waithaka J, Kamanga J (2015) Finding space for wildlife beyond national parks and reducing conflict through community-based conservation: the Kenya experience. PARKS 21:51–62

Young TP, Okello B, Kinyua D, Palmers T (1997) KLEE: a long-term multi-species herbivore exclusion experiment in Laikipia, Kenya. Afr J Range Forage Sci 14(3):94–102

Building Resilience of Urban Ecosystems and Communities to Sea-Level Rise: Jamaica Bay, New York City

5

A. Saleem Khan, Kytt MacManus, Jane Mills, Malgosia Madajewicz, and Laxmi Ramasubramanian

Contents

Abstract

Climate change-induced sea-level rise (SLR) and coastal flooding pose serious threats to low-lying urban coastal regions. Densely populated cities with fragile

---

A. Saleem Khan (✉) · K. MacManus · J. Mills
Center for International Earth Science Information Network (CIESIN), The Earth Institute, Columbia University, New York, NY, USA
e-mail: asaleemkhan.cc@gmail.com; kmacmanu@ciesin.columbia.edu; jmills@ciesin.columbia.edu

M. Madajewicz
Center for Climate Systems Research, The Earth Institute, Columbia University, New York, NY, USA
e-mail: malgosia.madajewicz@gmail.com

L. Ramasubramanian
Department of Urban Policy and Planning and The Institute for Sustainable Cities, Hunter College, CUNY, New York, NY, USA
e-mail: laxmi.hunter@gmail.com

W. Leal Filho (ed.), *Handbook of Climate Change Resilience*,
https://doi.org/10.1007/978-3-319-93336-8_29

urban ecosystems face the brunt of rising sea levels and other coastal disasters. Building urban climate change resilience to SLR reduces vulnerability and enhances the urban ecosystem and the community's potential to maintain both social and ecological functions to adapt to the rising sea levels. Jamaica Bay in New York City provides an example of social and ecological interactions in an urban area that address challenges of SLR. It offers an opportunity to study different approaches that communities adopt to build resilience, while placing ecosystems at the center of adaptation initiatives. COREDAR- (a capacity building tool) based SLR and flood risk information assortment and GIS-based SLR and flood risk assessments were carried out. The findings of the study in this chapter (1) describe the potential areas and population that are at risk to the predicted impacts of SLR in the Jamaica Bay region and (2) present empirical case studies on climate resilience efforts taken by a range of stakeholders from city governments, research and academic institutions, civil society, to others from the Jamaica Bay region. The chapter illustrates different approaches to integrate adaptation efforts in the social and ecological systems to improve adaptation to SLR in urban communities.

Keywords

Urban climate change · Sea-level rise · Ecosystems · Communities · Resilience · Adaptation · Jamaica Bay · New York City

Introduction

Global sea levels have risen throughout the twentieth century. These rises will almost certainly accelerate through the twenty-first century and beyond because of global climate change, but their magnitude remains uncertain (Nicholls and Cazenave 2010). The Fifth Assessment Report (AR5) of Intergovernmental Panel on Climate Change (IPCC) has projected sea-level rise (SLR) for all four Representative Concentration Pathways (RCPs), based on CMIP5 climate projections. Global mean SLR for 2081–2100 relative to 1986–2005 will likely be in the range of 0.26–0.55 m for RCP2.6, 0.32–0.63 m for RCP4.5, 0.33–0.63 m for RCP6.0, and 0.45–0.82 m for RCP8.5 (Church et al. 2013; IPCC 2013; Ramachandran et al. 2017). However, the effects of climate change (SLR) on coasts are not uniform, but vary considerably from region to region and over a range of temporal scales (Nicholls et al. 2007). The most serious physical impacts of SLR are as follows: (1) coastal erosion, (2) inundation and displacement of wetlands and lowlands (in particular, coastal wetlands including salt marshes are projected to be negatively affected by SLR especially where they are constrained on their landward side, or starved of sediment) (Breitmeier et al. 2009), (3) increased coastal storm flooding and damage, (4) increased salinity of estuaries and aquifers (Barth and Titus 1984; Marfai and King 2008), etc.

Cities are home to over half of the world's people and are at the forefront of the climate change issue. Importantly, the rising sea levels lead to increased inundation of low-lying areas and put global cities and urban coastal communities at greater risk

(Khan 2017). For coastal cities, enhanced SLR and storm surges affect inhabitants and essential infrastructure, property, and ecosystems (Rosenzweig et al. 2011). According to the United Nations, coastal flooding affects ten million people each year, a number expected to increase exponentially due to climate change. IPCC (2013) states that about 65% of the world's cities with populations of over five million are located in the LECZ (low elevation coastal zone). SLR is not uniform across regions; some communities in places such as subsiding deltas are highly vulnerable (IPCC 2007). A study by the Organisation for Economic Co-operation and Development (OECD 2007) found that the populations of Mumbai, Guangzhou, Shanghai, Miami, Ho Chi Minh City, Kolkata, New York City, Osaka-Kobe, Alexandria, and New Orleans will be most exposed to surge-induced flooding in the event of SLR.

One of the possible measures to address the challenges of SLR at the city level is by building resilience of urban ecosystems and communities as an adaptation strategy. "Resilience" can be defined as the capacity of social, economic, and environmental systems to cope with a hazardous event, trend, or disturbance, responding or reorganizing in ways that maintain their essential function, identity, and structure, while also maintaining the capacity for adaptation, learning, and transformation (IPCC 2014). To understand resilience we need to understand how physical, ecological, and social systems interact (Branco and Waldman 2016). Socio-ecological resilience theory understands systems as constantly changing in nonlinear ways; it is therefore a highly relevant approach for dealing with future climate uncertainties (Rodin 2014; Tyler and Moench 2012). Nevertheless, urban resilience refers to the ability of an urban system and all its constituent socio-ecological and socio-technical networks across temporal and spatial scales to maintain or rapidly return to desired functions in the face of a disturbance, to adapt to change, and to quickly transform systems that limit current or future adaptive capacity (Meerow et al. 2016). The literature on resilience in complex adaptive systems emphasizes the integration of social agents and institutions along with biophysical elements as components of socio-ecological systems (Folke 2006; Folke et al. 2002; Gunderson and Holling 2002; Tyler and Moench 2012). Institutions for collective action and governance can also be designed to strengthen resilience by supporting ecosystem restoration and sustainability (Adger et al. 2005; Folke et al. 2005; Ostrom 1990; Tompkins and Adger 2004; Tyler and Moench 2012).

There is broad consensus that (1) cities must become resilient to a wider range of shocks and stresses and become better prepared to address climate change and (2) efforts to foster climate change resilience must be bundled with efforts to promote urban development and sustainability (Leichenko 2011). The starting point in managing risks and building long-term resilience is for a city to understand its exposure and sensitivity to a given set of impacts and develop responsive policies and investments that address these vulnerabilities (The World Bank 2011). To the extent that low elevation coastal populations are at risk from SLR stronger storms and other seaward hazards induced by climate change, it is important to assess the size of these populations and how they are distributed (McGranahan et al. 2007) and how the ecological system, present in the midst of social system, is impacted by rising sea levels in the urban context. Thus, building resilience and adapting to climate change is increasingly a high priority for cities. At the same time, cities are well positioned to propose and implement thoughtful development policies to address climate change and promote resilience.

In this context, this chapter focuses on SLR impacts in the low elevation coastal zones of urban areas with a question of how coastal cities can be transformed into systems in the changing climate that are more resilient and adaptive to the rising sea levels. Jamaica Bay in New York City is our focus for ideas about global coastal resilience to changing climate, and it has been taken to assess the impacts of SLR and explore various strategies taken to build socio-ecological resilience in the bay area.

Jamaica Bay, New York City

Jamaica Bay is a tidal lagoon on the south shore of western Long Island, New York, bounded by residential areas of Brooklyn, Queens, and Rockaway Beach and John F. Kennedy (JFK) International Airport (Brand et al. 2018). The geographical coordinates of Jamaica Bay are 41° N, 74° W (Fig. 1a). The Jamaica Bay unit is the largest of the three administrative units (Jamaica Bay, Sandy Hook, and Staten Island) and is one of the largest expanses of open space in the region, consisting of over 19,000 acres of land, bay, and ocean waters within two boroughs of New York, i.e., Brooklyn and Queens (US Department of Interior 2013). Jamaica Bay Wildlife Refuge (JBWR) is located at the center of the bay area and has been protected since 1972 as part of the Jamaica Bay Unit of Gateway National Recreation Area (GNRA) administered by the National Park Service (Tanacredi and Badger 1995; Hartig et al. 2002). Social science literature about Jamaica Bay has used various scales to define study areas. The various boundaries represent specific agency boundaries, realistic or imagined ecosystem boundaries, or boundaries drawn because of particular scientific expertise (Allred et al. 2016). The bay area units surrounding the Jamaica Bay Wildlife Refuge including (1) Seagate, Coney Island; (2) Brighton Beach; (3) Sheepshead Bay, Gerritsen Beach, and Manhattan Beach; (4) Georgetown, Marine Park, Bergen, and Mill Basin; (5) Canarsie; (6) Far Rockaway, Bayswater; (7) Hammels, Arverne, and Edgemere; and (8) Breezy Point, Belle Harbor, Rockaway Park, and Broad Channel are demarcated for this study (Fig. 1b).

The bay is shallow with numerous salt marsh islands and tidal mudflats that vary in extent with the lunar cycle (Black 1981; Burger 1981; Brand et al. 2018). Water depths in the bay average 16 ft with depths up to 40 ft in the deepest portions of the dredged channels and basins. Some of the islands in the bay have upland communities including open field, shrub thicket, developing woodlands, and beach grass dune (DOS 1992). It is also designated as a Significant Coastal Fish and Wildlife Habitat by the New York State Department of State (NYC-DEP 2016). The bay estuary contains diverse wetland ecosystems and 330 different wildlife species (Van Hooreweghe 2012). It's a critical stopover area along the Eastern Flyway migration route and is one of the best bird-watching locations in

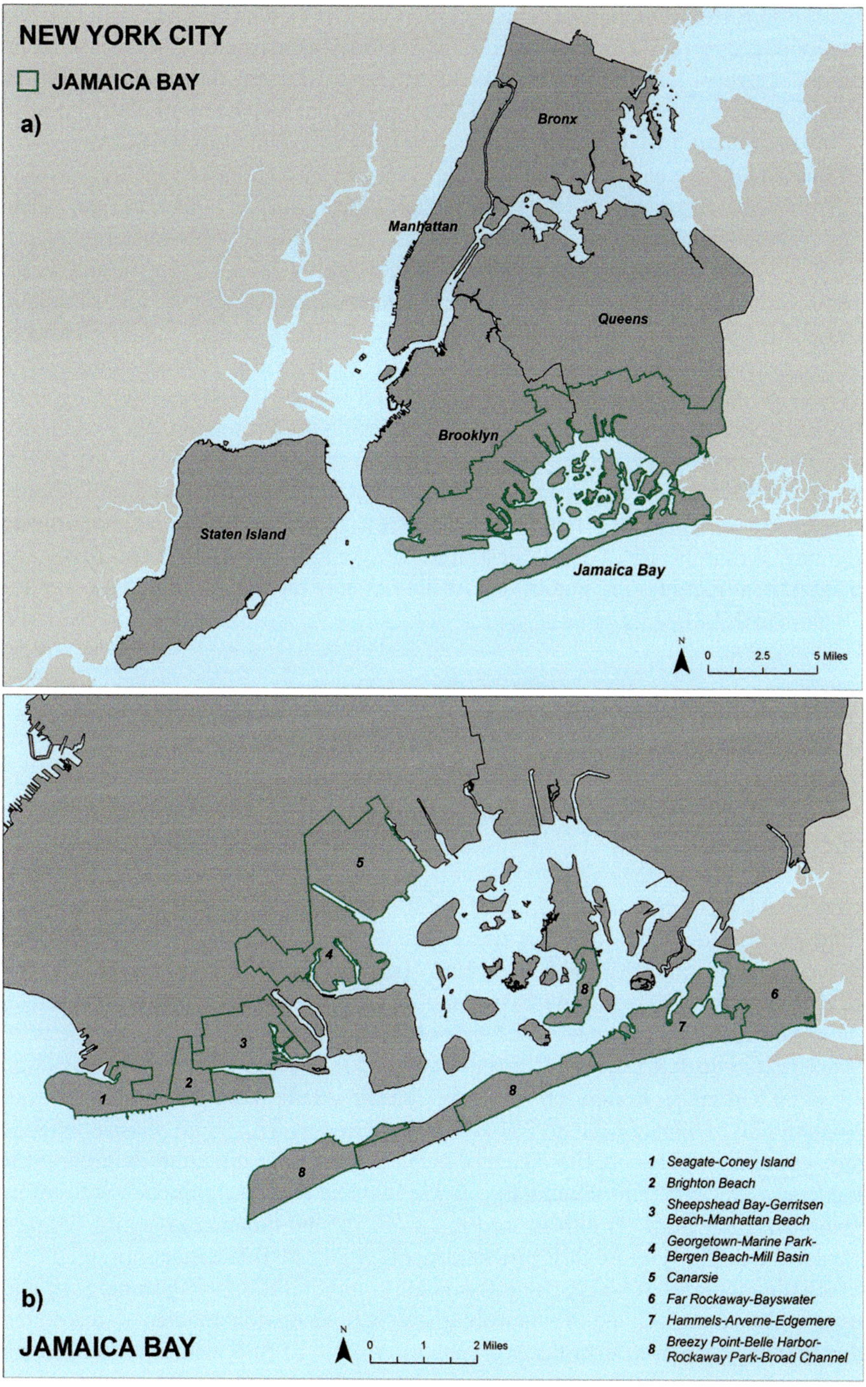

Fig. 1 (**a**) Boundary of New York City (NYC); (**b**) study area of Jamaica Bay region, NYC (Source: CIESIN, The Earth Institute, Columbia University, NY)

the Western Hemisphere (National Park Service 2004). On the other hand, the bay is embedded within a heavily urbanized region with extremely high population densities. According to 2016 US Census Bureau estimates, there were 2,629,150 people residing in Brooklyn and 2,333,054 in Queens as part of 8537, 673 people residing in New York City (NYC Department of City Planning 2018). As a result of human activity and development, the ecologically rich habitat of Jamaica Bay has changed considerably (NYC-DEP 2016). In terms of total area, the urban environment is the dominant feature of the modern-day landscape. In the last century, urban expansion has resulted in the filling of tidal salt marsh and freshwater wetlands, loss of all freshwater riparian habitat areas, and the loss of upland grassland, shrub land, and forest habitat (NYC-DEP 2007). Salt marshes covered an estimated 6550 ha in 1900 (Englebright 1975). By 1970, only around 1620 ha of salt marshes remained. Although protected since 1972, the remaining tidal wetlands within the boundaries of Gateway National Recreation Area were still shrinking in part due to ongoing SLR and other factors (Gorntiz et al. 2002). In October of 2012, Hurricane Sandy inflicted significant damage and loss of life in the Jamaica Bay region (ERG 2016). Despite profound alterations to the bay, it is still one of the largest areas of open space within New York City, a significant natural area within one of the nation's most populous urban centers (Waldman 2008).

Methodological Approach to Address Sea-Level Rise and Floods at Jamaica Bay

COREDAR-Based Sea-Level Rise and Flood Risk Information Assortment

COREDAR (Communicating Risk of sea-level rise and Engaging stakeholDers in framing community-based Adaptation stRategies), a capacity building tool for SLR risk communication and urban community-based adaptation, has been used in this study (Khan et al. 2015; UNAI 2016; State of the Planet 2016). The tool is based on the framework that was developed following the IPCC Fifth Assessment Report, Working Group I Report on Climate Change 2013: The Physical Science Basis; Working Group II Report on Climate Change 2014: Impacts, Adaptation and Vulnerability; Climate Change 2014: Synthesis Report with an emphasize on SLR and urban CBA (Khan 2017). The tool helps to gather holistic information on SLR risk communication and urban CBA in a systematic stepwise approach with information ranging from (1) climate and the profile of the urban coastal city, (2) past sea-level trends, (3) future SLR projections, (4) predicted SLR impacts, (5) predicted SLR vulnerabilities, (6) SLR risk information communication, (7) framing SLR and urban CBA strategies, and (8) mainstreaming SLR adaptation policies. In this study, this tool has been used to collect various information of SLR with relevance to the Jamaica Bay region of New York City. Each step offers a checklist for scientific and research communities (SLR science); local communities, NGOs, and private and public partners (SLR and society); and policy-planners and decision-makers (SLR policy) to involve and contribute to the SLR and urban CBA decision-making process (Khan 2017).

GIS-Based Sea-Level Rise and Flood Risk Analysis

This section describes the processes and data used to produce estimates of flooded area and population at risk for the 100- and 500-year floodplains in the years 2020 and 2050. (1) ***Data source:*** In this study, data such as (a) Neighborhood Tabulation Areas, Department of City Planning (2013); (b) 2010 Census Blocks Department of City Planning (2013); (c) SLR maps of 2020 and 2050, 100–500-year floodplain, Mayor's Office of Long-Term Planning and Sustainability (2013); and (d) 2010 Census Summary File 1, US Census Bureau (2011) are used. (2) ***Flood scenario extent preparation:*** Flood scenario extent feature classes were first projected to a local NYC coordinate system (NAD 1983 StatePlane New York/Long Island FIPS 3104 ft), and then each flood extent was dissolved into a single feature class. For cartographic purposes the extents were clipped to the census block boundaries from the NYC Department of City Planning. (3) ***Census block boundary preparation:*** The 2010 census population was downloaded in ASCII format from Summary File 1 of the US Census. It was then converted into text format and joined to the census block boundaries using a common identifier (a concatenation of state, county, tract, and block geoids). Next the area in square meters was calculated for each census block using the Calculate Field tool in ArcGIS. Subsequently, each census block was assigned to a neighborhood defined by the Neighborhood Tabulation Areas feature class from the NYC Department of City Planning. This was accomplished by (a) converting the census blocks to points, (b) using the Spatial Join tool in ArcGIS to assign a neighborhood name to all census block points contained within each neighborhood boundary, and (c) joining the neighborhood names back to the census block boundaries using a common identifier. (4) ***Analysis:*** Estimating the population at risk and flooded area in each of the four flood scenarios was a multi-step process. The census block boundaries were clipped to each flood extent to delineate flooded areas. The flooded area (square meters) of each census block within the flood extent was determined using the Calculate Field tool in ArcGIS. Next, the population within the flooded area of the census block was calculated using the formula below. This calculation relies on the assumption that the population in a census block is evenly distributed (Doxsey-Whitfield et al. 2015), and therefore the proportion of population at risk would be equal to the proportion of flooded area:

$$P_{\text{atrisk}} = P_{\text{total}} * \frac{A_{\text{flooded}}}{A_{\text{total}}}$$

where P is population and A is area. Finally the neighborhood names along with borough name were used to summarize the flooded area and population at risk for each flood scenario. The schematic representation of the methodology adopted in this study is outlined below (Fig. 2).

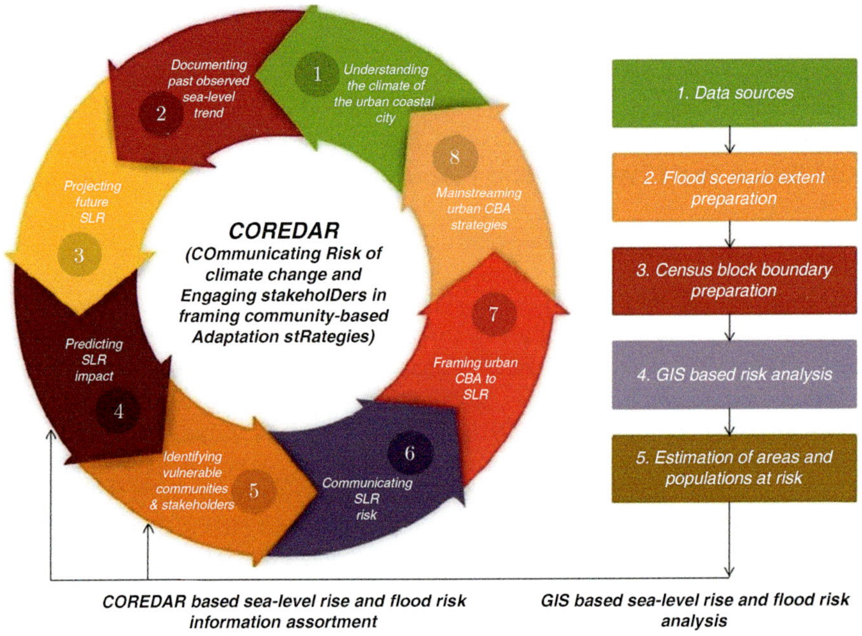

Fig. 2 Schematic representation of the methodological approach adopted (Source: Khan 2017 and CIESIN, The Earth Institute, Columbia University, NYC, NY)

Predicted Impact of Sea-Level Rise and Floods at Jamaica Bay

The rate of local SLR in New York City is around 2.84 mm/year with a 95% confidence interval of +/– 0.09 mm/year based on monthly mean sea-level data from 1856 to 2016 which is equivalent to a change of 0.93 ft in 100 years, based on the tide gauge data from Battery Park in Manhattan (NOAA 2018). The New York City Panel on Climate Change (NPCC2) has projected SLR in four time slices as the 2020s, 2050s, 2080s, and 2100 for NYC. Based on 24 GCMs and RCPs, NPCC has estimated SLR for NYC as 10 in. (0.25 m) in the 2020s, 30 in. (0.76 m) in the 2050s, 58 in. (1.47 m) in the 2080s, and 75 in. (1.91 m) by 2100 under the high estimate (Horton et al. 2015; Gornitz et al. 2017). These estimated increases in sea level could accelerate the loss of wetlands and uplands from wave–driven erosion and flooding. Warming associated with climate change also is predicted to increase the frequency and severity of major storms and hurricanes. These storms, in conjunction with high sea levels, have the capacity to cause extensive change to the present morphology of Jamaica Bay, including the Rockaway barrier system and inlet (Waldman 2008). Using the static approach, the NPCC2 created specific map products such as GIS shape files of the future 100–year flood extent for the 2020s, 2050s, 2080s, and 2100 based on FEMA's Preliminary FIRMs. These estimates illustrate that Queens is the

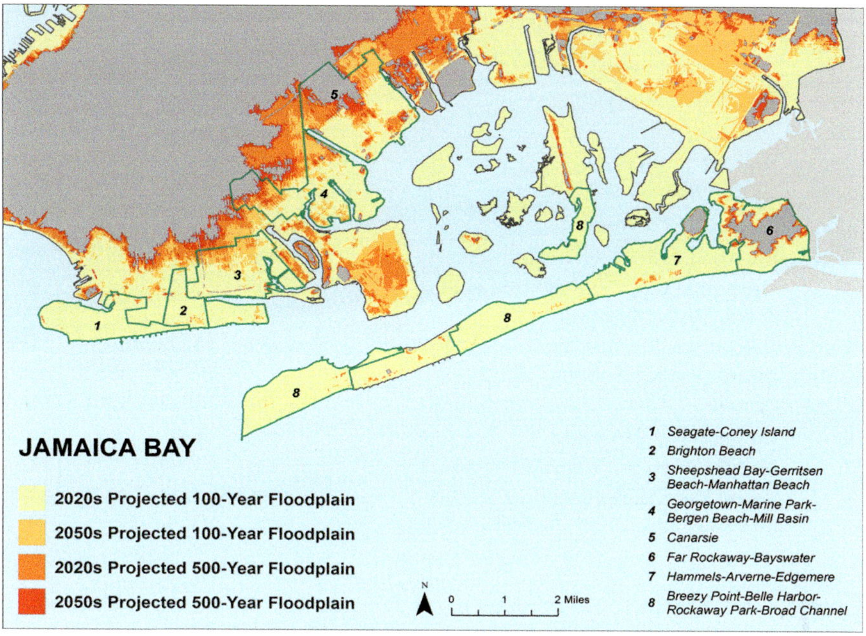

Fig. 3 Predicted areas at risks of Jamaica Bay region to 100- and 500-year floods of the 2020s and 2050s (Source: CIESIN, The Earth Institute, Columbia University, NY)

borough with the most affected land area, followed by Brooklyn, Staten Island, the Bronx, and Manhattan (Patrick et al. 2015). For example, large portions of land subject to flooding in Queens are part of the Gateway National Recreation Area, including the uninhabited intertidal salt marshes of the Jamaica Bay Wildlife Refuge and the parks of Fort Tilden, Breezy Point Tip, and Jacob Riis.

Rather than adopting the NPCC2 approach, in this study, the map products (Fig. 3) of predicted impact of SLR specific to the neighborhoods of Jamaica Bay region are created for SLR projections such as 10 in. (0.25 m) and 30 in. (0.76 m) of the future 100-year and 500-year floods of 2020 and 2050. Breezy Point, Belle Harbor, Rockaway Park, and Broad Channel; Sheepshead Bay, Gerritsen Beach, and Manhattan Beach; and Hammels, Arverne, and Edgemere are three neighborhoods of Jamaica Bay that are at high risk when compared to Far Rockaway, Bayswater, and Brighton Beach and other regions (in terms of acres of areas) to 100-year and 500-year floods of the 2020s and 2050s (Table 1). Likewise, map products (Fig. 4) are created for populations of Jamaica Bay region at risk to future 100-year and 500-year floods of the 2020s and 2050s. Population at Sheepshead Bay, Gerritsen Beach, and Manhattan Beach and Canarsie and Hammels, Arverne, and Edgemere are at high risk when compared to Breezy Point, Belle Harbor, Rockaway Park, and Broad Channel and Far Rockaway, Bayswater, and other regions (in terms of population) to 100-year and 500-year floods of the 2020s and 2050s (Table 2). As a whole, the Sheepshead Bay, Gerritsen Beach, and Manhattan Beach and Hammels,

Table 1 Estimation of areas at risks in Jamaica Bay region to 100- and 500-year floods of the 2020s and 2050s

S.No	Name of the neighborhood of Jamaica Bay	Total area (acres)	100-year flood (acres of area at risk of flooding)		500-year flood (acres of area at risk of flooding)	
			2020	2050	2020	2050
1.	Seagate, Coney Island	889	871	879	888	888
2.	Brighton Beach	394	368	389	393	393
3.	Sheepshead Bay, Gerritsen Beach, Manhattan Beach	1454	1093	1327	1430	1433
4.	Georgetown, Marine Park, Bergen Beach, Mill Basin	1592	891	1202	1370	1434
5.	Canarsie	1883	762	1045	1364	1507
6.	Far Rockaway, Bayswater	1248	614	679	769	829
7.	Hammels, Arverne, Edgemere	1421	1267	1294	1321	1326
8.	Breezy Point, Belle Harbor, Rockaway Park, Broad Channel	2283	2255	2279	2283	2283

Arverne, and Edgemere neighborhoods of Jamaica Bay are at high risk (both in terms of areas and population) when compared to the Far Rockaway, Bayswater, neighborhood and other regions of Jamaica Bay (both in terms of areas and population) to 100-year and 500-year floods of the 2020s and 2050s.

Case Studies on Socio-ecological Resilience to Sea-Level Rise and Floods at Jamaica Bay

Hurricane Sandy has brought public attention to the climate hazards of the New York area. It demonstrated the system's vulnerability to a whole host of weather-related threats, ranging from storm surge and SLR to heavy downpours – threats that are expected to worsen as the climate changes (Plan NYC 2013). It was a wake-up call for studying hurricane effects on the resilience and stability of coastal wetlands along the Atlantic coast, especially coastal areas along the northeast Atlantic seaboard where a dense human population resides. Over the last decade, there has been an active debate on the best ways to protect areas such as Jamaica Bay from storms. Hurricane Sandy only highlighted the need to provide better information. There have been a number of initiatives taken by various stakeholders of New York City in general and Jamaica Bay in particular, and one such example is the NYC Special Initiative for Rebuilding and Resiliency (SIRR). However, given the effect of accelerating SLR and the possibility of more intense hurricanes due to projected climate change (Wang et al. 2017), it is necessary to build the socio-ecological resilience of the Jamaica Bay region. One possible approach involves building sea walls (or flood walls) and other "gray" structures that will work to stop storm surge and strong waves caused by coastal storms. This is often referred to as "shoreline armoring." A second

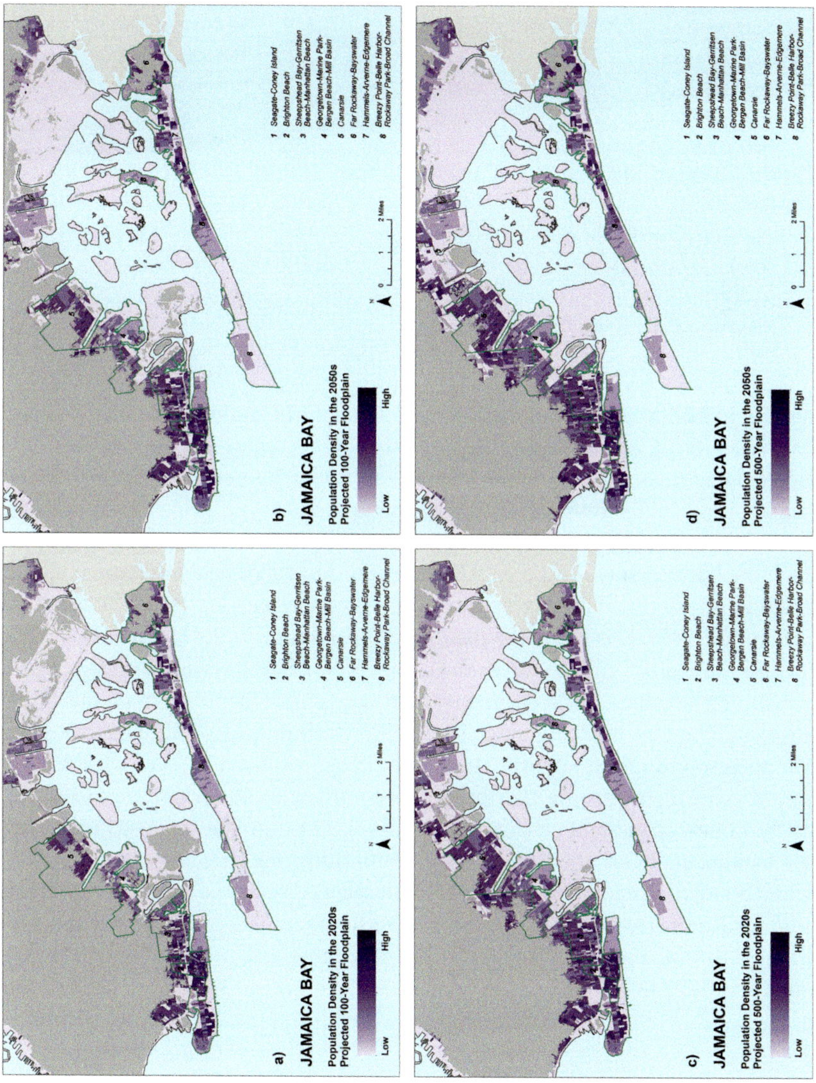

Fig. 4 Predicted population densities at risks of Jamaica Bay region to 100- and 500-year floods of the 2020s and 2050s (Source: CIESIN, The Earth Institute, Columbia University, NY)

Table 2 Estimation of population densities at risks in Jamaica Bay region to 100- and 500-year floods of the 2020s and 2050s

S.No	Name of the neighborhood of Jamaica Bay	Total population	100-year flood (population at risk)		500-year flood (population at risk)	
			2020	2050	2020	2050
1.	Seagate, Coney Island	31,965	31,819	31,868	31,963	31,965
2.	Brighton Beach	35,547	32,768	35,182	35,547	35,547
3.	Sheepshead Bay, Gerritsen Beach, Manhattan Beach	64,518	43,779	57,724	64,094	64,155
4.	Georgetown, Marine Park, Bergen Beach, Mill Basin	45,231	17,806	28,862	35,216	37,720
5.	Canarsie	83,693	33,412	46,204	64,314	70,744
6.	Far Rockaway, Bayswater	50,058	23,454	25,799	28,961	31,161
7.	Hammels, Arverne, Edgemere	36,885	34,933	36,061	36,651	36,885
8.	Breezy Point, Belle Harbor, Rockaway Park, Broad Channel	28,018	27,251	28,009	28,018	28,018

approach is to build "green" infrastructure such as dunes and marshes that will also protect coastal areas and provide habitat as well as recreational opportunities. The "green" approaches are sometimes referred to as "living shorelines" (ERG 2016). A National Oceanic and Atmospheric Administration (NOAA) Coastal Ocean Climate Application program award evaluated the potential impacts of a living shoreline approach on Jamaica Bay (Orton et al. 2017). The results of this work are available in an online interactive mapping application and accompanying technical report at http://adaptmap.info/. On the other hand, local communities in Jamaica Bay have identified a number of projects to increase their resiliency to coastal storms. Resilience planning engages decision-makers and vulnerable population in managing climate change and in implementing specific activities that can build understanding on how to respond (ADB 2014). The communities of Jamaica Bay with partner agencies and organizations are charting a path to urban coastal resilience that others around the country can build on (Fitzpatrick 2014). The case studies presented in this chapter stemmed out of the larger impetus to study the effects and impacts of Hurricane Sandy, focusing on coastal SLR and related themes of adaptation, mitigation, and resilience. However, these case studies do not reflect on the newer analyses and findings of the predicted impact of sea-level rise and floods at Jamaica Bay described in this chapter.

Adaptation to Coastal Flooding in New York City Neighborhoods

Adaptation to coastal flooding due to storms is a high priority for city governments on the densely populated northeastern coast of the United States (The City of New York 2011). The concern increased sharply after Hurricane Sandy. New York City has been at the forefront of engaging with climate scientists to understand the city's vulnerability to climate risks in general and to coastal flooding in particular and to develop adaptation strategies. Initial efforts to address the increasing coastal flood risk were primarily focused on engineered solutions (The City of New York 2013). More recently, the city has engaged in efforts to improve adaptation of residential communities.

This case study complements the efforts to improve adaptation at the neighborhood level by documenting the recovery from Hurricane Sandy in two coastal neighborhoods in New York City, which were most devastated by the storm. The study demonstrates sources of vulnerability that need to be addressed as well as factors that facilitated the recovery, which illustrate how adaptation can be improved. This study focuses on the Rockaways neighborhoods of the Jamaica Bay region. This neighborhood bore the brunt of Hurricane Sandy, experiencing similar intensities of impacts, and is still recovering from the storm. This case study adopted a mixed method approach, combining evidence from in-depth interviews of community leaders, people involved in the recovery, and residents with data from surveys of residents in the two neighborhoods.

The evidence suggests that low- and middle-income homeowners are least recovered from Hurricane Sandy and most vulnerable to future coastal flooding. This evidence is contrary to the common assumption in the literature on social vulnerability that low-income populations are most vulnerable to climate risks, and it illustrates the importance of documenting vulnerabilities that are specific to each climate risk and context in order to inform policy decisions (Madajewicz and Coirolo 2018). The main loss caused by the storm, whose cost was borne by residents of the affected areas, was damage to people's homes. Rebuilding was the component of recovery for which there was least assistance available. Fifty-two percent of the homeowners who responded to the surveys had flood insurance. Eighty-seven percent sustained flood damage. Insurance did not cover the full cost of damages for almost anyone. Low- to middle-income homeowners spent on average $30,000 on the recovery out of their own pockets, over and above any money recovered from insurance or from other assistance. For the homeowner at the 25th percentile of the income distribution, which is $42,500 in the survey sample, the amount was 71% of their gross annual income. Higher income homeowners spent a similar amount, $28,000, which is about 32% of gross annual household income for the homeowner with median annual household income. Renters, who are in the same income range as the low- to middle-income homeowners, spent on average $3400 out of their own pockets. Low- to middle-income homeowners received on average $9000 in financial assistance and $24,000 in insurance payments, while wealthier homeowners received on average $19,000 in financial assistance and $21,000 in insurance payments. Renters received $3700 on average in financial assistance. After these expenditures, low- and middle-income homeowners are far less resilient to future storms than they were before Sandy.

Limited access to several types of information and knowledge was a major source of vulnerability in coastal neighborhoods after Hurricane Sandy. Poor understanding of storm warnings resulted in residents remaining in their homes. Recovery was slower where distribution of information about treating mold and about sources of assistance was limited. Widespread lack of awareness about future flood risk and how homeowners can reduce damage from future flooding contributed to prevalent rebuilding that retained the same risks, which affected homes before Hurricane Sandy. For example, 80% of survey respondents had to replace their boilers and/or hot water heaters after the storm. Yet only 17% raised any of their electric utilities and the heights to which these utilities were raised were not informed by estimates of future flood risk. Progress on adaptation requires improved understanding at the local, neighborhood level of future flood risk, storm warnings, and actions that residents can take to adapt, in order to inform the resident's decisions about the costs they want to incur to reduce their expected future damages from coastal flooding and whether or not they want to relocate.

Neighborhoods that had stronger community organizations recovered more quickly and completely. Social networks as well as volunteers, community groups, and nonprofit, non-governmental organizations (NGOs) had the presence on the ground that enabled them to collect precise information about who needed what and where and to address those needs quickly. Community organizations were the principal sources of help with rebuilding homes. The rebuilding work ground to a halt when many homes still needed assistance because the NGOs depleted their funding. Public/private partnerships could enhance the capacity of local and volunteer networks and organizations to deliver well-targeted assistance that would improve adaptation to future storms. Partnerships could combine local information and capacity to meet needs rapidly, which are the strengths of the community organizations, and the resources and technical expertise of the public sector (Madajewicz and Coirolo 2018).

Building Community Resilience Capacity in Jamaica Bay

When Hurricane Sandy lashed into New York, most New York City residents thought they were well prepared to weather the hurricane and its after-effects. The negative consequences of the storm disproportionately impacted its most vulnerable residents: the working poor living in immigrant enclaves, those individuals without access to private transportation, public housing residents, and others who were already living in challenging circumstances before the storm eventuated. In 2014, approximately 18 months after the event, researchers from the Institute for Sustainable Cities (ISC) at Hunter College, with support from the Science and Resilience Institute at Jamaica Bay, and the Rockefeller Foundation undertook a qualitative study in several neighborhoods in and around Jamaica Bay to understand how individuals in neighborhoods and communities were navigating the recovery.

The ISC study used qualitative and ethnographic field research methods to connect with these vulnerable populations. The data collection was organized in two levels; a team of graduate students compiled a database of over 400 direct service providers and community-based organizations working in the Jamaica Bay area. This was no small challenge, when considering the reality that Jamaica Bay itself extends 31 square miles and the population in the greater Jamaica Bay watershed approximates about one million people. The service providers shared their localized and specialized knowledge about the geographical and the social context of different neighborhoods and communities. The research team selected eight neighborhood areas including Gerritsen Beach, Canarsie, Idlewild, Broad Channel, Far Rockaway, Rockaway Park, Arverne, and Breezy Point to conduct focused interviews with residents, engaging them to reflect on their post-Sandy experiences.

In encapsulating the lived experiences of vulnerable populations, the research study was startled to document how a singular event such as Hurricane Sandy served as a disruptive event that negatively impacted long-term resilience. The picture of community resilience that gathered after synthesizing interviews with residents, service providers, and professional experts is a complex and sometimes a counter-intuitive narrative. For some individuals and in some neighborhoods, resilience was a state of mind in part, because of an indomitable individual spirit, coupled with a neighborhood that was already connected and engaged. In these resilient neighborhoods liked Broad Channel or Breezy Point, residents spoke English, had stable jobs, and owned their homes. They had lived in Jamaica Bay long enough to understand the risks that came with living on the coast, like the effect that tides have on SLR. A mixture of stoicism and a determination to rebuild and stay in a place indicated to the researchers that the benefits of staying in that neighborhood far outweighed perceived "temporary" inconveniences created by the storm.

At the same time, in some neighborhoods like Canarsie, newer New Yorkers who had moved there because of low rents in full/partial basement apartments were displaced for several months. The absence of stable and affordable housing options made renters vulnerable as soon as their places of residence were damaged. The research reaffirmed what became widely known – that both private and public supplies of housing were affected as a result of the storm; the research further elucidated the disproportionate burden borne by renters. It also documented that the instability in housing access created additional problems for vulnerable populations. However, one of the more important findings of the ethnographic work was the absence of key informants who were scattered because of the housing and transportation challenges. This is one of the main challenges researchers encountered in understanding community resilience practices in the low-income neighborhoods around Jamaica Bay (Ramasubramanian et al. 2016a).

In documenting the lived experiences of everyday people, the research team recognized that community (or neighborhood) resilience relies heavily on preserving affordable housing stock in safe and usable condition after a natural disaster. Likewise, maintaining effective public transportation networks after disaster events is an important cornerstone of community resilience because it provides access to jobs and much needed income support. Housing, transportation, and jobs then are the cornerstones of community resilience. The post-Sandy research in Jamaica Bay also educated the research team in recognizing the value of micro-neighborhood research. It was a preliminary step in understanding how to integrate the lived experiences of individuals and groups to make recommendations to state and non-state actors that would enhance community resilience. In this context, community resilience includes the experiences of individuals and groups and takes into the account the natural environment, the built environment, and the network of civil society and government organizations that provide infrastructural and resource support for that community.

Importantly, the field work resulted in two broad categories of recommendations: one related to education and the second related to engagement. (a) Educating the general public about the issues related to climate change and consequent impact propensity for more frequent and more powerful storms, rising sea levels, numerous flooding events, and so on is critical to creating community resilience in coastal neighborhoods. This education should be available to young people as well as older adults and include both long-term residents and newcomers to the area. (b) The theme of engagement captures the complexity and the challenges of promoting community resilience. The research team recommends that community resilience activities become incorporated as part of a larger neighborhood-based development strategy that emphasizes opportunities for people in the community to come together for vocational education, job training, or learning about ways to revitalize their neighborhoods through civic action. The need for a physical location to host these types of engagement activities that was accessible to neighborhood residents was also recognized (Ramasubramanian et al. 2016b).

In conclusion, the ISC qualitative research about Jamaica Bay communities, approximately 18 months after Hurricane Sandy made landfall, revealed that community resilience among poor communities was present, although fragile and in need of external inputs of resources and expertise. Additional research is needed to investigate the ways in which such support can be effectively delivered to these communities.

Conclusion

Accelerated SLR is a major long-term outcome of climate change leading to increased inundation of low-lying areas which will subject global cities and urban coastal communities to greater risk (Khan 2017). It plays havoc with coastal cities, submerging some areas, and making others far more vulnerable to storm surges, or adversely impacting key infrastructure (Rosenzweig et al. 2011). Building resilience and adapting to climate change threats like SLR are increasingly high priorities for urban coastal cities (Romero Lankao and Dodman 2011; Khan 2017). The starting point in managing risks and building long-term resilience is for a city to understand its exposure and sensitivity to a given set of impacts and develop responsive strategies (The World Bank 2011). Building socio-ecological resilience is one of the significant ways to address SLR as an adaptation strategy. Social and ecological resilience to climate change are inextricably linked and should be considered as integrated socio-ecological systems (Wasterman et al. 2012). Thus, building socio-ecological resilience of densely populated urban coastal regions warrants urgent attention, and cities can take major steps toward building urban climate resilience to SLR. However, there is no single action that will make a city resilient to climate change (SLR). Resilience is instead achieved through a number of actions, building upon each other over time. These actions would be enhanced and progressed as peoples and institutions learn from past experiences and apply it to future decisions (ADB 2014). Furthermore, building resilience requires not only robust decision-making by those in positions of formal authority but also a strong web of institutional and social relationships that can provide a safety net for vulnerable populations (The World Bank 2011). Jamaica Bay in New York City is a paradigmatic example of environmental vulnerability, particularly given future SLR (Seavitt 2015). Nevertheless, the New York metropolitan region is a classic example of a complex socio-ecological system (Cadenasso et al. 2007; McPhearson et al. 2014). In particular, Jamaica Bay is an ultimate socio-ecological system that exhibits interactions to physical and ecological systems and human or social systems (Binder et al. 2013; Allred et al. 2016). It has the potential to be transformed as an ecological, infrastructural, and community asset – an anchor of the region's resiliency (Seavitt 2015). The bay provides an incredible opportunity for exploring the performance of nature-based features, particularly salt marshes, maritime forests, and dunes. The bay is a dynamic ecological entity – a sandy, shifting terrain geologically capable of functioning resiliently during and after disturbance events (Seavitt 2015). In this study, COREDAR- (a capacity building tool) based SLR and flood risk information assortment using GIS-based SLR and flood risk assessments of the Jamaica Bay region of New York City (1) provide important information about neighborhood areas and population of Jamaica Bay region that are at risk for SLR projections such as 10 in. and 30 in. of the future 100-year and 500-year floods of the 2020s and 2050s and (2) offer experiences and lessons learned from case studies emphasizing the interaction of socio-ecological systems as a powerful mechanism to enhance resilience since they bring together stakeholders from different sectors, build on their strengths, and craft capacity that would not otherwise exist (Thunder Bay 2015). However, generating and sharing knowledge is critical to build resilience of ecosystems and communities to SLR. As outlined in this chapter, the information gleaned and the case studies highlighted with Jamaica Bay could help spur coastal cities to build resilience of urban ecosystems and communities to SLR. Importantly, it provides access to new information and creates the opportunity to reflect, learn, and act. Thus, the fate of Jamaica Bay serves as a wake-up call to build socio-ecological resilience of other coastal wetlands facing the intertwined effects of SLR and human-induced stresses (Hartig and Gorntiz 2001).

Acknowledgment The authors are grateful to the Department of State, Government of the United States of America, and Government of India for funding this study (COREDAR) through Fulbright-Nehru Postdoctoral Research Program (2015–2016).

References

ADB (2014) Urban climate change resilience: a synopsis. Asian Development Bank, Philippines https://www.adb.org/sites/default/files/publication/149164/urban-climate-change-resilience-synopsis.pdf. Accessed 05 Mar 2018

Adger WN, Hughes TP, Folke C, Carpenter SR, Rockstro MJ (2005) Social-ecological resilience to coastal disasters. Science 309(5737):1036–1039. https://doi.org/10.1126/science.1112122

Allred S, DuBois B, Bunting-Howarth K, Tidball K, Solecki WD (2016) Social-ecological system transformation in Jamaica Bay. In: Sanderson EW, Solecki WD, Waldman JR, Parris AS (eds) Prospects for resilience. Island Press, Washington, DC

Barth MC, Titus JG (1984) Greenhouse effect and sea level rise. Van Nostrand Reinhold, New York

Binder CR, Hinkel J, Bots PWG, Pahl-Wostl C (2013) Comparison of frameworks for analyzing socio-ecological systems. Ecol Soc 18(4):26

Black FR (1981) Jamaica Bay: a history, Cultural resources management study No. 3. US Department of Interior, National Park Service, Washington, DC

Branco B, Waldman JR (2016) Resilience practice in urban watersheds. In: Sanderson EW, William DS, Waldman RJ, Parris AS (eds) Prospects for resilience: insights from New York City's Jamaica Bay. Island Press, Washington, Covelo, London, p 23

Brand CJ, Windingstad RM, Siegfried LM, Duncan RM, Cook RM (2018) Avian morbidity and mortality from botulism, aspergillosis, and salmonellosis at Jamaica Bay Wildlife Refuge, New York, USA. Colon Water bird 11(2):284–292

Breitmeier H, Kuhn J, Schwindenhammer S (2009) Analyzing urban adaptation strategies to climate change: A comparison of the coastal cities of Dhaka, Lagos and Hamburg. Contribution to the panel "Regieren im Klimawandel" section,, Regierungssystem und Regieren in der Bundesrepublik Deutschland", DVPW-Kongress 21.-25.09.2009, Kiel, Germany. https://www.unigiessen.de/fbz/fb03/institute/ifp/personen/schwindenhammer/dateien/paper1. Accessed 05 Mar 2018

Burger J (1981) Movements of juvenile Herring Gulls hatched at Jamaica Bay Refuge, New York. J Field Ornithol 52:285–229

Cadenasso M, Steward T, Pickett A, Schwarz K (2007) Spatial heterogeneity in urban ecosystems: reconceptualizing land cover and a framework for classification. Front Ecol Environ 5:80–88

Church JA, Clark PU, Cazenave A, Gregory JM, Jevrejeva S, Levermann A, Merrifield MA, Milne GA, Nerem RS, Nunn PD, Payne AJ, Pfeffer WT, Stammer D, Unnikrishnan AS (2013) Sea level change. In: Stocker TF, Qin D, Plattner GK, Tignor M, Allen SK, Boschung J, Nauels A, Xia Y, Bex V, Midgley PM (eds) Climate change 2013: The physical science basis. contribution of working group I to the fifth assessment report of the intergovernmental panel on climate change. Cambridge University Press, Cambridge, UK, NY

DOS (1992) Coastal fish and wildlife habitat rating form. Department of State, New York. https://www.dos.ny.gov/opd/programs/consistency/Habitats/nyc/Jamaica_Bay.pdf. Accessed 05 Mar 2018

Doxsey-Whitfield E, MacManus K, Adamo S, Pistolesi L, Squires J, Borkovska O, Baptista S (2015) Taking advantage of the improved availability of census data: a first look at the gridded population of the World, version 4. Papers in Applied Geography 1(3):226–234

Englebright S (1975) Jamaica Bay: a case study of geo-environmental stresses. A guidebook to field excursions. New York State Geological Association, Hempstead

ERG (2016) Hurricane Sandy and the value of trade-offs in coastal restoration and protection. Eastern Research Group, Inc, Lexington https://coast.noaa.gov/data/digitalcoast/pdf/sandy-report.pdf. Accessed 05 Mar 2018

Fitzpatrick M (2014) As sea level rises in Jamaica Bay, New York, Tidal flooding moves from occasional to chronic. Union of concerned scientist. https://blog.ucsusa.org/melanie-fitzpatrick/sea-level-rise-in-jamaica-bay-new-york-688. Accessed 05 Mar 2018

Folke C (2006) Resilience: The emergence of a perspective for social-ecological systems analysis. Global Environ Chang 16:253–267. https://doi.org/10.1016/j.gloenvcha.2006.04.002

Folke C, Carpenter S, Elmqvist T, Gunderson L, Holling CS, Walker B, Svedin U (2002) Resilience and sustainable development: Building adaptive capacity in a world of transformations. Environmental Advisory Council, Stockholm

Folke C, Hahn T, Olsson P, Norberg J (2005) Adaptive governance of social-ecological systems. Annu Rev Environ Resour 30:441–473. https://doi.org/10.1146/annurev.energy.30.050504.144511

Gornitz V, Horton R, Bader DA, Orton P, Rosenzweig C (2017) Coping with higher sea levels and increased coastal flooding in New York City. In: Walter LF, Keenan JM (eds) Climate change adaptation in North America: fostering resilience and the regional capacity to adapt. Springer, Hamburg, pp 209–223

Gorntiz V, Couch S, Hartig EK (2002) Impacts of sea level rise in the New York City metropolitan area. Glob Planet Chang 32:61–88
Gunderson LH, Holling CS (2002) Panarchy: understanding transformations in human and natural systems. Island Press, Washington, DC
Hartig EK, Gorntiz V (2001) The vanishing marshes of Jamaica Bay: sea level rise or environmental degradation? Science Briefs, NASA GISS
Hartig EK, Gornitz V, Kolker A, Mushacke F, Fallon D (2002) Anthropogenic and climate- change impacts on salt marshes of Jamaica Bay, NYC. Wetlands 22(1):71–89
Horton R, Little C, Gornitz V, Bader D, Oppenheimer M (2015) Chapter 2: Sea level rise and coastal storms. In: New York city panel on climate change 2015 report. Building the knowledge base for climate resiliency. Annals of New York Academy of Sciences. New York Academy of Science, New York, pp 36–44
IPCC (2007) Assessing key vulnerabilities and the risk from climate change. In: Parry ML, Canziani OF, Palutikof JP, van der Linden PJ, Hanson CE (eds) Impact, adaptation and vulnerability, contribution of working group II to the fourth assessment report of the intergovernmental panel on climate change. Cambridge University Press, Cambridge, UK, NY, p 976
IPCC (2013) Chapter 5. Coastal systems and low-lying areas. In: Poh Poh W, Iñigo JL (eds) Impacts, adaptation and vulnerability. Contribution of working group II to the fifth assessment report of the intergovernmental panel on climate change. Cambridge University Press, Cambridge, UK, pp 1–28
IPCC (2014) Glossary – Annex II. Climate Change 2014: Impacts, adaptation and vulnerability. Intergovernmental panel on climate change, fifth assessment report. Cambridge University Press, New York, USA
Khan AS (2017) COREDAR for cities: developing a capacity building tool for sea-level rise risk communication and urban community-based adaptation. In: Leal Filho W, Keenan JM (eds) Climate change adaptation in North America: Fostering resilience and the regional capacity to adapt. Climate Change Management, Springer, Berlin, pp 137–157
Khan AS, Chen RS, De Sherbinin A, Ramachandran A, Palanivelu K (2015) COREDAR framework-communicating risk of sea-level rise and engaging stakeholders in framing community-based adaptation strategies. American Geophysical Union (AGU), fall meeting, 16 Dec 2015, San Francisco, California, USA. https://agu.confex.com/agu/fm15/webprogram/Paper78074.html. Accessed 05 Mar 2018
Leichenko R (2011) Climate change and urban resilience. Curr Opin Environ Sust 3(3):164–168
Madajewicz M, Coirolo C (2018) Adaptation to coastal flooding in New York City neighborhoods: A summary. National Oceanic and Atmospheric Administration (NOAA) Coastal and Ocean Climate, NOAA Regional Integrated Science and Assessment (RISA) grant that is funding the Consortium for Climate Risk in the Urban Northeast (CCRUN)
Marfai MA, King L (2008) Potential vulnerability implications of coastal inundation due to sea level rise for the coastal zone of Semarang city, Indonesia. Environ Geol 54:1235–1245
McGranahan G, Balk D, Anderson B (2007) The rising tide: assessing the risks of climate change and human settlements in low elevation coastal zones. Environ Urban 19(1):17–37
McPhearson T, Hamstead ZA, Kremer P (2014) Urban ecosystem services for resilience planning and management in New York City. Ambio 43:502–515
Meerow S, Newell JP, Stults M (2016) Defining urban resilience: a review. Landscape Urban Plan 147:38–49
National Park Service (2004) The evolving legacy of Jamaica Bay. Gateway National Recreation Area, and Jamaica Bay Institute http://www.ciesin.org/jamaicabay/stakeholder/LegacyofJB_110503_NPS.pdf. Accessed 05 Mar 2018
Nicholls RJ, Cazenave A (2010) Sea-level rise and its impact on coastal zones. Science 328(5985):1517–1520
Nicholls R, Wong P, Burkett V, Codignotto J, Hay J, McLean R, Ragoonaden S, Woodroffe C (2007) Coastal systems and low-lying areas. In: Parry P, Canziani O, Palutikof J, Van der Linden P, Hanson C (eds) Climate change 2007: impacts, adaptation and vulnerability, contribution of working group II to the fourth assessment report of the intergovernmental panel on climate change. Cambridge University Press, Cambridge, UK, pp 315–356

NOAA (2018) Mean sea level trend, Station ID 8518750. The Battery, New York Tides and Current https://tidesandcurrents.noaa.gov/sltrends/sltrends_station.shtml?stnid=8518750. Accessed 05 March 2018

NYC Department of City Planning (2018) Current estimates of New York City's population for July 2016. http://www1.nyc.gov/site/planning/data-maps/nyc-population/current-future-populations.page#collapse3 Accessed 20 Mar 2018

NYC-DEP (2007) The Jamaica Bay Ecosystem. In: Draft Jamaica Bay watershed protection plan, New York City, New York, USA http://www.nyc.gov/html/dep/pdf/jamaica_bay/volume-1_3-1-07_part-2.pdf. Accessed 05 Mar 2018

NYC-DEP (2016) Jamaica Bay watershed protection plan 2016 Update. New York City department of environment protection, New York City, New York. http://www.nyc.gov/html/dep/pdf/jamaica_bay/jbwpp_update_10012016.pdf. Accessed 05 Mar 2018

OECD (2007) The analysis focused on the exposure of population and assets to a 1 in 100 year surge-induced flood event (assuming no defences) rather than "risk" from coastal flooding. http://www.rrojasdatabank.info/statewc08093.3.pdf. Accessed 05 Mar 2018

Orton P, MacManus K, Sanderson E, Mills J, Giampieri M, Fisher K, Yetman G, Doxsey-Whitfield E, Hugens D, Wu Z, Yin L, Georgas N, Blumberg A (2017) Quantifying the value and communicating the protective services of nature-based flood mitigation using flood risk assessment: technical report, draft 2.1. http://adaptmap.info/jamaicabay/technical_report.pdf. Accessed 22 Mar 2018

Ostrom E (1990) Governing the commons: the evolution of institutions for collective action. Cambridge University Press, Cambridge, UK

Patrick L, Solecki W, Jacob KH, Kunreuther H, Nordenson G (2015) Chapter 3: static coastal flood mapping. In: New York City Panel on Climate Change 2015 Report. Building the knowledge base for climate resiliency. Annals of New York Academy of Sciences. New York Academy of Science, New York, pp 45–55

Plan NYC (2013) A stronger and more resilient New York. NYC Special Initiative for rebuilding and resiliency. http://s-media.nyc.gov/agencies/sirr/SIRR_spreads_Lo_Res.pdf. Accessed 05 Mar 2018

Ramachandran A, Khan AS, Prasnnahvenkatesh R, Palanivelu K, Jayanthi N (2017) Projection of climate change induced sea level rise for the coasts of Tamil Nadu and Puducherry, India using SimCLIM: a first step towards planning adaptation policies. J Coast Conserv 21:731–742

Ramasubramanian L, Menser M, Rieser E et al (2016a) Neighborhood and community perspectives of resilience in the Jamaica Bay Watershed. In: Sanderson E, Solecki W, Waldman J, Parris A (eds) Prospects for resilience: insights from New York City's Jamaica Bay. Island Press, Washington DC, pp 117–137

Ramasubramanian L, Menser M, Rieser E et al (2016b) Strategies for Community Resilience for the Jamaica Bay Watershed. In: Sanderson E, Solecki W, Waldman J, Parris A (eds) Prospects for resilience: insights from New York city's Jamaica Bay. Island Press, Washington DC, p 241

Rodin J (2014) The resilience dividend: being strong in a world where things go wrong. Public Affairs, New York

Romero Lankao P, Dodman D (2011) Cities in transition: transforming urban centers from hotbeds of GHG emissions and vulnerability to seedbeds of sustainability and resilience. Curr Opin Environ Sustain 3(3):113–120

Rosenzweig C, Solecki WD, Hammer SA, Mehrotra S (2011) Climate change and cities. First assessment report of the urban climate change research network, Center for climate systems research, Earth Institute, Columbia University. Cambridge University Press, New York City, NY

Seavitt C (2015). Experimental research studio: Jamaica Bay. Urban Omnibus, A publication of the architectural league of New York https://urbanomnibus.net/2015/01/experimental-research-studio-jamaica-bay/ Accessed 06 Mar 2018

State of the Planet (2016) Helping communities respond to climate change. Earth Institute, Columbia University, New York, USA. http://blogs.ei.columbia.edu/2016/02/23/newcommunications-tool-to-help-communities-respond-to-climate-change/. Accessed 05 Mar 2018

Tanacredi JT, Badger CJ (1995) Gateway, a visitor's companion. Stockpile Books, Mechanicsburg

The City of New York (2011) PlaNYC: A Greener. Update, Greater New York
The City of New York (2013) A stronger, more resilient New York. The City of New York, Mayor Michael R. Bloomberg. PlaNYC: A Greener, Greater New York Update
The World Bank (2011) Guide to climate change adaptation in cities. The World Bank Group. http://siteresources.worldbank.org/INTURBANDEVELOPMENT/Resources/336387-1318995974398/GuideClimChangeAdaptCities.pdf. Accessed 05 March 2018
Thunder Bay (2015) Climate ready city. City of Thunder Bay Climate adaptation strategy. https://climatereadycity.com/wpcontent/uploads/2015/12/Earthcare_Climate_Ready_City_Web1.pdf. Accessed 05 Mar 2018
Tompkins EL, Adger WN (2004) Does adaptive management of natural resources enhance resilience to climate change? Ecol Soc 9(2):10 http://www.ecologyandsociety.org/vol9/iss2/art10. Accessed 05 Mar 2018
Tyler S, Moench M (2012) A framework for urban climate resilience. Clim Dev 4(4):311–326
UNAI (2016) New tool helps communities better understand climate change. United Nations Academic Impact, United Nations Headquarters, New York, USA https://academicimpact.un.org/content/new-tool-helps-communities-better-understand-climate-change. Accessed 05 Mar 2018
US Department of Interior (2013) Draft general management plan environmental impact statement July 2013. A new vision for a great Urban National Park, Gateway National Recreation Area. National Park Service, US Department of Interior, USA https://www.nps.gov/gate/learn/management/upload/GNRA-GMP-EIS-Executive-Summary-July-2013–1.pdf. Accessed 05 Mar 2018
Van Hooreweghe KL (2012) The creeks, beaches, and bay of the Jamaica Bay Estuary: The importance of place in cultivating relationships to nature. CUNY Academic Works, City University of New York, NY, USA. https://academicworks.cuny.edu/cgi/viewcontent.cgi?referer=https://www.google.co.in/&httpsredir=1&article=2843&context=gc_etds. Accessed 05 Mar 2018
Waldman J (2008) Research opportunities in the natural and social sciences at the Jamaica Bay unit of gateway national recreation area. National Park Service Jamaica Bay Institute, Brooklyn, NY, USA
Wang H, Chen Q, Hu K, Snedden AG, Hartig EK, Couvillion BR, Johnson CL, Orton PM (2017) Numerical modeling of the effects of hurricane Sandy and potential future hurricanes on spatial patterns of salt marsh morphology in Jamaica Bay, New York City US geological survey open-file report 2017–1016, p 43
Wasterman K, Oleson LLK, Haris AR (2012) Building socio-ecological resilience to climate change through community-based coastal conservation and development: experiences in southern Madagascar, Western Indian Ocean. J Mar Sci 11(1):87–97

Risk Perception and Action to Reduce the Impact of Floods in the Czech Republic

6

Mohan Kumar Bera and Petr Daněk

Contents

Abstract

In the Czech Republic, the increasing impacts of floods in the late twentieth century led local communities and governments to question the usefulness of conventional ways in reducing the risk of disaster. This chapter aims to understand how changes in risk perception and in the disaster management paradigm have influenced the strategies local communities and government use to reduce the risk of floods. It finds that the perception of risk has been changed by coordination between villagers and local governments, the acceptability of local leadership, social capital and social network, community resilience, a sense of community, and by changes in insurance policies. Villagers trust the local government's efforts to reduce the impacts of floods, and the local government cannot overlook the people's voice in disaster management measures. Clearly, both risk

M. K. Bera (✉)
Institute of Economic Growth, New Delhi, India
e-mail: mohan.bera@gmail.com

P. Daněk
Department of Geography, Masaryk University, Brno, Czech Republic
e-mail: danek@sci.muni.cz

W. Leal Filho (ed.), *Handbook of Climate Change Resilience*,
https://doi.org/10.1007/978-3-319-93336-8_31

perception and consciousness of self-responsibility towards society influence people in the Czech Republic to engage in reducing the risk of disaster.

Keywords
Flood · Disaster · Risk perception · Disaster reduction · Czech Republic

Introduction

Since the mid-1990s, Central European countries have been struck by disastrous floods on an unprecedented scale and frequency. The increasing frequency of floods seems to be related to climatic variations (Brázdil et al. 2006), but also to changing patterns of land use. It has, therefore, become crucial to adopt measures and strategies to prevent flood-related risks. In the Czech Republic, floods in 1997, 2002, 2006, 2007, and 2013 created "crisis situations" as defined by Act No. 240/2000 Coll (Government of the Czech Republic 2000). Such crises are managed by the Integrated Rescue System (IRS), which guides the activities of various bodies at the municipality, district (municipalities with extended competences), provincial (kraj), and state levels. All hierarchical levels share the responsibility for managing crises, but the local municipality has final responsibility.

The damage and loss caused by floods and low participation in property insurance schemes have impacted people's risk perception and coping mechanisms and raised awareness of flood-related risks and – in some places – involvement in actions aimed at minimizing flood impact. A new discourse on flood risk management has emerged. The factors of the new discourse are both environmental (such as climatic variations and changing land use) and social (such as increased risk perception and the resulting change in the behavior of people and institutions).

This chapter aims to understand whether the new approach has changed local communities' perception of disaster risk and increased their resilience. It aims to understand also how the new approach has impacted the collective identity of people living in localities exposed to the risk of flood and their willingness to work collaboratively. This chapter discusses the new discourse's impact and paradoxes, and its influence over communities in flood-prone areas and over local governments' decision and implementation strategies. To understand the new approach to managing flood risk, researchers interviewed residents and authorities in two flood-prone municipalities.

The rest of this chapter is organized as follows. Section "Floods and Flood Management Regimes in the Czech Republic" briefly introduces the history of floods in the Czech Republic. Section "Risk Perception and Its Components" provides a theoretical discussion of the risk perception. Section "Research Methodology" outlines the research methodology. Section "Data Analysis" presents the research results, which are discussed in section "Discussion." Section "Conclusion" concludes the chapter.

Floods and Flood Management Regimes in the Czech Republic

Statistical records on floods in the Czech Republic are available since the seventeenth century (Brázdil et al. 2006). There were 102 flood events recorded between 1671 and 1870 (5.0 events per decade) and 95 floods between 1871 and 2009 (6.9 events per decade) (Brázdil et al. 2011). However, the frequency of floods decreased significantly in the second half of the twentieth century (Vávra et al. 2015). There were no floods exceeding 20 years return period on any of the major rivers in the Czech Republic after 1954; the last exceeding 20 years return period occurred in the Morava River basin in 1938, the Labe River basin in 1940, and in the Vltava River basin in 1954 (Brázdil et al. 2011). This was attributed to the success of intense technical measures taken to prevent floods – mainly the construction of large dams – and interpreted ideologically as the success of socialist engineering.

Building large dams constituted the major flood prevention strategy since the late nineteenth century – 588 reservoirs, including 123 large dams, were built from 1900 to 1997 (Ministry of Agriculture and Ministry of the Environment 2006). Between 1930 and 1991, a complex, extensive system of large dams was built on the Vltava River in Prague and its tributaries upstream. Since the frequency of floods dropped in the second half of the twentieth century (Vávra et al. 2015), it was believed that these dams served to prevent floods.

Disaster management used to be considered the government's responsibility, and citizens had only a limited role. However, disastrous floods in the late twentieth and early twenty-first centuries changed the perception of risk and, in turn, questioned the relevance of conventional strategies of flood risk reduction (Vávra et al. 2015). In 1989, communism ended, the responsibility of centralized flood management was transferred to the municipality (Čamrová and Viktorová 2006), and the protectionism of the communist system changed into the individualism of a liberal economy. Citizens were expected to take responsibility for their actions and face the consequences of their decision-making (Čamrová and Viktorová 2006), but continued to rely on the state from time to time. In 1997, floods on the Morava River claimed 50 lives, made evident that solely structural measures cannot provide sufficient prevention, and triggered a change in the approach to flood management. In June 2000, the Czech government adopted a new approach to flood management (Act No. 240/2000 Coll). In 2002, the Czech capital of Prague was extensively flooded by the overflowing Vltava River – despite the extensive network of dams. In 2004, the Czech Republic joined the EU. The EU Directive (European Union 2007) reinforced the new flood management approach, which aims to combine the continuing improvement of structural measures with the individual responsibility of people (Ministry of Agriculture 2007). The new integrated system of flood prevention specifies responsibilities for each institution and ways of effective communication between institutions, and between institutions and the population (Ministry of the Environment 2012).

Flood-related property insurance is an important factor of the risk-reduction strategy of individuals but, after the 1997 floods, insurance companies raised the cost of flood insurance so steeply that many households could not afford insurance any longer (Duží et al. 2017). Households who live on high ground or have good flood insurance rarely have a strong interest in contributing to the local government that reduces the risk of floods (Duží et al. 2017). But less well-off households and/or those in risky locations close to the river willingly cooperate with local government and other institutions in measures and strategies that prevent floods. This new cleavage follows both social-economic and geographical cleavages.

Risk Perception and Its Components

Risk indicates the likelihood of experiencing injury, damage, or loss (Slovic 1999), and perception of risk changes with the type and probability of disaster and people's judgement of such disaster. Risk can be classified into "dread risk" – lack of control over, and dread of, potential catastrophic events and their consequences – and "unknown risk," a perception of disasters that are unknown, have not been experienced, and have unobservable consequences. In the social science approach, risk is a subjective feeling; there is no "objective risk" (Slovic 1999). The phenomenon of risk is socially constructed around the perception of disaster (Slovic 1999), and awareness of risks is reflected in people's behavior and activities (Fischhoff et al. 1978).

Perception is the process of collecting, selecting, and interpreting information about the environment (Wachinger et al. 2013). Risk perception is a subjective image of a likely phenomenon developed through direct and indirect sources of information (Wachinger et al. 2013). It can also be modified and changed by the socioeconomic and cultural environment. Risk perception is an intuitive judgement of individuals or groups about risk in the context of limited and uncertain information (Raaijmakers et al. 2008). It is based on characteristics that result from awareness, worry, and preparedness (Slovic 2000).

Awareness, in the context of risks, is the knowledge of, and consciousness about, the existence of potential disaster. Awareness is developed through individual experience and external awareness programs. Worry depends on an individual's awareness of risk. Many people worry about risks and insist that their safety be ensured. Awareness can raise (or lower) worry and lead to higher (or lower) preparedness (Raaijmakers et al. 2008). Preparedness increases people's capability to reduce the negative consequences of risk. Preparedness depends on social, economic, technical, and institutional factors. A better prepared community is least worried about risk (Raaijmakers et al. 2008).

Risk perception, awareness, and preparedness are closely related. Higher risk perception leads to better preparedness and vice versa (Wachinger et al. 2013), because of experience and motivation, trust and responsibility, and individual ability. To understand the acceptance, rejection, or ignorance of risk or adaptation to it, Starr (1969) proposed the concept of "revealed preference," based on "risk-benefit tradeoffs." Despite their awareness of the threat of flood, many people live in flood-prone areas because they perceive that the benefits of living there outweigh the potential of negative consequences. They perceive the loss of livelihood related with their potential migration to a safer place or disconnection with friends, family, and neighbors to be a more real threat than disaster (Wachinger et al. 2013). Livelihood security is an important factor in the decision-making of people living in flood-prone areas. People in frequently flood-affected areas judge the threat through direct lived experience and indirect sources of information (Wachinger et al. 2013).

"Trust" is an important component of risk perception. Trust can reduce risk if the individual has knowledge of disasters and their possible consequences. It can help in reducing uncertainty and in enabling an individual to deal with complex, unfamiliar threats. If an individual trusts the authorities and flood protection measures, such perception could reduce their risk perception of natural hazards and, in turn, their willingness to prepare for disaster (Grothmann and Reusswig 2006). In the Netherlands, trust in structural protection measures and in government preventive programs reduced the intent to prepare for disaster preparedness (Terpstra 2011). In contrast, mistrust and lack of resources influenced the government to make individuals and communities responsible for adopting preventive measures (Armas and Avram 2009). Moreover, if individuals and communities trust public authorities, they heed the authorities' advice (Wachinger et al. 2013).

Jóhannesdóttir and Gísladóttir (2010) explain the relation between risk perception and vulnerability. When communities evaluate risk and vulnerability to disaster, it helps them to prioritize strategies of preventing and reducing risk and decide which would be most effective. In disaster management, vulnerability is a set of physical, socioeconomic, political, and psychological conditions that help people survive disaster (Cutter et al. 2003). The factors of risk perception are social (Cutter et al. 2003) and geographical (Brody et al. 2004). People living in low-lying areas perceive the risk of flood to be higher (Brody et al. 2008). The probability and severity of disaster also determines vulnerability.

However, as the frequency of disaster increases, understanding vulnerability is inadequate. When individuals and communities perceive threat and uncertainty to increase, they participate in government strategies of reducing the risk of disaster (Terpstra and Gutteling 2008). People and the government must collaborate to reduce the risk of disaster. Such participation in risk reduction strategies and decision-making processes improves peoples' awareness of the risk (Stanghellini and Collentine 2008), willingness to collaborate with the government, and trust in its strategies.

Until recently, The Czech Republic government flood prevention strategies relied heavily on building dams, dykes, and levees, and led people in flood-prone areas to believe that physical protection measures can contain flood risk and trust government strategies, which affected risk acceptance. However, it is impossible to achieve "zero risk" (Motoyoshi 2006) and, thus, the government must consider endangered communities' opinion and risk perception in preparing flood prevention strategies. This led, in recent years, to adoption of integrated flood risk management strategies, which aspire to accept the strategies and perception of people who are at risk (Motoyoshi 2006). However, the government has adopted a flood prevention approach that pays inadequate attention to flood preparedness and networking among different institutions in flood risk reduction.

Research Methodology

With the aim to assess the changes in the risk perception after the legislation of the Crisis Management Act (Act No. 240/2000 Coll.) at the local level of municipality, two case studies were conducted in municipalities in the flood-affected villages in the Morava River and Odra River basins. The first municipality, Podhradí nad Dyjí, is situated on the banks of the Dyje river (the major tributary of the Morava river). The village was flooded in 2002, when 300 years return period on the Dyje River was exceeded, and then twice in 2006. During the devastating floods in June 2006, the 500 years return period was exceeded. Podhadí nad Dyjí is a small, remote municipality near the Austria/Czech Republic state border. Almost all the employment opportunities in the region are in the tourist industry, forestry, and agriculture, and the skilled and educated working-age population are forced to migrate either temporarily or permanently. Weekend cottages or second homes of affluent city dwellers outnumber the homes of permanent residents in the municipality. The second municipality, Jeseník nad Odrou, is located near the confluence of the Odra and Luha rivers. A destructive flood afflicted the municipality in 2009. Six lives were lost, 150 family houses and 21 public buildings were damaged during the flood (AFP 2009). Jeseník nad Odrou is centrally located, on the railway connecting major agglomerations of the country. Most houses in the municipality are occupied by households whose working-age members either commute to nearby towns on daily basis or are employed in small local businesses.

A total of 15 in-depth interviews was conducted in both municipalities. The interviews sought to understand the risk perception of the affected population and other stakeholders and to solicit their opinion on mitigation strategies. In each municipality, researchers conducted interviews with households affected by the recent floods (13 respondents), a local government representative (one respondent), and experts (two respondents). I selected households by permanent residency and ownership of property in the village. Appointments were made before the interviews, which were conducted in the Czech language. Although the researchers specifically interviewed each household head, all other family members participated actively in the conversation. Each interview lasted an hour on average. A series of interviews was also conducted with the mayor of the municipality, who was later sent an additional unstructured questionnaire. Expert respondents included a meteorologist from the Czech Hydro Meteorological Institute (CHMI) and an environmental activist working in the Czech Republic. The recorded interviews provided the main source of data, which is analyzed in the following section.

Data Analysis

As per the law (Law No.458/1992), and in cooperation with the River Basin Authority, the CHMI provides a flood forecasting and warning service (Kubát 1999). Provincial and district authorities declare flood warnings and flood emergencies at the regional level and the municipal authority at the local level. Sirens, the traditional means of warning, are installed in all villages, and tested regularly. Usually, the municipal authority texts a message to the mobile phones of all villagers registered for the service. Villagers can also receive flood warnings from provincial authorities on mobile apps and, sometimes, receive warnings even before the municipality office does. Flood risk information is sufficient, but villagers lack the technical knowledge to analyze it.

Warnings are disseminated by public media such as TV or internet but, largely, residents trust early warnings issued by the municipal authority because its information is place-specific. Individuals accept early warnings depending on their age and prior experience of floods. Usually, older citizens trust their neighbors and community more than the government early warning system. The recent frequent floods were so destructive that some people are still apprehensive. And frequent unnecessary warnings may tempt villagers to migrate out of the municipality permanently, so the local government pays attention to warnings only if there is a real emergency. As the mayor mentioned, "It is too much! I do not want to lose my villagers" (Respondent No.7, 2016).

Villagers in Podhradí nad Dyjí and Jeseník nad Odrou have been living along the river bank for generations. The price of land increases with the distance from the river bank. Young couples stay on cheaply available land on the river bank, despite their worry about floods, and economic incapability influence many villagers to accept the risk. Their memory tends to consolidate minor flood events, and people refer only to destructive flood events. Sometimes, the garden near the river bank gets flooded during heavy rains, but destructive flood events are beyond the everyday experience. Villagers affected badly never forget the event. One respondent said, "People were shouting for help, we were not prepared, it was the night of horror" (Respondent No. 3, 2016). Dykes cannot always prevent damage during major floods. However, villagers who live near river banks trust and support the government for their work on structural measures at the river bank. In contrast, villagers who live at higher altitudes do not trust these construction measures, because of their failure to protect the village from floods.

Flood risk perception varies with day and night, weekday and weekend, and season. Most floods in the Czech Republic occurred between spring and summer, but winter floods were most detrimental. During summer, the municipality flood committee cleans the river bank and removes debris along the river to avoid unwanted incidents. However, villagers do not consider the post-winter season risky until the municipality issues an early warning. During the summer vacation in academic institutions, children and youth have time to help their parents in cleaning, repairing, relocating, and reconstructing their houses after a flood disaster. Floods during weekdays are considered destructive because most working people in villages are at work in cities nearby, and only the elderly and children are at home to deal with an emergency.

Elderly people are more endangered than younger people because they are chronically ill or their health conditions limit their movement but do not worry about floods more than younger people, because they consider floods an unusual experience, an extreme situation, or an accident. They presume that a destructive flood similar to that of 1997 will not occur again during their lifetime. During flood events, the elderly usually enjoy staying with their friends and neighbors for a long time. The emotional attachment to the place, and to friends and neighbors, influences their decision to continue living in flood-prone areas. Their inability to pay expenses for newly built houses at safer places is another reason for not moving out of risky zones. They would rather spend money to repair their damaged houses than shift to safer places. They also feel that other villagers consider it their responsibility to help the elderly during an emergency.

The elderly are dependent on their pension. Although paltry, it suffices. The absence of flood insurance facilities – not old age – is troublesome. Most problems after major flood disasters stem from their inability, or limited ability, to access insurance benefits because of the increasing price of insurance and restrictions of insurance companies. Elderly villagers tend to have different experiences of floods because of their age-old practices. Very often, they keep cash in their homes, mainly under their beds, or in the locker/drawer attached to their beds. Strong flood waters wash away household materials, along with money. Thus, it is only their bank savings that helps them to recover from flood impacts. Close relatives living in safer places constantly worry about the elderly members of their family, but they don't earn enough to accommodate them. However, the close bonding of elderly villagers with their children and grandchildren helps them forget their difficulty and loss.

During a flood disaster, the absence of male members in a house does not pose a problem; what is important is an individual's capability to deal with disaster. Women deal with disaster better if they are self-dependent and educated and if family members contribute equally. There is no gender bias during floods. One respondent said, "It is not a matter of gender, but supporting each other during an emergency is important" (Respondent No. 3, 2016). Households having three female children did not have any difficulties in managing an emergency. Village women are financially independent and share resources to recover from flood impacts. However, women having small children prefer to stay at safer places in the village.

Post-communist agricultural activities and dependency have changed villagers' lives. Now, they are less dependent on agriculture and, therefore, less interested in protecting their farmland from flooding. Private companies have taken over land to cultivate. People are largely dependent on non-farming activities and fishing. Tourism has become another source of income. Villagers work in industry in towns and cities nearby. Therefore, loss of crops due to disasters does not greatly affect their economic activities.

Risk perception of floods varies with location and distance from the river bank. One respondent had said, "I do not live in an endangered area, so I do not buy flood insurance" (Respondent No. 2, 2016). Only villagers who live near the river bank – conscious of the impact of past floods and of flood damage – worry about potential floods and shift to higher altitudes or safer places. Geographical location influences experience of floods and perception of risk, as do housing structure and building material. Single-storey houses are at greater risk than two-storey houses. Floods

Table 1 Background and factors of purchasing flood insurance

Background / Factors	Elderly household living at low lying area	Young couple living at low lying area	Young couple shifted at high land
Experience of flood	More than one time	More than one time	One time
Average family member	2	4	3
Source of earning	Pension	Private job and consultancy	Private job
Bank loan	No	No	Yes (for new house)
Dread of flood	Yes	Yes	Yes, but not now
Attachment with landscape	Yes	Yes	Not important
Alternative house	No	Yes	No
Trust towards the government	Yes	Yes	Yes
Received relief and compensation	Yes	Yes	Yes
Participation in community activities	Not much	Yes	Yes
Change in the structure of house	Not much	Yes	Not important
Old flood insurance	Yes	No	No
Received benefit of insurance	Not much	Yes	Not important
New flood insurance	May be for a bed room or not	Yes	Not important
Renewal of flood insurance	Irregular	Regularly	Not important

Source: Authors

damage the ground floor of most houses, so families living in two-storey houses shift their movables to the first floor. The water level touched the roof of houses, but people could save their lives. One respondent said, "Saving life is more important than attachment to household materials" (Respondent No. 2, 2016).

Another respondent said, "Government will not protect us, we are protected by flood insurance" (Respondent No. 1, 2016). Floods damage property; they hardly affect income-generating activities. Flood insurance changes people's risk perception and activities (Table 1), but is not compulsory. Purchasing insurance depends on perception of risk and capability to purchase; villagers who live at higher altitudes are not interested in flood insurance, and it has become too expensive for many households living near river banks and on flood plains. The major floods in 1997 complicated the issue of insurance. Most insurance policies are old and do not properly mention the names of beneficiaries or the benefits for different types of flood. Therefore, holders of old flood insurance policies do not receive adequate benefits.

But fear and dread, their sense that disaster is inevitable, and the severity of losses suffered in previous floods prompt villagers in the Czech Republic to buy flood insurance (Zaleskiewicz et al. 2002; Raška 2015). Residents are particularly anxious if their homes are within or near a flood risk zone. Our research corroborates the findings of Pynn and Ljung (1999), whose study of the Grand Forks flood of 1997 demonstrates that anxiety and the experience of previous floods are important factors that motivate people to buy insurance. The decision to purchase insurance, despite its high cost, is aimed partly at reducing fear and creating a sense of security. Flood insurance is now regarded as a mitigation measure that enhances the confidence of households living in flood risk zones.

During an emergency, the government's priority is to save lives. There is another responsibility: assess flood loss and damage. Inadequate resources limit the local government's management of emergencies. The municipality prefers to build dykes on private land, but few are willing to forgo their land; most prefer that the government rejuvenate and maintain the river's natural flow. Now, villagers are more conscious about sustainable non-structural flood maintenance measures.

Most villagers perceive government assistance inadequate to cope with loss and damage, and spend individual savings and labor to offset loss and damage. They realize that individuals cannot cope with such emergency situations, and families help each other to relocate, shift household movables, and fill sand bags. During an emergency, neighbors of affected families – though not affected themselves – consider it their responsibility to help. People forget previous disputes and participate in community activities, and the sense of community grows. These households have lived closely for generations and are connected through extended family relationships and social networks. Assistance from friends, relatives, neighbors, and people from neighboring villages, even distant Moravian villages, revitalizes them and helps them cope with disaster.

Another challenge for the local government is the distribution of relief and aid. Damage to houses and loss of household goods severely inconveniences villagers in flood risk zones. The local government is responsible for rebuilding confidence and stopping people from migrating. People trust and depend on the government to provide support and safety during emergencies. Instead of expecting popular support of their decisions, local governments should focus on collaborative work and participation and enhance people's resilience and preparedness.

Discussion

Flood impact tends to change community activities and strategies that depend on risk perception. The trade-offs of perceived risk and community activities are largely matters of cultural and individual choices. Households in the Czech Republic are aware of flood risk, but their lack of experience of floods limits their worry. A flood becomes disastrous only when it is beyond everyday experience. It influences people at risk to adopt strategies based on capability and risk-benefit trade-offs. So, risk perception is the combined outcome of awareness, preparedness, and coping capacity (Table 2).

Table 2 Factors of worry, awareness, preparedness, and coping capacity

	Factors of worry	Factor of awareness	Factors of preparedness	Factors of coping capacity
Physical factor	Geographical location of house	Geographical location of house	Geographical location of house	Type of housing
	Frequency of floods	Damages and losses	Damages and losses	Frequency of floods
	Season and time		Season and time	Season of flood
Socio-cultural factors	Security of lives and properties	Security of lives and properties	Security of lives and properties	Security of lives and properties
	Age	Knowledge about flood risk zone	Age	Age
	Prior experiences	Prior experience of flood risk	Prior experience of flood risk	Gender
	Level of awareness	Academic qualification	Accessibility of early warning	Community solidarity
	Technical know-how	Knowledge of modern technology	Fear	Enhancement of community bonding
	Safety for children	Alternative place to migrate	Social network	Safety of the young and old members
	Social network	Watching weather bulletin on TV	Gender	Extended family
	Community bonding	Access to internet	Knowledge about flood risk zone	Alternative safe place to migrate
	Physical ability	Social network	Dependent family member	Social network
	Sickness			
Economic factor	Flood insurance	Flood insurance	Flood insurance	Purchasing flood insurance
	Livelihood security	Livelihood security	Economic activity and resource holding	Dependency on primary source of livelihood
Institutional factor	Leadership of the mayors	Prohibition of active flood plains	Trust in the government and physical measures	Prohibition of active flood plains
	Regular updates about floods and early warning	Regular updates about floods and early warning	Regular updates about floods and early warning	Changing landscape or increasing the height of levees
	Controlling flow of river		Installing loud speakers for emergency	Protection of forests
	Trust in the government		Evacuation route	Trust and cooperation with the government

Source: Authors

In the past couple of decades, the rural landscape has changed drastically (Brázdil et al. 2011). Developmental activities, such as bridges built to improve the villagers' commute, have disturbed the natural flow of rivers. Regulation of streams, such as dykes built to stop water from overflowing, and the changing landscape have played an important role in increasing flood events in villages (Brázdil 2006). Villagers now prefer to rejuvenate the natural flow of rivers and other, sustainable solutions.

Disaster vulnerability is directly connected with geographical location (Cutter et al. 2003), which influences risk perception. Most villagers in risk zones are poor and vulnerable to disaster. Researchers (Stojanov et al. 2015) have focused on the risk perception of survivors of the 1997 floods in the Morava region. These survivors had taken the risk of floods seriously and raised the ground floor of their newly built houses.

Now, the main question is: Is the increased height of the ground floor a result of higher risk perception or a trend in the village? Flood survivors in Bangladesh build houses above the average flood level (Momtaz and Shameem 2016); this practice is a century old. Higher consciousness and risk perception of floods have helped them to formulate strategies to cope with disasters. Villagers have adopted different measures to protect their houses from future floods (Stojanov et al. 2015). However, even though villagers in Podhradí nad Dyjí have experienced three floods since 2002, few have increased the height of the ground floor, because the investment is higher. In flood risk areas, building houses that are safe from floods is expensive, and people who cannot afford flood insurance cannot afford building such houses either. However, measures are being instituted to protect the existing houses, because a house represents the investment of a good sum of money.

Many affected people live in the flood risk zone because they are financially incapable of moving to safer places or because the activities they depend on for their livelihood compel them to live there. Many villagers living near river banks can relocate, but do not, because they are attached to the place. In contrast, elderly people at risk cannot afford the rent on safer houses or land and, therefore, are compelled to accept the risk. Households who have flood insurance ignore the risk. Young couples avoid risk to ensure the safety of younger family members and move to safer places. Destructive disaster events influence people's decisions, but acceptance and risk avoidance depends on socioeconomic and cultural factors. Czech households live at risk because they have no alternative, in line with previous findings (Duží et al. 2017).

The findings of this chapter corroborate the arguments of Figueiredo et al. (2008) that there is high consensus towards structural solutions to ensure disaster reduction, and government regulation of active flood risk zones helps to reduce potential loss. The Water Act (Act No. 254/2001 Coll.) also prohibits citizens to engage in construction activities at flood zone (Government of the Czech Republic 2001). Non-structural and structural solutions are required in equal measure to reduce flood impacts; structural measures alone are not enough (Rasid and Haider 2002), though these enhance people's confidence. Villagers do not trust dykes when flood water marks are visible on the wall, and no dyke can be built up to the water mark. Therefore, villagers consider permanent evacuation from flood risk zones the best way to avoid future floods, better even than flood insurance.

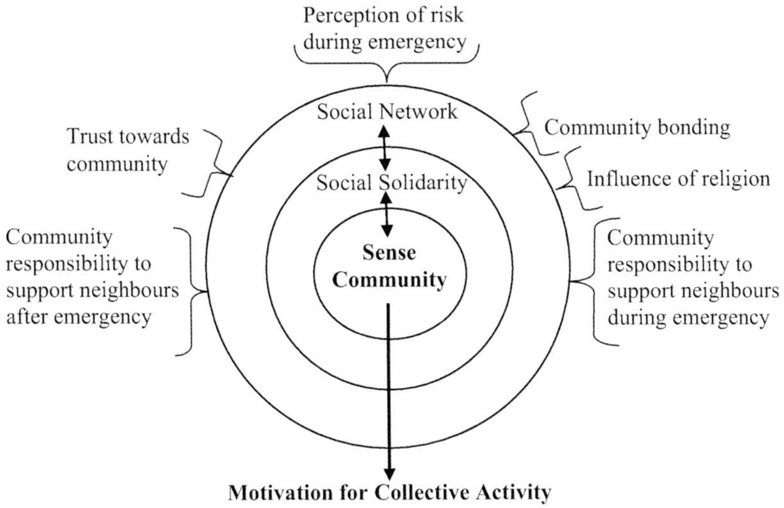

Fig. 1 Sense of community to collective activity. (Source: Authors)

None of the aforementioned measures is in itself fully adequate to deal with a disaster; options are considered because the people lack faith in the efficiency of adopted measures. It has been discovered that few residents have the capacity to deal with floods on their own, and in this our research corroborates the findings of Svahn (2013) in the flood-affected areas of Hungary. Therefore, people rely on community activities for successful disaster management (Fig. 1). The evidence from the flood-affected villages in the Czech Republic supports the findings on disaster resilience of survivors of the 2011 Van earthquake in Turkey (Doğulu et al. 2016).

Experience changes the perception of risk. Slovic (1987) argues that experience of disaster changes views and attitudes towards it. Although people were unaffected, awareness of uncontrollable and inequitable catastrophic events can make them worry about future risks. The research on flood-affected villages in the Czech Republic supports the findings of Klemešová and Andráško (2015) about the motivations of affected people in disaster reduction. Villagers think that such floods will not recur in the future. It is considered as a "horror night."

The state has invited the mayor of Jeseník nad Odrou village to participate in making disaster reduction plans. This shows that the centralized governance system in the Czech Republic is slowly accepting the community-centric flood risk reduction approach. The perception of risk is changed by coordination between villagers and local government, acceptability of local leadership, social capital and social network, community resilience, and sense of community. Slovic (1987) suggests that laymen perceive risk differently from experts and that this difference in perception affects the communication of risk. Czech citizens cannot follow the complex analysis of flood forecasting and trust the local government's early warning system and efforts to reduce flood impact. The local government cannot overlook people's needs and demand in disaster reduction. Therefore, risk communication between people and local government becomes crucial in disaster risk reduction.

Conclusion

Destructive flood events are beyond prior experience. Experience of flood risk varies with location of house and physical measures at the river bank, number of dependent family members, damaged properties and house, loss of lives, and difficulties in receiving insurance benefits. It is not only risk perception but also consciousness of self-responsibility towards society that influence people to engage in reducing the risk of disaster. Households help one another and support government personnel during such emergencies, but the trend of collective activities is very low. Religion does not have importance during any emergency, but religious belief combines and enhances the unity that influences risk perception and activities in disaster reduction. Villagers living near river basins are more worried about the increasing frequency of floods. Social networks of friends and neighbors and trust in government policies help villagers to deal with such emergencies. A strong sense of community helps villagers to remain connected with each other. The sense of collectivity during emergency is very strong within social networks. Social networks enhance people's ability to engage in collaborative activities to cope with disaster. Media intervention is also needed to educate people about the sense of community that would enhance the resilience to cope with potential risks.

The municipality's efficiency in distributing relief and aid material, resolving disagreements and disputes in emergency management, and processing reconstruction measures has increased trust in the government and changed the risk perception of survivors. But the top-down approach to disaster management has made people dependent on the government, reduced their preparedness and resilience, and overburdened the local government. It cannot implement disaster prevention measures without the consent of people at the grassroots. Government negligence of people's opinion on disaster reduction has distanced them from such activities. The paradigm of disaster management is changing, as is peoples' attitude and willingness to reduce impacts. Therefore, the Czech people must be willing to contribute to reducing the risk of disaster so that the government can enhance their resilience to cope with flood risk.

References

AFP (2009) Six dead in Czech floods: emergency services. Staff Writers Prague (AFP). http://www.terradaily.com/reports/Six_dead_in_Czech_floods_emergency_services_999.html. Accessed 12 May 2016

Armas I, Avram E (2009) Perception of flood risk in Danube Delta, Romania. Nat Hazards 50(2):269–287

Brázdil R (2006) Climate change and losses through natural disaster: some remarks from experience in the Czech Republic. Climate change and disaster losses workshop: understanding and attributing trends and projections. Munich Re

Brázdil R et al (2006) Historical and recent floods in the Czech Republic: causes, seasonality, trends, impacts. In: Schanze J et al (eds) Flood risk management: hazards, vulnerability and mitigation measures: NATO science series, vol 67. Springer, Dordrecht, pp 247–259
Brázdil R, Řezníčková L, Valášek H et al (2011) Fluctuations of floods of the river Morava (Czech Republic) in the 1691–2009 period: interactions of natural and anthropogenic factors. Hydrol Sci J 56(3):468–485
Brody SD, Highfield W, Alston L (2004) Location matter? Measuring environmental perceptions of creeks in two San Antonio watersheds. Environ Behav 36(2):229–250
Brody SD, Zahran S, Vedlitz A et al (2008) Examining the relationship between physical vulnerability and public perceptions of global climate change in the United States. Environ Behav 40(1):72–95
Čamrová L, Viktorová D (2006) Policy-making decisions under the thumb of disasters – a case of the floods in the Czech Republic. Curr Polit Econ Russia East Cent Eur 21(3):203–204
Cutter SL, Carolina S, Boruff BJ et al (2003) Social vulnerability to environmental hazards. Soc Sci Q 84(2):242–261
Doğulu C, Karanci AN, Ikizer G (2016) How do survivors perceive community resilience? The case of the 2011 earthquakes in Van, Turkey. Int J Disaster Risk Reduct 16:108–114
Duží B, Vikhrov D, Kelman I et al (2017) Household measures for river flood risk reduction in the Czech Republic. J Flood Risk Manage 10(2):253–266
European Union (2007) Directive 2007/60/EC on the assessment and management of flood risks. Official Journal European Union, p 27
Figueiredo E, Valente S, Coelho C et al (2008) Coping with risk: analysis on the importance of integrating social perceptions on flood risk into management mechanisms – the case of the municipality of Águeda, Portugal. J Risk Res 12(5):581–602
Fischhoff B, Slovic P, Lichtenstein S et al (1978) How safe is safe enough? A psychometric study towards technological risks and benefits. Policy Sci 9(2):127–152
Government of the Czech Republic (2000) Act No. 240/2000 Coll. Crisis Act
Government of the Czech Republic (2001) Act No. 254/2001 coll. (Water Act)
Grothmann T, Reusswig F (2006) People at risk of flooding: why some residents take precautionary action while others do not. Nat Hazard 38(1):101–120
Jóhannesdóttir G, Gísladóttir G (2010) People living under threat of volcanic hazard in southern Iceland: vulnerability and risk perception. Nat Hazards Earth Syst Sci 10(2):407–420
Klemešová K, Andráško I (2015) Perception of floods – towards more effective flood management in the Czech Republic. Carpathian J Earth Environ Sci 10(2):199–208
Kubát J (1999) 1997/1998 Floods in the Czech Republic: hydrological evaluation. In: Marsalek J et al (eds) Flood issues in contemporary water management: NATO science series. Springer, Dordrecht, pp 25–39
Ministry of Agriculture (2007) Plán hlavních povodí České republiky, schválen usnesením vlády České republiky ze dne 23. května 2007 č. 562
Ministry of Agriculture and Ministry of the Environment (2006) Plán hlavních povodí České republiky – 2. pracovní verze
Ministry of the Environment (2012) Předběžné Vyhodnocení Povodňových Rizik, V České Republice, 2011, Implementace Směrnice Evropského Parlamentu A Rady 2007/60/ES, O Vyhodnocování a Zvládání Povodňových Rizik, verze 5.0
Momtaz S, Shameem M (2016) Experiencing climate change in Bangladesh: vulnerability and adaptation in coastal regions. Academic, Oxford
Motoyoshi T (2006) Public perception of flood risk and community-based disaster preparedness. In: Ikeda S et al (eds) A better integrated management of disaster risks: toward resilient society to emerging disaster risks in mega-cities. TERRAPUB, Tokyo, pp 121–134
Pynn R, Ljung GM (1999) Flood insurance: a survey of Grand Forks, North Dakota, homeowners. Appl Behav Sci Rev 7(2):171–180
Raaijmakers R, Krywkow J, Veen A (2008) Flood risk perceptions and spatial multi-criteria analysis: an exploratory research for hazard mitigation. Nat Hazards 46(3):307–322

Rasid H, Haider W (2002) Floodplain residents' preferences for non-structural flood alleviation measures in the Red River basin, Manitoba, Canada. Water Int 27(1):132–151
Raška P (2015) Flood risk perception in Central-Eastern European members states of the EU: a review. Nat Hazards 79(3):2163–2179
Slovic P (1987) Perception of risk. Science 236:280–285
Slovic P (1999) Trust, emotion, sex, politics, and science: surveying the risk-assessment battlefield. Risk Anal 19(4):689–701
Slovic P (2000) The perception of risk. Earthscan, London
Stanghellini LPS, Collentine D (2008) Stakeholder discourse and water management-implementation of the participatory model CATCH in a Northern Italian Alpine sub-catchment. Hydrol Earth Syst Sci 12(2):317–331
Starr C (1969) Social benefit versus technological risk. Science 165(3899):1232
Stojanov R, Duží B, Daněk T et al (2015) Adaptation to the impacts of climate extremes in Central Europe: a case study in a rural area in the Czech Republic. Sustainability 7(9):12758–12786
Svahn C (2013) Risk perception and communication: a study on how people living in the Tisza river basin, Hungary perceive the risk of floods and how the flood risk communication between authorities and the public could be improved. MA thesis, Stockholm University, Stockholm
Terpstra T (2011) Emotions, trust, and perceived risk: affective and cognitive routes to flood preparedness. Risk Anal 31(10):1658–1675
Terpstra T, Gutteling JM (2008) Households' perceived responsibilities in flood risk management in the Netherlands. Int J Water Resour Dev 24(4):555–565
Vávra J, Lapka M, Cudlínová E et al (2015) Local perception of floods in the Czech Republic and recent changes in state flood management strategies. J Flood Risk Manage 10(2):1–15
Wachinger G, Renn O, Begg C et al (2013) The risk perception paradox-implications for governance and communication of natural hazards. Risk Anal 33(6):1049–1065
Zaleskiewicz T, Piskorz Z, Borkowska A (2002) Fear or money? Decisions on insuring oneself against flooding. Risk Decis Policy 7(3):221–233

Water Management as a Means for Climate Change Adaptation and Sustainable Development

7

Ghrmawit Haile

Contents

Abstract

The earth's total water resources reserve is said to be limited, becoming depleted and a more precious commodity over time, due to different stressors or drivers which include climate change and variability impacts. Moreover, it is becoming clear that no one is sure how the future climate will unfold. However, as the impacts of climate change and variability on social and natural systems grow and are projected to persist unless there are substantial and sustained reductions in greenhouse gas emissions which, together with adaptation, can limit the climate

G. Haile (✉)
Ministry of Environment Forest and Climate Change and Addis Ababa University, Addis Ababa, Ethiopia
e-mail: h.ghrma@gmail.com

W. Leal Filho (ed.), *Handbook of Climate Change Resilience*,
https://doi.org/10.1007/978-3-319-93336-8_32

change risk, the need to adapt is evident. Even if it seems clear that climate change adaptation and actions toward resilience are important for sustainable development, adaptation faces many constraints, particularly in low-income settings. Climate data, scenarios and impact models are insufficient for supporting adaptation, particularly as they relate to food systems and rural livelihoods; the adaptation response to date has been limited, fragmented and divorced from national planning processes, with limited engagement with local expertise, and adaptation policies and programs are too narrowly focused on explicit responses to climate change rather than responses to climate variability or broader development issues.

The stress on water resources is further compounded by improper management and handling of the resource while the demand for water, food and energy is expected to rise by 30–50% in the next two decades. This paper discusses how water resources need to be managed as a means of climate change adaptation for sustainable development. It also discusses the ever increasing competing and conflicting demands for water under the climate change and variability–driven impacts on the water resources.

Keywords
Water resources management · Climate change adaptation · Sustainable development

Introduction

Earth's water is found as saltwater and freshwater. A large amount of the water is saline water which accounts for about 97%, and only about 3% is freshwater, found as glaciers and ice caps, groundwater and surface water (Du Plessis 2017). Mostly water resources refer to freshwater which appears as surface water (rivers, lakes and reservoirs) and groundwater – the part which is useful or potentially useful to humanity and the ecosystem. Previously water was taken as a nature-given abundant resource. However, due to different stressors or drivers, the earth's total water resources reserve is said to be limited, becoming depleted and a more precious commodity over time. Different drivers such as the accelerated pace of population growth, increasing rates of urbanization, expanding middle-class lifestyles and unsustainable production and consumption patterns (Hoff 2011; van Vuuren et al. 2014; World Economic Forum 2011) fueled by climate change, variability and land use change are putting serious stress in general on the overall natural resources and in particular on the water resources. The projected increase of the global mean surface temperature by the end of the twenty-first century (2081–2100) relative to 1986–2005 is *likely* to be 0.3–1.7 °C under Representative Concentration Pathway (RCP) 2.6, 1.1–2.6 °C under RCP4.5, 1.4–3.1 °C under RCP6.0 and 2.6–4.8 °C under RCP8.5 (Pachauri 2014). The increase in the mean surface temperature and variation in precipitation together with the change and transformation in land use influence the evaporation, surface runoff and subsurface infiltration which in turn determine the surface and groundwater distribution and availability within a sub-watershed or watershed.

Water is a key development ingredient, in particular in developing countries whose development mainly depends on natural resources capital. Water scarcity due to different drivers including climate change and variability challenges socio-economic development, human livelihoods and ecosystem well-being. The stress on water resources is further compounded by improper management and handling of the water resource while the demand for water, food and energy is expected to rise by 30–50% in the next two decades (Bizikova et al. 2013).

Therefore, in an era highly threatened by climate challenge complexity and uncertainty, managing the water resources capital effectively and coherently would contribute to building climate change and variability adaptive capacity and development sustainability. It is prudent and well timed to devise a management system which considers competing demands and conflicting interests in a holistic and inclusive manner. This may require rational policies and effective regulatory frameworks, holistic plans and developmental interventions taking into account climate change adaptation mainstreaming at all scales.

Water resources management is a series/sequential intervention applied at different scales, ideally to manage the resource for the most beneficial and sustainable utilization. Generally, the major interventions are:

- Policies and relevant legal and regulatory frameworks (such as laws, regulations, institutions, standards)
- Plans at different scales
- Developmental intervention or implementation (such as programs and projects of different types and scales)

Policies together with relevant legal and regulatory frameworks being at the top of the governance system have a significant role in guiding the overall planning, implementation of the developmental interventions of different scales and decision-making processes throughout the water resources management landscape. Almost all interventions by nations ideally are intended for or are designed for optimum utilization of the water resources; the outcome highly depends on the overall development governance system of nations and virtually may or may not result in the intended desirable outcome. Together with the implementation of the different key water resources management interventions at different scales, periodic and systemic performance monitoring and tracking which inform back to the system are very important for continuous learning and improvement. The water resources management intervention as a means for climate change adaptation needs to be mainstreamed and executed at all scales. Furthermore, for coherence and realization of the desired outcome, effective coordination and synergy within and between public organs at different scales in a water resources sector governance structure would be necessary. Moreover, continuous demand-driven capacity development intervention based on gap analysis is important at institutional, systemic and human resources levels.

Water Resources Depletion: Major Drivers

Different directly or indirectly interlinked drivers account for impacts on water resources, such as the accelerated pace of global population growth, climate change and variability, rapid urbanization, consumption–production patterns, the trend in changing land use patterns, and national policies and related legal and regulatory frameworks. The accelerated pace of the world population increase is resulting in growing demand for water for domestic supplies, increased food/agricultural and energy production and processing of increased industrial products or commodities. The global fast urbanization has put stress on water due to the rising standard of living related to the increase in the middle-income segment of the global population and the overall consumption–production pattern. The increase in global population together with the increased consumption pattern results in increased waste which also pollutes the water resources (both surface and ground) and needs additional water for appropriate disposal and treatment. All these impacts fueled by impacts from climate change and extreme events are now affecting every continent, disrupting national economies and affecting lives, and costing people, communities and countries dearly today and even more tomorrow, SDG 13 (Development et al. 2016). Clearly identifying the key drivers of the stresses on the water resources capital and prioritizing them as per the degree and severity of the stress would help in devising an appropriate and contextual management approach which could address broad issues and adaptation responses that can complement and enhance the handling of significant co-benefits and synergies. Furthermore, identifying and addressing the drivers would help in dealing with the interconnection the water resources have with different developmental interventions and guide the water resources management system setting, in particular the formulation of the policies and relevant legal and regulatory frameworks. Moreover, identifying the drivers would help in prioritizing the subsequent plans and interventions which would likely make a significant contribution to sustainable development.

In most cases, the stress from different drivers on the water resources would be intensified and compounded by the situation of the water resources capital and the existing management system. In managing the water resources as a means for climate adaptation, nations and relevant institutional decision-makers need to be farsighted and think of transforming the overall economy through building adaptive capacity and sustainable development. This requires a shift from incremental adaptation actions by just taking action on specific impacts of climate change and variability to addressing the major drivers of the challenges for development sustainability – that is, a shift from responding to or focusing on particular symptoms to identifying and addressing the underlining sources or causes of the challenges in a holistic manner, and eventually building a resilient economy and ecosystem.

In improving the water resources management system, applying climate change projections and hydrologic models for the area of concern (at regional, national or river basin level) would be very important to simulate the impacts of different drivers such as land use/land cover (LULC) change, and climate change and climate variability (CCCV), on the water resources. The simulation result would be helpful

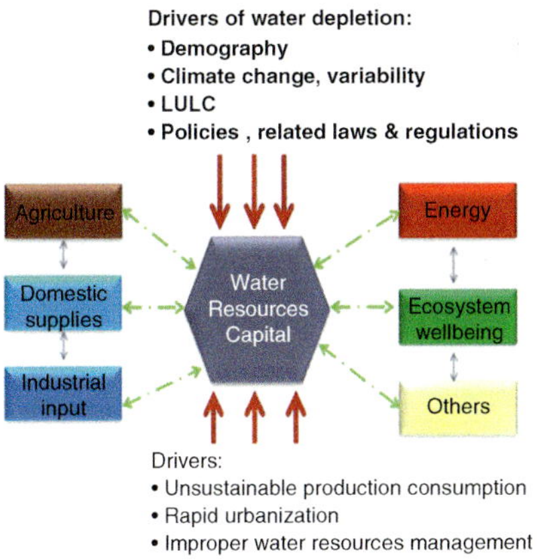

Box 1: Key points:

Water resources management requires clear inderstanding of the major drivers of the challenges on the resource, and

The need for holistic water Resources management demand for;

- *addressing key drivers, and*
- *considering interconnections and conflicting demands in the resource allocation.*

Fig. 1 Water resources are key inputs into different sectors and highly affected by different drivers

and important in designing a water resources management system that would contribute to building climate adaptive capacity (Fig. 1).

Climate Change Impacts on Water Resources and Development

Climate change intensifies existing risks and creates new risks for natural and human systems. The risks are unevenly distributed due to different reasons such as geographic position and level of development. In addition, the risks are severely felt by disadvantaged people and communities in countries at all stages of development. Each of the last three decades has been successively warmer at the earth's surface than any preceding decade since 1850. The period from 1983 to 2012 was likely the warmest 30-year period of the last 1400 years in the northern hemisphere. Evidence of observed climate change impacts is strongest and most comprehensive for natural systems. In many regions, changing precipitation or melting snow and ice are altering hydrologic systems, affecting water resources in terms of quantity and quality, IPCC 2014: *Climate Change 2014*.

Climate change is one of the different drivers impacting the water resources. Extreme events make a significant contribution to water depletion both in quantity (affecting the availability) and quality. Climate change causes incidents of flooding and droughts which are directly related to water resources quantity and quality. The increase in temperature will increase transpiration. The demand for water by the biotic system and overall ecosystem also increases as the temperature rises.

Climate change over the twenty-first century is projected to reduce renewable surface water and groundwater resources in most dry subtropical regions, intensifying competition for water among sectors. In presently dry regions, the frequency of droughts will *likely* increase by the end of the twenty-first century under RCP8.5. In contrast, water resources are projected to increase at high latitudes. Globally, in all

RCPs, it is *likely* that the area encompassed by monsoon systems will increase, monsoon precipitation is *likely* to intensify and El Niño–Southern Oscillation (ENSO)–related precipitation variability on regional scales will *likely* intensify. The interaction of increased temperature; increased sediment, nutrient and pollutant loadings from heavy rainfall; increased concentrations of pollutants during droughts; and disruption of treatment facilities during floods will reduce raw water quality and pose risks to drinking water quality. Observational records and climate projections provide abundant evidence that freshwater resources are vulnerable and have the potential to be strongly impacted by climate change, with wide-ranging consequences for human societies and ecosystems, IPCC 2014: *Climate Change 2014*. Current water management practices may not be robust enough to cope with the impacts of climate change on water supply reliability, flood risk, health, agriculture, energy and aquatic ecosystems.

From a developmental perspective, the level and degree of vulnerability to climate change and variability depend on the level of development and overall available capacity of nations. From a poverty perspective, climate change impacts are projected to slow down economic growth, make poverty reduction complicated and more difficult, further erode food security, and prolong existing and create new poverty traps, the latter particularly in urban areas and emerging hot spots of hunger. International dimensions such as trade and relations among states are also important for understanding the risks of climate change at regional scales. Climate change is projected to increase the displacement of people. Populations that lack the resources for planned migration experience higher exposure to extreme weather events, particularly in developing countries with low incomes. Climate change can indirectly increase risks of violent conflicts by amplifying well-documented drivers of these conflicts such as poverty and economic shocks (Pachauri 2014).

Water Resources Management as a Means for Adaptation

The need to adapt is increasingly recognized, from the community level to the regional and national government level to the donor community level, yet adaptation faces many constraints, particularly in low-income settings. Four key concerns about adaptation emerge:

- Climate data, scenarios and impact models are insufficient for supporting adaptation, particularly as they relate to food systems and rural livelihoods.
- The adaptation response to date has been limited, fragmented and divorced from national planning processes, with limited engagement with local expertise.
- Adaptation policies and programs are too narrowly focused on explicit responses to climate change rather than responses to climate variability or broader development issues (Adenle et al. 2017).

In an uncertain situation where no one is sure how the climate will unfold and where the timing and magnitude of the climate change variability impacts are unknown, it would be wise to manage the water resources with continuous dynamism, learning from the process and continuously adapting to climate change and variability adverse impacts. For an "improved" water resources management system that considers the complexity of the sector, and that would cope with the climate change challenges to ensure sustainable development, interventions need to be made at the following scales:

- Policy and related legal and regulatory frameworks
- Planning level (planning at the national/mega level guided by the water resources policies and related legal and regulatory frameworks, and subnational, sectoral or grassroots planning guided by the national/mega water-related plans)
- Developmental intervention/implementation level
- Water resources management system monitoring and tracking

Effective Water Resources Policy, Legal and Regulatory Frameworks as a Means for Climate Change Adaptation and Sustainable Development

To have an effective water resources management system that considers the complexity, uncertainty and competition for development demands for the resource under the climate change adverse impacts, the role of "effective" policies and related legal and regulatory frameworks is indispensable. Therefore, this would entail countries developing or checking/reviewing their existing water resources management policies and overall legal and regulatory frameworks with the intention to come up with coherent and effective water resources policy and relevant legal and regulatory frameworks. Effective policy and legal frameworks guide and regulate the nation's water resources developmental plans, interventions and decision-making process in a challenging climate change situation. In a situation highly challenged by climate change impacts, it is highly unlikely that business-as-usual policies and regulatory frameworks will result in the desired climate change adaptive capacity and sustainable development outcome. Effective water resources management instruments which guide development sustainability and promote climate change adaptation are important. From the top, effective policies are very instrumental in guiding the implementation of plans and subsequently achieving the national vision. This would be particularly important for countries which manage/run their water resources development activities on a silo/sectoral basis. Silo/sectoral management, without considering the competing demands and conflicting interests of different sectors (e.g., water for the agriculture sector, water for the energy sector, water for domestic and industrial supplies) under challenging climate change and variability circumstances needs particular consideration. The policies, related laws and regulatory frameworks being at the top of the governance system need particular emphasis, because (i) well designed, coherent and effective national policies and (ii) relevant regulatory frameworks which read and are based on the actual situation and reality are highly likely to contribute significantly to development sustainability. Moreover, since water resources, like other natural resources, operate in an interconnected manner, effective water resources management policy would have co-benefits with other developmental interventions. Any intervention that uses water operates in an interconnected manner and intervention in one sector may unintentionally affect the others either in a positive or negative way.

For (i) the policy to be based on the actual national water resources situation and (ii) the legal and regulatory frameworks to be comprehensive so as to assist in addressing a wide range of water-related issues in the country in synergy with the country's vision, development pathway and priorities, it would be advisable for the policy, legal and regulatory frameworks review/development processes to be inclusive and participatory for all stakeholders – bottom-up, top-down, and both ways. This would contribute to the productivity and sustainability of the policy and related legal instruments. Therefore, in setting the water resources policy it would be advisable to work in a synergic and holistic manner.

In addition, for the policy to link with and to guide the subsequent planning process, the regulatory framework needs to properly translate and promote the policy. This will help subnational sectoral plans to contribute in achieving the national vision and development sustainability. For effective implementation of the water-related climate change adaptation interventions or responses at different scales (regional, national, local institutions/communities), designing comprehensive and coherent enabling regulatory frameworks (such as laws, regulations, standards, effective institutions, suitable governance structures) is essential. These legal instruments need to devolve from the national water resources management policy and link with the actual sociocultural and behavioral context.

Effective Water Resources Planning as a Means for Climate Change Adaptation and Sustainable Development

Guided by the national policies and relevant legal and regulatory frameworks for the water resources planning process and plans at different scales – national, subnational/regional, grassroots or district level (depending on the political/institutional hierarchy of a nation) – each planning level and plan has specific issues to consider or address. The national planning process housed at the national scale needs to be entirely based on the national water resources policy and the national development pathway in harmony with the regional agreement and aiming toward the global goal if there is any. National plans need to reflect and be based on the country's vision and approach to achieving sustainable development in accordance with the nation's priorities, national circumstances and available resources. The plans at all scales need to align with, supplement and promote the national water resources policy. The plan at the national scale, which is mega level, needs to utterly consider the interconnections and conflicting demands for water allocation for the macro plans of different sectors (agriculture, energy, domestic, industrial supplies, ecosystem, others) in the context of climate change and variability projections and potential impacts on the water resources. Here water resources optimization and trade-off reduction/ minimization exercises need to be worked out in a synergic manner because fragmented and siloed/sectoral handling where each and every sector closes its doors and plans for water without having data/information on the availability level of the resource, without considering the interconnection with other sector demands which are also planning for the same resource, may not show the broader picture and demand for the resource. Silo or sectoral plans eventually may end up being counterproductive or maladaptive in the medium to long term if not earlier. Therefore, to take care of any co-benefits, synergies or trade-offs between and within water-related sectoral plans to optimize the trade-offs and any competing intervention, sectors need to work in synergy and in a holistic manner in setting the mega plans at national level.

At this level of planning to understand the hydrologic process and trends and to accommodate planning for climate change and projected potential impacts on the water resources, it is advisable either to develop models for climate projection or to use available climate projection data and information from reliable sources for the area or basin under concern. Mostly, plans at this level of planning span the medium to long term, with the duration varying from country to country; therefore, devoting the necessary resources and importance would be prudent. It would also be advisable and beneficial for the planning process to be participatory and inclusive for all relevant actors or stakeholders.

The water resources planning at a lower scale – basically action plans – is a denomination of the national mega plan. Depending on the political structure of countries, this scale plan could be a sectoral, regional, subregional or district plan. Usually, plans at this scale are done every year to devolve the mega plan to the ground and for budgetary purposes of the national treasury. This level of water resources plans are plans for actual intervention actions mostly at the grassroots level. This level of the planning process, being the vehicle which realizes the national policies and national mega plans in action, needs to be designed and communicated in a clear, logical and contextual manner involving the local beneficiaries/communities and with understanding of the ecosystems under consideration. For those involved directly or indirectly in the planning process and responsible for the realization of the plan to be inclusive, transparent and informative, particularly to the community that ideally will benefit from the plan, would help in the effective implementation of the plan. Moreover, inclusiveness also helps in bracing ownership and accountability at all levels and scales. In addition, to improve water resources sustainability, this level of planning needs to recognize and consider the local sociocultural context (culture, behavior, different interests, settings) and expectations. Indigenous groups and the local and traditional knowledge systems and practices need to be well considered and taken care of.

Effective Water Resources Management of Developmental Interventions as a Means of Climate Change Adaptation for Sustainable Development

Water resources management of developmental interventions design and implementation translates the plans at the different scales into action. For different developmental interventions to contribute toward achievement of the national vision and development of sustainability, the design and implementation need to recognize and consider that climate change and climate variability have a significant effect on water resources infrastructure operation and sustainability. Climate has a significant effect on water infrastructure design and system operation compared to the business-as-usual situation. Extreme events may result in sudden water resources system failure or may require measure/serious design and operational modification from the original design for normal performance. Both higher and lower precipitation which may result either in flooding or drought significantly affect water resources system operation. Therefore, to accommodate design for climate change and extreme variability, water system infrastructures would be most effective if the design considers the context of expected, plausible future climate conditions. In doing this the role of climate projection is very significant. However, climate change is characterized by uncertainty and complexity; considering climate projection data in the water infrastructure design may not be an easy task in terms of the required resources (skill, innovation, finance) as compared to the business-as-usual design and implementation. This may be particularly challenging for vulnerable poor nations which may have limitations in skilled personnel and knowledge in the sector and financial capacity to implement infrastructures designed with consideration of climate projections, mostly presumed to be huge and sophisticated as compared to the business-as-usual case. Furthermore, climate change uncertainty and complexity complicate the issue of considering climate projection in water resources infrastructure design. Here the idea or issue of designing for climate change may need partnering and collaboration for innovation, dissemination of knowledge and best practices both internally within a nation and regionally and globally between nations so as to strengthen the sustainable management of the water resources.

Practically, the policies, related laws, regulations and plans at different scales are translated to the ground in the different socioecological environments to suit local contexts in a country through the implementation of developmental interventions of different scales and types. In the design, implementation and operation of the developmental interventions, one-size-fits-all does not work. Hence, guided by the national water resources management policy and higher plans, understanding of the local sociocultural context and behavior, expectations, different interests and settings would be highly important. Indeed, the effectiveness and rationality of the policies and all the plans are measured and evaluated by the degree of social acceptance, and the extent to benefit the local community and safeguard the environment these developmental interventions will result in fulfilling promised or planned developmental interventions (Fig. 2).

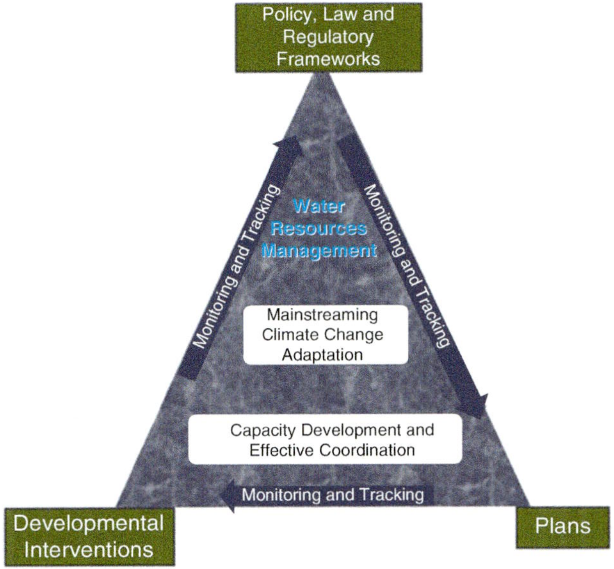

Box2: Major water resources management levels

1. Policy, Legal and Regulatory Frameworks:
- need to link with and guide the overall water resources management and decision-making process:
-need to be comprehensive, address national and regional issues, demands and benefits-
-mostly, of medium to long-term duration,

2. Planning
-National /mega plans mostly for medium term
-Sub-national, sectoral/district mostly annual span

3. Developmental interventions/Implementation:
- Such as programs, projects

Management System Continuous Monitoring and Tracking that information back to decision making process for learning and accountability

Capacity Development and Effective Coordination for coherence and complementarities

Mainstreaming Climate Change adaptation at all levels and processes.

Fig. 2 Major water resources management levels

Effective Coordination and Accountability Across Governance Scales and Capacity Development

To realize national plans and ultimately achieve the national vision, "effective" coordination, "improved" communication and synergy are very important across horizontal and vertical governance structure at all scales.

Capacity development enhances the ability of national institutions to effectively:

- Formulate rational policies, laws and related regulatory frameworks which guide sustainable water resources management
- Prepare plans and budgets in a holistic and coherent manner
- Implement, monitor and track developmental intervention implementation
- Report

For improved water resources management in the context of climate change and variability, the challenging situation demands well performing or "competent" institutions, systems and human resources throughout the landscape of the system for successful implementation of the water resources management initiatives. With this intention the implementation of a capacity-building initiative enables public institutions to develop competencies and skills that can make the system more effective and sustainable. The capacity development intervention is not one-size-fits-all. Hence, it has to be context specific, aiming at identifying the available capacities and capacity gaps at institutional, systemic and human resources levels for the "effective" implementation of water management system initiatives throughout the landscape of the water resources management stratum. Therefore, the capacity development intervention needs to be based on an exhaustive assessment of the existing water resources management system, climate change and variability impacts on the water resources and the existing coping mechanisms and strategies and existing decision-making processes. The capacity development intervention needs to work toward transforming and strengthening the overall water resources management system at all governance scales, particularly in policy formulation, planning, budget preparation and implementation of developmental activities. Capacity development with regular monitoring and tracking of system performance geared toward achieving institutional goals and periodic reporting are critical in contributing to the realization of the climate adaptive capacity.

Mainstreaming Water Resources Management for Climate Change Adaptation

Adaptation to climate change is not an independent, isolated sector or intervention. Climate change affects the natural resources and most if not all development interventions and sectors. Hence, climate change adaptation need to be mainstreamed and implemented at all levels of the water resources developmental interventions, starting from the policy intervention at the top to the planning and

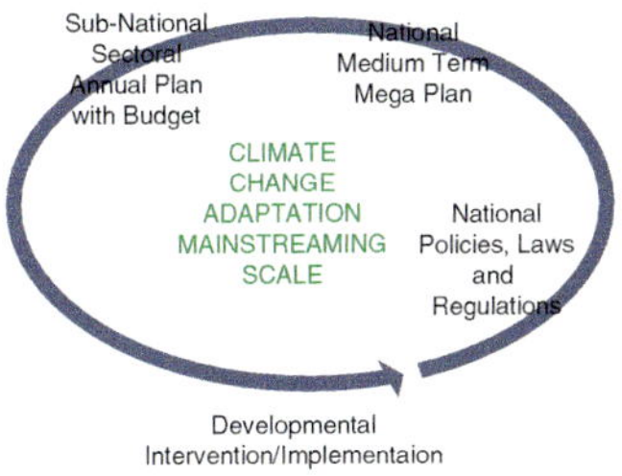

Box 3: Climate Change Adaption Mainstreaming at:

- **Policy, Laws and Regulations,**
- **All Planning levels, and**
- **Developmental Interventions.**

Fig. 3 Climate change adaptation mainstreaming

developmental intervention implementation at the ground. Literally mainstreaming and integrating the water resources management system as a means for climate change adaptation into policy formulation at all levels of planning, implementation of actions and the decision-making process are greening the water resources development and hence contributing in greening the economy.

For effective water resources management as one way of adapting to climate change and variability, in mainstreaming climate change adaptation throughout the water resources management landscape, the role of climate projection information would be significant and useful for effective decision-making processes.

The soft and hard developmental intervention implementation at the local or ground levels are the actual actions or vehicles for the implementation of the plans at different scales. These interventions need to concretely mainstream the designed "effective" water resources management system interventions as climate change adaptation actions for the development to be green and sustainable and contribute toward building resilient sustainable development. Interventions which mainstream climate change adaptation and better performance under the uncertainty of climate change and variability impacts and that have broad-based multiplier effects would be beneficial (Fig. 3).

Water Resources Management System Performance Monitoring and Tracking

Monitoring and tracking of the outcome of water resources management interventions as a means of climate change adaptation are important to evaluate or find out how effective the interventions are and decide on the areas that need to be scaled up or need modification or further improvement. Monitoring and tracking of the water resources management system outcome requires developing measurable outcome indicators and setting a baseline from which progress can be measured and tracked. This will help in evaluating the achievement or performance of the water resources management system in building climate adaptive capacity in the water resources. However, despite the importance of monitoring and tracking outcome, it could be challenging to monitor and track the "effective" water resources management as a means for climate change adaptation and sustainable development in setting the outcome indicators, estimating the existing level of adaptive capacity as a baseline and measuring the achievements in building adaptive capacity in the water resources. The currently available knowledge to monitor and track adaptation outcome is limited and fragmented, particularly at the global level (Gagnon-Lebrun and Agrawala 2007; Preston et al. 2009; Berrang-Ford et al. 2011; Ford et al. 2011). But, the water resources management system could also be monitored and tracked against the designed system processes focusing on the processes and objectives of the water resources management interventions at the different scales of the governance structure.

From the available knowledge, weather and climate data documentation and experienced disasters/impacts, nations could design water resources management systems that serve as a means of climate adaptation. However, due to the uncertain and unknown future climatic and socioeconomic conditions, climate extreme events and related disasters may disrupt the water resources system. Such incidents, which are typical characteristics of climate change, should not lead to a conclusion, because the system for building adaptive capacity in the water resources sector is influenced by many other factors beyond the water resources management system. Indeed, such occurrences should be taken as opportunities to continuously learn, adjust and build a robust water resources management system that serves as a means for climate change adaptation. Furthermore, it is worth noting that the results of an "effective" or "successful" water resources management system that serves as a means for climate change adaptation and sustainable development may not be perceived in a short time period.

Therefore, depending on the available knowledge in the country, it would be advisable to assess the extent of climate change extreme events, impacts, the degree of vulnerability, existing water resources management systems, decision-making processes, policies, laws and regulatory frameworks, and come up with what to monitor and track. Monitoring and tracking of the designed water resources management system periodically and in a systemic way are seriously important for the eventual success of the management system through continuous learning and review. However, caution is needed with regard to what to monitor and track. For the monitoring and tracking system to serve both for continuous improvement of the management process through learning by doing and for accountability purposes, the overall water resources management monitoring and tracking system needs to inform back to the decision-making process on a continual periodic basis.

Conclusion

Water scarcity due to different drivers including climate change and variability challenges affects, socio-economic development and human livelihood, and ecosystem well-being. The stress is further compounded by improper management and handling of the water resources. The roles of policies, plans, and developmental interventions are significant in managing the water resources. This requires rational policies, comprehensive and broad-based laws and regulatory frameworks, holistic plans and developmental interventions considering climate change adaptation mainstreaming at all scales of the governance system. The policies, the plans at different levels and the developmental interventions need to read, promote and link with each other. If any one aspect is missed or not performing as designed, the system as a whole may fail to perform as desired; therefore, proper and effective delivery is required for every intervention. In the formulation of the water resources management system there is a need for a major shift to a more holistic and interconnected approach at all scales of governance and intervention, and it is needed to come up with well informed, environmentally sound policies that address the threats already faced while preparing for the future projected challenges. In doing so, the impacts of different drivers on the water resources need to be well assessed and addressed contextually.

In improving the water resources management system, applying hydrologic models for the area of concern would be very important to simulate the impacts of different drivers on the water resources. The simulation result would be helpful and important in designing a water resources management system and infrastructure that would contribute to building climate change adaptive capacity.

Inclusiveness and transparency in the process – particularly (i) with the community that ideally is to benefit from the interventions and (ii) with all stakeholders who directly or indirectly have stakes or concerns in the policy formulation, planning and developmental intervention implementation – would help in the effective realization of the plans to build climate change adaptation and sustainable development. Moreover, inclusiveness helps in bracing ownership in the community and accountability at all governance levels and scales. At all scales, recognition and consideration of local sociocultural contexts and expectations are very important. Indigenous groups and the local and traditional knowledge systems and practices need to be well taken care of.

References

Adenle AA, Ford JD, Morton J, Twomlow S, Alverson K, Cattaneo A et al (2017) Managing climate change risks in Africa – a global perspective. Ecol Econ 141:190–201. https://doi.org/10.1016/j.ecolecon.2017.06.004

Berrang-Ford L et al. (2011) Global Environmental Change 21

Bizikova L, Roy D, Swanson D, Venema HD, McCandless M (2013) The water–energy–food security nexus: towards a practical planning and decision-support framework for landscape investment and risk management. International Institute for Sustainable Development, Winnipeg. www.iisd.org/pdf/2013/wef_nexus_2013.pdf

Development B, Group E, Paper W (2016) Sustainable development and climate change: implementing the 2030 agenda and the Paris agreement, (April)

Du Plessis A (2017) Global water availability, distribution and use. In: Du Plessis A (ed) Freshwater challenges of South Africa and its Upper Vaal River: current state and outlook. Springer, Cham, pp 3–11. https://doi.org/10.1007/978-3-319-49502-6_1

Gagnon-Lebrun, Agrawala (2007) Implementing adaptation in developed countries: an analysis of progress and trends. Frédéric Gagnon-Lebrun and Shardul Agrawala. Climate Policy

Hoff H (2011) Understanding the Nexus. Background Paper for the Bonn 2011 Conference: The Water, Energy and Food Security Nexus, Stockholm Environment Institute (SEI), Stockholm, Sweden

IPCC [Intergovernmental Panel on Climate Change] (2014) Climate change 2014: synthesis report. Contribution of Working Groups I, II and III to the fifth assessment report of the Intergovernmental Panel on Climate Change [Core Writing Team, Pachauri RK, Meyer LA (eds)]. IPCC, Geneva. http://www.ipcc.ch/report/ar5/syr/

Pachauri RK (2014) Climate change 2014 synthesis report

Preston et al. (2009) Putting vulnerability to climate change on the map: A review of approaches, benefits, and risks. Sustainability Science 6:177–202. https://doi.org/10.1007/s11625-011-0129-1

van Vuuren DP, Kriegler E, O'Neill BC et al. (2014) Climatic Change 122:373. https://doi.org/10.1007/s10584-013-0906-1

World Economic Forum - Geneva Switzerland (2011) Klaus Schwab World Economic Forum

Urbanization, Climate Linked Water Vulnerability as Impediments to Gender Equality: A Case Study of Delhi, India

8

Jagriti Kher, Savita Aggarwal, Geeta Punhani, and Sakshi Saini

Contents

J. Kher (✉) · S. Aggarwal · G. Punhani · S. Saini
Department of Development Communication and Extension, Institute of Home Economics, University of Delhi, New Delhi, India
e-mail: jagriti28@gmail.com; savitaaggarwal@gmail.com; spunhani@hotmail.com; aatishkaa@gmail.com

W. Leal Filho (ed.), *Handbook of Climate Change Resilience*,
https://doi.org/10.1007/978-3-319-93336-8_33

Abstract

Poor urban and rural women bear the brunt of climate change as the dwindling availability of water and other natural resources makes their lives full of drudgery. They trudge longer distances and head load water for the family and struggle with poor water quality and unsafe sanitation. Climatic changes and the non-climatic drivers such as rapid urbanization and high rate of population growth will further confound the scenario and make the lives of poor women harder.

The present study has been conducted to assess the vulnerability of poor women residing in slums of Delhi, the capital city of India, to water- and climate-linked stresses using quantitative and qualitative approaches. An index called CVI-WH was used to quantify vulnerability of the slum women; the qualitative study was done using various participatory approaches. The study has shown very high vulnerability of the slum women to climate-linked water stresses as reflected by high CVI-WH values ranging between 0.62 and 0.67 across different regions.

Therefore, if the quality of life of poor women has to be improved, it is extremely important to enhance the adaptive capacity of women to face climatic stresses and to invest in water- and sanitation-related infrastructure.

Keywords

Water · Climate change · Vulnerability · Slums · Poverty · Urbanization · Climate Vulnerability Index (CVI)

Introduction

Climate change has become a global phenomenon, the impacts of climate change are already being felt across the globe, and the variability in climate is likely to increase. The world mean temperature has already risen by 0.74 °C in the last century and is further projected to increase by 0.3–4.8 °C by 2100 (IPCC 2013). In India, the mean temperature has risen by 0.56 °C between over the period (1901–2007), and it is projected that the temperature may rise between 2.2° and 5.5 °C by the turn of the century (MoEF 2012; Kothawale et al. 2010; Kumar et al. 2013; Dholakia et al. 2015). This, in turn, will lead to changes in all aspects of the hydrological cycle. It is projected that rainfall will become more variable and uncertain leading to increased frequency of droughts and floods (IPCC 2013). This would have an impact on the spatial and temporal availability of water in many parts of the world. Since women are the key procurers and managers of natural resources, especially water and fuel, such environmental stresses due to increased frequency and intensity of droughts, floods, heat episodes, and deforestation will further enhance hardships of poor women and will make their lives more drudgerous.

Water and Poverty

Water is an important natural resource, which pervades all aspects of human development. There is a two-way relationship between income poverty and deprivation in access to water. Globally, almost two-thirds of the population, lacking access to safe water, lives on less than 2 dollars a day (1.25$ being the poverty line threshold). The Multidimensional Poverty Index, which indicates human deprivations in a decent standard of living, has also used access to water and safe sanitation as its components to ensure a decent quality of life (UNDP 2010). The Sustainable Development Goals to be achieved by 2030 also focus on ensuring equitable access to safe and affordable drinking water and sanitation facilities. However, both these basic necessities are still a luxury for large chunk of the global population. According to WHO/UNICEF Joint Monitoring Program for Water Supply and Sanitation, 2017, as many as 884 million people in the world still do not get their drinking water from improved sources, and 2.3 billion people lack access to basic sanitation facilities. This is apparent by the fact that 892 million people in the world still practice open defecation (WHO/UNICEF, JMP 2017). It is well known that poor access to safe water and lack of sanitation facilities in developing countries multiply the work burdens of women as they are generally responsible for cooking, household management, hygiene of children, and sanitation.

Water and Women

Women are the primary stakeholders in water accession and management. Their role as water managers for the family is undisputed in most of the developing world. The studies conducted on the time spent on household tasks by women over a period of 40 years across different countries confirm the anecdotal evidence that the task of water collection and management is largely shouldered by women (Desai et al. 2010; Sorenson et al. 2011; WHO/UNICEF 2012; Pickering and Davis 2012; Karim et al. 2013; Arora 2014). Rural, semi-urban women spend considerable time in procuring, transporting, and storing water from different water bodies such as rivers, lakes, wells, or community taps. Women carry water in different containers on their heads or arms and often have to travel long distances. They pay heavily in terms of physical (energy), temporal (time), affective (negative feelings and attitudes), and cognitive costs (foregoing opportunities for formal education, training, and intellectual development) to meet the needs of fresh water, fuel, and fodder as well as other natural resources required by the family. These costs magnify in times of sickness of family members, disasters, and environmental stresses, such as droughts or floods since women are the primary caregivers for the family (UNDP 2011). An analysis of data from 25 countries in sub-Saharan Africa (2006–2009), representing 48% of the region's population, reveals that women bear the primary responsibility of water collection. In all the households without water on the premises 62% women were responsible for water collection as compared to only 29% men. While the women spent a combined total of at least 16 million hours every day collecting drinking water, the men spent only six million hours followed by children who spend about four million hours (WHO/UNICEF, JMP 2012). In order to explore gender differences in water collection, a cross-sectional survey was conducted covering 602 households spread across 15 villages in Makondo Parish in Uganda. The results clearly revealed that the burden of water collection was mostly borne by women and children (Asaba et al. 2014).

An analysis of water collection among 24 sub-Saharan countries revealed that adult females were primarily responsible for water collection ranging from 46% in Liberia to 90% in Côte d' Ivoire. Across countries, a significantly higher number of female children (62%) were shouldering the responsibility of water collection as compared to the (38%) male children (Graham et al. 2016). To assess the impact of water insecurity on the families especially women, a study conducted in Kenya highlighted that water insecurity was harmful to women and their families in five different domains including physical health. It was highlighted that a large majority of women (women 78%) were responsible for water collection spending 4.5 h per week, which was also calorically very expensive (Collins et al. 2017). A study conducted by WHO/UNICEF, Joint Monitoring Program, 2017, for water supply and sanitation across 61 developing countries has documented that women (73%) shoulder the major bulk of water collection load. Women are twice as likely as men to fetch water for the household (WHO/UNICEF 2017).

The Indian scenario is even bleaker and likely to worsen because of climatic stresses, variability, and extremes. The time spent by women in fetching water in rural India represents nearly 22% of their working days representing a significant unproductive component of their work time (Mkhize et al. 2012). An assessment covering 183 villages in 28 districts of nine drought-affected states of India revealed that the scarcity of water during drought conditions compelled women to spend extra hours and trudge longer distances to fetch water for the family. The women across these states were spending 2 to 6 h each day walking several kilometers to fetch water. The study also highlighted three- to sixfold increase in time spent by women in queuing up and fetching water since the distance to the source had increased forcing women to travel farther for water collection (UNICEF 2016). Carrying water over long distances is not only a time-consuming task but also leads to various health hazards for women (Geere et al. 2010; Asaba et al. 2014).

Access to Water: Impact on Education

It is not only women but also children especially female children who often share the temporal and physical burden of carrying water, often leading to very large gender gaps in school attendance in many countries. A study conducted in Ethiopia assessing the perception of girls about their domestic tasks including fetching water found that all young girls felt that water collection limited their ability to participate and succeed in school (Abebaw et al. 2010). Another study conducted in Africa has shown that one-third of girls and boys in Ghana and 8% of girls and 3% of boys in Malawi reported to arrive late for school because of water collection activities (Porter et al. 2012). Research conducted in Ethiopia and Bangladesh has shown that as water sources are being depleted by increased frequency of extreme events and rapidly rising population, girls are spending greater time collecting water for drinking and other household purposes. Fetching water could take up to 6 h a day in the future, whereas previously it had taken around 2 h (Swarup et al. 2011). A study conducted in the least developed countries in the Asia-Pacific has shown that women and girls generally have the primary responsibility for collecting and managing water, spending up to 6 h per day, traveling as far as 6 km to collect water. Often girls are forced to drop out of school to assist in this household task (Brown Kourtnii 2012). Over 98% of the respondents in Nakuru, in Kenya, reported that their children had to walk long distances to fetch water for the family and did not attend school on several days, negatively impacting their mental and physical development due to lost opportunities of education (Jonah et al. 2015). A survey was conducted in eight secondary schools in two watershed areas in Gujarat and Rajasthan to assess the student's perceptions about groundwater scarcity and its impact on their educational opportunities. School absenteeism was found to be gender differentiated as female students were missing schools more frequently as compared to male children. The study highlighted that the scarcity of groundwater leads to an enhanced burden of household work including water collection on female children, thus limiting their educational and economic opportunities (Kookana et al. 2016). It has been estimated that women and girls across the world collectively spend about 200 million hours every day collecting water (WHO/UNICEF, JMP 2010) that could be spent in undertaking income-generating activities and contributing to the formal economy. Thus fetching water takes several hours and shortens the time that girls have for schooling, thus impacting their overall development. Carrying water from distant places not only causes physical disorders in women but also reduces their opportunities for formal education, income generation, community and political participation, skill development, leisure, recreation, and training. Although developing countries including India have made great strides in terms of all-round development including water and sanitation works, such development is not able to keep pace with the demographic changes, thus diluting the efforts. Women in these regions continue to spend massive amount of time and energy in fulfilling the water requirements of the family. Translating the losses borne by girls and women in terms of (woman) hours showed a loss of human capital and reduced ability of the household to capitalize fully on its other resources. This is depicted by an estimate of women's contribution of 150 million workdays per year on water collection for the household, which is equivalent to a loss of whopping 10 billion rupees per year (RFSTE 2005).

Climatic changes coupled with demographic and technological factors will further confound this scenario, since climate change has a significant impact on the availability of water (Bates et al. 2008). It is estimated that the demand for water in India will rise by 20–40% in the next 20 years due to rapid urbanization, modernization, and associated changes in lifestyle (Planning Commission 2002). The per capita availability of water in India has drastically reduced from 5177 m^3 in 1951 to 1654 m^3in 2007and is likely to go down to as low as 1100 m^3/year by 2050, leading to exacerbation of the water crisis (Krishna and De 2010; MOWR 2011). It is important to highlight that research has shown that India is not poor in terms of water resources; it lacks the ability to efficiently capture and effectively utilize the available resources for maximum benefit. Due to large spatial and temporal variability of rainfall, conservation of water through proper storage is of utmost importance. The decisively lower level of storage capacity is an important area of concern. The per capita storage in the country is about 225 m^3 which is far below the storage capacity of Russia (6103 m^3), Australia (4733 m^3), and China (1111 m^3) (MOWR 2011; MoEF 2012). Thus the lack of water storage facilities enhances the level of vulnerability of the Indian population especially women who are the prime procurers and mangers of water.

Impact of Improved Services on Women

There is substantial evidence to show that access to water at the household level helps women save time which can be put to several income-generating or productive tasks such as agricultural production, food preparation, technical training, and education, all of which may contribute to growth and improving their quality of life. The women may also have increased time for rest and leisure, become healthier and happier, gain confidence, and attain a greater sense of personal dignity (Ray 2007; Mkhize et al. 2012; Van Houweling et al. 2012; Crow et al. 2013). Freed-up time may also be devoted to child nutrition and health needs including visits to health centers (Sorenson et al. 2011; Koolwal and van de Walle 2013).

Studies have also shown that provision of water at or near the household has a positive effect on the attendance of girls and their enrolment in school.

A study conducted in Tanzania has depicted a very strong relationship between school attendance and distance of water source from home. It was found that when the source of water was 15 min or less from the home, the school attendance of girls was 12% higher as compared to homes where water source was an hour or more away. This was not the case with boys since their attendance rates in school was far less sensitive to distance of water source from home, highlighting the integral role that girls play in fetching water (Hanchate and Dyson 2004). In Morocco, as a result of rural water supply and sanitation project, attendance of girls in school increased by 20% over 4 years. It also reduced the time spent in water collection by 50–90% (World Bank 2003). In Ghana four rounds of Demographic and Health Survey indicated that 50% reduction in the time to haul water on an average would increase the proportion of girls 5 to 15 years attending school by 2.4% points with stronger effect for rural communities than in the urban areas (Nauges and Strand 2013).

Trends of Urbanization

Due to massive increase in urbanization, managing urban areas has become one of the most important development challenges of the twenty-first century to which India is no exception. The world's population is becoming increasingly urban as almost half of the world's population (54%) resides in the urban areas and is expected to increase to 66% by 2050. It is projected that largest urban growth will take place in three developing countries including India. The rate of India's urbanization is 1.1%, which is much higher as compared to global average of 0.9% (UNDESA 2014; UN-Habitat 2016). Trends suggest that the urban population in India is likely to grow from 410 million in 2014 to 814 million in 2050 constituting almost 50% of the total population (International Organization for Migration Report 2015). Of this, a considerable number will reside in slums and slum-like settlements giving India a character of a giant slum to the country (Sankhe et al. 2010). The 2-3-4-5 phenomenon has often been used to describe the character of Indian demography. This implies that the overall population of India has grown at an average rate of 2%, the urban areas at 3%, the big cities at 4%, and the slum population at 5% (USAID 2007), indicating the phenomenal growth in the slum population. This segment of the society faces a high level of vulnerability due to limited access to basic services such as housing facilities with reliable water supply and sanitation.

Thus in the coming years, problems of climate change coupled with urbanization and modernization will put an increased pressure on water as well as other resources. The decreased availability of overall water supply in the future is further likely to affect the quality of life of people and is bound to impact the lives of women who are responsible for procurement and management of water. Poor women living in the rural and urban areas will suffer the most because of their limited adaptive capacity.

The primary objective of the present study was to assess the vulnerability of poor urban women residing in slums and related settlements to climate-linked water stresses using a combination of qualitative and quantitative approaches. The other objective of the study was to create evidence of the need to focus on sound adaptation strategies, which could enable women greater participation in socioeconomic development and lead to greater gender equity and equality in societies.

Methodology

As per the Census of India 2011, about 17.4% of urban households in India live in slums. This study has been conducted in Delhi, the capital city of India, since it has more than considerable population (16%) living in slums (Census of India 2011). These numbers continue to grow at a fast pace. Since the capital city represents the highest level of infrastructure and socioeconomic development, this study would provide a show window to the vulnerability of poor women and their families living in impoverished areas of other cities and towns in India and other developing countries facing similar problems.

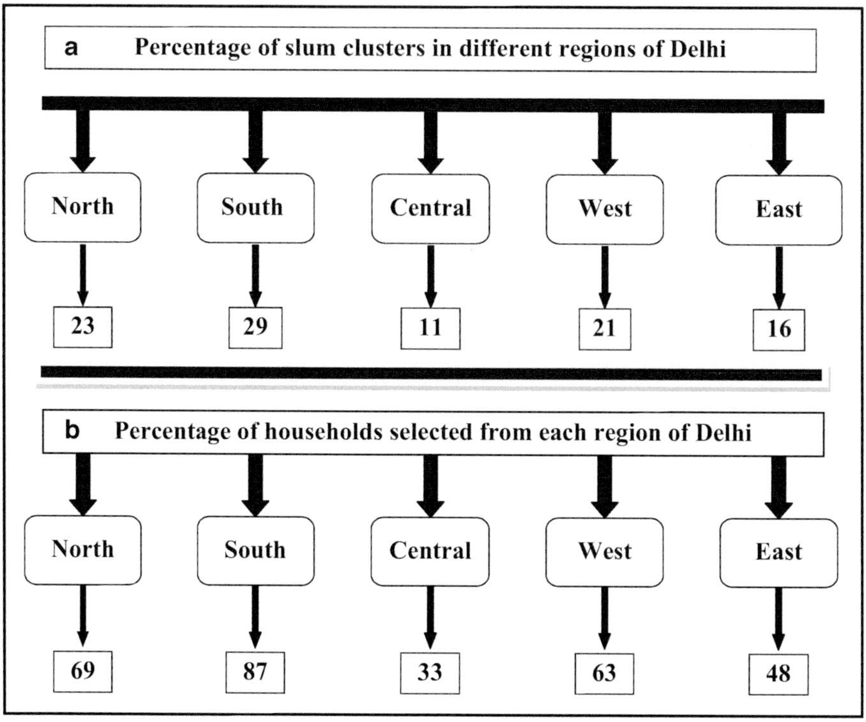

Fig. 1 (**a**) Percentage of slum clusters across regions of Delhi. (**b**) Percentage of households selected from each region. (Source: Based on Fig. 1a; Ward wise Data of Delhi Slums, Census of India 2001)

The study was conducted across 300 households in five main regions in the National Capital Territory of Delhi, namely, North, South, Central, East, and West. The number of households selected from each region was propionate to the percentage of slum clusters in that region (Fig. 1a).

A two-stage sampling technique was used. In the primary stage, slums were selected randomly from each of the five regions of National Capital Territory of Delhi. Circular systematic sampling was used for selecting the households from the selected slums/JJ* clusters. The number of households covered from each region was proportionate to the number of slum population in the five regions of Delhi. Out of 300 households, varying number of households were selected as per the distribution of slums across different regions (Fig. 1b).

The quantitative assessment was done by a primary survey using an interview-administered questionnaire focusing on water-related vulnerability of slum families especially women living in Delhi, the capital city of India. This was accomplished by collecting gender disaggregated data on the mode of water accession, distance of water source from home, impact of water collection on income generation and education of children and skill development opportunities on women, use of newer methods or technologies for water accession, and storage and coping strategies to deal with water supply shortage. Data was collected over a period of 1 year in the five regions of Delhi. Several visits were made to the slums to establish rapport with the women, observe their water management practices, and later collect data. The data was later used to quantify Climate Vulnerability Index for Water at the Household (CVI-WH) level to indicate the extent of vulnerability.

The CVI-WH is based on Water Poverty Index (WPI) and Climate Vulnerability Index (CVI) developed between 2002 and 2005 by a group of hydrologists, water practitioners, and social scientists (Lawrence et al. 2002; Sullivan and Meigh 2005).

CVI-WH has been adapted to capture the water- and climate-related vulnerability in India at different temporal and spatial scales keeping in mind suitability of indicators used as well as feasibility of data availability. CVI-WH is characterized by six components, namely, *resources* (aims to capture the overall availability of water), *access* (includes the access of the population to safe drinking water at a reasonable distance), *human capacity* (the ability of people to manage water), *use* (consumption of water in different sectors), *environment* (reflect the state of the environment), and *geospatial characteristics* (geographical characteristics of the location that make it vulnerable to climatic stresses and extremes). The computation of CVI-WH has been done using normalization method; the index values of CVI-WH as well all its components ranged from zero (least vulnerable) to one (most vulnerable). For the detailed methodology of CVI-WH, refer to (MoEF 2012).

The qualitative data was captured by a number of Participatory Learning and Action (PLA) tools such as focus group discussions, resource maps, seasonal calendars, and in-depth interviews. Guidelines for conducting focus group discussions and in-depth interviews were made to get deeper insights into the problems faced by women and focused on water-related issues, mode of accession, the time spent by different family members on water collection, and hardships faced and missed opportunities for women and children due to their involvement in water collection. These exercises were conducted with a group of 8–10 women in an open area available in the slum or in the compound of the house of one of the respondents; in some areas, men and adolescent girls and boys also participated in the activity. Data triangulation was done to get a clearer picture of the issues under study. To get deeper insights about various issues, the findings from qualitative study have been presented along with quantitative data analysis.

Results

Demographic Profile of the Sample

Age Group

Majority of the women (almost 90%) were below 49 years of age, of which about 54% were in the age group of 20–34 years and 35% in the age group of 35–49 years. Only about 10.8% of women were aged 50 years and above (Table 1). This was

Table 1 Age group of the respondents ($n = 300$)

Regions of Delhi (percentages)	20–34 years	35–49 years	50 years and above
North	50	39.4	10.5
South	32.9	50.5	16.4
East	62.5	29.1	8.33
West	53.9	30.1	15.8
Central	72.7	24.2	3.03
Delhi (average)	54.4	34.6	10.8

Source: Primary Survey

Table 2 Religion of the respondents

Regions of Delhi (percentages)	Hindu	Muslims
North	98.6	1.32
South	89	10.9
East	77	22.9
West	85.7	14.2
Central	90.9	9.09
Delhi (average)	88.24	11.68

Source: Primary Survey

because more young people had migrated to cities for jobs, while the parents stayed back in a more familiar environment of the village looking after land and property.

Religion

A large majority (88%) of the respondents belonged to Hindu religion, and the rest were Muslim. This composition was comparable to the religion-wise breakup at the national level. It was also found that Central and North Delhi had higher number of Hindus (91–99%), whereas West and East Delhi had higher number of Muslims (14.2–22.9%) as compared to other regions of Delhi. The other religions were not represented by the slum population (Table 2).

Family Size

Only 30% of the respondents had less than five members in their family; 52% had five to seven members. Overall, 82% of families had up to seven members in their family (Table 3). Across all the regions, a very small fraction of respondents (almost 4%) had a family size of more than 10 members.

Family Type

The data highlighted the trend of nuclear families in the slums of Delhi since on an average 76% of the families were nuclear, the rest being either extended or joint families. The nuclear families across different regions ranged from 72 to 83% (Table 4). Since a large majority of the families had migrated from rural areas, the older members (parents, in-laws) of their families often stayed back in the village. Small size of houses in the slums was another reason cited by respondents for having nuclear families.

Table 3 Number of family members per household

Regions of Delhi	Less than 5 members	5–7 members	8–10 members	More than 10 members
North	22.3	61.8	11.8	3.9
South	18.6	58.2	19.7	3.2
East	29.1	54.1	10.4	6.2
West	34.9	36.5	22.2	6.3
Central	42.4	51.5	6	0
Delhi (average)	29.46	52.42	14.02	3.92

Source: Primary Survey

Table 4 Family type

Regions of Delhi (percentages)	Joint	Nuclear
North	27.6	72.3
South	25.2	74.7
East	16.6	83.3
West	23	76.1
Central	24.2	75.7
Delhi (average)	23.32	76.42

Source: Primary Survey

Educational Level

The educational attainments of slum women were much below the national average literacy rate. It was startling to note that more than three-fourths of the women residing in the slum areas had never attended any kind of school and were totally illiterate as compared to 65% literacy rate of women in India. On an average, only 18% of women had education up to primary level and 4% up to secondary level, and less than 2% of women had any form of higher education (Fig. 2). These values are in sharp contrast to the female literacy values of the whole of Delhi placed at 77% (NSSO, 2004–2005). There were hardly any differences in educational level across different regions of Delhi. This clearly depicted the dismal state of education of women in slums of Delhi. Focus group discussions revealed that most women who reported studying up to primary level were also either semiliterate or illiterate as they had dropped out of school in first, second, or third grades and thereafter had never tried to read or write. They had attended school in their respective villages and had dropped out due to a variety of reasons such as lack of finances, lack of interest in studies, family-related problems, household work, or marriage.

Monthly Family Income

There was a lot of skepticism in the families in reporting their monthly income. It is important to note that there may have been some underrepresentation while reporting the income by the respondents. Many women said their husbands were part-time employees, daily wagers, or entrepreneurs operating petty business; their incomes

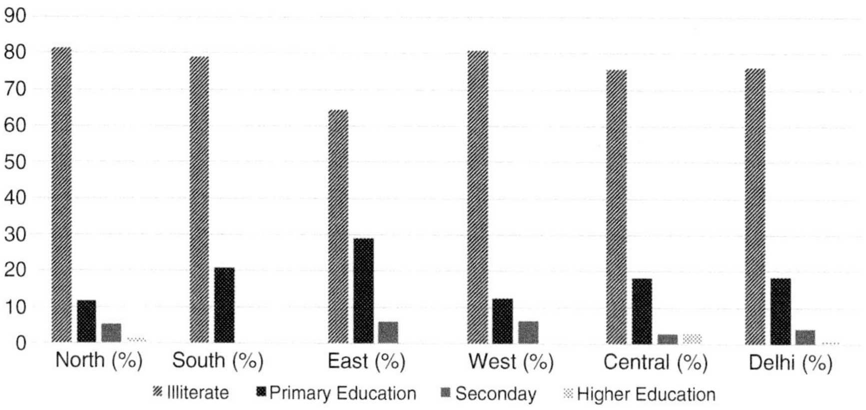

Fig. 2 Educational level of the respondents. (Source: Primary Survey)

were never fixed. Majority of the families (90.5%) had reported monthly income up to INR* 5000/- (90 US $). The families were therefore eligible to get the BPL (below poverty line) cards. As many as 45% of the families earned a monthly income of Rs 3001–4000/- followed closely by 28% of the families who were placed in the income bracket of INR 2001–3000/- (Fig. 3). Only about 10% of the families earned a monthly income above five thousand Indian rupees. There were considerable variations across the different regions in terms of income distribution. In South Delhi, a much larger number of people earned more than INR 5000 as compared to other regions. One of the reasons for this was, higher salaries were being paid to workers in South Delhi since this region comprised of posh colonies of Delhi. The per capita annual income of people in the slums across different regions ranged from 8 to 10 thousand rupees. By comparison, the average per capita annual income for Delhi as a whole was much higher, placed at 36 thousand INR (Economic Survey 2003).

Women's Occupation

It was quite surprising to find that despite low levels of income, only a quarter of the women respondents were engaged in any form of income-generating activities and the rest were homemakers. The women reported that it was not easy for them to find suitable jobs since they were preoccupied with household responsibilities and had little time to spare for working outside the home. The women needed to look after small children since most of them were in nuclear family setups. Among those who worked outside the home for earning a livelihood, about half of them worked as domestic helpers or petty sellers and the other half as daily wagers in factories or construction sites (Table 5). Maximum employment was seen in South and West Delhi perhaps due to greater availability of jobs there. During the FGDs, women reported that the male members were engaged in a variety of occupations and worked as rickshaw pullers, auto drivers, sellers, ear cleaners, or wage workers.

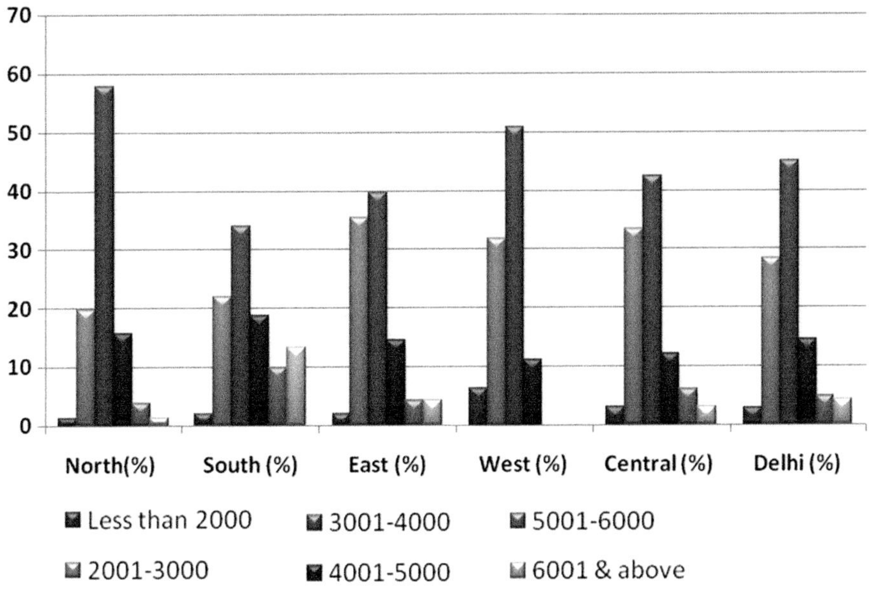

Fig. 3 Monthly Income families (INR). (Source: Primary Survey)

Table 5 Occupation of the women respondents

Regions of Delhi (percentages)	Housewife	Employed
North	84.2	15.7
South	68	32
East	89.5	10.4
West	61.9	38
Central	78.7	21.2
Delhi (average)	76.4	23.4

Source: Primary Survey

Water Accession and Management at the Household Level

Source of Water in the Household

A small number of families (7%) had installed illegal water connections at home from the community point using plastic pipes with taps and therefore did not have to walk to the water source located at community point. These families however lived in fear of civic authorities since the water connections were illegal. The rest of the families, which constituted a large majority, collected water from a community point. Majority of the respondents (73%) used piped water supply from the taps installed at community points, while a smaller (11%) drew water from the tube wells followed by hand pumps (3.5%), both installed at community point. A small number

Table 6 Available sources of water for drinking for domestic use

Regions of Delhi (percentages)	Piped water supply in the house	Piped water community point	Hand pump (house)	Hand pump (community)	Motorized hand pump (house)	Tube well community point	Tanker
North	18.4	61.84	0	11.84	3.95	3.95	
South	3.3	53.8	0	1.1	2.2	29.67	5.49
East	6.25	89.5	2.08	0	2.08	0	
West	0	73.02	0	1.59	0	23.8	1.59
Central	9.09	87.88	0	3.03	0	0	
Delhi	7.41	73.21	0.42	3.51	1.65	11.48	3.54

Source: Primary Survey

of families (5%) filled up water from the tanker sent by the *Delhi Jal Board* or the waterworks authority of the local government. A few families (1.6%) had installed motorized hand pumps in the house either individually or in collaboration with other families so that the cost could be shared, and a very few (0.4%) had hand pumps within the house (Table 6). Overall 97% of the families had access to safe source of water (tap, hand pump, or tube well), which is almost the same as for whole of Delhi.

It was found that most families were using more than one water source to meet their requirements depending upon the ease of availability, the quality of water, and the use it has to be put to. As an example, some families were using tube well water to meet their sanitation requirements and tanker water which was of better quality to meet their cooking and drinking requirements.

There were some variations across different regions in terms of source of water supply. These differences can be explained due to variation in population density, tap-to-user ratio because of differences in the number of installed taps and hand pumps, timings, regularity, and pressure of water supply. At some places, people had broken down the water pipelines, dug up holes in the ground to collect water, or used plastic pipes on the broken connection to withdraw water. Sometimes people had to buy water from water vendors (small-scale independent water providers) by paying substantial amounts of money.

Distance of Water Source from Home

The water source could be reached from home within 5 min time for almost 30% of the families and in 5–15 min for another 34% of the families. A considerable number of families (27%) reported that the water source was farther away and they had to travel more than 15 min from their homes to reach the water source (Table 7). These figures of safe location of water were in sharp contrast to Delhi, which was placed at 92% (Census of India 2001).

There was variation across different regions of Delhi as only 4% of the families in East Delhi reported that they had to walk more than 15 min to the water source as compared to

Table 7 Time taken to reach the water source from the house

Regions of Delhi (percentages)	Within house	Less than 5 min	5–15 min	More than 15 min
North	22.3	25	42	10.53
South	5.49	29.6	39.5	25.27
East	10.4	45.8	39.5	4.17
West	0	6.3	30.16	63.49
Central	9.09	39.3	18.18	33.33
Delhi (average)	9.46	29.2	33.87	27.36

Source: Primary Survey

63% in case of West Delhi. This is because of high population density, poor water supply, and high tap-to-user ratio. A small number of women (9.5%) did not have to spend any time in travel as they had installed water connections (illegally) at home.

Responsibility of Water Collection

Almost 95% of women were involved in water collection. More than 78% of water collection task was performed by the daughters-in-law followed by mothers-in-law (10.4%). The daughters-in-law involvement ranged from 63.6% to 85.4% across different regions of Delhi. The findings of the study are well supported by a number of research studies (mentioned earlier) conducted across the globe, which highlight a clear link between women and water accession and management. The children, especially girls (9.5%), also helped their mothers in fetching water (Fig. 4). Girls were frequently asked to skip school to help their mothers in water collection or to look after the young siblings while their mothers went for water collection especially in areas facing water shortages. This finding is also well supported by the studies conducted across a number of countries, which reveal high dropout rates among girls to fulfill the family requirements of water. The participation of men (3.6%) in water collection was found to be very low. The qualitative assessment using focus group discussions highlighted that men fetched water for the household only in some special situations as when their wives were away to their relatives or maternal home and were unwell and the older women were not physically fit to carry water. The study therefore has shown that water management in the slum households continued to be a deeply gendered task allocated almost exclusively to females. In fact in many instances, the women themselves said that if men would start collecting water, then who would earn money for the family?

Time Spent in Water Collection

More than 90% of women reported spending up to 4 h every day in water collection for meeting the household needs across different regions of Delhi. The timings of water supply were very erratic and kept fluctuating, causing hardships and forcing

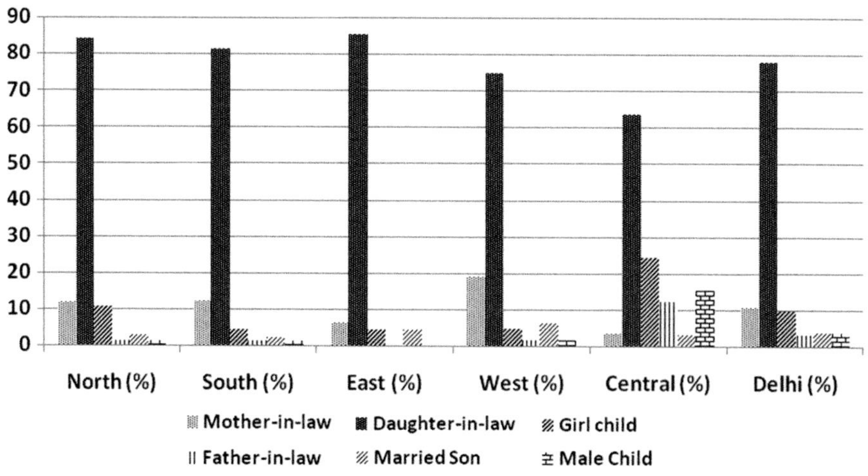

Fig. 4 Persons normally responsible for water collection. (Source: Primary Survey)

women to spent even greater amounts of time and energy in water collection. About (2%–31%) of women were spending more than 4 h in water collection across different regions in Delhi (Fig. 5). The seasonal calendars revealed that the availability of water was a problem in the slums throughout the year. However, the worst times were the months from April to July. During this time, the families had to suffer a lot due to highly inadequate supply of water. Women had to wake up early in the morning to queue up for filling water. During the summer months, they spent much more time in water collection and cut down time spent on household chores, childcare, leisure, sleep, as well as on income- generating activities. They also had no option but to involve their children in water collection as well. By comparison, the winter months from November to February were less problematic in terms of water scarcity and lower demand for water.

Number of Trips to the Water Source

The women reported making several trips every day to collect water in buckets and jerry cans of about 15–20 l capacity. A large majority of the women (36.3–66.6%) across different regions were making more than six trips per day for collecting water reflecting the vulnerability of women to inadequate system of water supply (Fig. 6). Almost 42% of the respondents made up to six trips per day to collect water for the family. The resource maps revealed the uneven distribution of the location of water supply sources causing even greater hardships to women whose houses were located away. In-depth interviews with the women leaders of the area highlight the increasing population of the slum areas over the years without a proportionate increase in the water supply sources leading to overcrowding, long queues, and fights over water collection. When water was in short supply, especially during the summer

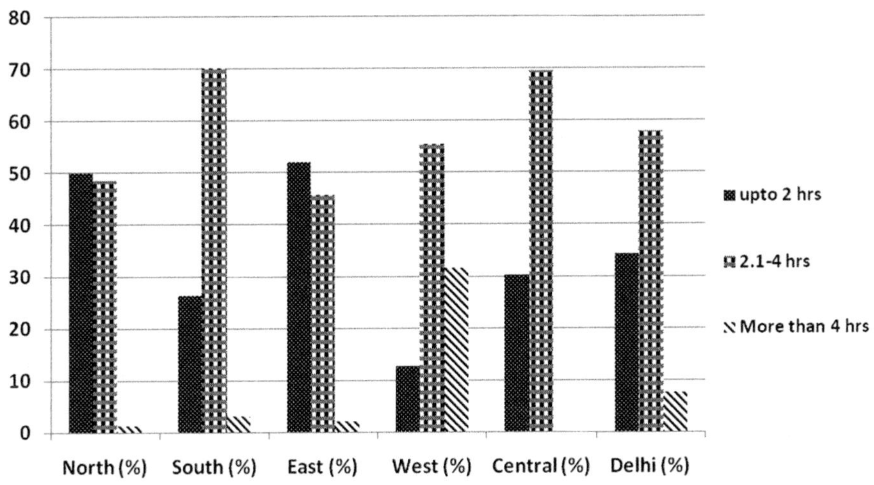

Fig. 5 Time spent by slum women in water collection. (Source: Primary Survey)

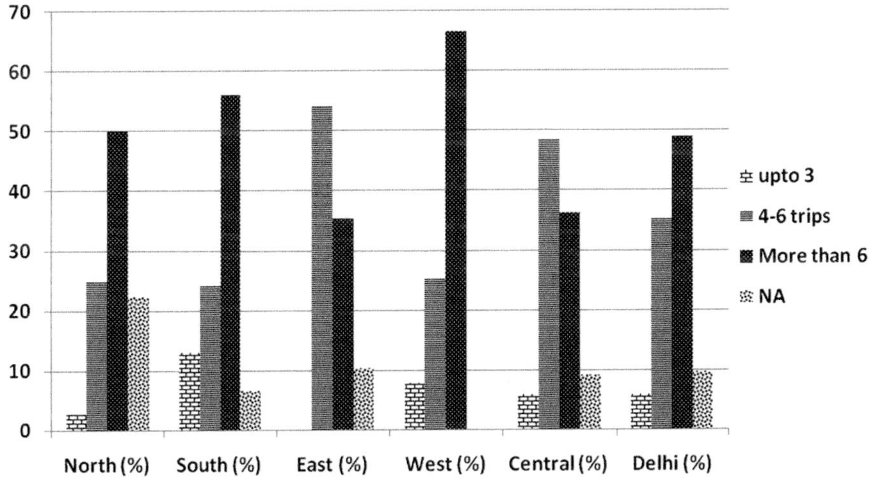

Fig. 6 Number of trips made for water collection per day. (Source: Primary Survey)

months, there was a corresponding increase in the number of trips for water collection. A few trips had to be made just to find out whether the water tanker had arrived or the water pump had been started by the in charge as well as to place the cans/buckets in the queue. In times of water shortage, the women had to depend on alternate sources of water supply (taps, hand pumps) located in the adjoining colonies. During such periods, their waiting and collection time increased substantially causing them even greater hardship. Thus the vulnerability of slum dwellers especially of women and children increased tremendously during such environmentally stressful periods.

Frequency and Time of Water Supply

The tap water supply at community point was available two times a day in the morning and evening for 1 to 2 h as stated by majority of respondents (73%), while the rest (26.7%) reported receiving water supply throughout the day. Round-the-clock availability of water ranged from 5.4% in South Delhi to 45% in Central Delhi. The women however pointed out that the water supply from the hand pumps if present was available all the time, bringing some relief to the families. The tankers of the city water corporation had no fixed time at all and, despite a fixed schedule on paper, would arrive at any time of the day or week. Due to this, the residents had considerable problems especially during periods of water shortage as they had to be continuously alert for filling up water. About 62% said that the timings of water supply were not convenient to them, since they clashed with their working hours. A smaller number (less than 40%) said the timings were convenient.

Water Quality

About 68% of the respondents were satisfied with the quality of water supplied to them, while the rest were not. Those who were not satisfied reported multiple problems such as muddy water (25%); hard water (23%); unpleasant smell, taste, and color (14%); and health-related problems due to poor quality of water (6%). It is important to mention that few women from East Delhi also reported presence of worms and insects in the water supply.

Problems Faced in Water Collection

Multiple problems were reported by the respondents during water collection such as very long queues, low pressure of water, and too much consumption of time as well as the occurrence of regular fights and conflicts. As many as 78% of the respondents found water collection to be a time-consuming task, whereas 44% said that the long queues was one of the major problems faced during water collection. About 70% of the respondents reported fights and conflicts at the time of filling water (Fig. 7). All these lead to great physical hardship for the women, tensions in the community, and mental stress. While majority of the conflicts were resolved by the people themselves, on some occasions the police had to be called in. Moreover, some families said they have been doing it all their life and had got used to it. A small percentage (6%) also reported incidents of eve teasing during water collection. Due to these numerous issues, almost 90% of the respondents reported body pain and general tiredness as a result of filling and carrying water every day. The variations across the different regions in terms of problems experienced by people in water accession and collection reflected the differences in water supply infrastructure and demographic characteristics.

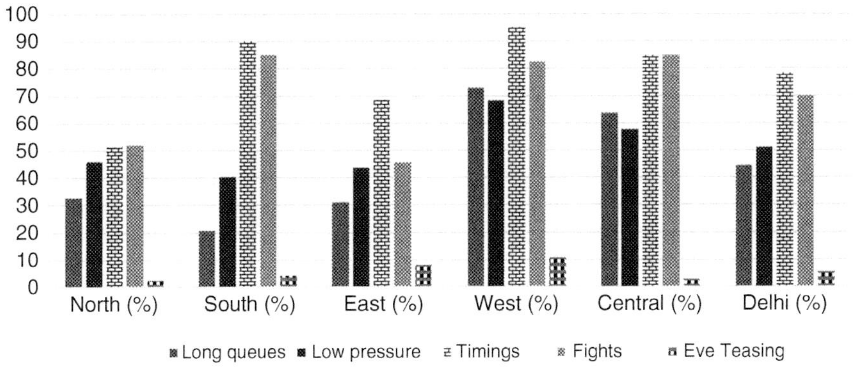

Fig. 7 Problems faced in water accession. (Source: Primary Survey)

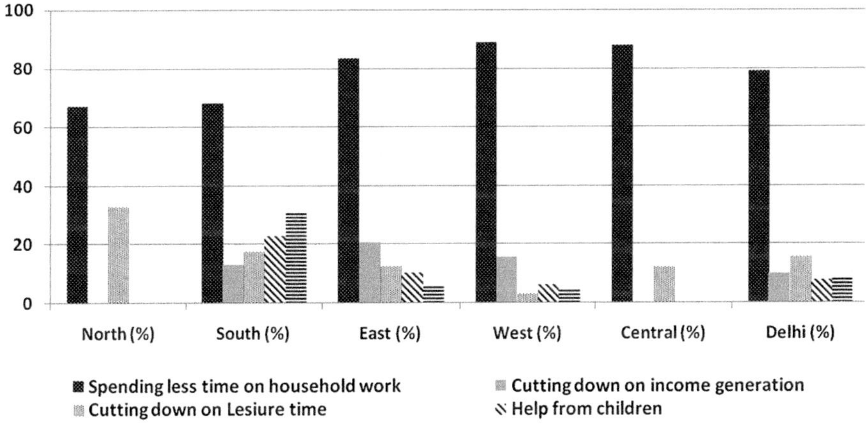

Fig. 8 Coping strategies to cope with water shortage. (Source: Primary Survey)

Coping Strategies of Women During Water Stress Periods

Such times were very common in the lives of these poor women such as during the summer months from April to June when there is dwindling of water supply due to peak demand and lowered water availability. The women adopted several strategies to tide over periods of environmental and water stress. Almost 80% of the respondents reported that they spent less time on household work, cut down on leisure time (16%), and sought help of children (8%) and help of other family members (8.3%). Of the almost 25% of women who were engaged in paid employment, 10% of women reported cutting down time on income-generating activities (Fig. 8). There were some differences across regions in the coping strategies practiced by women. Whereas 67% of women from North Delhi said that they cut down time spent on household work, as many as 88% reported the same from West and Central Delhi. Except in North and Central region, the help of both male and female children was taken across the remaining regions. A relatively higher number of families in South Delhi took the help of children in water accession since the water sources were few and far off and the women alone could not manage the water collection task.

Availability of Sanitation Facilities

Toilet Facilities

Only a small number of women (4%) had access to sanitary (pit) latrines. Majority of the women (77%) were using community toilets though not exclusively. Of these, 45% of women reported practicing open defecation sometimes. By comparison, almost 62% of the population of whole of Delhi had access to sanitary toilets (Census of India 2001).

The women complained that there was a dearth of sanitation facilities as each slum had only one toilet block and accessing it in the peak hours was almost impossible. Community toilets were often very dirty, so the families preferred to use open areas for defecation. Moreover, people were forced to use open areas at night after 10 P.M. since the toilets were locked due to security reasons. Many women left home before dawn or after nightfall to maintain privacy. It was reported that women have to wait until dark to defecate and urinate in the open and they tend to drink less during the day, resulting in serious health problems. The findings of the study are well supported by a number of researches conducted in different parts of the developing world where women adopt practices such as inadequate food and water intake to avoid defecating during nighttime (Fisher 2006; McFarlane 2008; O'Reilly 2010; Truelove 2011; Khanna and Das 2016).

In case of males, a higher proportion (76%) were using community toilets, and when the community toilets were not available due to the abovementioned reasons, 71% of them were practicing open defecation. There was a large variation across different regions of Delhi with regard to use of community toilets or open defecation. Whereas 97% of women used the community toilets in Central Delhi, only 50% of women did so in the South Delhi. This was because of differences in population density, availability of and access to useable toilets, their upkeep and level of cleanliness, as well as the availability of open areas for defecation in and around the slum. The practice of open defecation made the areas very unhygienic and filthy. The findings of the study is supported by a report by Water Aid (2017) that reports India has the highest number of people (723 million) without access to toilets. Women and girls are among the worst hit with 350 million of them lacking access to basic sanitation. According to a study by the Water and Sanitation Program (WSP) in India, the economic impacts of inadequate sanitation at the national level are estimated at 53.8 billion US$ (World Bank 2003; Water and

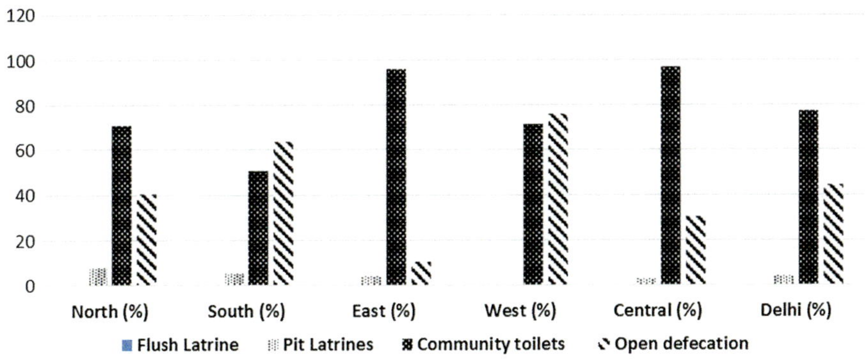

Fig. 9 Defecation pattern of the community (women). (Source: Primary Survey)

Sanitation Programme 2011). A study has also reported that whereas only 31% of the population in India had access to sanitary toilets in 2008, a much higher number (45%) owned mobile phones. These statistics not only indicate lack of knowledge and awareness in people regarding good hygiene and sanitation facilities (UNU-INWEH 2010) but also the difficulties involved in building sanitary toilets.

Bathroom Facilities

A very large number of families (84%) did not have any bathrooms in the house and had to manage with a makeshift bathroom without proper drainage or door. By comparison, almost 71% of households in the city of Delhi had bathroom facilities within the house (Census of India 2001). Across different regions of Delhi, only 5% of the slum families had bathrooms with drainage facility in their houses, whereas 10% used the community bathrooms. The rest used makeshift bathroom created from their living space. There were considerable variations across the regions of Delhi in terms of the availability of proper bathrooms ranging from a mere 1.5% in West Delhi to 6.2% in East Delhi. Similarly, the use of community bathrooms were much higher in West and East Delhi (20–27%) in sharp contrast to nil usage in Central and South Delhi and mere 2% in the slums of North Delhi (Fig. 10). This was because of differences in the availability as well as management of community bathrooms across the different regions (Fig. 9). During the focus group discussions, the women complained that because of absence of drainage facilities, there was flooding in the slums even after light rains leading to breeding of mosquitoes and a high incidence of diseases. The use of makeshift bathrooms led to severe drainage problems in the areas, leading to formation of sludge and creation of breeding areas for flies and mosquitoes depicting poor state of environment across all the slums in Delhi.

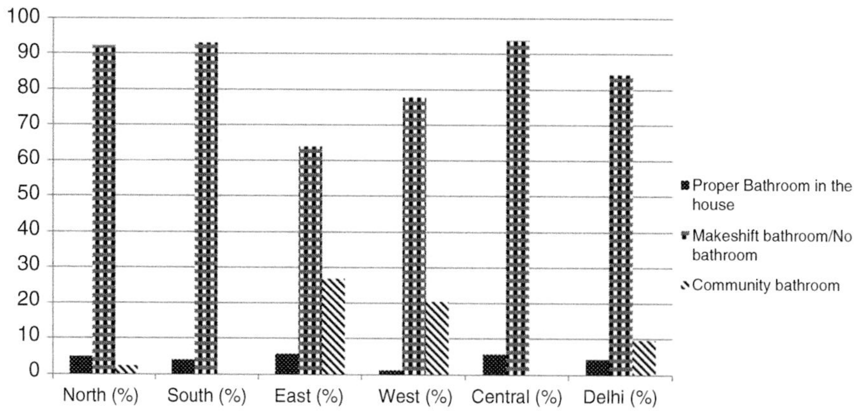

Fig. 10 Bathroom facility at home. (Source: Primary Survey)

Per Capita Water Usage by Families

On an average the per capita water consumption per day across Delhi slums was 37.6 l. The values ranged from 36.76 l per capita per day in East Delhi to 40 l in the case of North Delhi slums, significantly lower as compared to 57.98 l of water consumption per capita in the entire city of Delhi. It is important to highlight that safe water and sanitation are essential for healthy living, dignity, empowerment, and prosperity and have been declared by UN as fundamental human rights of every person.

It can be inferred that water and sanitation facilities in the slums in Delhi were dismal. Since women are closely related to water and allied activities, the results indicate their high vulnerability, and changing climate in the near future will make the scenario even worse. This is well supported by Climate Survey, 2017, of the India Meteorological Department (IMD) which reports extreme weather events, natural disasters, and a failure of climate change adaptation as three of the environmental risks that made it to the top ten risks facing humankind. A comparison of the income level (people below poverty line), literacy of women, and state of water and sanitation facilities in the slums with the whole of Delhi highlights stark differences between the two pointing to the vulnerability of women since they are at the core of household activities (Fig. 11).

Quantification of Vulnerability

In order to assess the vulnerability of families to climatic and water stresses at the household level across different regions of Delhi, the index values of CVI-WH were computed. The values collected by the primary survey were converted into index values, to generate the CVI-WH values across different regions of Delhi. Though the

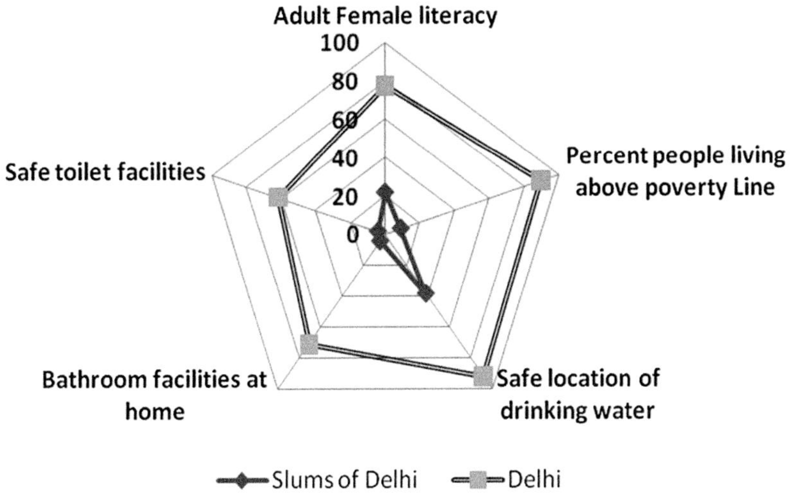

Fig. 11 Differences in Human capacity and state of environment of slums and whole of Delhi. (Source: Slums data-primary survey: Delhi-Secondary sources)

slum areas were exposed to the same climatic variability and extremes as the whole of Delhi, the results of this study have shown that the slums across all the regions in Delhi were extremely vulnerable and had high CVI-WH values ranging from 0.62 in case of North Delhi to 0.67 in case of West Delhi (the higher the value, the greater the vulnerability) with an average vulnerability index of 0.63 across the regions. By comparison, Delhi as a whole faced much lower level of vulnerability with its CVI-WH value at 0.36 (Fig. 12). The slums were extremely vulnerable in terms of poor state of environment due to poor sanitation facilities followed by limited human capacity and limited use and access to water.

The vulnerability of slum women of Delhi was found to be comparable to the general city dwellers of the most vulnerable states in India such as Jharkhand, Madhya Pradesh, Bihar, and Orissa (for state level values, refer to MoEF 2012) (Fig. 13). The high vulnerability of urban poor in Delhi suggests that the situation in urban slums and rural areas of vulnerable states could be much worse since Delhi being the capital city of India presumably represents the highest level of development.

The situation is not likely to improve considering the high levels of urbanization, population growth reducing the per capita availability of water, climatic changes, and no significant changes in the gendered distribution of household responsibilities. Looking at trends of rapidly rising urban population, India is expected to add 300 million new urban residents by 2050. In India, urban areas already contribute more than 60% of GDP, and an extra 300 million new urban residents will contribute to GDP growth of the country but will create additional pressure on the already stressed or limited resources of the cities (UN Habitat 2016). Trends have shown that greater access to jobs, educational opportunities, and health-care facilities is

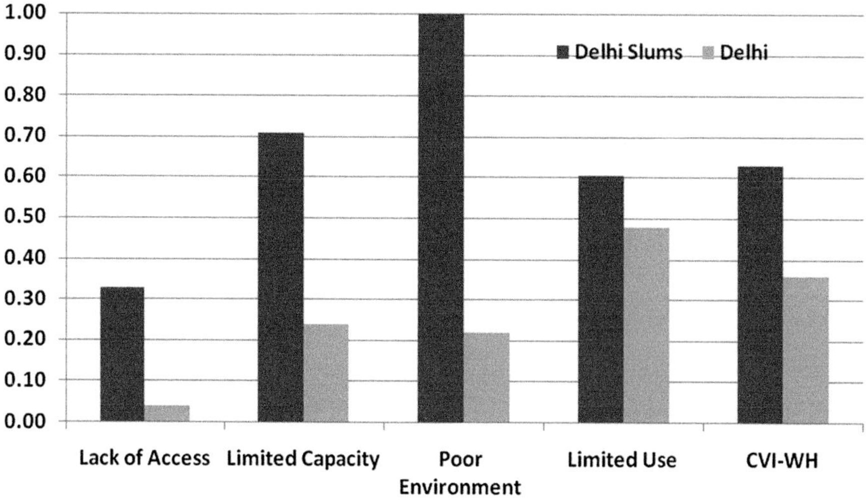

Fig. 12 Index values of CVI-WH and its sub-components across slums in different regions of Delhi. (Source: Authors calculation)

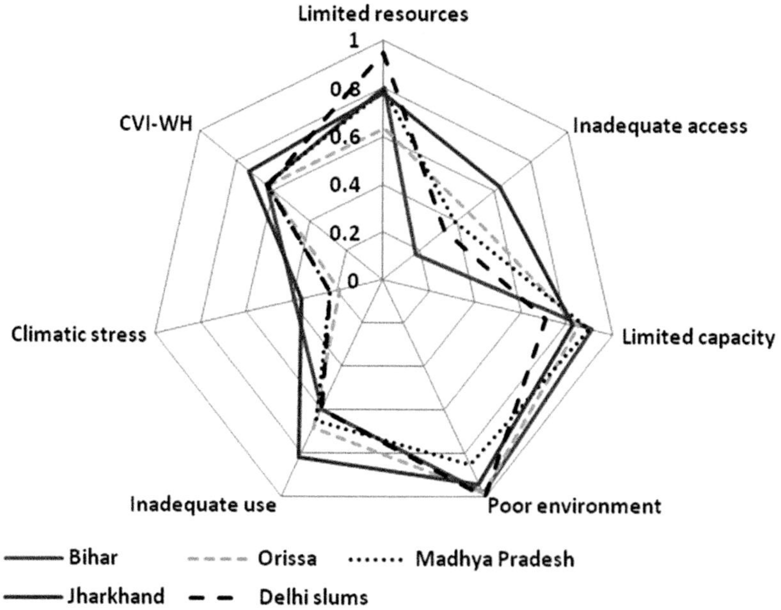

Fig. 13 Comparison of CVI-WH values of slums in Delhi with most vulnerable states in India (Source: Authors calculations)

attracting more people to the cities, thereby compounding the problem. All these factors make the slum population especially women and children highly vulnerable to environmental stresses related to water, sanitation, as well as climatic extremes.

Conclusion

The results of the study using both qualitative and quantitative techniques revealed high vulnerability of the slum dwellers especially women and children to climate- and water-linked stresses because of their prime role in water accession and management (Ballesteros 2010; Duflo et al. 2012; Yasmin 2016). The quantification of CVI-WH at the micro level can help to identify the population most at risk to water- and climate-related stresses and can provide evidence to trigger targeted action. The qualitative tools were instrumental in highlighting the problems and vulnerabilities of the poor slum women to climate- and water-related stresses. The number of slum dwellers is rising at a tremendous pace since as many as 55 million new slum dwellers have been added to the global population since 2000 (Chopra et al. 2016). On one hand, increased income, employment opportunities, and multifaceted facilities are attracting more and more people to the cities; on the other hand, the cities do not have enough infrastructure and resources to support this migrant population. The result is the ever-increasing number of slums as well as the population in existing slums. Without access to basic housing and water and sanitation facilities and the reducing per capita availability of water, the slum population especially women and children are likely to be highly vulnerable to water-related stresses which will be compounded by climatic changes and extremes.

To meet the goals of gender equality in society, it is crucial to pay special attention to the provision of clean water and sanitation, which are important needs of women. On the one hand, it is necessary to integrate housing of economically weaker sections of society into the smart city model equipped with water- and sanitation-related infrastructure; on the other hand, it is important to build adaptive capacity of poorer sections and most vulnerable groups such as women to face challenges posed by climatic and non-climatic drivers. This can be done by enhancing human capacity especially of women through more education, awareness, knowledge, and infusion of new technology. Such efforts have the potential to address the gender gaps in society in different domains and enable women to use their time more productively and equip them to lead climate-resilient lives.

Note: This study has been conducted as part of India's second national communication to the UNFCCC and was funded by the Ministry of Environment and Forests, Government of India.

References

Abebaw D, Tadesse F, Mogues T (2010) Access to improve water source and satisfaction with services: evidence from rural Ethiopia. International Food Policy Research Institute, Washington, DC

Hanchate A, Dyson T (2004) Prospects of food demand and supply. In: Cassen R, Visaria L, Dyson T (eds) Twenty first century India: population, economy, human development, and the environment. Oxford University Press, New York

Arora D (2014) Gender differences in time poverty in rural Mozambique. Department of Economics, University of Utah, Salt Lake City

Asaba RB, Fagan Honor G, Kabonesa C, Mugumaya F (2014) Women and access to water in rural Uganda: a review. J Gend Water 3(1):1–8. The University of Pennsylvania
Bates BC, Kundzewicz ZW, Wu S, Palutikof JP (eds) (2008) Climate change and water. Technical paper of the Intergovernmental Panel on Climate Change. IPCC Secretariat, Geneva, 210 pp
Ballesteros MM (2010) Linking Poverty and the Environment: Evidence from Slums in Philippine Cities. Discussion Paper Series NO. 2010–33. Philippine Institute for Development Studies
Brown Kourtnii S (2012) Women and water security. Asia foundation. Available from http://asiafoundation.org/in-asia/2012/03/21/women-and-water-security. Accessed 25 Feb 2016
Census of India (2001) Available via http://www.censusindia.gov.in/Tables_Published/H-Series/houselist_main.html
Census of India (2011) Cited in compendium government of India Ministry of Statistics & Programme implementation central statistics. Government of India
Chopra R, Sivakumar DK, Chopra R (2016) GeoSlum – Empowering local authorities for slum improvement through an open source Geospatial Platform. Indian Institute for Human Settlements, Bangalore
Collins MS, Miller DJ, Mbullo P et al (2017) The Multi-Faceted consequences of water insecurity for pregnant and Postpartum Kenyan women of Mixed HIV status. J Water Health 14(6):1028–1031
Crow B, Davies J, Patersonc S, Milesb J (2013) Using GPS and recall to understand water collection in Kenyan informal settlements. Dev Pract 38(1):43–60. https://doi.org/10.1080/02508060.2013.752315
Desai SB, Dubey A, Joshi LB, Sen M, Shariff A, Vanneman R (2010) Human development in India. Challenges for a society in transition. Oxford University Press, Oxford
Dholakia HH, Mishra V, Garg A (2015) Predicted increases in heat related mortality under climate change in urban India. Indian Institute of Management, Ahmadabad
Duflo E, Galiani S, Mobarak M (2012) Improving Access to Urban Services for the Poor: Open Issues and a Framework for a Future Research Agenda. Cambridge
Economic survey (2003) Income and poverty: Government of India, Economic Survey 2003
Fisher J (2006) For her, it is the big issue: putting women at the center of water supply, sanitation and hygiene. WSSCC and WEDC, Geneva
Geere J-AL, Hunte PR, Paul J (2010) Domestic water carrying and its implications for health: a review and mixed methods pilot study in Limpopo province, South Africa. Environ Health 9(52):1–13. https://doi.org/10.1186/1476-069X-9-52
Graham JP, Hirai M, Kim SS (2016) An analysis of water collection labor among women and children in 24 sub-Saharan African countries. PLoS One 11(6):0155981. https://doi.org/10.1371/journal.pone.0155981
Intergovernmental Panel on Climate change, IPCC (2013) The physical science basis. Working group I contribution to the fifth assessment report of the intergovernmental panel on climate change. New York, Cambridge
International Organization for Migration (2015) World migration report 2015. Urban migration trends, challenges and opportunities in India. International Organization for Migration (IOM)
Jonah C, Maitho T, Omware Q (2015) Water access and sustainable rural livelihoods: a case of Elementaita division in Nakuru County, Kenya. Int J Sci Technol Soc 3(1):9–23
Karim RkM, Emmelin M, Resurreccion, Wamala S (2013) Water, women and marital violence in a Bangladesh village. Global water Forum. Available via http://www.globalwaterforum.org/2013/03/25/water-women-and-marital-violence-in-a-bangladesh-village/?pdf=6365. Accessed 15 Oct 2016
Kookana RS, Maheshwari B, Dillon P et al (2016) Groundwater scarcity impact on inclusiveness and women empowerment: insights from school absenteeism of female students in two watersheds in India. Int J Incl Educ 20(11):1155–1171. https://doi.org/10.1080/13603116.2016.1155664
Koolwal G, van de Walle D (2013) Access to water, women's work, and child outcome. Econ Dev Cult Chang 61(2):369–405. https://doi.org/10.1086/668280
Kothawale DR, Munot AA, Kumar KK (2010) Surface air temperature variability over India during 1901–2007, and its association with ENSO. Clim Res 42:89–104

Khanna T, Das M (2016) Why gender matters in the solution towards safe sanitation? Reflections from rural India. Global Public Health, 11(10): 1185–1201
Krishna S, De A (2010) Women managing water: inspiring stories from South Asia. SaciWATERs, Hyderabad
Kumar P, Wiltshire A, Mathison C, Asharaf S et al (2013) Downscaled climate change projections with uncertainty assessment over India using a high resolution multi-model approach. Sci Total Environ 1:468–469. https://doi.org/10.1016/j.scitotenv.2013.01.051
Lawrence P, Meigh JR, Sullivan CA (2002) The water poverty index: an international comparison. Keele Economics Department Working Paper, Staffordshire. Available via http://www.keele.ac.uk/depts/ec/wpapers/kerp0219.pdf. Accessed on 29 July 2015
McFarlane C (2008) Sanitation in Mumbai's informal settlements: state, 'slum' and infrastructure. Environ Plan A 40(1):88–107
Ministry of Environment and Forests, Government of India (MoEF) (2012) India second national communication to the United Nations Framework Convention on Climate Change. MoEF, Government of India, New Delhi
Ministry of Water Resources, Government of India (2011) Restructuring the central water commission. MOWR, vol I. Government of India
Mkhize BH, Nag S, Nagarajan KH, Pradhan CK, Singh S (2012) Access to water, time allocation and income in rural India. National Council of Applied Economic. Research, Working Papers, New Delhi
Nauges C, Strand J (2013) Water hauling and girls school attendance. Some evidence from Ghana. Policy research working paper. The World Bank
O'Reilly K (2010) Combining sanitation and women's participation in water supply: an example from Rajasthan. Dev Pract 20(1):45–56
Pickering AJ, Davis J (2012) Freshwater availability and water fetching distance affect child health in sub-Saharan Africa. Environ Sci Technol 46(4):2391–2397. https://doi.org/10.1021/es203177
Planning Commission (2002) Report of the committee on India vision 2020. Planning Commission, Government of India, New Delhi
Porter G, Hampshire K, Abane A, Munthali A, Robson E, Mashiri M et al (2012) Child porterage and Africa's transport gap: evidence from Ghana, Malawi and South Africa. World Dev 40:2136–2154
Ray I (2007) Women, water and development. Annu Rev Environ Resour 32:421–449
RFSTE/Navdanya (2005) Women and water: a report by Research Foundation for Science, Technology and Ecology for National Commission for Women. Navdanya/RFSTE, New Delhi
Sankhe S, Vital I, Dobbs R, Mohan A, Gulati A, Ablet J et al (2010) India's urban awakening: building inclusive cities, sustaining economic growth. McKinsey Global Institute (MGI), Mumbai
Sorenson BS, Morssink C, Campos AP (2011) Safe access to safe water in low income countries: water fetching in current times. Soc Sci Med 72(9):1522–1526. https://doi.org/10.1016/j.socscimed.2011.03.010
Sullivan C, Meigh J (2005) Targeting attention on local vulnerabilities using an integrated index approach: the example of climate vulnerability index. Water Sci Technol 51(5):69–78
Swarup A et al (2011) Weathering the storm: adolescent girls and climate change. Plan International, London
Truelove J (2011) (Re-) conceptualizing water inequality in Delhi, India through a feminist political ecology framework. Geoforum 42:143–152
UNDESA (2014) World urbanization prospects: the 2014 revision, highlights (ST/ESA/SER.A/352). Department of Economic and Social Affairs, Population Division, New York
UN-Habitat (2016) World cities report. Urbanization & development: emerging futures. United Nations Human Settlements Programme (UN-Habitat), Kenya
UNICEF (2016) When coping crumbles. Drought in India 2015–16. A Rapid assessment of the impact of drought on children and women in India. UNICEF, New Delhi

United Nations Development Programme (2010) Human development report 2010, the real wealth of nations: pathways to human development. UNDP, New York

United Nations Development Programme (2011) Human development report 2011: sustainability and equity: a better future for all. United Nations Development Programme, New York

UNU-INWEH (2010) Mobile telephones more common than toilets in India. Available via https://news.un.org/en/story/2010/04/335332-mobile-telephones-more-common-toilets-india-un-report-finds. Accessed on 3 July 2017

USAID (2007) Health of the urban poor in India. Issues, challenges and the way forward. Urban Health Resource Centre, New Delhi

Van Houweling VE, Hall PR, Diop SA, Davis J, Seiss M (2012) The role of productive water use in women's livelihoods: evidence from rural. Senegal Water Altern 5(3):658–677

Water and Sanitation Programme (WSP) (2011) Flagship report economic impacts of inadequate sanitation in India. Water and Sanitation Programme

WHO/UNICEF (2017) Safely managed drinking water – thematic report on drinking water 2017. WHO/UNICEF, New York

WHO/UNICEF Joint monitoring programme for water supply and sanitation (2010) Progress on sanitation and drinking water 2010 update. World Health Organization and UNICEF, New York

WHO/UNICEF Joint monitoring programme for water supply and sanitation (2012) Progress on drinking water and sanitation 2012 update. Geneva World Health Organization, New York

WHO/UNICEF Joint monitoring programme for water supply and sanitation (2017) Progress on drinking water, sanitation and hygiene, 2017. Update and SDG baselines. UNICEF, New York

World Bank (2003) Implementation completion report on a loan in the amount of US$ 10 million equivalent to the Kingdom of Morocco for a rural water supply and sanitation project. World Bank, Washington, DC, Report No. 25917

Yasmin S (2016) Climate Change Effect in Physical and Natural Capital: A Study on Urban Poor in Dhaka City of Bangladesh. J Geogr Nat Disast 6:170. https://doi.org/10.4172/2167-0587.1000170

Vulnerability of Uganda's Electricity Sector to Climate Change: An Integrated Systems Analysis

9

Vignesh Sridharan, Eunice Pereira Ramos, Constantinos Taliotis, Mark Howells, Paul Basudde, and Isaac V. Kinhonhi

Contents

Abstract

Hydropower contributed to about 86% of Uganda's total electricity generation in 2016 (UBOS, 2016). With more than 2000 MW of investments in the pipeline, within the next decade (Platts 2016), this technology is expected to play a critical role in Uganda's transition to a higher consumption level in the multi-tier

V. Sridharan (✉) · E. P. Ramos · C. Taliotis · M. Howells
Unit of Energy Systems Analysis, KTH – Royal Institute of Technology, Stockholm, Sweden
e-mail: vsri@kth.se; epramos@kth.se; taliotis@kth.se; mark.howells@energy.kth.se

P. Basudde
Sector Planning and Policy Analysis Department, Ministry of Energy and Mineral Development (MEMD), Kampala, Uganda
e-mail: p.asudde@energy.go.ug; pbasudde@yahoo.com

I. V. Kinhonhi
Electricity Regulatory Authority (ERA), Kampala, Uganda
e-mail: i.Kinhonhi@era.or.ug

W. Leal Filho (ed.), *Handbook of Climate Change Resilience*,
https://doi.org/10.1007/978-3-319-93336-8_45

framework for measuring energy access (MEMD 2015). Competition for water sources is a common challenge among its users. In this case, hydropower infrastructure is not an exception, and water allocation is frequently prioritized to supply domestic and agriculture sectors. With Uganda's population expected to double by 2050 compared to 2015 levels (UNDESA 2017), the competition for water among the different sectors is only expected to increase. In addition to this, climatic variables, like precipitation and temperature, introduce a high variability in the availability of surface water (Maslin and Austin 2012). Hence, before locking down on major infrastructure decisions as is the case of large-scale hydropower plants (>100 MW), it is prudent to take into consideration the cross-sectorial dependencies, trade-offs, and potential impacts of climate variability. This study develops a methodology based on the established Climate, Land, Energy and Water strategies (CLEWs) framework (Howells et al. 2013) to assess the vulnerability of the electricity sector to climate change by also considering minimum environmental flows in major Ugandan rivers. This assessment utilizes the cost of electricity generation as an indicative metric to compare conditions of different hydropower output, in light of changing climates and hypothetical environmental flow constraints. It concludes that irrespective of the climate, if key environmental services have to be maintained, there will be a reduction in hydropower generation in the country, and proper adaptation measures need to be taken to avoid disruptions in power supply.

Keywords
Energy systems analysis · Climate change · Hydropower · Uganda · Hydrology · CLEWs · Environmental flow regulations

Introduction

With only 20% of the country electrified (IEA 2017), Uganda has a significant challenge moving forward to meet electricity access targets of the 2030 Agenda for Sustainable Development (UN 2015). About 87% of Uganda's Total Primary Energy Consumption (TPEC) comes from traditional biomass (MEMD 2014), with more than 90% of the households using firewood and charcoal for daily cooking and heating purposes. As the country transitions to use cleaner forms of energy, more people are expected to switch to using electricity for their daily needs, according to Uganda's second National Development Plan (NDP II 2015). In addition to higher household electricity consumption, ambitious irrigation plans (MWE 2011) and increased household water consumption (MWE 2013) are expected to increase electricity demand in the country. NDP II (2015) takes into consideration the need for higher electricity generation capacity in the country to meet this expected increase in demand.

Among the available sources of electricity generation in Uganda, hydropower, with its high potential (MEMD 2017), can become the backbone of the country's economic development. Hydropower plants produce a significant share (86%) of electricity generated in the country (UBOS 2016a); they constitute about 77% of the total installed generation capacity in 2016 (MEMD 2017). With more than 2000 MW of planned capacity expansion in the pipeline (Platts 2016), hydropower is expected to play a substantial role in Uganda moving to a higher tier in the World Bank's Multi-Tier Framework for Measuring Energy Access (ESMAP 2007; MEMD 2015).

However, the source of water for these existing and forthcoming large hydropower plants (>100 MW) is also used and often prioritized for other sectors, namely, agriculture and domestic water supply. Less than 1% of the cultivatable land in Uganda is irrigated (MWE 2011). With an expected, at least, fivefold increase in the irrigated agricultural land by 2035 (compared to 2011 level) (MWE 2011) to ensure self-sufficiency and food security, water use in the agricultural sector is expected to increase, resulting in prioritized water allocation. In 2010, about 28% of water used for irrigation was from surface water sources (UBOS 2011), and, according to the irrigation master plan, this share is expected to increase as more cultivatable land near surface water sources is expected to be brought under irrigation. As per capita purchasing power increases, water consumption in households is also expected to rise until it reaches the phase, where the consumption levels reach the flat section of the environmental Kuznets curve (Katz 2015), thus adding to the competition. It must be noted that different sections of the Ugandan society will fit at varying points on the curve, based on their GDP per capita.

In parallel to the discussion on competition for the same sources of water supply, investments for improving and scaling up energy and water infrastructure have impacts on the ecosystem: starting from altering river flows and changing the land cover to involuntary displacement/resettlement of communities, to name a few (Roy 1999). Large hydropower plants often require big dams (for plants with reservoirs) or a diversion of flow from the main river (for some run-of-river-type plants). Such infrastructure buildup may result in water flow reduction and diversion in rivers, leading to potentially irreversible damage to the local ecosystem (Anderson et al. 2015). To address this issue, a discussion was stimulated at the annual river symposium in Brisbane (2007), with the aim to maintain minimum flow rates in rivers, irrespective of future diversions and infrastructure buildup. This discussion led to the Brisbane Declaration (2007) which stipulates that "[e]nvironmental flows describe the quantity, timing, and quality of water flows required to sustain freshwater and estuarine ecosystems and the human livelihoods and well-being that depend on these ecosystems." The declaration encourages river commissions across the world to make sure new and large infrastructure projects do not create any detrimental impacts on river flows.

Expansion planning for investments in the power sector, without considering the interlinkages and cross-sectoral impacts on land use, water resource systems, and the ecosystem as a whole, can result in lopsided policies, as illustrated by Howells et al. (2013) for the case of Mauritius. The study, using a multi-model ensemble, explores the options of using bagasse to produce ethanol instead of using it traditionally in combined heat and power (CHP) plants to reduce greenhouse gas (GHG) emissions. Rasul (2016) describes the need for a nexus approach to meet the water and energy targets of the United Nations Sustainable Developments Goals (SDGs) from a South Asian perspective. FAO (2014) discuss the water-energy-food nexus in the context of the Sustainable Energy for All (SE4ALL) Initiative. Thus, multi-sectoral interdependencies should be taken into consideration while planning long-term investments in energy infrastructure.

Amidst the discussion on the competition for sources of water supply, the resilience of the power sector to climate change introduces high uncertainty in the generation of electricity, and hence high risk in the ability to meet demands. Climatic impacts on the energy system are multifold (Ebinger and Vergara 2011; Gerlak et al. 2018); the following list calls attention to a few important impacts:

1. Fluctuations in energy demand due to changes in atmospheric temperature (Staffell and Pfenninger 2018) and variations in season and spatial water availabilities.
2. Fluctuations in energy supply: erratic changes in climatic variables like precipitation and temperature can result in unexpected irregularities in electricity generation from hydropower and inland, water-cooled thermal power plants. Similar effects are noticeable in electricity generation from wind and solar power plants (van Vliet et al. 2016). Production of liquid biofuels is directly related to crop production and hence water consumption; climate affects the availability of the water for crops (Tirado et al. 2010).
3. Impact on energy infrastructure: sudden and extreme weather events could lead to infrastructure damage (damage of hydropower spillways due to flash floods, damage to electricity transmission infrastructure).

For countries with a large share of hydropower in the electricity generation mix, like Uganda, the stakes are high when drier weather conditions persist. De Lucena et al. (2010) describe the impacts of climate variability in Brazil, where hydropower plants constitute more than 80% of the installed electricity generation capacity; they suggest diversifying the mix to reduce damages. Conway et al. (2017) demonstrate that the case in many sub-Saharan African countries with a significant share of hydropower is no different; they correlate a reduction in GDP to lower hydropower production due to drought in Kenya and Zambia. Lower availability of water in the Zambezi River has led to frequent blackouts in Zimbabwe and Zambia (NBC News 2015), resulting in a reduction of industrial outputs (Hamududu and Killingtveit 2016).

Variability in hydropower generation due to climate change, resulting in fluctuations in the cost of electricity generation, might have far-reaching impacts on the economy, which are only exacerbated by the high share of hydropower in the generation mix. The focus of this chapter is to assess the vulnerability of a hydro-dominated electric power system, such as in Uganda, to climate change.

There have been recent studies that focus on analyzing the impact of climate change on different sectors in Uganda: agriculture, water resources, and energy, to name a few. Twagiramaria et al. (2018) discuss the impact of climate on agricultural practices in Ugandan highlands and suggest some adaptation measures to climate proof the crop production. Taylor et al. (2014) discuss the impact of climate change on water supply for domestic and agricultural purposes but do not consider the country's irrigation master plan and the possible impact of hydropower on the water resources. Van Vliet et al. (2016) discuss the impacts of climate change on the power sector on a global scale, using a hydrological-electricity modeling framework; this study is extensive in its assessment of hydropower and climate scenarios on a continental and global scale, but does not capture the detail required to provide insights on a country level. The recent electricity sector master plan of the Eastern African Power Pool (EAPP) discusses the impact of climate on electricity infrastructure but does not take into consideration the water and land-use sectors and national plans for irrigation (EAPP master plan 2016). Nevertheless, the EAPP master plan is an excellent source for technological and fuel price projections, which are critical for evaluating the resilience of the power sector.

The purpose of this study is to assess the vulnerability of Uganda's electricity sector to climate change, taking into consideration the interlinkages between energy, land use, water resources, and ecosystem services. This study complements and builds on the "Climate, Land, Energy and Water strategies" (CLEWs) framework developed by Howells et al. (2013). Since hydropower and electricity sector expansion planning is the focus, a long-term electricity sector planning tool is used in this study. The Open Source energy MOdelling SYStem (OSeMOSYS), a least-cost, optimization-based, long-term, energy systems model, is chosen owing to its simplicity and open-source code which allows for easy modifications to suit the modeling needs (Howells et al. 2011). In addition to OSeMOSYS, a water balance model having the capability to represent land use and take into consideration long-term climate variability is used to model water supply and demand components in the analysis. The Water Evaluation And Planning (WEAP) system is chosen for this purpose. The capability of WEAP models to represent energy, water, and land-use sectors in a user-friendly manner is utilized in this study. Different methods exist to link and iterate data between models. In this assessment, we soft link the two models to preserve the technical detail in the different sectors that are represented in each of them, along with the possibility to exchange data between them. The study uses the cost of electricity generation in the country as an indicative metric to assess the climatic impacts.

The Methods and Tools section of the chapter briefly presents the different tools, the model structure, and the information flow. Key results from selected scenarios are explored in the Results and Discussion section. The chapter concludes with a summary of key outcomes in the Conclusion section and suggests future steps and improvements to the model.

Methods and Tools

The work presented in this chapter is based on the soft linking of two modeling frameworks: one model for infrastructure expansion in the electricity sector and one that deals with the management of water resource systems in Uganda also accounting for land-use changes. The following sections summarize these tools, the underlying assumptions and data used in this exercise.

Water Resource Management

This section details the methods used to represent water resources management in Uganda. It includes a subsection describing the method used to select climate scenarios for this study, followed by a section on maintaining minimum environmental flows in rivers so as to preserve/safeguard ecosystem services.

Water Management Framework

Table 1 summarizes the existing, under-construction, and planned hydropower capacity in Uganda in 2016 (Platts 2016); the large hydropower plants are highlighted in the table. Bujagali, Nalubaale, and Kiira, the existing large hydropower plants, constitute about 70% of total installed electricity generation capacity in Uganda. These hydropower plants are located at the outlet of Lake Victoria, close to the city of Jinja (see Fig. 1).

Although 45% of Lake Victoria's surface area lies inside Ugandan borders, more than 80% of the catchment area that drains into it lies outside Uganda, in Tanzania, Kenya, Burundi, and Rwanda (Prado et al. 1991). Therefore, the management of water resources in the catchments that fall outside Ugandan borders plus the natural availability of water dictated by the water balance in Lake Victoria play a significant role in defining the flow of the Victoria Nile River that reaches the hydropower plants located over its course.

As illustrated in Fig. 1, all future, large, hydropower plants are situated downstream of existing hydropower plants on Victoria Nile River, between Lake Victoria and Lake Albert. All of these have capacity over 100 MW. Catchments around Victoria Nile is where a significant share of cultivatable land is expected to be brought under irrigation (MWE 2011). Hence, the catchments that drain into rivers inside Ugandan borders needed to be represented in detail to capture the different systems that compete for the same sources of water supply.

A water balance of the catchments that drain into Lake Victoria and ones that drain into rivers inside Uganda is developed individually in the Water Evaluation And Planning (WEAP) system, a widely used integrated water management framework, which has been fine-tuned over the past 20 years (Yates et al. 2005).

Since the focus of this study is to assess the vulnerability of the power system inside Uganda including the other relevant sectors, a greater spatial detail is used to represent the catchments that drain into Ugandan rivers (like Victoria Nile River) than for the catchments that drain into Lake Victoria (that lie in Tanzania, Kenya, Burundi, and Rwanda).

A lumped catchment-type water balance model is developed for the catchments that drain into Lake Victoria. The catchments are represented by two hydrological units, representing nonirrigated and irrigated areas, and are used to represent areas of different land covers within the catchment. The nonirrigated node is further

Table 1 Hydropower in Uganda (Operating-Plants in Operation, Construction-Plants under construction, Licensed-Licensed Projects (Allowed to Start Construction), and Feasibility-Projects with Permits under Feasibility)

Name	Type	Capacity (MW)	Status	Commission dates
Nalubaale	large Hydro	180	Operating	1954
Kiira	large Hydro	200	Operating	2000
Bujagali	large Hydro	250	Operating	2012
Mobuku 1 (KML)	Small Hydro	5	Operating	1990
Mobuku 3 (KCCL)	Small Hydro	10	Operating	1990
Ishasha	Small Hydro	6.5	Operating	2011
Bugoye	Small Hydro	13	Operating	2009
Mpanga	Small Hydro	18	Operating	2011
Hydromax(Kabalega)	Small Hydro	9.9	Operating	2013
Nyagak I	Small Hydro	3.5	Operating	2013
Rwimi	Small Hydro	5.5	Operating	2017
Siti 1	Small Hydro	5.0	Operating	2017
Lubilia	Small Hydro	5.4	Operating	2018
Nyamwamba	Small Hydro	9.2	Operating	2018
Muvumbe	Small Hydro	6.5	Operating	2017
Isimba	large Hydro	183.0	Construction	Sept 2018
Nkusi	Small Hydro	9.0	Construction	May 2018
Waki	Small Hydro	4.8	Construction	Oct 2019
siti 2	Small Hydro	16.0	Construction	Mar 2019
Mahoma	Small Hydro	2.7	Construction	Oct 2018
Agago/Achwa II	Small Hydro	42.0	Construction	Sept 2018
Kikagati	Small Hydro	16	Licensed	2019
Kakaka	Small Hydro	5	Licensed	2019
Kyambura	Small Hydro	7.6	Licensed	2019
Nyamagasani I	Small Hydro	15	Licensed	2019
Nyamagasani II	Small Hydro	5	Licensed	2019
Nyamabuye	Small Hydro	7	Licensed	2019
Ndugutu	Small Hydro	5.1	Licensed	2019
Bukinda	Small Hydro	6.5	Licensed	2020
Kiba	large Hydro	330	Feasibility	2025
Oriang	large Hydro	392	Feasibility	2026
Nsongezi	Small Hydro	35	Feasibility	2020
Muzizi	Small Hydro	44.7	Feasibility	2020
Muyembe	Small Hydro	7	Feasibility	2019
Nyabuhuka	Small Hydro	3.2	Feasibility	2019
Sironko	Small Hydro	7	Feasibility	2019
Kabeywa	Small Hydro	12	Feasibility	2019
Bukwa	Small Hydro	9	Feasibility	2019
Keere	Small Hydro	6.3	Feasibility	2020
Ngoromwo	Small Hydro	8	Feasibility	2021

(*continued*)

Table 1 (continued)

Name	Type	Capacity (MW)	Status	Commission dates
Cresta	Small Hydro	5	Feasibility	2019
Kanyampara	Small Hydro	4.8	Feasibility	2019
Achwa III	Small Hydro	13	Feasibility	2020
Achwa IV	Small Hydro	10.5	Feasibility	2020
Achwa V	Small Hydro	5	Feasibility	2021

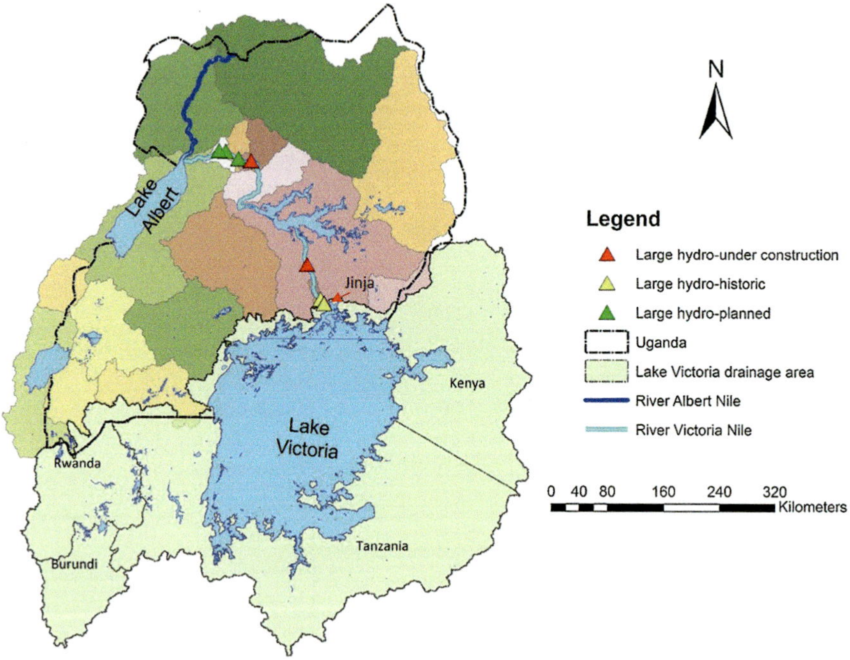

Fig. 1 Study area, catchments inside Uganda and Lake Victoria sub-basin

disaggregated into grassland, forestland, wetland, cultivated land, water bodies, and built-up land. The irrigated land represents solely the agricultural area for cultivation of irrigated crops, with the latter being specific to the region covered by the catchment. Cultivated agricultural land in the Victoria Nile catchments is predominantly rainfed, and there is a significant push to irrigate and increase crop yield. It is expected that irrigation water withdrawals in the region will increase in the future, resulting in varying runoff into Lake Victoria, hence the need to differentiate irrigated and nonirrigated irrigated areas. The lumped catchments that drain into Lake Victoria are represented in Fig. 1 under a single color.

A semi-distributed (sub-basin level) type balance is developed to represent catchments that drain into Ugandan Rivers. The model takes into consideration the following:

- Detailed land cover maps (RCMRD 2014)
- Distinct groundwater potential for 112 districts in Uganda
- Catchment areas delineated from digital elevation maps (DEM) (Lehner et al. 2008)
- Agricultural land-use classifications (MWE 2013)
- National irrigation master plan (MWE 2011)
- Population distribution (UBOS 2016b) and population projections (UNDESA 2017)
- Historical information on precipitation and ambient air temperature (Sheffield et al. 2006)

The developed WEAP models have a monthly temporal resolution, and the model period extends until 2050. Similar to the lumped catchment approach followed to represent the drainage basin of Lake Victoria, each catchment inside the Ugandan border is classified into irrigated and nonirrigated area nodes. The nodes in each catchment are classified based on the latest land cover maps obtained from the Regional Centre for Mapping Resources For Development (RCMRD 2014). The share of the irrigated area in the total cultivated agricultural land in Uganda is less than 1% (Wanyama Joshua et al. 2017). An irrigation master plan was developed by the Ministry of Water and Environment (MWE 2011); it discusses the need to increase the share of land under irrigation and lays out a plan for it. According to the plan, about 70% and 20% of the irrigable land, which are close and far off from surface water sources, respectively, will be irrigated by 2035. This study utilizes district-specific irrigation expansion from the master plan as base data. Along with land use, also other water uses in the agricultural sector were represented in the water model.

The large hydropower plants in Uganda are represented in the model with site-specific information. In the Victoria Nile catchments, covered in this study, these correspond to a total existing installed capacity of 630 MW, which represent 70% of the total existing electricity generation capacity of the country in 2016. The small or mini hydropower plants account for about 67.5 MW of current installed electricity generation capacity in the country, and the planned additions amount to about 380 MW (MEMD 2017). Due to non-availability of site-specific power plant information, we considered only a few of these small power plants (in the WEAP model) for which we had all the technical specifications. All these small-scale plants are run-of-the-river (ROR) type and do not have any reservoir capacity; hence they are characterized by seasonal variation in electricity generation. However, the entire hydropower capacity is represented in the energy systems model, both large and smaller infrastructures, as it is critical to represent all the available generation capacity in the country for long-term infrastructure planning. In the water model, this is expected to create little impact as they have low turbine flows and do not have any reservoir, thus not affecting the flow in the river to a significant degree.

It must be taken into consideration that the water balance model prioritizes allocation of water for agricultural and domestic household consumption over hydropower generation. The model also considers the losses present in transporting water from the source to demand centers. Moreover, all the scenarios including the baseline consider the same irrigation expansion plan and water allocation priorities for household water use. These two competing uses of water are considered in the same manner across all the scenarios to capture the climatic impacts on the power system effectively.

Figure 2 illustrates the calibration that is obtained for the flow out of Lake Victoria for the years 1952–1990. The blue and orange lines refer to the modeled and observed actual outflows from Lake Victoria, respectively, measured at Jinja. The historical climatic information used in the calibration process is obtained from Li et al. (2010).

Choice of Future Climates

Climatic data is a critical component of any water resource assessment (Silberstein 2006); hence, it is of paramount importance to use site-specific data to model river flows representative of the natural flow pattern. However, relevant, site-specific data is not easy to obtain, especially if the information from climate data centers is not easily accessible, which is the case in this study. A global dataset of meteorological forcing variables for land surface modeling, developed by the Princeton Land Surface Hydrology Research Group, is used to define the baseline (Sheffield et al. 2006). A spatial resolution of $0.5^{\circ} \times 0.5^{\circ}$ and a monthly temporal resolution are used for this study. All the $0.5^{\circ} \times 0.5^{\circ}$ grid points, which fell under each of the studied catchments, were averaged to obtain one set of monthly temperature, precipitation, humidity, and wind speed data. Climatic data is used by WEAP to calculate reference evapotranspiration (ET_{ref}), while crop coefficients (K_C) obtained from the Food and Agricultural Organization (FAO) (Allen et al. 1977) are used to calculate crop and land cover-specific evapotranspiration (ET_C). For the reference scenario, climate data for the periods of 1951–2000 was cycled into the future, therefore assuming that the climate in 2001–2050 was similar to the period 1951–2000.

For defining the climate change scenarios, we considered two categories of emission scenarios: one medium (A1B) and one high (A2), based on the Special Report on Emission Scenarios (SRES) from the IPCC's Fourth Assessment Report (AR4) (IPCC 2007), and one medium (RCP 4.5) and one high (RCP 8.5) emission scenario based on the Representative Concentration Pathways (RCP) from the IPCC's Fifth Assessment Report (AR5) (IPCC 2013). The Coupled Model Intercomparison Projects, CMIP3 and CMIP5, which contributed to IPCC AR4 and AR5 reports, respectively, consist of outputs from many general circulation models (GCMs), 22 from AR4 and 23 from AR5, respectively. Hence, two emission scenarios from each of the abovementioned GCMS resulted in a set of 90 climate model-emission scenario combinations. These combinations were ranked based on their Climate Moisture Index (CMI), which is an indicator related to precipitation and evapotranspiration projections and is used to rank the combinations to identify the wettest and driest scenarios (Willmott and Feddema 1992).

The Climate Moisture Index (CMI) is a measure of aridity in the region. The index values vary between -1 and $+1$, with lower values representing conditions that are arider. A CMI value greater than zero indicates that precipitation rates are higher than potential evapotranspiration rates. CMI is often used as an indicator of

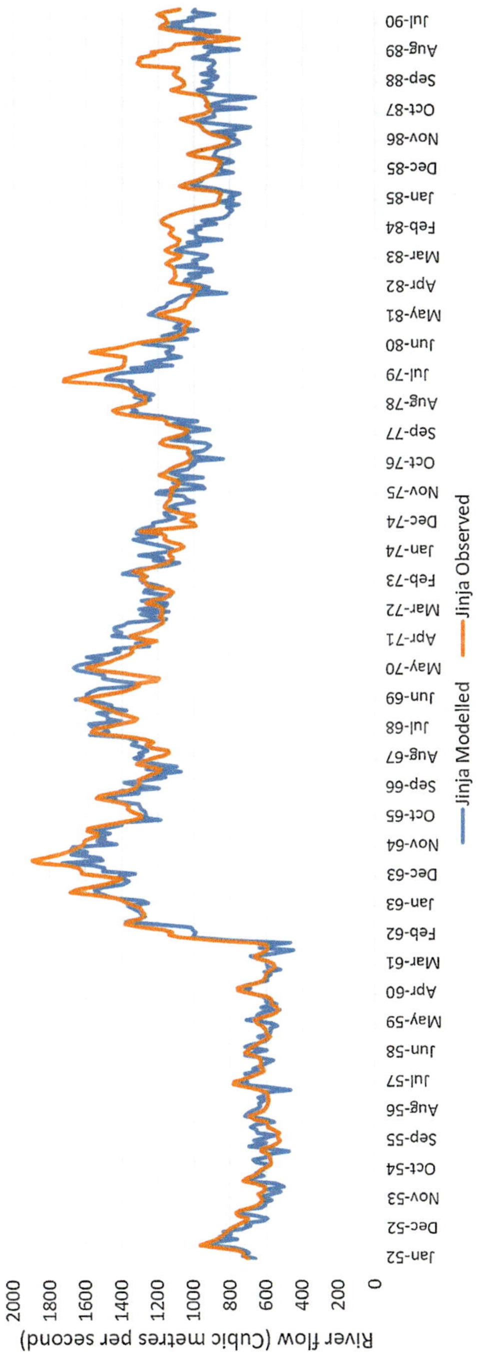

Fig. 2 Flow calibration for Lake Victoria outflow at Jinja (1952–1990)

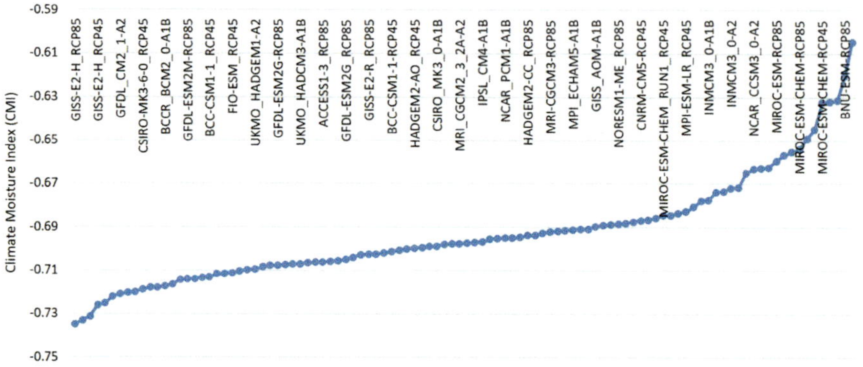

Fig. 3 Average Climate Moisture Index (2010–2050)

crop water demand and surface runoff (Cervigni et al. 2016). A CMI value for the Nile River basin, average over the period 2010–2050, is used for this study (Boehlert et al. 2016). Figure 3 illustrates the CMI value for all the combinations.

It can be noticed that the least arid combination is the RCP 4.5 scenario from the BNU-ESM (GCM) model (Ji et al. 2014) and the most arid combination is the RCP 8.5 scenario from the GISS-E2H (GCM) model (Schmidt et al. 2014). In addition to the scenarios selected above, for each of the chosen GCMs, an alternative RCP is chosen to analyze the variability that a different emission pathway from the same GCM will induce. Hence, five climate scenarios are analyzed as part of this exercise:

(a) Historical recycled climate (reference/baseline climate) – *ref*
(b) The RCP4.5 scenario from BNU-ESM (wettest) – *bnu45*
(c) The RCP8.5 scenario from BNU-ESM – *bnu85*
(d) The RCP4.5 scenario from GISS-E2H – *giss45*
(e) The RCP8.5 scenario from GISS-E2H (driest) – *giss85*

Projections of all the climatic variables for each scenario, including the baseline, are introduced into the WEAP model to estimate the surface runoff at each catchment node.

Maintaining Ecosystem Services: Estimating Environmental Flow (EFRs) With Uganda and its neighboring transboundary countries, in the Nile Basin, increasing the share of agricultural land under irrigation and the share of hydropower in their electricity generation mix, there is a high imperative to estimate and implement necessary EFRs on their shared rivers to maintain ecosystem services in the region. The Nile Basin Initiative's "Strategy for Management of Environmental Flows in the Nile Basin" emphasizes the need to take adaptive action before locking into large energy and irrigation infrastructure (NBI 2016). Over

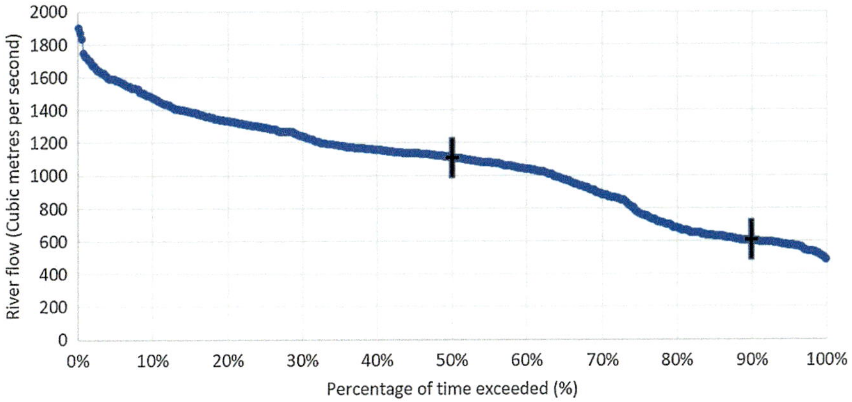

Fig. 4 Flow duration curve, Victoria Nile at Jinja (1952–1990)

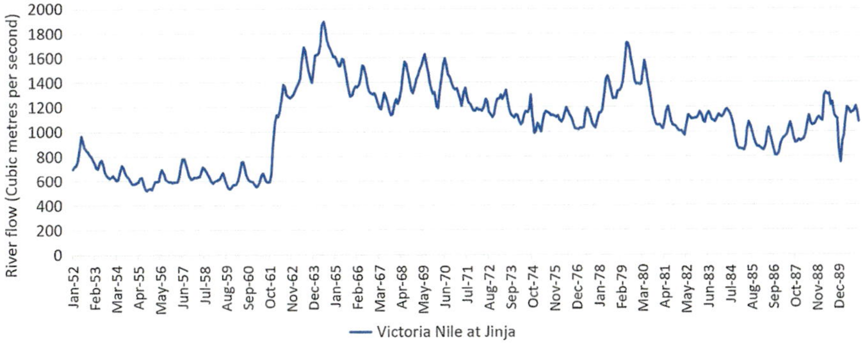

Fig. 5 Absolute river flows, Victoria Nile at Jinja (1952–1990)

200 different methods exist to calculate minimum environmental flows, classified under various categories ranging from purely hydrological to holistic methodologies involving multiple parameters (Tharme 2003). For this study, five different EFR methodologies implemented in separate case studies over different hydrological regimes are considered: Smakhtin et al. (2004), Tennant (1976), Tessmann (1979), and two new methods developed by Pastor et al. (2014). The chosen 5 EFR frameworks have been validated at a local level across 11 different basins across the globe. Since the electricity sector and, especially, large hydropower plants are the prime focus, the EFRs were implemented for the section of the Victoria Nile River starting from Lake Victoria's outlet until the point where it reaches Lake Albert. Note that all the existing and planned hydropower infrastructure is situated in this section of the river. Historical monthly flow information at Jinja, a gauging station located near the start of Victoria Nile, for the period 1950–1990 is used to calculate the EFRs. Figures 4 and 5 illustrate a flow duration curve (FDC) showing the Q_{50} and Q_{90} levels and the absolute flow,

Table 2 Minimum environmental flow methods analyzed in this study. MAF (the mean annual flow), MMF (the mean monthly flow), Q_{90} (where the flow exceeded 90% of the period of assessment), and Q_{50} (where the flow exceeded 50% of the period of assessment). HFRs, IFRs, and LFRs are used for high-, intermediate-, and low-flow requirements, respectively (Pastor et al. 2014)

	EFR methodologies				
Characteristics	Smakhtin	Tennant	Q90_Q50	Tessmann	Variable monthly flow
Determination of low-flow months	MMF≤MAF	MMF≤MAF	MMF≤MAF	MMF≤0.4*MAF	MMF≤0.4*MAF
Low-flow requirements (LFRs)	Q_{90}	0.2* MAF	Q_{90}	MMF	0.6*MMF
Determination of high-flow months	MMF>MAF	MMF>MAF	MMF>MAF	MMF>0.4*MAF and 0.4*MMF>0.4*MAF	MMF>0.8*MAF
High-flow requirements (HFRs)	0 to 0.2*MAF[a]	0.4*MAF	Q_{50}	0.4* MMF	0.3* MMF
Determination of intermediate-flow months	–	–	–	MMF>0.4*MAF and 0.4*MMF ≤ 0.4*MAF	MMF>0.4*MAF and MMF≤0.8*MAF
Intermediate-flow requirements (IFRs)	–	–	–	0.4* MAF	0.45*MAF

[a]If Q_{90}>30%MAF, HFRs = 0. If Q_{90}<30% and Q_{90}>20%, HFRs = 7%MAF. If Q_{90}<20% and Q_{90}>10%, HFRs = 15%MAF. If Q_{90}<10%, HFRs = 20%MAF

respectively, from the Jinja gauge. Q_{50} and Q_{90} (flow quantile) markers correspond to the river flow that has exceeded 50 and 90 percent of the time, respectively, between 1952 and 1990.

Table 2 details different methods and the respective criteria used to arrive at the EFRs (Pastor et al. 2014). The river flows are classified into low, intermediate, and high depending on the type of EFR method as shown in Table 2.

EFRs calculated using all the methods mentioned above for Victoria Nile River are shown in Fig. 6. Also presented are the mean monthly flow (MMF) for the period 1950–1990 and the average value that resulted from the application of the different EFR methods.

As shown in Fig. 6, on average and for the period of measurements, the MMF and MAF for Victoria Nile River at Jinja (black and brown lines, respectively, on the chart) is above 1000 m^3/s. The red line represents the average of the five EFRs. From the analyzed frameworks, three methods are chosen: one based on the Tennant method (LFR) to represent a low-flow requirement where a minimum flow is

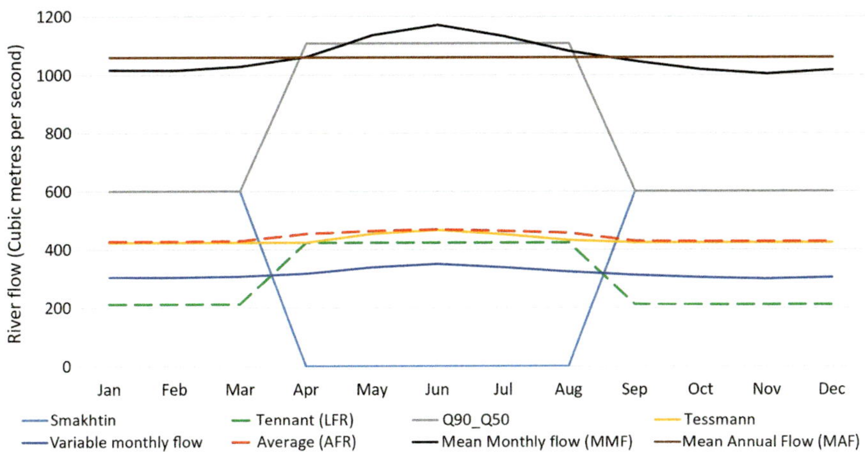

Fig. 6 Minimum environmental flows for Victoria Nile River

maintained across all the months in a year, an average of all the EFRs (AFR) to represent a high-flow requirement, and a case where there is no minimum flow requirement in the rivers (NFR). The chosen three adequately represent the spectrum of maintaining varying levels of ecosystem services.

The choice is also partially guided by the fact that with rapid electrification plans in place to improve electricity access in the region, it is considered unlikely that the environmental policies will tend toward higher EFRs suggested by the Q_{90}_Q_{50} method; it will render almost all the planned hydropower plants without water for electricity generation. The EFRs are implemented, in the model, in all sections of Victoria Nile, where a diversion is created for the run-of-river hydropower plants. Ideally, the EFR has to be calculated for each section of the river where the power plants are located, but due to lack of available data (flow information) at each plant site, the EFR calculated for Victoria Nile at Jinja is used for the stretch of the river until Lake Albert, which is being modeled in the four successive catchments.

Long-Term Electricity Sector Expansion Planning

Uganda's latest National Development Plan (NDP) mandates to improve electricity access in the country from 14% in 2015 to 30% by 2030 (NDP II 2015). Though there has been some delay in the construction of hydropower plants mentioned in the NDP (UEGCL 2018), Uganda is expected to double its electricity generation capacity by 2020. In this study, an expansion model of Uganda's power sector is developed using an open-source, linear, techno-economic optimization framework – OSeMOSYS (Howells et al. 2011). OSeMOSYS, written in GNU MathProg and also available in GAMS and Python (Makhorin 2008), is a dynamic, bottom-up,

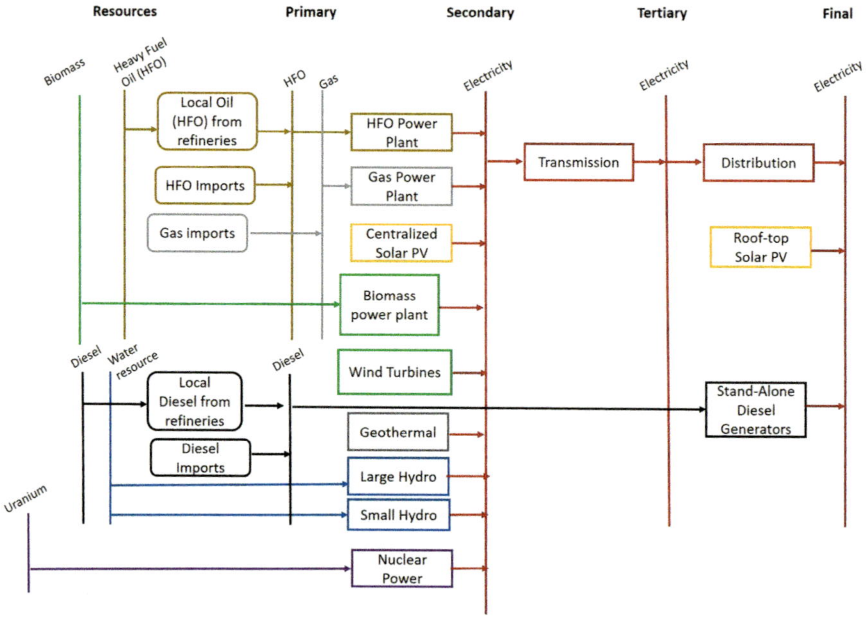

Fig. 7 Reference Energy System (RES)

multi-year energy system model applying linear optimization techniques. It is used to develop pathways for medium- and long-term energy planning on national, continental, and global levels (Cervigni et al. 2016; Pinto de Moura et al. 2017; Taliotis et al. 2014, 2016). Effectively, this section explores the different assumptions considered in the reference/baseline scenario of this exercise. In the electricity system model developed for Uganda, all electricity generation options, except for large hydropower plants, are grouped based on the fuel used. For example, all the heavy fuel oil- (HFO) based power plants are grouped into one option and represented with the same techno-economic characteristics, the same in the case for Solar PV, wind turbines, and the other technologies. Figure 7 illustrates the Reference Energy System (RES) – the schematic representation of energy flows following a bottom-up resource-to-use approach – of the Ugandan model, from the supply on the left-hand side to demand on the right-hand side.

The electricity demand projections used in this exercise are obtained from an assessment carried out by (AF Consult 2015a) for the Ministry of Energy and Mineral Development (MEMD) in Uganda for the year 2015. Three different demand scenarios are considered: high, low, and base as illustrated in Fig. 8. In 2050, the electricity demand in the high demand scenario is expected to be 35% and 14% higher than the low and base demand scenarios, respectively.

The yearly electricity demand is classified into domestic, industrial, and commercial sectors based on the information obtained from MEMD (2017). The sectorial demand shares are kept constant until the end of the modeling period. Transmission

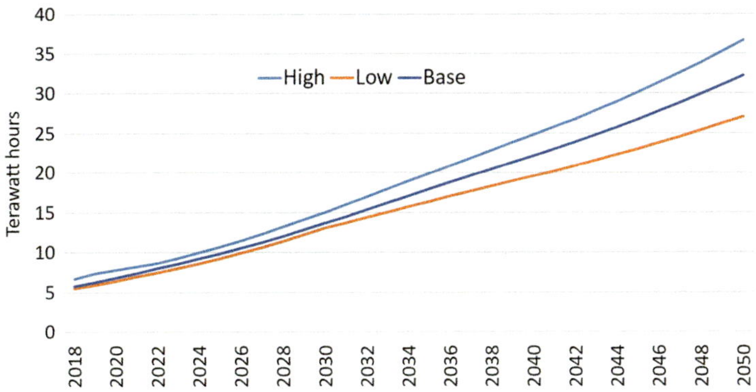

Fig. 8 Final electricity demands for the electricity demand scenarios investigated in this study

and distribution losses in the current system are estimated based on a demand and supply balance obtained from the MEMD. For 2014, the losses were 3.6% on transmission grid and 21.3% on the distribution grid (MEMD 2015, 2017). It is assumed that until the end of the modeling period (2050), these will reduce to 2% and 5%, respectively.

The techno-economic parameters for the electricity generation technologies are obtained from the latest edition of the EAPP master plan (2016) and AF Consult (2015b). The large hydropower plants have site-specific capital costs derived from the estimated total cost of the respective projects. Fossil fuel prices and their projections for this study are obtained from the EAPP master plan (2016). A discount rate of 10% is applied to the model. Although the national policies extend only up to 2040 (NDP II 2015), the model period of this analysis has been extended until 2050 since the climate scenarios analyzed as part of this assessment exhibit high variability in between 2040 and 2050, as illustrated in Fig. 10.

The temporal resolution of the model is more detailed than the water management model; it has 48-year splits. Each year is split into 12-month periods; additionally, each month is classified into four dayparts (morning, day, peak, and night), this division led to 48 splits in a year. The share of electricity load falling under each of the computed time steps is calculated and input to the model.

It must be mentioned that this model has a perfect foresight into the future and computes an optimal generation mix for the entire modeling period using a least-cost, optimization-based approach. The OSeMOSYS code used in this study can be accessed from the GitHub portal and corresponds to version 2017_11_08.

Model Soft Linking and Methodological Flow Process

Figure 9 illustrates the methodological flow of this assessment. The precipitation and evapotranspiration values for the different climate projections are used to calculate the CMI, which is used as a metric to choose the climate scenarios for the analysis. Climatic

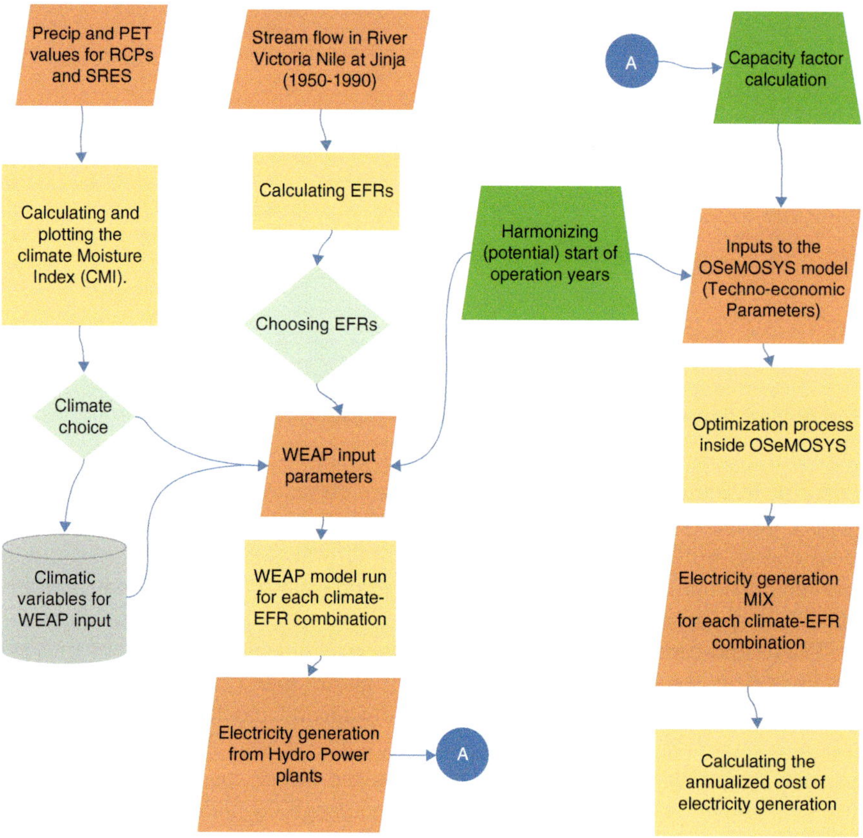

Fig. 9 Flowchart methodology

parameters, for the chosen climates, required as input to the WEAP model, are obtained from the CMIP databases. In parallel, historical river flows at Jinja are used to calculate minimum environmental flows based on the selected EFR frameworks. During the model development phase, the potential start dates for future power plants are harmonized between the OSeMOSYS and WEAP models. The minimum environmental flows and climate scenarios are implemented in WEAP simultaneously to estimate the maximum hydropower generation for each scenario combination. Five climate scenarios and 3 EFRs were analyzed, resulting in 15 different combinations. Each combination is explored with the water model and the power sector model. Hydropower generation from each of the large plants, as obtained from the water model runs on a monthly basis, is used to estimate capacity factors of the respective power plants, to be used in the OSeMOSYS model. However, the power pool model has a better temporal resolution; it has four splits in each month. Hence, to maintain coherence, the four-time slices of each month in the OSeMOSYS model are made to associate with the same hydro capacity factor to reflect the modality of the WEAP model.

Results and Discussion

The following subsections explore key findings from this study. The impact of climate on the flow out of Lake Victoria is discussed as the hydropower generation from historical power plants is affected by it. After that, the electricity generation mix of the reference scenario is explained illustrating the generation options available in Uganda. The discussion is taken forward by analyzing the impacts of climate and enforcing minimum environmental flows on Victoria Nile in a mutually exclusive manner. The section concludes with an assessment of the combined impacts of environmental flows and changes in the climate.

Climatic Impact on Lake Victoria Outflow

Figure 10 illustrates the impact of different climate projections on Lake Victoria's outflow. It should be noted that a wet/dry climate scenario chosen based on their CMI values is only cumulatively wet/dry, but not necessarily wet/dry across all time steps of the period of the analysis. This characteristic is explained by the inter-annual and seasonal variability of Lake Victoria outflows in the modeled period. Indirectly, the outflow reflects the impact of climate on the catchments that drain into Lake Victoria and the lake's water balance.

It is interesting to observe how the reference climate differs from the cumulatively wetter and drier climates. According to the data sources, it is expected to be relatively wet in the initial years of the modeling period and becomes progressively drier toward the second half, with some wet spells toward the end. The outflow of Lake Victoria, measured at Jinja, is critical to hydropower generation, as the contribution of drainage from Victoria Nile catchments is minimal, at least for the existing power plants: Bujagali, Nalubaale, and Kiira. Hence, proper management of the lake and its drainage area is essential for higher levels of energy security in

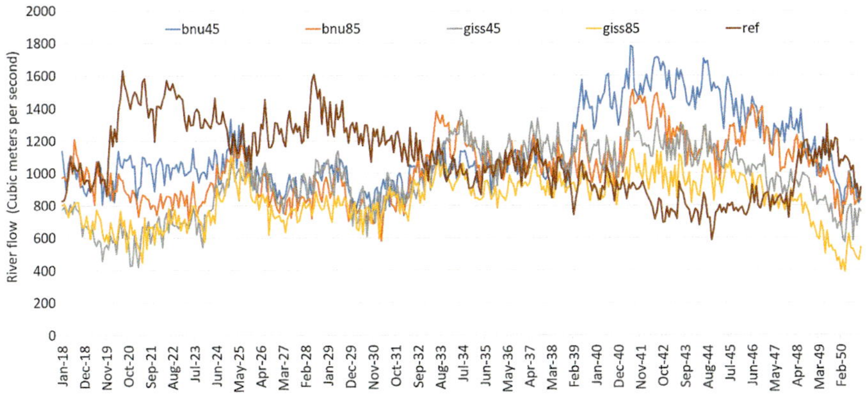

Fig. 10 Lake Victoria outflow, at Jinja, for different climate scenarios (2018–2050)

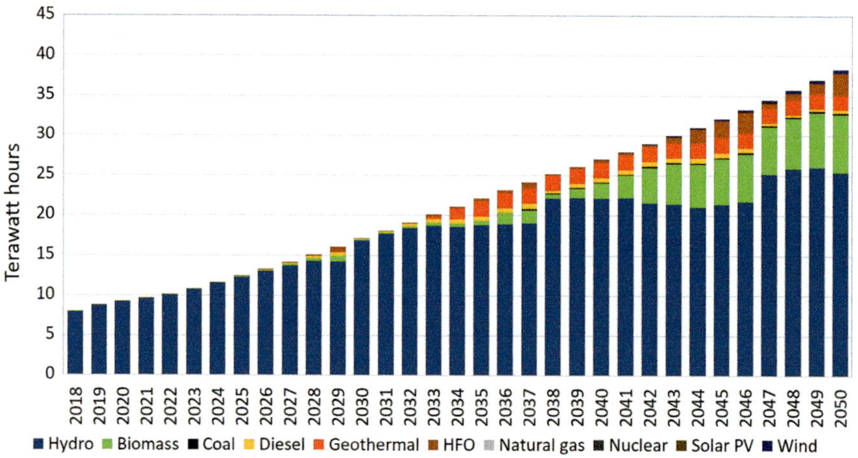

Fig. 11 Electricity generation mix, reference climate, no EFR

Uganda; this calls for higher levels of transboundary cooperation between Uganda, Rwanda, Burundi, Kenya, and Tanzania. Cumulative (2018–2050) hydropower generation from the three historical large hydropower plants is approximately 143 TWh for the baseline scenario (ref). The generation in the wettest (bnu45) and driest (giss85) climates are expected to be 2.6% higher and 11.6% lower, respectively, compared to the baseline.

Electricity Generation Mix in Uganda

Despite analyzing three demand scenarios and running the OSeMOSYS model to generate electricity sector expansion plans for each of them, only the implications for the high demand scenario are presented in this chapter.

Electricity generation mix for the reference climate scenario without considering any EFRs is illustrated in Fig. 11. The dominance of hydropower is noticeable throughout the model period. Reduction in the share of hydropower in the total generation mix can be noticed post-2030; this could be due to low flows in Victoria Nile or can also be attributed to the need for extra capacity in the power system to meet higher demand. Biomass, geothermal, and HFO-based power plants are expected to play a vital role by contributing a more significant share to the generation mix. Geothermal capacity in the country, though in its nascent stages of development, is restricted to a maximum capacity of 250 MW as in the latest electricity Master Plan of the Eastern African Power Pool (EAPP master plan 2016). Oil production from the oil fields in Lake Albert region is expected to commence in 2020, and with the East African countries sharing the cost of the pipeline, locally sourced HFO is expected to play a supporting role for the power system laden with hydropower (MEMD 2016). There is approximately 85 MW (Platts 2016) of biomass-fired power plants in the system, all of them use bagasse from the sugar

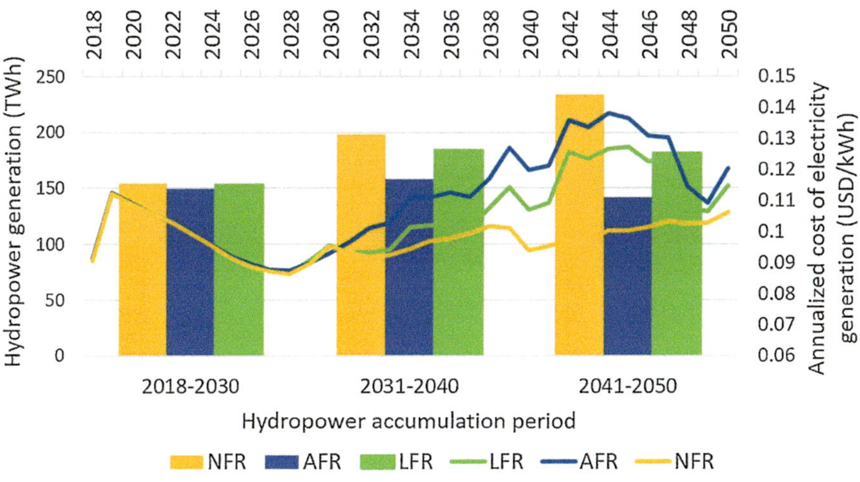

Fig. 12 Impact of EFRs on hydropower generation (bar graph, primary Y-axis), annualized cost of electricity generation (line graph, secondary Y-axis)

industry as fuel, that are used predominantly for self-consumption by the industrial sector (sugar works) and contribute excess to the grid. With 233,170 hectares of cultivatable land planned to be irrigated by 2035 (MWE 2011), the availability of inexpensive biomass residue/bagasse available for power generation is expected to increase and play a key role to strengthen the power system.

Wind and solar power are also expected to contribute a minor share (2% in 2050) to the total generation. In spite of the model considering a reduction in Solar PV and wind power costs over the modeling period, the technologies do not seem to be cost-effective to warrant higher penetration in the generation mix. Similarly, despite the availability of nuclear power options, it is not part of the cost-optimal energy mix until 2050 in the reference climate scenario with no EFR.

Impact of Minimum Environmental Flows on the Power System

The impact of implementing minimum environmental flows (on River Victoria Nile) on electricity generation is illustrated in Fig. 12. The annualized cost of electricity generation and hydropower generation in the different EFR scenarios explain the economic impacts on the power system. Hydropower generation is accumulated over three different time periods. The only induced change in the scenarios are the capacity factors of the hydropower plants taking into consideration the different EFRs; the OSeMOSYS model computes an optimal generation mix based on power plant availabilities.

In the period 2030–2050, the effect of enforcing higher environmental flows on the cost of electricity can be clearly noticed. The annualized cost of electricity generation is calculated exogenously from the total expenditure (capital, fixed, and variable) on electricity infrastructure that contributes to the generation mix. It includes all the power plants – both grid-connected and stand-alone systems – transmission and distribution (T&D) infrastructure, and fuel prices. In the high EFR scenario, consumer expenditure is expected to be approximately 17% higher than the scenario with no EFR, cumulatively for the period 2018–2050, a difference corresponding to about 11 billion USD_{2015}. It is assumed that there will not be any government subsidy in electricity pricing. Consumers are expected to pay 10% (6.5 billion USD_{2015}) more in the low EFR scenario in comparison to the no EFR case. There is a significant similarity in the cost of electricity for the initial years (until 2027); this is because the reference climate is wet in those years, resulting in higher Victoria Nile outflows. Moreover, the hydropower plants are working at part load capacity, since the demand is not high enough for them to generate at full load. Hence, even a stricter environmental flow regulation does not affect hydropower and electricity cost. It must be taken into consideration that the results are for a model with perfect foresight. In reality, there is no perfect foresight, and a dry spell can result in higher electricity prices and can render some infrastructure redundant.

Impact of Climate on the Power System

The impact of climate on the power system without enforcing any minimum environmental flows is shown in Fig. 13. The cumulatively driest climate (giss85) results in higher costs compared to other climates, approximately resulting in 3.6% – 2.3 billion USD_{2015} – higher consumer expenditure when compared to the reference climate. Under the cumulatively wettest climate (bnu45), consumer expenditure is expected to reduce by 2.5% – 1.6 billion USD_{2015} – in comparison to the reference scenario.

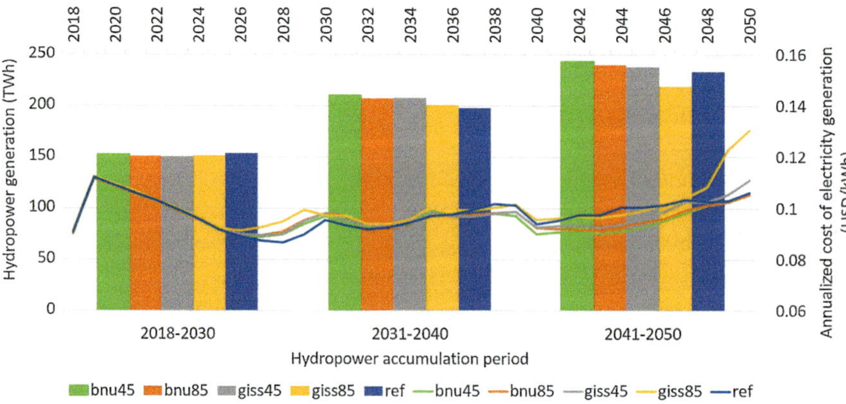

Fig. 13 Impact of climate on hydropower generation (bar graph, primary Y-axis) and the annualized cost of electricity generation (line graph, secondary Y-axis)

The generation mix for the different climates, without any EFR, does not exhibit significant deviations from each other except for the cumulatively driest climate, where the reduction in hydropower generation toward the end of the modeling period results in more significant fossil fuel usage and as such higher cost of electricity.

However, having no restriction on the river flow for environmental and ecosystem services could result in a much more significant impact in the region under a dry climate. This being said, a wet climate could result in lesser or no damages; but, even a cumulatively wet climate could have a few dry year spells, and not maintaining minimum flows could alter critical processes like sediment trapping and recycling of soil nutrients (Kiwango et al. 2015). The effects of not maintaining minimum environmental flows in rivers span over many categories ranging from entertainment and tourism to loss of habitat for endangered species and land-use change. The stretch of Victoria Nile between Lake Victoria and Lake Kyoga is popular for white water rafting but is under threat due to the construction of Isimba dam (International Rivers 2016).

Impact of Both Climate and Environmental Flows on the Power System

Maintaining a minimum environmental flow in the rivers, irrespective of the climate, is mandatory to sustain ecosystem services. When the climate and EFRs are considered together, the impact on the cost of electricity generation is quite noticeable through out the modeling period. Two EFRs, a low and high option, were analyzed along with the five different climates. Figure 14 illustrates only the impact of introducing a high EFR. The cumulatively driest climate (giss85) is expected to cost the consumer 8.2% more than in the reference climate (ref) and the cumulatively wettest climate (bnu45) 9.6% less compared to the reference. The cost of electricity generation is similar for the reference and the two wet scenarios in the initial years until 2024, whereas the cost is significantly higher in the drier climates.

The cumulative impact of all the scenarios and their electricity generation mix can be observed in Fig. 15. For the cumulatively driest climate (giss85), high EFR scenario (AFR) is expected to cost consumers about 80 billion USD_{2015}, between 2018 and 2050, 6 billion USD_{2015} more than the scenario with the Low EFR (LFR), whereas in the cumulatively wettest climate (bnu45), the AFR and LFR scenarios are expected to cost 65 billion USD_{2015} and 63 billion USD_{2015}, respectively. The yellow line corresponds to the cumulative expenditure in the reference climate scenario with no EFR.

It is interesting to note that nuclear power generation options are being considered only in the cumulatively drier scenarios (giss45, giss85) and only when EFRs are considered; the optimization results indicate that investments in nuclear power plants will occur toward the end of the modeling period. This situation is expected to change as the Ugandan Government is discussing with the International Atomic Energy Agency to conduct feasibility studies for the development of nuclear

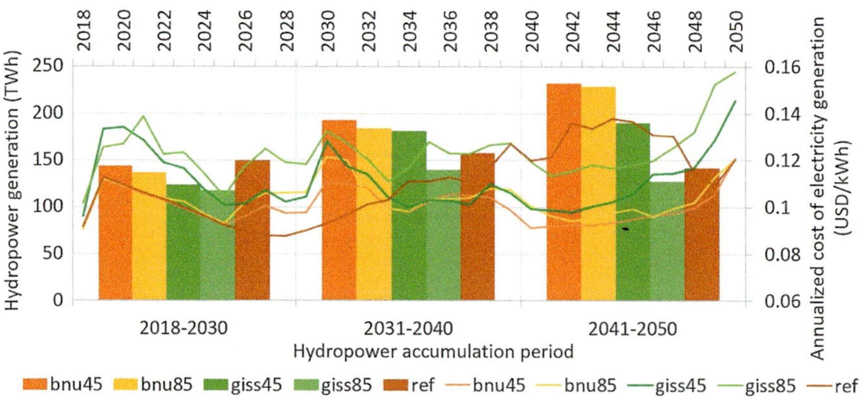

Fig. 14 Impact of high EFRs (AFR) on hydropower generation (bar graph, primary Y-axis) and the annualized cost of electricity generation (line graph, secondary Y-axis) under different climates

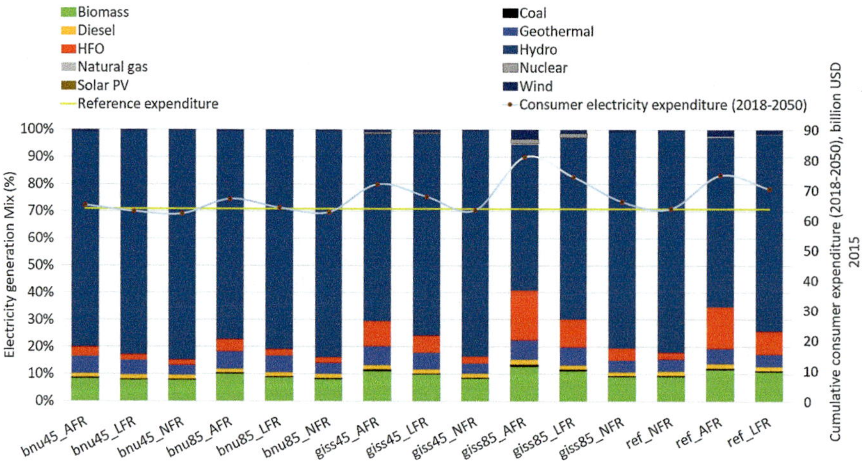

Fig. 15 Cumulative electricity expenditure (2018–2050) and electricity generation mix for each scenario combination

power plants and investigation of uranium exploration possibilities (Uganda State House 2018).

The vulnerability of electricity infrastructure, while maintaining minimum environmental flows in the rivers, is quite evident from the variations in the cost of electricity generation across the 15 different scenario combinations. In addition to the primary objective of improving electricity access in the country, Uganda aspires to be a net electricity exporter in the region by trading electricity with neighboring Kenya and Tanzania, at least to start with. If the hydro infrastructure is built without considering potential climatic changes and other competing uses, an unexpected drier year could fail to deliver on power purchase agreements; since all the large hydro facilities are the run-of-the-river type, there is no possibility to store water in reservoirs as a buffer.

Uncertainty around climate change is inevitable; despite the advancements in climate science, it is difficult to assign probabilities for the climate to unfold in a particular pattern. Moreover, even if the climate is uniformly wet, this study concludes that if the environmental flows need to be maintained, a reduction in electricity generation is expected. This is assuming the technical specifications of the existing and planned hydropower infrastructure in Uganda remains the same.

Whereas, from an environmental point of view, the potential long-term advantages of maintaining a minimum flow in rivers may have a value that could outweigh the impact of variations in the cost of electricity generation. However, establishing regulations to implement minimum flows is not straightforward; robust policies need to be put in place to ensure their maintenance.

Conclusion

In countries with a large share of hydropower in their electricity generation mix, like Uganda, there is an increasing need to ensure that investments made in large infrastructure are judicious and prudent. The case of hydropower in the power system is unique because it is dependent on a resource that is shared with other sectors and is the most affected by climate patterns, among the sources of electricity generation in Uganda. Methods implemented in this study explore the climatic impacts on a hydro-dominant power system considering the allocation of water resources for maintaining environmental services while safeguarding the supply of other users such as the agricultural and household sectors. The results indicate that assuming a minimum environmental flow is maintained on Victoria Nile, irrespective of the climate, there could be periods of fluctuation in hydropower generation resulting in volatile electricity costs, though, cumulatively, the costs are lower for a wetter climate and vice versa for a drier climate. In addition to the cost of electricity generation to consumers, it is expected that a reduction in electricity generation from hydropower plants will lead to delayed payback times for power plant operators; this is of importance in Uganda, where there are large privately financed independent power plants like Bujagali (Meyer et al. 2018). Moreover, this work also contributes to expand the knowledge on the implications of introducing minimum environmental flows when planning the development of the electricity generation sector. This study does not intend to predict the cost of electricity generation in Uganda rather to gather insights regarding risks faced by the electricity system under different climate futures.

It must be observed that this study does not take into consideration the sensitivities in fuel prices and final electricity demand, and various other geopolitical factors, which only indicate a higher uncertainty on the power system impacts. With that being said, numerous adaptation options could be considered. Diversifying the electricity generation mix to increase the share of non-hydro renewables like geothermal, wind, and solar technologies has the potential to offset the climatic impacts and overreliance on specific technologies. Though the National Development Plan indicates a high geothermal potential (1500 MW) (NDP II 2015) in the country, this assessment considers only candidate power plants that are either under construction or planned, according to the latest figures from the MEMD, which are lower than the NDP estimates. However, it remains interesting to wait and observe how the geothermal potential in the country will be expanded to reach its maximum potential. Biomass-based electricity could serve as a potential alternative; a recent study estimates the total theoretical potential of agricultural residues in Uganda will be 148 PJ/year (Okello et al. 2013). Assuming a conservative efficiency estimate of 10%, approximately 4 TWh of electricity demand could be met each year from only the agricultural biomass residue, not including the forest and animal residue. In addition to the technological options, some mitigative measures like interlinking the Ugandan power system to hydropower in neighboring Democratic Republic of the Congo (DRC), where the Congo Basin is expected to have low vulnerability to climate change (Cervigni et al. 2016), could be an alternative, provided the DRC develops its hydro capacity. Similarly, power system interlinkages within the EAPP might help mitigate the climate effects of basins with high vulnerability to climate change.

In parallel to the electricity supply options, transmission and distribution (T&D) infrastructure is critical to improving the electricity access in Uganda. The need to establish T&D systems is all the more critical in the case of hydropower plants as the resources are far from demand centers, as in the case of Uganda.

The power sector model, developed for this study, could be improved to perform a limited foresight analysis to account for myopic decision-making. This study does not quantify the indirect monetary benefits of implementing minimum environmental flows, which could potentially outweigh any increase in the cost of electricity generation.

References

AF Consult (2015a) Demand forecast report -study on integrating nuclear power in generation capacity plan 2015–2040. Ministry of Energy and Mineral Development, Kampala

AF Consult (2015b) Generation plan report -study on integrating nuclear power in generation capacity plan 2015–2040. Ministry of Energy and Mineral Development, Kampala

Allen RG, Pereira L, Raes D, Smith M (1977) FAO irrigation and drainage paper no. 56 crop evapotranspiration

Anderson D, Moggridge H, Warren P, Shucksmith J (2015) The impacts of 'run-of-river' hydropower on the physical and ecological condition of rivers. Water Environ J 29:268–276. https://doi.org/10.1111/wej.12101

Boehlert B, Strzepek KM, Groves D, Hewitson B, Jack C (2016) Climate change projections in Africa-Chapter 3. In: Enhancing the climate resilience of Africa's infrastructure: the power and water sectors. The World Bank, Washington DC, p 219

Brisbane Declaration (2007) The declaration – 2007 – International River symposium, Brisbane

Cervigni R, Liden R, Neumann JE, Strzepek KM (2016) Enhancing the climate resilience of Africa's infrastructure: the power and water sectors. Africa development forum series. The World Bank, Washington, DC

Conway D, Dalin C, Landman WA, Osborn TJ (2017) Hydropower plans in eastern and southern Africa increase risk of concurrent climate-related electricity supply disruption. Nat Energy 2:946–953. https://doi.org/10.1038/s41560-017-0037-4

de Lucena AFP, Schaeffer R, Szklo AS (2010) Least-cost adaptation options for global climate change impacts on the Brazilian electric power system. Glob Environ Change 20:342–350. https://doi.org/10.1016/j.gloenvcha.2010.01.004

EAPP master plan, 2016. EAPP regional power system master plan (2016) (Regional master plan for electricity infrastructure). Eastern Africa power pool (EAPP), EA Energy Analyses, Energinet.dk, Addis Ababa

Ebinger J, Vergara W (2011) Climate impacts on energy systems: key issues for energy sector adaptation (No. 60051). The World Bank, Washington, DC

ESMAP (2007) Technical and economic assessment of off-grid, mini-grid and grid electrification technologies. Energy Sector Management Assistance Program, The World Bank, Washington, DC

FAO (2014) The water-energy-food nexus at FAO: concept note. Food and Agriculture Organization of the United Nations, Rome

Gerlak AK, Weston J, McMahan B, Murray RL, Mills-Novoa M (2018) Climate risk management and the electricity sector. Clim Risk Manag 19:12–22. https://doi.org/10.1016/j.crm.2017.12.003

Hamududu BH, Killingtveit Å (2016) Hydropower production in future climate scenarios; the case for the Zambezi River. Energies 9:502. https://doi.org/10.3390/en9070502

Howells M, Rogner H, Strachan N, Heaps C, Huntington H, Kypreos S, Hughes A, Silveira S, DeCarolis J, Bazillian M, Roehrl A (2011) OSeMOSYS: the open source energy modeling system. Energy Policy Sustain Biofuels 39:5850–5870. https://doi.org/10.1016/j.enpol.2011.06.033

Howells M, Hermann S, Welsch M, Bazilian M, Segerström R, Alfstad T, Gielen D, Rogner H, Fischer G, van Velthuizen H, Wiberg D, Young C, Roehrl RA, Mueller A, Steduto P, Ramma I (2013) Integrated analysis of climate change, land-use, energy and water strategies. Nat Clim Chang 3:621–626. https://doi.org/10.1038/nclimate1789

IEA (2017) Energy access outlook 2017: from poverty to prosperity. IEA, Paris

International Rivers (2016) Uganda weighs energy needs against tourism industry | marketplace [WWW document]. Int Rivers. https://www.internationalrivers.org/resources/11223. Accessed 7 Mar 2018

IPCC (2007) In: Core Writing Team, Pachauri RK, Reisinger A (eds) Climate change 2007: synthesis report. Contribution of working groups I, II and III to the fourth assessment report of the intergovernmental panel on climate change. Assessment report no. 4. IPCC, Geneva

IPCC (2013) Climate change 2013: the physical science basis. Contribution of working group I to the fifth assessment report of the intergovernmental panel on climate change. IPCC/Cambridge University Press, Cambridge, UK/New York

Ji D, Wang L, Feng J, Wu Q, Cheng H, Zhang Q, Yang J, Dong W, Dai Y, Gong D, Zhang R-H, Wang X, Liu J, Moore JC, Chen D, Zhou M (2014) Description and basic evaluation of Beijing Normal University Earth System Model (BNU-ESM) version 1. Geosci Model Dev 7:2039–2064. https://doi.org/10.5194/gmd-7-2039-2014

Joshua W, Herbert S, Isaya K, John KA, Noble B, Ahamada Z, Oker ET, Charles M, Nicholas K, Kato KR, Banaga MD, Lubwama KF (2017) Irrigation development in Uganda: constraints, lessons learned, and future perspectives. J Irrig Drain Eng 143:04017003. https://doi.org/10.1061/(ASCE)IR.1943-4774.0001159

Katz D (2015) Water use and economic growth: reconsidering the environmental Kuznets curve relationship. J Clean Prod Sustain Dev Energy Water Environ Syst 88:205–213. https://doi.org/10.1016/j.jclepro.2014.08.017

Kiwango H, Njau KN, Wolanski E (2015) The need to enforce minimum environmental flow requirements in Tanzania to preserve estuaries: case study of the mangrove-fringed Wami River estuary. Hydrol Curr Res 6:1–10. https://doi.org/10.4172/2157-7587.1000205

Lehner B, Verdin K, Jarvis A (2008) New global hydrography derived from spaceborne elevation data. EOS Trans Am Geophys Union 89:93–94. https://doi.org/10.1029/2008EO100001

Li H, Sheffield J, Wood EF (2010) Bias correction of monthly precipitation and temperature fields from intergovernmental panel on climate change AR4 models using equidistant quantile matching. J Geophys Res Atmos 115:D10101. https://doi.org/10.1029/2009JD012882

Makhorin A (2008) Modeling language GNU MathProg

MEMD (2014) Uganda energy balance

MEMD (2015) Uganda's sustainable energy for all (SE4All) initiative action agenda
MEMD (2016) Petroleum production licences to Tullow Uganda Operations Pty Limited (Tullow) and total E&P Uganda B.V. (Total). Kampala, Uganda
MEMD (2017) Communications with the Ministry of Energy and Mineral Development (MEM), Uganda
Meyer R, Eberhard A, Gratwick K (2018) Uganda's power sector reform: there and back again? Energy Sustain Dev 43:75–89. https://doi.org/10.1016/j.esd.2017.11.001
MWE (2011) A national irrigation master plan for Uganda (2010–2035) (Master plan). Ministry of Water and Environment, PEM consult, Kampala
MWE (2013) Uganda national water resources assessment. Ministry of Water and Environment (MWE), Kampala
NBC News (2015) Drought-caused blackouts batter Zambia, Zimbabwe economies [WWW document]. NBC News. https://www.nbcnews.com/business/energy/drought-caused-blackouts-batter-zambia-zimbabwe-economies-n454581. Accessed 9 Apr 2018)
NBI (2016) Strategy for management of environmental flows in the Nile Basin | Nile Information System
NDP II (2015) Second national development plan (NDPII)-vision 2040
Okello C, Pindozzi S, Faugno S, Boccia L (2013) Bioenergy potential of agricultural and forest residues in Uganda. Biomass Bioenergy 56:515–525. https://doi.org/10.1016/j.biombioe.2013.06.003
Pastor AV, Ludwig F, Biemans H, Hoff H, Kabat P (2014) Accounting for environmental flow requirements in global water assessments. Hydrol Earth Syst Sci 18:5041–5059. https://doi.org/10.5194/hess-18-5041-2014
Pinto de Moura GN, Loureiro Legey LF, Balderrama GP, Howells M (2017) South America power integration, Bolivian electricity export potential and bargaining power: An OSeMOSYS SAMBA approach. Energy Strategy Rev 17:27–36. https://doi.org/10.1016/j.esr.2017.06.002
Platts (2016) Platts- world electric power plants database [WWW document]. World Electric Power Plants Database. http://www.platts.com/Products/worldelectricpowerplantsdatabase. Accessed 1 Mar 2016
Prado J, Beare RJ, Siwo Mbuga J, Oluka LE (1991) A catalogue of fishing methods and gear used in Lake Victoria
Rasul G (2016) Managing the food, water, and energy nexus for achieving the sustainable development goals in South Asia. Environ Dev 18:14–25. https://doi.org/10.1016/j.envdev.2015.12.001
RCMRD (2014) Uganda landcover 2014 scheme I & II
Roy A (1999) The greater common good. India Book Distributor, Bombay
Schmidt GA, Kelley M, Nazarenko L, Ruedy R, Russell GL, Aleinov I, Bauer M, Bauer SE, Bhat MK, Bleck R, Canuto V, Chen Y-H, Cheng Y, Clune TL, Del Genio A, de Fainchtein R, Faluvegi G, Hansen JE, Healy RJ, Kiang NY, Koch D, Lacis AA, LeGrande AN, Lerner J, Lo KK, Matthews EE, Menon S, Miller RL, Oinas V, Oloso AO, Perlwitz JP, Puma MJ, Putman WM, Rind D, Romanou A, Sato M, Shindell DT, Sun S, Syed RA, Tausnev N, Tsigaridis K, Unger N, Voulgarakis A, Yao M-S, Zhang J (2014) Configuration and assessment of the GISS ModelE2 contributions to the CMIP5 archive. J Adv Model Earth Syst 6:141–184. https://doi.org/10.1002/2013MS000265
Sheffield J, Goteti G, Wood EF (2006) Development of a 50-year high-resolution global dataset of meteorological forcings for land surface modeling. J Clim 19:3088–3111. https://doi.org/10.1175/JCLI3790.1
Silberstein RP (2006) Hydrological models are so good, do we still need data? Environ Model Softw Integr Model Clim Terr Fluvial Syst 21:1340–1352. https://doi.org/10.1016/j.envsoft.2005.04.019
Smakhtin V, Revenga C, Döll P (2004) A pilot global assessment of environmental water requirements and scarcity. Water Int 29:307–317. https://doi.org/10.1080/02508060408691785
Staffell I, Pfenninger S (2018) The increasing impact of weather on electricity supply and demand. Energy 145:65–78. https://doi.org/10.1016/j.energy.2017.12.051

Taliotis C, Howells M, Bazilian M, Rogner H, Welsch M (2014) Energy security prospects in Cyprus and Israel: a focus on natural gas. Int J Sustain Energy Plan Manag 3:5–20

Taliotis C, Shivakumar A, Ramos E, Howells M, Mentis D, Sridharan V, Broad O, Mofor L (2016) An indicative analysis of investment opportunities in the African electricity supply sector – using TEMBA (The electricity model base for Africa). Energy Sustain Dev 31:50–66. https://doi.org/10.1016/j.esd.2015.12.001

Taylor T, Markandya A, Droogers P, Rugumayo A (2014) Impacts of climate change in Uganda on the water sector – future water. Climate Change Unit, Ministry of Water and Environment, Uganda Climate & Development Knowledge Network (CDKN). Kampala, Uganda

Tennant D (1976) Instream flow regimes for fish, wildlife, recreation and related environmental resources. Fisheries 1:6

Tessmann SA (1979) Environmental use sector: reconnaissance elements of the western Dakotas region of South Dakota study. Water Resources Institute, South Dakota State University, Brookings

Tharme RE (2003) A global perspective on environmental flow assessment: emerging trends in the development and application of environmental flow methodologies for rivers. River Res Appl 19:397–441. https://doi.org/10.1002/rra.736

Tirado MC, Cohen MJ, Aberman N, Meerman J, Thompson B (2010) Addressing the challenges of climate change and biofuel production for food and nutrition security. Food Res Int Clim Change Food Sci 43:1729–1744. https://doi.org/10.1016/j.foodres.2010.03.010

Twagiramaria F, Tolo CU, Zinyengere N (2018) Chapter 4 – adaptation to and coping strategies for climate change and variability by rural farmers in Kigezi Highlands, Uganda. In: Beyond agricultural impacts. Academic, London, pp 55–75. https://doi.org/10.1016/B978-0-12-812624-0.00004-1

UBOS (2011) Uganda census of agriculture 2008/2009: crop area and production report, vol IV. Uganda Bureau of Statistics, Kampala

UBOS (2016a) Uganda Bureau of Statistics 2016 statistical abstract, statistical abstract, Kampala

UBOS (2016b) National population and housing census 2014 (National census). Uganda Bureau of Statistics (UBOS), Kampala

UEGCL (2018) Karuma project: great strides despite some challenges – Uganda Electricity Generation Company Ltd (UEGCL) [WWW document]. https://uegcl.com/corporate-affairs/news/43-karuma-project-great-strides-despite-some-challenges.html. Accessed 2 Mar 2018

Uganda State House (2018) IAEA gives node as Uganda pushes for nuclear power use | State House Uganda [WWW document]. http://www.statehouse.go.ug/media/news/2018/01/17/iaea-gives-node-uganda-pushes-nuclear-power-use. Accessed 8 Mar 2018

UN (2015) Sustainable development goals – United Nations [WWW document]. UN Sustainable Development. https://sustainabledevelopment.un.org/sdgs. Accessed 22 Feb 2016

UNDESA (2017) World population prospects – population division – United Nations. United Nations, New York

van Vliet MTH, Wiberg D, Leduc S, Riahi K (2016) Power-generation system vulnerability and adaptation to changes in climate and water resources. Nat Clim Chang 6:375–380. https://doi.org/10.1038/nclimate2903

Willmott CJ, Feddema JJ (1992) A more rational climatic moisture index*. Prof Geogr 44:84–88. https://doi.org/10.1111/j.0033-0124.1992.00084.x

Yates D, Sieber J, Purkey D, Huber-Lee A (2005) WEAP21 – a demand-, priority-, and preference-driven water planning model. Water Int 30:487–500. https://doi.org/10.1080/02508060508691893

Managing the Increasing Heat Stress in Rural Areas

10

Adithya Pradyumna, Ramkumar Bendapudi, Dipak Zade, Marcella D'Souza, and Premsagar Tasgaonkar

Contents

Abstract

Increasing temperatures are likely to impact human health. An increase in severe heat wave days and heat mortalities has been observed in India over the past few decades. At present, there is little evidence on the heat exposure, impact, and the adaptation measures in the rural context. The present study examines

A. Pradyumna (✉)
Watershed Organisation Trust (WOTR), Pune, India

Society for Community Health Awareness Research and Action (SOCHARA), Bangalore, India
e-mail: adithya@sochara.org

R. Bendapudi · P. Tasgaonkar
Watershed Organisation Trust (WOTR), Pune, India
e-mail: rbendapudi@yahoo.com; premsagar.tasgaonkar@wotr.org.in

D. Zade · M. D'Souza
The WOTR-Centre for Resilience Studies (W-CReS), Pune, India
e-mail: dipak.zade@wotr.org.in; marcella.dsouza@gmail.com

W. Leal Filho (ed.), *Handbook of Climate Change Resilience*,
https://doi.org/10.1007/978-3-319-93336-8_46

vulnerability of rural communities to heat stress in the semiarid villages in Maharashtra state of India.

The study was conducted in five villages of Jalna and Yavatmal districts of Maharashtra. Household survey covering 20% of the households was conducted in Jalna during 2016 summer months. Twenty data loggers were installed to measure the indoor temperatures in Yavatmal.

Exposure to heat in various circumstances, both outdoors and indoors, was reported. Age, gender, wealth, and pre-existing health conditions were significantly associated with occurrence of heat-related symptoms (HRS). Exposure factors such as working outdoors during midday, roofing material, and indoor ventilation were significantly associated with occurrence of HRS. The indoor temperature in houses with tin roofs was found to be higher as compared to cement-roofed houses.

Existing coping strategies appear to be inadequate to protect people from outdoor and indoor heat stress. People from poorer households reported socioeconomic and livelihood challenges in adopting coping strategies as well. A long-term and locally appropriate strategy in terms of knowledge about HRS and infrastructure and access to timely medical facilities is needed. Development of effective surveillance mechanism and a comprehensive state-level heat action plan are needed to prevent and monitor heat mortalities in the future.

Keywords

Heat stress · Heat-related symptoms · Differential vulnerability · Rural health · Semiarid regions

Background

The risk of heat-related illnesses and deaths is likely to increase due to rising temperatures (Intergovernmental Panel on Climate Change 2014). In India, there has been an increasing trend of heat wave-related deaths in the last few decades. A report by National Disaster Management Authority highlights that between the periods 1992 and 2015, about 22,562 heat-related deaths were reported (Government of India 2016). In fact, heat stroke was the second major cause of deaths from natural disasters in India (accounting for 15% of those deaths) during the decade 2001–2012 (Paul and Bhatia 2016). Most studies on the health impacts of heat have been conducted within the context of officially declared heat waves. It has been found that significant mortality and morbidity can be attributable to heat exposure. The effect could be up to 43% increase in deaths due to heat wave in urban India (Azhar et al. 2014).

Future climate projections for India indicate that heat waves will likely be more intense, have longer durations, and occur more often and earlier in the year. Intensification of heat waves will also lead to increased mortality rates (Dholakia et al. 2015; Murari et al. 2015). It was also found that death attributable to extreme heat was about the same as moderately high temperatures (Gasparrini et al. 2015).

Increasing heat exposure is linked to occupational health risks and negatively impacts work productivity (Dash and Kjellstrom 2011; Kjellstrom et al. 2009).

Studies on occupational heat stress impacts in India found that decreased productivity was reported most commonly among outdoor/semi-outdoor occupations with high workload (e.g., brick manufacturing, metal fabrication, construction, etc.), whereas productivity losses were reported less frequently among indoor workers (Venugopal et al. 2015).

An association between high temperature and noninfectious disease mortality was found in a study conducted in rural western India. Men in working age involved in outdoor activities like agricultural and industrial workers were more vulnerable to heat (Ingole et al. 2015).

In Gujarat and Rajasthan, workers in the rural and semi-urban industries of ceramics, pottery, iron works, and stone quarry were found to be vulnerable (Nag et al. 2009). In the rural context, it was found that among the rice farm workers in West Bengal, high workplace heat exposure caused heat strain and reduced work productivity (Sahu et al. 2013).

Apart from decreased work productivity, increased temperatures also have other social impacts. A 21-year rural longitudinal survey conducted in Pakistan showed that men move out of the village due to extreme heat stress and that the landless and asset-less poor are more likely to do so (Mueller et al. 2014). A large-scale study conducted in Thailand found that working under heat stress conditions is associated with worse overall health and psychological distress (Tawatsupa et al. 2010).

Apart from outdoor heat stress, exposure to hot indoor conditions is also a concern. In a study assessing the impact of heat wave events on dwellings (Quinn et al. 2014), it was found that a substantial fraction of houses exceeded dangerous heat thresholds during extreme events. Older adults staying indoors during hot indoor temperature conditions are at a risk of significant detriment of physical functions (Lindemann et al. 2017). There is a need to improve awareness regarding management of indoor heat stress. The use of passive building design strategies such as shading, thermal mass, and internal air movement has been suggested to mitigate the overheating of dwellings and to postpone the use of active systems (Din and Brotas 2017). It has also been demonstrated in an experimental study that simple and effective hybrid passive cooling system can significantly help in reducing thermal loads of roofs (Ponni and Baskar 2015).

A recent study from urban India identified exposure (geographic location, housing characteristics, and occupational and behavioral factors), susceptibility (age, pre-existing health status, and socioeconomic factors), and adaptive capacity (access to health services and information, coping mechanisms, and societal factors (infrastructure, information, and social capital) determine vulnerability to heat (Tran et al. 2013). Pre-existing conditions, such as cardiovascular diseases, may be exacerbated by heat stress (Khan et al. 2014). However, there is little evidence of the heat experience, impact of heat exposure, and adaptation measures to heat and heat waves in the rural context.

Most of the above studies examine vulnerability to heat stress due to work exposure (occupational hazard) and focused on urban areas. Very few studies have attempted to understand the factors contributing to vulnerability of the communities.

In this context, this study examines vulnerability of rural communities to heat stress in villages located in semiarid regions of Maharashtra state in India. An exploratory qualitative study was conducted to inform this survey (Mhaskar et al. 2016). The specific objectives addressed by this study are:

(i) To quantify heat-related symptoms and illnesses in the rural communities during summer time
(ii) To identify the categories of the rural population that are most affected by heat stress
(iii) To understand exposure to outdoor and indoor heat stress
(iv) To identify factors contributing to vulnerability to heat stress
(v) To examine various existing strategies being used to cope with heat stress

Methods

Study Location

The study was conducted in Jalna and Yavatmal districts of Maharashtra. Jalna is located in the central part of Maharashtra state in northern Marathwada region. The district has a sub-tropical climate with average annual rainfall ranging between 650 and 750 mm. The district experiences years of drought with annual rainfall recording as low as 400 mm. The hot dry summer season is from March to June. During summer, the maximum day temperature ranges between 42 °C and 43 °C (Government of India 2018). The summer months are dry, with relative humidity generally between 20% and 25% in the afternoon.

The blocks of Jafrabad and Bhokardan in Jalna district were selected as there was anecdotal evidence of few deaths due to heat stress in 2014. Exploratory visits were undertaken during the months of April and May 2015 to select the villages. The factors considered for selecting the villages were sparse vegetative cover, lack of access to water, and remoteness. Three villages were purposively selected for the study, namely, *Adha* and *Sindi* in the Jafrabad block and *Goshegaon* in the Bhokardan block. In *Goshegaon* village, the local government authorities had supplied drinking water in tankers during summer of 2015. Village *Adha* had a primary health subcenter in the village, whereas the nearest government healthcare system for *Sindi* and *Goshegaon* villages is located at a distance of 15 km and 7 km, respectively.

For understanding indoor temperatures, *Sonurli* and *Ekhlara* villages located in the Ralegaon block of Yavatmal district were selected for measuring indoor temperatures using data loggers. According to the India Meteorological Department, when the temperature reaches 45 °C, it is considered as heat wave condition. In both Jalna and Yavatmal districts, the maximum summer temperature ranges between 42 °C and 43 °C, thus nearing the heat wave conditions. Figure 1 shows the location of Jalna and Yavatmal districts and of the study villages.

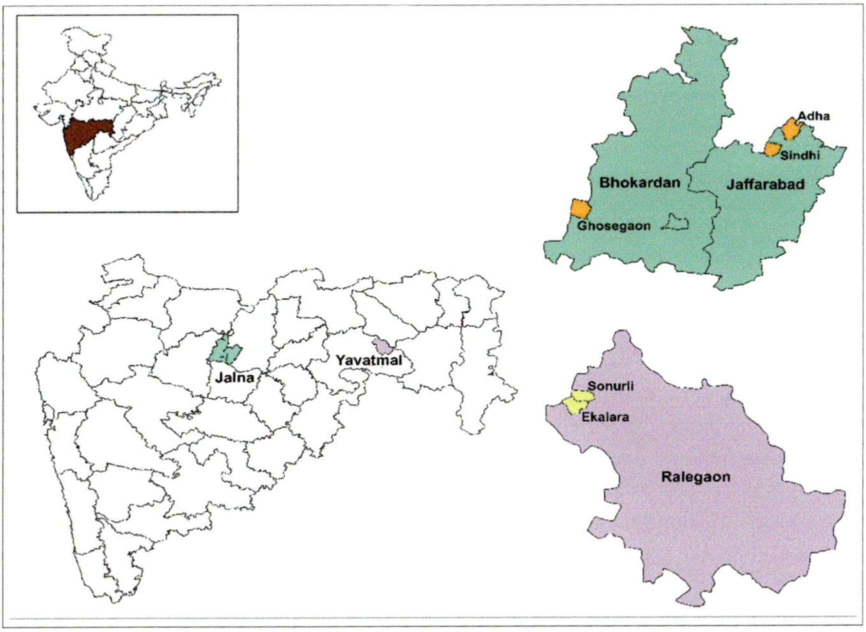

Fig. 1 Location map of Jalna and Yavatmal districts

Sample Survey

Households in each village were categorized according to socioeconomic criteria based on a participatory wealth ranking exercise. Wealth ranking exercise is a participatory tool for classifying the households based on the indicators related to land ownership, asset ownership, food security, migration, and sources of income. The households are classified as very poor, poor, middle class, and better off.

From each of the three villages, a sample of 20% of total households was selected for survey. Stratified random sampling was used for selecting the respondent households. Accordingly, the sample households were selected in proportion to the number of households in the respective wealth categories. In total, 215 households contributed to the detailed household survey (Table 1).

Information collected from the respondents included the following: living conditions (housing structure, access to water for drinking and domestic use, access to electricity), work profile (type of livelihood activities, exposure to outdoor heat), health problems (pre-existing health conditions, self-reported heat-related symptoms, access to medical facilities, and sources of information on preventive measures), and coping strategies used to manage heat exposure and impacts. This list was informed by the literature and also based on our exploratory qualitative study conducted prior to this survey (Mhaskar et al. 2016).

Table 1 Socioeconomic status of sampled households

Wealth category of household	Total households in the study area	Households included in the study
Very poor	226 (21.9%)	46 (21.4%)
Poor	420 (40.7%)	87 (40.5%)
Middle class	241 (23.3%)	51 (23.7%)
Better off	146 (14.1%)	31 (14.4%)
Total	**1033 (100%)**	**215 (100%)**

Monitoring of Indoor and Outdoor Temperatures

The selection of household for installation of data loggers was made based on types of roofing. Twenty indoor digital temperature data loggers were installed, inside the houses in the room where maximum time was spent by household members. Twelve of these loggers were installed in houses with tin/galvanized sheet roof structure, 7 loggers in houses with cement concrete roofs, and 1 in a tiled roof structure. Each data logger had a unique serial number embedded within its firmware, allowing for tracking of deployed loggers. The data loggers and the automated weather station were set to record temperatures at a 10-min interval, allowing for a maximum monitoring period of summer months.

An automated weather station was also installed in the village for measuring the location-specific outdoor temperature during the summer months.

Data Collection

A structured questionnaire was developed and pretested. Individual interviews were conducted in the month of May 2016, which is the peak summer period in Maharashtra. An adult household member was selected as the respondent who provided information regarding all other members of the family (in the context of exposure to heat and heat-related symptoms).

The types of heat-related symptoms (HRS) included in the survey are as follows: small blisters or pimples, dry mouth, fatigue, leg cramps, heavy sweating, intense thirst, rapid heartbeat, headache, leg swelling, paranoid feeling, swelling of face, fever, diarrhea, vomiting, hallucinations, and fainting. The occurrences of these were recorded with a recall of 2 months (the hottest months – April and May, during the year 2016). Hallucinations and fainting were considered as severe HRS, whereas the rest were considered as mild HRS for analysis. This list was informed from literature and also from the exploratory qualitative study conducted prior to this survey.

Sample Characteristics

The sampled households comprised of 1224 individuals in total. Out of these, there were 671 male members and 553 female members (accounting for 55% and 45% of total individuals, respectively). About 64% of the total household members were in the age groups of 15–30 years and 31–59 years.

Table 2 Sociodemographic characteristics of the study population

Variable	Number of individuals (% of total, $n = 1224$)
Gender	
Male	671 (54.8)
Female	553 (45.2)
Age	
0–4	104 (8.5)
5–14	212 (17.3)
15–30	450 (36.8)
31–59	330 (27.0)
60+	128 (10.5)
Wealth ranking	
Very poor	223 (18.2)
Poor	448 (36.6)
Middle class	335 (27.4)
Better off	218 (17.8)
Caste category	
Scheduled caste	276 (22.8)
Scheduled tribe	57 (4.7)
Nomadic and denotified nomadic tribes	44 (3.6)
Vimukta Jati nomadic tribes	13 (1.1)
Other backward classes	66 (5.4)
Forward caste	757 (62.4)
Education	
Illiterate	286 (25.6)
Primary school	210 (18.5)
Secondary school	450 (39.8)
High school	122 (10.8)
Graduate and above	54 (4.8)
Summer occupation	
Non-income generating activities	205 (22.2)
Farming	211 (22.9)
Agricultural and nonagricultural labor	342 (37.0)
MGNREGS work	113 (12.2)
Others	52 (5.6)

About 18% of persons interviewed belonged to households that were very poor and 37% to households that were poor (based on wealth ranking). The majority of the households belonged to forward caste category (62% of total sample households) followed by scheduled caste category (23%). Illiteracy is prevalent among 25.6% of individuals covered by the survey. About 37% of individuals reported wage labor (agricultural and nonagricultural) as the major summer occupation (Table 2). About 23% of the individuals indicated farming as an important livelihood source. Non-income generating activities such as household chores, education, and other such activities were indicated by 22% of the total individuals from the sample.

Data Analysis

Descriptive statistics and cross-tabulations were first used to understand the data. The occurrence of at least one HRS in the individual was the health outcome of interest. This health outcome was cross-tabulated against every other variables (all of which relate to susceptibility, exposure, and coping strategies), through which odds ratios (OR) were calculated on the lines of multinomial logistic regression. Confidence intervals (CI) and p-values for the odds ratios were also reported to help with the interpretation. All the ORs presented are unadjusted.

For analyzing the temperature data generated from each data logger, Hoboware software was used. The software allowed the temperatures to be plotted on a continuous graph covering the monitoring period.

Results and Discussion

Heat-Related Symptoms Among the Households

Among the heat-related symptoms experienced by the household members, headaches, heavy sweating, fatigue, intense thirst, dry mouth, and small blisters/rash were found to be the most commonly reported symptoms (Graph 1).

Of all individuals, 46.2% experienced at least one HRS during the study period. On average, at least two individuals were affected per household (Table 3). Individuals from 64.5% households experienced fever, and 21.5% households experienced diarrhea during the study period.

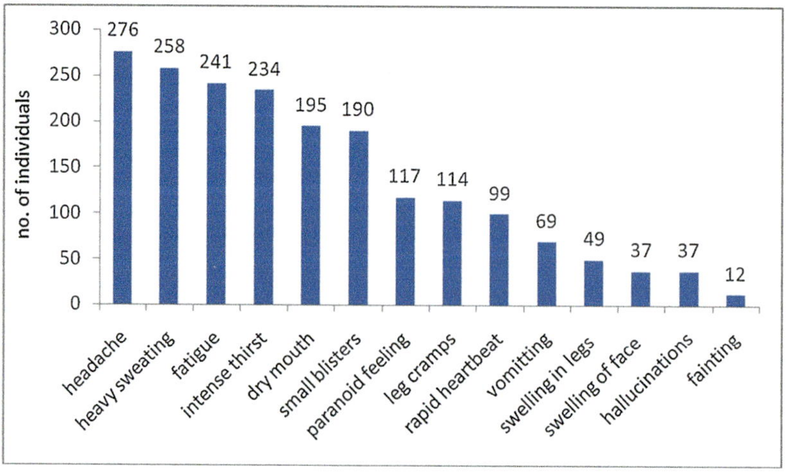

Graph 1 Frequency of reported heat-related symptoms

Table 3 Occurrence of heat-related symptoms (HRS) at individual level

Particulars	Number of individuals ($N = 1224$)
Total individuals experiencing at least one HRS (%)	566 (46.2)
Average individuals affected in each household	2.64
Heat stress categories	
No HRS (%)	658 (53.8)
Mild HRS (%)	523 (42.7)
Severe HRS (%)	43 (3.5)

Figures in parenthesis indicate percentage to total

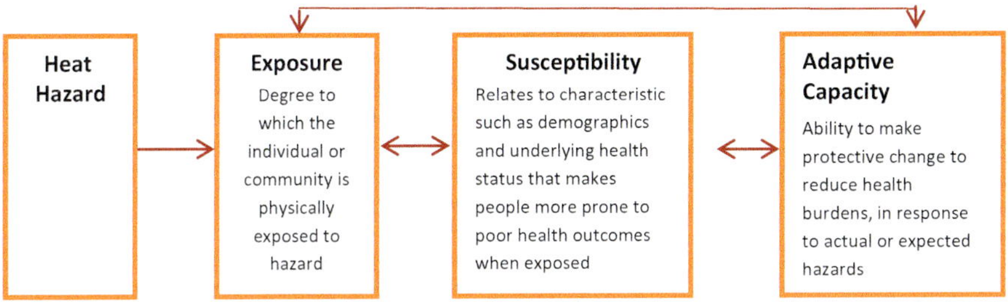

Fig. 2 Framework for assessing heat vulnerability. (Adapted from Tran et al. 2013)

Vulnerability to Heat Stress

The analysis is based on the framework given by Tran et al. wherein heat vulnerability is conceptualized as a function of interacting biophysical and socioeconomic determinants that can be broken down into heat hazard probability as well as factors associated with population exposure, susceptibility, and adaptive capacity (Tran et al. 2013) (Fig. 2).

Susceptibility Factors

Susceptibility factors included age, gender, and pre-existing health conditions. The analysis of wealth status has also been discussed here for convenience, though it relates to all dimensions of vulnerability through various pathways.

Elderly population (>60 years old) were more affected as compared to children under 4 years (OR 0.16; CI 0.09–0.29) and those between 5 and 14 years (OR 0.23; CI 0.15–0.30) and those under 30 years of age (OR 0.55; CI 0.37–0.82). Only adults between 30 and 59 years of age were more affected (OR 1.86; CI 1.22–2.85) than the elderly (Table 4). This is also the age group where higher percent of individuals (71.2%) reported heat-related symptoms. It has been reported in literature that elderly and children (Li et al. 2015; Lundgren et al. 2013; McGeehin and Mirabelli 2001; Oudin Åström et al. 2011) are more susceptible to heat stresses. Our finding indicates that "exposure" might be driving the health impact in our study region, rather than "susceptibility."

Table 4 Occurrence of HRS in relation to demographic variables (univariate analysis)

Parameter	Odds ratio (unadjusted)	Lower CI	Upper CI	*p*-value
Age				
0–4	**0.16**	0.09	0.29	<**0.001**
5–14	**0.23**	0.15	0.37	<**0.001**
15–30	**0.55**	0.37	0.82	**0.003**
31–59	**1.86**	1.22	2.85	**0.004**
60+ (ref)	**1**			
Gender				
Male (ref)				
Female	**0.75**	0.6	0.94	**0.012**
Pre-existing health conditions				
None (ref)				
At least one pre-existing health condition	**6.34**	4.16	9.66	<**0.001**
Wealth categories				
Very poor (ref)				
Poor	**1.07**	0.78	1.48	**0.676**
Middle class	**0.67**	0.48	0.94	**0.021**
Better off	**0.52**	0.36	0.77	<**0.001**

Both men and women reported suffering from HRS. However, the proportion of men reporting HRS (49.5%) was relatively more as compared to women (42%). The unadjusted odds ratio for this comparison was 0.75 (CI 0.6, 0.94). This could be due to a greater proportion of men working outdoors during the day (76% vs. 71%, respectively), performing strenuous work (41% vs. 36%, respectively), and use of protective clothing by women, and not due to physiological susceptibility among men.

Those belonging to middle-class category (OR 0.67; CI 0.48–0.94) and better off families (OR 0.52; CI 0.36–0.77) were less affected as compared to the very poor. Higher proportion of individuals from the "very poor" and "poor" categories reported at least one HRS (about 51% and 53% of total individuals in respective categories) as compared to the individuals in the "better off" category (only 35%). Our finding corroborates previous literature which have indicated that people with low socioeconomic status have been reported to be more affected by heat stress (Harlan et al. 2006; Li et al. 2015). The mechanisms through which poverty might make individuals more susceptible include general health condition and working and living conditions.

Among the 1224 total individuals in the sample households, 157 individuals (about 13% of the sample) reported to be suffering from at least one pre-existing health condition (ranging from asthma to cancer). Those with at least one pre-existing chronic health condition reported relatively high HRS (OR of 6.34, CI 4.16, 9.66) (Table 4). Individuals with pre-existing health conditions have been

Photo 1 Villagers working at MGNREGS sites during summers. (Source: WOTR)

reported to be more susceptible to heat stress (Li et al. 2015; McGeehin and Mirabelli 2001).

The majority of the sample population in the study area belonged to general and scheduled caste categories. The difference in proportion of individuals reported HRS among the two social categories was not statistically significant.

Exposure Factors

Exposure can be influenced by hazard factors, amplifying factors, and protective factors (Tran et al. 2013). In this study, the hazard factors are the outdoor and indoor temperatures; the amplifying factors include outdoor work during peak heat hours, type of occupation, and roofing material; and the protective factors include the coping strategies being employed. In this section, the key amplifying factors are discussed.

Exposure to heat was reported both outdoors and indoors. While the working adult population was exposed to direct sunlight outdoors, the elderly and children were exposed to hot indoor environments during midday and afternoons (Mhaskar et al. 2016). Those from the poorer sections and from the lower castes were disproportionately exposed based on their daily activities and roofing material used for the houses (Photo 1).

Exposure due to Livelihood Activities

One of the amplifying factors affecting exposure was the type of occupation an individual is engaged in. Majority of the individuals from the sample households (72%) were engaged in outdoor livelihood activities during summer months such as farming on own land, wage labor (agriculture, non-farm), and in the

Mahatma Gandhi National Rural Employment Guarantee Scheme (MGNREGS), a government employment scheme.

Those engaged in wage labor (whether privately employed or through the governmental MGNREGS scheme) were more affected as compared to those engaged in activities such as household chores and education. Those engaged in farming (OR 2.05, CI 1.38–3.03), labor activities (agriculture and non-agriculture) (OR 3.35; CI 2.33–4.80), and MGNREGS (OR 3.71; CI 2.28–6.04) were more affected as compared to those performing non-income generating activities (Table 5). Occupation (especially physically intense occupations) is a known risk factor for heat-related illness (Centers for Disease Control and Prevention 2011; Jackson and Rosenberg 2010). People perceived those working outdoors (farm laborers) and the elderly and young children stay indoors in heat-trapping tin houses to be more exposed to heat. It was also perceived that those working outdoors during daytime, especially women laborers, are most vulnerable to health impacts of heat exposure. Exposure during travel to work was also reported by many local respondents (Mhaskar et al. 2016).

The work day was divided into four segments – morning (4 am to 11 am), midday (11 am to 3 pm), afternoon (3 pm to 5 pm), and evening (5 pm to 7 pm). Of these, midday and afternoon together were considered as "peak heat hours" based on local people's opinion (Mhaskar et al. 2016). The total number of individuals exposed to peak heat hours for 1.5–6 h is 428 (55%) (Table 6). Majority of these individuals were engaged in physically strenuous work such as wage labor (47.9%) and farming (20%). About 14.7% worked as MGNREGS laborers during peak heat hours

Table 5 Occurrence of HRS in relation to other variables (univariate analysis)

Parameter	Odds ratio (Unadjusted)	Lower CI	Upper CI	*p*-value
Summer occupation				
Non-income generating activities (ref)				
Farming	**2.05**	1.38	3.03	**<0.001**
Agricultural and nonagricultural labor	**3.35**	2.33	4.80	**<0.001**
MGNREGS labor	**3.71**	2.28	6.04	**<0.001**
Others	**1.95**	1.06	3.61	**0.03**
Minutes spent performing activities outdoors during peak heat hours				
0 (ref)				
1–90 min	**1.56**	1.00	2.44	**0.050**
91–180 min	**2.88**	1.90	4.35	**<0.001**
181–360 min	**1.75**	1.14	2.69	**0.010**
Exposure				
Type of roof				
Tin (ref)				
Cement slab	**0.69**	0.52	0.92	**0.011**
Others	**1.96**	1.22	3.13	**0.005**

Table 6 Time spent outdoors during peak heat hours

Time spent outdoors during peak heat hours for work or activities	No. of individuals
0	195 (25.4)
<1.5 h	145 (18.9)
1.5–3 h	255 (33.2)
3–6 h	173 (22.5)
Total individual responses	**768 (100)**

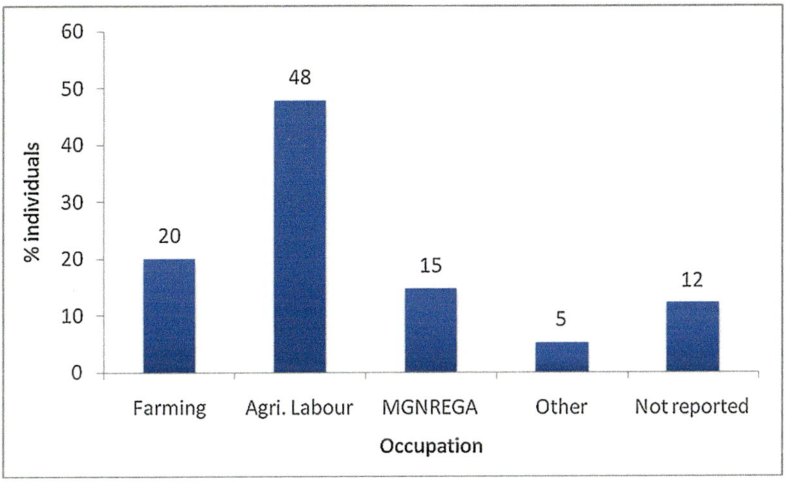

Graph 2 Occupation of individuals working outdoors for more than 1.5 h during peak heat hours ($N = 428$)

(Graph 2). Respondents from over 83% households reported regularly using synthetic clothing during summer, which may add to heat-related discomfort.

Occurrence of HRS was relatively much higher in individuals spending more time outdoors during peak heat hours (Table 5).

Exposure During Other Household Chores

Apart from the livelihood activities, other household activities could also result in outdoor exposure to heat. Over 96% households collect firewood from field and forest. In 43.7% households, only women take the responsibility for collection of firewood. About 84% of households indicated spending more than 4 h each week on this activity. The preferred time for this activity was in the morning, early evening, and evening (these are times with relatively less heat outdoors (Mhaskar et al. 2016)).

On an average, during summer, the time taken for collecting water was 71 min (against 40 min during other seasons). In over 93% households, the women were primarily responsible for fetching water. But during summer, in some households men also support women in fetching water. Similar to the practice of collecting firewood, water too is collected during morning, early evening, and evening hours.

Table 7 Roofing material

	Total individuals living in those houses	Total individuals who experienced at least one HRS	Proportion of individuals experiencing at least one HRS
Tin	887	412	47.0
Cement slab	260	99	38.1
Others	82	52	63.4

Photo 2 Tin-roofed households in the villages. (Source: WOTR)

Type of Roofing

Another amplifying factor of exposure is the type of shelter or dwellings. A large proportion of households (about 74% of sample households) were living in tin-roofed houses. Tin roofs get heated quickly during daytime as compared to cement slab roofs, but they also cool up rapidly as compared to the cement roofs. About 47% of respondents residing in tin-roofed houses indicated at least one HRS as compared to 38% living in cement slab houses (Table 7). Cross ventilation helps in dissipating heat. However, only 13% of the households reported having cross ventilation in all rooms. Those living in households with cement roof were less affected as compared to those with tin roof (OR 0.69, CI 0.52–0.92) (Photo 2).

Without adequate rest and recovery time from heat, people become more vulnerable to heat stress (Khan et al. 2014). Poor design of houses with inadequate ventilation (lack of windows) could exacerbate the condition due to high daytime and nighttime temperatures. The elderly were also perceived as vulnerable due to their physiological condition, with several reported cases of fatigue, dehydration, and loss of consciousness, despite majority of them staying back at home (Mhaskar et al. 2016).

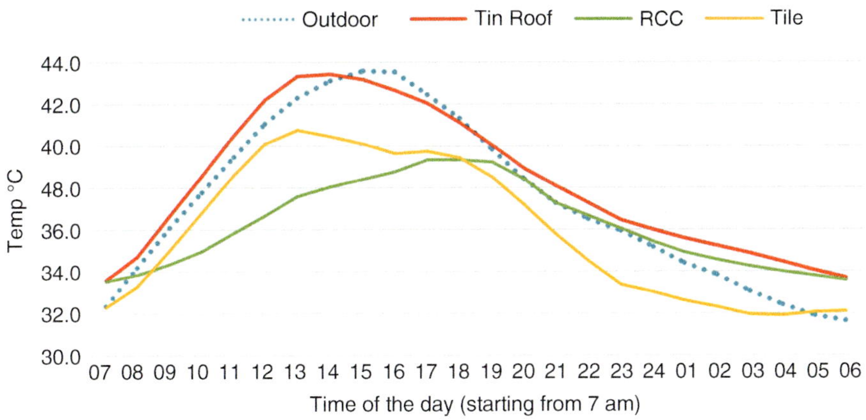

Graph 3 Average indoor and outdoor temperatures during the whole day (period of 10 May–5-June 2016)

Monitoring of Indoor Temperatures

Graph 3 depicts the indoor temperatures of the houses in Sonurli village recorded using data loggers for the period of 10th of May to 5th of June 2017. On average, as the day progressed, the indoor temperature increased and peaked at around 2 pm. Among the three different roofing types, tin roofs get heated the most (having the highest diurnal maximum and minimum temperatures). During the peak heat hours (11 am–5 pm), the indoor temperature in tin-roofed houses was also more than the outdoor temperatures. Elderly, children, and women who rest indoors during morning and afternoon hours may be at risk for heat stress. Local elderly people felt indoors were "intolerable" after 12 noon. Some of them reportedly rest just outside their houses under the shade of a tree or a portico (Mhaskar et al. 2016).

During the day the indoor temperatures recorded in cement-roofed houses were consistently less than the outdoor temperatures, but during the night the temperatures in cement houses were more than the outside temperatures. These houses heated up relatively slowly, but they also cool down slowly. These characteristics of houses and their respective indoor temperatures have important implications for health vulnerability, especially with cooler nighttime temperatures being important for recovering from daytime heat exposure.

Coping Strategies and Adaptation Measures

Individuals, families, and the community took various steps to cope with heat and its impacts. These include changing work timings, resting under trees, resting between working hours, planned reduction in work, using fans indoors, staying hydrated, eating appropriate foods, and using protective clothing. These address the immediate problem and are mostly employed during hot days or following them (Mhaskar et al. 2016).

The different strategies used to cool the houses were as follows: use ceiling fans, use water coolers, spread layer of crop residue over the roof, or a combination

Table 8 Indoor and outdoor temperature difference when coping strategies employed by households

	Temperature difference (indoor to outdoor) (°C) when different coping strategies are employed						
Time of the day	Tin roof +fan	Tin roof+fan +layer of crop residue	Tin roof +cooler	Tin roof +cooler+crop residue layer	Cement slab +fan	Cement slab +Cooler	Tile +cooler
11 am	2.8	−3.3	−3.4	−0.5	−5.4	−4.7	−0.9
12 pm	3.5	−4.2	−4.3	−1.1	−6.8	−5.6	−1.0
1 pm	4.0	−4.8	−5.1	−2.2	−7.4	−6.1	−1.6
2 pm	3.9	−4.7	−5.8	−4.1	−7.3	−6.9	−2.7
3 pm	3.4	−4.4	−5.9	−5.2	−6.5	−7.3	−3.5
4 pm	2.9	−3.7	−5.4	−6.1	−5.4	−7.1	−3.9
5 pm	2.4	−2.5	−4.0	−4.7	−3.4	−5.2	−2.7

(Time period – 10 May to 6 June)

of these (Mhaskar et al. 2016). Table 8 shows the difference between the indoor and outdoor temperature when different coping strategies were employed by the households. It indicates that irrespective of the coping strategy employed, houses with tin roof get more heated as compared to houses with cement roof. Even after employing some kind of coping strategy, the temperature inside tin-roofed households was higher by over 4 °C as compared to outdoors. In those houses where water cooler was used, the indoor temperatures were found to be somewhat lower as compared to outside (in cases of both tin- and cement-roofed houses, respectively). Thus using water coolers may provide some relief. However, many times erratic electricity supply and inadequate water availability prevent the households from using these coolers. In addition, coolers were found in less than 4% of study households. There is no clear indication about what might be the best option for roofing and other housing-related aspects. The choice of the roof material also depends on the economic condition of the household. It might be a combination of things including using hay or crop residue over the roof and improving indoor ventilation. Respondents from poorer families reported that tin roofs were useful during the rainy season as compared to thatched roof (Mhaskar et al. 2016) indicating the need for holistic understanding before making knee-jerk recommendations on addressing the challenge of heat stress.

Finding time to rest was found to be a challenge for women having responsibilities at home and in the fields. Those who work on their own land stated that they rest for an hour or more at home during the afternoons. People prefer to sit outdoors under the shade of trees. It was also observed that some house, especially ones which did not have trees around, have erected a thatched roof propped up by sticks outside the front of the house. Traditional cots were also kept outside the houses in the shade during summer months. Some shared that they rest for a day or two when they are affected by HRS.

About 25.6% households reported no specific measures being used. Almost all families reported increased consumption of water for drinking during summer.

Photo 3 Elderly woman resting outside her house during a summer afternoon. (Source: WOTR)

Almost 60% of families change their cooking schedules, with most of them (49.8%) cooking earlier in the day during summer months. In majority of households, male and female members slept outdoors and changed their sleeping pattern (waking up earlier in summers) (Photo 3).

Respondents reported drinking more water. Other simple strategies such as seeking shade, staying indoors, wearing protective clothing, changing work schedule, and avoiding outdoor activity received less affirmative responses (performed only sometimes), which indicates limited scope to adopt such measures (Graph 4). Almost 89% and 79% respondents reported that the water bodies and tree cover in their respective villages were poor or very poor.

A large proportion of households (44%) approached private doctors for HRS. The average distance between villages and their preferred healthcare facilities was 9.5 km. Most people reported that the doctor they consulted was qualified (94.6%) (though this did not adequately corroborate with our earlier qualitative enquiries with local health practitioners). While individuals provided various reasons for inconvenience of visiting healthcare providers, poor transport facilities (reported by 65.7% households) was the most important, followed by distance to health center (41.4%) and cost of utilizing healthcare (31.4%).

About 53% of the study households reported having received information on heat stress from at least one reliable source like TV, radio, newspaper, or a medical professional. Majority of respondents (72% of sample households) felt it was important to have access to weather information. About 23% households received information only through word of mouth (Table 9). Other sources included information from friends, TV, family members, and doctors.

In recent years, heat action plans have been prepared at state level (e.g., Uttar Pradesh, Andhra Pradesh, Orissa, and Telengana), city level (e.g., Ahmedabad, Nagpur),

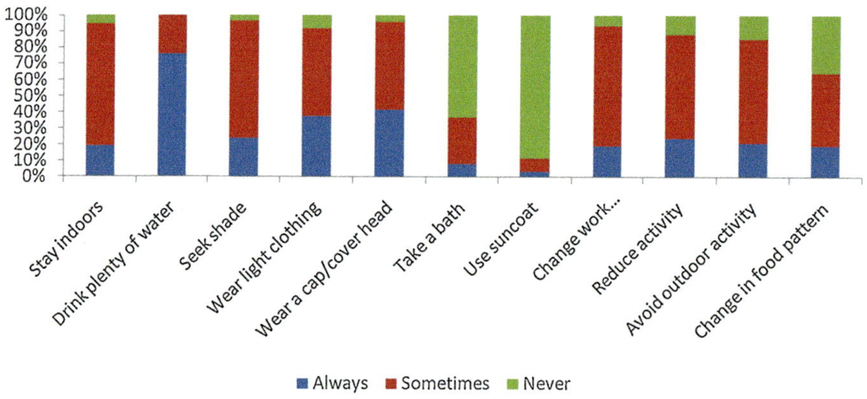

Graph 4 Coping strategies employed by respondents during very hot weather (In %, $N = 215$)

Table 9 Reported sources for information on heat stress

Type of source	Number	Percentage ($N = 215$)
At least one reliable source of information (TV, radio, newspaper, medical professional)	114	53.0
Hearsay	49	22.8
Didn't hear any information	52	24.2

and district level (e.g., Hazaribaug in Jharkhand state). Broadly, these action plans aim to build public awareness and community outreach, develop early warning system and institutional mechanism for inter-agency cooperation, build capacity for healthcare professionals, and promote adaptive measures. Their effectiveness on the ground needs to be assessed.

Conclusion

Heat-related symptoms were highly prevalent in the area. The common HRS were found to be headache, heavy sweating, and fatigue, which are typically of mild or moderate nature. It could be useful to use HRS as indicators of heat stress in an area to intervene before they become severe. However, it is important to note that these symptoms are not specific to heat stress, and so there is scope for underreporting and over-reporting. Though heat was considered a problem locally, its priority was relatively low for local people (Mhaskar et al. 2016).

Findings indicate that age, gender, wealth, and pre-existing health conditions were significantly associated with occurrence of HRS. Working men and women (31–59-year category) had the highest proportion of affected individuals compared to all other age groups. Though a greater proportion of men were affected as compared to women, the dimensions of vulnerability and exposure need further study to understand the mechanisms through which more men were found to be affected. Local people, in our qualitative study, attributed the perceived poorer health resilience in the current generation to change in diets over the years (Mhaskar et al. 2016).

The identified exposure factors of working outdoors during midday, roofing material of the households, and indoor ventilation were significantly associated with occurrence of HRS. A smaller proportion of women reported experiencing HRS as compared to men. The type of livelihoods and housing structures influenced exposure to heat stress. It was found that individuals performing physically intensive tasks were more vulnerable to heat stress and so were individuals residing in tin-roofed houses. Indoor temperature in tin-roofed houses was found to be generally high as compared to other roofing types – cement or tiled. During the peak heat hours, temperatures inside tin-roofed houses were more than the outdoor temperatures. This has implications for elderly and young children who may rest inside these houses during those times. The use of coolers reduced indoor temperatures, but the availability of electricity and water as well as the funds to purchase coolers is a challenge in rural areas. A very small proportion of houses (<4%) had these facilities. In addition, relatively high nighttime temperatures inside tin- and cement-roofed houses are of health concern for persons resting indoors.

Further in-depth studies are needed to monitor the indoor temperature of various housing structures (considering not only roofing but also ventilation, location of windows, type of walls, and flooring) and the effect on individuals. This would provide insights for improving the housing designs that can better handle peak heat conditions. There is a need for policy studies as well, as during the time of the study subsidies were being provided for purchase of tin roofs for houses. There is a need for health assessment before the launching of health-sensitive policies.

Existing coping strategies were found to be inadequate to protect people from indoor and outdoor heat-related stresses. Changing the work timing and reducing work on hot days was a challenge for many laborers working in other people's farms. Those from poor families and woman-headed households also reported having to work without rest and even during periods of illness. Sleeping outdoors at night was also relatively common, despite other reported health risks such as mosquitoes (Mhaskar et al. 2016). A major challenge in rural areas in the context of heat stress is accessibility to well-equipped health centers. Though accessing healthcare was found to be a relatively common coping strategy, poorer families found healthcare to be expensive and avoided accessing it (Mhaskar et al. 2016). A more long-term strategy for competence on HRS, infrastructure, and accessibility to timely medical facilities is needed. For this, investments in upgrading of rural health infrastructure to handle heat stress-related incidences should be considered in all areas with high temperatures. In addition, there is a role for improved awareness on heat stress and associated precautionary measures.

In the future, population exposed to heat waves is projected to increase. Hence, there is a need for pre-emptive strategies to ensure that people in areas where heat waves are not yet a phenomenon are adequately supported to reduce their vulnerability. Improving health systems will benefit not just in the context of heat-related illnesses but for all illnesses, and so it is a no-regret intervention. Heat stress symptoms are easily recognized and can be used for early identification and prevention of more serious impacts such as heat stroke. Such measures will help protect individuals as well as the livelihoods. Effective planning through development of surveillance mechanism to monitor heat-related mortalities and morbidity could help in mitigating and avoiding heat-related stresses and deaths in the future. Housing designs should be improved to facilitate adequate ventilation and reduce the adverse impacts of tin roofs. The government can play an effective role here through existing housing schemes. Our study aims at understanding vulnerability within few villages in a semiarid area. It may be useful to compare our findings to villages in other agroclimatic regions to get a better understanding of relative vulnerability. However, different areas have different temperature-health response relationships, which make it difficult to compare.

At present there are heat action plans for some states (Government of Andhra Pradesh 2016; Odisha State Disaster Management Authority 2018) and for few cities (Ahmedabad Municipal Corporation et al. 2017; District Disaster Management Authority, Hazaribagh 2016). Maharashtra does not have a state-level heat action plan. Therefore, priority should be given to develop a comprehensive state-level heat action plan for Maharashtra, which addresses the needs and contexts of both urban and rural communities to prepare and prevent heat-related illness and deaths.

Cross-References

▶ Agro-met Services and Farmer Responsiveness to Advisories: Implications for Climate Smart Agriculture
▶ Assessment of the Degree of Households' Vulnerability to Climate Variability in Nigeria
▶ Building Resilience to Climate Change: Water Stewardship in Rainfed Agrarian Villages in Maharashtra, India
▶ Climate Change, Vulnerability, and Adaption Under the Small Farming Households of Konso Community, Southern Ethiopia
▶ Climate Resilience in African Coastal Areas: Scaling Up Institutional Capabilities in the Niger Delta Region
▶ Discussion with Farmers in Southwestern Nigeria on Climate Change and Agriculture
▶ Impacts of Climate Change on Health: Evidences from Multi-stakeholders in the Western Region of Cameroon

References

Ahmedabad Municipal Corporation, Natural Resources Defense Council, Indian Metereological Department, Indian Institute of Public Health, Gandhinagar, Public Health Foundation of India, Mount Sinai School of Medicine, & Climate and Development Knowledge Network (2017) Ahmedabad heat action plan 2017 - Guide to extreme heat planning in Ahmedabad, India.

India: NDRC. Retrieved from https://www.nrdc.org/sites/default/files/ahmedabad-heat-action-plan-2017.pdf

Azhar GS, Mavalankar D, Nori-Sarma A, Rajiva A, Dutta P, Jaiswal A, Sheffield P, Knowlton K, Hess JJ (2014) Heat-related mortality in India: excess all-cause mortality associated with the 2010 Ahmedabad heat wave. PLoS One 9. https://doi.org/10.1371/journal.pone.0091831

Centers for Disease Control and Prevention (2011) Extreme heat and your health: warning signs and symptoms of heat illness [WWW document]. CDC Center for Disease Control Prevention. Available online at https://www.cdc.gov/disasters/extremeheat/warning.html

Dash SK, Kjellstrom T (2011) Workplace heat stress in the context of rising temperature in India. Curr Sci 101:496–503

Dholakia HH, Mishra V, Garg A (2015) Predicted increases in heat related morbidity under climate change in Urban India (No. W.P. No. 2015-05-02). Indian Institute of Management, Ahmedabad, India

Din A, Brotas L (2017) Assessment of climate change on UK dwelling indoor comfort. Energy Procedia 122:21–26. https://doi.org/10.1016/j.egypro.2017.07.296. CISBAT 2017 International Conference Future Buildings & Districts – Energy Efficiency from Nano to Urban Scale

District Disaster Management Authority, Hazaribagh (2016) Heat Wave Action Plan-2016 Hazaribagh. Hazaribagh, Jharkhand: District Disaster Management Authority. Retrieved from https://cdn.s3waas.gov.in/s3ed265bc903a5a097f61d3ec064d96d2e/uploads/2018/05/2018052643.pdf

Gasparrini A, Guo Y, Hashizume M (2015) Mortality risk attributable to high and low ambient temperature: a multicountry observational study. Lancet 14:464–465. https://doi.org/10.1016/S0140-6736(14)62114-0

Government of India (2018) Demography | Jalna [Internet]. Jalna. Available online at https://jalna.gov.in/about-district/demography/. Accessed 01 May 2018

Government of Andhra Pradesh (2016) Heat wave action plan of Andhra Pradesh. Hyderabad, India: Government of Andhra Pradesh. Retrieved from www.imdhyderabad.gov.in/apsite/andhra.pdf.

Harlan SL, Brazel AJ, Prashad L, Stefanov WL, Larsen L (2006) Neighborhood microclimates and vulnerability to heat stress. Soc Sci Med 63:2847–2863. https://doi.org/10.1016/j.socscimed.2006.07.030

Ingole V, Rocklöv J, Juvekar S, Schumann B (2015) Impact of heat and cold on total and cause-specific mortality in Vadu HDSS – a rural setting in Western India. Int J Environ Res Public Health 12:15298–15308. https://doi.org/10.3390/ijerph121214980

Intergovernmental Panel on Climate Change (2014) Human health: impacts, adaptation, and co-benefits. In: Climate change 2014: impacts, adaptation, and vulnerability. Part A: global and sectoral aspects. Contribution of Working Group II to the Fifth Assessment Report of the Intergovernmental Panel on Climate Change. https://doi.org/10.1186/1471-2334-9-160

Jackson LL, Rosenberg HR (2010) Preventing heat-related illness among agricultural workers. J Agromedicine 15:200–215. https://doi.org/10.1080/1059924X.2010.487021

Khan F, Malik S, Rehman A (2014) Sheltering from a gathering storm: temperature resilience in Pakistan. ISET-International, Boulder

Kjellstrom T, Holmer I, Lemke B (2009) Workplace heat stress, health and productivity-an increasing challenge for low and middle-income countries during climate change. Glob Health Action 2:1–6. https://doi.org/10.3402/gha.v2i0.2047

Li M, Gu S, Bi P, Yang J, Liu Q (2015) Heat waves and morbidity: current knowledge and further direction-a comprehensive literature review. Int J Environ Res Public Health 12:5256–5283. https://doi.org/10.3390/ijerph120505256

Lindemann U, Stotz A, Beyer N, Oksa J, Skelton D, Becker C, Rapp K, Klenk J (2017) Effect of indoor temperature on physical performance in older adults during days with normal temperature and heat waves. Int J Environ Res Public Health 14:186. https://doi.org/10.3390/ijerph14020186

Lundgren K, Kuklane K, Gao C, Holmér I (2013) Effects of heat stress on working populations when facing climate change. Ind Health 51:3–15. https://doi.org/10.2486/indhealth.2012-0089

McGeehin M, Mirabelli M (2001) The potential impacts of climate variability and change on temperature-related morbidity and mortality in the United States. Environ Health Perspect 109(Suppl):185–189. https://doi.org/10.2307/3435008

Mhaskar B, Pradyumna A, Shinde Y, Kadam Y (2016) Heat stress and human health: vulnerability of rural communities in dry semi arid areas of India. Presented at the adaptation futures 2016, Rotterdam

Mueller V, Gray C, Kosec K (2014) Heat stress increases long-term human migration in rural Pakistan. Nat Clim Chang 4:182–185. https://doi.org/10.1038/nclimate2103

Murari KK, Ghosh S, Patwardhan A, Daly E, Salvi K (2015) Intensification of future severe heat waves in India and their effect on heat stress and mortality. Reg Environ Chang 15:569–579. https://doi.org/10.1007/s10113-014-0660-6

Nag PK, Nag A, Sekhar P, Pandit S (2009) Vulnerability to heat stress: scenario in Western India. National Institute of Occupational Health, Ahmedabad

National Disaster Management Authority (2016) Guidelines for preparation of action plan – prevention and management of heat-wave. Government of India, New Delhi. Available at: https://ndma.gov.in/images/guidelines/guidelines-heat-wave.pdf ; Accessed on 29.07.2018

Odisha State Disaster Management Authority (OSDMA) (2018) Heat Action Plan for Odisha. Odisha, India: OSDMA. Retrieved from www.osdma.org/Download/heat-wave-action-plan.pdf

Oudin Åström D, Bertil F, Joacim R (2011) Heat wave impact on morbidity and mortality in the elderly population: a review of recent studies. Maturitas 69:99–105. https://doi.org/10.1016/j.maturitas.2011.03.008

Paul S, Bhatia V (2016) Heat stroke – emerging as one of the biggest natural calamity in India. Int J Med Res Prof 2:15–20

Ponni M, Baskar B (2015) A study on indoor temperature and comfort temperature. Int J Eng Sci Invent 4:7–14

Quinn A, Tamerius JD, Perzanowski M, Jacobson JS, Goldstein I, Acosta L, Shaman J (2014) Predicting indoor heat exposure risk during extreme heat events. Sci Total Environ 490:686–693. https://doi.org/10.1016/j.scitotenv.2014.05.039

Sahu S, Sett M, Kjellstrom T (2013) Heat exposure, cardiovascular stress and work productivity in rice harvesters in India: implications for a climate change future. Ind Health 51:424–431. https://doi.org/10.2486/indhealth.2013-0006

Tawatsupa B, Lim L-Y, Kjellstrom T, Seubsman S, Sleigh A (2010) The association between overall health, psychological distress, and occupational heat stress among a large national cohort of 40,913 Thai workers. Glob Health Action 3:5034. https://doi.org/10.3402/gha.v3i0.5034

Tran KV, Azhar GS, Nair R, Knowlton K, Jaiswal A, Sheffield P, Mavalankar D, Hess J (2013) A cross-sectional, randomized cluster sample survey of household vulnerability to extreme heat among slum dwellers in Ahmedabad, India. Int J Environ Res Public Health 10:2515–2543. https://doi.org/10.3390/ijerph10062515

Venugopal V, Chinnadurai JS, Lucas RAI, Kjellstrom T (2015) Occupational heat stress profiles in selected workplaces in India. Int J Environ Res Public Health 13:1–13. https://doi.org/10.3390/ijerph13010089

Resettlement and Relocation Options for Coastal Communities

11

Peni Hausia Havea, Sarah L. Hemstock, and
Helene Jacot Des Combes

Contents

Abstract

With a nation highly vulnerable, with a risk index of 28.57% and ranked second worldwide by the World Risk Report to be affected by disasters caused by natural hazards and to the effects of climate change, many people are ramping up plans to address these projected events. As a result, it is a necessity for Tonga to strengthen its response to climate change, continue to be more resilient, and, most importantly, address migration and human rights in the context of climate change (e.g., climigration). Here, a concurrent convergence parallel triangulation design

P. H. Havea (✉) · H. J. D. Combes
PaCE-SD, Lower Campus, Laucala, USP, Suva, Fiji
e-mail: ilaisiaimoana@yahoo.com; helene.descombe@usp.ac.fj

S. L. Hemstock
Bishop Grosseteste University, Lincoln, UK
e-mail: sarah.hemstock@bishopg.ac.uk

W. Leal Filho (ed.), *Handbook of Climate Change Resilience*,
https://doi.org/10.1007/978-3-319-93336-8_50

of 460 residents from 5 coastal communities, Kanokupolu, ‘Ahau, Tukutonga, Popua, and Manuka in Tongatapu, Tonga, was used – to provide a better understanding of whether people in these communities needed a resettlement and relocation options or not. The quantitative analysis revealed that based on this sample, people in this chapter would choose Australia and New Zealand as the two most preferred countries for climigration. The same factors were explored qualitatively and found Australia to be the preferment country for climigration. Based on this result, it is recommended that a relocation and/or climigration policy should be adopted for Tonga as part of its resilient Tonga by 2035 and beyond.

Keywords
Climigration · Climate change · Relocation · Resettlement · Tonga

Introduction

Global climate change is known as one of the major risk factors for migration. Importantly, according to the International Organization for Migration (IOM) (Laczko and Aghazarm 2009), it is estimated that about 200 million to 1 billion people worldwide will be forced to move or to migrate by 2050 if nothing can be done to minimize climigration (migration due to climate change) (Laczko and Aghazarm 2014). So, while climigration is discussed globally, it is essential, however, not to view it as a failure of communities to adapt to climate change. Rather, it is one of the ways in which people have chosen to adapt (Kelpsaite and Mach 2015; Shearer 2012) because although some people may choose to live and die in their own communities or countries because their spiritual connection to their lands is strong (Havea et al. 2017, 2018; McNamara and Gibson 2009; Mortreux and Barnett 2009), studies have shown that climate adaptation in the form of relocation and/or resettlement seen as opportunity to give them a more sustainable and resilient life now and in the future (De Sherbinin et al. 2011; Farbotko and Lazrus 2012; Gibson et al. 2013; Kelpsaite and Mach 2015; Scheffran et al. 2012; Taufatofua 2011; Yamamoto and Esteban 2017).

In the Pacific, Fiji, Papua New Guinea (Fig. 1a) (Connell 2016; Luetz and Havea 2018), Nauru, Kiribati (Wyett 2014), and Tuvalu (Fig. 1b) already experienced migration within the region, or to New Zealand and Australia, although it is mostly due to other factors (e.g., economic situation, access to better education, and health) (Fig. 1) (Campbell et al. 2016; Milan et al. 2016; Oakes et al. 2016). As a result, it is a necessity that a climigration policy and a climate strategic framework for migration should be developed for the people in the Pacific in order to ensure that building of safety and resilience Pacific Islanders by 2030 and beyond is achievable (Ash and Campbell 2016; Pacific Community et al. 2016). This can be done by developing a national adaptation plan that would guide countries to relocate further inland

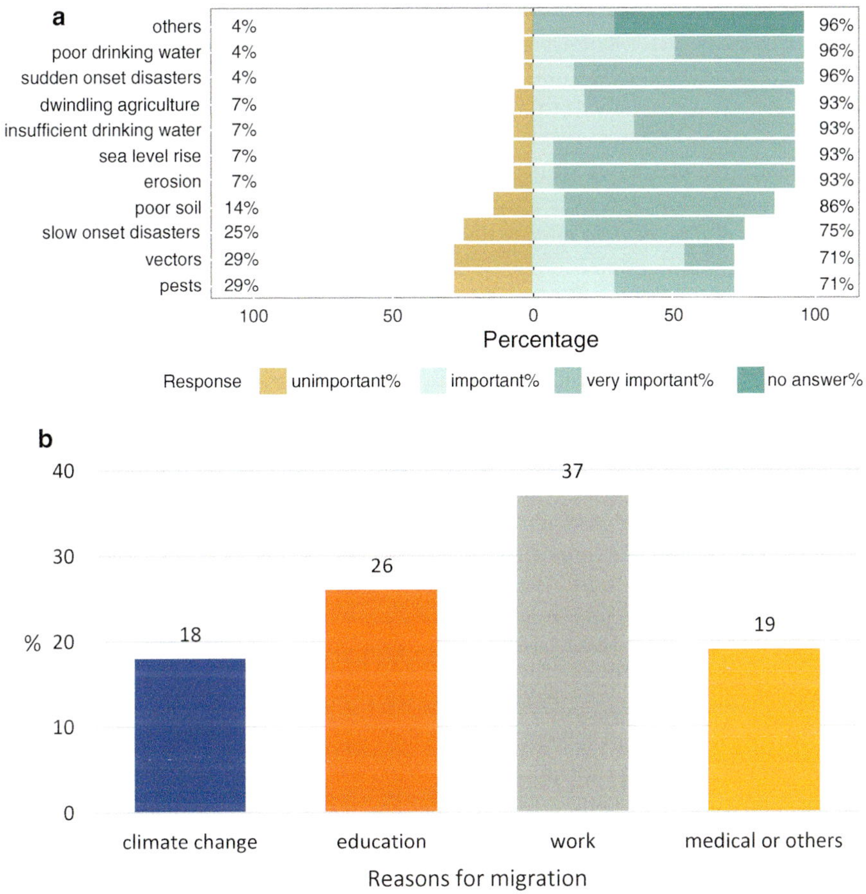

Fig. 1 Climate change a risk factor for forced migration in the Pacific. (**a**) Environment-related factors for forced migration in Papua New Guinea. (Source: Luetz and Havea (2018)). (**b**) Reasons for migration for Nauru, Kiribati, and Tuvalu to Fiji, New Zealand, and Australia. (Source: Milan et al. (2016), Campbell et al. (2016), and Oakes et al. (2016))

(internal migration) and/or to resettle in other countries (external migration) as part of their climate adaptation regime for sustainable growth and development (Pacific Community et al. 2016).

In Tonga, migration due to climate change has been viewed as a negative form of coping mechanism and adaptation strategy like in the other Pacific islands by the neighboring country, Australia (Ash and Campbell 2016) and theoretically by New Zealand (Dastgheib 2015). If this is to be continued in the future, then climigration as a long-term adaptation solution may be difficult for climate refugee whose last resort in response to an unbearable circumstance is to relocate or resettle (De Sherbinin et al. 2011; Tacoli 2009). With a nation highly vulnerable to disasters caused by naturals hazards and to the effects of climate change, there are plans to propose climigration and platform for disaster displacement framework (Berchin et al. 2017; Herrmann 2017; Kolmannskog 2012; Lübken 2012; Mayer 2016; Nunn 2013, 2016; Pacific Community (SPC) 2017; Platform on Disaster Displacement 2016; United Nation University (Institute for Environment and Human Security) 2016; Williams 2008; Yamamoto and Esteban 2017) for the Government of Tonga.

One of these plans is to request for Tonga to address gaps in its national climate change and hazard acts and specifically draft a policy where people who are displaced in the context of climate change and disasters have a right to relocate or migrate. These policies need to be written and agreed upon in a form which can be regionally and internationally negotiated (Barnett 2017; Birk and Rasmussen 2014; Campbell et al. 2014; Connell 2016; Donner 2015; Gemenne and Blocher 2017; Kelman 2015; Ketola 2015; Kutzbach 2014; Locke 2009; Nash 2018; Radio New Zealand 2018; Shearer 2012; Wyett 2014; Yamamoto and Esteban 2017). With climate change an enabling factor for promoting migration to New Zealand and/or Australia, public perception on relocation and resettlement or climigration for the affected communities in Tonga is significant. Reflecting this, the chapter aims to provide a better understanding of whether people in five coastal communities – Kanokupolu, ‘Ahau, Tukutonga, Popua, and Manuka – needed resettlement and relocation options or not.

Method

Methodology

A mixed-method approach named concurrent convergence parallel triangulation design (Creswell and Plano Clark 2011) was used on 460 Tongans aged 15 to 75 from 5 coastal communities in Tongatapu, Tonga: Kanokupolu, ‘Ahau, Tukutonga, Popua, and Manuka. It was concurrent because both the quantitative and the qualitative studies were conducted in the same time frame. Since they were merged during the design, the analysis, and reporting, the chapter followed convergence design. Parallel refers to an equally weighted study without any emphasis to either the quantitative or the qualitative aspect of the studies. Then triangulation indicates that a multi-methodological approach was used (Creswell and Plano Clark 2007).

Research assistants (RA) were recruited from these local communities, to assist in the data collection process, and the incentive in the form of a gift of $1 Tongan pa‘anga (TOP) per survey paper was paid to the RA from the project to show appreciation of their time and effort, since this chapter was conducted in Tongans. Information on climigration was collected via self-administered questionnaire ($n = 460$), individual in-depth interview ($n = 24$), focus group discussion ($n = 28$), and informant interview ($n = 12$). Participants were selected using a random sampling framework named census block householding using the Statistic Department of the Government of Tonga instruction (Tonga Department of Statistic 2011, 2013a, b, 2016). Then a simple randomization technique was used to recruit participants for the interviews and focus group discussion (‘Ata’ata 2012;

Clark and Ivankova 2016; Creswell and Plano Clark 2011; Nastasi et al. 2007; Peck and Devore 2012; Sullivan 2013).

Data Analytical Strategy

The data analysis used a convergence model (Creswell 2013; Creswell and Plano Clark 2011). For the quantitative study, the data was analyzed using frequency analysis, cross-tabulation, and correlation. The results of this analysis were then explored qualitatively using thematic analytical strategy by taking one piece of information to compare iteratively with the statements of the participants. In doing so, it has allowed the chapter to provide a better understanding about why and where the participants chose to relocate and/or resettle to as part of its Tonga's climate resilient policy by 2035 and beyond. The data analysis was conducted using SPSS and R.

Study Limitations

The main limitation of this chapter is that climigration was a small component of the assessment since the main project (Ph.D. Thesis) was mainly focused on assessing the perceived impacts of climate change on Tongans aged 15–75 livelihoods, health, and well-being in these five coastal communities. In the view of this limitation, it is recommended that another research study investigate further the results on climigration from this chapter.

Results

Study Population

This chapter population is a sample of 460 women and men aged 15–75 from Kanokupolu, ʻAhau, Tukutonga, Popua, and Manuka in Tongatapu, Tonga. Of all the 460 participants, 53% were female and 47% were male. The distribution of age and gender across the pyramid is almost comparable with the distribution of the total population pyramid in Tonga (Tonga Department of Statistic 2013a, b) indicating this sample would provide a good representation of how displacement in the context of climate change and disasters caused by natural hazard would affect the general population. Based on this sample, the average age was 37.25 with a mode of 27 years old (Fig. 2).

Climigration and Well-Being

This theme connotes migration due to climate change and improving people's overall well-being and happiness (Helliwell et al. 2018) about moving and relocate

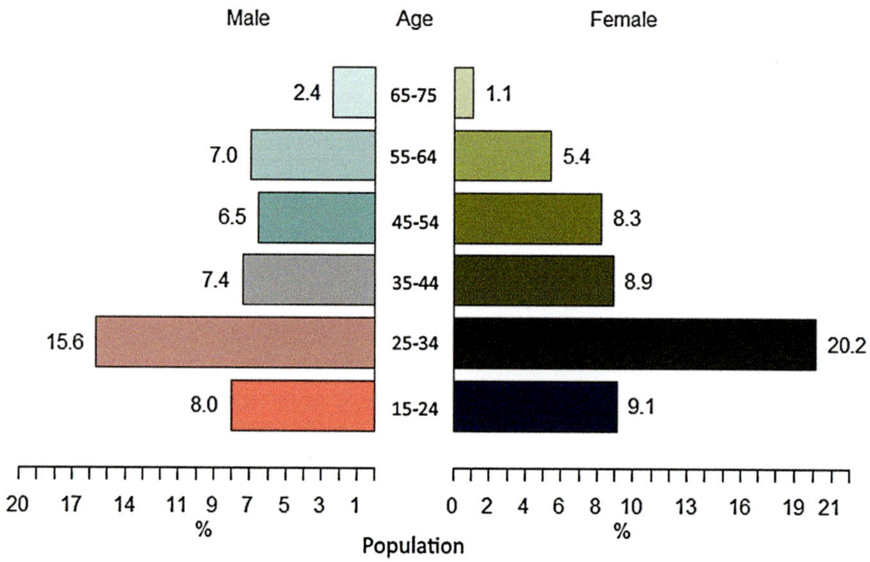

Fig. 2 Study population pyramid for the five studied coastal communities in Tongatapu, Tonga. (Source: Authors)

or resettle in another country. Based on this chapter, 78.5% (361) of the participants perceived that migration may lead to improving their overall well-being and the well-being of their family household in their communities. Most importantly, the chapter revealed that those who opted for this climate change-induced migration are young people (Table 1).

As complemented by one of the female participants from Kanokupolu, when she was asked to reflect on the question if she had to choose to relocate or to resettle overseas, in which country she would prefer to do so and why? She replied:

> I would prefer to relocate our families in Australia. The reasons are because one of my children is there, I have been travelling on and off several times, I am familiar with the culture there and to me as an individual, living in Tonga is the same as living in Australia. The weather in the part of Australia that I've been travelling to is almost the same as what we have here in Tonga. They also have a lot of Christians and I am happy to migrate there instead of any other country.

Climigration and Options for Coastal Communities to Relocate or Resettle in Australia, New Zealand, and Other Countries in the Region and Around the World

This theme denotes options for the participants in this chapter to choose any country in the world, for them to relocate and/or to resettle in response to the effects of

Table 1 Climigration associated with improving people's well-being

		Age						
		15–24	25–34	35–44	45–54	55–64	65–75	Total
By self-resilience via mitigation, adaptability, evacuation, migration, and relocation/resettlement would improve my overall well-being and the well-being of my family household	Very unimportant	2	5	2	3	4	0	16
	Unimportant	3	7	5	7	2	0	24
	Neutral	15	17	11	10	4	2	59
	Important	19	35	17	18	20	4	113
	Very important	40	101	40	30	27	10	248
Total		79	165	75	68	57	16	460

Source: Authors

climate change. So, as people in this chapter have indicated that migration is significant to their overall well-being, of the 460 participants, 46.9% (216) have shown that Australia is the best country to meet their expectations for migration (Table 2). As stated explicitly by a male participant from Tukutonga, when he was asked to which country he would like to relocate or resettle to and why? He said: "I would prefer to relocate to Australia. I prefer Australia because I have the opportunity to travel and work there. I have been in New Zealand and when I compared them, I would rather stay in Australia."

Another male participant from Tukutonga mentioned the same affinity to Australia and said: "There is a country that I have envisioned of relocating to, but of course our leaders always have the final sayings, regardless whether we agree to it or not, but rather we will be relocating. I think Australia would be the best to relocate to on the basis that the weather is quite the same with us so that we can do agriculture to support our livelihoods." Then lastly, a young man from 'Ahau put it in such a diplomatic manner. He says: "To me, this is important, but there is no other country that is safer than the one Jesus has been recommended, which is heaven. But on earth, I would recommend Australia. So, that we improve working together as stakeholders and our partnership better than before. But God has the final says."

However, although Australia seems to be the first destination for many people in this chapter, New Zealand is another country that was chosen by 32.6% (150) of the participants who thought it is also an excellent place to live. As stated by a male participant from Manuka: "There is an old saying that goes like this: worshipping God is wisdom or making sense of the reality. This means if there is something happens (e.g. meaning if Tonga will be destroyed by earthquake, tsunami, cyclone or submerged by rising sea level), and we need to relocate, we must go ahead and

Table 2 Countries where people chose for migration due to climate change and/or hazards

		Age						
		15–24	25–34	35–44	45–54	55–64	65–75	Total
If a massive relocation or resettlement is to be recommended for Tonga in the future, which country would you prefer for relocation/ resettlement?	Australia	31	85	34	31	25	10	216
	New Zealand	30	50	21	20	24	5	150
	America	7	11	11	6	4	0	39
	Fiji	2	1	0	2	1	0	6
	Britain	1	5	1	1	1	0	9
	Tonga (e.g., Vava'u, 'Eua, Tongatapu)	5	10	4	5	1	1	26
	Hawaii	0	1	0	0	1	0	2
	England	0	1	0	0	0	0	1
	Italy	0	0	0	1	0	0	1
	Africa	1	0	0	0	0	0	1
	Japan	1	0	2	0	0	0	3
	India	0	0	1	1	0	0	2
	China	0	0	0	1	0	0	1
	Trinidad and Tobago	1	0	0	0	0	0	1
	Solomon Islands	0	0	1	0	0	0	1
	Canada	0	1	0	0	0	0	1
Total		79	165	75	68	57	16	460

Source: Authors
Note: These countries were freely chosen by the participants in this chapter. It was not a pre-established list of countries to choose from in this chapter

relocate. Like, if I was a kid, and we were told to relocate, but I didn't do so because we rather die here, I suppose that God doesn't want that to happen, as it is more like of a suicide attempt. God wanted you to relocate, so to carry out his work somewhere else. So, if anything like that happens to us here in Tonga, I would relocate to New Zealand. First, the air is clean and fresh over there. Second, I have many relatives there too."

This perception was supported by another female participant from Popua, who stated: "If we were told to relocate overseas, I would prefer to relocate to New Zealand. My reason for this is that the air is clean as well as the environment is neat," and another woman from the same community supported this by saying: "If the King for some reasons wanted us to relocate overseas, I would choose to relocate to New Zealand because it is closer to Tonga."

A woman from Popua has given her testimony for the strong and spiritual connection to her land stated:

> To be honest I have kids overseas who are residing in Hawaii, United States of America, one in New Zealand as well as my brothers and relatives and my sisters in Australia. But you know, there is no place like my home (meaning Tonga). Even nowadays, my kids from

> overseas are trying their best to convince me to come for holidays, but I told them, no way my garden will die out, as well as my animals. I would rather just be happy to see myself at home and stay here. Although I don't have everything, I like it here so much. But if for some reasons, we are forced to relocate overseas for whatever reasons that would be, I would follow where ever my kids would want me to stay. I'm talking about my kids at home (meaning in Tonga in Popua) right now. If they wanted to migrate, I would just follow them overseas. You know the tsunamis warning we have had just recently, they ran away from our home and relocate temporarily on a high point spot, but for me, I just stayed home. And then, when my older son heard about it, he called me up and say, Mom, if there is any tsunami warning again, please just get on the vehicle and drive with the kids and find a good spot safe to relocate for the time being, and then go back home afterwards.

Interestingly, an old man from Manuka went further into this issue on migration and discussed his goal for the future by saying:

> God created Tonga and this universe, and we shouldn't complain because we can't do anything like that. God created us to stay here in Tonga until we die. Because we have a lot of people who are staying here now in this community (so when the seawater comes), they escape to the bush and stay there because there are lots of reclamation going on now. But is like this, if we are planning to reclaim the land up to the sky, it doesn't really matter because if Jesus wanted to destroy it, no one can stop it. You see, Jesus, through his father (meaning their Christian God) control seawater, air and everything else in this universe. So, my point is to try to be honest with God and worship Him (God) in the right way and work hard.

The remaining 20.5% (94) were divided among the USA, Fiji, the UK, Hawaii, Italy, Africa, Japan, India, China, Trinidad and Tobago, Solomon Islands, and Canada. Interestingly, although many people have chosen to migrate, others are confined and chose to stay in Tonga until they die. Therefore, since this chapter is the first of its kind to investigate the platform for disaster and climate change displacement framework for a Talanoa (national dialogue) in Tonga, the government should address climigration and population mobility as part of their campaign to achieve resilience of Tonga by 2035 and beyond. Most importantly, all participants in this chapter have thought about, and chosen, a place or a country to move and stay. So, if this is the case, then it is a necessity for the Government of Tonga to draft a climigration legal framework or acts and have them reviewed, modified, and do a global consultation before seeking endorsement and approval first by the Government of Australia, New Zealand, and then the other countries.

Discussion

Resettlement and Relocation Options for Coastal Communities

Eight in ten of those whose livelihoods, health, and well-being were affected in these five coastal communities in Tonga were most likely to migrate to New Zealand and/or Australia inter alia due to the negative effects of climate change and disaster caused by natural hazards in order to improve their overall happiness and well-being. Since displacement in the context of climate change and disasters caused by natural hazards and/or climigration in Tonga is unprecedented, an unprecedented partnership – collaboration between international organizations, private sectors, civil society, academia, and governments – is needed, to inform the global dialogue in the United Nations Framework Convention on Climate Change (UNFCCC), Sendai Framework, and the Pacific Platform on Disaster Displacement (PDD) framework, including the Framework for Resilient Development in the Pacific (FRDP). This is significant for the people of Tonga because climate change is expected to continue and is essential to give the Tongans as many options as possible as stated by Lord Fusitu'a, a Tongan Nobel and Member of Parliament of Tonga, to the UN General Assembly in New York in 2016.

He stated:

> The main message was two-fold. Number one; because we bear the major brunt of climate change although we bear practically, no responsibility for the emissions, we have to engage with international bodies and with metropolitan foreign governments to work towards implementation and enforcement ratification as soon as possible of things like the Paris Agreement. So that the causes of climate change and therefore climate change-induced migration are brought down, as well as ensuring that on our end we have as much adaptation as possible so that we are able to institute preventative measures so that we are ready for whatever the effects of climate change may be, within a comprehensive policy, giving our citizens as many options as possible (Radio New Zealand 2018).

The same perception perceived by the participant in this chapter. As illustrated by the above-stated excerpts, this theme emerged because all the participants in this chapter perceived they have a right for climigration and/or innovative capabilities to relocate or resettle since most of their families and relatives have already experienced the benefits of remittances among other positive influences of being relocated, living, and working in the first world countries (Fig. 3). This is why one of the male participants from Tukutonga highlighted the same positive impact of climigration

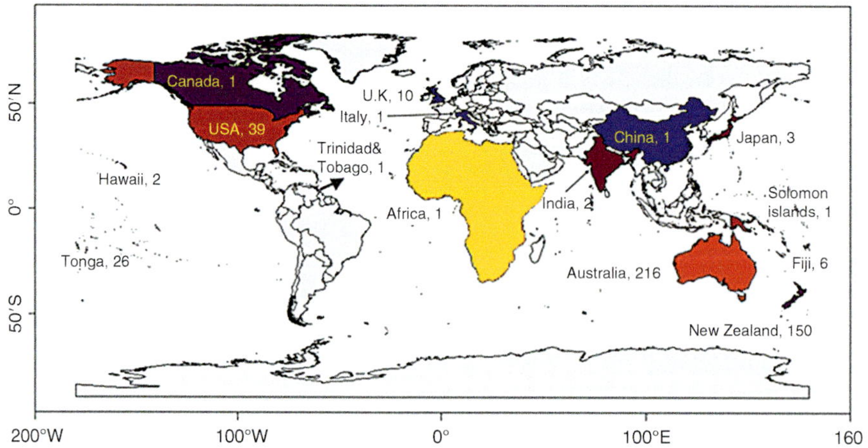

Fig. 3 Continents around the world where people in the five coastal communities indicated that they wanted relocation and resettlement. (Source: Authors)

and stated: "I would prefer to relocate to Australia. I prefer Australia because I had the opportunity to travel and work there. I have been in New Zealand and when I compared them, I would rather prefer to stay in Australia."

Based on these results, it is vital not to view climigration as simply a failure of these five coastal communities to adapt to the negative effects of climate change; rather it should be perceived as one of the ways in which people in these communities choose to adapt to the changing environment caused by climate change and/or disasters caused by natural hazards.

Migration out of the country still remains a last-resort solution to the adaptation challenge. However, although it is not the same as retreat or short-distance migration, but the fact that this chapter shows that Tonga is thinking ahead of these issues, as stated in the Tongan proverb of the "Tā e Lango kei Mama'o" (cutting the slip while at a distance) framework toward climigration, is significant now and in the future. Most importantly, the fact that the people thought about it and seem ready to relocate abroad also comes from the fact that they have family members already living abroad and that in that case, relocating would not mean to go into an unknown territory but would build on existing family ties in the chosen foreign country.

As clearly indicated in this world map, while 94.3% (434) of the participants have chosen to relocate and/or resettle in the region and other countries around the world, 5.7% (26) of the participants choose to stay in Tonga until they die (Fig. 3). As a result, the Government of Tonga should reinforce its climigration policy, as a conceivable way or as one of the several adaptation strategies and a coping strategy of enhancing adaptation to climate change that would be beneficial for both parties for those who choose to migrate, as well as for those, who vie to stay. Those who vie to stay may have the opportunity to relocate and/or resettle to a better and higher land areas in Tonga as well as get significant support from the government.

The reason is that the climate change and migration nexus is very complex for Tonga and because migration is triggered by other factors, especially economic reasons, and that climate change is rarely the only or even the major factor to migrate at the moment, although it may change in the future. And with a low adaptive capacity due to low levels of human, financial, and natural resources and limited institutional and technological capability, to exercise climigration even inland or moving out of the country may look good on papers, but in practice, the government could not afford it, since migration and reestablishment of even a single community is costly (Charan et al. 2017).

Therefore, for climigration to be worked for Tonga and the Pacific, there is a need to develop a framework, where a platform for people who are displaced in the context of climate change and disaster will be addressed appropriately. This can be done by calling on all Pacific Island Countries (PICs) to develop a platform for a human right discussion and a Pacific climigration framework in response to the migratory impacts of climate change and hazards in the region. Learning from the climigration approach that was used in Arctic Alaska in 2016 (Hamilton et al. 2016), the Pacific countries can develop a suitable approach by putting in more preventative measures to make either planned or unplanned migration unavoidable now and in the future, as already done by Kiribati, according to its previous president, Anote Tong, who developed education in Kiribati so the people could move with dignity and become able citizens in the country they migrate to, instead of being refugees (Office of Te Beretitenti 2013).

Conclusion

Nine persons in 10 in Tonga perceived that they would prefer to migrate to 14 countries (Australia, New Zealand, America, Fiji, Britain, Hawaii, Italy, Africa, Japan, India, China, Trinidad and Tobago, Solomon Islands, Canada), as a result of impacts from climate change and hazards. Therefore, based on this sample, there is no doubt that climate change had threatened human security including people's livelihoods, health, and well-being in these coastal communities, thus leading them to consider the options to relocate or remain in Tonga until they die. Among all the countries in the world to relocate to, most of the participants have chosen Australia and New Zealand. There are several reasons that people chose these two countries for climigration. First, it is closer to Tonga. This is significant for them because once the resettlement will be done, they can also come back and visit their families who resided in Tonga. Second is that people perceived that it would be easier for them to get a job and so to help to support them financially. The other reasons are that, based on this chapter, the presence of family member and a similar environment and weather are also important factors to influence their choice.

As a result, based on this chapter, the Government of Tonga has a case to pursue to help those who will be displaced because of climate change and disaster. This could be achieved by developing a legal framework that would allow Tongans to exercise climigration in the future. For those who chose to stay, this could be a rights-based approach that supports the resilience and adaptation strategies for them, thereby reducing the effects of climate change and the propensity to engage in climigration. For those who would migrate, a rights-based approach to averting the migratory outcomes and challenges that such migration will create in another country would also be helpful.

To support this goal, this chapter is the first to provide evidence which decision-makers in Tonga may use to plan for a climigration policy in its climate change act or migration act. In doing so, not only may Tonga contribute to achieving its climate change policy named resilience Tonga by 2035 and beyond, but it will also help people in the affected communities to improve their capacity to live a sustainable and resilient life now and the future. Based on the results of this chapter, two policy proposals are recommended for Tonga:

1. A climigration policy for the people of Tonga who wanted to migrate overseas and/or relocate in-country (e.g., Tongatapu, Vava'u, 'Eua) due to climate change and hazards
2. A climigration policy to ensure safety and guarantee the protection of those who would be willing to relocate from overseas or the region and resettle in Tonga in the future

References

'Ata'ata MF (2012) Kingdom of Tonga national accounts statistics 2010–2011. Statistic Department of the Government of Tonga, Nuku'alofa

Ash J, Campbell J (2016) Climate change and migration: the case of the Pacific Islands and Australia. J Pac Stud 36(1):53–72

Barnett J (2017) The dilemmas of normalising losses from climate change: towards hope for Pacific atoll countries. Asia Pac Viewpoint 58(1):3–13

Berchin II, Valduga IB, Garcia J, de Andrade Guerra JBSO (2017) Climate change and forced migrations: an effort towards recognizing climate refugees. Geoforum 84:147–150

Birk T, Rasmussen K (2014) Migration from atolls as climate change adaptation: current practices, barriers and options in Solomon Islands. Nat Res Forum 38(1):1–13

Campbell J, Bedford R, Bedford R (2014) Migration and climate change in oceania. In: Piguet E, Laczko F (eds) People on the move in a changing climate: the regional impact of environmental change on migration. Springer Netherlands, Dordrecht, pp 177–204

Campbell J, Oakes R, Milan A (2016) Nauru: climate change and migration – relationships between household vulnerability, human mobility and climate change. United Nations University Institute for Environment and Human Security, Bonn

Charan D, Kaur M, Singh P (2017) Customary land and climate change induced relocation – a case study of Vunidogoloa Village, Vanua Levu, Fiji. In: Filho WL (ed) Climate change adaptation in pacific countries: fostering resilience and improving the quality of life. Springer International Publishing, Heidelberg, pp 19–33

Clark VLP, Ivankova NV (2016) Mixed method research: a guide to the field. Sage, London

Connell J (2016) Last days in the Carteret Islands? Climate change, livelihoods and migration on coral atolls. Asia Pac Viewpoint 57(1):3–15

Creswell JW (2013) Qualitative inquiry and research design: choosing among five approaches. Sage, Los Angeles

Creswell JW, Plano Clark VL (2007) Designing and conducting mixed methods research. Sage, Thousand Oaks

Creswell JW, Plano Clark VL (2011) Designing and conducting mixed methods research. Sage, Thousand Oaks

Dastgheib S (2015) Kiribati climate change refugee told he must leave New Zealand [Online]. Stuff, Wellington. https://www.stuff.co.nz/national/72276281/kiribati-climate-change-refugee-told-he-must-leave-new-zealand. Accessed 11 Mar 2018

De Sherbinin A, Castro M, Gemenne F, Cernea MM, Adamo S et al (2011) Preparing for resettlement associated with climate change. Science 334(6055):456–457

Donner SD (2015) The legacy of migration in response to climate stress: learning from the Gilbertese resettlement in the Solomon Islands. Nat Res Forum 39(3–4):191–201

Farbotko C, Lazrus H (2012) The first climate refugees? Contesting global narratives of climate change in Tuvalu. Glob Environ Chang 22(2):382–390

Gemenne F, Blocher J (2017) How can migration serve adaptation to climate change? Challenges to fleshing out a policy ideal. Geogr J 183(4):336–347

Gibson J, McKenzie D, Rohorua H (2013) Development impacts of seasonal and temporary migration: a review of evidence from the Pacific and Southeast Asia. Asia Pac Pol Stud 1(1):18–32

Hamilton LC, Saito K, Loring PA, Lammers RB, Huntington HP (2016) Climigration? Population and climate change in Arctic Alaska. Popul Environ 38(2):115–133

Havea PH, Hemstock SL, Combes HJD (2017) Preparing for better livelihoods, health and well-being – a key to climate change adaptation. In: Leal Filho W (ed) Climate change adaptation in pacific countries: fostering resilience and improving the quality of life. Springer International Publishing, Cham, pp 87–99

Havea PH, Hemstock SL, Jacot de Combes H (2018) "God and Tonga Are My Inheritance!" – climate change impact on perceived spirituality, adaptation and lessons learnt from Kanokupolu,

'Ahau, Tukutonga, Popua and Manuka in Tongatapu, Tonga. In: Leal Filho W (ed) Climate change impacts and adaptation strategies for coastal communities – climate change management. Springer, Cham

Helliwell JF, Layard R, Sachs JD (2018) World happiness report 2018. Sustainable Development Solutions Network, New York

Herrmann V (2017) America's first climate change refugees: victimization, distancing, and disempowerment in journalistic storytelling. Energy Res Soc Sci 31:205–214

Kelman I (2015) Difficult decisions: migration from small island developing states under climate change. Earth's Future 3(4):133–142

Kelpsaite L, Mach E (2015) Migration as adaptation? A comparative analysis of policy framework on the environment and development in MECLEP countries. Migration, environment and climate change: policy brief series, vol 1, no 5. International Organization for Migration, Geneva. ISSN 2410-4930

Ketola T (2015) Climigration: how to plan climate migration by learning from history? Global Environment. 8(2):410–445

Kolmannskog V (2012) Climate change, environmental displacement and international law. J Int Dev 24(8):1071–1081

Kutzbach M (2014) Migration and climate change. Growth Change 45(1):159–161

Laczko F, Aghazarm C (eds) (2009) Migration, environment and climate change: assessing the evidence. International Organisation for Migration, Geneva

Laczko F, Aghazarm C (2014) Introduction. In: Piguet E, Laczko F (eds) People on the move in a changing climate. The regional impact of environmental change on migration. Global migration issues. Springer, Dordrecht

Locke JT (2009) Climate change-induced migration in the Pacific Region: sudden crisis and long-term developments. Geogr J 175(3):171–180

Lübken U (2012) Chasing a ghost? Environmental change and migration in history. Glob Environ 5(9):4–24(21)

Luetz J, Havea PH (2018) "We're not refugees, and we'll stay here until we die!" – Climate change adaptation and migration experiences gathered from the Tulun and Nissan Atolls of Bougainville, Papua New Guinea. In: Leal Filho W (ed) Climate change impacts and adaptation strategies to coastal communities. Springer, Cham

Mayer B (2016) Migration in the UNFCCC workstream on loss and damage: an assessment of alternative framings and conceivable responses. Transnational Environ Law 6(1):107–129

McNamara KE, Gibson C (2009) 'We do not want to leave our land': Pacific ambassadors at the United Nations resist the category of 'climate refugees'. Geoforum 40(3):475–483

Milan A, Oakes R, Campbell J (2016) Tuvalu: climate change and migration: relationships between household vulnerability, human mobility and climate change. United Nations University Institute for Environment and Human Security, Bonn

Mortreux C, Barnett J (2009) Climate change, migration and adaptation in Funafuti, Tuvalu. Glob Environ Chang 19(1):105–112

Nash SL (2018) From Cancun to Paris: an era of policy making on climate change and migration. Global Pol 9(1):53–63

Nastasi BK, Hitchcock J, Sarkar S, Burkholder G, Varjas K et al (2007) Mixed methods in intervention research: theory to adaptation. J Mixed Methods Res 1(2):164–182

Nunn PD (2013) The end of the Pacific? Effects of sea level rise on Pacific Island livelihoods. Singap J Trop Geogr 34(2):143–171

Nunn PD (2016) Sea levels, shorelines and settlements on Pacific reef islands. Archaeol Ocean 51(2):91–98

Oakes R, Milan A, Campbell J (2016) Kiribati: climate change and migration – relationships between household vulnerability, human mobility and climate change. United Nations University Institute for Environment and Human Security, Bonn

Office of Te Beretitenti (2013) National framework for climate change and climate change adaptation. Office of Te Beretitenti, Kiribati

Pacific Community (SPC) (2017) Pacific platform for action on gender equality and women's human rights 2018–2030. SPC, Suva

Pacific Community, Secretariat of the Pacific Regional Environment Programme, Pacific Islands Forum Secretariat, United Nations Development Programme, United Nations Office for Disaster Risk Reduction et al (2016) Framework for resilience development in the Pacific: an integrated approach to address climate change and disaster risk management (FRDP) 2017–2030. The Pacific Community, Geoscience Division, Suva

Peck R, Devore JL (2012) Statistics: the exploration and analysis of data. Brooks/Cole Cengage Learning, Boston

Platform on Disaster Displacement (2016) Strategic framework 2016–2019. Platform on Disaster Displacement, Geneva

Radio New Zealand (2018) Tongan noble MP wants focus on climate change migration [Online]. Radio New Zealand, Wellington. https://www.radionz.co.nz/international/programmes/datelinepacific/audio/201817626/tongan-noble-mp-wants-focus-on-climate-change-migration. Accessed 31 Jan 2018

Scheffran J, Marmer E, Sow P (2012) Migration as a contribution to resilience and innovation in climate adaptation: Social networks and co-development in Northwest Africa. Appl Geogr 33(1):119–127

Shearer C (2012) The political ecology of climate adaptation assistance: Alaska Natives, displacement, and relocation. University of California at Santa Barbara

Sullivan M (2013) Statistics: informed decisions using data. Pearson, New York

Tacoli C (2009) Crisis or adaptation? Migration and climate change in the context of high mobility. Environ Urban 21(2):513–525

Taufatofua P (2011) Migration, remittance and development Tonga. FAO, Tonga

Tonga Department of Statistic (2011) Tonga national population and housing census 2011: preliminary results. Statistic Department of Tonga, Nuku'alofa

Tonga Department of Statistic (2013a) Tonga 2011 census of population and housing, volume 2: analytical report. Government of Tonga, Nuku'alofa

Tonga Department of Statistic (2013b) Tonga 2011 census of population and housing: Volume 1 – basic tables and administrative report. Tonga Department of Statistics, Nuku'alofa

Tonga Department of Statistic (2016) Tonga – national population and housing census 2011. Tonga Department of Statistics, Tonga

United Nation University (Institute for Environment and Human Security) (2016) World risk report 2016: focus (logistic and infrastructure). UNU, Tokyo

Williams A (2008) Turning the tide: recognizing climate change refugees in international law. Law Policy 30(4):502–529

Wyett K (2014) Escaping a rising tide: sea level rise and migration in Kiribati. Asia Pacific Policy Stud 1(1):171–185

Yamamoto L, Esteban M (2017) Migration as an adaptation strategy for Atoll Island States. Int Migr 55(2):144–158

Building Resilience to Climate Change: Water Stewardship in Rainfed Agrarian Villages in Maharashtra, India

12

Eshwer Kale and Marcella D'Souza

Contents

Abstract

In rain-dependent drylands of India, unseasonal rainfall and decrease in precipitation cause losses in agriculture-related livelihoods and water scarcity for humans and livestock. Traditional knowledge and local water governance practices of rural communities are unable to cope with these newer risks and losses. Hence, it is necessary to equip people with knowledge and tools to take informed decisions at

E. Kale (✉) · M. D'Souza
The WOTR-Centre for Resilience Studies (W-CReS), Pune, India
e-mail: eshwer.kale@wotr.org.in; marcella.dsouza@gmail.com

W. Leal Filho (ed.), *Handbook of Climate Change Resilience*,
https://doi.org/10.1007/978-3-319-93336-8_51

the farm, enterprise, and community levels that enhance resilience and adaptive capacities. This chapter is based on lessons learnt and observations from implementing a Water Stewardship Initiative (WSI) that seeks to facilitate a cognitive and organizational shift by bringing science, policy, and governance together at the level of practice and community action. This initiative was launched by the Watershed Organisation Trust, a not-for-profit organization, in 100 villages in Maharashtra State, India. In the WSI, the approach adopted is of "co-production of knowledge for behavioral and institutional change" toward building the communities' knowledge and capacity to effectively face varying weather conditions. This chapter presents the approach, key concepts, and a process applied in the WSI and highlights its potential to be taken forward in other similar regions. It has the potential to influence state policies in the water sector.

Keywords

Behavioral change · Equity · Drought · Governance · Resilience · Sustainability · Water budget · Water security · Water stewardship

Introduction

In the rain-dependent drylands of India, increasing droughts and climate variability are observed to cause a drastic fall in agricultural production and acute water scarcity for domestic use. Models indicate that by 2025, in the absence of appropriate action to save water, the combination of population growth, economic growth, and climate change will create scarcity far and wide (Roger 2008). As a result of the rapid reduction in India's groundwater, the per capita freshwater availability is lowered to 1545 cubic meters per year, as against the 5000 cubic meters per capita per year when India become independent in 1947 (Singh and Rahal 2013). This is far lower than the 1700 cubic meters that is the threshold for water stress (Khurana and Sen 2012; Prakash et al. 2013). Hence it is necessary to equip communities with the knowledge and tools to take informed decisions at farm, enterprise, and community levels that build adaptive capacities and enhance their resilience.

Key Challenges for the Water Sector and Rainfed Agriculture

Maharashtra, the third largest state in India, has a total geographical area of 30.76 million hectares (mha) of which 52% is prone to drought (GSDA 2014). In groundwater-dependent arid and semiarid regions, water resources required to meet the multiple demands, such as livelihoods, food, and domestic use for an increasing population, are at risk due to the increasing area under agriculture and irrigation (dependent on groundwater extraction) and changes in the rainfall regime. Users are barely literate about water and its judicious use; they consider groundwater a private good. The lack of appropriate and effective governance mechanisms for water resources has established the rule of capture resulting in anarchy in groundwater management (Shah 2009).

The general understanding in India is that water beneath the land belongs to the owners of land (Ballabh 2008), which is well described by Ramaswamy Iyer (2003), "water is attached like a chattel, to the land property." This has led to mismanagement of groundwater resources particularly by the better-off in rural areas, putting responsibility on the state to provide water for all, while it negates the responsibility of communities to manage their locale specific resource. Market price fluctuation for farm produce and a shift to high external input commercial crops worsens the groundwater crisis and farmers' woes. It is not just the economic and productive uses of water, i.e., livelihoods that are under threat, but water for domestic and sanitation needs and particularly the work load of women.

Groundwater: The Backbone of Arid and Semiarid Regions

In the 1950s and 1960s, the share of surface water to groundwater in irrigation was around 60% and 30%, respectively (Shankar et al. 2011). By the end of the last decade, groundwater replaced surface water irrigation as the main source (ibid.). A report of the "Expert Group on Groundwater Management and Ownership" of the Planning Commission in 2007 states that in 2004, 28% of India's blocks showed alarmingly high levels of groundwater extraction (Planning Commission 2011) due to its use in irrigation. Today, irrigation in India is highly dependent on groundwater sources with nearly 90% of the rural water needs met by groundwater sources (Prakash et al. 2013), making it the backbone of India's agriculture and drinking water security (Shah 2013).

A recent report of the Groundwater Surveys and Development Agency (GSDA), a leading state agency in Maharashtra, indicates that since the past 5 years in many regions of the state, the water level by the month of October is drastically reduced. The report points out that of the 353 blocks, in 230 blocks (10,167 villages), the water level has dropped by over 1 m; in 2587 of the 10,167 villages, the groundwater level has decreased by over 3 m. The report also warns that 7139 villages in 132 blocks may face water scarcity for drinking water during 2018 (GSDA 2017). The CGWB report (2017) states that in the last 10 years, groundwater levels have fallen in around 50% area of Maharashtra (TOI 2018).

Impacts of Climate Change on Groundwater Resources

In water resources management today, climate change and global warming are of growing concern as India is one of the largest and most important regions of high overall human vulnerability (Thow and Blois 2008). The climate change phenomenon is directly related to patterns of the monsoon behavior and change in temperature, as global warming is likely to intensify, accelerate, or enhance the global hydrological cycle (IPCC 2008). Data shows that the hydrological cycle is already being impacted (Dragoni and Sukhija 2008). Precipitation forecasts for India suggest higher but more variable rainfall, except in the drier parts, where rainfall could decrease. While the intensity of rainfall may increase, the actual number of rainy days may decrease. The changing patterns of rainfall and runoff are expected to significantly impact groundwater recharge and availability (Zbigniew et al. 2009), thus in water-scarce years, farmers and other users will mostly depend on groundwater options to compensate for inadequate rainfall and surface water supplies. The scenarios also indicate that after good rainfall years, the dry period of low or no rainfall which is currently of 2–3 years may increase to 4–5 years, and with the average global temperature rising, the rate of evaporation of surface water will also increase.

Therefore, in the rain-dependent drylands of India within which much of Maharashtra falls, erratic rainfall and drought will result in a drastic drop in agricultural production and acute water scarcity for drinking and livelihood needs. Hence it is necessary to prepare and equip rural communities and farmers with knowledge, skills, and the means to address these challenges. This chapter is based on experiences in 100 villages where the Water Stewardship Initiative (WSI) is implemented by the Watershed Organisation Trust (WOTR) with an aim to facilitate a cognitive and organizational shift by bringing science, governance, and policy together at the level of practice and community action. The approach adopted in WSI is of "co-production of knowledge and learning for behavioral and institutional change" toward building community knowledge and enhancing their capacities to effectively deal with varying weather conditions.

The Theoretical Framework Adopted in the Water Stewardship Initiative

As discussed in the above section, at the center of this initiative is the motivation and mobilization of individuals and community for behavioral change around water-use practices. The initiative is designed to address the drivers of the individual's behavior and how these can be used to enable the primary actors to take appropriate measures for water use that is economically efficient, socially fair, and environmentally sustainable.

Existing Frameworks and Models of Explaining Human Behavior

In existing literature and analysis of behavioral change practices, the "Theory of Planned Behavior" proposed by Ajzen and Fishbein, the "Social Cognitive Theory" of Albert Bandura, "Institutional Analysis and Development Framework (IAD)" of Elinor Ostrom, and "Governance Dynamic Framework (GDF)" of Subodh Wagle are important and widely used frameworks that analyze the behavioral underpinnings of individuals and of the community as a whole.

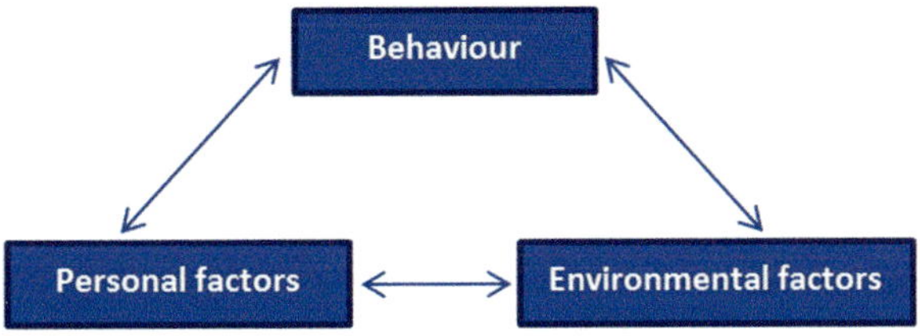

Fig. 1 The social cognitive theory model. (Adapted from Bandura 1986)

Theory of Planned Behavior (TPB)

The TPB proposes the "intention" of an individual, driven by beliefs and values of the outcome of the action, and also of opinions of society, i.e., general social pressure, that trigger one's (individual) behavior (Fishbein and Ajzen 1975). When seeking behavioral change, the intention, i.e., the personal gain, is seen as most important; hence it is necessary to present information to an individual such that it addresses both the outcome of the intention and the subjective norms that support the desired behavior (Sommer 2011). However, the individual must have the confidence of being able to achieve the desired behavior which is manifested by a perceived control over opportunities, resources, and skills needed for change. In short, the theory states that the attitude toward behavior, subjective norms, and perceived behavioral control together shape an individual's behavioral intentions and actions.

Social Cognitive Theory

The "social cognitive theory" proposes that external and environmental factors shape an individual's behavior rather than inner forces. As depicted in Fig. 1, the external situation in which an individual is located (environmental factor) drives a person's behavior, which is also influenced by individual traits and motivational forces (personal factors) (Bandura 1999).

To enhance levels of self-efficacy (personal factors), the individual's confidence needs to be boosted by provisioning of resources and support. However this theory proposes the need for incentives, rewards, and an encouraging environment for shaping good behavior.

Institutional Analysis and Development (IAD) Framework

The IAD framework developed by Elinor Ostrom (2011) helps to understand how actors/stakeholders behave in particular action situations and how the dynamics of their interactions and behavior are shaped by rules and norms in the given context. Applied and tested in a range of situations, it systematically analyzes how individuals behave in the structure of situations faced and determines how rules and attributes of the surrounding environment and local community affect these situations over time (Smajgl et al. 2009). While this theory acknowledges that in simple situations individuals may act rationally, it posits that in complex circumstances, individuals lack complete knowledge and means to act in a fully economically rational manner (Ostrom et al. 1992; Polski and Ostrom 1999).

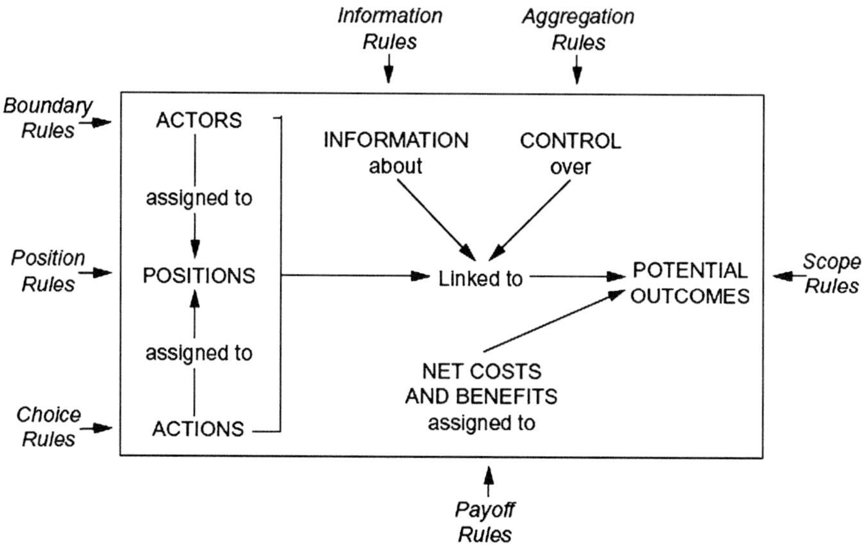

Fig. 2 Rules as exogenous variables directly affecting the elements of an action situation. (Adapted from Ostrom 2011)

Central to the IAD framework is the "action arena" characterized by the (a) physical and (b) social context and (c) institutional arrangements (Polski and Ostrom 1999). These three attributes contextualize the action arena where underlying factors affecting the action output are addressed. According to the IAD framework, as depicted in Fig. 2, all action situations include seven elements: (1) *participants* who have specific (2) *positions* that decide between various (3) *actions* based on (4) *information* they have about how the actions are (5) *linked* to possible (6) *outcomes* and the ensuring (7) *costs and benefits* (Ostrom 2011). These seven types of working rules can affect the structure of an action situation (actor's behavior). As illustrated in Fig. 2, the cumulative effects of these seven types of rules affect an action situation while the individual takes decisions and acts.

Governance Dynamic Framework (GDF)

GDF presents a schema of interactions between behaviors of actors involved and the relevant public policies and development interventions (Wagle 2011). The framework explains that the misalignment between the goals of policies or programs and actors' behaviors mainly occurs because actors are not fully prepared to take appropriate actions. Thus preparedness of the actor to adopt expected behavior/actions is at the center of GDF (Kale 2018). Preparedness of actors comprises of four preparedness elements (PEs), viz., (a) awareness, (b) vision, (c) willingness, and (d) abilities, which are influenced by innate factors related to the actors. These innate factors related to the actors that shape the preparedness elements are here referred to as preparedness determinants (or PDs) which include (i) internal norms (unwritten, informal rules), (ii) interests (tangible and intangible), (iii) information and knowledge the actors have, (iv) formal rules of the governing agencies (where the actors function), (v) resources and capabilities of the actors, (vi) pressures and enticements from other actors, and (vii) environmental factors (that make the expected behavior more convenient). Thus, the GDF presents an extensive set of determinants which shape the individual's behavior and recommends that, while making any efforts toward changing the actor's behavior, the preparedness determinists have to be well addressed.

Behavioral Change Approach Adopted in WSI

The understanding behind development of the WSI is that rural farming households in water scarce regions are driven by market pulls and the need for income from an increasingly uncertain agriculture. Access to water resources holds the key to success, particularly as low rainfall, delayed monsoons, prolonged dry spells during monsoons and droughts are on the increase. Water in rural areas is considered a private resource; those who can capture such a resource draw maximum benefits, an idea which unintentionally is supported by some government schemes. Excessive extraction of groundwater by a few causes inequity, yet people tolerate it considering this their "fate." In a nutshell, the different dimensions of groundwater and its availability (the invisible resource) are barely discussed at village level.

Hence to bring people together to manage water resources responsibly, it is necessary to demystify knowledge through providing scientific information in a manner that is understandable by the village community. Technical tools for preparing sound water budgets, bringing various village groups (based on land and water accessibility) to actively engage in a realistic understanding of their context and preparation of plans, are easily accepted by people. Besides capacity building and trainings, workshops and discussions around real-life situations in their villages get people actively engaged. But most of all, putting the village community at the center, not as recipients of generosity but as key players and "stewards" with responsibility and authority to develop and use their resources, is an important attitude required in the WSI.

Contribution of WSI to the SDGs and Policy Process

The WSI provides important contributions for achieving the Sustainable Development Goals – SDG 6, 12, 13, and 16 – and in the context of climate change where 1.5–2 °C temperature rise is expected by the end of the twenty-first century (IPCC 2014). The water budget process discussed here becomes useful to build the community's capabilities to adapt to adverse conditions and reduce threats while obtaining water for agriculture and drinking needs.

Overview of the Water Stewardship Initiative

This section provides an overview of experiences of local water governance in India and, in particular, the WSI that the Watershed Organisation Trust (WOTR) has implemented in 100 villages in Maharashtra.

Water Governance: The Concept and Experiences

The concept of people managing water resources is not new in India. Groundwater depletion and the consequent misery have triggered attempts to create a new social order of "community-based natural resource management" (CBNRM). The argument behind CBNRM is that there exists a certain kind of cooperation and community solidarity within village communities that sustains with endogenous or exogenous stimuli when the expected outcomes are for the benefit of the concerned group (Lopez-Gunn and Cortina 2006). These ideas are echoed in experiments by social entrepreneurs like Anna Hazare and Popat Rao Pawar, the best-known examples of community revival through participatory watershed and water resource management in Maharashtra (World Bank 2010). Besides these, there are numerous successful participatory watershed development projects where NGOs put community at center. The works of Hardevsingh Jadeja in Rajsamadhiala village near Rajkot; Swaminarayan Sampraday and Swadhyaya Pariwar in Saurashtra, Gujarat; Rajendra Singh of Tarun Bharat Sangh in Rajasthan; and the late Vilasrao Salunke's Gram Gaurav Pratishthan in Maharashtra are few pioneering voluntary, self-regulated community-based experiments toward sustainable use of groundwater (Shah 2009). These experiments went beyond just harvesting rainwater to invoking visions of its sustainable and judicious use. However, most of such successful initiatives are characterized by the presence of charismatic leadership and/or are location specific; therefore do not offer a process-based approach/model that can be upscaled. The Andhra Pradesh Farmer-Managed Groundwater Systems (APFAMGS) project stands apart. The APFAMGS experiment indicates that carefully designed community-based approaches hold significant promise for addressing groundwater use, especially in hard-rock aquifers (World Bank 2010). These successful initiatives, however, challenge the underlying assumptions in the discourse and practice of groundwater management in India that "legal and policy reform is a necessary first condition for attempting groundwater management in the country" (ibid.).

At the international level, Integrated Water Resource Management (IWRM) is promoted since the 1990s. For equitably maximizing economic and social welfare without compromising the sustainability of vital ecosystems, IWRM promotes a coordinated development and management of water, land, and related resources (UNEP-DHI 2009; Harsh 2012). In many developed countries, the IWRM framework is practiced by allocating water through regulation and incentive mechanisms such as water pricing (Shah and Prakash 2010) which increases the efficiency of water use and allows for better maintenance of the water-related infrastructure. However, in countries like India, the challenge in applying IWRM practices is mainly in formalizing the existing informal water economy (Shah and Koppen 2016) where the number of water users is large and water resource management is compartmentalized into surface and groundwater managed by departments and agencies. As the proportion of the population in India which does not have access to the most basic facilities of safe and adequate drinking water and sanitation is large, putting a price on water for these sections is not justifiable and possible (NWM 2010). For promoting IWRM practices, the financial, infrastructure, and human capacities of countries to fulfill the responsibility of water governance matters a lot, which are, however, inadequate in India (Azhonia et al. 2017). Although in many countries the IWRM approach is not very successful (ADB 2007), it has provided an important framework for integrated water management and a tool for establishing multi-stakeholder dialogue.

At the global level, systematic large-scale experiences in community-based groundwater management come principally from Mexico and Spain, which have adopted community-based models as central in their official groundwater governance and management policy. In these countries, groundwater user associations or aquifer management councils are statutorily promoted as the institutions for collective and participatory self-governance of aquifers. While these community institutions have been successful in some aspects, they have not been able to achieve the basic objective of demand management to ensure sustainable use of groundwater resources (Shah 2009). The assessment of such experiences for possible application in India indicates that merely substituting the community by the state cannot work unless groundwater management interventions are structured to serve the basic interests of and management by the users while taking into account the socioeconomic realities of the ecosystem needs of each particular groundwater setting (Kale 2018). Therefore, while it is clear that the circumstances of groundwater use in most settings of India necessitate the adoption of community-based interventions, the growing and explosive demand on groundwater use is the cause of overexploitation of aquifers, and its management is a challenge for both people and the government.

Rationale and the Objectives of the Water Stewardship Initiative

Considering the above discussion, there is an urgent need to move beyond supply-side enhancement of participatory watershed development (WSD) toward the demand-side management and improvement in water governance at the local level. This is the underpinning of the Water Stewardship Initiative. In these 100 villages, soil and water conservation structures and/or WSD is ongoing or has been completed, and while water availability and agriculture productivity show a marked increase (Vani et al. 2007), these villages also experience water shortage for irrigation and domestic use in summer. It's because of the absence of effective regulations on water use, the shift to cash crops, and increased lifting of groundwater for agriculture particularly by the better endowed farmers. Watershed Committees have not been proactive and are not confident enough to set up water-use norms and practices. Unequal water access in villages has reached to an extent where better-off farmers have water available in their farm ponds promoted through government schemes, even in the dry season and drought periods, while in the same period, the lesser endowed farmers and women struggle to fetch water for their basic needs from public tankers (Kale 2017). It is assumed that the state is responsible for meeting the various water needs of the people, while individuals and communities do not have any responsibility which reinforces a dependency syndrome. Community action which is the key to village development and progress is lacking in most villages. At another level, healthy dialogue and cooperation between government officials and village communities does not occur besides the implementation of government schemes, e.g., in the water context, those who have water resources and can construct farm ponds and apply for micro-irrigation, both of which are beneficial to the resource rich. Moreover, water extraction for agriculture generally lacks relevant scientific information and knowledge of the agroclimatic and hydrogeological conditions that are essential for planning appropriate water use.

The Water Stewardship Approach

To address the challenges discussed above, WOTR designed an approach to "Water Stewardship" that brings key actors into a dialogue to achieve mutual cooperation, to develop a plan of action and implement the same. The "Water Stewardship" approach treats all water users as important stakeholders – "water stewards" – with the potential to be "good water managers" who also protect the resource base and ensure its sustainable, judicious, and efficient use (AWS 2014). This concept considers the needs of every individual, their livelihoods and livestock, along with the ecosystem. The use comes with a responsibility and accountability to oneself and the community for its proper management. The different water users are not passive beneficiaries who may exploit the resource, but custodians who also have the responsibility to protect and manage it for the generations to come.

In WSI, the water resource is seen as public trust rather than private entity. Hence this approach sees the necessity to bring the different users/stakeholders together on one platform to discuss, dialogue, be informed, and come to a consensus in preparing a plan and executing the same. The water stewardship approach has three levels of important actors: (1) the primary stakeholders, i.e., all households within a given geography of a village who are the water users for different purposes; (2) the secondary stakeholders are the neighboring villages that affect or are affected by the water availability of a particular village, e.g., the downstream and upstream villagers; and (3) the tertiary stakeholders are decision-makers, administrators (government officials), water experts, donor agencies, and civil society agencies, who influence the water-related activities at the village level. These diverse stakeholders need to come on one platform for a common understanding and collective planning for the sustainable management of this precious resource. Hence, the WSI requires not just a set of actions to be carried out by the primary and secondary stakeholders but also needs to be informed by sound scientific knowledge so as to make informed decisions. Consensus building through dialogue and workshops (stakeholder engagements) is important where perspectives, aspirations, and conflicts between the different groups are encouraged to surface and sustainable solutions are sought.

This chapter presents the experience of WSI implemented in 100 villages in Maharashtra, India. To find pathways to address the critical balance between supply and demand-side management and achieve water security in varying weather situations, a scalable WSI model is implemented in a semiarid region, regularly affected by drought and heavily dependent on groundwater for agriculture and domestic use. These villages are spread over five blocks in three districts (refer to

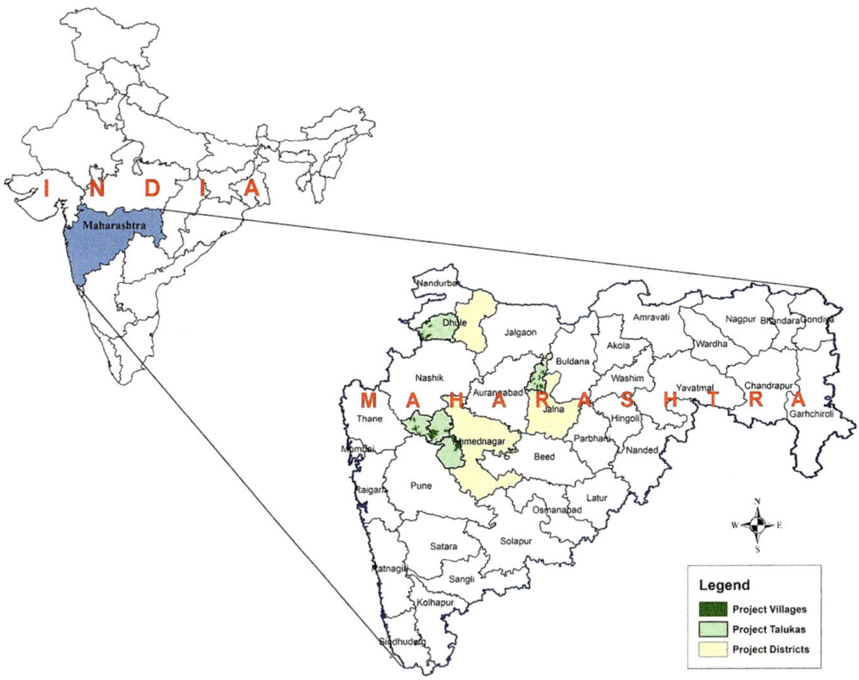

Fig. 3 Locations map of the project villages of WSI

Fig. 3). In this region, both extreme weather events and slow-onset disasters (low rainfall, droughts, and drought-like conditions) and increase in annual and summer temperature affect income from agriculture and increase pressure on water requirement for living, livelihoods (agriculture and livestock), and nature.

The goal of the WSI is to ensure that water at the local level is managed in a responsible manner and is socially equitable, environmentally sustainable, and economically efficient in climate-varying conditions.

Specific Objectives of the WSI

In line with the goal, the five specific objectives set for the WSI are:

1. Building skills and capacities of villagers to prepare the Village Annual Water Budget that ensures water availability for domestic use and that minimizes crop losses.
2. All households of the village are motivated and facilitated to prepare plans for harvesting rainwater (in the initial years) and for increasing water productivity ("more crop per drop") through the adoption of water-efficient technologies and practices.

3. The village – either the Grampanchayat or its subcommittee – is motivated, organized, and guided to formulate norms for water management and to implement effective water governance arrangements such that it ensures equitable access to and availability of water for domestic use for all local inhabitants throughout the year.
4. To promote a common understanding and trust between different stakeholders of the water sector, i.e., the primary, secondary, and tertiary stakeholders.
5. Develop workable operational guidelines to upscale the initiative in similar conditions (arid and semiarid regions) with appropriate contextual modifications.

Key Components and Processes Adopted in the WSI

This section provides the details of different components, activities, and processes adopted and executed in the WSI that are essential for achieving the desired outcomes. While implementing these components, conscious effort was made to achieve quantifiable targets as well as qualitative richness to ensure the motivation as well as skills and capacity enhancement of the primary stakeholders.

Institutional Development and Capacity Building

As institutions are the "pillars of governance," in the WSI, the Village Water Management Team (VWMT) is responsible for water stewardship. *Jal Sevaks* support and guide the VWMT toward meeting the goal of sustainable water management.

Jal Sevaks

Jal Sevaks (water volunteers) are crucial to this initiative. They play the important role of organizer, facilitator, and motivator to VWMTs for the preparation and implementation of water stewardship plans. *Jal Sevaks* are committed and motivated rural youth having a passion for the progress of their villages and are capacitated to handle various challenges in the water sector at the level of village and cluster of villages. They organize villagers of all land and water ownership categories to (1) form Village Water Management Team (VWMT), (2) conduct a baseline inventory of the existing soil and water harvesting structures with detailed measurements, (3) facilitate and guide the VWMT to prepare water stewardship plans, and (4) motivate villagers to implement the same. The *Jal Sevaks* promote water literacy and its management.

However for effectiveness (achieving good impacts and outcomes) and for addressing conflicts and challenges, the *Jal Sevaks* require continuous skill enhancement (Kale et al. 2018) and handholding through interactions and on-field experience, particularly to engage with their own community. As water management is based on water available in the year, i.e., related to climate variability, bimonthly

Photo 1 Exposure visit of *Jal Sevaks* to learn water budget preparation. (Photo Credit: Mohan Dhuldhar, WOTR)

training workshops equip the *Jal Sevaks* to prepare weather-varying water budgets. As they have an intense time-consuming role extending to a number of villages, financial incentives motivate the *Jal Sevaks* to give dedicated time and effort toward the WSI (Photo 1).

Thus, by engaging rural youth (*Jal Sevaks*) and building their capabilities, WSI has resulted in developing a cadre of motivated and skilled water stewards who can take forward water management practices in villages.

The Village Water Management Team (VWMT)

As discussed earlier, the water stewardship approach goes beyond treating water users as passive "beneficiaries" or "target" groups and looks on them as "water managers." Provided with adequate information and knowledge of the different aspects of water as a "good" essential for life, as a "resource" meant for all living creatures and for the ecosystem, and as a resource that needs to be used and protected, the VWMT is motivated to understand their responsibility of governing the water resource effectively. All users are considered stewards of their respective water resource; therefore they necessarily need to adopt good water management practices. To sensitize this diverse set of village-level stakeholders, representatives of different categories of households within a village, e.g., landownership groups, landless, women, and other water-dependent livelihood groups, are selected to form the VWMT. This team is responsible for the preparation and execution of village water stewardship plans that engage all villagers.

VWMTs are appropriately capacitated to lead and be responsible for water governance in the village. Their roles and responsibilities include to create a positive environment for adoption of water management, ensure that the Gramsabha (village general body meeting) takes the appropriate and timely resolutions required to meet good water management practices, monitor the local rainfall and collect groundwater data, and implement the water stewardship plan in the village, which includes (1) half yearly water budgets, (2) water-saving/conservation plan, (3) water-use efficiency plan, (4) setting the necessary rules/norms to implement these plans, and (5) approach different government departments, NGOs, and corporates to obtain the necessary support for convergence to realize the water stewardship plans.

Bringing Diverse Stakeholders Together Through an Engagement Process

For good water management, the diverse stakeholders who have a shared problem and use the same resources at various levels (within a village, clusters of villages, and aquifer) need to be engaged in focused dialogue. During such events, sharing of scientific information of the cluster of villages or even of the block is useful to develop a common understanding of the landscape, the water resources, and the climate variability – all of which affect the livelihood of the particular group of villages. In such a forum, the stakeholders understand the link of one village to another and how together they can take decisions to improve crop production while using the available water sustainably. Stakeholders need to meet at least every half year to plan and act accordingly, based on water availability. Bringing the respective government officials, experts, and facilitating agencies (tertiary stakeholders) to these engagements further strengthens the WSI. Stakeholder engagement workshops foster free and healthy discussions and interaction between participants and help uncover the various aspects and surface queries particularly the sticky problems. Relevant scientific information presented during such discussions provides a good basis for understanding the problematic aspects of water management.

Preparation and Implementation of Water Budgets

Once the VWMT is motivated to take concrete action for improving their village water health status and address related problems, they are guided to prepare the village water budget. Most of the existing water accounting and budgeting tools are either at river basin or at a macro-watershed level where its application faces many difficulties (Kale and Sathe 2017). This water-budgeting tool presents a practical and sound method at village and micro-watershed level for approximately 1000–2000 ha. The tool and methodology is described in detail in the Manual for Water Stewardship (Kale et al. 2018) based on WOTR's experience in water budgeting of over 10 years. The tool is designed for agrarian villages to facilitate and mobilize them to prepare appropriate crop plans based on the available water including rainfall decrease and also ensure water for domestic use and livestock for all households throughout the year. The water budget preparation process follows three steps: (1) calculation of the total water availability, (2) calculation of the water requirement for different uses, and (3) planning the use of the balance water for different crops.

Preparation of the Village Water Budget

The exercise of water budget preparation is done twice during the agriculture year: around the month of March, the first water budget is prepared which calculates the annual requirement, i.e., for domestic use and livestock and water use for agriculture in the monsoon, winter, and summer seasons. This helps to assess the amount of water that can be harvested within the village from structures earlier constructed and to identify those that require maintenance. The second water budget is prepared in the first week of October (post-monsoon), to precisely calculate the available water at the end of the monsoon and to plan its use for all needs particularly for agriculture in the winter and summer seasons. Currently water for ecosystem services is not factored in the water budget tool applied in the WSI, as estimates of water requirements for ecosystem services in the Indian context is not yet available.

Based on the water available for agriculture including perennial horticulture, villagers discuss and decide the types of crops to be sown and the area (number of hectares) that can be cultivated. Often, a water budget results in deficit water because people cultivate a greater area than planned; besides they may also cultivate water guzzling crops. An understanding of the water available provides them the opportunity to revise their crop plans so as to minimize the deficit and thus crop loss.

The water-budgeting exercise may result in either surplus water in the village, after meeting all water requirements, or a deficit in meeting these requirements. The results of the exercise highlights to the VWMT and villagers the overall picture of the water-use plans for agriculture. It gives them the opportunity to understand their water resources and the potential and different water needs of the members of all categories of households of their village. Through this exercise in most villages, people become aware that the area of agriculture planned is greater than the water available – which would otherwise lead to crop and investment loss. They are also aware that they often compromise the water needs for domestic use and for livestock. The exercise is thus an eye opener. As a first reaction, the VWMT and villages make efforts to revise the total water required for crops by reducing the area under cultivation; however, even after this revision, the deficit remains. To address the deficit, the water stewardship tool provides three sub-plans which focus on supply- and demand-side measures, as well as governance mechanisms: (1) additional water-harvesting plan, (2) water-saving plan, and (3) preparation of social governance rules to ensure the efficient water use.

The Water-Harvesting Plan

To address water deficit in the budget, a water-harvesting plan gives an opportunity to inhabitants to increase the water-harvesting potential within the village either by repairing defunct water-harvesting structures or, where appropriate, undertaking new soil and water conservation measures.

Water-Saving Plan

Besides addressing the supply side, it is important that water-saving measures are adopted through water-use efficiency practices in irrigation along with soil quality-enhancing practices. Water-saving plans enable villagers to assess the existing water-use practices and determine the initiatives that can increase water saving. Water-saving plans include the use of micro-irrigation (drip, sprinklers), mulching that reduces evaporation, and organic and vermicompost that increases the biomass content of soil. The VWMT motivates its farmers to adopt these practices and assists in linking them with the different government schemes.

Water Governance (Social) Rules

For implementation of crop plans based on the water-budgeting exercise and the sustainable, equitable, and efficient regulation of water use, it helps when villagers themselves formulate the rules to control their own behavior and practices. Initiated by the VWMT, the rules are decided upon at the Grampanchayat level and ratified at the Gramsabha, so that it becomes obligatory to all in the village. Here are some examples of rules prepared: (1) no direct lifting/pumping of water from surface water bodies such as check dams and percolation tanks, (2) government norms for the depth (not more than 200 ft) for borewells to be complied with, and (3) ban or limit on the area under water guzzling crops. Thus, the water stewardship plan is a comprehensive tool that motivates villagers for water governance as well as helps them to implement and monitor their actions.

Execution of the Water Stewardship Plans

The effectiveness of the processes and transparency in its execution determines the level of success of the WSI. These are essential to achieve the expected impacts.

Clarity on the water stewardship plans and water-budgeting processes are shared in the Gramsabha by the VWMT, and only after open discussions, it is ratified by the Gramsabha. All the necessary social norms required for ensuring sustainable and judicious water use in the village and execution of the water budget plan are discussed and finalized by the Gramsabha. The final water budget details are displayed on a notice board located at a public place within the village.

In our experience of its implementation, during the pre-monsoon period (March to May 2016–2017 and 2017–2018), the VWMT and *Jal Sevaks* sensitized and mobilized villagers for contributing *shramdaan* (sweat equity) to repair the soil and water conservation structures. In few villages people constructed new stone bunds, continuous contour trenches (CCT), and water absorption trenches (WAT). In other villages, besides *shramdaan*, villages were able to access government schemes. After preparing the water conservation plans, the VWMT presented the same to the concerned district government officials, which initiated a dialogue between the government authorities and the proactive local community. Thus, preparing and executing water stewardship plans create a space for people to work together for water management in a way that generates a positive relationship with the respective government departments and both contribute to responsible water management.

Piloting Key Principles of the Maharashtra Groundwater (Development and Management) Act 2009

The Maharashtra Groundwater (Development and Management) Act 2009 (henceforth referred to as the 2009 Act) (GoM 2013) came into force in the State in December 2013. The broad objective of the 2009 Act is to strike a balance between protecting drinking water sources and promoting the optimum utilization of groundwater for irrigation. For ensuring the supply of adequate drinking water and sustainable groundwater use, the main approach set in the law is to create an institutional structure for community participation, which was absent in the provisions of the earlier 1993 Groundwater Act of the state (Phansalkar and Kher 2006). To achieve the abovementioned broad objective of the 2009 Act, two new governing agencies, i.e., the State Groundwater Authority and the Watershed Water Resource Committee (WWRC), are proposed. Limiting the depth of wells, securing drinking water availability throughout the year, aquifer-based groundwater use plans, and prospective crop plans based on groundwater budgets are few important measures proposed in the State's 2009 Groundwater Act (Kale 2018). These broad strategies are further articulated for three types of areas based on water stress: (i) notified area, (ii) scarcity area – within the notified area, and (iii) non-notified area (GoM 2013).

The 2009 Act is well intended and focuses on institutional building with remedial measures; however it is somewhat vague on provisions regarding allocation of necessary resources and the capabilities required to effectively carry out functions. The WWRC is considered an important governing agency at the local level, and has been given major functions and authority such as granting permission for sinking wells, implementing prospective crop plans; yet, the proper process and system for establishing transparency and accountability in its functioning is not elaborated in the 2009 Act. However, the effective implementation of provisions proposed in the Act regarding preparation and implementation of prospective crop plans, limiting the depth of wells and mandatory permissions for new wells, has potential to change the status quo in the area of groundwater overexploitation in the state.

In the WSI, key principles of the 2009 Act are tested on a pilot basis with the intent to operationalize the Act.

Prospective Crops Plans

The Act mandates preparation and implementation of prospective crop plans based on the available water stock. In the WSI, in all project villages, capacities and skills of the VWMT and villagers have been developed to prepare and execute water-budgeting plans, which include both supply- and demand-side measures and ensure drinking water availability. These plans are prepared and executed by the VWMTs.

Institutional Governance at Aquifer Level

The Act proposes formation of the WWRC comprising of 11 villages or more. This scale is proposed considering an aquifer boundary where villages share the common aquifer. They come together to benefit from the aquifer by managing it in

Photo 2 Aquifer management workshop at Bhokardan block with 3D maps (CDVI). (Photo Credit: Eshwer Kale, WOTR)

a sustainable manner. To explore the practical possibilities, aquifer-based management is piloted in the Bhokardan block of Jalna district, where representatives from 14 villages share a common aquifer work together. Water resource literacy was initiated by applying the tool called CoDriVE-Visual Integrator (CDVI) which resulted in a three-dimensional model of the landscape above and the shallow aquifer characteristics below surface level. A group comprising of key persons from these 14 villages formed the aquifer-based committee. They were capacitated with motivational exercises and activities to plan water use at the aquifer level (Photo 2).

Rules for Water Use

The Act bans sinking deep well (below 60 m) in notified areas and has a provision to levy a cess on extracting groundwater below 60 m in non-notified areas. Currently in the WSI, communities formulate their own rules rather than have these externally enforced. However, it is acknowledged that strict policy rules at the state level create a supportive policy environment for motivating villagers to frame Grampanchayat and aquifer-level rules. In the project, through the intensive stakeholder engagement process at the cluster of villages' level, and by informing villagers of the key provisions of the 2009 Act, the VWMTs were motivated to frame rules considering their local context. In more than 80 villages, the VWMTs have framed such rules that have been ratified by the Grampanchayat. The rules framed are mainly regarding the ban on directly lifting of water from water-harvesting structures, putting a ceiling on the depth of borewells, and/or the ban on borewell drilling in the village and a ban on growing water-intensive crops.

Registration of Groundwater Abstraction Structures

Registration of all wells is a precondition for applying the regulations. In the WSI, in sample villages the process of recording the details of all water-harvesting structures and wells, including detailed well inventory, has been completed.

Results and Outcome of the WSI

In project villages, the execution of different components of WSI resulted in improving the overall level of water management, specifically in terms of supply- and demand-side water management and better institutional environment to govern it. These impacts are observed as changes in the behavior of people (water stewards) regarding water-use practices, access, and crop management.

Changes in Behavior of the Water Stewards

Feedback from the primary stakeholders of different regions shows that discussions and information received during the stakeholder engagement workshops have changed the perspective of many villagers; they have come to understand new aspects they were not aware of. As shared by some participants, the stakeholder engagement workshops provided them an opportunity to deliberate and discuss on water issues as a "shared problem," leaving aside all other differences and dynamics of the village. They learnt how to calculate the water budget and to use water efficiently. Many shared that they now understand water as a common property and that everyone has a right to it; therefore water should be used judiciously. Mrs. Meera Ramesh Shinde, a VWMT member from Lingewadi village, Jalna, describes how the situation has changed in her village.

Mrs. Meera Ramesh Shinde (Lingewadi, Jalna)

"Earlier, for fetching drinking water, no private well owners allowed others to draw water from their wells. After exposure and learning about this in stakeholder engagement workshops and discussions of the same within the village, some well owners now switch on their private pumps to allow others to take water for home use. As water is needed for all, even for the animals and birds, we have made special water-troughs for the animals in the remote hilly area where monkeys, wild boar, and deer face difficulties in finding water in summer. They now frequently come and drink water from these troughs."

Although formulation and enforcement of rules regarding water use at village level is challenging, in some villages, the inhabitants have taken the initiative. Mr. Kisan Sakharam Icche, VWMT members from Kotha Jahangir who is also the village Sarpanch, explains how the stakeholder engagement workshops motivated their VWMT for setting village-level rules.

Kisan Icche (Sarpanch, Kotha Jahangir, Jalna)
"Getting motivated through Stakeholder Engagement workshops, we passed a resolution in the Gramsabha on the ban on drilling new borewells and changing crops in the rabi (winter) season in the village. We tried to convince every irrigated farmer to use sprinkler and drip instead of flood irrigation and this year (2017-18), almost 60 to 70% farmers installed drips and sprinklers. Because of the rule, not a single bore well was drilled during this year in the village." 

In Kotha Jahangir, the Gramsabha passed a resolution to ban the drilling of borewells deeper than 150 ft, and farmers have to take permission from the Grampanchayat before drilling a borewell. They have also passed a resolution in the Grampanchayat to ban sand extraction from the river bed. Such locally appropriate rules are made in more than 80 villages. The stakeholder engagement workshops have succeeded to some extent in building a common understanding and consensus, changing perspectives and motivating villagers to work together toward the efficient and judicious use of water.

Changes in Local Water Governance

In all 100 villages, VWMTs are formed and members are ratified by the Grampanchayat. Most of the VWMT members have participated in different stakeholder engagement workshops, where they have prepared water stewardship plans comprising of *Jal-Arogya takta* (water health chart), water budget, water-harvesting, and water-saving plans. In many villages, the VWMT succeeded in mobilizing their people for *shramdaan* to repair water-harvesting structures. The exercise of preparing governance rules at the village level is an important contribution of the WSI, particularly as state agencies are still struggling to enforce external regulations on villages for water management. The VWMTs in most of villages have compiled with their water stewardship plans and have submitted the same to district authorities. This is an important exercise where government authorities also experience an encouraging atmosphere at village level to implement their schemes. Villagers too have found a space to interact and dialogue with government authorities.

24 *Jal Sevaks* facilitated and motivated the VWMTs in these 100 villages. They have collected timely and relevant data regarding soil and water-harvesting structures and other socioeconomic details of the villages. In all these villages, water budgets are displayed on boards in prominent locations where the VWMTs regularly update the water budget for the current season. In all the villages, two manually operated rain gauge units are installed, where the VWMT members monitor the local rainfall. These activities have resulted in creating important spaces for villagers to discuss and share their interest and concern regarding water management practices.

Benefits of Increase in Water Availability and Governance

Between the 100 villages, they annually harvested a total of approximately 38 billion liters of water stored in different water-harvesting structures and brought this under governance through water budgets and local rules. In more than 60 villages, people provided *shramdaan* to repair water-harvesting structures or construct new ones (61 sand-bag dams across streams) and implemented the soil- and water-harvesting project of the government or other funding sources – a convergence initiative. Through these efforts, 8.95 billion liters of additional water-harvesting potential has been created. More than 2000 farmers have adopted practices of micro-irrigation (drips, sprinklers, and mulching) and, through this, saved 3.24 billion liters of water.

The WSI has promoted a community-embedded scientific method of water budgeting at the village level with appropriate governance practices, which enables villagers to effectively handle water stress especially in times of low rainfall. The latter is increasingly becoming a norm (however, if the climate event is very severe, external assistance is required). Ultimately, water security achieved by this exercise leads to an increased and sustained production of crops and livestock as well as water for households and livelihoods.

Stakeholder engagement workshops brought together the hundreds of "Water Stewards," researchers, scientists, and administrators and resulted in an important forum where different aspects of water scarcity and climate change, impacts on livelihoods, and ways to adapt and address these crises were discussed. This helped villagers to understand how climate change impacts them and helped prepare them to better adapt to climate variability. These various actions and interventions work toward building resilience of villagers and the farming community to adapt to climate variability and build their resilience.

Recommendations

Lessons learnt from the experience of the Water Stewardship Initiative can contribute singularly to the sensitive but crucial issue of groundwater use and governance:

1. State governments, donors, NGOs, and even villagers almost exclusively consider addressing drought through supply-side enhancement, e.g., watershed development and soil and water conservation. While this is important, it is the demand-side management and water-use efficiency that are more important, essential, sensitive, and challenging.
2. Villages cannot work in isolation along administrative boundaries to manage their water resources, especially groundwater. The shared aquifer resource lends itself to bringing the related villages together, to work not just on water management but also crop planning and management, which can be further incentivized by promoting market linkages.
3. Promoting water stewardship and building the capacities of communities for water budgeting and effective governance is a medium term (5–8 years) process. In a project mode of generally 2–3 years, sustainability of the efforts put in is threatened. Given the fact that villages face constant change from exogenous

factors, a medium-term commitment with progressively tapering external inputs and involvement is necessary to achieve a sustainable breakthrough.
4. As water for agriculture is the predominant use, women are generally left out. However, women in rural areas are also farmers, and water budgeting also includes water for domestic use; hence it is necessary that women are adequately represented in the VWMT and in all its activities. Including them in the various components of the WSI will bring in a rich diversity of perspectives for managing water, besides addressing gender concerns.
5. Building the skills of local educated youth in conducting water literacy, water budgeting and crop planning is the real need of hour. It motivates youth to be actively engaged in productive and managerial roles and can become a source of income. The *Jal Sevaks* are very useful for this purpose and can contribute meaningfully to the progress and development of their villages.

Conclusion

The phenomenon of climate change being little understood by rural communities, the hydrogeological frame within which each aquifer is set and the unknown quantity of water in the aquifers provide an occasion, reason, and platform that can bring people within a village and aquifer to work together. Empowering communities with scientific knowledge and information and treating them as "stewards" rather than as "beneficiaries/target population and opportunists" place responsibility on them which, when given proper guidance and skill enhancement, they are willing to assume, as demonstrated. They are open to facts and learning new ideas together, discuss their concerns and aspirations openly, dialogue and address conflicts, and work together to find solutions to manage groundwater more effectively and fairly.

Enhancing the skills of the local people, particularly the youth to manage their water resources scientifically while also improving agriculture productivity (through crop-water management) and income through better market access will further incentivize people for sustainable and equitable groundwater management. However, when this is backed by policy and law (e.g., the 2009 Act), there are greater chances of continuity of effective water management practices on the ground. Moreover, as clusters of villages work together, space is created for dissemination of good management practices to neighboring villages and beyond.

The Water Stewardship Initiative responds to a growing social, economic, and ecological demand. It outlines a pathway to ensuring more equitable, efficient, and sustainable use of water for life, livelihoods, and nature in a changing climate that tends toward increasing water stress and scarcity. The WSI has relevance to and offers important lessons to shape the existing legislative framework that seeks to regulate water use. There is, therefore, an urgent need to widely promote the Water Stewardship Initiative in water-stressed, arid, and semiarid regions.

References

ADB (2007) Institutional options for improving water management in India: the potential role of river basin organization. Publications stock No. 110507. Asian Development Bank, New Delhi

AWS (2014) The AWS International Water Stewardship Standard. Alliance for Water Stewardship. Available at http://a4ws.org/wp-content/uploads/2017/04/AWS-Standard-Full-v-1.0-English.pdf

Azhonia A et al (2017) Adapting water management to climate change: institutional involvement, inter-institutional networks and barriers in India. Glob Environ Chang 44(2017):144–157. Elsevier

Ballabh V (2008) Governance of water: institutional alternatives and political economy. Sage, New Delhi

Bandura A (1986) Social foundations of thought & action: a social cognitive theory, 1st edn. Prentice-Hall, Englewood Cliffs

Bandura A (1999) Social cognitive theory: an argentic perspective. Asian J Soc Psychol 2:21–24. Available https://www.uky.edu/~eushe2/Bandura/Bandura1999AJSP.pdf

Dragoni W, Sukhija B (2008) Climate change and groundwater. Geological Society, London

Fishbein M, Ajzen I (1975) Belief, attitude, intention and behavior: an introduction to theory and research. Addison-Wesley Publication Company, Philippines. Available at http://people.umass.edu/aizen/f%26a1975.html

GoM (2013) Maharashtra Groundwater (Development and Management) Act, 2009. Government of Maharashtra. Available at http://bombayhighcourt.nic.in/libweb/acts/Stateact/2013acts/2013.26.PDF

GSDA (2014) Report on the dynamic groundwater resources of Maharashtra (2011–2012). Groundwater Surveys and Development Agency, GoM and Central Ground Water Board, Central Region, Government of India, Nagpur. Available at https://gsda.maharashtra.gov.in/GWSpdf/Talukawise_GWA2011-12.pdf

GSDA (2017) Comparative study of water levels in Oct 2017. Report submitted by Groundwater Survey and Development Agency to Government of Maharashtra, Water and Sanitation Department

Harsh J (2012) IWRM and IRBM concepts envisioned in Indian water policies. Curr Sci 102(7):986–990

IPCC (2008) Climate change and water, IPCC technical paper VI. Available at https://www.ipcc.ch/pdf/technical-papers/climate-change-water-en.pdf

IPCC (2014) Climate change 2013: the physical science basis (eds: Stocker TF et al). Cambridge University Press, Cambridge, UK

Iyer R (2003) Water: perspectives, issues and concerns. Sage, New Delhi

Kale E (2017) Problematic uses and practices of farm ponds in Maharashtra. Econ Polit Wkly LII(3):20

Kale E (2018) Governing the misgoverned. PhD thesis, Tata Institute of Social Sciences, Mumbai

Kale E, Sathe M (2017) Groundwater budgeting. Econ Polit Wkly 52(42–43):4

Kale E et al (2018) Promoting efficient, judicious and sustainable water use: manual on developing water stewardship. Manual under publication, W-CReS, Watershed Organisation Trust

Khurana I, Sen R (2012) Drinking water quality in rural India: issues and approaches. WaterAid. Available at http://www.wateraid.org/documents/plugin_documents/drinking_water.pdf

Lopez-Gunn E, Cortina L (2006) Is self-regulation a myth? Case study on Spanish groundwater user associations and the role of higher-level authorities. Hydrogeol J 14:361–379. https://doi.org/10.1007/s10040-005-0014-z

NWM (2010) Draft guidelines for integrated water resources development and management. National Water Mission, Government of India, Central Water Commission, Basic Planning and Management Organization, New Delhi

Ostrom E (2011) Background on the institutional analysis and development frameworks. Policy Stud J 39(1):7

Ostrom E et al (1992) Covenants with and without a sword – self-governance is possible. Am Polit Sci Rev 86(2):404–417

Planning Commission (2011) Twelfth five year plan (2012–2017): Faster, more inclusive and sustainable growth, sage publication, India. Available at http://planningcommission.gov.in/plans/planrel/fiveyr/12th/pdf/12fyp_vol1.pdf
Phansalkar S, Kher V (2006) A decade of the Maharashtra groundwater legislation: analysis of the implementation process. Law Environ Dev J 2(1):67. Available at http://www.lead-journal.org/content/06067.pdf
Polski M, Ostrom E (1999) An institutional framework for policy analysis and design. Working paper No.W98–27, The Vincent and Elinor Ostrom workshop in political theory and policy analysis. Indiana University, Bloomington. Available at https://mason.gmu.edu/~mpolski/documents/PolskiOstromIAD.pdf
Prakash A et al (2013) Water in India: situation and prospects. UNICEF, FAO, SaciWATERs
Roger P (2008) Facing freshwater crises. Sci Am. Available at http://ohm.ecce.admu.edu.ph/wiki/pub/Main/ResearchProjects/Facing_the_Freshwater_Crisis_Scientific_American.pdf
Shah T (2009) Taming the anarchy: groundwater governance in South Asia. International Water Management Institute (IWMI), Colombo
Shah (2013) Water: Towards a paradigm shift in the twelfth plan, economic and political weekly, January 19, 2013 XLVIIII No 3
Shah T, Koppen B (2016) The precept and practice of integrated water resources management (IWRM) in India. In: Narain V, Narayanamoorthy A (eds) Indian water policy at the crossroads: resources, technology and reforms. Global issues in water policy, vol 16. Springer, Cham
Shah A, Prakash A (2010) Integrated water resource management in India: from critique to constructive engagement. Central Research Institute for Dryland Agriculture, Hyderabad, pp 333–346
Shankar V et al (2011) India's groundwater challenge and the way forward. Econ Polit Wkly XLVI(2):37
Singh Y, Rahal A (2013) Climate change-impact on agriculture and adaptive and mitigative measures. Kurukshetra, Ministry of Rural Development, Government of India, p 39
Smajgl A et al (2009) Outback institutions: an application of the Institutional Analysis and Development (IAD) framework to four case studies in Australia's outback. DKCRC report 31. Desert Knowledge Cooperative Research Centre, Alice Springs
Sommer L (2011) The theory of planned behaviour and the impact of past behaviour. Int Bus Econ Res J 10(1):91. Available at https://clutejournals.com/index.php/IBER/article/download/930/914
Thow A, Blois M (2008) Climate change and human vulnerability: mapping emerging trends and risk hotspots for humanitarian actors. Discussion paper, Maplecro. Available at https://reliefweb.int/sites/reliefweb.int/files/resources/E5549F80673D1E38C12574B0004B698F-care_mar2008.pdf. Accessed on 12 Apr 2018
TOI (2018) Times of India. Dated 22 March 2018. News available at https://timesofindia.indiatimes.com/city/nagpur/groundwater-level-falling-in-almost-50-area-ofmaharashtra/articleshow/63404707.cms. Accessed on 15 May 2018
UNEP-DHI (2009) Integrated water resources management in action. World Water Assessment Programme, DHI Water Policy, UNEP-DHI Centre for Water and Environment
Vani S et al (2007) Best-bet options for integrated watershed management. In: Proceedings of the comprehensive assessment of watershed programs in India. ICRISAT, Hyderabad
Wagle S (2011) Policy and governance: perspective and analytical framework for management of urban sewage. Available at http://gangapedia.iitk.ac.in/sites/default/files/Second%20Set%20of%20Report/009_PLG_Poiic%20and_Governance_Perspective.pdf
World Bank (2010) Deep wells and prudence: towards pragmatic action for addressing groundwater overexploitation in India. The International Bank for Reconstruction and Development/The World Bank, Washington, DC
Zbigniew W et al (2009) Will groundwater ease freshwater stress under climate change? Hydrol Sci J 54 (4):665. https://doi.org/10.1623/hysj.54.4.665. Special issue: Groundwater and Climate in Africa

Resilience in Climate Stressed Environment Through Water Grabbing

13

Makarius C. S. Lalika

Contents

Abstract

Climate change (CC) is currently considered as the most severe and devastating environmental catastrophe facing the globe. CC is the latest environmental driver of global environmental change causing rainfall shortage and water stress across the globe. This chapter reports a study carried out on water grabbing as solution for food production and water shortages in CC-stricken areas along the Pangani River Basin (PRB) in Tanzania. Results indicated that the mean yields before

M. C. S. Lalika (✉)
Department of Geography and Environmental Studies, Solomon Mahlangu College of Science and Education, Sokoine University of Agriculture, Morogoro, Tanzania
e-mail: makarius.lalika@yahoo.com; lalika_2mc@sua.ac.tz

W. Leal Filho (ed.), *Handbook of Climate Change Resilience*,
https://doi.org/10.1007/978-3-319-93336-8_68

water grabbing were statistically ($P < 0.001$) higher than yields after water grabbing. Drivers for water grabbing include climate change and variability in some parts of the world, human population dynamics, change of consumption patterns, economic growth, and technological advancement. The effects of CC and water grabbing include loss of some aquatic flora and fauna, water use conflicts, and poor crop production. It is recommended that smallholder farmers have to adapt to drought-resistant crops and short-term seed varieties and carry out groundwater research and rainwater harvesting. More research should be carried out so as to know the link between CC and the decline of rainfall and water flow. Moreover, efforts should be made to conserve the Kilimanjaro Mountain to restore the vanishing ice cape.

Keywords
Ecosystem services · Land grabbing · Population growth · Climate change · Water

Introduction

Water resource depletion and rising demand on limited water supplies result in putting at risk ecosystem services (ES) thereby creating water use conflicts and significant deterioration of water and aquatic life (UNDP 2006; URT 2002). The problem is exacerbated by climate change (CC), climate variability, and competition for water uses, notably for large-scale agriculture. Commercial pressure on land and water and its consequence on small farmers are among the major issues being discussed at local, national, and international scale (Cotula et al. 2009, 2011; Hall 2011; Rulli et al. 2012; World Bank 2010). For instance, water has been the prime target of the continued pressures exerted by developed countries and international commercial firms in their acquisitions of land on the African continent (Cotula and Vermeulen 2009a, b; GRAIN 2008; White and Dasgupta 2010; Borras et al. 2012).

Throughout history, human development has depended on water and the potential of water availability as a productive resource (UNDP 2006). Water for life and livelihoods are two of the foundations for human development. Yet, in recent years (World Bank 2010), large section of local communities in the African continent has not benefitted significantly from water-related ES due to the way large-scale foreign firms are established and implemented. In different parts of Tanzania, for instance, water sources and their associated ES are deteriorating due to the establishment of large-scale plantations. The establishment of *Jatropha*, sugarcane, aloe vera, oil palm, *Croton megalocarpus*, and white sorghum (Sulle and Nelson 2009; Gordon-Maclean et al. 2008) has resulted in exclusive rights for land and water which led to what is popularly known as "water grabbing." Excessive water withdrawals and abstraction resulted in water stress and scarcity thereby affecting downstream smallholder irrigators (Lankford 2005a, b; Lankford and Mwaruvanda 2005; Sotthewes 2008; Franks et al. 2011). The situation has been accelerated with the impacts of CC and climate variability in some parts of Tanzania (Lalika et al. 2017; Lalika 2017).

Since mid-1990s, water grabbing has been the order of the day along the Pangani River Basin (PRB). The presence of the mushrooming foreign and private companies mainly dealing with flower irrigation is causing water stress and scarcity; hence water use conflicts along the PRB (Mbonile 2005; Mbonile 2001; Ngana et al. 2010; Lalika et al. 2011). This situation has been worsened by policy failures to enforce clauses and conditions stipulated in signed investment contracts (Sulle and Nelson 2009; Matondi 2010; Komakech and van der Zaag 2011). Despite the presence of well-developed institutional structures and policy frameworks on water, environment, and land management in the country including the National Water Policy (URT 2002), Land Policy (URT 1995), Land Act of 1999 (URT 1999a), Village Land Act of 1999 (URT 1999b), and Environmental Policy (URT 1997), smallholder farmers in the PRB are still affected by water stress and water shortages.

Policy failures to regulate allocation and enhance equitable water distribution have been exacerbated by the influx of human population along the PRB (Mbonile 1999a, b, 2001). The repercussions of population increase may reflect the now classical economic theory of population growth and resource scarcity of Malthus (1798). Malthus in his famous publication on population growth feared that the interaction between people and resources (water) would mean that the combined effects of population growth and increasing demand on a fixed water resource base would have consequences on water stress at an extraordinary scale. His views hold water in today's modern life and water scarcity situation along the PRB.

Water scarcity has been exaggerated by the current impact of CC change, and smallholder farmers have resorted to water grabbing as solution for irrigated agriculture. Nevertheless, there are little information on studies about the empirical evidence on the role of water grabbing as solution in climate stressed areas and water shortage problems along the PRB. The current chapter is an attempt to fill this existing information gap.

The objectives of this chapter were (i) to determine the effect of water grabbing on crop yield along PRB, (ii) to identify drivers for water grabbing in PRB, (iii) to analyze population dynamics in PRB, and (iv) to review and analyze land and water deals in Tanzania. Findings of this chapter and subsequent recommendation are useful to land use for policy makers for making decisions on different land uses and allocation. In addition, the policy recommendations of this chapter could be instrumental for other stakeholders aiming to achieve a more equal water distribution along the PRB.

Materials and Methods

Description of the Study Area

Location

The research for this chapter was conducted in four villages, namely, Patanumbe and Valesca villages in Meru District, Arusha Region; Mawala village in Hai District; and Ngasinyi villages in Moshi Rural District, Kilimanjaro Region along the PRB,

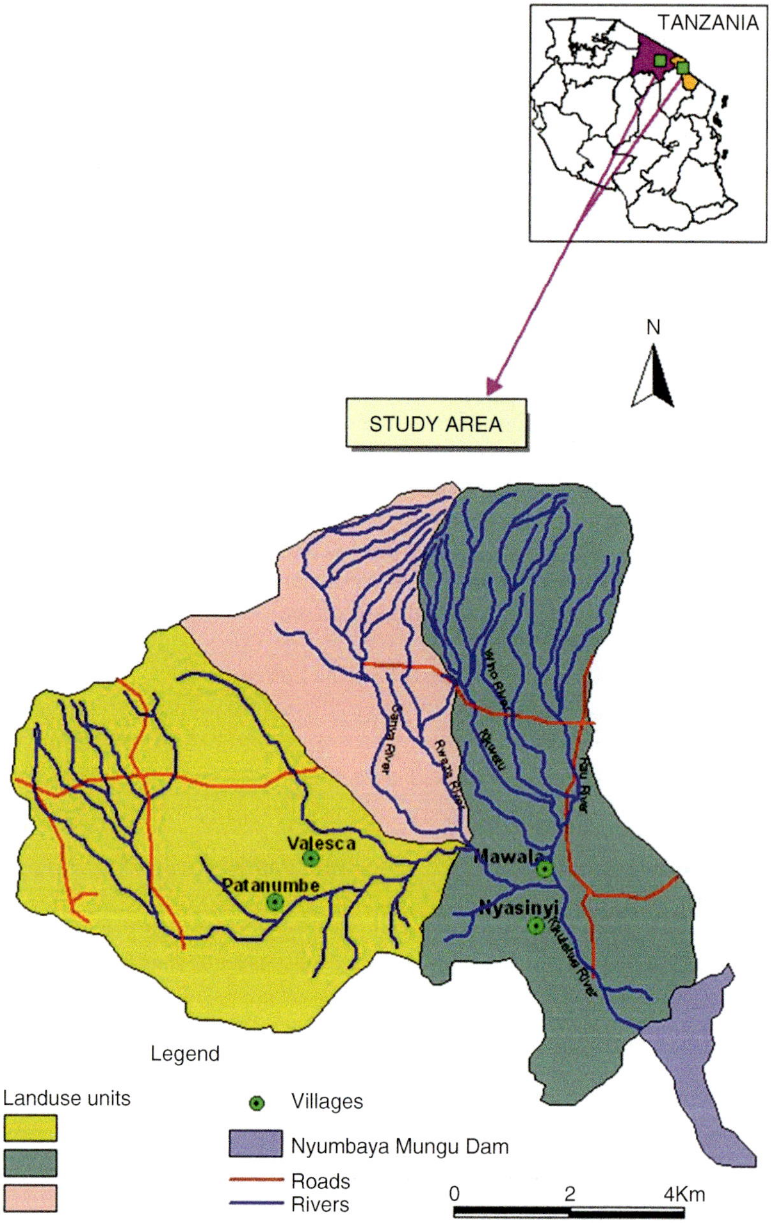

Fig. 1 Location of the study area along Pangani River Basin, Tanzania

Tanzania (Fig. 1). The PRB extends from the northern highlands to the northeastern coast of Tanzania. It lies between latitude 03° 05′ 00″ and 06° 06′ 00″ south and longitude 36° 45′ 36″ and 39° 36′ 00″ east.

Hydrology and Drainage Pattern

The hydrology and drainage pattern in the PRB catchment vary considerably. The PRB comprises of several sub-catchments of widely different characteristics. The Pangani River (PR), which is also referred to (in other publications) as the Pangani Mainstem, rises as a series of several small streams and springs on the southern sides of the Africa's highest peak, Mt. Kilimanjaro, and on Mt. Meru (IUCN and PBWO 2008; IUCN 2007). These streams (Nduruma, Tengeru, Sanya, Malala, etc.) create the Kikuletwa and Ruvu Rivers (Himo, Muraini, etc.) which drain further downstream into the Nyumba ya Mungu (NyM) dam (IUCN and PBWO 2008, 2011; IUCN 2007). The NyM dam has created a man-made water reservoir of ecological and economic importance along PRB. The overflow of the dam (outlet) is known as the Pangani River Mainstem and flows for 432 km before emptying into the Indian Ocean at the Pangani estuary.

The NyM reservoir is the largest water body in the PRB and was constructed in 1965 to enhance river flows for hydropower generation. It was later incorporated into irrigation plans (Mulungu 1997; Ndomba et al. 2008). Besides the power station at the outlet of this dam, other hydropower plants in the PRB are located near Hale and New Pangani Falls. Water released from the NyM dam is essential for supporting ecosystem services downstream. These include nutrient cycling at Kirua Swamp and sedimentation and support of enhancement of ecological processes (e.g., hindering salt water intrusion and coastal erosion) at the estuary mouth in Pangani Town (Ndomba et al. 2008; Shaghude 2006; Sotthewes 2008; Valimba 2005, 2007). Other river tributaries draining in the PRB are Mkomazi and Luengera from the Pare and Usambara Mountain ranges, respectively.

Forest and Vegetation Types

Forest and vegetation in PRB range from forests on mountain slopes to semiarid grasslands (IUCN 2003). The main vegetation types include forests, woodlands, and bushland, along with grassland thicket and plantation forest (Turpie et al. 2005). Plantation forests have replaced natural forests in the highlands, and the larger part of the lowlands is composed of woodland, bushland, grassland, and thicket. Forests perform vital hydrological functions in the PRB including the regulation of runoff, prevention of soil erosion, water storage, and improvement of water quality (IUCN 2003; Msuya 2010). According to IUCN (2003), dominant forest types in PRB include *mangrove forests* (located at the confluence of the Pangani River and the Indian Ocean and protecting the coastlines and soft sediment shorelines from erosion, trapping sediments, and recycling nutrients), *East African coastal forests* (containing remarkable levels of biodiversity and endemism), *Afromontane forests* (playing a key part in hydrological functions), and *riverine forests* (controlling erosion along the river banks). Research and previous studies on forest health conducted in the PRB show that between 1952 and 1982, catchment forests in the PRB declined at a fairly high rate of 3.8% of forest cover per year, while farmlands and settlements increased dramatically by 83% of forest cover per year (Kaoneka 1993; Lambrechts et al. 2002; Newmark 1998).

Climate

Variations in the local climate in the PRB are mostly related to topography. The flatter, lower-lying southwestern half of the basin is arid and hot, while the mountain ranges along the northern and southeastern catchment boundaries have cooler, wetter conditions. The high altitude slopes above the forest line on Mt. Meru and Mt. Kilimanjaro have an afro-alpine climate and receive more than 2500 mm of rainfall per year. Mean annual rainfall increases in a southerly direction along the mountain ranges and varies from about 650 mm per year in the north and south Pare Mountains to 800 mm per year in the western Usambara Mountains and 2000 mm per year in the eastern Usambara Mountains.

Population and Economic Activities

The PRB has an estimated 4.5 million people (data from 2007), and population densities vary between highlands and lowlands. About 90% of the basin's population resides in the highlands with some 900 people per km^2, while lowland densities were around 65 people per km^2 (IUCN 2003). The main causes of forest degradation and deforestation include encroachment for settlement and agriculture as well as increasing demand of forest products (mainly timber and fuel wood) (IUCN 2003). In terms of human population, PRB is a densely populated area in Tanzania, posing serious challenges to sustainable watershed management (Msuya 2010).

Data Collection

Sampling Procedure

A purposive sampling procedure was used where four villages were earmarked for the questionnaire survey (two in Arusha and Kilimanjaro regions, respectively). The decision on the location of the villages was based on their proximity to rivers and the reliance of the local communities on water for irrigation. Based on these two criteria, the main target for this survey was smallholder irrigators. Within each village, respondents were selected using a table of random numbers that corresponded to the household numbers in the village register. Household heads were the target for interviews; however, wherever the head of the household was not around, any household member within that particular household who of 18 years or above was picked for interviews. According to Tanzania regulations and laws, any one at 18 years or above is regarded as mature person. I adapted the 10% sampling intensity giving a total of 170 respondents were interviewed (Table 1).

Data Collection Methods

Quantitative and Qualitative Data

During data collection, both quantitative and qualitative research approaches were used to collect primary and secondary data. Structured questionnaires were used as the main tools to collect primary (quantitative) data. Questionnaire items comprised of questions mainly on drivers/factors of water grabbing, socioeconomic activities, types of water investors/grabbers, types of watershed services affected by water grabbing, crop yield before and

Table 1 Sampled and interviewed respondents

Region	Village	Total households	Sample size
Kilimanjaro	Mawala	330	33
	Ngasinyi	440	44
Arusha	Valesca	500	50
	Patanumbe	430	43
	Total	**1700**	**170**

after the coming of investors, and environmental and socioeconomic effects of water grabbing, to name just a few. Also, different methods for collecting qualitative data were used. Among others, these methods include group focus discussions, face-to-face interviews, and informal and formal interviews. Secondary data were searched and reviewed relevant literatures on irrigated agriculture and water use conflicts in the study area.

Data on Population Census

Data on population dynamics and increase in PRB were collected from the national library for Tanzania Bureau of Statistics in Dar es Salaam. These data were from the National Population Census which is normally conducted after every 10 years. Therefore, I collated population census reports for 6 years, i.e., 1957, 1967, 1978, 1988, 2002, and 2012. Thereafter, relevant information on PRB were extracted and later converted from analogue to digital format (Table 2).

Data on Water and Land Deals

Secondary data on land and water policy issues were reviewed from relevant document on land and water grabbing literatures. For policy issues, documents such as National Water Policy (URT 2002), Land Policy (URT 1995), Land Act of 1999 (URT 1999a), Village Land Act of 1999 (URT 1999b), and Environmental Policy (URT 1997) were reviewed. On land and water grabbing, reviewed document includes Biofuels in Tanzania: Status, Opportunities and Challenges (Mshandete 2011); *Biofuels, Land Access and Rural Livelihoods in Tanzania* (Sulle and Nelson 2009); Foreign land acquisitions in Tanzania: Global ideology and local perspectives (Larsen 2002); Biofuel Industry Study in Tanzania (Gordon-Maclean et al. 2008); and Accumulation by land dispossession and labour devaluation in Tanzania (HakiArdhi 2010).

Data Analysis

Quantitative and Qualitative Data

The 170 structured questionnaires were coded, entered, and cleaned for final analyses. The Statistical Package for Social Sciences (SPSS) version 20.0 was used to analyze quantitative data collected through the structured questionnaires. Later on, multiple response analysis was carried out to obtain frequency and percentages of responses from smallholder farmers. On the other hand, qualitative data were analyzed with the help of participants during group focus discussions through dialogue and intensive debates.

Table 2 Population dynamics in Pangani River Basin, Tanzania

	Years and population					
Region	1957	1967	1978	1988	2002	2012
Kilimanjaro	476,530	650,533	902,437	1,108,699	1,376,700	1,640,090
Tanga	375,923	769,504	1,037,767	1,283,636	1,636,280	2,045,210
Arusha	399,866	601,515	926,223	1,351,675	1,288,090	1,694,310

Population Census Data

Data on population census for five window periods, i.e., 1957–1967, 1968–1978, 1978–1988, 1988–1998, and 2002–2012, were scrutinized, rearranged, and grouped in a single table (Table 10).

With regard to statistical tests of crop yield before and after water grabbing, a two-tailed t-test was used.

Null hypothesis: There is no significant difference in yield before and after water grabbing.

$$Ho : \mu_1 = \mu_2, (\text{i.e.}, \mu_1 - \mu_2 = 0)$$

Alternative hypothesis: There is significant different in yield before and after water grabbing.

$$H_1 : \mu_1 \neq \mu_2, (\text{i.e.}, \mu_1 - \mu_2 \neq 0)$$

To test and compare the means for the net loss in yield before and after water grabbing between villages and within villages, the following hypothesis was used:

"There is no significant difference in net loss in yield before and after water grabbing."

One-way analysis of variance (ANOVA) was applied to test this hypothesis. Thereafter, the Duncan multiple range test was applied to separate the means.

Results

Effect of Water Grabbing on Crop Yield Along PRB

The results in Table 3 indicate that mean yields and their respective standard deviations (std) before water grabbing were higher than yields after water grabbing in all villages. This implies water loss due to water grabbing had negative effects on crop yield. Similarly, findings in Table 3 indicated that standard deviations for yield before water grabbing were higher than their means only in two villages (i.e., Mawala and Patanumbe).

Furthermore, a two-tailed statistical test was applied to test the significance of the crop yield before and after water grabbing. The two-tailed t-test was preferred

Table 3 Maize means yield before and after water grabbing (kg/ha/season) in PRB, Tanzania

Village	Maize Yield (kg/ha/season)	Mean ± Std	Std error mean
Mawala	Before water grabbing	1479.89 ± 2931.02	544.28
	After water grabbing	355.17 ± 214.88	39.90
Patanumbe	Before water grabbing	846.30 ± 1097.75	167.41
	After water grabbing	252.36 ± 150.66	22.98
Valesca	Before water grabbing	1255.31 ± 663.56	95.78
	After water grabbing	462.76 ± 271.68	39.21
Ngasinyi	Before water grabbing	1067.54 ± 503.82	77.74
	After water grabbing	480.00 ± 237.95	36.72

Table 4 Statistical test for crop yield before and after water grabbing (kg/ha) in PRB, Tanzania

Village	Mean ± Std	Std error mean	t-value	Sig (two tailed)
Mawala	746.14 ± 1374.82	108.02	6.91	0.000***
Patanumbe	593.93 ± 1060.42	161.71	3.67	0.001***
Valesca	792.54 ± 527.69	76.17	10.41	0.000***
Ngasinyi	587.54 ± 375.96	58.01	10.13	0.000***

***Significance at $P < 0.001$

Table 5 Net loss in crop yield in PRB, Tanzania

				95% Confidence interval for mean	
Village	N	Mean ± Std	Std error	Lower bound	Upper bound
Mawala	29	485.06 ± 704.22	130.77	217.19	752.93
Patanumbe	43	562.54 ± 1044.38	159.27	241.13	883.95
Valesca	48	795.52 ± 528.13	76.23	642.17	948.87
Ngasinyi	42	587.54 ± 375.96	58.01	470.38	704.70
Total	162	624.18 ± 708.38	55.66	514.27	734.09

because the assessment was from the same respondents for the two scenarios (i.e., yields before and after water grabbing). Results are summarized in Table 4 below.

As denoted in Table 4, the mean yield before and after water grabbing was statistically significant ($P < 0.001$). Thus, the null hypothesis that "there was no significant difference in mean yield before and after water grabbing" was rejected.

Results from one-way ANOVA on the net loss in crop yield are displayed in Table 5. While the standard deviation for the net loss in Mawala and Patanumbe villages was greater than their respective means, the situation was opposite in Valesca, and Ngasinyi villages means were greater than their respective standard deviations.

Results on statistical test about the net loss between and within villages before and after water grabbing are displayed in Table 6.

It was revealed that $p = 0.225$. Therefore, $p > 0.05$ implying that there was no significant difference in yield net loss between villages and within villages. Based on these findings, the null hypothesis is hereby accepted.

Table 6 One-way ANOVA statistical test for net loss between and within villages in PRB, Tanzania

Villages	Sum of squares	df	Mean square	F	Sig.
Between villages	2190236.13	3	730078.71	1.47	0.225**NS**
Within villages	78600914.59	158	497474.14		
Total	80791150.71	161			

NS Non statistical significance ($P > 0.05$)

Table 7 Responses on local drivers for water grabbing in the PRB, Tanzania

Driver	Frequency	Percentage
Poor water governance	160	29.6
Corruption	150	27.7
Lack of transparencies in contracts/agreements	107	19.8
Inadequate environmental impact assessments	64	11.8
Lack of integrated planning	57	10.5
Top-down management approaches	3	0.6

Drivers for Water Grabbing

Drivers at (in PRB) Local Scale

Land and water grabbing are currently seen as the major problem to many poor families who rely solely on smallholder agriculture. In the PRB, we identified a number of local drivers for water grabbing (Table 7). According to perceptions from smallholder irrigators, they include poor water governance (29.6%), corruption (27.7%), lack of transparency (19.8%), inadequate environmental impact assessment (11.8), lack of integrated planning (10.5), and top-down management approaches (0.3%).

It was found that poor water governance in the PRB is related to inadequate administration in water conservation authorities, insufficient approaches in information delivery, policy isolation in conservation issues, and funding constraints (Charbit 2011), to name just a few.

As indicted in Table 7, corruption (27.7%) was identified as one of the drivers for water grabbing in the study villages. It was reported that some of unfaithful government leaders at national and local levels conspired with investors over investment contracts at the detrimental of smallholder farmers. Consequently, the situation led to water use conflicts and unnecessary loss of human properties and chaos. Furthermore, lack of transparency (19.8%) was identified as one of the drivers for water grabbing along the PRB. It was reported that lack of transparent and democratic decision process accounted a lot for the loss of land with ample water to investors. The absence of negotiation with the local communities contributes a lot to resource use conflicts (Mbonile 2005), and this was evident in this study along the PRB. Findings by the World Bank (2010), Cotula et al. (2009), and Rulli et al. (2013)

Table 8 Responses on global drivers for water grabbing

Driver	Frequency	Percentage
Climate change and variability	153	28.5
Global and local population growth	136	25.4
Shifting consumption patterns	134	25.0
Economic growth	113	21.1

reported similar sentiments that many land deals were done with inadequate local population involvement, inadequate compensation, and without clear plans for new jobs and ecological integrity. Other contributing drivers identified in the study villages include inadequate environmental impact assessments (11.8%), lack of integrated planning (10.5%), and top-down management approaches.

Drivers at Global Scale

Table 8 reveals drivers for water grabbing at international scale. Local communities' perceptions are that commercial firms that involve in water grabbing are new forms of neocolonialism.

Findings in Table 8 show that climate change (28.5%) plays a key role in global water grabbing. Climate change and persistent climate variability influence water (Boko et al. 2007) in some areas thus forcing multinational companies to shift their investments abroad.

Consequently, these investments affect water availability, accessibility, normal water circle, and daily economic activities of local communities in the area of destination.

It was also found that an increase of global and local human population (25.4%) is to blame for the current water grabbing in the PRB. During field excursion, it was found that different economic undertakings range from irrigation and ranching to mining. Some of these socioeconomic activities were carried out by either foreign investors or Tanzanians but from different areas of the country. Furthermore, it was revealed that population increase in PRB was either natural (i.e., natural birth) or artificial increase (through population immigration). Therefore, the number of human population outpaced the land's carrying capacity, thereby, resulting into completion for land and water resources. Confiscation of water resources and water withdrawals along the PRB is mainly for irrigated agriculture. The situation results in irrigation water scarcity to the downstream smallholder irrigators, social unrest, and resource use conflicts (Mbonile 2005; The Oakland Institute 2011).

The results of shifting consumer patterns (25%) are presented in Table 8. The combination of drivers such as climate change, economic growth, and global and local population growth has contributed to the shift of consumer preferences and change in production systems. These attributes when combined with the declining natural resource base result in poverty thereby exposing the rural poor in vulnerable situation. The situation is also evident in the PRB where change in consumer preference across the globe has driven some foreign firms to invest abroad. Changes in consumer behavior necessitate a deliberate transformation in the approach to

Table 9 Responses on ecosystem services affected by water grabbing in the PRB, Tanzania

Ecosystem service	Frequency	Percentage
Disappearance of fish species	104	29.3
Water for HEP production	63	17.7
Water for livestock uses	62	17.5
Disappearance of birds	55	15.5
Water for domestic use	37	10.4
Disappearance of aquatic plants/trees	17	4.8
Reduced food crop harvest	9	2.5

development and a significant increase of investment in agriculture in the developing world. For quite some time, foreign firms along the PRB are dealing with the production of sugarcane, *Jatropha*, flowers, vegetables, and horticultural crops instead of cereals (e.g., wheat and rice) and coffee that used to be grown in the previous years.

Results of economic growth (21.1%) as one of the global drivers for water grabbing are presented in Table 8. It is believed that once the economy flourish with super profit, it needs to be reinvested to generate more profit. The ideal area for this reinvestment is in developing countries (the African continent) where there is ample fertile and virgin land.

Other Ecosystem Services Affected by Water Grabbing

Table 9 presents findings of other ecosystem services that are affected by the appropriation (water grabbing) of water resources in the PRB.

Excessive water (abstraction) grabbing (29.3%) causes water fluctuation, change of river flow regime, and affects the river health. Water decrease affects fish breeding and supplies of tilapia species. In recent years, fishing activities have become seasonal due to water fluctuation accompanied by fish disappearance. Some parts of the river were no longer perennial; some had fast-flowing water, shallow areas, and deeper runs with fish species adapted to living in fast-flowing streams (e.g., *Garra dembeensis*) (PBWO and IUCN 2008). Some of fish species endemic to the PRB include the *Oreochromis pangani*. Mwamila et al. (2008) found that the reduction in water flow affected negatively fish productivity, whereas higher flows result into swamp and floodplain formation. Normally, flooding conditions are necessary for fish spawning, feeding, and growth of young fish species.

Table 9 reveals that hydroelectric power production (17.7%) is another crucial ES along the PRB which is affected by water grabbing. The NyM dam is the largest water body potential for hydropower production. Other hydropower plants affected by water grabbing along the PRB include Hale and Pangani. Other ES along the PRB includes water for livestock uses (17.5%), disappearance of water birds (15.5%), water for domestic use (10.4%), and disappearance of aquatic plants/trees (4.8%). Aquatic tree affected by reduction of water flow includes the fig trees (*Ficus* ssp.) which are adapted to surviving in dry areas due to their ability to reserve/store water and mangrove tree species. Other ES that is affected by shortage of water flow is

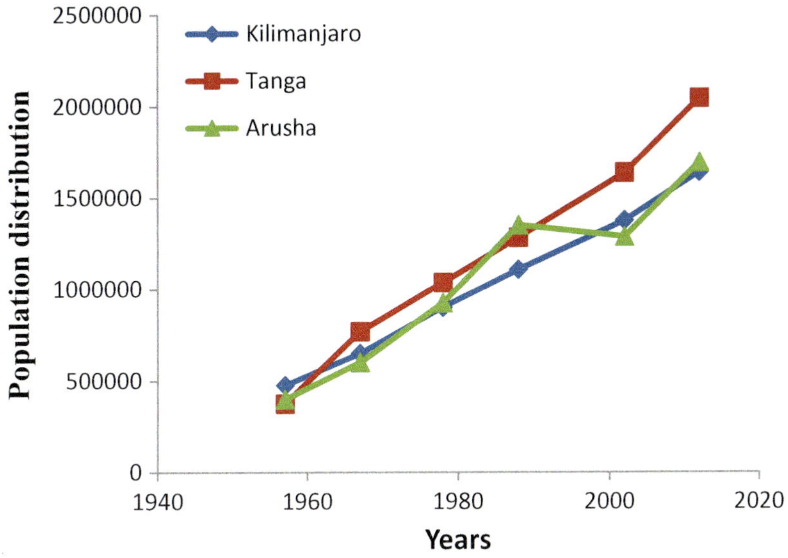

Fig. 2 Population dynamics for 1957–2012 along Pangani River Basin, Tanzania

food crop harvest (2.5%) as elaborated in section "Effect of Water Grabbing on Crop Yield Along PRB" in Tables 3, 4, and 5.

Population Dynamics in PRB

Figure 2 shows the trend of population increase along PRB. Overall, Tanga Region had many people than Arusha and Kilimanjaro (Table 2). This is also testified by the alignment of the trend lines in Fig. 2.

According to Fig. 2, population census 1957–2012 population increase was consistent in Tanga and Kilimanjaro regions. The dramatic fall of population trend in Arusha Region in 2000 is due to the split of administrative units where Manya Region was chopped out from Arusha Region. The increase of population trend is also reflected in Table 10 where population has increased over time.

Land and Water Deals in Tanzania

Table 11 summarizes examples of land acquired by foreign firms for biofuel plantations in Tanzania by the end of 2010. During literature review, it was realized that majority of investments were geared toward biofuel production for filling the void of energy demand. During the documentary review, it was found that majority of foreign firms engaged on large-scale plantations of *Jatropha*, sugarcane, aloe vera, oil palm, flowers, white sorghum, and *Croton megalocarpus*.

Table 10 Population changes along the Pangani River Basin, Tanzania, between 1952 and 2012

	Population per region		
Years	Kilimanjaro	Tanga	Arusha
1957	476,530	375,923	399,866
1967	650,533	769,504	601,515
1978	902,437	1,037,767	926,223
1988	1,108,699	1,283,636	1,351,675
2002	1,376,700	1,636,280	1,288,090
2012	1,640,090	2,045,210	1,694,310

Believers of the foreign investments assert that these investments would have positive outcomes to agricultural production, add value to local products and markets, and improve social services such as road infrastructure, health facilities, clean water supply, and education (HakiArdhi 2010). Experience, nevertheless, indicates that problems tied with foreign investments outweigh the perceived benefits.

Discussion

Effect of Water Grabbing on Crop Yield Along

Investment in irrigated agriculture is one of the main means of achieving sustained crop yield. It is through sustained crop yield where food security and community welfare can be ensured at household level. On the contrary, irrigated agriculture is faced with water grabbing by foreign investments (Table 3) something which leads to loss of crop yield. Sustainable agriculture and crop yield in the study villages are curtailed by foreign companies (located upstream) who abstract water for flower irrigation. Policy failures to enforce signed agreements (e.g., use of groundwater) are a testimony on how the policy framework hasn't answered problems of water shortages faced by smallholder irrigators.

The influence of water grabbing on crop yield indicated in Table 4 is another indicator of the plight of water grabbing to irrigated agriculture. Results of statistical significance on mean yield before water grabbing ($P < 0.001$) confirm how bad water grabbing is. Given the current situation of the failure of rainfed agriculture due to climate change and climate variability in PRB, irrigated agriculture would have been the solution for boosting crop yield and community livelihoods through increased income and provision of ES (i.e., food). Similar observation is echoed by FAO (2002) that increased crop yield has extra benefits than just income increase. Sustainable irrigation creates on-farm employment, ensures food security, and lowers food prices. Irrigated agriculture reduces poverty as well, because the poor normally spend 60–70% of their income on food (FAO 2002).

Table 11 Land grabbing status in Tanzania by 2010

S/N	Investor	Crop	Location	Land requested (hectares)	Land acquired (hectares)	Project status
1	FELISA	Oil palm	Kigoma	5000	4258	Land dispute in court for extra 350 ha obtained from 2 villages. No EIA done
2	BioShape	*Jatropha*	Kilwa, Lindi	82,000	34,000	400 ha pilot farm planted. Integrity of EIA questioned
3	Sun biofuel	*Jatropha*	Kisarawe, coast	50,000	8211	8211 ha of land formerly belonging to 12 villages transferred to general land; derivative title being finalized
4	SEKAB BT	Sugarcane	Bagamoyo, coast	24,500	22,500	Seed cane planted, and irrigation reservoir constructed
5	SEKAB BT	Sugarcane	Rufiji, coast	400,000	0	In land acquisition process
6	Diligent Tanzania ltd.	*Jatropha*	Arusha; Babati, Manyara; Handeni, Tanga; Singida; Monduli, Arusha	n/a	n/a	Contracted over 4000 farmers
		Croton megalocarpus		n/a	n/a	Collecting seeds from natural and planted forests
7	Donesta ltd. & Savannah biofuels ltd.	*Jatropha*	Dodoma	n/a	2000	200 ha planted
8	Trinity consultants/ bioenergy TZ ltd.	*Jatropha*	Bagamoyo, coast	30,000	16,000	Surveying land to be granted

(*continued*)

Table 11 (continued)

S/N	Investor	Crop	Location	Land requested (hectares)	Land acquired (hectares)	Project status
9	Shanta Estates ltd.	*Jatropha*	Bagamoyo, coast	n/a	14,500	Agreement with villagers signed
10	Tanzania biodiesel plant ltd.	Oil palm	Bagamoyo, coast	25,000	16,000	Land not surveyed; land granted by district but not by TIC
11	Clean power TZ ltd.	Oil palm	Bagamoyo, coast	n/a	3500	Project abandoned after realized high cost of doing land use plans
12	Agriculture bio-energy Tanzania	White sorghum	Bagamoyo, coast	n/a	25,000	Land request approved but asked to do land use plans
13	ZAGA	*Jatropha*	Kisarawe, coast	n/a	n/a	Applied for land
14	African green oils	Oil palm	Rufiji, coast	n/a	860	Planted 360 ha and financing land use plans in 7 villages
15	InfEnergy co. ltd.	Oil palm	Kilombero	n/a	5818	Land lease pending. Cultivating rice while growing oil palm
16	Massive	*Jatropha* and Pangamia	Lindi	n/a	50,000	
17	JCJ co. ltd.	*Jatropha*	Mwanza Mara Shinyanga Tabora	n/a	n/a	Aimed to sensitize local communities, but project abandoned due to alleged lack of government support
18	ABERC	*Croton megalocarpus*	Biharamulo, Kagera	n/a	20,000	No operational progress due to lack of funds
19	Prokon BV	*Jatropha*	Mpanda, Rukwa	n/a	10,000	Contract farming with 2000 smallholders; does not own

(*continued*)

Table 11 (continued)

S/N	Investor	Crop	Location	Land requested (hectares)	Land acquired (hectares)	Project status
						any plantation land
20	Mitsubishi corporation	*Jatropha*	Arusha Dar es Salaam coast	n/a	n/a	Looking for land in these regions
21	Kapunga Rice project	*Jatropha*	Mbarali, Mbeya	n/a	50,000	Planned to replant rice with *Jatropha*; president recently ordered that rice cultivation patterns not be changed
22	DI oils Tanzania ltd.	*Jatropha*			n/a	Abandoned plans for Tanzania
23	Kikuletwa farm	*Jatropha* and Aloe vera	Kilimanjaro	n/a	400	Growing *Jatropha*

Source: Adapted from HakiArdhi (2010), Gordon-Maclean et al. (2008) and Sulle and Nelson (2009)

Local and Global Drivers for Water Grabbing in PRB

Local in PRB

Water grabbing exists in PRB because local communities (majority of whom are small scale irrigators) have lost access to fertile land and ample water they used previously to support their living. Opinions from respondents suggest that poor water governance (Table 7) is the main driver for water grabbing in the study villages. Village and irrigation officers do not enforce the political, social, economic, and administrative procedure articulated in the bylaws guiding water use and distribution. Majority of villagers don't participate fully during contract negotiations and decision making. Lack of full local community's involvement leads to poor and unfair decisions at the detrimental of local communities. In the study villages, for instance, smallholder irrigators have no access to fertile land and water that they previously used to have the right to. In Arusha, large tracts of land are appropriated by foreign investors after signing with the government leaders. Local communities were poorly informed of the actual benefits and consequences of these land deals. As things stand, they have little rights to stop the land acquisition and claim the land back (Sulle and Nelson 2009). Furthermore, foreign companies located upstream divert large quantities of water (than the amount portrayed in the signed contracts) at the expense of downstream smallholder irrigators. This affects irrigation activities, provision of other ES (e.g., production of hydropower electricity), and the enhancement of ecological integrity along the river basin.

Corrupt leaders and lack of transparency are also key factors for water grabbing in the study villages. Responses from local smallholder irrigators revealed in Table 7 testify this existence. Decisions on who should be given a fertile land (and water) are corrupt oriented and are with low transparency and accountability. Findings from informal discussion with smallholder irrigators indicated that rampant corruption and absence of open village meetings are key to water grabbing in PRB. Consequently, once investment contracts are signed, environmental issues are not given special priority thus affecting the sustainable provision of ES and conservation of aquatic diversity within the river basins.

Inadequate environmental impact assessment, lack of integrated planning, and top-down management approaches are associated with the current level of water grabbing and water use conflict in the study villages. Large-scale investments on agriculture require a careful assessment of its environmental consequences before its initiation. FDIs of *Jatropha*, sugarcane, and flower plantations in Kilimanjaro and Arusha regions do not adhere to the recommendation from scientists about adverse environmental impacts. In turn, there have been cases of water pollution caused by agrochemicals and pesticides (Hellar and Kishimba 2005; Hellar-Kihampa 2011; 2013) thereby affecting aquatic species, ES, and human health. Sometimes, majority of these kind of investments are often negotiated at the highest level (e.g., national or regional level) neglecting the views of local leaders (at village levels) where investment are to take place. Moreover, important decisions on agricultural investment plans (poor planning) or integrated water resource management (IWRM) are always decided at ministerial level thereby leading to chaos and conflicts during implementation process. For example, uprooting natural forest in favor of large-scale plantation (e.g., sugarcane and *Jatropha*) affects biodiversity, water flow, river regime, nutrient cycling, and the other ES necessary for human well-being.

Global Drivers

Climate change and climate variability have already caused effects on production of food at global scale especially in developing countries. These negative effects are felt more on provisioning ES where significant reduction of maize, wheat, and rice yield has been reported (Howden and O'Leary 1997; Hoogenboom 2000; Gbetibouo and Hassan 2005; Challinor and Wheeler 2008). To feed the growing population in the areas hit by climate change and climate variability and also fill the gap of global food shortages and demands, investments are carried out in developing countries (including in the PRB Tanzania). However, the way these investments are carried out (i.e., excessive water abstraction) seems to have detrimental effects to the majority of native population and provision of ES. Water grabbing is depriving smallholder farmers of irrigation water thereby affecting the local economy.

Climate variability has accelerated the droughts, compromised rainfed agriculture and hence food shortages thereby necessitating irrigated agriculture. Land grabbing in the PRB is not just rush for fertile soil, rather it aims at fertile soil with ample water for irrigated agriculture. Thus the motive behind land appropriation is to carry out water irrigation because rainfall is not reliable for rainfed agriculture. These views concur with the argument by Rulli et al. (2012) who asserted that irrigated agriculture will remain the largest user of water where it accounts for more than 80% of use in developing countries.

Population increase has an implication on the future of natural resources, utilization, and conservation (Rulli et al. 2012). This is also revealed by the responses on Table 8 where respondents in PRB indicated that the global and local population growth enhance water and land grabbing. According to Mbonile (2005), global and local population growth result in increased food demands and changes in food preferences thereby exerting pressure on the global land and water resources to satisfy those growing human needs. The global land and water resources are, therefore, under severe demands to satisfy the needs of growing human population (Gleick 2000; HakiArdhi 2010; Molden and de Fraiture 2010; Rulli et al. 2012). In recent years, population increase has been accompanied with consumer preference on certain food types. Unfortunately, some of the preferred food types are not locally grown in European countries and in the USA, thus necessitating overseas plunder through water and land grabbing (World Bank 2010; White and Dasgupta 2010; Borras et al. 2012). The change of food preferences is sometimes induced by health reasons (high blood pressure and diabetes), lifestyle, indicator of income increase, or just effects of globalization. These changes influence types of manufacturing and processing industries, energy required, and raw materials as well. Thus, crossing the boundaries for water and land grabbing is a strategy to fulfill the consumers' requirements (Rulli et al. 2012) at the expense of ES at the area of destination.

As the economies tend to flourish, the need for reinvestments arises for capital accumulation (Harvey 2003). This in turn necessitates the exportation of surplus value (financial) for investment abroad. Therefore, economic growth (Table 8) is also a driver believed to have fueled water and land grabbing globally in the PRB. Developed countries like the Gulf States, China, the Netherlands, the USA, India, and South Korea are at the forefront of new investments in farmland abroad either for food or biofuel (von Braun and Meinzen-Dick 2009). In economic terms, financial capital is reinvested where production costs are much lower and where land and water are more abundant. Unfortunately, these foreign-funded biofuel and food plantations in PRB are causing environmental and socioeconomic disaster through watershed degradation, forest and biodiversity abuse, water use conflicts and ecological, significant reduction of ES (water), and sustainability uncertainty along the PRB.

Other Ecosystem Services Affected by Water Grabbing

Rivers require a certain quantity of water and quality of riparian vegetation for ecological integrity and to support aquatic life (Baker et al. 2006; Kediziora 2010; Randhir and Ekness 2013) including fish species. The current water abstraction in the PRB affects availability of ES, and low water flow is a problem to aquatic plants and ecological integrity. As indicated in Table 9, opinions from respondents show that low water flow affects existence of some fish species and aquatic plants. Low water level affects also hydroelectric power production at NyM dam, Hale, and New Pangani falls power plants. Increase of FDIs upstream consumes a lot of water intended for downstream smallholder irrigators and for electricity generation. Low contribution of these three electricity plants to the national electricity network affects the country's economy. Furthermore, reduced water flow and water level affects water bird breeding grounds, the bird watching sub-economy, and the tourist industry as well. Therefore, excessive water diversions (by investors) affect the bird watching tourism along PRB due to low water in the breeding grounds. Thus, efforts need to be done in order to improve water availability for domestic uses and restore aquatic plants and water tree species and water availability for crop-irrigated agriculture. Sustainability of these ES will ensure the welfare of communities residing along PRB.

Population Dynamics in the PRB

The population dynamics in the PRB (Table 10) according to population census between 1957 and 2012 is relatively in increasing trends. The increase has a clear implication to the supply of ES especially water for domestic and irrigated agriculture and more land for agriculture and settlement. But the timing of this increase is coinciding with influx of foreign investments looking for fertile land and ample water for investments. Substantial amount of water is abstracted by foreign companies (large-scale flower irrigators) that located upstream at the detrimental of smallholder farmers located far downstream.

Population growth along the PRB has resulted in increased number of irrigation farms, irrigation canals, and other human activities that rely on water use. As a result, water is not enough to satisfy human demands, and this water shortage is causing chaos and water use conflicts. Mbonile (2001, 2005) reported a number of water use conflicts along the PRB. They include conflict between communities and conservation organizations, upstream and downstream water users, hydroelectricity plants and other water users, farmers and livestock keeper, rural dwellers and urban communities, and communities and river basin authorities.

Land and Water Deals in Tanzania

Results on land and water deals presented in Table 11 are just indicative examples representing few cases of the major problem prevailing in the country. These examples focus much on biofuel and food crop investments. Other land deals in Tanzania (which are not in the table) focus on carbon sequestration projects under the umbrella of reducing emissions from deforestation and forest degradation (REDD) and clean development mechanism (CDM) initiatives.

Land and water grabbing highlighted in this study not only reveal a sign of poor knowledge of local communities over their resources, but it indicates the extent of corruption among our leaders. Some of the leaders responsible for safeguarding natural resources conspire with investors in order to violate regulations, guideline, and laws that guide proper land and water utilization. In some case, even clauses stipulated clearly in the 2002 National Water Policy, 1995 Land Policy, 1999 Land

Use Act, and 1999 Village Land Use Act are bypassed in favor of foreign investments. Land and water grabbing in Tanzania are an example of the validity of the applicability of classical economic theories (Harvey 2003) on wealth appropriation through "primitive accumulation."

Primitive accumulation in the PRB involves depriving poor smallholder farmers of their fertile land in the name of boosting the local economy through capital investment. Letting fertile land and water to foreign firms (Table 11) is a testimony of the operationalization of the Marxist theory of accumulation by dispossession (Harvey 2003). In the PRB, there are transfers of factors of production (land and water) from the majority (poor Tanzanians) to the few (foreign companies). What is actually going on is the transfer of sources of income from smallholder farmers to few who claim themselves as investors. In economic terms, this situation is like taking away sources of income from the majority (smallholder farmers) to the minority without creating new investment. Water and land grabbing in the PRB are like "the conversion of various forms of property rights into exclusive private property rights" (Larsen 2012). In a nutshell, land and water grabbing in Tanzania are another form of capital accumulation through neocolonialism dubbed as "accumulation by deprivation." It is a new neocolonial approach in disguise aimed at plundering natural wealth from developing countries.

Conclusion

Foreign investments have potential contribution to the development especially where the central government and private sector lack sufficient capital and skilled labor. Nevertheless, the current chapter has demonstrated that these foreign investments are exploitative in nature and influence water grabbing along the PRB. The significant influence of water grabbing on crop yield before water grabbing is an indicator of the negative effects of foreign investments to smallholder irrigators. This testifies that the perceived positive expectations from these investments are outweighed by the actual negative outcomes.

This chapter has revealed that drivers for land and water grabbing displayed are influenced by policy failures, poor governance, and lack of commitment among officers responsible for laws and regulations enforcement. Therefore, these negative outcomes and failure to oversee their responsibilities call for enforcement of policies, laws, and bylaws in order to offset the drivers and negative effects of land and water grabbing. Contracts signed between the government and foreign investors don't adhere to clauses outlined in national policies. Furthermore, these contracts are violated and increasing corruption sums up the problems facing land and water deals. Adjusting land laws, policies, by-laws and enforcing them to benefit smallholder farmers would bring about a win-win situation between villagers and foreign investors. This would also remove features of neocolonialism (e.g., land alienation, forced labor, bad working conditions, and discrimination) which are witnessed in majority of large-scale plantations along the PRB.

Investments involving land and water in Tanzania have so far created more problems (conflicts) than the expected benefits (such as employment and community welfare). Majority of foreign investments has jeopardized the security of tenure and the national interest as well. Lack of transparency and accountability in contract negotiations has created loopholes for corruption. Based on the above results and discussion, I recommend enforcement of Land Policy, Land Use Act, and the Village Land Act; implementation of investments aimed at improving food security; transparency in land acquisition; and investments that aim at ensuring the social security and remove vulnerability to water grabbing problems. Furthermore, I recommend foreign investments to assist local communities to explore groundwater, rainwater harvesting, and construction of water reservoirs.

Acknowledgment The financial support for this work came from the Government of Belgium (PhD scholarship grant number 09TAN/5917) through the Belgium Technical Cooperation (BTC).

References

Baker ME, Weller DE, Jordan TE (2006) Improved methods for quantifying potential nutrient interception by riparian buffers. Landsc Ecol 21:1327–1345

Boko M, Niang I, Nyong A, Vogel C, Githeko A, Medany M, Osman-Elasha B, Tabo R, Yanda P (2007) Africa. In: Parry ML, Canziani OF, Palutikof JP, van der Linden PJ, Hanson CE (eds) Climate change 2007: impacts, adaptation and vulnerability. Contribution of working group II to the fourth assessment report of the intergovernmental panel on climate change. Cambridge University Press, Cambridge, UK, pp 433–467

Borras SM, Kay C, Go'mez S, Wilkinson J (2012) Land grabbing and global capitalist accumulation: key features in Latin America. Can J Develop Stud 33:402–416

Challinor AJ, Wheeler TR (2008) Crop yield reduction in the tropics under climate change: processes and uncertainties. Agric Forestry Meteorol 148:343–356

Charbit C (2011) Governance of public policies in decentralized contexts: the multi-level approach. In: OECD regional development working papers, 2011/04. OECD Publishing, Paris

Cotula L, Vermeulen S (2009a) Land grabs in Africa: can the deals work for development? Briefing. International Institute for Environment and Development (IIED), London

Cotula L, Vermeulen S (2009b) Deal or no deal: the outlook for agricultural land investment in Africa. Int Aff 85(6):1233–1247

Cotula L, Vermeulen S, Leonard R, Keeley J (2009) Land grab or development opportunity? Agricultural investment and international land deals in Africa. International Institute for Environment and Development, Food and Agricultural Organization of the United Nations, and International Fund for Agricultural Development, London/Rome

Cotula L, Vermeulen S, Mathieu P, Toulmin C (2011) Agricultural investment and international land deals: evidence from a multi-country study in Africa. Food Security 3:99–113

FAO (2002) Investments in land and water. In: Proceedings of the regional consultation, Bangkok, Thailand, 3–5 October 2001. RAP Publication 2002/09

Franks T, Bdliya H, Mbuya L (2011) Water governance and river basin management: comparative experiences from Nigeria and Tanzania. Int J River Basin Manage 9(2):93–101. https://doi.org/10.1080/15715124.2011.591292

Gbetibouo G, Hassan R (2005) Economic impact of climate change on major south African field crops: a Ricardian approach. Global Planet Change 47:143–152

Gleick PH (2000) The changing water paradigm: a look at twenty-first century water resources development. Water Int 25(1):127–138

Gordon-Maclean A, Laizer J, Harrison P, Shemdoe R (2008) Biofuel industry study, Tanzania: an assessment of the current situation. A report submitted to the World Wide Fund for Nature (WWF) Tanzania Country Office, Dar es Salaam. 95pp

GRAIN (2008) Seized! The 2008 land grab for food and financial security, GRAIN, http://www.grain.org/go/landgrab

HakiArdhi (2010) Accumulation by land dispossession and labour devaluation in Tanzania: the case of biofuel and forestry investments in Kilwa and Kilolo, Tanzania. Research report submitted to land rights and resources institute, 62pp

Hall H (2011) Land grabbing in southern Africa: the many faces of the investor rush. Rev Afr Polit Econ 38(128):193–214. https://doi.org/10.1080/03056244.2011.582753

Harvey D (2003) The new imperialism. Oxford University Press, New York 265pp

Hellar-Kihampa H (2011) Pesticides residues four rivers running through an intensive agricultural area, Kilimanjaro Tanzania. J Appl Sci Environ Manag 15:307–316

Hellar-Kihampa H (2013) Micro-contaminants analysis of water and sediments from Pangani River Basin, Tanzania. Thesis for Award of PhD Degree of the University of Antwerp, Belgium. 203pp

Hellar H, Kishimba MA (2005) Pesticides residues in water from TPC sugarcane plantations and enviros, Kilimanjaro, Tanzania. Tanzan J Sci 31:13–22

Hoogenboom G (2000) Contribution of agrometeorology to the simulation of crop production and its applications. Agric For Meteorol 103:137–57

Howden SM, O'Leary GJ (1997) Evaluating options to reduce greenhouse gas emissions from an Australian temperate wheat cropping system. Environ Model Soft 12:169–176

IUCN (2003) Pangani Basin: a situation analysis. IUCN Eastern Africa Programme, Nairobi, 104pp

IUCN (2007) Pangani river system. State of the basin report – 2007 Tanzania. Submitted to Pangani Basin Water Office and IUCN Eastern Africa Regional Office. 51pp

IUCN, PBWO (2008) Scenario report: the analysis of water allocation scenarios for the Pangani River Basin. Moshi: The World Conservation Union (IUCN) and Pangani Basin Water Office (PBWO)

Kaoneka ARS (1993) Land use in the West Usambara Mountain: analysis of ecological and socio-economic aspects with special reference to forestry. Doctor of science thesis, Department of Forestry, Agricultural University of Norway, Report No. 2. 78pp

Kediziora A (2010) Landscape management practices for maintenance and enhancement of ecosystem services in countryside. Ecohydrol Hydrobiol 10:133–152

Komakech HC, van der Zaag P (2011) Understanding the emergence and functioning of river committees in a catchment of the Pangani basin, Tanzania. Water Alternatives 4(2):197–222

Lalika MCS (2017) Potential for payments for watershed services and climate change adaptation in Pangani River Basin, Tanzania. PhD thesis. Submitted for the degree of doctor in environmental science at the University of Antwerpen, Belgium, 268pp

Lalika MCS, de Deckere E, Ngaga YM (2011) Payment for water services as basis for natural resource management: experience from Pangani Basin Tanzania. In: Anderson K, Okeyo-Owur JB, Hezron M (eds) Towards implementation of payment for environmental services (PES). BOD Publishers, London, pp 176–195

Lalika MCS, Meire P, Ngaga YM, Sanga GJ (2017) Willingness to pay for watershed conservation: are we applying the right paradigm? Ecohydrol Hydrobiol 17(1):33–45

Lambrechts C, Woodley B, Hemp A, Hemp C, Nyiti P (2002) Aerial survey of the threats to Mount Kilimanjaro forests. United Nation Development Programme (UNDP), Dar es Salaam, 48pp

Lankford, Bruce A (2005a) Rural infrastructure to contribute to African Agricultural Development: The case of irrigation. Report for the commission for Africa. Overseas Development Group, Norwich, UK. www.uea.ac.uk/dev/faculty/lankford/cfa_irrig_may05.pdf. Accessed 30 June 2013

Lankford Bruce A (2005b) Water resources management: finding space in scarcity. Scarcity and the politics of allocation workshop, 6–7 June, University of Sussex, Brighton, UK

Lankford Bruce A, Mwaruvanda W (2005) A framework to integrate formal and informal water rights in river basin management. In: African water laws: plural legislative frameworks for rural water management in Africa, 26–28 January, Johannesburg, South Africa

Larsen S (2012) Foreign land acquisition in Tanzania: global ideology, global perspectives. Master MSc dissertation submitted in partial fulfillment of MSc in human geography. University of Oslo Norway. 93pp

Malthus T (1798) An essay on the principle of population. J. Johnson, St. Paul's Church-Yard 134pp

Matondi P (2010) Agro-investments in Zimbabwe at a time of redistributive land reforms. Paper presented at the regional workshop on commercialization of land and "land grabbing" in Southern Africa hosted by the Institute for Poverty, Land and Agrarian Studies (PLAAS), University of the Western Cape, Cape Town, 24–25 Mar 2010

Mbonile MJ (1999a) Population dynamics on the slopes of Mount Kilimanjaro. In: Mount Kilimanjaro: land and environmental management. French Institute for Research in Africa, Nairobi, pp 18–23

Mbonile MJ (1999b) Population mobility and migration in Mount Kilimanjaro. In: Mount Kilimanjaro: land and environmental management. French Institute for Research in Africa, Nairobi, pp 24–33

Mbonile MJ (2001) Rural population mobility in the Pangani basin, Tanzania. In: Ngana JO (ed) Water resources management in the Pangani basin: challenges and opportunities. University of Dar es Salaam Press, Dar es Salaam, pp 14–27

Mbonile MJ (2005) Population, migration, and water conflicts in the Pangani river basin, Tanzania. In: Population, health, environment, and conflict. ECSP report from Africa. Issue no 2

Molden D, de Fraiture C (2010) Comprehensive assessment of water management in agriculture. Agric Water Manag 97(4):493–578

Mshandete AM (2011) Biofuels in Tanzania: status, opportunities and challenges. J Appl Biosci 40:2677–2705

Msuya TS (2010) Developing integrated institutional framework for sustainable watershed management in Pangani River Basin, Tanzania. Thesis for Award of PhD degree of Sokoine University of Agriculture, Morogoro, Tanzania, 359pp

Mulungu DMM (1997) Application of linear models for inflow forecasting to Nyumba ya Mungu Reservoir in Pangani River basin Tanzania. Unpublished Master's thesis, Department of Water Resources Engineering, University of Dares Salaam, Tanzania, 120pp

Mwamila TB, Kimwaga RJ, Mtalo FW (2008) Eco-hydrology of the Pangani River downstream of Nyumba ya Mungu reservoir, Tanzania. Phys Chem Earth 33:695–700

Ndomba PM, Felix W, Mtalo FW, Killingtveit A (2008) A guided SWAT model application on sediment yield modeling in Pangani River basin: lesson learnt. J Urban Environ Eng 2(2):53–62

Newmark WD (1998) Forest area, fragmentation, and loss in the Eastern Arc Mountains: implications for the conservation of biological diversity. J East Afr Nat Hist 87:1–8

Ngana J, Notter B, Messerli P, Wiesmann U, Mbeyale G, Msuya T, Chitiki A (2010) Managing water resources in dynamic settings: a multi-level, multi-stakeholder perspective. In: Hurni H, Wiesmann U (eds; with an international group of co-editors). Global change and sustainable development: a synthesis of regional experiences from research partnerships. Perspectives of the Swiss National Centre of Competence in Research (NCCR) North-South, University of Bern, vol 5. Geographica Bernensia, Bern, pp 91–106

PBWO and IUCN (2011) Ground water assessment of the Pangani Basin/Tanzania. Final report submitted to Pangani Basin Water Office. 168pp

Randhir TO, Ekness P (2013) Water quality change and habitat potential in riparian. Ecohydrol Hydrobiol 13:192–200

Rulli MC, Savioria A, D'Odorico P (2012) Global land and water grabbing. PNAS 110:4. www.pnas.org/cgi/doi/10.1073/pnas.1213163110. Accessed Nov 2013

Shaghude YW (2006) Review of water resource exploitation and land use pressure in Pangani River basin Tanzania. Western Ind Ocean Mar Sci Assoc J 5(2):195–207

Sotthewes W (2008) Forcing on the salinity distribution in Pangani estuary, Tanzania. MSc thesis submitted in partial fulfillment of Masters of Science at Delft University of Technology, The Netherlands, 82pp

Sulle E, Nelson F (2009) Biofuels, land access and rural livelihoods in Tanzania. International Institute for Environment and Development (IIED), London

The Oakland Institute (2011) Understanding land investment deals in África; Land grabs leave Africa thirsty. http://polarisinstitute.org/files/OI_brief_land_grabs_leave_africa_thirsty_1.pdf. Accessed Dec 2013

Turpie J, Ngaga YM, Karanja F (2005) Catchment ecosystems and downstream water: the value of water resources in the Pangani Basin, Tanzania, Lao PDR. IUCN water, nature and economics technical paper No. 7, IUCN the World conservation union, ecosystems and livelihoods group Asia 108pp. http://www.waterandnature.org. (site visited on June 20, 203)

UNDP (2006) Human development report 2006. United Nations Development Programme beyond scarcity: power, poverty and the global water crisis, 440pp

URT (1995) National land policy. Ministry of land and human settlements development. United Republic of Tanzania, Dar es Salaam, 49pp

URT (1997) National environment policy, division of environment, Vice – President's Office (VPO). United Republic of Tanzania, Dar es Salaam, 44pp

URT (1999a) 1999 Land Act. National Land Policy. Ministry of Land and Human Settlements Development. United Republic of Tanzania, Dar es Salaam

URT (1999b) 199 Village land act. National land policy. Ministry of Land and Human Settlements Development. United Republic of Tanzania, Dar es Salaam

URT (2002) National water policy. Ministry of water and livestock development. United Republic of Tanzania, Dar es Salaam, 49pp

Valimba P (2005) Rainfall variability in southern Africa, its influences on stream flow variations and its relationships with climatic variations, Unpublished PhD thesis, Rhodes University

Valimba P (2007) Spatial variation of hydrological floods during the short rains in Northeast Tanzania. In: Proceedings of the international conference on climate and water, Helsinki (Finland), 3–6 Sep 2007

von Braun J, Meinzen-Dick R (2009) "Land grabbing" by foreign investors in developing countries – Risk and opportunities. In: International food policy research institute, Policy brief no. 13. www.ifpri.org/sites/default/files/publications/bp013all.pdf Accessed March 2015

White B, Dasgupta A (2010) Agrofuels capitalism: a view from political economy. J Peasant Stud 37(4):593–607

World Bank (2010) Rising global interest in farmland: can it yield sustainable and equitable benefits? The World Bank, Washington, DC

Impacts of Some Climatic Factors on Soil Quality of Tropical Acid-Sand Soils

14

Nkem J. Nwosu and Paul B. Okon

Contents

Abstract

There have been changes in yield of agricultural crops over the past decades, and these have been attributed in part to changes in environmental conditions including the soil. These changes have been induced by climatic factors of rainfall,

N. J. Nwosu (✉)
Department of Agronomy, University of Ibadan, Ibadan, Oyo State, Nigeria
e-mail: nkemj.nwosu@gmail.com

P. B. Okon
Department of Soil Science, University of Calabar, Calabar, Nigeria
e-mail: pbokon@yahoo.com

W. Leal Filho (ed.), *Handbook of Climate Change Resilience*,
https://doi.org/10.1007/978-3-319-93336-8_72

temperature, relative humidity, and solar radiation acting singly or in combination to promote or impede beneficial soil reactions and processes. The impact of some climatic factors on soil quality of acid-sands was evaluated to assess the effect of variability in climate on soil physical and chemical properties. Annual variation in climatic factors of rainfall, air temperature, relative humidity, solar radiation, and wind patterns induce seasonal changes that influences agriculture and soil quality in the tropics. Two profile pits was described prior to routine soil physical and chemical analysis. Information on the climatic conditions of the area were obtained for the past 10 years and current year. Results from soil analysis were used to deduce the soil quality using a modified version of Lal's soil quality index (SQI). A Pearson correlation was established to ascertain the relations between soil quality and climatic variability of the area over the past decade using ORIGINPRO v.8. The results of correlation studies showed that there was no significant relationship between the soil quality and variability of climatic parameters over the study period. The result also showed that soil quality merely decreased with an r-value of -0.19 ($r^2 = 0.0361$) as the year advanced. Therefore, the climate variability had no impact on soil quality of Calabar acid-sands.

Keywords
Soil quality · Climate variability · Soil quality index · Correlation · Sustainability

Introduction

Climate is the average condition of weather in a place, taken over a long period of time (usually about two to three decades). In a more general sense, the term "climate variability" denotes the inherent characteristics of climate which manifests itself in changes of climate with time. Climate variability is an overarching driver affecting numerous soil quality issues. Agriculture is highly sensitive to variation of climatic factors (temperature, rainfall, humidity, etc.). The forces that shape our climate are also critical to farm productivity. Human activities have already changed atmospheric characteristics such as temperature, rainfall, levels of carbon-dioxide (CO_2), and ground level zone. While food production may benefit from a warmer climate, the increased potential for droughts, flood, and heat waves have posed challenges for farmers due to their adverse effect on soil quality required for optimum crop production (IPCC 2007a).

Soil quality is the capacity of soils within landscapes to sustain biological productivity, maintain environmental quality, and promote plant and animal health (Nuttall 2007). The variation in climatic factors like rainfall, temperature, wind, sunlight, and relative humidity (RH) pose great effect on the quality of tropical soils (FAO 2008). The Intergovernmental Panel on Climate Change (2007b) (in EEA 2017) and the Australian Department of Primary Industries (2007) reported that the relationship between climate change and soil is complex but can generally be described in terms of soil quality and soil health. Climate is considered by many to be the most important factor in determining the properties of many soils. Moisture and temperature are the two main aspects of climate most important in controlling soil properties. The Ohio State University (2007) reported that projected climate change and variability may exacerbate the problem of desertification, drought and soil degradation in addition to soil fertility and microbial conditions. Hence, it is essential to identify the various impacts of the variations in climates on the quality of tropical soils in order to devise and implement coping (mitigation) and adaptive strategies on how to maintain soil quality required and/or necessary for optimum crop production and agricultural operations. The objective of this study therefore is to assess the impact of climate variability on soil physical and chemical indicator properties of soil quality.

Materials and Methods

Site Description and Soil Sampling

The study was conducted at the University of Calabar Teaching and Research Farm (UNICAL T and R farm) in Calabar, Cross River State. The area is geographically situated between latitude 4°N of the equator and 8°E of the Greenwich meridian. Sampling was carried out during the late rainy season (October, 2011), and two soil profile pit were sunk and fully described. One profile sited at the lower slope, and the other at the upper slope of the study area. However, climate data was also obtained from the Nigerian Meteorological Agency as well as from the Department of Geography and Regional Planning, University of Calabar.

Soil Profile Sampling

Two profile pits was sunk one at the upper slope (Profile A) located at Lat. 04°57′03″N and Long. 008°21′20″E. Profile A was 1.3 m deep while profile B (Lat. 04°56′11″N and Long. 008°20′17″E) was 1.5 m deep. More so, each profile comprised of four distinct horizons and was fully described. The soil description sheet was used to obtain the horizon's morphological features while the Munsell color chart was employed in color description. However, four samples were taken from each profile with the aid of a hand trowel from the lower horizon to the surface. In addition, four core samples were obtained from each profile for bulk density and saturated hydraulic conductivity determinations.

Laboratory Analysis

Particle size distribution was determined using the Bouyoucos hydrometer method (Gee and Or 2002). Bulk density was determined using core method (Grossman and Reinsch 2002). Saturated hydraulic conductivity (Ksat) was determined by the constant head method (Singh 1982). The soil pH was determined with the pH meter using glass electrode in a 1:1 soil to water ratio (Udo et al. 2009). Total nitrogen (TN) was determined by Kjehdahl digestion method (Bremmer 1996). Organic carbon was determined using the Walkey-Black wet oxidation method (Nelson and Sommers 1982). Organic matter was calculated by using the van Bemmelen's correction factor of 1.724. Available phosphorus was determined with the aid of a spectrophotometer using Mehlich III as extractant (Udo et al. 2009). Exchangeable bases were determined using neutral NH_4OAC leachate. Exchangeable Ca and Mg were determined by EDTA versanate titration method (McLean 1982). Exchangeable Na and K were determined by the flame photometer method. Exchangeable acidity was determined by leaching the soil with 1 N KCl and titrating with 0.05 N NaOH (McLean 1982). Crusting hazard which is a useful concept for land use planning as it is indicative of soil's vulnerability to crusting was mathematically computed using the expression as devised by Gicheru and Nyamai (2007):

$$R = OM/(Silt + Clay) \quad (1)$$

where:

R = Crust risk
OM = Organic matter

Evaluation Technique

The soils of Unical Training and Research Farm were evaluated using the sustainability index developed by Lal (1994). In this model, ten physicochemical properties were selected and their cumulative weighting factors were computed and subsequently compared with the sustainability index as reported by Lal (1994). The soils were evaluated using the soil quality index model developed by Lal (1994). The SQI model is a function of soil organic carbon (SOC), cation exchange capacity (CEC), pH, and water content (WC) and water stable aggregates (WSA).

The model is, however, presented as:

$$SQ = f(SOC, CEC, pH, WC, WSA) \quad (2)$$

where:

SQ = Soil quality

Data Analysis

Soil data were subjected to correlation analysis using Origin Pro. software version 8.1 to ascertain the relationship between soil quality and climate variability.

Table 1 Climate data report of the study area

Year	Temperature (°C)		Rainfall (mm)		Relative humidity (%)	
	UNICAL	NIMET	UNICAL	NIMET	UNICAL	NIMET
2000	26.9	–	2666.4	–	85	–
2001	26.8	–	3073.6	–	84	–
2002	26.7	–	2691.4	–	83	–
2003	26.4	–	2113.7	–	86	–
2004	27.6	–	1728.6	–	85	–
2005	27.4	–	2533.8	–	83	–
2006	26.8	27.0	2951.8	2646.7	84	85.2
2007	26.0	27.0	2417.5	3428.2	84	84.3
2008	26.7	27.1	1931.3	3125.2	82	84.5
2009	26.6	27.4	1401.1	2526.9	82	86.2
2010	26.7	27.5	2003.7	3071.7	82	85.8
2011	27.8	–	3055.3	–	86	–
Mean	26.8	27.2	2380.7	2959.7	84	85.2

Source: UNICAL (2011) and NIMET (2011)

Results and Discussions

Climate Data of the Study Area

The 11 years climate data obtained from UNICAL showed a mean temperature of 26.8 °C while mean rainfall and relative humidity figures were 2319.4 mm and 84%, respectively. The 5 years climate statistics obtained from the NIMET, Calabar, showed a mean temperature of 27.2 °C while rainfall and relative humidity figures were 2959.7 mm and 85.2%, respectively (Table 1). Furthermore, the study area is characterized by tropical rainforest vegetation which is evident in the low air temperature and high rainfall and relative humidity figures obtained. Isirimah (2001) reported that these vegetations tend to play significant role on soil temperature because of the insulating properties of plant cover which tends to reduce the daily variations in soil temperature thereby creating a conducive microclimate for organic matter decomposition and enhancing beneficial soil microbial activities in the soil medium. This was similar to the report of Pietikainen et al. (2005) where respiration rate increase at all temperature conditions and fungi were more adapted to low-temperature conditions than bacteria.

Gicheru and Nyamai (2007) suggested that areas with high rainfall range of 2500–3500 mm possessing low porosity and considerable low infiltration rates tend to have high run-offs leading to low soil and environmental quality and also tends to be susceptible to crusting. Nevertheless, aside from the vegetation of the area, the climate of the area is also influenced by the altitude, slope, and presence of ocean currents obtained from surrounding rivers and streams (Isirimah 2001). However, such climatic conditions makes room for the cultivation of vegetables like cucumber; okra; amaranthus (green); pumpkin; plantation crops like oil palm and rubber; arable crops like cassava, yam, water yam, etc.

Physico-Chemical Properties of Profile A at UNICAL T & R Farm

The particle density of the profile pit varied between 2.44 and 2.56 g cm^{-3} with a mean value of 2.50 g cm^{-3}. The mean values of the various soil separates – sand, silt, and clay were 75.3%, 2.0%, and 22.7%, respectively, giving a soil textural class of sandy clay loam. The saturated hydraulic conductivity (Ksat) showed a range of 0.0002–0.008 cm^3 s^{-1} and a mean value of 0.0078 cm^3 s^{-1}. The low values of Ksat obtained may be associated to low water transmission (hydraulic restriction) probably due to fewer mobile regions due to crusting (Bigger and Nelson 1976) and/or higher clay contents that constrict voids and which can make an area prone to flooding during rainy season, especially in lower physiographies (Ezeaku 2010). Ezeaku and Anikwe (2006) also reported that the cause of high water transmission (flow) in soils may be, probably due to increases in bioturbation (e.g., burrowing activities of animals and roots movement) that result to higher biopores and cross-sectional areas that contribute to flow and such soils could be regarded as hydraulically conductive.

The bulk density (B.D) ranged from 1.46 to 1.76 g cm^{-3} with an average mean value of 1.59 g cm^{-3}. From the principle that a good soil quality has the least soil degradation rate (SDR), Lal (1994) reported that soils with B.D range of 1.5–1.6 tend to possess a "severe" SDR. The high bulk density observed in the area may be associated to low soil porosity and seasonal flooding of soils by erosion that leads to surface crusting and compaction (Lal 1990, 1994; Isirimah et al. 2003).

Total porosity varied between 28% and 43% indicating that the soil was slightly porous with a mean value of 34.8% (Table 2).

Also, the color of the profile samples were dark brown (7.5 YR 3/2) for horizon A_1 – Ap: 0–25 cm; brown (7.5 YR 4/3) for horizon A_1 – Bt: 25–42 cm; dark yellowish brown (10 YR 3/4) for horizon A_1 – BC: 42–62 cm; and strong brown (7.5 YR 4/6) for horizon A_1 – C: 62–130 cm. The crust risk of the soils varied from 0.02 to 0.09 with a mean value of 0.05. Gicheru and Nyamai (2007) reported that soils whose crust risk range between 0 and 5 tend to possess a high crusting hazard which entails that the soils are highly susceptible to crusting. Furthermore, the pH of the profile samples varied between 5.1 and 6.0 with a mean value of 5.7. This value according to Schoeneberger et al. (2002) is slightly acidic. Lal (1994) reported that optimum pH for most agricultural crops falls between 6.0 and 7.0, and nutrients are more available at pH of about 6.5. Soils with pH range of 5.4–5.8 tend to possess moderate limitation in terms of soil degradation ratio with a relative weight factor (RWF) of 3 (Lal 1994).

The organic carbon content of the profile sample, range from 0.4% to 0.9% with a mean value of 0.7%. Islam and Well (2000) reported a near value in a natural forest in Bangladesh. Soils with organic carbon content range of 0.5–1.0 tend to possess a severe soil degradation rating (SDR) with an RWF of 4 (Lal 1994). Invariably, the general low levels of soil organic matter (SOM) could be attributed to management practices involving burning, continuous cultivation with reduced fallow period and scanty vegetation coverage of the land mass, an indication of low carbon stock available (Ezeaku 2010).

Table 2 Physical properties of Profile A at UNICAL T & R Farm

Sample	Sand (%)	Silt (%)	Clay (%)	Textural class	Color code	Color name	Bulk density (g cm^{-3})	Particle density (g cm^{-3})	Porosity (%)	Hydraulic conductivity (cm^3 s^{-1})	Crust risk
A: 0–25 cm	82.3	4.0	13.7	Loamy sand	7.5 YR 3/2	Dark brown	1.48	2.44	33.0	0.007	0.09
A: 24–42 cm	76.3	1.0	22.7	Sandy clay loam	7.5 YR 4/3	Brown	1.76	2.44	28.0	0.002	0.05
A: 42–62 cm	75.3	1.0	23.7	Sandy clay loam	10 YR 3/4	Dark yellowish brown	1.67	2.56	35.0	0.008	0.04
A: 62–130 cm	67.3	2.0	30.7	Sandy clay loam	7.5 YR 4/6	Strong brown	1.46	2.54	43.0	0.0002	0.02
Mean	75.3	2.0	22.7	Sandy clay loam			1.59	2.50	34.8	0.004	0.05
S.D	6.2	1.4	7.0				0.14	0.1	6.2	0.004	0.03
C.V (%)	8.2	70.0	30.8				8.8	4.0	17.8	100.0	60.0
Min	67.3	1.0	33.7				1.46	2.44	28.0	0.0002	0.02
Max	82.3	4.0	30.7				1.76	2.56	43.0	0.008	0.09
Range	15.0	3.0	17.0				0.3	0.12	15.0	0.0078	0.07

Total nitrogen followed a similar trend as soil organic matter since soil nitrogen constitute the bulk of total N for tropical soils (Noma et al. 2005). Soil total nitrogen varied between 0.03% and 0.08% with an average mean value of 0.06%. Rankine and Fairhurst (1999) reported similar value for production of oil palm but Enwezor et al. (1989) suggested that soils with TN values less than 0.1% possess low total nitrogen content and that the cause of N deficiency in the soils may be related to intense leaching and erosion due to rainfall with high mineralization rate. The available phosphorus varied from 5.5 to 11.7 mg kg^{-1} with a mean value of 8.5 mg kg^{-1}. This value according to Lal (1994) possesses no limitation in terms of soil degradation with an RWF value of 1.

Also, the following means were obtained for Ca^{2+} (4.0–8.2), Mg^{+} (0.8–4.0), K^{+} (0.1–0.11), and Na^{+} (0.07–0.08); 6.2, 2.1, 0.11, and 0.075 cmol kg^{-1}, respectively. Critical values of 2.0, 0.4, and 0.20 cmol kg^{-1} for Ca, Mg, and K, respectively was reported by Adeoye and Agboola (1985). In comparison, the mean values obtained in the study location are higher and hence disagrees with these critical limits. Therefore, it may be reasonable to say that the nutrients are not deficient and may not be limiting crop performances in the study area. Exchangeable H^{+} varied from 0.68 to 0.96 cmol kg^{-1} with a mean value of 0.77 cmol kg^{-1}. ECEC and base saturation had a mean value of 9.3 cmol kg^{-1} and 91.6%, respectively (Table 3).

Physico-Chemical Properties of Profile B at UNICAL T & R Farm

The particle density of the profile pit varied between 2.41 and 2.62 g cm^{-3} with a mean value of 2.50 g cm^{-3}. The mean value of the soil separates – sand, silt, and clay were 82.3%, 3.3%, and 14.5%, respectively, denoting a soil textural class of loamy sand. Saturated hydraulic conductivity varied from 0.002 to 0.026 cm^3 s^{-1} with a mean value of 0.01 cm^3 s^{-1}. Bulk density ranged from 1.43 to 1.72 g cm^{-3} with an average of 1.60 g cm^{-3}. Lal (1994) reported that soils with Ksat less than 0.8 tend to possess no limitation in SDR with an RWF of 1, while soils with bulk density range of 1.5–1.6 pose severe limitation to soil degradation with a relative weighting factor of 4. Total porosity varied between 33% and 41% with a mean value of 27.1% indicating very slight porosity. The color of the profile samples were black (7.5 YR 2.5/1) for horizon B_2 – Ap: 0–24 cm; very dark gray (7.5 YR 3/1) for B_2 – A: 24–60 cm; brown (7.5 YR 4/4) for horizon B_2 – B: 60–85 cm; and dark yellowish brown (10 YR 4/6) for horizon B_2 – C: 85–150 cm. The crust risk of the soils varied from 0.01 to 0.07 with a mean value of 0.03 (Table 4). This value according to Gicheru and Nyamai (2007) tends to possess a high crusting hazard which inherently entails that the soils are highly susceptible to crusting. Furthermore, the pH of the soil profile samples varied from 5.4 to 6.4 with an average value of 6.0. This value is slightly acidic according to Schoeneberger et al. (2002). Lal (1994) reported that such soils possess slight limitation with a relative weighting factor of 2. The soil organic carbon content ranged between 0.14% and 0.62% with a mean value of 0.3%. Lal (1994) reported that such soils possess extreme limitations with an RWF of 5.

Table 3 Chemical properties of profile Pit 1 at UNICAL T & R Farm

Sample	pH (H_2O)	Organic carbon (%)	Total N (%)	Available P (mg kg^{-1})	Exchangeable cations (cmolkg^{-1})				Exchangeable H^+ (cmol kg^{-1})	ECEC (cmol kg^{-1})	Base saturation (%)
					Ca^{++}	Mg^{++}	K^+	Na^+			
A: 0–25 cm	5.1	0.90	0.08	11.7	5.6	2.6	0.10	0.07	0.96	9.33	89.7
A: 24–42 cm	6.0	0.74	0.07	9.5	7.0	0.8	0.11	0.07	0.72	8.7	91.7
A: 42–62 cm	5.9	0.63	0.05	5.5	4.0	4.0	0.10	0.08	0.72	8.9	91.9
A: 62–130 cm	5.8	0.40	0.03	7.3	8.2	1.0	0.11	0.08	0.68	70.07	93.2
Mean	5.7	0.7	0.06	8.5	6.2	2.1	0.11	0.075	0.77	9.3	91.6
S.D	0.4	0.2	0.02	2.7	1.8	1.5	0.008	0.05	0.1	0.6	1.4
C.V (%)	7.0	28.6	33.3	31.8	29.0	71.4	7.3	66.6	13.0	6.5	1.5
Min	5.1	0.4	0.03	5.5	4.0	0.8	0.10	0.07	0.68	8.7	89.7
Max	6.0	0.9	0.08	11.7	8.2	4.0	0.11	0.08	0.96	10.07	93.2
Range	0.9	0.5	0.05	6.2	4.2	3.2	0.01	0.01	0.28	1.37	3.5

Table 4 Physical properties of profile Pit 2 at UNICAL T & R Farm

Sample	Sand (%)	Silt (%)	Clay (%)	Textural class	Color code	Color name	Bulk density (g cm^{-3})	Particle density (g cm^{-3})	Porosity (%)	Hydraulic conductivity (cm^3 s^{-1})	Crust risk
B: 0–24 cm	84.3	4.0	11.7	Loamy sand	7.5 YR 2.5/1	Black	1.43	2.42	41.0	0.026	0.07
B: 24–60 cm	85.3	4.0	10.7	Loamy sand	7.5 YR 3/1	Very dark brown	1.72	2.62	34.0	0.005	0.02
B: 60–85 cm	87.3	3.0	9.7	Sand	7.5 YR 4/4	Brown	1.65	2.46	33.0	0.005	0.02
B: 85–150 cm	72.3	2.0	25.7	Sandy clay loam	10 YR 4/6	Dark yellowish brown	1.58	2.41	34.0	0.002	0.01
Mean	82.3	3.3	14.5	Loamy sand			1.60	2.50	27.1	0.01	0.03
S.D	6.8	1.0	7.5				0.1	0.1	10.4	0.01	0.03
C.V (%)	8.3	30.3	51.7				6.3	4.0	38.4	100.0	100.0
Min	72.3	2.0	9.7				1.43	2.41	33.0	0.002	0.01
Max	87.3	4.0	25.7				1.72	2.62	41.0	0.026	0.07
Range	15.0	2.0	16.0				0.29	0.21	8.0	0.024	0.06

Total nitrogen ranged between 0.01% and 0.05% with a mean value of 0.02%. This value according to Enwezor et al. (1989) possesses low TN content which could be due to high mineralization accompanied with intense leaching. The available P varied between 5.1 and 35.4 mg kg^{-1} with a mean value of 20.6 mg kg^{-1}. Enwezor et al. (1989) suggested a critical limit of 8–12 mg kg^{-1} for available P of tropical soils implying that the soil was not deficient in available P. However, Lal (1994) reported that such soils possess slight limitation with an RWF of 2. More so, the following means were obtained for Ca^{2+} (7.4–9.0), Mg^{+} (0–4.8), K^{+} (0.09–0.12), and Na^{+} (0.07–0.1); 7.8, 3.1, 0.11, and 0.08 cmol kg^{-1}, respectively. Adeoye and Agboola (1985) suggested critical values of 2.0, 0.4, and 0.2 cmol kg^{-1} for Ca, Mg, and K implying that the soils in the study location had excess of Ca and Mg but slightly deficient in K. Total acidity (exchangeable H^{+}) varied between 0.028 and 0.56 cmol kg^{-1} with a mean value of 0.44 cmol kg^{-1}. ECEC and base saturation had a mean value of 11.5 cmol kg^{-1} and 96.2%, respectively (Table 5).

Soil Properties Under Influence of Climate Variability at the Study Area

Significant shifts in the regional climate variability pattern took place during the last decade in the area as shown in Table 1. Physical properties are grossly static in their response to change in climatic conditions, and though human activities tend to induce a more significant change in soil physical properties, it is rather the chemical constituents of the soil that are more susceptible (or dynamic) to changes in climatic conditions on a daily, monthly, and yearly basis (Lal 2001). Hence, more emphasis will be laid on the dynamic properties (chemical properties) of the soil with little emphasis on the physical nature of the soil (Table 6). Also, it is understandable that nutrient concentrations are predominant in soils at a depth of 0–30 cm. Hence, predominant soil analyses are carried out at such depth (Jones and Case 1990). The physical property of the soil that has shown significant shift in the study area includes texture, bulk density, particle density, and porosity. On the other hand, soil chemical properties that have shown wide change in the location under the same management and land use pattern includes pH; SOC (soil organic carbon content); TN; available P; exchangeable Ca^{++}, K^{+}, Mg^{++}, and Na^{+}; acidity; ECEC; and base saturation (Table 7).

Sustainability of the Soils at UNICAL T and R Farm (Study Area)

The cumulative rating index of the soil is 29 as observed in Table 9. The sustainability of the soils of the study area was evaluated using ten physico-chemical properties, and the cumulative weighting factors of these parameters were then computed and compared with the sustainability index of Lal (1994) showed that such soils with cumulative rating ranging from 25 to 30 show sustainability with high input (Table 10).

Table 5 Chemical properties of profile Pit 2 at UNICAL T & R Farm

Sample	pH (H_2O)	Organic carbon (%)	Total N (%)	Available P (mg kg^{-1})	Exchangeable cations (cmol kg^{-1})				Exchangeable H^+ (cmol kg^{-1})	ECEC (cmol kg^{-1})	Base saturation (%)
					Ca^{++}	Mg^{++}	K^+	Na^+			
B: 0–24 cm	6.1	0.62	0.05	19.9	7.4	3.8	0.11	0.08	0.52	11.91	95.6
B: 24–60 cm	5.4	0.20	0.01	22.1	7.4	3.8	0.12	0.10	0.4	11.82	96.6
B: 60–85 cm	6.4	0.14	0.01	35.4	9.0	0.0	0.11	0.08	0.28	9.47	97.0
B: 85–150 cm	6.2	0.18	0.01	5.1	7.4	4.8	0.09	0.07	0.56	12.92	95.7
Mean	6.0	0.3	0.02	20.6	7.8	3.1	0.11	0.08	0.44	11.5	96.2
S.D	0.4	0.2	0.02	12.4	0.8	2.1	0.1	0.01	0.1	1.5	0.7
C.V (%)	6.7	66.7	100.0	60.2	10.3	67.7	90.9	12.5	22.7	13.0	0.7
Min	5.4	0.14	0.01	5.1	7.4	0.0	0.09	0.07	0.28	9.47	95.6
Max	6.4	0.62	0.05	35.4	9.0	4.8	0.12	0.10	0.56	12.92	97.0
Range	1.0	0.48	0.04	30.3	1.6	4.8	0.03	0.03	0.28	3.45	1.4

Table 6 Soil physical properties under influence of climate variability at the study area

Year	Climate variability			Soil physical properties							Source
	Temp. (°C)	Rainfall (mm)	RH (%)	Sand (%)	Silt (%)	Clay (%)	Textural class	Bulk density (g/cm^3)	Particle density (g/cm^3)	Total porosity (%)	
2001	26.8	3073.6	84	81.0	5.0	14.0	Sandy loam	–	2.4	–	Okoh (2001)
2002	26.7	2691.4	83	66.86	5.03	16.66	Sandy loam	–	2.4	–	Etukakpan (2002)
2004	27.6	1728.6	85	84.0	8.0	8.0	Sandy loam	–	2.45	–	Chijoke (2004)
2005	27.4	2533.8	83	84.0	3.3	9.4	Sandy loam	–	2.43	–	Edemumoh (2005)
2007	26.0	2417.5	84	67.0	2.0	30.0	Sandy clay loam	1.42	2.45	42.0	Peters (2007)
2008	26.7	1931.3	82	84.6	8.9	10.7	Sandy loam	–	2.44	–	Iyabo (2008)
2010	26.7	2003.7	82	76.7	6.0	17.3	Sandy loam	1.43	2.39	40.2	Udo (2010)
2011	27.8	3055.3	86	82.3	4.0	13.7	Loamy sand	1.48	2.44	33.0	Field data (2011)

N/B: *RH* relative humidity

Table 7 Soil chemical properties under influence of climate variability at the study area

	Climate variability			Soil chemical properties											
Year	Temp. (°C)	Rainfall (mm)	RH (%)	pH	Organic carbon (%)	TN (%)	Available P (mg kg^{-1})	Ca^{++} (cmol kg^{-1})	Mg^{+} (cmol kg^{-1})	K^{+} (cmol kg^{-1})	Na^{+} (cmol kg^{-1})	EA (cmol kg^{-1})	ECEC (cmol kg^{-1})	BS (%)	Source
2001	26.8	3073.6	84	5.7	0.52	1.20	12.7	1.40	0.32	0.12	0.07	0.91	2.8	67.7	Okoh (2001)
2002	26.7	2691.4	83	4.9	0.16	0.76	46.5	3.5	2.6	0.09	0.07	3.40	9.6	63.8	Etukakpan (2002)
2004	27.6	1728.6	85	5.1	1.22	0.10	9.0	7.2	0.11	0.2	3.4	2.75	13.66	79.9	Chijoke (2004)
2005	27.4	2533.8	83	4.8	1.10	0.20	NR	NR	NR	NR	NR	NR	NR	NR	Edemumoh (2005)
2007	26.0	2417.5	84	5.0	1.44	0.12	45.8	1.4	1.0	0.07	0.08	2.16	4.71	54.1	Peters (2007)
2008	26.7	1931.3	82	4.7	0.90	0.06	66.0	1.7	0.6	N.R	0.06	4.80	3.62	87	Iyabo (2008)
2010	26.7	2003.7	82	4.8	2.31	0.33	73.0	1.6	2.8	0.09	0.07	21.0	25.56	17.84	Udo (2010)
2011	27.8	3055.3	86	5.1	0.90	0.08	11.7	5.6	2.6	0.10	0.07	0.96	9.33	89.7	Field data (2011)

NR Not reported
N.B: *RH* relative humidity, *ECEC* exchangeable cation exchange capacity, *BS* base saturation, *EA* exchangeable acidity

$$SQ = f^{10}(\text{physical,chemical,and biological indicators})$$

i.e., soil quality as a function of any ten easily measured and indexed physical, chemical, and biological indicators of soil quality. Inherently, the ten selected properties (including particle density, texture, total porosity, pH, total N, available P, exchangeable K, ECEC, base saturation, and organic carbon) indicated sustainability with high input.

Soil Quality Evaluation

The soil quality of UNICAL T and R Farm was evaluated using the soil quality model (Eq. 2) developed by Lal (1994). The model was developed as a function of soil organic carbon (SOC), cation exchange capacity (CEC), pH, water content (WC), and water stable aggregates (WSA). Thomas (2008) modified the model by replacing WSA, CEC, WC, with sodium adsorption ratio, texture, and particle density. Nevertheless, in this study, sodium adsorption ratio (SAR) was replaced with total porosity (TP). Furthermore, the reason for doubling the summation is to increase the index to 10. Although this has been a much deliberated subject, however, since soils possess both native and dynamic qualities, the native or inherent quality depicting the soils natural ability to function. Hence, doubling the values of the soil properties was to establish an index that would hold true even when the dynamic properties of soil changes to influence soil quality conditions in varying tropical environments. Nakajima et al. (2015) and Mandal et al. (2010) reported that there is seldom a universal method or tool for the assessment of soil quality, the selected properties used in this study were considered appropriate indicators for sustainability of the soil in the area as past studies conducted in the area have shown that they greatly influence the capacity of the soil to perform within the confines of its ecosystem and as such, are highly sensitive to the outcome. Albiet reviewed studies have shown the use of only organic matter, soil texture, and soil pH in the determination of soil quality, and this study added total porosity and particle density in order to cover a wider range of indices for adequately explaining the ability of a given tropical soil to function within its ecosystem boundaries. Also, the indicators used conformed to the rationale as documented by Andrews et al. (2004). The model used is thus given below;

$$SQ = f\left(2\sum[\text{SOC,pH, T.P,Texture,Particle density}]\right) \tag{3}$$

The weighting factor for each indicator according to Lal (1994) was computed using the average of the two profiles and is presented as follows: SOC: 4, pH: 4, TP: 1, texture: 4 and particle density: 2.

The critical rating index for soil physical and chemical properties (Table 8) was used to compute rating index of soils in the study area (Table 9). The cumulative

Table 8 Suggested critical levels of some soil physic-chemical indicators

Limitation	Relative weighting factor	Particle density (mg/m^3)	Texture	Total porosity (%)	Soil pH	Total N (%)	Phosphorus (mg/kg)	Potassium (cmol kg^{-1})	ECEC (cmol kg^{-1})	Base saturation (%)	Organic carbon (%)
None	1	<2.4	Loam	>20	6.0–7.0	<0.20	>20	>0.6	>20	>70	5–10
Slight	2	2.4–2.6	Silt loam, silt, silt clay loam	18–20	5.8–6.0	0.18–0.20	15–20	0.4–0.6	15–20	60.70	3–5
Moderate	3	2.6–2.8	Clay loam, sandy loan	15–18	5.4–5.8	0.15–0.18	10–15	0.2–0.4	10–15	50–60	1–3
Severe	4	2.8–3.0	Silt clay, loamy sand	10–15	5.0–5.4	0.12–0.15	5–10	0.1–0.2	5–10	40–50	0.5–1.0
Extreme	5	>3.0	Clay, sand	<10	<5.0	<0.12	<5	<0.1	<5	<40	<0.5

Source: Modified by Paul Okon from Lal (1994) and Ezeaku (2010)

Table 9 Critical levels of physicochemical indicators of soils at UNICAL T & R Farm

Soil property/indicator	Weighting factor	Limitation
Particle density	2	Slight
Texture	4	Severe
pH	1	None
Organic carbon	4	Severe
Total nitrogen	5	Extreme
Available phosphorus	3	Moderate
Exchangeable potassium	4	Severe
Total porosity	1	None
ECEC	4	Severe
Base saturation	1	None
Cumulative rating index (CRI)	**29**	

Table 10 Sustainability of a land use in relation to cumulative rating

Sustainability	Cumulative rating index
Highly sustainable	<20
Sustainable	20–25
Sustainable with high input	25–30
Sustainable with another land use	30–40
Unsustainable	>40

Source: Lal (1994)

index rating is the sum of all weighting factors (Table 10). This is equivalent to soil quality (SQ). Hence,

$$SQ = 4 + 4 + 1 + 4 + 2 = 15 \times 2 = 30$$

However, according to Lal's (1994) classification of sustainability which depicts soil groups with cumulative rating index (CRI) below 20 as highly sustainable, CRI of 20–25 as sustainable, CRI of 25–30 as sustainable with high input, CRI of 30–35 as sustainable with another land, while soils with CRI > 40 can be termed unsustainable; UNICAL T and R Farm is said to be sustainable with another land since the CRI or SQI (soil quality index) is 30. This could be attributed to the cropping history of then area where the soil has been periodically subjected to arable farming with little or no measure to conserve the soil against erosion and nutrient losses.

Influence of Climate Variability on Soil Quality of UNICAL T & R Farm

The slight shift in climate condition on a yearly basis has without doubt resulted in slight changes in the soil dynamic properties. This change was observed to be negative according to the result of Pearson Correlation (r) analysis (Table 11).

Table 11 The correlation (r) matrix of soil quality with weather variables and time (years)

	Soil quality	RH	Rainfall	Temp	Year
Soil quality	1.00				
RH	0.044	1.00			
Rainfall	−0.139	0.406	1.00		
Temp	0.373	0.546	0.097	1.00	
Year	−0.197	0.028	−0.198	0.120	1.00

Number of samples = 8
RH relative humidity, *Temp* air temperature

In 2001, temperature, rainfall, and relative humidity readings were 26.8 °C, 3073.6 mm, and 84%, respectively. Okoh (2001) reported a soil quality index (SQI) of 32 based on results of soil analysis, showing sustainability with another land with a CRI value of 31 (Lal 1994).

In 2002, temperature, rainfall, and relative humidity were 26.7 °C, 2691.4 mm, and 83%, respectively (UNICAL 2011) while Etukakpan (2002) reported a SQI of 32 but with sustainability with high input (CRI = 30). In 2004, temperature, rainfall, and RH readings were 27.6 °C, 1728.6 mm, and 85%, respectively (UNICAL 2011) as Chijoke (2004) reported a SQI of 34 with CRI value of 31 showing sustainability with another land use. In 2005, temperature, rainfall, and RH readings were 27.4 °C, 2533.8 mm, and 83%, respectively (UNICAL 2011) as Edemumoh (2005) reported a SQI of 30. In 2007, temperature, rainfall, and RH values were 26.0 °C, 2417.5 mm, and 84%, respectively (UNICAL 2011), while the SQI was observed to be 26 with a CRI value of 31 showing sustainability with another land.

Furthermore, in 2008, temperature, rainfall, and RH values were 26.7 °C, 1931.3 mm, and 82%, respectively (UNICAL 2011), while SQI was 38 (Iyabo 2008) with a CRI of 29 showing sustainability with high input. In 2010, temperature, rainfall, and RH values were 26.7 °C, 2003.7 mm, and 82%, respectively (UNICAL 2011), while SQI was 26 with a CRI of 22 showing sustainable land (Udo 2010). However, in 2011 (current year), temperature, rainfall, and RH readings were 27.8° C, 3055.3 mm, and 86%, respectively (UNICAL 2011) with a SQI of 32 and a CRI of 29 giving sustainability of land with high input.

Nevertheless, based on the findings, the slight decrease in SQI of the area can be attributed to increased temperature and rainfall distribution all year round, leading to leaching, erosion, etc. Hence, it is safe to say that climate variability of the area has negatively affected soil quality of the area in the past 10 years. Moreover, the slight variation in climate over time has resulted in increase in soil pH, exchangeable cations, and base saturation with corresponding decrease in available phosphorus, organic carbon, total nitrogen, and clay content of soil.

However, the soil quality reduced as the year progressed from 2001 to 2011 (Table 12), again, this relationship was very weak and not statistically significant. As such it could be vehemently said that within the period of 10 years, there was no direct impact of rainfall, temperature, and relative humidity on the soil quality of acid-sands soil of Calabar. Soil quality varied independent of the climatic variability.

Table 12 Soil quality index over the period of 10 years under study

Year	Climatic parameters/variables Relative humidity (%)	Annual rainfall (mm)	Temp (°C)	Soil quality index
2001	84	3073.6	26.8	32
2002	83	2691.4	26.7	32
2004	85	1728.6	27.6	34
2005	83	2533.8	27.4	30
2007	84	2417.5	26.0	26
2008	82	1931.3	26.7	38
2010	82	2003.7	26.7	26
2011	86	3055.3	27.8	30

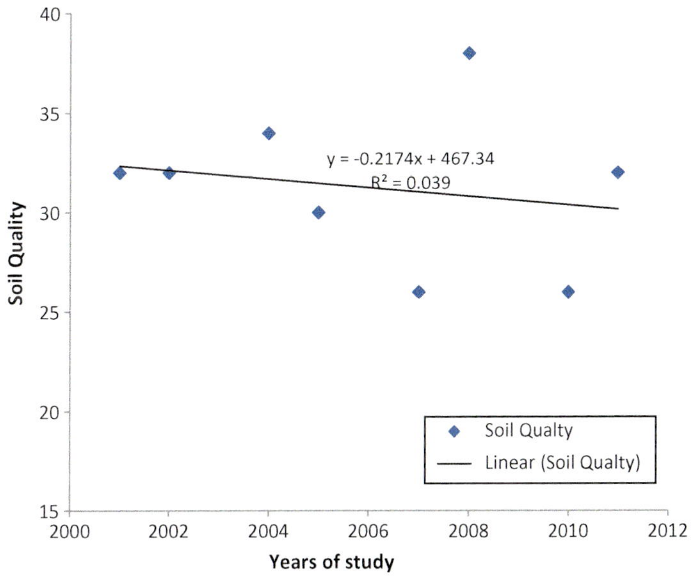

Fig. 1 Influence of time on soil quality in the University of Calabar Research farm

Figure 1 shows the weak relationship between soil quality and time. In the figure, the coefficient of determination R^2 was as low as 0.039.

Conclusion

Tackling climate change requires focus on to agriculture. More importantly, soil is a major factor in our response to tackling climate change, and as such, tackling climate change cannot be done without a better understanding of our soils. However, achieving climate change in practice still poses great challenges especially to smallholder famers in sub-Sahara Africa. The study was designed to ascertain the impact of some climatic factors on soil quality of Calabar acid-sand, a case study of UNICAL T and R farm. The results showed that spatial variability of soils under climate variability and their soil property vulnerability potentials or degradation rate can be evaluated. The results of the sustainability index based on current soil analysis showed that the soil was sustainable with high input due to the cropping history, while results from Lal's (1994) soil quality index was 32 showing sustainability with another land use. However, the study also exclusively observed that the slight variations in climatic condition over the past 11 years at the study area negatively affected the soil quality as expressed in a Pearson's Correlation (r) value of −0.19.

Recommendations

In view of the study area's climatic and soil conditions, the following recommendations could be made:

1. The low nutrient contents (organic carbon, total N, available P) observed in the location could be increased through organic farming – mulching and returning crop residues (trash farming) to the soils. Appropriate fertilizers could also be used.
2. Farmers need early warning signals and monitoring tools to help them assess the status of their soils since soils in the area are vulnerable to erosion and by the time degradation becomes visible and irreversible, it might be too late or very expensive to reverse it.
3. Planting of tree covers and use of cover crops can be established to improve microclimate of the area in order to reduce effects of temperature and rainfall extremes on soil conditions.
4. However, the use of fertilizers both organic and inorganic should not be done in a simple habitual manner or for-so-called insurance purposes. Rather, soil testing and other diagnostic tools should be used to determine the true needs for added nutrient.

Cross-References

- ▶ Climate Change and Coastal Resilience in Nigeria
- ▶ Climate Change Awareness and Adaptation Measures Among Farmers in Cross River and Akwa Ibom States of Nigeria
- ▶ Environmental Sustainability and Systems Thinking: A Foundation for More Effective Climate Policy
- ▶ Impacts of Climate Change on Agriculture, Livelihoods, and Women in Nile Delta, Egypt
- ▶ Impacts of Climate Change on Food Security: An Appraisal

- Impact of Climate Change on Growing Season in Nigeria: Seasonal Rainfall Prediction (SRP) as Assessment and Adaptation Tool
- Microbial Inoculants for Improving Carbon Sequestration in Agroecosystems to Mitigate Climate Change
- Sustainability of Climate Change Adaptation Measures in Rivers State, South-South, Nigeria

References

Adeoye GO, Agboola AA (1985) Critical levels for soil pH, available P, K, Zn and Mn and maize ear-leaf content of P, Cu and Mn in sedimentary soils of South-West Nigeria. Fertil Res 6:65–75

ADPI (2007) Soil health for Victoria's agriculture – context, terminology and concepts. Primary Industries Research Victoria, Bendigo

Andrews SS, Karlen DL, Cambardella CA (2004) The soil management assessment framework: a quantitative soil quality evaluation method. Soil Sci Soc Am J 68:1945–1962

Bigger, J. W. and D. K. Nielson (1976) Spatial variability of the leaching characteristics of a field soil, Water Resources Research, 12:78–84

Bremmer JM (1996) Nitrogen – total. In: Sparks DL (ed) Methods of soil analysis, Part 3, chemical method. Soil Science Society of America book series, no. 5, 2nd edn. Soil Science Society of America, Madison, pp 1085–1121

Chijoke IA (2004) The effects of different rates of NPK fertilizers on the soil properties and growth of aquatic bitter leaf grown in Calabar, South-Eastern Nigeria. Unpublished B. Agric Project, Department of Soil Science, University of Calabar, Calabar. 46pp

Edemumoh IE (2005) Review of organic matter and total nitrogen in soils of Cross River State. Unpublished B. Agric Project, Department of Soil Science, University of Calabar, Calabar. 22pp

EEA (2017) Climate change mitigation and adaption – soil and climate change. EEA report no 2, European Environment Agency Publications, Copenhagen, Denmark

Enwezor WO, Udo EJ, Usoroh NJ, Ayotade KA, Adepetu JA, Chude OV, Udegbe CI (1989) Fertilizer use and management practices for crops in Nigeria. Series 2. Bobina Publishers, Ibadan, p 18

Etukakpan GA (2002) Influence of termites on soil properties. Unpublished B. Agric Project, Department of Soil Science, University of Calabar, Calabar. 43pp

Ezeaku PI (2010) An evaluation of the spatial variability of soils of similar lithology under different land use. Types and degradation risks in a Savannah Agro-ecology of Nigeria. A paper presented at college on soil physics March, 2010. The Abdus Salam International Centre for Theoretical Physics Publication. 26pp

Ezeaku, P. I. and M. A. N. Anikwe (2006) A model for description of water and solute movement in soil-water restrictive horizons across two landscapes in South-East Nigeria, Journal of Soil Science 171(6):492–500 (USA)

FAO (2008) Food and Agricultural Organization 2008 yearbook. FAO, Rome

Gee GW, Or D (2002) Particle size analysis. In: Dane J, Topp GC (eds) Methods of soil analysis, Part 4: physical methods. Soil Science Society of America, Madison, pp 255–294

Gicheru PT, Nyamai M (2007) Crusting hazard in Kenyan soils. Abstract 0.42. 4th African Soil Science Society international conference, 7–13 January 2007, Accra, Ghana. www.assland.org

Grossman RB, Reinsch TG (2002) Bulk density and linear extensibility. In: Dane JH, Topp GC (eds) Methods of soil analysis Part 4. Physical methods. Soil Science Society of America book series no 5. ASA and Soil Science Society of America, Madison, pp 201–228

IPCC (2007a) Chapter 4, Ecosystems, their properties, goods and services. In: Parry ML, Canziani OF, Palutikof JP, van der Linden PJ, Hanson CE (eds) Climate change 2007: impacts, adaptation and vulnerability. Contribution of Working Group II to the Fourth Assessment

Report of the Intergovernmental Panel on Climate Change. Cambridge University Press, Cambridge, UK, pp 7–22

IPCC (2007b) Summary for policy makers. In: Parry ML, Canziani OF, Palutikof JP, Van der Linden PJ, Hanson CE (eds) Climate change 2007: impacts, adaptation and vulnerability. Contribution of Working Group II to their Fourth Assessment Report of the IPCC. Cambridge University Press, Cambridge, UK

Isirimah NO (2001) Introduction to climatology and biogeography for agriculture. Nichdano Publishers. Port Harcourt, Nigeria, 96pp

Isirimah NO, Igwe C, Ali CM (2003) Important ions in soil environment. In: Isirimah NO, Dickson AA, Igwe C (eds) Introductory soil chemistry and biology for agriculture and biotechnology. OSIA Int. Publishers Ltd. Port Harcourt, Nigeria, pp 34–97

Islam KR, Well RR (2000) Land use effects on soil quality in a tropical forest ecosystem of Bangladesh. Agric Ecosyst Environ 79(2000):9–16

Iyabo OA (2008) Effect of cultivation on chemical and biological properties of acid sand soil under continuous gardening in Calabar. Unpublished B. Agric Project, Department of Soil Science, University of Calabar, Calabar. 66pp

Jones JB, Case VB (1990) Sampling, handling and analyzing plant tissue samples. In: Westerman RL (ed) Soil testing and plant analysis, 3rd edn. Soil Science Society of America, Madison, pp 389–427

Lal R (1990) Tropical soils: distribution, properties and management. Resour Manag Optim 7:39–52

Lal R (1994) Methods and guidelines for assessing sustainability use of soil and water resources in the tropics. Soil Manag Support Serv Tech Monogr 21:1–78

Lal, R (2001) Potential of desertification control to sequester carbon and mitigate the greenhouse effect, Climate change 51, 35–72

Mandal D, Singh R, Dhyani SK, Dhyani BL (2010) Landscape and land use effects on soil resources in a Himalayan watershed. Catena 81:203–208

McLean EV (1982) Aluminium. In: Page AL, Miller RH, Keeney DR (eds) Methods of soil analysis, Part 2. American Society of Agronomy, Madison, pp 978–998

Nakajima T, Lal R, Jiang S (2015) Soil quality index of a Crosby silt loam in central Ohio. Soil Tillage Res 146:323–328

Nelson DW, Sommers LE (1982) Total carbon, organic carbon and organic matter. In: Page AL, Miller RH, Keeney DR (eds) Methods of soil analysis, Part 2. American Society of Agronomy, Madison, pp 539–579

Nigerian Meteorological Agency (2011) Climate data. NIMET, Calabar

Noma SS, Ojanuga AG, Ibrahim SA, Iliya MA (2005) Detailed soil survey of Sokoto-Rima flood plain at Sokoto. In: Managing oil resources for food security and sustainable management. Proceedings 29th annual conference. Soil Science Society of Nigeria. In: Uzoho BU, Oti NN, Ngwuta A (2007) Fertility rates under land use types on soils of similar lithology. J Am Sci 3(4):20–29

Nuttall M (2007) Climate change: identifying the impacts on soil and soil health. Department of Primary Industries, Agriculture Victoria Publications, State of Victoria, Australia

Ohio State University (2007) Projected climate change and future expectations. Presentation, USA

Okoh AC (2001) Evaluation of coastal plain sand soils supporting pineapple in Cross River State. Unpublished B. Agric project, Department of Soil Science, University of Calabar, Calabar. 40pp

Peters, K. C. (2007) Effects of different soil cover plants on sediment yield on soil loss in Calabar South Eastern Nigeria. Unpublished B. Agric. Project, Department of Soil Science, University of Calabar, Calabar, 66pp

Pietikainen J, Pettersson M, Baath E (2005) Comparison of temperature effects on soil respiration and bacterial and fungal growth rates. FEMS Microbiol Ecol 52(1):49–58

Rankine I, Fairhurst T (1999) Mature: field handbook. Oil palm series. International Plant Nutrition Institute Production, Georgia, 50pp

Schoeneberger PJ, Wysocki DA, Benham EC, Broderson WD (2002) Field book for describing and sampling soils (ver. 1.1). Natural Resources Conservation Service, USDA, National Soil Survey Centre, Lincoln

Singh RA (1982) Soil physical analysis. Kalyani Publishers, New Delhi, p 71

Thomas I (2008) Evaluation of soil quality of Cross River National Park. Unpublished B. Agric Project, Department of Soil Science, University of Calabar, Calabar. 64pp

Udo IE (2010) Amelioration of acid sands of Southern Nigeria using liquid pig manure (LPM) and lime ($Ca(OH)_2$). Unpublished B. Agric Project, Department of Soil Science, University of Calabar, Calabar. 71pp

Udo EJ, Ibia TO, Ogunwale JA, Ano AO, Esu IE (2009) Manual of soil, plant and water analysis. Sibon Books Ltd., Lagos. 183pp

University of Calabar (2011) Climate data. Department of Geography and Regional Planning, Calabar

Bioremediation Effects of Nitrogen Fixing Trees on Nutrients and Heavy Metals in Spent Engine Oil Polluted Soil

15

Bamidele Adanikin and Joshua Kayode

Contents

Abstract

Soil contamination by spent engine oil is a growing concern in many countries, especially in some less developed countries such as Nigeria. Greenhouse experiment was conducted for 4 months using *Leucaena leucocephala* to determine its effect on the nutrients (N, P, K, organic carbon) and heavy metals (Zn, Ni, Pb, Cu) of soil polluted with spent engine oil [5% (w/v)]. Soil without spent engine oil was used as control. Bioaccumulation of the nutrients and heavy metals in *Leucaena leucocephala* was also determined. Reductions in the nutrients were observed in the soils during the first to third months of bioremediation. However, in the fourth month, 33.33% and 16.67% increase in nitrogen were recorded in the control and polluted soils, respectively. Reduction in Zn, Ni, Pb, and Cu was 41%, 48.39%, 61.60%, and 52.72%, respectively, in the polluted soil, and bioaccumulation of Zn, Ni, Pb, and Cu in *Leucaena leucocephala* planted in the polluted soil increased by 73.41%, 85.46%, 3366.04%, and 125.53% in the plant biomass. This chapter has shown that *Leucaena leucocephala* is effective in bioremediation of heavy metals in spent engine oil-polluted soil.

B. Adanikin (✉) · J. Kayode
Department of Plant Science and Biotechnology, Ekiti State University, Ado Ekiti, Ekiti State, Nigeria
e-mail: adanikindele4dove@yahoo.com; joshua.kayode@eksu.edu.ng

W. Leal Filho (ed.), *Handbook of Climate Change Resilience*,
https://doi.org/10.1007/978-3-319-93336-8_73

Keywords
Bioremediation · Spent engine oil · Heavy metals · *Leucaena leucocephala*

Introduction

Soil contamination by spent engine oil from automobiles is a growing concern in many countries, especially in Asia and Africa. This is due to the increase in quantity of engine oil and other engine fluids from servicing and subsequently draining spent oil from automobiles and generator engines due to the increase in population, roadway, railway, generators, and vehicles (Baladincz et al. 2008; Agamathua et al. 2010). In some less developed countries, such as Nigeria, up to 80 million liters of spent engine oil are generated and discharged into the environment without any form of processing. This may pose a great threat to the environment and human being at large (Adelowo et al. 2006).

Various studies have confirmed the deleterious effects of spent engine oil in our environment. Agbogidi and Ejemete (2005) noted that oil in soil has deleterious effects on biological, chemical, and physical properties of the soil. Ekundayo et al. (2001) reported that the germination of seeds planted in crude oil-polluted soil area was delayed, and percentage germination was significantly affected. Kayode et al. (2009) reported the inhibitions of radicle and plumule growths as well as the reduction in the number of leaves in seedlings of *Vigna unguiculata* and *Zea mays* in spent engine oil-polluted soil.

Phytoremediation is a promising strategy for in situ removal of many contaminants (Greenberg 2006; Pilon-Smith 2005; Pulford and Watson 2003). According to Palmroth et al. (2002), root exudates from plants do help to degrade toxic organic chemicals and act as substrates for soil microorganisms in the soil which directly results in increased rate of biodegradation of the organic contaminants. Different types of plants have been found useful for phytotreatment of soil contaminated by hydrocarbons. *Mirabilis jalapa* was found to reduce concentration of total petroleum hydrocarbon (TPH) by 41.61–63.2% in 127 days (Peng et al. 2009). Studies by researchers have shown that *Jatropha curcas* is a potential plant for remediation of heavy metal-contaminated soil due to its bioaccumulation potential (Jamil et al. 2009; Mangkoedihardjo and Surahmaida 2008).

Leucaena leucocephala has numerous inherent characteristics that can be exploited to augment phytoremediation and lower the cost of soil regeneration. The species can survive in harsh environmental conditions with the exception of heavily frosted conditions and occurs in a wide range of ecological settings. It is fast growing, capable of reaching maturity in 6–7 months to produce a vast amount of seeds that can germinate into numerous seedlings to carry on further remediation of the polluted site. It can produce large quantities of phytomass that can accumulate heavy metals and can repeatedly be harvested to regenerate a polluted area through phytoextraction (Jamilu et al. 2017). Therefore the objective of this chapter is to investigate the potential of *Leucaena leucocephala* in enhancing the nutrients and reducing the heavy metals of spent engine oil-polluted soil.

Materials and Methods

The research was carried out at the greenhouse of the Department of Plant Science and Biotechnology, Faculty of Science, Ekiti State University, Ado Ekiti, Nigeria.

Collection of Soil Samples and Plant Materials: Top soil was collected from the research farm of the Faculty of Agricultural Sciences, Ekiti State University, Ado Ekiti, Nigeria. Spent engine oil was sourced from mechanic workshops in Ado Ekiti, Ekiti State. *Leucaena leucocephala* (LL) were nursed from seeds at the greenhouse of the Department of Plant Science and Biotechnology of the university.

Experimental Procedures: Ten (10) planting pots were used for the experiment; each of the planting pots was filled with soil (10 kg). The planting pots were divided into two groups (A and B) making five (5) pots in each group. Each pot in group A was polluted with 500 ml spent engine oil, while pots in group B were not polluted with Spent engine oil they served as the control. The polluted soil samples were left for seven (7) days without watering for the spent engine oil to sink into the soil. The unpolluted soil samples were also left for 7 days without wetting. At 3 weeks of growth, a seedling of *Leucaena leucocephala* was transplanted from the nursery into each pot in both group A and B on the same day. Watering of pots was done in alternate days with 500 ml of water per pot.

Sample Collection and Analysis: After 7 days, representative samples were taken from each pot to determine the physicochemical properties of the soil (polluted and not polluted), and samples of plant seedlings harvested at 3 weeks of growth were also taken for analysis [total organic carbon (TOC), nitrogen (N), phosphorus (P) and potassium (K), zinc (Zn), nickel (Ni), lead (Pb), and copper (Cu)] before transplanting into each pot. Representative soil samples were collected every month in each of the pots and analyzed for N, P, K, TOC, Zn, Ni, Pb, and Cu contents. At the end of the experiment, plants were harvested from each pot; the roots were washed with distilled water to remove soil particles and were analyzed for N, P, K, TOC, Zn, Ni, Pb, and Cu.

Soil Analysis: Soil samples were collected, air-dried, crushed, and sieved through 2 mm sieve. Soil organic C and total N were determined by dichromate oxidation (Nelson and Sommers 1982) and Kjeldahl digestion–distillation procedure (Bremner 1996), respectively. The total metal content was determined using an atomic absorption spectrophotometer (AAS, GBC Avanta Ver. 1.33, Australia) after digesting 0.5 g of dried soil samples with 15 ml of HNO_3, H_2SO_4, and $HClO_4$ in 5:1:1 ratio at 80 °C in a microwave digestion system (MDS-2000) and filtered through Whatman No. 42 filter paper followed by dilution up to 50 ml with triple distilled water (Baker and Amacher 1982).

Plant Analysis: Harvested plant samples were washed with water followed by $CaCl_2$ solution and distilled water and oven-dried at 70 °C up to constant weight. The whole plants were milled to pass through a 2 mm sieve. The milled plant material (0.5 g) was digested with a mixture of concentrated HCL/HNO_3 (4:1 v/v) and analyzed for metals using AAS (Allen et al. 1986).

Accumulation percentage (AP) was used to compare performance of various plants:

$$AP = \frac{\text{Level of heavy metals in treated plants at harvest}}{\text{Level of heavy metals in plants before planting}}) \times 100$$

Bioaccumulation ratio (BR) was calculated by using the procedure by Cai and Lena (2003) and expressed as:

$$BR = \frac{\text{Concentration in Plant}}{\text{Cocentration in soil}}$$

Percentage reduction (PR) and increase (PI) of metals in soil were also determined as follows:

$$PR = \frac{\text{Metal level before transplanting} - \text{Metal level after harvest}}{\text{Metal before transplanting}}) \times 100$$

$$PI = \frac{\text{Metal level after harvest} - \text{Metal level before transplanting}}{\text{Metal before transplanting}}) \times 100$$

Statistical Analysis: Experimental data collected was subjected to one-way analysis of variance (ANOVA) using SAS (2002) to compare variations in nutrients and heavy metals in each treatment. Significance was indicated at $p \leq 0.05$; Duncan's multiple range test was used to separate the means.

Results and Discussion

Table 1 shows that N and TOC in soil had increased by 19% and 336%, respectively, in the polluted soil before LL was planted. This was in agreement with previous studies of Liu et al. (2007) and Okonokhua et al. (2007). These increases in N and TOC have been attributed to input arising from mineral elements in the spent engine oil (Liu et al. 2007; Okonokhua et al. 2007; Nwite and Alu 2015). Values recorded for available P (0.33 ± 0.003 mg/g) and K (0.21 ± 0.005 mg/g) in polluted soil were lower than P (0.43 ± 0.008 mg/g) and K (0.28 ± 0.001 mg/g) recorded in the control. Decrease in P and K in soil after pollution with spent engine oil has been reported as due to the inability of microbes to transform organic matter which led to low mineralization of the elements in the previous study by Ogboghodo et al. (2004).

As shown in Table 1, values of metals, Zn (28.00 ± 0.2 mg/kg), Ni (14.63 ± 0.38 mg/kg), Pb (14.63 ± 0.38 mg/kg), and Cu (12.25 ± 0.35 mg/kg), were lower in the unpolluted soil compared to 73.17 ± 1.07 mg/kg, 26.30 ± 1.10 mg/kg, 57.35 ± 0.65 mg/kg, and 15.10 ± 0.10 mg/kg recorded in polluted soil. The spent engine oil in this research was not analyzed; however, previous studies have established solute contamination in spent engine oil which may cause metal pollution when introduced to soil (Okonokhua et al. 2007; Yong 2001).

Table 1 Physicochemical properties of control (B) and polluted soil (A) after 7 days and *L. leucocephala* used for the research

Parameters	LL	B	A
N (mg/g)	57.2 ± 0.3	2.1 ± 0.2	2.5 ± 0.4
P (mg/g)	0.58 ± 0.003	0.43 ± 0.008	0.33 ± 0.004
K (mg/g)	3.91 ± 0.005	0.28 ± 0.001	0.21 ± 0.005
TOC (mg/g)	414.8 ± 6.2	22.5 ± 0.3	98.3 ± 0.5
Zn (mg/kg)	12.7 ± 0.1	28.00 ± 0.2	73.17 ± 1.07
Ni (mg/kg)	5.53 ± 0.08	14.63 ± 0.38	26.30 ± 1.10
Pb (mg/kg)	0.3 ± 0.09	14.63 ± 0.38	57.35 ± 0.65
Cu (mg/kg)	4.09 ± 0.04	12.25 ± 0.35	15.10 ± 0.10

Table 2 Nutrients (mg/g) and heavy metal content (mg/kg) in control and polluted soil in the first, second, third, and fourth months of bioremediation

		Nutrients				Heavy metals			
Treatments	Months	N	P	K	TOC	Zn	Ni	Pb	Cu
A	First	2.80a	0.031b	0.198a	175.25a	64.67a	23.52a	49.45a	14.21
	Second	1.35b	0.037b	0.197b	38.95b	54.67b	22.37a	38.68b	13.57b
	Third	1.20c	0.028c	0.181c	27.80c	46.79c	20.45b	32.60c	10.16c
	Fourth	1.40b	0.020d	0.166d	9.05d	42.26d	14.21	22.38d	7.195d
B	First	1.80e	0.038e	0.241e	126.95e	24.62e	12.65e	9.21e	10.23e
	Second	1.80e	0.038e	0.247e	12.50f	24.42e	11.95f	8.22f	10.60e
	Third	1.20f	0.030f	0.221f	11.15 g	22.11f	10.11 g	8.05f	9.23f
	Fourth	1.60e	0.025 g	0.181 g	2.45 h	20.30 g	9.83 g	4.18 g	5.66 g

N, P, K, and TOC in LL were 57.2 ± 0.3 mg/g, 0.58 ± 0.003 mg/g, 3.91 ± 0.005 mg/g, and 414.8 ± 6.2 mg/g, respectively. Metals, Zn, Ni, Pb, and Cu, were also recorded in the LL but in smaller amount than in the soil.

Monthly values of nutrients and metals in the control and polluted soils were presented in Table 2. In the first month of bioremediation, values recorded for N and TOC in treatment B, 1.80 mg/g and 175.25 mg/g, respectively, were lower compared to A. The higher values recorded for N and TOC in treatment A after the first month of bioremediation may be due to the addition from mineral elements which is contained in spent engine oil (Liu et al. 2007). Reduction in values of N was recorded in both treatments (A and B) in the second and third months of bioremediation. In the fourth month, 33.33% and 16.67% increases in N were recorded in treatments B and A, respectively. Previous studies have reported that nodulation in nitrogen-fixing tree may take up to 3 months after planting (Aganga and Tshwenyane 2003; Schneider et al. 2013). Saraswat and Rai (2011) also reported that the nodulation of LL in soil may take longer time as the level of pollution increases.

Lower value of N (1.40 mg/g) recorded in treatment A compared in the fourth month may be due to delayed nodulation as a result of the pollution. The study by Binkley and Menyailo (2005) has indicated that the ability of nitrogen-fixing trees to improve TOC in the soil is associated with their nitrogen-fixing capability. The

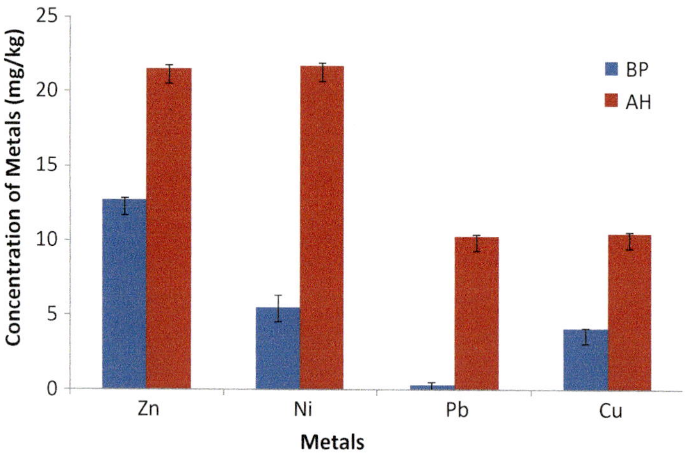

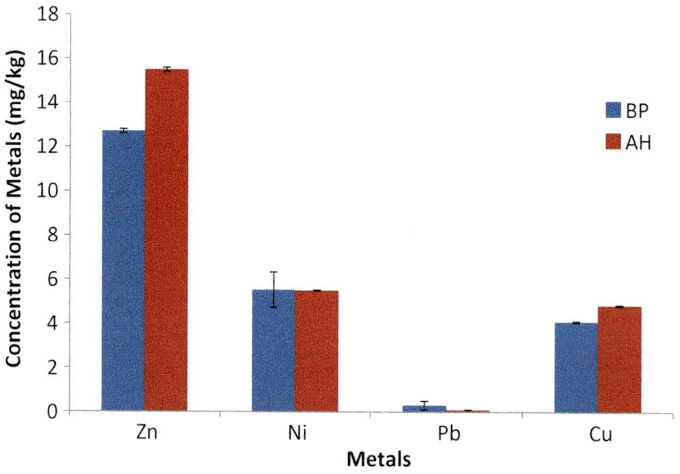

Fig. 1 Metals in *L. leucocephala* biomass in Group A; before planting (BP); after harvest (AH). Metals in *L. leucocephala* biomass in Group B; before planting (BP); after harvest (AH)

reduction observed in these elements (P, K, and TOC) may also be due to delayed nitrogen-fixing ability in LL.

Metals (Zn, Ni, Pb, Cu) in the polluted soil were observed to be reduced in each month of bioremediation. In treatment A, significant reductions ($p \geq 0.05$) was observed in all the months of bioremediation. This agrees with the previous studies that LL has the ability to sequestrate metals from polluted soils (Ma et al. 2006; Tawfix 2008). However, values recorded in treatment A in the fourth month of bioremediation (Zn (42.26 kmg/g), Ni (14.21 kmg/g), Pb (22.38 kmg/g), and Cu (7.20 kmg/g)) were still higher than that recorded in treatment B (20.30 kmg/g, 9.83 kmg/g, 4.19kmg/g, and 5.66 kmg/g, respectively). This may have suggested that a longer period of time is needed for effective bioremediation of these metals which were initially higher in the polluted soil.

Figure 1 shows that Pb was the highest bioaccumulated metal in LL biomass from the polluted soil. Pb increased by 3366.04% in LL biomass after harvest. Juson et al. (2016) have reported the ability of LL to bioaccumulate Pb in its tissues from polluted soils. This also agrees with Yitao et al. (2014) that high concentration of Pb can be bioaccumulated by LL in its aboveground parts. Cu, Ni, and Zn also increased in plant biomass after harvest (125.53%, 85.46%, and 73.41%, respectively). Studies have reported the ability of LL to accumulate Zn and Cu in its roots and shoots (Gupta et al. 2000; Schneider et al. 2013).

Bioaccumulation factor is used to evaluate the efficiency of a biota to efficiently remediate a polluted soil. Bioaccumulation factor higher than 1 has been considered most efficient in bioremediation of polluted soils (Garcial et al. 2009; Kalac and Svoboda 2000). The bioaccumulation factors for Zn, Ni, Pb, and Cu were 0.51, 1.53, 0.46, and 1.46, respectively, in LL. Bioaccumulation factor (>1) recorded for Ni and Cu in LL may suggest that this plant tends to be more prolific in the uptake of these metals from spent engine oil-polluted soil.

Conclusions

The chapter shows that LL has the potential to bioremediate heavy metals from spent engine oil-polluted soil. The reduction observed in N both in the control and polluted soil indicates that longer period of time (more than 3 months) may be needed for the nitrogen-fixing ability of LL to attain its full potential. The bioaccumulation factor >1 recorded for Ni and Cu in LL suggests that this tree is most efficient in remediating these metals from polluted soil.

References

Adelowo OO, Alagbe SO, Ayandele AA (2006) Time-dependent stability of used engine oil degradation by cultures of *Pseudomonas fragi* and *Achromobacter aerogenes*. Afr J Biotechnol 5(24):2476–2479

Agamuthua P, Abioyea OP, Abdul Aziz A (2010) Phytoremediation of soil contaminated with used lubricating oil using Jatropha curcas. J Hazard Mater 179:891–894

Aganga AA, Tshwenyane SO (2003) Lucerne, Lablab and Leucaena leucocephala forages: production and utilization for livestock production. Pak J Nutr 2:46–53

Agbogidi OM, Ejemete OR (2005) An assessment of the effects of crude oil pollution on soil properties, germination and growth of *Gambaya albida* (L.). UNISWA Res J Agric Sci Technol 8(2):148–155

Allen SE, Grimshaw HM, Rowland AP (1986) Chemical analysis. In: Moore PD, Chapman SB (eds) Methods in plant ecology, 2nd edn. Blackwell Scientific Publication, Oxford/London, pp 285–344

Baker DE, Amacher MC (1982) Nickel, copper, zinc and cadmium. Methods of soil analysis, No. 9, Part II, Madison, pp 331–333

Baladincz J, Szabo L, Nagy G, Hancsok J (2008) Possibilities for processing of usedlubricating oils – part 1. MOL Sci Mag 3:81–86

Binkley D, Menyailo O (eds) (2005) Tree species effects on soils: implication for global change. NATO Science Series, IV: Earth and Environmental Sciences 55:155–164

Bremner JM (1996) Nitrogen-total. In: Sparks DL (ed) Methods of soil analysis, Part 3 chemical methods. SSSA Inc., ASA Inc, Madison, pp 1085–1122
Ekundayo EO, Emede TO, Osayande DI (2001) Effects of crude oil spillage on growth and yield of maize (*Zea mays* L.) in soils of Mid-western Nigeria. Plant Foods Hum Nutr 56:313–324
GarcíaI DM, Martín F, Simón M, Dorronsoro C (2009) Mobility of arsenic and heavy metals in a sandy-loam textured and carbonated soil. Pedosphere 19:166–175
Greenberg BM (2006) Development and field test of multi-process phytoremediation system for decontamination of soil. Can Reclam Spring/summer(1):27–29
Gupta M, Kumari A, Yunus M (2000) Effect of fly-ash on metal composition and physiological responses in *Leucaena leucocephala* (lamk.) de. wit. Environ Monit Assess 61:399–406
Jamil S, Abhilash PC, Singh N, Sharma PN (2009) Jatropha curcas: a potential crop for phytoremediation of coal fly ash. J Hazard Mater 172:269–227
Jamilu ES, Muhammed N, Ingvar B, Oryem-Origa H (2017) Phytoremediation potential of *Leucaena leucocephala* (Lam.) de wit. For heavy metals polluted and heavy metals degraded environments. In: Phytoremediation potentials of bioenergy plants. Springer, Singapore, pp 159–209
Juson AE, Maria-Kariza MM, Johnny AC (2016) Accumulation and distribution of heavy metals in *Leucaena leucocephala* (Lam.) and *Bougamvillea Spectabilis* Willd. Plant systems. J of Exp Biol Agric Sci 4(1):01–06. ISSN No. 2320-8694
Kalac P, Svoboda L (2000) A review of trace element concentrations in edible mushrooms. Food Chem 69:273–281
Kayode j, Olowoyo O, Oyedeji A (2009) The effects of used engine oil pollution on the growth of early seedling performance of *Vigna unguiculta* and *Zea mays*. Res J Soil Biol 1(1):15–19
Liu MS, Luo YP, Su ZY (2007) Heavy metal concentrations in soils and plant accumulation in a restored manganese mine land in Guangxi, South, China. Environ Pollut 147:168–175
Ma Y, Dickinson NM, Wong MH (2006) Beneficial effects of earthworms and arbuscular mycorrhizal fungi on establishment of leguminous shrubs on Pb/Zn mine tailings. Soil Biol Biochem 38:1403–1412
Mangkoedihardjo S, Surahmaida A (2008) Jatropha curcas L for phytoremediation of lead and cadmium polluted soil. World Appl Sci J 4(4):519–522
Nelson DW, Sommers LE (1982) Total carbon, organic carbon and organic matter. In: Page AL et al (eds) Methods of soil analysis: part 2. Chemical and microbiological properties, ASA monograph number 9. ASA SSSA, Madison, pp 539–579
Nwite JN, Alu MO (2015) Effect of different levels of spent engine oil on soil porperties, grain yield of maize and its heavy metal uptake in Abakaliki, Southeastern Nigeria. J Soil Sci Environ Manag 5(4):4451
Ogboghodo AE, Inaga EK, Osemwota O, Chokor JU (2004) An assessment of the effects of crude oil pollution on soil properties, germination and growth of maize (*Zea mays* L.) using two types Forcades light and Escravos light. J Environ Monitor Asses 96:142–152
Okonokhua BO, Ikhajiagbe B, Anoliefo GO, Emede JO (2007) The effect of spent engine oil on soil properties and growth of maize (*Zea mays* L.). J Appl Sci Environ Manag 11(3):147–152
Palmroth MRT, Pichtel J, Puhakka JA (2002) Phytoremediation of subarctic soil contaminated with diesel fuel. Bioresource Technol 84:221–2281
Peng S, Zhou Q, Cai Z, Zhang Z (2009) Phytoremediation of petroleum contaminated soils by Mirabilis jalapa L. in a greenhouse plot experiment. J Hazard Mater 168(2–3):1490–1496
Pilon-Smith E (2005) phytoremediation. Annu Rev Plant Biol 56:15–39
Pulford ID, Watson C (2003) Phytoremediation of heavy metal contaminated land by trees-a review. Environ Int 4:529–540
Saraswat S, Rai JPN (2011) Prospective application of *Leucaena leucocephala* for phytoextraction of Cd and Zn and nitrogen fixation in metal polluted soils. Int J Phytoremediation 13(3): 271–288
SAS (2002) Statistical analysis software guide for personal computers. cary, NC27513, USA: release 9. 1 SAS Institute Inc. Scale comparison with water hyacinth. Bioresour Technol 163: 82–91

Schneider J, Labory CRG, Rangela WM, Alves E, Guilherme LRG (2013) Anatomy and ultra-structure alterations of Leucaena leucocephala (Lay.) inoculated with mycorrhizal fungi in response to arsenic-contaminated soil. J Hazard Mater 262:1245–1258

Tawfik KM (2008) A monitory field study at El Saaf-Helwan faba bean farms irrigated by industrial waste water and polluted water with sewage. J Appl Sci Res 4:492–499

Yitao XU, Qixian F, Jixiu W, Mingrui LI, Fangdong Z, Xiao Y (2014) Heavy metal uptake characteristics of three plants grown in lead/zinc mine tailings. Environ Sci Technol 4(2):118–138

Yong RN (2001) Geoenvironmental engineering: contaminated soils, pollutant fate and mitigation. CRC Press, Florida, 307 pp

Building Vulnerable Islander Resilience to Natural Hazard: A Participatory Approach

16

Riffat Mahmood and A. Q. M. Mahbub

Contents

Abstract

In last decade and so, disaster risk reduction community started paying attention on the notions of resilience; therefore, the purpose of this study is to provide an overview of human vulnerability and priority intervention issues based on vulnerability experience by which resilience can be promoted. In the context of natural hazards and disasters, coastal islands of Bangladesh are vulnerable where

R. Mahmood (✉)
Department of Geography and Environment, Faculty of Life and Earth Sciences, Jagannath University, Dhaka, Bangladesh
e-mail: riffat.mahmood.jnu@gmail.com

A. Q. M. Mahbub
Department of Geography and Environment, Faculty of Earth and Environmental Sciences, University of Dhaka, Dhaka, Bangladesh
e-mail: aqmahbub@yahoo.com

W. Leal Filho (ed.), *Handbook of Climate Change Resilience*,
https://doi.org/10.1007/978-3-319-93336-8_107

isolation itself works as an adverse factor regarding vulnerability so that this study chose Sandwip Island as the study area. The data collection methods included geospatial techniques and participatory rural appraisal (PRA) using 0–5 measuring scale followed by weighted matrix index (WMI) and in-depth interviews using semi-structured questionnaire. This research explored 8 major vulnerability issues and 17 urgent priority intervention issues for building vulnerable islander resilience. Authors identified that attempts in reducing vulnerability in inhabitation and livelihood is the starting point of being resilient, and it can be promoted further by undertaking interventions such as local infrastructure development and improvement in community facilities, i.e., long-term adaptation at community level, not only the post disaster responses. Finally, this research provides guideline of how resilience of coastal islanders can be promoted one step forward in the face of natural hazards and disasters.

Keywords
Bangladesh · Meghna Estuarine islands · Natural hazard · Participatory approach · Resilience · Vulnerability

Introduction

Greater numbers of people in recent times are more vulnerable to natural hazards since the impact and scope of disasters are increasing day by day through the combined effects of economic, social, demographic, ideological, and technological factors (Barman et al. 2012). It has already been reported by IPCC (2007) that Bangladesh is considered to be the most vulnerable to natural disasters which, coupled with an increasing population in the coastal zone, potentially increases vulnerability, risks, and losses (Munich 2012). Also, low economic strength, inadequate infrastructure, low level of social development, lack of institutional capacity, and higher dependency on natural resources make the coastal area more vulnerable to natural hazards and disasters. Therefore, the aim of the Department of Disaster Management of the Ministry of Disaster Management and Relief (MoDMR), Bangladesh, is to foster a paradigm shift in disaster management from conventional response and relief to a more comprehensive risk reduction culture which reflects the adoption of the "Hyogo Framework for Action 2005–2015" in national policies, emphasizing the need to reduce vulnerability and promote resilience (UNISDR 2005). Therefore, progression from the vulnerability of coastal community toward a state of climate resilience is a priority concern in Bangladesh as resilience is seen as contributing to sustainability and reducing vulnerability concerning with the interactions between people and nature, including vulnerability and disaster reduction (UNISDR 2002).

Most of the resilience models related to hazard involve environmental and engineering components of coastal vulnerability (Newton and Weichselgartner 2013). In these frameworks, the properties of resilient infrastructure reduce the probability of failures (Bruneau et al. 2003), yet these frameworks often fail to capture antecedent social factors and human dimensions that occur at the most local levels or to account for the vulnerability or resilience of the natural environment. Based on the extensive literature review on resilience in social system, numerous models have also been identified; each model developed in theoretical or conceptual isolation from the others; while there is some overlap in domains/content, differences are also present. But gap between linking theoretical thinking and local case studies have been seen (Zou and Thomalla 2008). Resilience is seen as contributing to sustainability and reducing vulnerability (Schulze and Mooney 1993; Peterson et al. 1998; Chapin et al. 2000), although clear guidance as to how resilience can be promoted is lacking.

It is already known that geographical location and hydro-morphological characteristics of Bangladesh along with the funnel-shaped coastal area influence it to be affected by tropical cyclone and storm surge, particularly in lower Meghna Estuary (Brammer 2014) where many coastal islands are situated (Brammer 2012). Inundation of these low-lying islands during cyclone surge causes adverse impact (World Bank 2010) as these small islands have limited human resource base with single urban center; in most case islanders are concentrated in the coastal zone (Newton and Weichselgartner 2013). Also, rapid geomorphological changes are taking place in Meghna Estuarine islands due to heavy sediment load and freshwater discharges. Remoteness and isolation of these islands from mainland itself works as an adverse factor to enhance livelihood vulnerability generating deeper poverty (FAO 2006). Therefore, this study attempts to develop a methodological framework for building an integrated participatory priority assessment of risk reduction issues of vulnerable islanders of Meghna Estuary of Bangladesh to the pathways of resilience to natural hazards through location-specific field studies.

An Overview of Resilience in Hazard Perspective

In 1970 the term "resilience" was first coined in ecology to describe systems that have the ability to recover and return to their original state after undergoing stress (Holling 1973; Surjan et al. 2011). Though these ideas developed from an ecosystem perspective, resilience became a concept of value for economic and social studies and disaster management as well; that refers to the ability of human communities to withstand external perturbations such as environmental variability or social, economic or political upheaval, and to recover from such perturbations (Timmerman 1981). The new notion of paying more attention to prevention and readiness of a society to external shocks makes resilience a concept more convergent in the context of disaster management (Keim 2008; Cutter 2013). In this context, resilience is defined as the ability of a system to prepare and plan for, absorb, recover from, or more successfully adapt to actual or potential adverse events (Berke and Campanella 2006; NRC 2012). Though, there is no specific definition of disaster resilience, yet there is some common agreement on its broad parameters, specifically the capacity

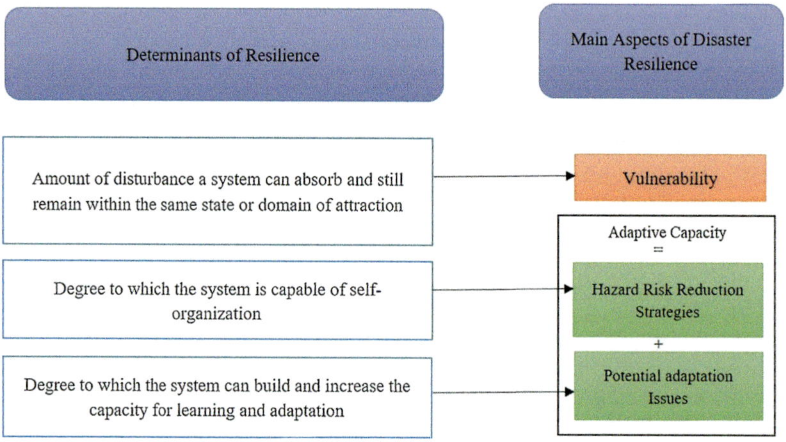

Fig. 1 Main aspects of disaster resilience

to reduce or avoid losses, contain the effects of disasters, and recover with minimal social disruptions after a hazard event (Klein et al. 2003a; Vale and Campanella 2005; Manyena 2006; Nelson et al. 2007; Turner 2010; Paton and Johnston 2006; Cutter et al. 2008; Leichenko 2011). It includes not only a capacity to return to the state (or multiple states) that existed before the disturbance but also to advance the state through learning and adaptation (UNISDR 2002; Klein et al. 2003b; Adger et al. 2003). Carpenter et al. (2001) mentioned three determinants of resilience by the Resilience Alliance which refers to the following main aspects of disaster resilience adopted in this study (Fig. 1).

Many scholars prioritized different aspects of disaster resilience; one is vulnerability (Twigg 2007) as a flip side of resilience (Khan 1974; Burton et al. 1993; Hewitt 1997; Twigg and Bhatt 1998; Weichselgartner 2001; IFRC 2006). Manyena (2006) considered vulnerability as an outcome when defined as the ability to cope with a hazard event and imbedded within vulnerability. Other aspect of disaster resilience is related to capacity to respond as to cope with or adapt in the context of hazard risk reduction, conceptualization of risk and disaster (Blaikie et al. 1994; Gallopın 2006). Twigg (2007) emphasized on adaptive capacity what community can do for themselves such as actions, or processes, undertaken toward achieving resilience (i.e., hazard risk reduction strategies) and how to strengthen their capacities (i.e., potential adaptation issues) while focusing on disaster resilience that makes the concept of adaptive capacity equivalent to the resilience (Handmer and Dovers 1996; Smit and Wandel 2006). So, there are two generic approaches to anticipatory adaptive capacity (Huq and Klein 2003): increasing the flexibility of potentially vulnerable systems that are managed by humans by allowing for mid-term adjustments in management practices, i.e., hazard risk reduction strategies, and improving public awareness and preparedness through informing the public about the risks and possible consequences of climate change, i.e., potential adaptation issues.

$$\text{Resilience} = \frac{\text{Adaptive capacity}}{\text{Vulnerability}}$$

Methodological Framework for Building Vulnerable Islander Resilience

The approach of this study aims to learn from experiences by the identification of vulnerability first as a starting point (i.e., assessment of vulnerability) and examining how households are able to cope with these vulnerabilities and become resilient (i.e., assessment of adaptive capacity). Those households that are able to bounce back to their condition in the pre-disaster period, or even to improve their situation, are considered resilient. In this study, households who resume levels of income and production in a timely manner after the disaster period are considered resilient; in contrast, those households who are collapsing or recovering but worse off than previously may be considered not resilient, i.e., vulnerable.

Assessing Vulnerability

Numerous frameworks, conceptual models, and vulnerability assessment techniques have been developed to advance both the theoretical underpinnings and practical applications of vulnerability (Gallopin 2006; Fussel 2007). In this study, vulnerability is viewed as a function of biophysical risk and social response and how these manifest itself locally or the hazardousness of place (Hewitt and Burton 1971). So, the assessment of vulnerability can be conducted through participatory rural appraisal (PRA) using a measuring scale from 0 to 5 where level 0 represents to a total lack of vulnerability and levels 1, 2, 3, 4, and 5 represent 20%, 40%, 60%, 80%, and 100% of vulnerability, respectively. In this study, one participant identified one vulnerability issue; this issue has to be recognized and accepted by all other participants and when accepted is weighted and agreed by all the participants. If anyone gives wrong information, the other participants would counter the information by which the cumulative information would be certified through the process. A semi-structured questionnaire survey based on the identified vulnerability issues in PRA session was then done to know about the extent of vulnerability.

Assessing Adaptive Capacity

Adaptive capacity in context of climate change resilience that is viewed in this study refers to the ability to adjust that takes place in natural or human systems in response to the actual or expected impacts of climate change, aimed at moderating harm or exploiting beneficial opportunities. Assessing adaptive capacity of resilient households was done through semi-structured questionnaire based on generic hazard risk reduction strategies at individual level modified from Burton et al. 1993 (Table 1) as

Table 1 Generic strategies at individual level for preparing for climate change and reducing weather-related hazard risks adopted in this study. (Modified from Burton et al. 1993)

Strategy	Option	Option in level of individual
Choose change	Change location	Migration and relocation of exposed population (Hussein and Nelson 1999; Rayhan and Grote 2007; IOM 2009)
	Change use	Sustainable land use change accepting the hazard (UN-Habitat 2010; Bajracharya et al. 2011)
Reduce loss	Prevent effects	Human coping and adaptive response to reduce the impacts of a hazardous event when it occurs (Blaikie et al. 1994; Gallopín 2006; Smit and Wandel 2006; Twigg 2007)
	Modify event	
Accept loss	Share loss	Economic recovery by exploiting reserves of wealthy households (Morris et al. 2002; Carter et al. 2007) or informal credit and insurance market for poorer households (Coate and Ravallion 1993) Livelihood diversification through activity diversification (Evans and Ngau 1991; Ghosh and Bharadwaj 1992; Dercon and Krishnan 1996; Taylor and Wyatt 1996; Carter 1997; Chambers 1997) or migratory strategies in the context of slow onset disaster (Stark and Levhari 1982; Stark and Bloom 1985; Hugo 2010; Renaud et al. 2011)
	Bear loss	

there is no clear-cut boundary between adapting climate change and reducing weather-related hazard risks (Fussel and Klein 2002). Finally, participatory rural appraisal (PRA) was done using 0–5 measuring scale to rank their priority of adopted strategies where 0 represents no priority and 5 represents highest priority of adopted strategies.

Assessing High-Priority Intervention Issues

Finally, priority intervention issues were proposed to build vulnerable islander resilience based on vulnerability experiences as well as suggestion from resilient households to cope with those vulnerabilities. Selection of high-priority intervention issues was done by PRA session where "weighted matrix index (WMI)" was followed (Baker and Rapaport 2005). Younus and Harvey (2014) has presented how this method has shown significant direction through which local-level climatic impact needs to be identified through local participant's consensus responses in a practical way. This method has been methodologically proven as a good tool to use in developing countries (Chambers 2002). In this study, this weighting was done by consultation with climate change and coastal management experts who have gathered extensive experience on the study area as well as with local people having a certain level of education allowing them to assign weights to the proposed priority intervention issues for building vulnerable islander resilience. At first, they were requested to validate the proposed priority intervention issues identified on the basis of the vulnerability and adaptive capacity of the grassroot community, and then they were asked to weigh proposed priority intervention issues through a 1–20 weighing scale, where the number 20 signifies highest priority intervention and 1 represents lowest priority intervention. The scale was subdivided into four categories, e.g., very high (16.00–20.00), high (11.00–15.99), low (6.00–10.99), and very low (1.00–5.99). The values assigned by each expert were averaged to obtain the final weight of each priority intervention issues to build vulnerable islander resilience.

Building Vulnerable Islander Resilience with Special Reference to Sandwip Island, Bangladesh

Brief Introduction of Sandwip Island

The focus of this study is to explore the vulnerability and priority intervention issues in the context of natural hazard so that not only the hazards context but also demographic context are the important parameters of area selection of this study as changes in demographics have influenced past conditions of exposure and vulnerability to natural hazards (Newton and Weichselgartner 2013). After reviewing geophysical, hazard, and demographic context of Meghna Estuarine islands, Sandwip Island was selected as the study area of this research. Sandwip Island is the major part of Sandwip Upazila (administrative unit consisting of 150–200 villages) of Chittagong District. It is located from 22°22 N′ 91°26′E to 22°34′N 91°34′E (Fig. 2). Geographically, the island is bounded with the Bamuni River to the north, Meghna River and Hatia Island to the west, Sandwip Channel and Sitakunda and Mirsharai Upazila to the east, and Bay of Bengal to the south. This island is part of the Young Meghna Estuarine Floodplain (USAID 2006) where about 2000–3600 mm/year rainfall occurs on average (Sarkar and Choudhury 2003). Apart from the geophysical and climatic context, the demographic context of Sandwip Island is also very distinctive than that of whole country. There is a decreasing trend in population growth rate (2001–2011) among the hazard-prone Unions (smallest rural administrative and local government units in Bangladesh) of Sandwip Island (BBS 2001, 2011).

Materials and Methods

Inductive approach of research was followed where both qualitative and quantitative data were collected in order to fulfill the purpose of this study. A three stages field survey was done in accordance with methodological framework presented in Fig. 3. According to the available literature review and analysis of satellite images, it had been revealed that erosion is the most prominent and active natural hazard in Sandwip Island which happens all year round. Historically, Sandwip comprised of three types of Unions: (1) Unions with only erosion, (2) Unions with only accretion, and (3) Unions with no accretion and erosion (i.e., stable). Type erosion-prone Union can be subgrouped into two categories, i.e., Unions with riverbank erosion and Unions with coastal erosion with salt water intrusion. So, four Unions were selected

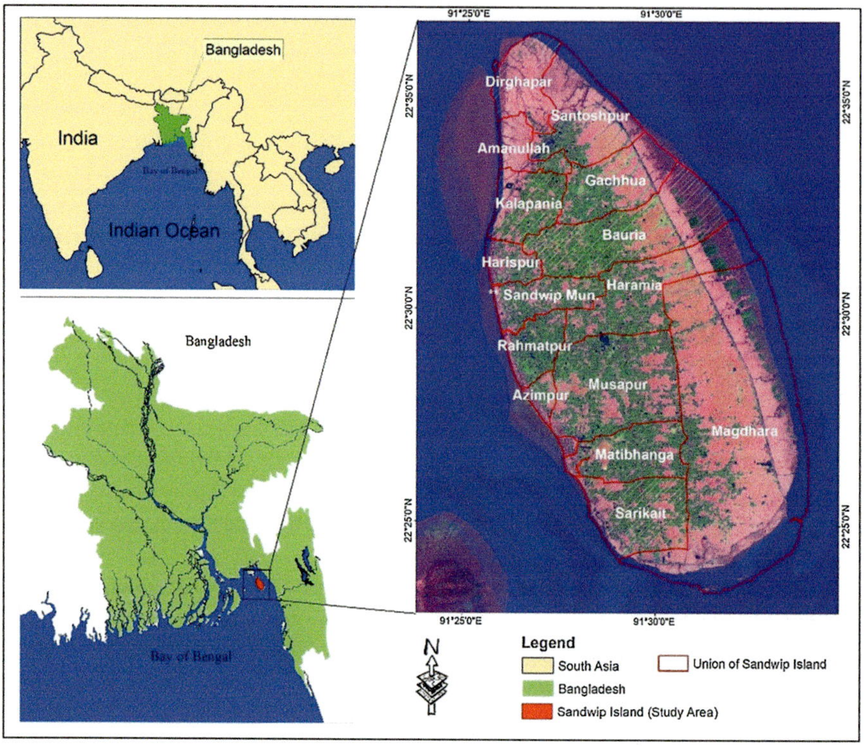

Fig. 2 Physical setting of the study area using satellite imagery of December 24, 2016 (Landsat 8 OLI image courtesy of the US Geological Survey), and spatial administrative boundary data (LGED, Government of Bangladesh)

for field survey that fall into these four categories. The Unions were Kalapania, Sarikait, Dirghapar (new), and Bauria.

In this study, first field survey consisted of four FGDs in four Unions to identify the major natural hazards in the last 10 years that affect local people's life and livelihood and one pilot survey to identify vulnerable and resilient households of survey Unions by asking past and present levels of income and productions in the last 10 years. Finally, four PRA sessions were done among vulnerable households in four Unions using 1–5 measuring scale to identify their vulnerabilities in the context of major natural hazards of the study area. In this field survey, also ground truthing data are collected for Participatory GIS (PGIS) for the major natural hazards of the study area.

The second field survey comprised of in-depth interviews of vulnerable and resilient households through questionnaire. Both vulnerable and resilient households were identified through pilot survey during the first field survey. Questionnaire for vulnerable households was intended to know the extent of identified vulnerabilities in first field survey, whereas questionnaire for resilient households was intended to

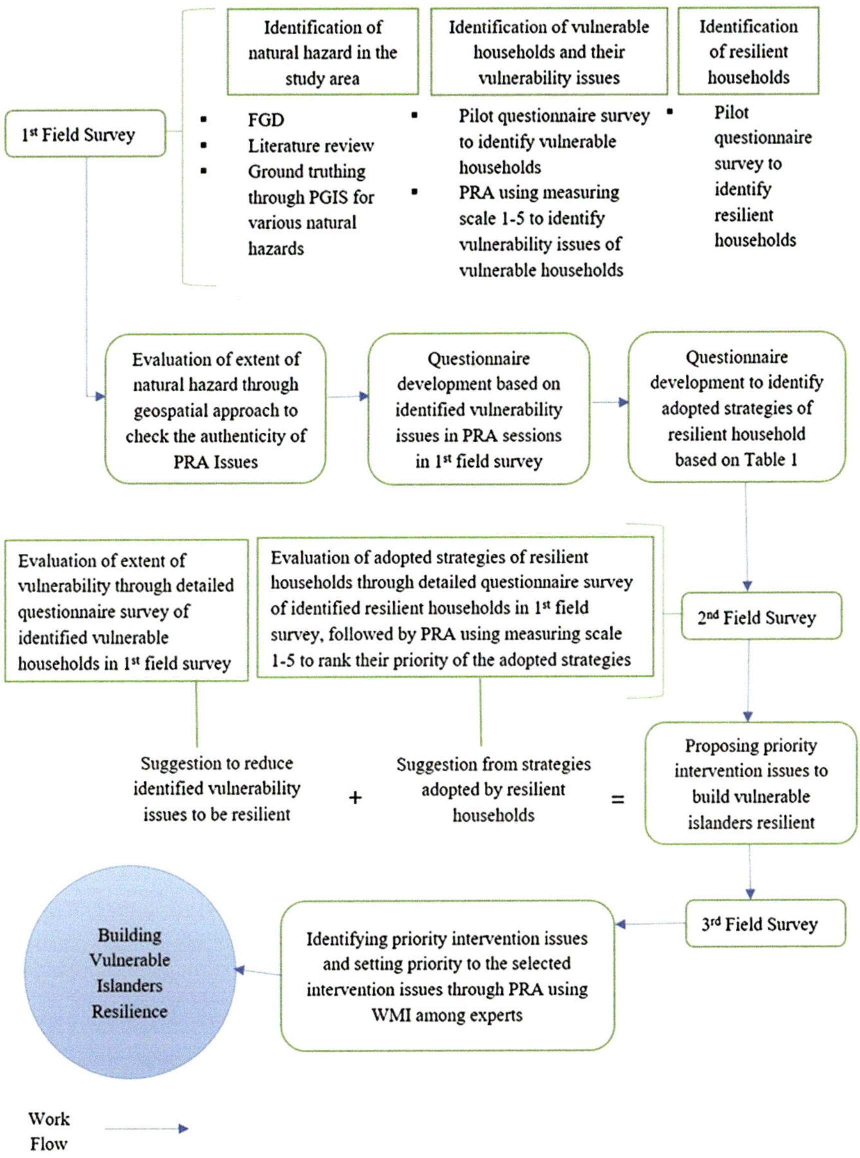

Fig. 3 The methodological framework for building vulnerable islander resilience adopted in this study

learn the strategies of resilient households based on vulnerability experiences. At the end of the second field survey, four PRA sessions were also conducted among resilient households in survey Unions using 1–5 measuring scale to rank the adopted strategies of resilient households.

Based on the suggestion to reduce identified vulnerabilities as well as experiences from the strategies adopted by resilient households, a list of priority intervention issues was proposed before the third field survey. In third field survey, one PRA session using "weighted matrix index (WMI)" was done where participants are environmentalist (02), university professors of related field (03), practitioners (02), and chairman of Union Parishad (3). They were asked first to validate the proposed priority intervention issues identified by the grassroot community, and then they were asked to weigh proposed priority intervention issues through a 1–20 weighing scale and selected "High Priority Intervention Issues" (WMI value, 16.00–20.00) for building vulnerable islander resilience.

In this study, for pilot survey Slovin's formula (1960) (Guilford and Frucher 1973) with 95% confidence level was used to determine sample size and obtained 398 dwelling households as a sample size, because total number of dwelling households in Sandwip Island is 54,480 (BBS 2011). Since Unions of Sandwip island was grouped into major three categories based on erosion and accretion activity, so the total households of the island had been grouped into three groups, e.g., (a) households in erosion-prone Union which cover 79.65% of the total households, (b) households in relatively stable Union which cover 14.02% of the total households, and (c) households in accretion-prone Union which cover 6.34% of the total households. So, 317 samples from erosion-prone Union (79.65% of 398), 56 samples from relatively stable Union (14.02% of 398), and 25 samples from accretion-prone Union (6.34% of 398) were collected accordingly (Table 2). For the convenience of the pilot survey, households were selected randomly from the "Household List" of the Union Council.

Among 398 dwelling households of the first field survey, 218 households were found vulnerable as they are collapsing but worse off in the last 10 years, whereas 180 households were considered as resilient as they resume income as well as production level in a timely manner after a disaster period in the last 10 years. In second field survey, in-depth interview was done among 218 vulnerable households to know the extent of their vulnerabilities and what they want from higher authority to reduce their vulnerabilities. Also, in-depth interview of 180 resilient households was conducted in second field survey to know their adopted strategies to cope with the vulnerabilities and become resilient.

Table 2 Sampling frame of the research

Subgroup of population	Sampled Union	Number of samples collected	Identified vulnerable households	Identified resilient households
Erosion prone (erosion>accretion)	Kalapania Sarikait	317	188	129
Relatively stable (insignificant erosion/accretion has been occurring)	Bauria	56	22	34
Accretion prone (accretion>erosion)	Dirghapar (new)	25	8	17
Total number of sample size		398	218	180

Result

Extent of Natural Hazard

Erosion: The strong seasonality of the freshwater input in Meghna Estuary near Sandwip Island has an impact on the water circulation (Brammer 2014) and salinity intrusion (Mimura 1999) and coastal erosion (Zhang et al. 2004) and subsequent loss of ecosystem services that impact human welfare (Newton and Weichselgartner 2013). High input of freshwater in Meghna Estuary accelerates shoreline erosion of this island causing 40% of land erosion in the last 200 years (Brammer 2014). Figure 4 depicts the areal change of Sandwip Island in the last 450 years which reveals continuous decrease of land of this island.

In the period of 1989–2017, 4368.22 hectares (ha) of land have been eroded, whereas 3418.01 ha land was accreted in Sandwip Island. The net land loss is 950.21 ha. Due to erosional activity, 1694.96 ha of agricultural land and 128.6 ha of settlement have been eroded in the last 28 years with the rate of 60.53 ha/year and 4.59 ha/year, respectively (Fig. 5).

Sea-level rise and flood risk: The contiguous area along the coast that is less than 10 m above sea level is one of the most vulnerable to sea-level rise (SLR) (McGranahan et al. 2007) defined as one of the consequences of climate change (IPCC 2007). As for Sandwip the aforesaid statement is very much pertinent because 47% of Sandwip Island is less than 6 m (Fig. 6) and 52% is between 6 and 10 m, whereas a few portions (0.53%) are 10 m above sea level (SRTM 2013). Due to low elevation of Sandwip Island, it may experience the largest relative increase in flood risk due to sea-level rise (Nicholls et al. 1999). According to USAID (2006), this island is prone to 0.3 to 0.9 m flooding with some areas prone to up to 1.83 m of flooded depth.

Salinity and salinity intrusion: There is paucity in depth survey on salinity and salinity intrusion in Meghna Estuarine islands. USAID (2006) claims that southern and southeastern part of Sandwip Island was moderately saline with some strongly

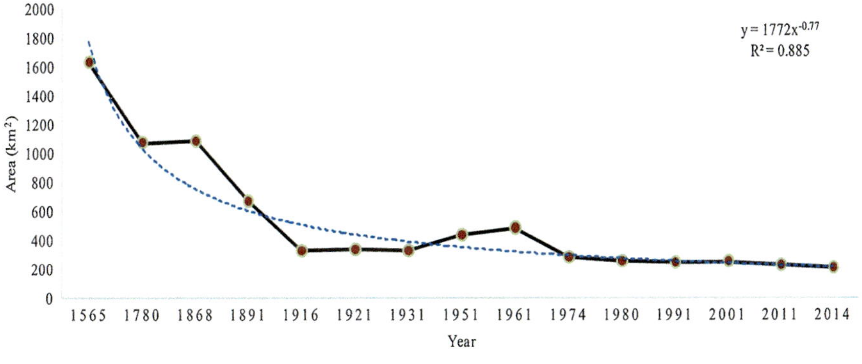

Fig. 4 Areal change of the Sandwip Island from the year 1565 to 2014. (Source: Humayun 2009 (Year from 1565 to 1961) and satellite image analysis by authors (Year from 1974 to 2014))

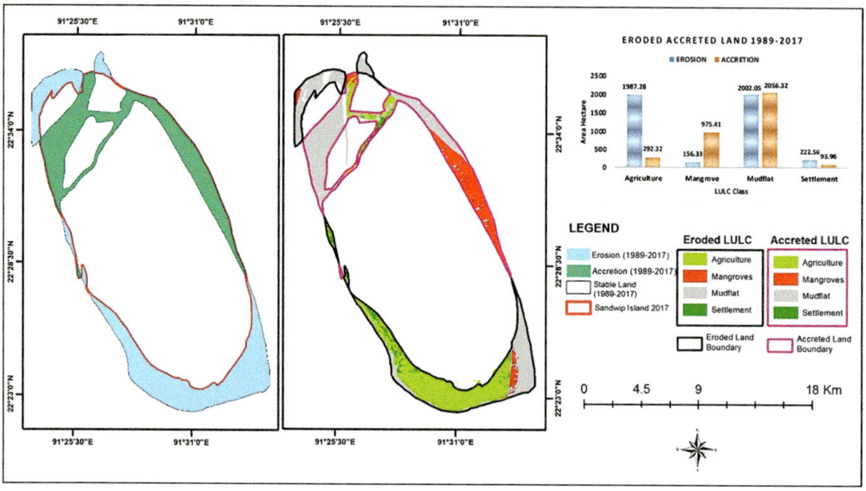

Fig. 5 Erosion and accretion of Sandwip Island in the period of 1989–2017 based on satellite image analysis (Landsat-5 and 8 image courtesy of the US Geological Survey) and ground truthing field survey data

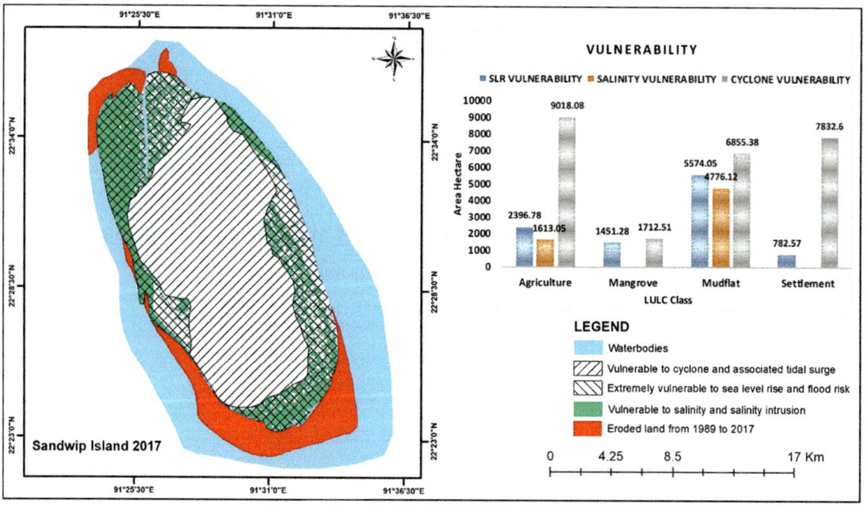

Fig. 6 Extent of natural hazard in Sandwip Island based on satellite image analysis (Landsat-5 and 8 image courtesy of the US Geological Survey), elevation data (SRTM Plus V3, 2013), and ground truthing field survey data

saline. From the field survey, it was evidently understood that southern and south-eastern part of this island are much prone to salinity intrusion due to low elevation as well as seasonal tidal flood because of proximity of salt water of the Bay of Bengal. During field survey, villagers of Sarikait Union mentioned that thin layer of salt crusts over agricultural land has made them unable of pursuing traditional agricultural purposes. The recent salinity concentration of groundwater of Sandwip Island ranges from 6083 to 8890 μS/cm (Glass 2013) which far exceeds the standard salinity level of 700 μS/cm that could reduce yields from salt-sensitive plants (Shahbaz and Ashraf 2013).

Cyclone: Vulnerability to mean sea-level rise is likely to increase extreme coastal high water levels; tropical cyclone with maximum wind speed (IPCC 2012) is a special issue for tropical small islands (Newton and Weichselgartner 2013) like Sandwip Island. In 1970, 1985, 1991, and 1997, the highest recorded cyclone storm surge heights (ranging between 0.90 and 13.64 meters) occurred in Meghna Estuary, particularly affecting Sandwip Island (Haider et al. 1991; Paul et al. 2002). According to Bay of Bengal Model (based on cyclone tracts, incorporating a 27 cm rise in sea level, a 10% increase in wind speed, and landfall of cyclones during high tide) recently upgraded and updated under the Comprehensive Disaster Management Program of Bangladesh, Sandwip is in high-risk zone to climate change scenario where tropical cyclones, storm, and tidal surges are occurring with a dead hit (UK DEFRA 2007) (Fig. 6).

Natural Hazard-Induced Vulnerability

There are 17 hazard-induced vulnerability issues (VI) that have been found through participatory rural appraisal (PRA) among vulnerable households using a measuring scale from 0 to 5 where level 0 represents to a total lack of vulnerability and levels 1, 2, 3, 4, and 5 represent 20%, 40%, 60%, 80%, and 100% of vulnerability. Among them top eight vulnerability issues have been termed as major vulnerability issues (vulnerability level, 3–5). It has been seen from Table 3 that multiple hazards are working as influencing factor of the top eight vulnerability issues in Sandwip Island which were explored further in the second stage of field survey, i.e., in-depth interviews.

All of the major vulnerabilities found in the field survey are associated with their inhabitation (i.e., settlement) and livelihood (i.e., farm activities/agriculture) which coincides with the Fig. 7 that represents total area of inhabitation and farm livelihood vulnerability to natural hazards operating in Sandwip Island (Table 4).

Adaptive Capacity to Natural Hazard

The adopted strategies of resilient households of Sandwip Island that were found during the second field survey are fallen into three categories according to Hazard Risk Reduction Strategy Framework of Burton et al. (1993), i.e., (1) *choose change*, relocation and migration of household, change in land use; (2) *reduce loss*, structural coping and adaptive response at household; and (3) *accept loss*, livelihood diversification through nonfarm activities and/or nonseasonal migration of family members, economic recovery strategies. Authors realized that all these options regarding hazard risk reduction strategies are mutually inclusive, related to each other (Table 5).

Table 3 Major vulnerability issues in Sandwip Island and their underlying factors

Major vulnerability issues		Shoreline erosion	Salinity and salinity intrusion	Cyclone and storm surge	High tide	SLR and associated flood	Variation in rainfall pattern
Homelessness		√		√	√	√	
Property loss		√		√	√		
Primary occupation loss	Agriculture	√	√		√	√	√
	Fishing	√		√	√		
Reduction in income and destruction of income source		√	√	√	√	√	√
Home damage				√	√	√	
Loss of working season		√	√	√	√	√	√
Less work opportunity		√	√	√	√	√	√

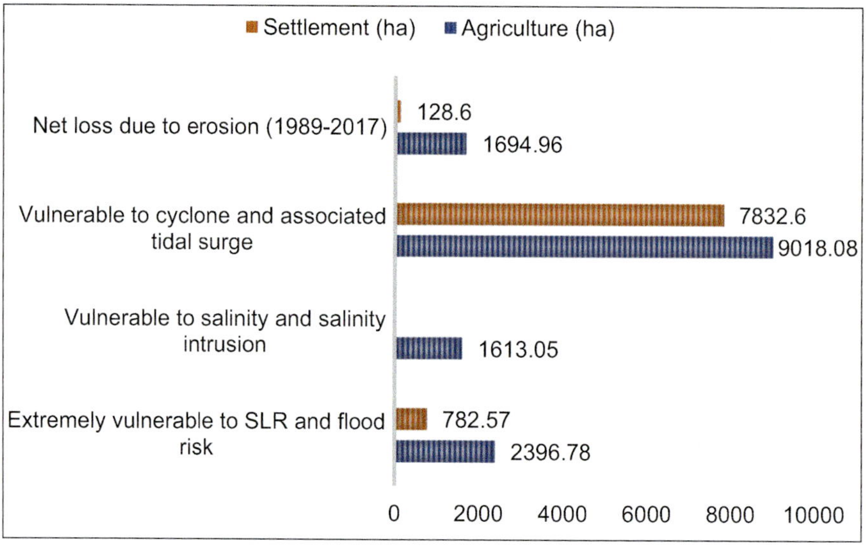

Fig. 7 Area of inhabitation and farm livelihood vulnerability to natural hazards in Sandwip Island

High-Priority Intervention Issues to Build Vulnerable Islander Resilience to Natural Hazard

A list of priority intervention issues was proposed to build vulnerable islander resilience based on suggestion to reduce identified vulnerabilities as well as suggestion from the adopted strategies of resilient households to cope with those vulnerabilities (Table 8).

In total 25 priority intervention issues were proposed; among them 17 intervention issues were categorized as high-priority intervention issues (mean weighted value (MWV), 16.00–20.00) which is further termed as "Urgent Priority Intervention

Table 4 Analysis of major vulnerability issues in Sandwip Island

Major vulnerability issues in Sandwip Island	Vulnerability level in 0–5 measuring scale *(Source: PRA)*	Extent of major vulnerabilities among vulnerable households *(Source: in-depth interviews)* (%)	Description of the extent of vulnerability
Homelessness	4.75	80.73	This phenomenon is very prominent in erosion-prone Unions of Sandwip Island. Authors saw that 749 and 337 families are living on the embankment of severely erosion-prone Unions, i.e., Kalapania and Sarikait, after losing their homes due to shoreline erosion
Property loss	4.75	86.70	More than four out of five households have lost their properties at least once in their lifetime
Primary occupation loss: Agriculture and animal husbandry	4.75	83.03	Damage and demolition of embankments due to shoreline erosion in the study area causes salinity intrusion leading to long-term damage to the productivity of agricultural lands and ponds. Most of these households are identified in natural hazard-prone (like erosion, salinity, waterlogging) areas, as natural hazards destroy or damage the environment upon which farm activities are heavily dependent
Primary occupation loss: Fishing	4.50	80.28	Respondents also stated that they have not got enough fish like before. The respondents assume that the increase of salinity in open water causes the destruction of breeding grounds of the fish, as according to them, most of the native fish species have become endangered
Reduction in income and destruction in income source	4.50	81.65	Natural hazards are the main causes of reduction in income among the farmers. Loss of agricultural land due to erosion is one of the most prominent vulnerabilities. Vast amount of agricultural lands is crusted by salt and is now unsuitable for cultivation. During the field

(*continued*)

Table 4 (continued)

Major vulnerability issues in Sandwip Island	Vulnerability level in 0–5 measuring scale *(Source: PRA)*	Extent of major vulnerabilities among vulnerable households *(Source: in-depth interviews)* (%)	Description of the extent of vulnerability
			survey, it has been observed that many inland areas were submerged during the tidal flood and became unusable for agricultural practices. Long-term waterlogging is also another reason for the decrease in agricultural production. Animal husbandry is declining due to deterioration of grasslands caused by shoreline erosion as well as salinity intrusion
Home damage	4.25	81.19	The main factors of home damage are cyclone and storm surge, shoreline erosion, and long-term waterlogging
Loss of working season	4.00	80.28	Loss of working days in an average due to shoreline erosion, flooding, cyclone, tidal surge, and storm surge is 7, 6, 4, 3, and 2, respectively, in Sandwip Island
Less of working opportunity	4.00	78.44	Decrease in area of this island due to shoreline erosion, destruction of income source of farm activities, and poor commutation with mainland create less working opportunity inside the island. More than three-fourth vulnerable households stated that they probably have to search for new employments outside the island due to adverse impact of natural disasters

Correlation coefficient between PRA and in-depth interview results, $r = 0.67$

Issues" (UPI) for building vulnerable islander resilience to natural hazards. These urgent priority interventions include:

UPI 1: Reconstructing embankment/building new embankments (MWV 20.00)
UPI 2: Building and maintaining sluice gate (MWV 20.00)
UPI 3: Introducing crop insurance for farmers (MWV 19.70)
UPI 4: Incorporating saline-tolerant crop varieties and studying crop suitability based on soil testing (MWV 19.40)

Table 5 Analysis of adopted strategies of resilient households of Sandwip Island

Strategies of resilient households	Context of the adopted strategies *(Source: PRA and in-depth interview)*	Extent of the adopted strategies among resilient households *(Source: in-depth interviews)* (%)	Rank of the priority of adopted strategies in 0–5 scale *(Source: PRA)*
Household relocation *(changing location within the same Union)* Permanent household migration *(changing location outside the Union)*	People choose change in household location (both relocation and migration) as a hazard risk reduction strategy to increase their resiliency to riverbank and coastal erosion (Table 6). Overwhelming majority of the poor households settled near the erosive coast so that they again suffer from bank erosion in the near future, whereas most of the nonpoor live away from the erosive coast to offset the risk the riverbank/ coastal erosion	80.55 (at least once in their lifetime)	4.75
Livelihood diversification through nonseasonal migration of family members	Nonseasonal migration of family members is common in upper and middle economic classes of households in Sandwip Island where the main cause of migration includes limitation of previous job opportunities due to natural hazards like bank erosion and salinity as well as opportunities to have better income. All migrated family members are engaged in off-farm activities (within Bangladesh or outside Bangladesh) which represents that households in hazard-prone area view off-farm activities as an economic insurance to natural hazards as these are less dependent on environment. Foreign remittance coming from family member plays an important role to upgrade their economic status as well as to increase resilience to natural hazard	91.11	5
Livelihood diversification through nonfarm and off-farm activities	Households prefer to pursue nonagricultural activities to diversify sources of livelihood as these activities are less dependent upon environment. Therefore, in	82.22	4.75

(*continued*)

Table 5 (continued)

Strategies of resilient households	Context of the adopted strategies *(Source: PRA and in-depth interview)*	Extent of the adopted strategies among resilient households *(Source: in-depth interviews)* (%)	Rank of the priority of adopted strategies in 0–5 scale *(Source: PRA)*
	the context of natural hazard, pursuing nonagricultural activities represents a risk minimization strategy to achieve basic household subsistence needs (Table 7). In Sandwip Island, tendency to engage in both off-farm and nonfarm activities rather than farm activities is increasing day by day		
Change in land use	Land use change in Sandwip Island includes conversion of nonproductive agricultural land, bare land, and pond into settlement (Fig. 8). But coping with changing environmental conditions by conversely land use change may increase vulnerability to disasters. But increase of settlement as well as homestead vegetation indicates sustainable land use change as most of the dwelling settlement is surrounded by good amount of branched tress which has been planted as a coping strategy to protect their house from cyclone, storm surge, etc. Many people of this island have lost their previous home due to bank erosion and bound to rebuild their home on agricultural land. Also increase in salinity intrusion causes agricultural land and ponds unproductive; hence people prefer to rebuild settlement there. Due to decrease in agricultural land, involvement in off-farm and nonfarm activities is increasing in this island, and pursuing such activities provides risk insurance to natural hazards and disaster addressed by many studies. Change in land use in Sandwip Island is, thus, viewed by people as a hazard risk	67.78	3.75

(*continued*)

Table 5 (continued)

Strategies of resilient households	Context of the adopted strategies *(Source: PRA and in-depth interview)*	Extent of the adopted strategies among resilient households *(Source: in-depth interviews)* (%)	Rank of the priority of adopted strategies in 0–5 scale *(Source: PRA)*
	reduction strategy in the context of natural hazards		
Structural coping and adaptive response at households	Most of the structural coping and adaptive response at household level are for natural hazards like flooding, cyclone, storm surge, tidal surge, etc. which include building *Pukka* and *Adha Pukka* house, strengthening the structure of the house, establishment of deep tube well in high area, improving sanitation facilities, raising the house floor, plantation of trees around the house, increase the security of preservation and storage areas, etc.	95.00	5
Economic recovery through informal credits and financial reserve	Poor households are better off than nonpoor households in economic recovery, because the major portion of the disaster loss of nonpoor households comes from the agricultural sector and there is no crop insurance available in rural areas of Bangladesh and the government does not compensate households for the crop loss. For poor households, the major portion of the disaster loss comes from the damage of the house. The government agencies and nongovernment organizations (NGOs) usually provide materials for house repairing after any natural disaster in Bangladesh So poor household depends on reliefs and informal credits in disaster economic recovery, whereas nonpoor households exploit financial reserve to recover disaster loss from agricultural sector that mostly come from the remittance by migrated family member	92.22	4.50

Correlation coefficient between PRA and in-depth interview results, r = 0.81

Table 6 Present migration status in Sandwip Island

Erosion prone	Kalapania	No in-migration occurred in the last 15–20 years
	Sarikait	
Relatively stable	Bauria	Families are migrating from the erosion-prone areas of Sandwip Every year 10–15 families are migrating here from other Unions of Sandwip
Accretion prone	Dirghapar (New)	Every year 20–25 families are migrating from Urirchar, Hatiya, and other parts of Sandwip who had been migrating from Dirghapar (old) when it was almost eroded

Table 7 Cross tabulation of sources of household income and household income of Sandwip Island (Field survey, 2017)

Sources of household income (%)	Household income (%)			Total (sources of household income)
	Decrease	Increase	Remain stable	
Farm	**42.46**	5.03	3.02	**50.51**
Off-farm	5.03	6.78	1.51	**13.32**
Nonfarm	7.29	**28.39**	0.50	**36.18**
Total (household income)	**54.78**	**40.20**	**5.03**	**100**

UPI 5: Banning fishing in spawning season and use of Current Jawl (MWV 19.20)
UPI 6: Easy access to bank/NGO/loans with low interest (MWV 19.10)
UPI 7: Promoting coastal afforestation (MWV 18.90)
UPI 8: Building climate fund in the Union Parishad level (MWV 18.80)
UPI 9 and 10: Giving emphasis on development work for climate vulnerable area and allocating budget for building islander resilience (MWV18.60)
UPI 11 and 12: Establishing vocational schools to create efficient manpower and export skilled manpower in foreign country (MWV 18.50)
UPI 13: Easy communication with mainland (MWV 18.30)
UPI 14: Permanent migration opportunities to better place (MWV 18.20)
UPI 15: Raising awareness among local people on sustainable land use (MWV 17.80)
UPI 16: Listing of priority relief by identifying actual losers (MWV 17.20)
UPI 17: Ensuring effective distribution of relief activities/goods to all of the affected area (MWV 17.20)

By reviewing the priority intervention issues, it was seen that experts recognize local infrastructure development (such as reconstruction and building of embankment, sluice gate, etc.) and improvements in community facilities (such as easy access to low interest loan, effective distribution of relief goods and activities, establishment of vocational schools, easy communication, etc.) as “Urgent Priority Intervention Issues.”

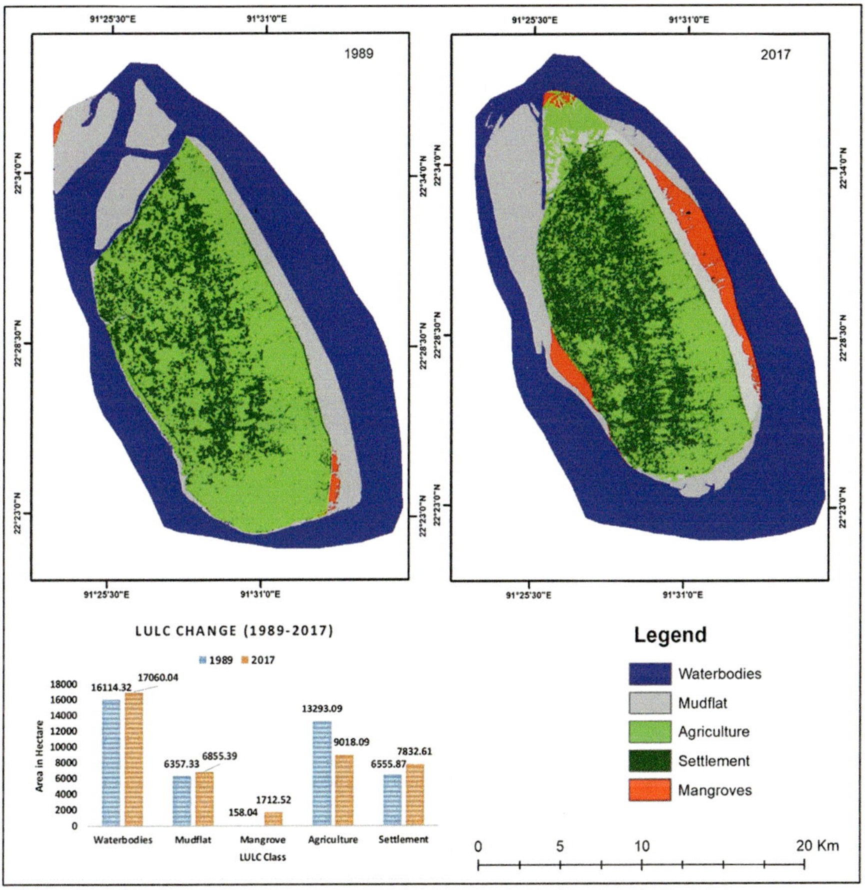

Fig. 8 Land use change in Sandwip Island in the period of 1989–2017

Discussion

Population and urban settlement patterns result in vulnerability hotspots (Newton et al. 2012) that are located disproportionately in coastal zones (Pelling and Blackburn 2013) which are subject to both hazards from land (e.g., river flooding, bank erosion, etc.) and from the sea (e.g., storm surge, salinity intrusion, etc.) (Newton and Weichselgartner 2013). Vulnerability to coastal hazards is multicausal, multidimensional, and nested in multiple scales so it is irrational to use "one-cause/one-effect" approach to address coastal vulnerability (Newton and Weichselgartner 2013). It has been seen from this research that Sandwip Island is very vulnerable because of its location, economy, and society; people are living here as a part of nature, extracting resources from nature more than anywhere else in the country. In reality, people of Sandwip Island are suffering from multiple natural hazards, and the associated hazard risks affect their physical infrastructure (houses, roads,

Table 8 Proposed priority intervention issues to build vulnerable islander resilience to natural hazard in Sandwip Island

Vulnerability issues of vulnerable households in Sandwip Island	Homelessness and property loss	Can be tackled by	1. Reconstructing embankment/ building new embankments 2. Building and maintaining sluice gate 3. Shifting house from outside of embankment to inside the embankment 4. Incorporating saline-tolerant crop varieties, studying crop suitability based on soil testing 5. Banning fishing in spawning season and use of Current Jawl (specialized fishing gears for catching small fish) 6. Raising boundary of ponds in order to prevent fish being carried away by flood water 7. Providing construction materials for temporary shelters 8. Incorporating modern low-cost techniques in building homes 9. Providing immediate cash/food for survival after disaster 10. Ensuring effective distribution of relief activities/goods to all of the affected area 11. Establishing vocational schools to create efficient manpower 12. Giving emphasis on development work for climate vulnerable area 13. Listing of priority relief by identifying actual losers 14. Building more multipurpose cyclone shelters with modern facilities	Proposed priority intervention issues to build vulnerable islander resilience to natural hazard
	Primary occupation loss: agriculture and animal husbandry			
	Primary occupation loss: fishing			
	Reduction in income and destruction in income source			
	Home damage			
	Loss of working season and less of working opportunity			

Adopted strategies by resilient household in Sandwip Island	Household relocation and permanent household migration	Can be well facilitated by	1. Permanent migration opportunities to better place 2. Easy communication with mainland 3. Establishing vocational schools to export skilled manpower in foreign country 4. Easy access to bank/NGO/loans with low interest to promote small business 5. Raising awareness among local people on sustainable land use 6. Promoting coastal afforestation 7. Introducing crop insurance for farmers 8. Introducing low-cost sanitation and alternative source of drinking 9. Easy access to bank/NGO/loans with low interest to promote economic recovery after disaster 10. Building climate fund in the Union Parishad level 11. Allocating budget for building islander resilience	
	Livelihood diversification through nonseasonal migration of family members			
	Livelihood diversification through nonfarm and off-farm activities			
	Change in land use			
	Structural coping and adaptive response at households			
	Economic recovery through informal credits and financial reserve			

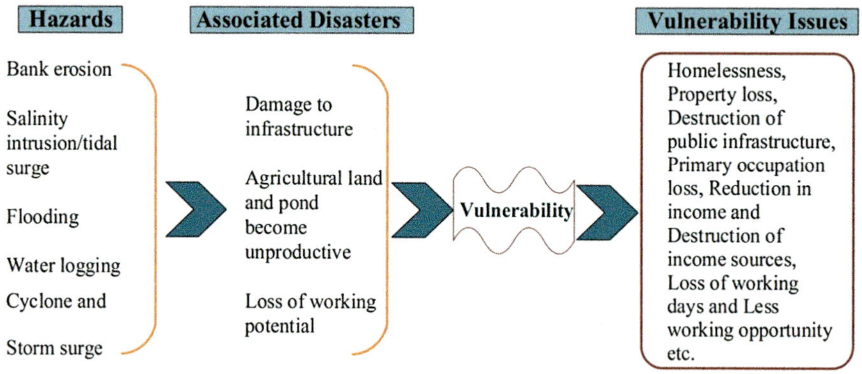

Fig. 9 Vulnerability issues in the context of natural hazard and disaster

communication, and shelters), health, working potentials, and their livelihood activities. Natural hazards make them vulnerable through making their lives more complicated, destroying properties, and limiting their livelihood options expressing the multidimensionality of natural disasters (Fig. 9).

In the case of the most common climate-related shocks, households generally experience a reduction in income and production which is largely related to a decline in weather-dependent activities, including rain-fed agriculture and livestock production. Vulnerability issues due to natural hazard and disaster identified in this study decline the standard of living of the costal islanders which has confirmed the findings of Akter and Mallick (2013). By reviewing the major vulnerability issues, it has been revealed that all the major vulnerability issues identified are related to living condition or livelihood option of the people of Sandwip Island which will help us to understand the answer of the question, "when do people feel vulnerable to natural hazards/disasters?" The answer could be "when they face problems in their living and livelihood." It is obvious that when people or communities face any problem in their inhabitations and livelihood, they want to mitigate it or adapt – which is the starting point of being resilient to natural hazard which is consistent with the findings of Twigg 2007; Manyena 2006 (Fig. 10).

Though natural hazards and disasters are responsible for the vulnerabilities of Sandwip people, lack of protective measure from the hazard is also the other dimension of vulnerability. To survive and cope with a disaster with minimum impact and damage, in this study, priority intervention issues have been identified by the grassroot community and selected as urgent priority intervention issues by giving high priority by experts. It was seen that both grassroot community and experts recognize local infrastructure development and improvements in community facilities as "Urgent Priority Intervention Issues (UPI)." Particular mention can be made of crop insurance and climatic fund at Union Parishad level which is consistent with Younus and Harvey (2014). Respondents also gave priority to permanent migration opportunity for poor communities. In Sandwip Island people prefer migration as the adaptation strategy to climatic stress because shoreline erosion,

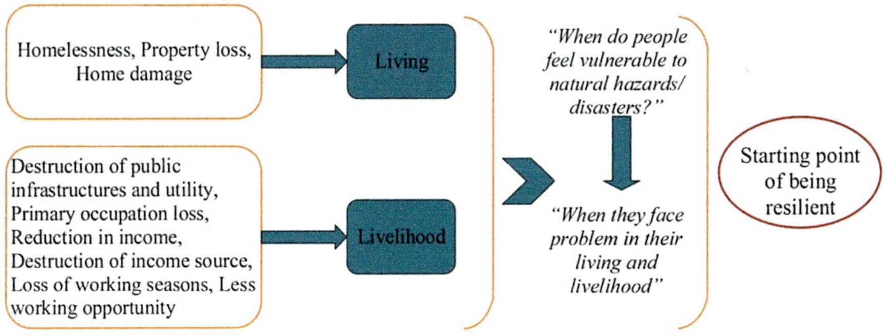

Fig. 10 Resilience complements vulnerability

salinity intrusion, increasing high tides, and long-term inundation cause problems in their inhabitation and limit livelihood options which is consistent with the statement of IOM (2009) so that Climate Change Strategy and Action Plan (CCSAP) and National Adaptation Programme of Action (NAPA) should pay special attention on "environmental migrant issue" in the islands of Meghna Estuary. Also, a separate allocation for "environmental migrants" needs to be established under the Bangladesh Climate Change Trust (BCCT). But priority intervention implementing factors are "no regret measures" (Newton and Weichselgartner 2013) such as funding, improvement of knowledge base, capacity building, public awareness, etc. which cannot be done without the help of government, NGO, or other organizations. The major constraints of implementing intervention issues are limited institutional and technical capacity, corruption, mismanagement, and nepotism of administration (Younus and Harvey 2014) eroding resilience of coastal societies and increasing their vulnerability (Glaser et al. 2010). In order to increase the resilience of coastal islanders, the government should develop adaptation policies in the context of natural hazard and disaster with the special attention of strengthening local Union Parishads consistent with the finding of Younus and Harvey (2014) and Biermann et al. (2012).

This paper focuses on the vulnerability issues in order to understand the resilience in the context of natural hazard as resilience is considered as an outcome when defined as the ability to bounce back or cope with a hazard event embedded within vulnerability (Manyena 2006). Authors argue that interventions in the context of natural hazard should be determined as priorities needed to be addressed on an urgent basis to reduce the effects of disasters with minimal social disruptions. By reviewing all the priority intervention issues, it has been seen that the inhabitants of Sandwip Island give priority to those interventions which are related to long-term adaptation at community level, not only in post disaster responses, so that implementing these priority interventions in the context of natural hazard can be said to be a part of adaptation to changing climatic scenario. Respondents also mentioned that it is necessary to improve island social-ecological systems response and adaptation capacities system coinciding with the findings of Bunce et al. (2009)

as recovery on a small island hit by a disaster can be difficult. As vulnerability and adaptation strategies are considered as one of the determinants of resilience according to the Resilience Alliance (Carpenter et al. 2001), this paper can provide guidelines for being one step forward to be resilient community to natural hazard. Respondents of grassroot community justify their priority intervention issues with the perception that these interventions at community level will help to reduce vulnerability or avoid loss due to natural disasters. So, it can be said that priority intervention issues based on vulnerability experiences of islanders identified in this paper coincide with the definition of resilience in hazards and disaster perspective (Berke and Campanella 2006; NRC 2011; Manyena 2006; Newton and Weichselgartner 2013). But the findings do not support any concluding remarks on how resilience to natural hazard and disaster of a community can be achieved fully. Further research is needed to assess how to achieve resilience fully by increasing the adaptive capacity of vulnerable islanders to natural hazards.

Conclusion

This study opens the opportunity of many future researches to identify fact and reality of resilience level of coastal islanders to natural hazard and disaster. Any research in this sector should concentrate on both social behaviors of human and physical parameters of natural hazards and disasters. Solely, concentrating only on the physical parameters of natural hazard and disaster will fail to recognize the complex human behavior to increase their resiliency level in the context of unusual climatic stress. Similarly, one should not consider only the social aspects ignoring the physical parameters. So, to understand the fact of changing environment and its impact upon living condition and livelihood, a comprehensive holistic approach is necessary here. To conduct field survey of this research, authors had to face difficulties as the study area was remote. It was difficult to carry on the field research as there were severe shortages of facilities to stay longer; heavy rain and storm interrupted the field survey many times. This study is hoped to be a creative one to understand the issues of vulnerability and priority interventions at community level to increase the resilience of vulnerable islanders in the context of natural hazard and disaster of Bangladesh.

References

Adger WN, Huq S, Brown K, Conway D (2003) Adaptation to climate change in the developing world. Prog Dev Stud 3:179–195

Akter S, Mallick B (2013) The poverty vulnerability resilience nexus: evidence from Bangladesh. Ecol Econ 96:114–124

Bajracharya B, Childs I, Hastings P (2011) Climate change adaptation through land use planning and disaster management: local government perspectives from Queensland. In: Paper presented at 17th Pacific rim real estate society conference on climate change and property: its impact now and later, Gold Coast, 16–19 Jan 2011

Baker D, Rapaport E (2005) The science of assessment: identifying and predicting environmental impacts. In: Hanna KS (ed) Environmental impact assessment: practice and participation. Oxford University Press, London
Barman SD, Majumder SC, Rahaman MZ, Sarker S (2012) Foundations of migration from the disaster consequences coastal area of Bangladesh. Dev Ctry Stud 2:22–29
BBS (2001) Bangladesh population and housing census 2001: socio-economic and demographic report. Bangladesh Bureau of Statistics, Ministry of Planning, Dhaka
BBS (2011) Bangladesh population and housing census 2011: socio-economic and demographic report. Bangladesh Bureau of Statistics, Ministry of Planning, Dhaka
Berke PR, Campanella TJ (2006) Planning for post disaster resiliency. Ann Am Acad Pol Soc Sci 604:192–207
Biermann F, Abbott K, Andresen S, Bäckstrand K, Bernstein S, Betsill MM, Bulkeley H, Cashore B, Clapp J, Folke C, Gupta A, Gupta J, Haas PM, Jordan A, Kanie N, Kluvánková-Oravská T, Lebel L, Liverman D, Meadowcroft J, Mitchell RB, Newell P, Oberthür S, Olsson L, Pattberg P, Sánchez-Rodríguez R, Schroeder H, Underdal A, Camargo Vieira S, Vogel C, Young OR, Brock A, Zondervan R (2012) Navigating the anthropocene: improving earth system governance. Science 335:1306–1307
Blaikie P, Cannon T, Davis I, Wisner B (1994) At risk: natural hazards, people's vulnerability and disasters, 1st edn. Routledge, London
Brammer H (2012) The physical geography of Bangladesh. The University Press Limited, Dhaka
Brammer H (2014) Bangladesh's dynamic coastal regions and sea level rise. Clim Risk Manag 1:51–62
Bruneau M, Chang S, Eguchi R, Lee G, O'Rourke T, Reinhorn A (2003) A framework to quantitatively assess and enhance the seismic resilience of communities. Earthquake Spectra 19:733–752
Bunce M, Mee L, Rodwell LD, Gibb R (2009) Collapse and recovery in a remote small island: a tale of adaptive cycles or downward spirals? Glob Environ Chang 19:213–226
Burton I, Kates R, White G (1993) The environment as hazard, 2nd edn. The Guilford Press, New York
Carpenter S, Walker B, Anderies JM, Abel N (2001) From metaphor to measurement: resilience of what to what? Ecosystems 4:65–781
Carter M (1997) Environment, technology, and the social articulation of risk in west African agriculture. Econ Dev Cult Chang 45:557–591
Carter M, Little P, Mogues T, Negatu W (2007) Poverty traps and natural disasters in Ethiopia and Honduras. World Dev 35:835–856
Chambers R (1997) Whose reality counts: putting the first last. Intermediate Technology Publications, London
Chambers R (2002) Relaxed and participatory appraisal: notes on practical approaches and methods for participants in PRA/PLA-related familiarization workshops. Participation Resource Centre, University of Sussex, Sussex
Chapin FS, Zavaleta ES, Eviner VT, Naylor RL, Vitousek PM, Raynolds HL, Hooper DU, Lavorel S, Sala OE, Hobbie SE, Mack MC, Diaz S (2000) Consequence of changing biodiversity. Nature 405:234–242
Coate S, Ravallion M (1993) Reciprocity without commitment: characterization and performance of informal insurance arrangements. J Dev Econ 40:1–24
Cutter SL (2013) Building disaster resilience: steps toward sustainability. Challenge Sustain 1:72–79
Cutter SL, Barnes L, Berry M, Burton C, Evans E, Tate E, Webb J (2008) A place-based model for understanding community resilience to natural disasters. Glob Environ Chang 18:598–606
Dercon S, Krishnan P (1996) Income portfolios in rural Ethiopia and Tanzania: choices and constraints. J Dev Stud 32:850–875
Evans HE, Ngau PM (1991) Rural-urban relations, household income diversification and agricultural productivity. Dev Chang 22:519–545
FAO (2006) Trends, poverty and coastal livelihoods in coastal fishing communities of Orissa state, India. FAO fisheries technical paper 490. Food and Agricultural Organization, Rome
Fussel HM (2007) Vulnerability: a generally applicable conceptual framework for climate change research. Glob Environ Chang 17:155–167

Fussel HM, Klein RJ (2002) Assessing vulnerability and adaptation to climate change: an evolution of conceptual thinking. UNDP expert group meeting on integrating disaster reduction and adaptation to climate change. United Nations Development Programme, Havana, 17–19 June 2002

Gallopin GC (2006) Linkages between vulnerability, resilience, and adaptive capacity. Glob Environ Chang 16:293–303

Ghosh J, Bharadwaj K (1992) Poverty and employment in India. In: Bernstein H, Crow B, Johnson H (eds) Rural livelihoods: crises and responses. Oxford University Press and The Open University, London

Glaser M, Baitoningsih W, Ferse SCA, Neil M, Deswandi R (2010) Whose sustainability? Top-down participation and emergent rules in marine protected area management in Indonesia. Mar Pol 34:1215–1225

Glass J (2013) Amid perfect storm of climate challenges, can aquaculture net food security gains in Bangladesh? New security beat the blog of the Wilson Center's Environmental Change and Security Program. https://www.newsecuritybeat.org/2013/10/perfect-storm-climate-challenges-aquaculture-net-food-security-gains-bangladesh/. Assessed 21 June 2015

Guilford JP, Frucher B (1973) Fundamental statistics in psychology and education. MC Graw-Hill, New York

Haider R, Rahman A, Huq S (1991) Cyclone 91: an environmental and perception study. Bangladesh Center for Advanced Studies, Dhaka

Handmer JW, Dovers SR (1996) A typology of resilience rethinking institutions for sustainable development. Ind Environ Crisis Q 9:482–511

Hewitt K (1997) Regions of risk: a geographical introduction to disasters. Longman, Harlow

Hewitt K, Burton I (1971) The hazardousness of a place: a regional ecology of damaging events. University of Toronto Press, Toronto

Holling CS (1973) Resilience and stability of ecological systems. Annu Rev Ecol Syst 4:1–23

Hugo GJ (2010) Climate change induced mobility and existing migration regime in Asia and Pacific. In: McAdam J (ed) Climate change and displacement: multidisciplinary perspectives. Hart Publishing Ltd, Oxford, pp 14–29

Humayun R (2009) Migration and settlement. In: Humayun R, Mohammad H (eds) Dwipannito. Barnarekha Press (In Bangla), Chittagong, pp 31–62

Huq S, Klein RJ (2003) Adaptation to climate change: why and how. https://www.scidev.net/global/policy-brief/adapting-to-climate-change-why-and-how.html. Assessed 7 Mar 2017

Hussein K, Nelson J (1999) Sustainable livelihoods and livelihood diversification. IDS working paper 69. Institute of Development Studies, Brighton

IFRC/RC (2006) World disaster report 2006: focus on neglected crisis. International Federation of Red Cross and Red Crescent Societies, Geneva

IOM (2009) Migration, environment and climate change: assessing the evidence. International Organization for Migration, Geneva

IPCC (2007) Climate change 2007: climate change impacts, adaptation and vulnerability. Cambridge University Press, Cambridge

IPCC (2012) Managing the risks of extreme events and disasters to advance climate change adaptation. Cambridge University Press, Cambridge

Keim ME (2008) Building human resilience: the role of public health preparedness and response as an adaptation to climate change. Am J Prev Med 35:508–516

Khan AA (1974) Perception of cyclone hazard and community response. Oriental Geographer 18:1–25

Klein RJ, Nicholls RJ, Thomall F (2003a) Resilience to natural hazards: how useful is this concept? Environ Hazards 5:35–45

Klein RJ, Nicholls RJ, Thomalla F (2003b) The resilience of coastal megacities to weather-related hazards: a review. In: Kremer A, Arnold M, Carlin A (eds) Building safer cities: the future of disaster risk. The World Bank, Washington, DC, pp 101–120

Leichenko R (2011) Climate change and urban resilience. Curr Opin Environ Sustain 3:164–168

Manyena SB (2006) The concept of resilience revisited. Disasters 30:433–450
McGranahan G, Balk D, Anderson B (2007) The rising tide: assessing the risks of climate change and human settlements in low elevation coastal zones. Environ Urban 19:17–37
Mimura N (1999) Vulnerability of island countries in the South Pacific to sea level rise and climate change. Clim Res 12:137–143
Morris S, Neidecker-Gonzales O, Carletto C, Munguia M, Medina J, Wodon Q (2002) Hurricane Mitch and the livelihoods of the rural poor in Honduras. World Dev 30:49–60
Munich RE (2012) Topics geo: natural catastrophes 2011. Munich Reinsurance, Munich
Nelson DR, Adger WN, Brown K (2007) Adaptation to environmental change: contributions of a resilience framework. Annu Rev Environ Resour 32:395–419
Newton A, Weichselgartner J (2013) Hotspots of coastal vulnerability: a DPSIR analysis to find societal pathways and responses. Estuar Coast Shelf Sci 140:123–133
Newton A, Carruthers TJB, Icely J (2012) The coastal syndromes and hotspots on the coast. Estuar Coast Shelf Sci 96:39–47
Nicholls RJ, Hoozemans FMJ, Marchand M (1999) Increasing flood risk and wetland losses due to global sea-level rise: regional and global analyses. Glob Environ Chang 9:69–87
NRC (2011) National earthquake resilience: research, implementation and outreach. National Academies Press, Washington, DC
NRC (2012) Disaster resilience: a national imperative. The National Academies Press, Washington, DC
Paton D, Johnston D (2006) Disaster resilience: an integrated approach. Charles C Thomas, Springfield
Paul A, Rahman MM, Mahbub AQM (2002) Role of shelter center for cyclone hazard mitigation in Cox's Bazar, Bangladesh. Chittagong Univ J Sci 26:113–123
Pelling M, Blackburn S (eds) (2013) Megacities and the coast: risk, resilience and transformation. Routledge, London
Peterson GD, Allen CR, Holling CS (1998) Ecological resilience, biodiversity and scale. Ecosystems 1:6–18
Rayhan I, Grote U (2007) Coping with floods: does rural-urban migration play any role for survival in rural Bangladesh? J Identity Migr Stud 1:82–98
Renaud FG, Dun O, Warner K, Bogardi J (2011) A decision framework for environmentally induced migration. Int Migr 49:5–29
Sarkar M, Choudhury E (2003) Knowledge portal on estuary development. Water Resources Planning Organization (WARPO) & The Center for Environmental and Geographic Information Services (CEGIS), Dhaka
Schulze ED, Mooney HA (1993) Biodiversity and ecosystem function. Springer, Berlin
Shahbaz M, Ashraf M (2013) Improving salinity tolerance in cereals. Crit Rev Plant Sci 32:237–249
Smit B, Wandel J (2006) Adaptation, adaptive capacity, and vulnerability. Glob Environ Chang 16:282–292
SRTM Worldwide Elevation Data (2013) https://lpdaac.usgs.gov. Assessed 1 Mar 2018
Stark O, Bloom DE (1985) The new economics of labor migration. Am Econ Rev 75:173–178
Stark O, Levhari D (1982) On migration and risk in less developed countries. Econ Dev Cult Chang 31:190–196
Surjan A, Sharma A, Shaw R (2011) Understanding urban resilience. In: Shaw R, Sharma A (eds) Climate and disaster resilience in cites. Community, environment and disaster risk management, vol 6. Emerald Group Publishing Limited, Bingley, pp 17–45
Taylor E, Wyatt T (1996) The shadow value of migrant remittances, income and inequality in a household-farm economy. J Dev Stud 32:899–912
Timmerman P (1981) Vulnerability, resilience and collapse of society, Number 1 in environmental monograph. Institute for Environmental Studies, University of Toronto, Toronto
Turner BL (2010) Vulnerability and resilience: coalescing or paralleling approaches to sustainability science? Glob Environ Chang 20:570–576

Twigg J (2007) Characteristics of a disaster-resilient community: a guidance note. United Kingdom Department for International Development (DFID) Disaster Risk Reduction Interagency Coordination Group. www.benfieldhrc.org/disaster_studies/projects/communitydrrindicators/community_drr_indicators_index.htm. Assessed 15 Dec 2017

Twigg J, Bhatt M (1998) Understanding vulnerability: south Asian perspective. Intermediate Technology Publications, London

UK DEFRA (2007) Investigating the impact of relative sea-level rise on coastal communities and their livelihoods in Bangladesh. Department for Environment, Food and Rural Affairs of United Kingdom, London

UN-HABITAT (2010) Land and natural disasters: guidance for practitioners. United Nations Human Settlements Programme, Nairobi

UNISDR (2002) Living with risk: a global review of disaster reduction initiatives. United Nations International Strategy for Disaster Reduction, Geneva

UNISDR (2005) Hyogo framework for action 2005–2015: building the resilience of nations and communities to disasters. United Nations International Strategy for Disaster Reduction, Geneva

USAID (2006) GIS-Bangladesh country almanac (BCA): a user friendly GIS tool for agricultural, forestry and natural resource management. https://pdf.usaid.gov/pdf_docs/pdaci949.pdf. Assessed 1 Jan 2016

Vale LJ, Campanella TJ (2005) The resilient city: how modern cities recover from disaster. Oxford University Press, New York

Weichselgartner J (2001) Disaster mitigation: the concept of vulnerability revisited. Disaster Prev Manag 10:85–94

World Bank (2010) Economics of adaptation to climate change: Bangladesh. The World Bank, Washington, DC

Younus M, Harvey N (2014) Economic consequences of failed autonomous adaptation to extreme floods: a case study from Bangladesh. Local Econ 29:22–37

Zhang KQ, Douglas BC, Leatherman SP (2004) Global warming and coastal erosion. Clim Chang 64:41–58

Zou L, Thomalla F (2008) A synthesis of insights on the causes of social vulnerability to coastal hazards in Southeast Asia. Stockholm Environment Institute, Stockholm

Climate Variability and Water Availability in Riparian Rural Communities of the Ebrié Lagoon in Côte d'Ivoire

17

Isimemen Osemwegie and Adjoua Nadège Boko-Koiadia

Contents

Abstract

Globally, droughts, sea level rise, and coastal water degradation are the main challenges facing coastal communities. Knowledge of their vulnerability is, therefore, a prerequisite to adequately respond to their immediate needs, especially in developing countries, in order to adapt to climate change as well as mitigate the damage already done. This regional vulnerability assessment of coastal communities examined climate change and its impacts on coastal waters and considered the adequacy of social conditions of communities to adapt to coastal water degradation. Results showed that medium hydrological droughts

I. Osemwegie (✉)
International University of Grand Bassam, Grand-Bassam, Côte d'Ivoire
e-mail: isiosemwegie@yahoo.com

A. N. Boko-Koiadia
Department of Sociology, University of Félix Houphouët-Boigny, Abidjan, Côte d'Ivoire
e-mail: juchriel@yahoo.fr

W. Leal Filho (ed.), *Handbook of Climate Change Resilience*,
https://doi.org/10.1007/978-3-319-93336-8_114

were frequent in past decades in this catchment and its impacts are relatively higher on water resources in urbanized areas where water demands are higher. These impacts are however superimposed on existing non-climatic stressors. Furthermore, coastal communities are sensitive to resource degradation with high dependence on coastal waters to satisfy daily water needs and sustenance. Communities demonstrated high willingness to participate in government-led programs geared toward restoration and conservation of coastal resources. Therefore, little coercion would be needed to instill the desired behavioral changes and sensitize these communities on the need to protect these declining water resources through sustainable and conservative water practices. The provision of pipe-borne water and improved sanitation facilities will be crucial elements in addressing the objectives of water resource conservation, livelihood enhancement, and minimization of risks associated with climate change.

Introduction

Twelve of the fifteen countries of the Economic Community of West African States (ECOWAS) lie on the coast of the South Atlantic Ocean and face challenges including, inter alia, sea level rise, increased storm surge, flooding, and erosions (Magadza 2000; Niang et al. 2014; Goussard and Ducrocq 2014). Large lagoons and water bodies separated from the Atlantic Ocean by islands are additional features along the coast from Côte d'Ivoire to the east of Nigeria (Diop et al. 2014).

This work, which focused on the Ivorian east coast, has been conducted for the West African Science Service Center for Climate Change and Adapted Land Use (WASCAL). Climate change impacts along the Ivorian east coast were projected in the context of coastal inundations (Jallow et al. 1999) and the economic loss of urban infrastructures (Hallegatte et al. 2013).

This chapter contributes to the scientific debate about the impacts of climate change on the quality of combined salt and freshwater systems along the land-sea interface and on the continued delivery of their beneficial ecosystem services. The objectives were to (1) examine the impacts of climate change on coastal water resources along the Ivorian east coast and (2) document and analyze water use by households in the context of degraded freshwater quality. While the focus of the overall study was the assessment of seasonal hydrologic interactions between the groundwater and surface water systems (Atlantic Ocean and the Ebrié Lagoon) along the Ivorian east coast (Osemwegie et al. 2017), this chapter focuses mainly on analyses of the local climate and the results of the qualitative data gathered through observations and a household survey among riparian rural communities of the Ebrié Lagoon catchment.

Water (too much or too little) the medium through which climate change will hurt humanity exhibits large spatial and temporal variations (Mooney 2016). In Côte d'Ivoire, rainfall events are more intense in the southern parts with an average annual of 2,500 mm compared to the northern part of the country with an annual average of less than 800 mm. This disparity is reflected in different vegetation patterns, from dense humid forest in the south to shrub savannah in the north (Konaté and Kampmann 2010). However, the vegetation along the coastal zone is savannah and croplands (Avenard et al. 1971). Provision of water for settlements along coastal zones is challenged with saltwater intrusion. In the Ivorian context, the majority of hand-dug wells in the coastal zones have been abandoned due to saltwater intrusion at one time or the other. Water for rural riparian communities needs to be supplied via long-distance pipelines from inland areas less prone to saltwater influence. But the majority of these communities are not served with pipelines and still depend on hand-dug wells and surface water bodies for domestic supply.

Water availability is precursory to sustainable development, serving as a springboard for the attainment of wider Sustainable Development Goals (SDGs). This makes coastal waters crucial elements in addressing livelihood enhancement and poverty alleviation. Poverty, more than just the lack of income, is often associated with the lack of clean water, substandard living conditions, and environmental degradation (Malik 1998). Coastal water systems, in addition to the provision of clean water, provide other ranges of ecosystem services which are highly beneficial to people, societies, and national economies (Millennium Ecosystem Assessment 2005). In Cote d'Ivoire, coastal aquatic ecosystem service product (ESP) has been estimated at US$ 151.23 million (Martínez et al. 2007).

The Ebrié Lagoon which is the largest estuary system in West Africa provides direct income for about 3000–4000 fishermen and their families, as well as indirect incomes to over 40,000 people employed or self-employed in fishing-related sectors, such as boat construction and fish sales (Dufour et al. 1985). Hydroelectricity power plants built on the tributaries of the Ebrié Lagoon generate about 12 kW-h of electricity annually (Dufour et al. 1985). In 2000, before the Ivorian civil war, about US$ 3 million was generated from aquaculture facilities in the country (Anonyme 2000). Other socioeconomic activities directly linked to the lagoon include the large international port in Abidjan, sand mining with a truckload of sand selling for about US$ 130, water transport with about 100,000 travelers monthly commuting via the lagoon (Noumé 2017), and recreational activities. Furthermore, mangroves growing wildly along the shores of the lagoon serve as fuelwood provider and an alternative source of income for numerous indigenous riparian communities (Osemwegie et al. 2016).

On the one hand, anthropogenic water pollution in the urban sector of the Ebrié Lagoon through lack of wastewater treatment in the Abidjan metropolis is well documented (Obrist et al. 2006; Diarra 2015). This pollution was largely due to the Ivorian civil war from 2000 to 2010 that had resulted in the doubling of the Abidjan population and the progressive failure of operational water treatment plants attributed to technical and manpower deficit, low maintenance, and inadequate investments. Also, discharge of industrial effluents and toxic substances has been mentioned as contributors to the pollution hazard of the lagoon (Scheren et al. 2004; UNEP 2015). Water pollution within the area of Abidjan compromises the quality and safety of water and fish for consumption (ibid., 2015). In hundreds of kilometers distance, however, degradation of the lagoon water has other reasons such as pollutants from agricultural runoffs which include herbicides, pesticides, and fertilizers (Adingra and Kouassi 2011), the use of toxic plants and chemicals for fishing (Anoh 2007), effluents and waste by-products from aquaculture ponds (Kouamé et al. 2018), and construction of traditional latrines on the lagoon (Fig. 8).

In this text, discussions on climate change and water availability along the Ivorian east coast on the basis of meteorological data analyses follow the presentation of the methodological approach. Afterward the views of the local communities regarding climate change and its impacts on their aquatic resources are presented and discussed. Lastly, assessments which were made of household water availability, local waste (water) disposal, the ecological benefits of the coastal waters, and extant community-based conservation strategies are reviewed.

Research Design and Methodology

This study was conducted in January 2016 by the first author and her research team in the Ebrié Lagoon catchment, a mixed watershed (cutting across urban, rural, and agricultural sectors) in southeast Cote d'Ivoire (Fig. 1). The team went by boat for field research to research zone (Fig. 1) where households in riparian villages (Kpass, Mopoyem, Boboury, Bodou, Toupah, Nigui-Assoko, Abraco, Agnéby, Azito, N'djem, M'badon, Gbéyo) were surveyed, while meteorological data were collected, compiled, and then analyzed in the office.

Meteorological Data

Data on precipitation and temperature from weather stations in Abidjan (the central urban sector of the catchment) and Bingerville (the eastern suburbs) was collected from the national meteorological service called SODEXAM (*Société d'Exploitation et de Developpement Aéroportuaire, Aéronautique et Météorologique*). Only precipitation data was available for Jacqueville (western rural areas). Abidjan, a large coastal city according to the classification scheme of Barragán and de Andrés (2015), lies between 4°08'W (Adiopodoumé bay) and 3°55'W (île Désirée) in close proximity to the bay of Bingerville (Dufour and Slepoukha 1975). The average of time series data was computed. Trends in the time series were defined using Kendall's tau nonparametric test (Kendall 1975), by calculating the probability of the existence of a trend at a 5% significance level. On the variety of rainfall patterns, dry and wet months are distinguished based on the Gaussen's ombrothermic diagram (Gaussen 1963).

Drought index was used as vulnerability indicator (quantitative measures intended to represent a characteristic or a parameter of interest, Cutter et al. 2008), since the response of hydrological systems vary markedly as a function of time (Pandey and Ramasastri 2001). Climate-water balance (i.e., precipitation minus evapotranspiration) was computed for 3-, 6-, and 24-month time scales using the Standard Precipitation Evapotranspiration Index (SPEI) version 1.6 (Vicente-Serrano et al. 2010) in statistical package R (R core team 2013) for the Abidjan and Bingerville. SPEI was not computed for Jacqueville on the basis of limited data. SPEI takes into account the role of antecedent conditions in quantifying drought

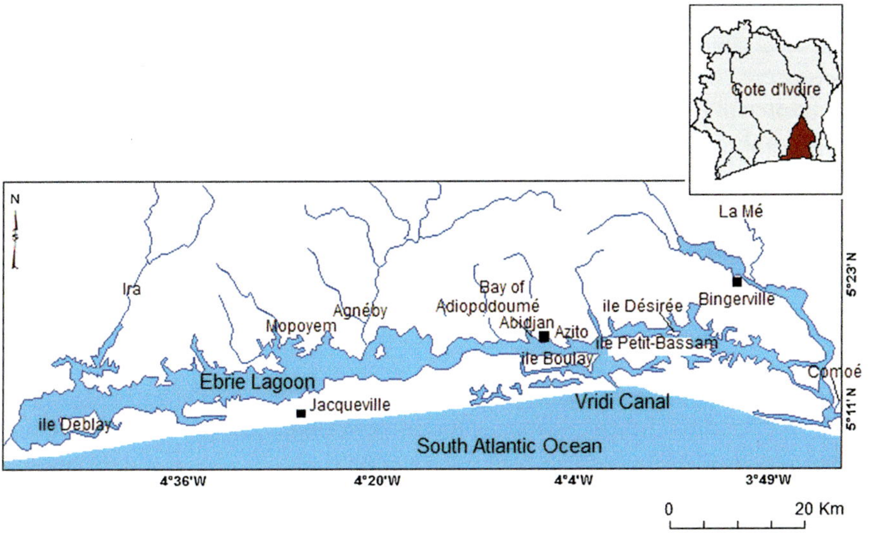

Fig. 1 Map of Cote d'Ivoire showing the study region and the weather stations (black boxes). The Agnéby-Mé coastal watershed is highlighted in red

severity (Potop and Možný 2011). PET was calculated from Thornthwaite's equation (Thornthwaite 1948) using the statistical package R. Results were fitted into log-logistic probability distribution to standardize (Vicente-Serrano et al. 2010), and values were accumulated to different time scales. Time scale functionally separates different drought levels. Drought, an aftermath of precipitation deficit (meteorological drought), can in the long term engender other drought types – hydrological, agricultural, and socioeconomic droughts (Wilhite and Glantz 1985; Mishra and Singh 2010). Meteorological drought (1–3 months) results from a shortage of precipitation, while hydrological drought (6 months) describes a deficiency in the volume of the water supply, which includes streamflow, reservoir storage, and/or groundwater table levels (Wilhite 2000). Agricultural drought usually relates to a shortage of available water for plant growth and is assessed as soil moisture deficit, while socioeconomic drought (24 months) occurs when the demand for an economic good exceeds supply as a result of weather-related shortfall in water supply (Keyantash and Dracup 2002; Mishra and Singh 2010). Positive SPEI implies greater than median precipitation, while values of −1 mark the onset of drought event. Count of the number of years with SPEI values lower (higher) than −1.0 (+1.0) was defined as either drought or wet periods.

Survey

The main objective of the survey was to assess the locals' perception of climate change and its impacts on their dependent water resources and underpin the vulnerability of these coastal communities to water degradation by virtue of available water sources for their household consumption. The duration of the interviews depended on the availability of the respondents. Thus, interviews lasted between 20 and 30 min including interruptions. The number of respondents considered for the survey was determined by us based on the normal approximation of binomial distribution (Dagnelie 1998). In this study, we considered a margin of error of 6.8% appropriate. The calculated sample size corresponded to 208 individuals, who were randomly selected from the surveyed coastal communities. The survey used a detailed questionnaire in French with ordinal and nominal questions which were later analyzed with R software. In some cases, it was necessary for research assistants to translate the French questions into local languages Ebrié, Adjoukrou, and Ahizi.

According to recommendations of Agresti (2002), Pearson's chi-squared test was used to determine if there was a significant association between age, gender, and professional occupation on the one hand and local perceptions of climate change impacts on the other, where its assumptions were satisfied (i.e., expected frequency count was greater than 0, with more than 80% at least 5 in each cell of the contingency table). Otherwise, Fisher's exact test was used at a significance level of 0.05.

While conducting the survey, the research team also paid ecological site visits to several other coastal rural communities along the Ivorian east coast, documented the local situation with photographs, and took water samples from coastal waters and other sources (Osemwegie et al. 2017).

Results and Discussions

Climate Variability and Local Water Availability

Results from the statistical analysis show clear longitudinal variation in annual precipitation amount along the Ivorian east coast (Fig. 2). This variation along the Ivorian coast is in concert with the findings of Le Loeuff and Marchal (1993). Precipitation is highest in the eastern sector of the catchment and decreases to the west: annual average (1970–2012) precipitation was 1,727 mm in Bingerville (eastern sector of the catchment), 1,704 mm (average for 1970–2014) in Abidjan (central/urban sector), and 1,520 mm (average for 2001–2013) for Jacqueville (western sector). The ombrothermic diagram (Fig. 3) showed that the eastern and central sectors of the catchment had five dry months (months with less than 100 mm of precipitation), while the western sector was drier with six dry months. Superposition of piezometer levels over the ombrothermic diagram (Fig. 3) shows a time lag of 1–3 months between peak rainfall event and groundwater recharge in Abidjan and Jacqueville. Contrariwise, groundwater in Bingerville showed simultaneous response to rain events. A likely explanation would be an increase in groundwater abstraction and/or underlying geology. In Abidjan, demand for water is usually

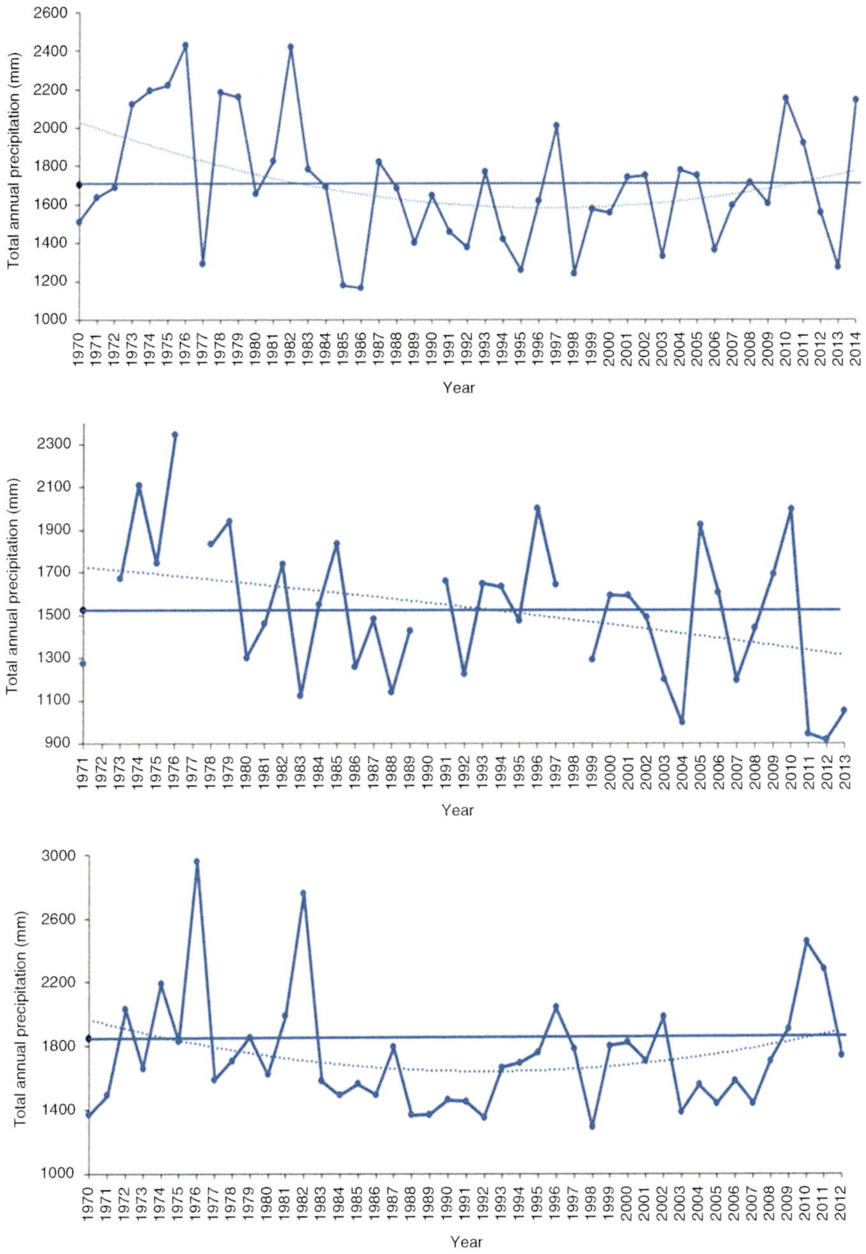

Fig. 2 Yearly trend in precipitation across the study area: Abidjan (top), Jacqueville (middle), and Bingerville (bottom). Thick lines represent moving average, while dashed lines are second-order polynomial fitted curves

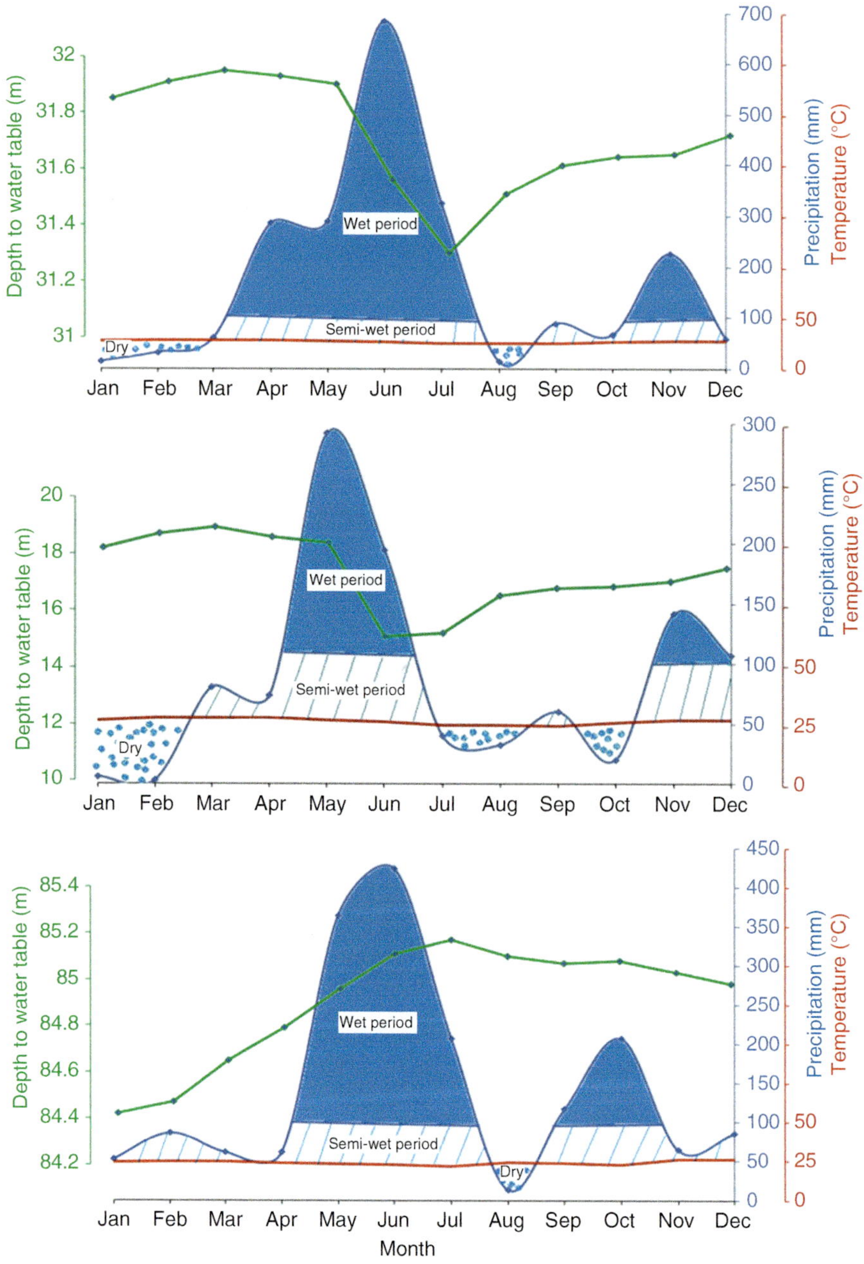

Fig. 3 1970–2014 Ombrothermic diagram superimposed over groundwater recharge history as shown by monthly water table level fluctuations in monitoring wells. (Data source: L'Office National d'Eau Portable (Ministry of water resources) in 2014). Abidjan (top), Jacqueville (middle), and Bingerville (bottom)

highest during the peak dry season and lowest during the peak rainy season (Saint-Vil 1983).

In Cote d'Ivoire, 1951–1968 was a period of precipitation surplus (Paturel et al. 1998) and continues into the 1970–1979 decade. For the Abidjan region, SPEI drought monitor (Fig. 4a) shows that the 1970–1979 decade was the wettest with over 40% of SPEI values above the wet line (+1.00). Extremely wet years (with total annual precipitation greater than 2000 mm) during this period were 1973–1976 and 1978–1979. The wettest period was 1976 with precipitation surplus of 42.7% (total annual precipitation of 2432.3 mm). Also evident from the SPEI index (Fig. 4a) is recurrent dry to extremely dry conditions in Abidjan. The 1980–1989 decade with 23% of the SPEI values below the drought line (−1.00) had the most persistent meteorological drought. Rupture, which is a downward trend in precipitation (values below the multi-temporal moving average of 1,704 mm for the observation period between 1970 and 2014), started in the year 1983. In this year 1983, earmarked as a year of intense drought, climate anomaly was observed over much of the globe due to the intense El Niño-Southern Oscillations (ENSO) event (Rasmusson and Wallace 1983). Precipitation deficit up to 31.5% was observed for 1986, the driest year. This is followed by the 1990–1999 decade and lastly the 2000–2010 decade. On the other hand, in Bingerville (Fig. 4b), the 1980–1989 decade was the wettest in this area with 23% of SPEI values above the wet line (+1.0). The 1990–1999 and 2000–2010 decades showed the most persistent meteorological drought, both with 18% of the SPEI values below the drought line (−1.0). Presently, the Bingerville area shows no sign of water stress. Recovery (precipitation amount above moving average) for the Abidjan and Bingerville areas is already underway since 2010 (Fig. 4a and b). On the contrary, for the Jacqueville area, a downward trend in precipitation began since the year 1993/1994 and continues until present (Fig. 2).

Precipitation deficit has far-reaching impacts on water availability, ephemeral surface water storage systems, and stream flows. Hydrological drought (mild to moderate) started since 1985 in Abidjan and worsened during the period of 1990–2000 (Fig. 4), consistent with the findings of Saley et al. (2009). Evidence of this drought registered in the Ebrié Lagoon catchment are dried and abandoned wells, receding surface water bodies and general lowering of piezometer levels. In Avagou, the village head informed us that the lake *gbodongbo* has shrunk in spatial extent by up to 10 m. In Gbéyo, we were informed that the lagoon has receded by up to 5 m. These impacts of drought inclusive of poor agricultural yields and socioeconomic impacts have been reported by past studies (Oga et al. 1998; Brou et al. 2005; Saley et al. 2009; Brou 2010). Streamflows have also been affected. For instance, mean annual discharge of the Comoé River, responsible for 65% of annual freshwater discharge to the Ebrié Lagoon (Guiral 1992), decreased from 331m^3/s in 1960–1970 to 161m^3/s in 1970–1980, 119m^3/s in 1980–1990 (Convention on Biodiversity 2009), and 83.8m^3/s in 1993 (Hauhouot 2002). In addition, the Agnéby River, also draining the lagoon, recorded an approximate flow deficit of 50% between 1955 and 2000 (Jallow et al. 1999; Kouakou et al. 2012).

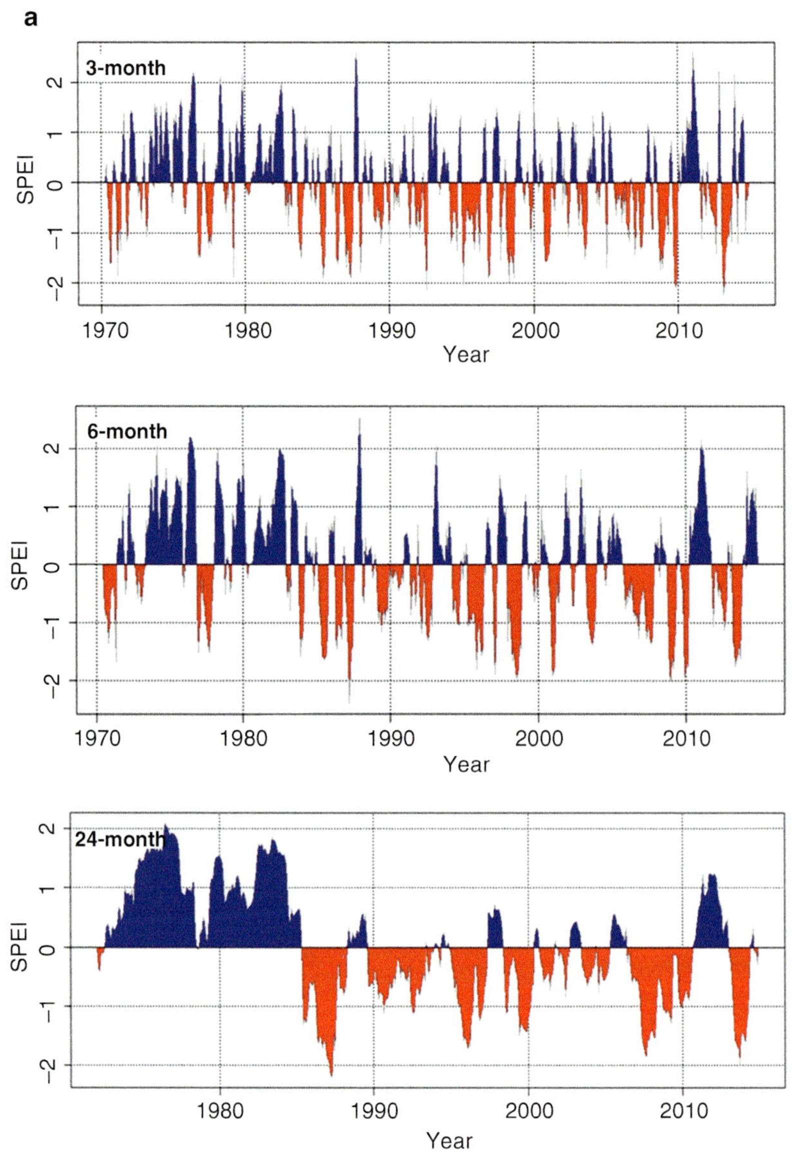

Fig. 4 (continued)

Climate variability along the Ivorian coastal zone has been attributed to natural causes such as variations in sea surface temperatures (SST) (Bigot et al. 2005), oceanic upwelling events, latitudinal dynamics of the ITCZ (Durand and Chantraine 1982), and coastal geomorphology. Direct relationship exists between coastline shape, progression of sea-breeze front, and rainfall amount on adjacent inland areas (Bakun 1978; Negri et al. 1994; Mazón and Pino 2013). Concave coastlines

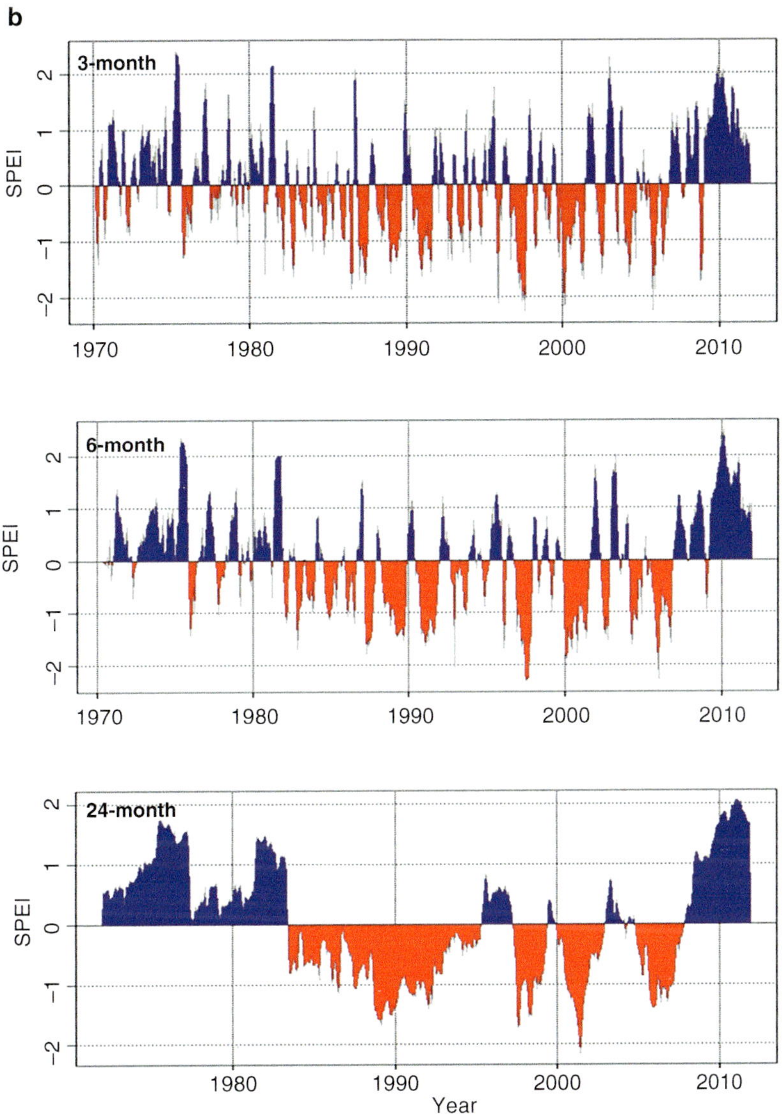

Fig. 4 (**a**) Temporal evolution of SPEI at different time scales for the Abidjan weather station. Time scale of 3, 6, and 24 months depicts meteorological, agricultural, and hydrological/socioeconomic droughts, respectively. (**b**) Temporal evolution of SPEI at different time scales for the Bingerville weather station. Time scale of 3, 6, and 24 months depicts meteorological, agricultural, and hydrological/socioeconomic droughts, respectively

diverge oceanic breeze, resulting in scarce precipitation on adjacent inland areas (Mazón and Pino 2013). The orientation of the Ivorian coastline is WSW-ENE between its western border in Tabou and Sassandra, E-W from Sassandra to Abidjan, and WNW-ESE from Abidjan to its eastern border (Wozazek 2001). The western and eastern borders with coastlines angles near to the horizontal (80°) receive relatively higher amount of rainfall (between 1700 and 2500 mm per annum) as compared to the near horizontal coastal areas. Rainfall divergence is reflected further inland by the V-shape delimitation (savannah intrusion) of the Guinea Savannah area commonly referred to as "the V-Baoulé" (Peltre 1978). Apart from natural climate variability, high deforestation rates are anthropogenic factor identified as influencing local climate (Brou et al. 1998); Côte d'Ivoire has lost 84% of its forest cover between 1960 and 2000 (Koné et al. 2014).

Although there is a progressive decrease in its water availability, Côte d'Ivoire does not suffer from physical water scarcity. It is endowed with a hydrographic network of four principal basins covering an area of 265,000 km^2. These are from west to east: Cavally (basin area, 15,000 km^2), Sassandra (75,000 km^2), Bandama (97,000 km^2), and Comoé (78,000 km^2) (Koffi 2016). The scarcity is rather economically induced from lack of investment in water infrastructures. Cote d'Ivoire was constantly a low water stress zone under the different climate change scenarios of Alcamo and Henrichs (2002). Total renewable water resources per capita (m^3/person/year) were 5,024 in 1997, 4,074 in 2007, and 3,385 in 2014 (FAO (Food and Agricultural Organization) 2015). Its present value of 3,385 m^3/person/year exceeds the benchmark of 1,700, indicative of water stress (Falkenmark 1989). This means that much more water per capita is still available than in, for example, the neighboring coastal countries: Ghana (2,050), Benin (2,426), Togo (2,012), or Nigeria (1,571). In spite of its abundant total renewable resources per capita, there is existential gap between water supply and demand in Côte d'Ivoire. The government in a public-private partnership (PPP) with SODECI (*Société de Distribution d'Eau de la Côte d'Ivoire*), the oldest and largest water PPP in the developing world, is charged with the operation of all water systems in the country since 1974 (Ménard and Clarke 2000; Marin et al. 2009). Despite its efforts at increasing production to meet booming population, there is still a daily water supply deficit of 225,000 m^3 (Global Water Intelligence 2014). Many rural coastal areas are not served by SODECI but rely on local water sources, such as wells, rivers, or water from the Ebrié Lagoon for their diverse household uses.

Local Perceptions of Regional Climate Change Impacts

In the research zone, which administratively belongs to Grands-Ponts, the population is rather homogenous. The languages are Ebrié, Alladian, Ahizi, and Avicam, and the main occupations are farming and fishing. Villages are usually located directly at the Ebrié Lagoon and display basic infrastructures, such as narrow earth roads and footpaths, electricity and potable water supply in some, and health centers and schools in some others. Typical local characteristics of the villages include small thatched roof houses, usually made of mud, built in clusters around one main narrow earth-covered access road. They also have meeting halls and small markets. There are specific market days of the week, in which farm and aquatic produces and other basic household items are sold.

Respondents of our survey were 107 men and 101 women, who were categorized by age into young (15–30 years old, $n = 71$), adults (31–60 years old, $n = 129$), and old (>60 years old, $n = 8$). Thus, the old age category was underrepresented in the random sample. Most households' counts in the studied communities (42% of the sample) were between 4 and 7 members. Smaller households made up 30% and households with more than 8 members 28% of our sample.

Our result shows that most of the respondents are aware of "climate change" and its impacts. Fifty-seven percent of respondents told us that in comparison to past decades, there is a progressive decrease in precipitation amount and frequency and 66% of the respondents reported relatively warmer temperatures. Regarding the perception of locals in relation to climate change, poll's results indicate a significant relationship between gender and the perception of climate change ($x^2 = 7.4393$, df = 3, p-value = 0.046). This significant link could be explained by the social division of labor and differential exposures to the effects of climate variability. This finding is consistent with those of another study (ADF (Seventh Forum for Africa's Development) 2010) that shows that climate change has disproportionate effect on gender. Women will be much more affected as they are responsible in the household for the provision of water, energy (fuelwood), and food and also because they do not have access to the same resources as men (Ngigi 2009). Statistical analysis to test for significant link between local views of climate change and age ($x^2 = 2.3183$, df = 6, p-value = 0.89) and profession ($x^2 = 13.515$, df = 12, p-value = 0.33) shows no relationship between the age of respondents or their profession and their views on climate change.

Forty-two percent of respondents attested to deteriorating coastal water quality and decline in quantity: 62% of them linked water quality-related problems to precipitation deficit during the warmer dry months of the year, while 33% of the respondents opined that water quality-related problems were mostly prevalent during the rainy season (see Fig. 5). The main water quality-related problem for these communities is mass fish mortality. Indeed, during the peak dry season in May 2013, there was a mass fish mortality event that affected thousands of tons of wild and farm fishes in the western part (Dabou and Jacqueville) of the Ebrié Lagoon (N'kaka 2013). This phenomenon occurred again in 2016 in the Marcory bay (Hermance 2016). During our ecological site visits, we found dead fishes floating in the lagoon around Petit-Bassam and Deblay Islands, during the dry season, and also in Azagny Island (Fig. 6), during the seasonal high tides. The species mainly affected were carp fishes. Reasons put forward by respondents include natural causes, owing to seasonal influx of seawater to the lagoon that results in the development of halocline and thermal stratification. Other possible causes mentioned are the putrefaction of aquatic weeds releasing toxic substances and the use of toxic poisonous pesticides by some fishermen to increase their catches. Informal talks with fishermen in the island village of Couvé revealed that the dead fishes had abnormally enlarged gallbladder, and prior to fish mortality events, fishes were seen gasping for air. This display usually indicates low oxygen levels. Dufour and Slepoukha (1975) have shown that during the dry season in the peripheral bays of the lagoon where water circulation and mixing are less significant, vertical stratification favors the formation of anoxic bottom layer.

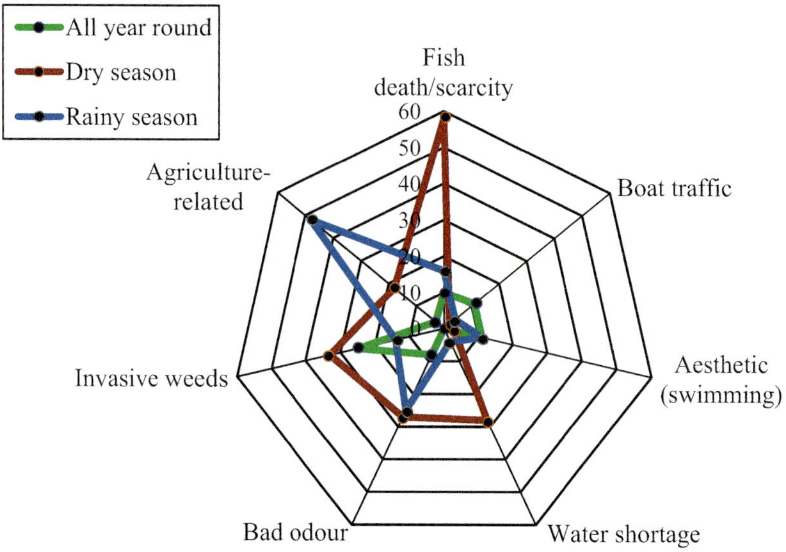

Fig. 5 Kiviat chart of local perception of coastal water degradation and likely period of occurrence

Fig. 6 Dead fish stuck in floating mats of *Eichhornia crassipes* (water hyacinth) in ile d'Azagny

The second most important problem for the riparian population is attributable to erratic rainfall patterns with consequent delays and uncertainties in planting seasons. Smallholder farmers in these coastal areas are dependent solely on rain-fed agriculture. In Abréby, a coconut farmer reported a decrease in plantation harvest from 3,000 coconuts/hectares to barely 300 coconuts/hectares within the past decade, hence relying heavily on fertilizers to boost production. In Avagou, a farmer also reported a decrease from ten jute bags of copra, dried flesh of coconut that are sold to cosmetic companies to about one and half jute bag only within the past 4 years. Past studies (PAM (Program Alimentaire Mondial) 2009; Sarr et al. 2012) have reported that climate variability will greatly increase the risk of famine and malnutrition. In essence, these studies indicate that of all human activities, agriculture remains the most influenced by climate and its variations. The intensification and frequency of extreme events due to climate variability will have increasing disastrous consequences for agricultural production and farmers income sources. Other problems highlighted include proliferation of invasive aquatic weeds (*Eichhornia Crassipes*) mostly in oligohaline (salinity less than 5 mg/L) areas of the lagoon. These heavy floating mats of aquatic plants obstruct fishing activities and render the affected areas of the lagoon non-navigable.

Household Water Availability and Waste (Water) Disposal

Poll's results illustrate that as water provision in our study region is only partly guaranteed by SODECI (26%), coastal waters in the form of groundwater and surface water are the primary source for household uses (15% from springs, rivers, and the lagoon, as well as 59% from wells). Thus, about 74% of households in our sample (Fig. 7) still lack access to clean, safe drinking water and depend mainly on natural coastal waters to meet their daily needs. In Gbéyo, Audoin Beugréto, Abraco, and ile d'Azagny also part of our study region, public water supply is completely nonexistent.

Coastal waters are easily accessible as 82% of respondents can fetch up to 10 l in less than 30 min. Irrespective of the water source, 49% of our respondents made direct use of water without any form of treatment. Locally adopted treatment methods include the addition of liquid household bleach (39%), boiling (32%), and filtration (21%), while others simply let the water stand and settle (decantation). In Audoin Beugréto, an informant told us that they regularly administer anthelmintic drug to children to help fight waterborne infections, while the adults take *kutuku*, locally fabricated gin, generally believed to be able to cleanse their body system. None of these methods can turn heavily contaminated water into safe drinking quality. But the communities under study were too poor to either invest in modern water purification infrastructures or pay for high-tech water treatment facilities. According to another informant in Gbéyo, water quality is not an issue for the people as it is believed that a crocodile, a spirit (locally called "genie") residing between the shared community well and the Ebrié Lagoon, ensures its good quality. Such belief is also widespread in Abraco, where it is believed that a "genie" makes use of water wells at night and ensures its protection, as a result of which wells are left uncovered.

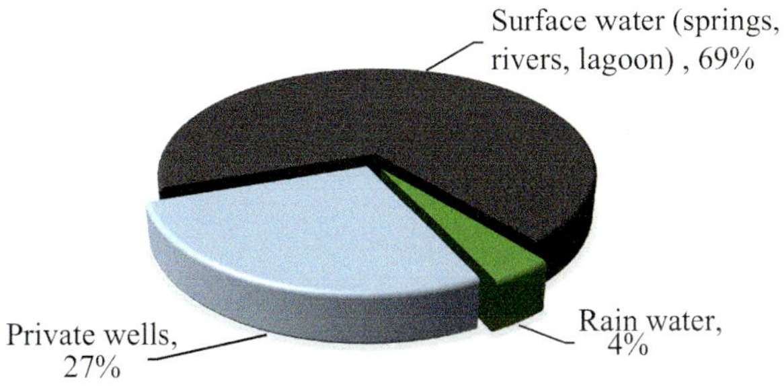

Fig. 7 Alternative domestic water sources for riverine communities

Fig. 8 A pit latrine directly on the Ebrié Lagoon (left) and stick on the beach to dig through the soft sand for open defecation

Faced with the lack of a functional local sewage system, respondents told us that they would dump their domestic effluents/wastewater directly into nature (56%) or surface water bodies (33%). Only a little percentage (11%) reported channeling their domestic effluents to public wastewater networks. Furthermore, sanitation coverage is low in these communities. About 48% of respondents have no access to basic sanitation system and still practice open defecation (Fig. 8). Many studies in West Africa have revealed that hand-dug wells in communities lacking basic sanitation facilities are unfit for drinking water source (e.g., Ernest 2013; Olayinka et al. 2014). Oga et al. (2007) have traced the origin of anthropogenic contaminants, e.g., nitrate in wells in central part of the catchment to agricultural runoffs and septic system effluents. Authors (Kiptum and Ndambuki 2012; Templeton et al. 2015) have shown good correlation between nitrate contamination in wells and closeness to a pit latrine and on the basis of this finding recommended a safe distance of 48 m between wells and pit latrines. In rural areas with low population densities, the provision of proper sanitation facility is a matter of planning and implementing better designs of latrines and wells, while in densely populated urban quarters, good design may be compromised due to sheer lack of space.

Ecological Benefits of Coastal Waters and Community-Based Conservation Strategies

Apart from its use as drinking water, other benefits of coastal waters are source of daily protein and income. In Abréby, fishermen revealed that about 5 years ago, an individual used to earn about 200,000 FCFA (368 USD) monthly from fishing, but due to the poor health of the waters, coupled with the prevailing drier conditions, this has greatly reduced, hence, the rural outmigration in search for greener pastures. Anchovies, shrimps, and crabs from the lagoon are roasted and used as seasoning in the preparation of local delicacies. Indigenes wash clothes and household utensils directly in surface waters. Other economic benefits are bivalve shell (*Corbula trigona*) harvesting and marketing. Other benefits are due to the presence of aquatic plants, which they opined serve as fish habitats, fuelwoods, local tidal gauge, medicine, and natural water filtration. In Adiapoto, about US$ 800 is generated weekly from sales of dried bivalve shells, which serve ornamental purposes in some other villages. Other benefits are due to the presence of aquatic plants, which they opined serve as fish habitats, fuelwoods, local tidal gauge, medicine, and natural water filtration. These aquatic plants support both climate change mitigation and adaptation strategies as has been discussed in detail by Osemwegie et al. (2016).

Community-based resource management systems are already in operation in some of these communities. As in some African countries, "sacred groves" and customary laws play a vital role in the sustainable use of natural resources and biodiversity conservation (Alohou et al. 2017), in these riparian communities. According to an informant in Abreby, it is a taboo to commercialize fish catches from the coastal lake Lason as it is believed that once this taboo is broken, all the fishes will disappear. In Toukouzou-Hozalem, with the predominance of the "Papa Nouveau" movement, it is customary not to fish on days of religious ceremonies and religious festivals. In Avagou, it is a taboo to fetch water from the community well at sunset (after 7 pm). This is tied to the belief that the well water will dry up. This practice controls household consumption and by so doing allows ample time for natural recharge. These customary laws already in operation should be strengthened and promoted as they help curb overexploitation and ensure the sustainable management of natural resources.

The local population has an adequate knowledge of the importance and ecological benefits of coastal water resources. A good percentage (63%) of them showed willingness to pay (WTP) and actively participate should there be a government-led restoration program for water quality improvement. They were willing to pay between US$ 1 (US$ 1 = 596.51 XOF FCFA) and whatever sum required for the cleanup, restoration, and conservation of this common pool resource. The Ebrié Lagoon was declared state property in 1998 (loi N° 98–750 of 23 December). Therefore, exclusive use rights are rather exceptional (as for areas of the harbor). Local fishermen communities enjoy use rights to the water body and its fishes and thus have a keen interest in the sustainable management of these water resources that ensure the provision of good quality water and fishes for its host communities. The conservation of public goods is an uphill task that requires collaborative efforts from relevant stakeholders, supplemented by some form of government enforcement for tangible results. Incoherent efforts to curb the pollution of the lagoon such as local initiatives of the local environmental committee in Anna (Bingerville) have met with little success, as they have limited influence over contaminant fluxes from other riparian communities.

Conclusion

This tropical Ebrié Lagoon catchment area demonstrates progressive decrease in precipitation amount from the east to the west. The SPEI index reveals persistent drought conditions in the catchment. The perceptions of the inhabitants of the study area to climate change correspond to results from meteorological data analyzed. However, the significant relationship between perceptions of locals to climate change and gender elicits the differing degree in exposure to the phenomenon. Anecdotal and empirical pieces of evidence of climate change such as dried up wells and retreating surface water bodies are already widespread in these communities and negatively impacting their sources of livelihood and compromising food security as a result of declining agricultural yields. Furthermore, sustainable development goal, SDG 11 goal to create sustainable cities and communities, cannot be attained in the absence of potable drinking water and improved sanitation facilities. The provision of potable water and other basic amenities would reduce the anthropogenic loads on coastal waters, stimulate societal development, foster economic growth, and enhance the success of poverty alleviation programs. As there exist no insurances for natural systems against anthropogenic disturbances and return to pristine conditions is not always guaranteed after the exceedance of its tolerant limits, man is ultimately invited to work with nature in a considerate and harmonious manner to ensure their mutual survival. This chapter has shown the importance of coastal waters for the rural riverine population along the Ebrié Lagoon catchment and the problems that undermine these resources. The involvement of the authorities and better consideration of coastal water issues could help local protection measures and facilitate the sustainable management of these natural resources.

Acknowledgments This study was conducted for the West African Service Center on Climate Change and Adapted Land Use (www.wascal.org) and funded by the German Federal Ministry of Education and Research.

References

Adingra AA, Kouassi AM (2011) Pollution en lagune Ebrié et ses impacts sur l'environnement et les populations riveraines. F Tech Doc Vulg 48–53

Agresti A (2002) Categorical data analysis, 2nd edn. Wiley, Hoboken

Alcamo J, Henrichs T (2002) Critical regions: a model-based estimation of world water resources sensitive to global changes. Aquat Sci 64(4):352–362

Alohou EC, Gbemavo DSJC, Mensah S, Ouinsavi C (2017) Fragmentation of forest ecosystems and connectivity between sacred groves and forest reserves in Southeastern Benin, West Africa. Trop Conserv Sci 10:1940082917731730

Anoh KP (2007) Impact environnemental et socio-économique de la pêche par empoisonnement en milieu littoral ivoirien Revue de Géographie Tropicale et d'Environnement 2:3–13

Anonyme (2000) Annuaire des statistiques de l'aquaculture et des pêches. Direction des productions halieutiques, Ministère de l'agriculture et des ressources animales, Abidjan

Avenard JM, Eldin M, Girard G, Sircoulon J, Touchebeuf P, Guillaumet JL, Adjanohoun E (1971) Le milieu naturel de la Côte d'Ivoire, vol 50. ORSTOM, Paris, p 391

Bakun A (1978) Guinea current upwelling. Nature 271:147–150

Barragán JM, de Andrés M (2015) Analysis and trends of the world's coastal cities and agglomerations, Ocean Coast. Manage 114:11–20

Bigot S, Brou TY, Oszwald J, Diedhiou A (2005) Facteurs de la variabilité pluviométrique en Côte d'Ivoire et relations avec certaines modifications environnementales. Science et changements planétaires/Sécheresse 16(1):5–13

Brou YT (2010) Risques climatiques, pressions foncières et agriculture en Côte d'Ivoire. IAHS-AISH Publ 340:320–326
Brou YT, Servat E, Paturel JE (1998) Contribution à l'analyse des inter-relations entre activités humaines et variabilité climatique: cas du Sud forestier ivoirien. Comptes Rendus de l'Académie des Sciences-Series IIA – Earth Planet Sci 327(12):833–838
Brou YT, Akindès F, Bigot S (2005) La variabilité climatique en Côte d'Ivoire: entre perceptions sociales et réponses agricoles. Cahiers Agric 14(6):533–540
Convention on Biodiversity (2009) Fourth National Report on Côte d'Ivoire (2005–2009). Available at http://www.cbd.int/doc/world/ci/ci-nr-04-fr.pdf
Cutter SL, Barnes L, Berry M, Burton C, Evans E, Tate E, Webb J (2008) A place-based model for understanding community resilience to natural disasters. Glob Environ Chang 18:598–606
Dagnelie P (1998) Statistiques théoriques et appliquées, 2nd edn. De Boeck et Larcier, Bruxelles
Diarra A (2015) La lagune Ebrié d'Abidjan À l'épreuve des pressions anthropiques. Un domaine lagunaire en Côte d'Ivoire fortement agressé par les actions humaines. EUE, Saarbrücken
Diop S, Barusseau JP, Descamps C (2014) The land/ocean interactions in the coastal zone of West and Central Africa. Springer, Cham
Dufour P, Slepoukha M (1975) l'oxygène dissous en lagune Ebrié: influences de l'hydroclimat et des pollutions. Documents Scientifiques, Centre de Recherches Océanographiques. Abidjan 6(2):75–118
Dufour P, Chantraine JM, Durand JR (1985) Impact of man on Ebrié lagoon ecosystem (Ivory Coast). In: Chao LN, Kirby-Smith WW (eds) Utilisation of coastal ecosystem: planning, pollution and productivity. Proceedings of an international symposium, Rio Grande, RS, Brazil. University of Rio Grande du Sul, Brazil, pp 467–484
Durand JR, Chantraine JM (1982) L'environnement climatique des lagunes ivoiriennes. Revue d'Hydrobiologie Tropicale 15(2):85–113
Ernest AA (2013) Effect of pit latrines on dug-well water quality – a case study of the Asankrangwa Community in the Wassa Amenfi West District of Ghana. Doctoral dissertation, Kwame Nkrumah University of Science and Technology
Falkenmark M (1989) The massive water scarcity now threatening Africa: why isn't it being addressed? Ambio 18(2):112–118
Food and Agricultural Organisation (2015) AQUASTAT Main Database, Food and Agriculture Organization of the United Nations (FAO). Accessed on 28 Oct 2016
Gaussen H (1963) Les cartes bioclimatiques et de la végétation. Principes directeurs et emploi de la couleur. Science du Sol 1:117–131
Global Water Intelligence (2014) Côte d'Ivoire reboots its water sector. GWI 15(3):10
Goussard J-J, Ducrocq M (2014) West African coastal area: challenges and outlook. In: The land/ ocean interactions in the coastal zone of West and Central Africa. Springer, Cham, pp 9–21
Guiral D (1992) L'instabilité physique, facteur d'organisation et de structuration d'un écosystème tropical saumâtre peu profond: la lagune Ebrié. Vie milieu 42:73–92
Hallegatte S, Green C, Nicholls RJ, Corfee-Morlot J (2013) Future flood losses in major coastal cities. Nat Clim Chang 3:802–806
Hauhouot C (2002) Les problèmes de l'aménagement de l'estuaire du fleuve Comoé à Grand-Bassam. Les Cahiers d'Outre-Mer Revue de géographie de Bordeaux 55(219):307–324
Hermance KN (2016) Alerte/Abidjan: Des poissons morts flottent sur la lagune Ebrié. Soir Info L'inter. Available from http://www.linfodrome.com/faits-et-mefaits/24577-alerte-abidjan-des-poissons-morts-flottent-sur-la-lagune-Ebrié. Accessed 16 Jan 2018
Jallow BP, Toure S, Barrow MMK, Mathie AA (1999) Coastal zone of The Gambia and the Abidjan region in Côte d'Ivoire: sea level rise vulnerability, response strategies, and adaptation options. Clim Res 12:129–136
Kendall M (1975) Multivariate analysis. Charles Griffin, London
Keyantash J, Dracup JA (2002) The quantification of drought: an evaluation of drought indices. Bull Am Meteorol Soc 83(8):1167–1180
Kiptum CK, Ndambuki JM (2012) Well water contamination by pit latrines: a case study of Langas. Int J Water Resour Environ Eng 4(2):35–43

Koffi (2016) La gestion durable de l'eau et le partenariat national en Cote d'Ivoire. In: Allah-Kouadio R, Cisse B, Gregoire LJ (eds) Developpement durable et Emergence de l'Afrique, Grandvaux Eds, Brinon-sur-sauldre, France
Konaté S, Kampmann D (eds) (2010) Biodiversity atlas of West Africa, vol 3. Côte d'Ivoire, Abidjan & Frankfurt/Main
Koné M, Kouadio YL, Neuba DF, Malan DF, Coulibaly L (2014) Évolution de la couverture forestière de la Côte d'Ivoire des années 1960 au début du 21eme siècle/[Evolution of the forest cover in Côte d'Ivoire since 1960 to the beginning of the 21st century]. Int J Innov Appl Studies 7(2):782
Kouakou KE, Goula BTA, Kouassi MA (2012) Analyze of climate variability and change impacts on hydro-climate parameters: case study of Côte d'Ivoire. Int J Sci Eng Res 3(2):1–8
Kouamé YP, Coulibaly JK, Benié RD, Berte S, Ouattara NI, Dosso M, Kouamelan EP (2018) Microbiological quality of effluents from a fishing station installed at the edge of the Lagoon Ebrie (Côte d'Ivoire). J Environ Sci Toxicol Food Technol 12(4):34–40
Le Loeuff P, Marchal E (1993) Géographie Littorale. In: Environnement et ressources aquatiques de Côte d'Ivoire. Tome 1. Le milieu marin. Le Lœuff, P., Marchal, E. and J.B. Amon Kothias (eds.), editions de l'ORSTOM, Paris, pp 15–22
Magadza CHD (2000) Climate change impacts and human settlements in Africa: prospects for adaptation. Environ Monit Assess 61(1):193–205
Malik S (1998) Rural poverty and land degradation: a reality check for the CGIAR. Report to the Technical Advisory Committee to the CGIAR. FAO, Rome
Marin P, Ouayoro E, Fall M, Verspyck R (2009) Partnering for water in Cote d'Ivoire: lessons from 50 years of successful private operation. No. 10529. World Bank Publications, Washington D.C., US
Martínez ML, Intralawan A, Vázquez G, Pérez-Maqueo O, Sutton P Landgrave R (2007) The coasts of our world: Ecological, economic and social importance. Ecological Economics, 63(2):254–272
Mazón J, Pino D (2013) The role of sea–land air thermal difference, shape of the coastline and sea surface temperature in the nocturnal offshore convection. Tellus A 65(1):1–13
Ménard C, Clarke G (2000) Reforming the water supply in Abidjan, Cote d'Ivoire: mild reform in a turbulent environment. Policy research working paper 2377. World Bank Publications. Washington DC
Millennium Ecosystem Assessment (2005) Ecosystems and human well-being: wetlands and water synthesis. World Resources Institute, Washington, DC
Mishra AK, Singh VP (2010) A review of drought concepts. J Hydrol 391(1–2):202–216
Mooney (2016) World Bank: the way climate change is really going to hurt us is through water. Energy & Environment, the Washington Post. Available from https://www.washingtonpost.com/news/energy-environment/wp/2016/05/03/. Accessed 12 June 2018
N'kaka N (2013) Dabou: Mort mystérieuse de plusieurs poissons: des familles touchees, les chiens soccombent, Abidjan touchee. Soir info. www.linfodrome.com/societe-culture/9427-dabou-mort-mysterieuse-de-plusieurs-poissons. Accessed 17 June 13
Negri AJ, Adler RF, Nelkin EJ, Huffman GJ (1994) Regional rainfall climatologies derived from Special Sensor Microwave Imager (SSM/I) data. Bull Am Meteorol Soc 75(7):1165–1182
Ngigi SN (2009) Climate change adaptation strategies: water resources management options for smallholder farming systems in sub-Saharan Africa. The Earth Institute at Columbia University, New York
Niang I, Ruppel OC, Abdrabo MA, Essel A, Lennard C, Padgham J, Urquhart P (2014) Africa. In: Barros VR, Field CB, Dokken DJ, Mastrandrea MD, Mach KJ, Bilir TE, Chatterjee M, Ebi KL, Estrada YO, Genova RC, Girma B, Kissel ES, Levy AN, MacCracken S, Mastrandrea PR, White LL (eds) Climate change: impacts, adaptation, and vulnerability. Part B: regional aspects. Contribution of Working Group II to the Fifth Assessment Report of the Intergovernmental Panel on Climate Change. Cambridge University Press, Cambridge, UK/New York, pp 1199–1265
Noumé A (2017) Transport lagunaire : une solution pour désengorger les routes. Cote d'Ivoire Economie No. 51. April 2017. Available from http://www.cotedivoire-economie.com/wp-content/uploads/2017/10/CI-ECO-N%C2%B051-leger-.pdf. Accessed August 25, 2018
Obrist B, Cissé G, Koné B, Dongo K, Grenado S, Tanner M (2006) Interconnected slums. Water sanitation and health in Abidjan, Cote d'Ivoire. Eur J Dev Res 18(2):319–336

Oga YMS (1998) Ressources en eaux souterraines dans la région du Grand Abidjan (Côte d'Ivoire): Approches hydrochimique et isotopique. Thése de doctorat, Université Paris Sud, Orsay, France, p. 240

Oga YMS, Sacchi E, Zuppi GM (2007) Origin and effects of nitrogen pollution in groundwater traced by $\delta^{15}N–NO_3$ and $\delta^{18}O–NO_3$: the case of Abidjan (Ivory Coast). Advances in isotope hydrology and its role in sustainable water resources management (IHS – 2007). Proceedings of a symposium Vienna, 1, 139–147

Olayinka OO, Adedeji OH, Sanni SS (2014) Water quality of hand-dug wells close to pit latrines: implications for sanitation and health. IOSR J Environ Sci Toxicol Food Tech 8:56–64

Osemwegie I, N'da HD, Stumpp C, Reichert B, Biemi J (2016) Mangrove Forest Characterization in Southeast Côte d'Ivoire. Open J Ecol 6:138–150

Osemwegie I, Stumpp C, Biemi J, Reichert B (2017) Multi-tracer assessment of seasonal water source changes of coastal water systems along the south-eastern Ivorian coast (West Africa). Submitted to the Hydrological Science Journal (HSJ), Taylor & Francis, UK

Pandey RP, Ramasastri KS (2001) Relationship between the common climatic parameters and average drought frequency. Hydrol Process 15(6):1019–1032

Paturel JE, Servat E, Delattre MO, Lubès-Niel H (1998) Analyse de séries pluviométriques de longue durée en Afrique de l'Ouest et Centrale non sahélienne dans un contexte de variabilité climatique. Hydrol Sci J 43(6):937–946

Peltre P (1978) Le "V-Baoulé" (Côte d'Ivoire centrale): héritage géomorphologique et paléoclimatique dans le tracé du contact forêt-savane. O.R.S.T.O.M, Paris

Potop V, Možný M (2011) The application a new drought index–standardized precipitation evapotranspiration index in the Czech Republic. Mikroklima a mezoklima krajinných structur a antropogenních prostředí 2, 2–14

Programme Alimentaire Mondial (2009) Côte d'Ivoire: Evaluation approfondie de la sécurité alimentaire des ménages ruraux en Côte d'Ivoire, rapport final. Available from http://www.ins.ci/n/templates/docss/secualim_ruraux.pdf. Accessed on 07.6.2018

Rasmusson EM, Wallace JM (1983) Meteorological aspects of the El Nino/Southern oscillation. Science 222:1195–1202

Saint-Vil J (1983) L'eau chez soi et l'eau au coin de la rue: les systèmes de distribution de l'eau à Abidjan. Cahiers ORSTOM 19(4):471–489

Saley MB, Tanoh R, Kouamé KF, Oga MS, Kouadio BH, Djagoua EV, Oularé S, Savane I (2009) Variabiltié spatio-temporelle de la pluviométrie et son impact sur les ressources en eaux souterraines: cas du district d'Abidjan (sud de la Côte d'Ivoire). In: 14e colloque International en évaluation environnementale, Niamey, 26–29

Sarr B, Atta S, Kafando L (2012) Le Sahel face aux changements climatiques: Enjeux pour un développement durable. Revue des indices climatiques utilisés dans les systèmes d'assurances agricoles indicielles en Afrique. Sècheresse 23:255–260

Scheren PAGM, Kroeze C, Janssen FJJG, Hordijk L Ptasinski KJ (2004) Integrated water pollution assessment of the Ebrié lagoon, Ivory Coast, West Africa. Journal of Marine Systems, 44 (1):1–17

Seventh Forum for Africa's Development (2010) Changements climatiques et développement durable en Afrique: Vue d'ensemble. Centre de conférences des Nations Unies • Addis-Abeba (Éthiopie)

Templeton MR, Hammoud AS, Butler AP, Braun L, Foucher JA, Grossmann J, Jourda JP (2015) Nitrate pollution of groundwater by pit latrines in developing countries. AIMS Environ Sci 2(2):302–313

Thornthwaite CW (1948) An approach toward a rational classification of climate. Geogr Rev 38(1):55–94

UNEP (2015) Côte d'Ivoire. Post-conflict environmental assessment. UNEP, Nairobi

Vicente-Serrano SM, Beguería S, López-Moreno JI (2010) A multiscalar drought index sensitive to global warming: the standardized precipitation evapotranspiration index. Journal of Climate, 23(7):1696–1718

Wilhite DA (2000) Chapter 1: Drought as a natural hazard: concepts and definitions. Drought Mitigation Center Faculty Publications, University of Nebraska, Lincoln 2000. Available from Publications 69. http://digitalcommons.unl.edu/droughtfacpub/69. Accessed on 18.1.2018

Wilhite DA, Glantz MH (1985) Understanding: the drought phenomenon: the role of definitions. Water Int 10(3):111–120

Wozazek S (2001) Die klastischen Sedimente von Süd-Elfenbeinküste: Provenanz, Umlagerungsprozesse und Entstehung des Goldvorkommens 'Belle Ville'. Fakultät Geo- und Biowissenschaften der Universität Stuttgart, Germany

Microbial Inoculants for Improving Carbon Sequestration in Agroecosystems to Mitigate Climate Change

18

Abeer Ahmed Qaed Ahmed, Kehinde Abraham Odelade, and Olubukola Oluranti Babalola

Contents

Abstract

The impacts of losing carbon from the terrestrial pool into the atmosphere have major consequences that affect many aspects of our planet in the long term and the short term. Agricultural lands could contribute in targeting these issues and

A. A. Q. Ahmed
Pharmacy Department, Al-Saeed University, Osaifrah, Taiz, Yemen

K. A. Odelade · O. O. Babalola (✉)
Food Security and Safety Niche Area, Faculty of Natural and Agricultural Sciences, North-West University, Mmabatho, Mafikeng, South Africa
e-mail: olubukola.babalola@nwu.ac.za

W. Leal Filho (ed.), *Handbook of Climate Change Resilience*,
https://doi.org/10.1007/978-3-319-93336-8_119

provide efficient solutions, such as decreasing the level of atmospheric carbon while increasing carbon levels in soil. The proper selection of suitable microbial inoculants that are able to sequester carbon into soils is very important in order to improve agricultural land's capability to sequester and store carbon. By achieving that, soil quality and properties would increase, and atmospheric carbon would be mitigated. In this review, we discussed the potential of using microorganisms as microbial inoculants to overcome the negative impacts of losing carbon from soils, by increasing carbon levels in soils. Soil microorganisms have the ability to affect the organic matter quantity and quality, which leads to affecting soil ecology and properties. Fungi and bacteria contribute differently in soil carbon sequestration by regulating multiple and different pathways inputs and losses of soil carbon such as possessing metabolic activities that can capture CO_2, the ability to sediment carbonates, the recalcitrant nature of their vegetative and products tissues, or the formation of stable forms that protect carbon in soil. More studies are needed to test the potential of certain microbial strains with carbon sequestration ability to improve soil quality and mitigate climate change.

Keywords

Agroecosystems · Bacteria · Carbon sequestration · Climate change · Fungi · Microbial inoculants · Global warming · Soil fertility · Soil health

Introduction

Soil is considered the largest carbon reservoir, as it contains more carbon than the atmosphere and vegetation combined (Averill et al. 2014; Tan et al. 2014). Soil organic matter is composed of organic fractions that contain decomposed animals and plants as well as microbial organisms, in addition to inorganic forms such as carbonates and lime (Chan 2008). Two types of components present in soil organic matter are active (35%) and passive (65%), in which active soil organic matter is living organisms as well as dead animals and plant residues that contain easily digested sugars and proteins; meanwhile passive soil organic matter is the components present in the soil that are hard to decompose by microbes (Hoorman and Islam 2010). All these different types of soil components affect soil ecology and properties.

Soil organic carbon originally comes from atmospheric CO_2 that is captured via a process called photosynthesis by plants and autotrophic organisms to turn atmospheric CO_2 into organic compounds (such as sugars and cellulose) which build up their biomass. The rate of atmospheric CO_2 that is captured through photosynthesis and enters terrestrial biomass is about 120 Pg carbon per year, half of which will be lost again from plants to the atmosphere by respiration, and the net production will be about 60 Pg per year (Fig. 1) (Amundson 2001; Janzen 2004). This amount will stay in a form of temporarily vegetative tissues. Afterward, other living organisms will feed on the plants, build up their biomass from carbon compounds, and then release some of the carbon as CO_2 to the atmosphere through respiration. After organisms die and decompose (e.g., plants and animals), carbon in the organic matter will return into the soil where it might stay until being broken down by soil microorganisms and released back to the atmosphere in the form of CO_2. Heterotrophic respiration, which is largely done by microorganisms in soil, and fire return an amount of carbon estimated to be ~60 Pg carbon per year back to the atmosphere in the form of CO_2 (Fig. 1) (Janzen 2004).

Soil contains about 2344–2500 Gt of carbon, of which about 1550 Gt of carbon is stored in organic forms and 950 Gt stored in inorganic forms (Fig. 1) (Lal 2004; Stockmann et al. 2013). There is a constant need to reduce carbon emissions and remove carbon from the atmosphere in order to fight against climate change and to overcome the negative impacts of losing carbon from the terrestrial pool to the atmosphere. High concentrations of atmospheric CO_2 ($\geq$400 ppm) lead to an increase in temperatures of more than 2–4 °C, causing changes in wet/dry cycles, heavy and intensive rainfall and storms, extreme frost and heat waves, and increased fire frequency. It also significantly influences environmental quality and water resources, crucially affects soil fertility and properties, and significantly abrupts changes in temperatures daily, seasonally, and interannually (Qafoku 2015). Due to these negative effects that are mainly caused by the large losses of soil carbon to the atmosphere, intensive studies should be undertaken to decrease carbon losses from soils as well as reduce atmospheric CO_2 levels.

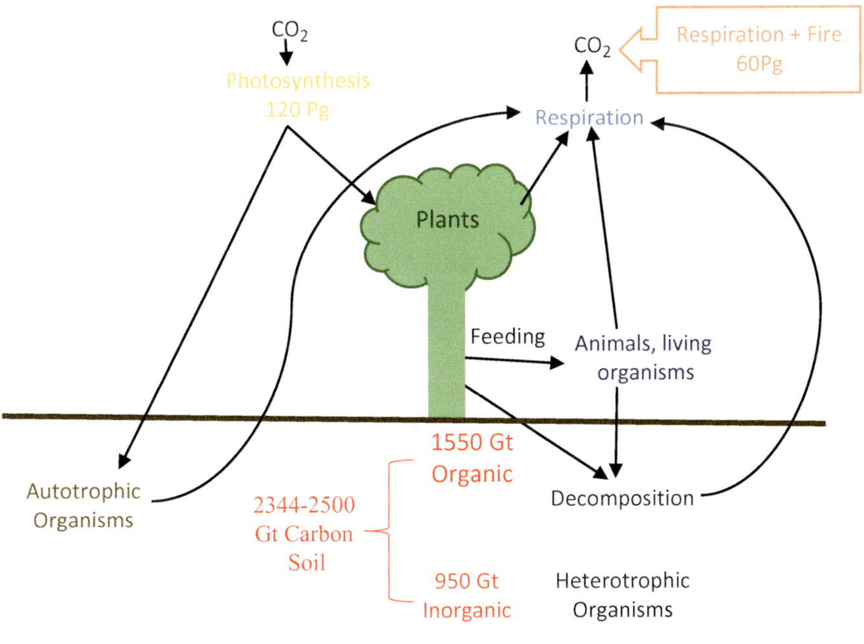

Fig. 1 Carbon sequestration and cycle in terrestrial ecosystems

The Impact of Losing Carbon from Soils on Agricultural and Climatic Conditions

Agricultural Conditions

Soil organic matter contributes significantly to soil's quality and fertility (Chan 2008), and it contains an amount of 58% of soil organic carbon (Stockmann et al. 2013). Thus, losing carbon from the soil will affect the soil's conditions and properties, which directly influence agricultural production. Soil organic matter increases the availability of nutrients in the soil through the decomposition of the organic matter, which will release phosphorus, nitrogen, and other important nutrients for plant health and growth (Chan 2008). Enough amount of organic matter in soil will ensure the availability of all nutrients that are important for plant health and growth. Losing organic matter from soils leads to decreasing the efficient amount of these nutrients, which would affect plant growth and health and directly leads to decreased plant productivity.

Soil organic matter enhances soil structure and physical properties such as water filtration, water-holding capacity, root growth, gaseous exchange, and ease of cultivation through soil aggregation formation (Chan 2008). Suitable soil structure and physical properties directly affect soil quality leading to better soil conditions. These conditions will contribute to optimize the soil ecology to provide all the required factors for plant's growth. Thus, losing or decreasing the organic matter in soil will disrupt these optimal conditions through disrupting the soil structure and physical properties, which will then affect plants. Soil organic matter significantly contributes in decreasing soil loss and runoff as well as reducing carbon leaching; therefore it will enhance carbon sequestration in soil (Kadlec et al. 2012). Hence, soil organic matter plays a major role in improving soil properties, either directly by immediate effect or indirectly through enhancing carbon sequestering into soil that would eventually affect soil properties. Improving soil structure and physical properties is critical for creating suitable soil ecological conditions for plant growth, health, and production.

Organic matter promotes soil biological health as it is considered an important food source for soil flora and fauna. It controls the type and number of soil inhabitants and thus contributes to other important soil functions such as nutrient availability and cycling, plant nutrient uptake, root growth, burrows creation, and suppression of crop diseases (Chan 2008). Sufficient organic matter in soil contributes to enhancing different soil biological activities of useful soil microbial organisms. The reduction of soil organic matter would influence the soil microbial organisms negatively, which would affect these critical soil biological activities and plant health. Therefore, efficient amount of organic matter in soil regulates various biological processes that contribute to nutrient cycling, as well as improving plant health and overcoming diseases.

Climatic Conditions

The loss of carbon from soils increases CO_2 levels in the atmosphere and leads to increasing global temperatures and causes global warming. Previous studies have focused on reducing carbon emissions in order to fight against climate change (Powlson et al. 2011; Shindell et al. 2012). Global temperature is reported to increase more than 2 °C by 2050 and about 3–5 °C in the coming 50–100 years (Ipcc 2007; Randers 2012). The increase in temperature correlates with the increase of the frequency of other environmental stresses such as heat waves, flood, and drought (Trenberth 2011; Jansen et al. 2014). These environmental stresses will affect agricultural lands afterward.

Drought affects plants by causing water stress in them, which could negatively affect their growth and health. Water is a necessity for plants. Therefore, the lack of water availability will damage plants by affecting important biological processes such as photosynthesis (Ghotbi-Ravandi et al. 2014) and then affect metabolisms and reduce plant growth and health.

Plant biochemical reactions are highly sensitive to high temperatures, and high temperature stress leads to affecting plants' metabolism, growth, and productivity (Hasanuzzaman et al. 2013). Strong impacts on plants are caused by heat waves through decreasing growth as well as shifting biomass allocation, and when heat waves combine with drought, it could be devastating for plants and lead to mortality (Teskey et al. 2014). Heat waves put plant under heat stress. This influences plants by affecting important biochemical reactions that are necessary for plant growth and health, which would limit plant yields. The continuous increasing of global temperature would contribute in decreasing the production of agricultural lands.

Flooding could affect agricultural production through changing the physiological and chemical properties by leaching organic matter and nutrients down the soils (Eni et al. 2011). Plants need adequate moisture in the soil to ensure optimized crop yields, and the decrease or increase of these levels will directly affect plant health and growth and will affect crop production eventually. The increase in global temperature could also affect agricultural lands through reducing the cultivated areas. The increase of temperature will melt the tundra, and then oceans would expand and decrease agricultural lands. It is predicted that ocean expansion in the next 40 years will cause sea level to raise 30 cm (Randers 2012). Thus, agricultural lands could face considerable reduction in size that definitely decreases agricultural production.

High temperatures could provide suitable conditions for pathogens growth and spread (Luck et al. 2011). In extreme weather, certain plant diseases such as powdery mildews and rusts could be sporadic and develop more quickly in warm conditions (West et al. 2015). Thus, controlling global temperature through sequestering carbon and reducing CO_2 mitigation could participate in controlling plant diseases subsequently. Maintaining plant health by controlling and limiting plants susceptibility for diseases would positively enhance crop yields and quality. Therefore, we reviewed the use of microbial inoculants as means to significantly enhance carbon sequestration in the soil, thereby mitigating against climate change.

Soil Carbon Sequestration

Carbon sequestration is a term used to describe a process that leads to increase soil organic carbon storage by removing CO_2 from the atmosphere and introducing it into soils. It is defined as the process of transferring atmospheric CO_2 into the soil of a land unit through unit plants, plant residues, and other organic solids in order to be retained in the unit as part of the soil organic matter (Olson 2013). Carbon could be stored in soils as stable long-lived forms of organic matter (Six et al. 2006; Powlson et al. 2011). Increasing the concentrations of hard-to-decompose organic forms in soil will lead to securely stored carbon in soils for long periods and to reduction of carbon emissions. The balance between carbon inputs and outputs determines soil carbon sequestration and the change in soil organic carbon levels (Chan 2008). The more carbon stored in the soil and less carbon lost indicates that this land has high ability to sequester carbon, and vice versa. Soil organic carbon levels are affected by management practices such as cultivation, overgrazing, and stubble burning or removal (Chan 2008). Therefore, poor management practices will affect carbon sequestration negatively and lead to depletion of soil organic matter in soils. Carbon levels and storage in soils result from the interaction between several ecological processes, which are photosynthesis, respiration, and decomposition (Ontl and Schulte 2012). Photosynthesis (by either plants or autotrophic organism) contributes to carbon sequestration process through increasing the amount of CO_2 input from the atmosphere to the terrestrial pool. Meanwhile respiration and decomposition contribute to losing carbon from the terrestrial pool back to the atmosphere. Due to this, studies should focus more on the processes that increase carbon sequestration and input in soils and decrease the level of carbon losses and emissions. Therefore, application of microbial inoculants is very paramount in this review.

Using microbial inoculants in agricultural practices could help to achieve desirable characteristics in soil. Bio-sequestration is the process that includes the natural capture and storage of CO_2 by photosynthetic organisms as well as soil microbes (Macreadie et al. 2014). Microbial inoculants could be used to increase the level of carbon inputs and decrease the levels of carbon outputs in the soils. Soil microbial communities have important roles in carbon sequestration and soil carbon emission (Fang et al. 2014). Choosing the right microorganism with the right mechanism for a specific land is very important in order to increase carbon sequestration in soils. Changes in microbial communities affect the soil organic matter cycling and storage due to the ability of soil microorganisms to regulate multiple pathways input and loss of soil carbon (Smith et al. 2014). The difference in microbial biomass contents and physiological characteristics between microbial communities can lead to different mechanisms and pathways in sequestering carbon into soils.

Microbial Inoculation in Agricultural Soils

Microbial inoculation can be defined as the application of cost-effective and eco-friendly formulated microbial inoculants that can be used as renewable plant nutrient sources and can be considered an alternative method to chemical pesticides and fertilizers (Babalola and Glick 2012; Selvakumar et al. 2012). This environmentally friendly approach plays a major role in maintaining sustainable agriculture by acting as natural biopesticides and bio-fertilizers, which reduce the negative effect of using mineral fertilizers and chemical pesticides on agricultural lands as well as agricultural crops, that might lead to environmental and health consequences.

Microbial inoculants provide agricultural soils with suitable conditions that ensure high protection and yield of crops. Agricultural lands with poor chemical and physical soil properties, under stressed or normal conditions, and with low soil organic matter, lead to a reduction in crop yield and quality. In poor lands, crop yield and quality were dramatically increased after using microbial inoculants (Tavasolee et al. 2011; Upadhyay et al. 2011; Wu et al. 2014a, b). Microbial inoculants increase the level of organic carbon, nitrogen, potassium, and soil respiration (Sangakkara et al. 2014); thus they could be used as an effective solution to restore soil quality and fertility in poor lands.

Microbial inoculants are very important in agricultural systems for many benefits that they can provide, such as promoting plant growth (Bashan et al. 2014), reducing heavy metals toxicity (Sessitsch et al. 2013; Wani and Khan 2013), enhancing drought tolerance of plants (Ortiz et al. 2015), alleviating plant growth stresses (Miransari 2010), protecting against different types of pathogens (Babalola 2010), and enhancing nutrient uptake (Mardukhi et al. 2011). In addition, microbial inoculants are considered by Pal and Pandey (2014) the key determination factor of soil quality. For all these benefits, microbial inoculants could be the best solution for improving agricultural lands and to solve unfavorable soil conditions that lead to directly and/or indirectly affect crop yield and quality.

Microbial Inoculants and Carbon Sequestration

Soil microorganisms contribute to carbon sequestration by different mechanisms such as possessing metabolic activities that capture atmospheric CO_2, the ability to sediment carbonates, forming recalcitrant vegetative tissues and products, and the ability to form stable forms such as soil aggregates that protect carbon soil organic forms. Total microbial biomass in soil is composed of 0.1–2% active, 42% dormant, and 56% dead microorganisms (Blagodatskaya and Kuzyakov 2013). Microorganisms contribute to soil physical characteristics and chemical contents through their living state as well as dead and dormant states. Different microbial communities vary in their metabolic activity states, which would lead to different effects on soil organic matter quantity and quality. The mycorrhizal and rhizobacterial type affects soil carbon independently and far more than the effects of soil clay contents, precipitation, temperature, and net primary production (Averill et al. 2014). Therefore, it is very important to study the possibility of using a specific microbial strain as microbial inoculant in order to sequester carbon in agricultural lands.

Bacterial Inoculants and Carbon Sequestration

Bacteria exist in large numbers in soil compared to fungi. Although previous reports had already discussed the ability of fungi to contribute more than bacteria in carbon sequestration (Six et al. 2006; Li et al. 2015), some reports showed that some bacterial strains have the ability to reduce CO_2 and could contribute to carbon sequestration and to mitigate atmospheric CO_2 levels (Han et al. 2013; Komala and Khun 2014; Nie et al. 2015). This indicates that bacteria could contribute to carbon sequestration through different pathways and metabolic activities.

Pseudomonas fluorescens is a plant growth-promoting bacterium that could be a useful tool for carbon sequestration and climate change mitigating (Nie et al. 2015). The results from this study indicate that this microbial inoculant increased plant productivity as well as having potential to mitigate high atmospheric CO_2 levels by increasing terrestrial carbon sequestration, particularly in high-CO_2 ecosystems. These findings could draw attention to the use of plant growth-promoting bacteria as microbial inoculants to sequester atmospheric CO_2.

Carbonate induction by some bacteria has the potential to capture atmospheric CO_2 and sequester carbon (Peng et al. 2010). *Bacillus mucilaginosus* produces carbonic anhydrase that first captures the atmospheric CO_2 and then fixes the atmospheric CO_2 through bacterial metabolism (Zhang et al. 2011). This process induces carbonate formation to form organic substances or carbonated minerals by fixing CO_2 from the atmosphere, which leads to decreased CO_2 levels and increased carbon sequestering in soils. In a recent study conducted by Xiao et al. (2015), further work was done to transcribe and translate carbonic anhydrase genes from *Bacillus mucilaginosus* genome into *Escherichia coli* to test the ability of this type of carbonic anhydrase to promote $CaCO_3$ precipitations. Results showed that the engineered bacteria produced carbonic anhydrase that helps in $CaCO_3$ formation in a relatively short time. In this regard, microbial strains that are able to produce carbonic anhydrase could be very useful microbial inoculants in the process of carbon sequestration and capturing atmospheric CO_2.

Bacterial strains with the ability to sediment carbonates such as *Bacillus pumilus* (Komala and Khun 2014), *Bacillus cereus* (Han et al. 2013), *Bacillus mucilaginosus* (Zhang et al. 2011), and *Bacillus pasteurii* (Stocks-Fischer et al. 1999) were reported to reduce CO_2 and could contribute to the carbon cycle and sequestration. These strains with other soil microorganisms that have the same ability could further be explored for use as microbial inoculants potential to sequester carbon in soils. The sequestration of CO_2 into sediment carbonates by bacteria could be a promising option in the quest to find solutions for increasing soil carbon sequestration and reducing carbon emissions.

Chen et al. (2012) genetically engineered *Synechococcus elongatus* PCC 7942 cyanobacteria to combine the biological CO_2 fixation with the chemical transformation of CO_2. This review also introduces the insight that CO_2 bio-mitigation could be enhanced through genetic engineering by CO_2 capturing or transformation via carbonic anhydrase, with biological fixation of the captured CO_2 to participate in biomass generation. Microorganisms could be genetically engineered to overcome low efficiency of CO_2 capturing and fixation. Then, these microbes could be further studied for their ability to be used as microbial inoculants for carbon sequestration in agricultural lands.

Fungal Inoculants and Carbon Sequestration

Fungi exist in a smaller population than bacteria, but they dominate the soil biomass, especially when soil is not disturbed (Hoorman et al. 2009). Fungi contribute more than bacteria to carbon sequestration and organic matter formation and stabilization in soil (Six et al. 2006; Li et al. 2015). This is due to that fungal tissues input large amount of carbon into soil and lead to the integration of more carbon into soil fungal biomass, having recalcitrant biomass tissues and by-products, and enhance carbon protection and stabilization by soil aggregate formation.

Effects of Fungal Mycelia on Carbon Sequestration

Fungi contribute to carbon accumulation and losses in soil through fungal biomass and by-products production as well as fungal necromass degradation. Carbon accumulation by fungi in soil varies according to difference between fungal species and fungal biomass content (Fernandez and Koide 2011; Ekblad et al. 2013). Fungal mycelium is one of the vegetative tissues which grow and spread into soils in order to provide the fungi with the required water and nutrients. The production of extrametrical mycelium could reach hundreds of kilograms per ha per year (Wallander et al. 2013). Since fungal mycelium is built up by carbon, the more fungal mycelia produced, the more carbon is captured in the soil. Therefore, fungal mycelia could be considered as a vital carbon sink to the soils.

Fungal necromass is the main microbial source of stable soil organic matter and contributes more than bacterial necromass (70.7% vs. 25.9%) to soil organic matter (Li et al. 2015). Thus, mycelial necromass could be a very important source of carbon storage in soil; and it could be an important source of carbon for living organisms as well. High chitin and nitrogen concentrations in fungal mycelium would contribute to the decomposition of mycelial necromass within weeks; meanwhile melanized cell wall mycelium is resistant to decomposition, which leads to long-term preservation of necromass in soil according to the concentration level of melanin in fungal tissues (Fernandez and Koide 2011; Drigo et al. 2012; Fernandez and Koide 2014; Koide et al. 2014). The ability of mycelial necromass to resist decomposition varies according to fungal species and the composition of fungal tissues. Thus, some fungal species could contribute to sequestering carbon more than others could. Fungal mycelia could grow into secure soil forms such as soil aggregates (Peng et al. 2013). Moreover, fungal necromass will remain protected from degradation for a long time. The more fungal necromass remains protected, the more carbon will be kept secured in soils.

Effects of Fungal Metabolites (Glomalin) on Carbon Sequestration

Glomalin is an insoluble hydrophobic glue-like proteinaceous substance (Peng et al. 2013), which is released from spore and hyphae of mycorrhizal fungi and *Scutellospora*, *Acaulospora*, *Entrophospora*, *Glomus*, and *Gigaspora* genera in order to protect the hyphae from drying out (Singh 2012; Pal and Pandey 2014). Glomalin affects carbon sequestration through two ways. First, the production of glomalin assists the formation of soil aggregates, which subsequently will affect carbon sequestration. The second way is due to the recalcitrant nature of glomalin, which makes it hard to decay, and it thus remains stable in soils for long periods.

Wu et al. (2014a, b) tested direct and indirect effects of different factors such as roots, mycorrhizal hyphae, and glomalin on aggregate stability and found that glomalin-related soil protein was the primary contributor to aggregate stability among all tested factors. This indicates that glomalin is the main factor that influences the carbon sequestration and storage in soil, either directly or indirectly by influencing other carbon sequestration factors. Glomalin attaches to soil particles and organic matter to form clumps. These clumps are stable structures that resist outside factors such as erosion, and at the same time, it has permeable characteristics that allow air, roots, and water to move inside these clumps. Soil particles and other organic and inorganic materials inside these clumps will remain protected for a long time, and this would increase the level of carbon in soils, which provides soil with good characteristics and gives soil better texture.

Since glomalin contains carbon and is abundant in soils, it could account for a considerable amount of the terrestrial carbon pool. Glomalin does not easily dissolve in water, and it resists microbial decomposition and could remain in soil for long periods, about 10–50 years (Wright and Upadhyaya 1999; Pal and Pandey 2014). The recalcitrance and complex nature of glomalin contribute significantly to carbon sequestration and storage in soil.

Effects of Soil Aggregation on Carbon Sequestration

Soil aggregation is a process that depends on providing glue-like materials by microorganisms to hold soil particles and other organic and inorganic materials together to form soil aggregates. Soil fauna and microorganisms, roots, environmental variables, and binding factors (such as glomalin) are the major factors that play a role in the formation and stabilization of aggregates (Pal and Pandey 2014). Root and fungal hyphae are considered the most important biotic factors in stabilization of soil aggregates (Singh et al. 2013). The soil aggregates form when fungal hyphae encircle soil particles and other solid materials, while glomalin acts as a gluing agent to stick particles together to form stable structured clumps, which are called aggregates. The cementing capacity of soil aggregates provides a stable uniform structure, which protects the organic and inorganic materials that are contained inside the soil aggregate. Stabilization of aggregates contributes to carbon sequestration through protecting carbonaceous compounds inside the aggregates from degradation, which keeps carbon stored and sequestered in soils for long periods.

Soil aggregates are important to maintain and improve soil fertility, especially when targeting soil physical properties (Rawlins et al. 2013; Pal and Pandey 2014). The soil aggregates help to overcome the unfavorable conditions of soils such as looseness, runoff, leaching and erosion. At the same time, soil aggregates increase other favorable conditions such as soil structure stability, nutrient supply capability, water-holding capacity, porosity, and aeration. These useful effects can work in concert to improve soil quality and fertility. The abundance and stability of soil aggregates are very important contributors in soil quality, as well as being critical for several properties of soils.

Method Used to Determine the Effects of Microbial Inoculants in Improving Carbon Sequestration, Soil Fertility, and Climate Mitigation

For nearly a century, soil organic matter (SOM) formation in conceptual and quantitative models has been depicted primarily as a function of plant inputs and their chemistry (Waksman 1936; Berg and McClaugherty 2003; Lehmann and Kleber 2015). As such, chemically diverse and stable SOM originates from the preservation of biochemically recalcitrant complex plant polymers, such as lignin derivatives and long-chain lipids (Waksman 1936; Berg and McClaugherty 2003). However, soil microbial communities are adept at decomposing a wide range of plant compounds and using the carbon (C) to synthesize their own biomass. The importance of soil microbes in processing plant inputs and synthesizing SOM is very paramount, because formation and stabilization of SOM is a great tool to determining how carbon is being controlled, level of soil fertility, and climate mitigation. Therefore, this review discussed the use of model soils to quantitatively assess whether microbial processing of simple C (i.e., low-molecular-weight) substrates alone, in the absence of complex plant compounds, can build significant amounts of chemically diverse, stable SOM. Since this model soils are initially C- and microbe-free, we eliminate the difficulties of isolating microbial residues that occur when using natural soils. We use a gradient of substrate-C inputs to represent different microbial-available energy in order to facilitate the development of diverging microbial communities and physiologies that are hypothesized to influence SOM chemistry and accumulation rates. The substrate gradient includes monomeric and dimeric sugars since they are abundant energy sources for microbial metabolism in natural soils (Fischer et al. 2010), but also because their rapid microbial uptake and intercellular breakdown should leave little to no unaltered substrate in the soil that would interfere with the detection of novel SOM molecules formed during the experiment (Baldock et al. 1989; Fischer et al. 2010). Further, we include a more recalcitrant lignin monomeric substrate, as well as plant-derived dissolved organic C (DOC), a natural analog to test whether substrate chemical diversity is requisite to generating SOM chemical diversity. We also compare two clay types (kaolinite or montmorillonite) to investigate the effects of clay mineralogy on SOM accumulation relative to microbial communities and substrate chemistry. We characterize the composition of SOM after 18 months of incubation using high-resolution molecular fingerprinting by pyrolysis-gas chromatography/mass spectrometry (py-GC/MS) to establish the chemical fingerprint of newly formed microbial residues (including cell wall and cytoplasmic materials, metabolites, and extracellular excretions). Our model soils accrue C concentrations between 1 and 1.4% C, with a chemical diversity and stability characteristic of natural soils. Substrate type has a stronger influence on SOM development than clay mineralogy. However, this effect appears to be an indirect consequence of diverging microbial communities, where different substrates select for distinct microbial communities, with microbial-SOM accumulation being greatest in soils where microbial abundances are highest and microbial biomass production is most efficient.

Method

Experimental Conditions and Sampling

Laboratory mesocosms consisted of 100 g dry wt of 33% kaolinite or montmorillonite clay mixed with quartz sand. The cation exchange capacity was 5.2 and 28.6 mg per 100 g soil for the kaolinite and the montmorillonite soil mixtures, respectively. Metal concentrations for the kaolinite mixture were 5.6% Al, 1% K, 0.9% Fe, 0.14% Ca, 0.11% Na, and 0.06% Mg. Montmorillonite mixture metal concentrations were 2.2% Al, 0.8% K, 2.3% Fe, 0.39% Ca, 0.14% Na, and 0.32% Mg. Soil mixtures received weekly additions of glucose, cellobiose, syringol, or switchgrass plant DOC (0.7 mg C per g dry soil) and biweekly additions of a multi-nutrient Hoagland's solution (0.023 mg N per g dry soil). The substrate-C (5 ml of 14,000 ppm C) and nutrient additions (0.5 ml of 4666 ppm N) were syringe-injected throughout the model soils to facilitate uniform substrate application. The rate of weekly C additions reflects the upper range of natural C inputs to soil from daily root exudations. Total C inputs from added substrates over the course of 15 months amounted to 46.9 mg C per g soil, and total inorganic N from nutrient solution was 1.56 mg N per g soil. To create DOC, we mixed 150 g switchgrass (*Panicum virgatum*) tissue (leaf, stem, and seed) with 2 l H_2O and shook for 24 h before filtering. The filtrate was then partially evaporated until the C concentration was 18 mg C per ml (C:N 20). We created the microbial soil inoculum in a soil slurry (1:100 w/v) using soil collected from a Michigan bromegrass (*Bromus inermis*) field at 0–7 cm depth. We then syringe-injected 100 μl of inoculum solution (~0.00001 g natural soil per g model soil) into the soil mixture. Each soil mesocosm treatment was replicated five times with five destructive harvests at 6, 9, 12, 15, and 18 months. During incubation, soils were maintained at 25 °C, 45% water-holding capacity, and a pH between 4.6 and 5. The pH of DOC-treated soils, however, increased to 7 after a few months of incubation, and therefore DOC biological data are minimally included in our analyses. During the 15-month incubation period, no soil mixing occurred in order to better simulate an undisturbed soil environment and to minimize aggregate destruction.

After 15 months, substrate and nutrient additions ceased, but soils were maintained for an additional 3 months under the previous incubation conditions with periodic hand mixing to facilitate microbial uptake of any remaining substrate or unstabilized microbial biomass. Microbial community activity was monitored weekly by CO_2 flux measurements for the first 9 months and then biweekly thereafter and by microbial hydrolytic and oxidative potential enzyme activities (Saiya-Cork et al. 2002) at each mesocosm harvest.

SOM Chemistry and Concentrations

SOM chemistry was assessed using pyrolysis-GC/MS (Grandy et al. 2009). Compound peaks were analyzed and identified with the Automated Mass Spectral Deconvolution and Identification Systems (AMDIS V 2.65) software and the National Institute of Standards and Technology (NIST) compound library and published literature (Grandy et al. 2009). Organic matter compounds are expressed as the % relative abundance of total sample peak area and classified based on origin (lipids, lignin derivatives, polysaccharides, proteins, nonprotein N-bearing, and phenolics). Identifiable compounds with potentially multiple origins were classified as unspecified. Aromatic compounds also with an unspecified origin were classified as aromatics. The SOM chemical diversity was determined based on the total number of identifiable compounds that were $>1\%$ relative abundance.

Total SOC concentrations were determined on an elemental analyzer (Costech ECS 4010). Total microbial biomass carbon (MBC) was determined by the difference in C concentration (using a Shimadzu TOC-L analyzer) in chloroform fumigated and unfumigated soil subsamples extracted with 0.5 M K_2SO_4. We calculated the total SOM conversion efficiency as the amount of total SOC relative to the total substrate-C added for each destructible harvest, where the total substrate added is the sum of the weekly substrate addition within the time period up until harvest.

SOM Stability

Determination of biological stability was performed in a separate 3-month incubation on soil subsamples harvested at 6 months. During this separate incubation experiment, soils received three applications of ^{13}C-labeled 1:1 glutamic acid/glucose 25 atom % mixture at 50 μg C per g soil. The labeled substrate enabled us to use a standard isotope mixing model (Ineson et al. 1995), to determine the amount of newly formed C vulnerable to decomposition by an active microbial community. During the 3-month incubation, soils were frequently hand-mixed to facilitate SOM mineralization and maintained under the same incubation conditions as the long-term incubation. We also assessed the chemical stability of the accumulated SOM with acid hydrolysis fractionation (Paul et al. 2001). After the full 18-month incubation, 1 g soil was treated with 50 ml 6 M HCl and heated at 100 °C for 18 h. After heating, the acid hydrolysable fraction was decanted, and the remaining nonacid hydrolysable (i.e., stable) residue was rinsed four times with DI-H_2O and isolated by centrifugation (Paul et al. 2001; Plante et al. 2011). The chemically stable

C was then determined based on the differences in C remaining in the nonacid hydrolysable residues and the initial C content of the sample.

Microbial Community Composition

Determination of differences in microbial community composition was performed from PLFA biomarkers, identified and quantified at 12 and 15 months (Guckert et al. 1985). We included a phosphate buffer in the single-phase solvent system (chloroform) to extract only viable microorganisms. Lipid extracts were fractionated on silicic acid columns to isolate and collect polar lipids. Polar lipids were methylated to fatty acid methyl esters with 0.2 M methanolic KOH, purified and then analyzed on a Varian 3800 GC-FID. Peaks were quantified using internal standards, and identification was based on retention time data with known standards. Fungal relative abundance was determined by the sum of polyenoic unsaturated fatty acids (18:2ω6 and 18:1ω9c); Gram-positive relative abundances by the sum of the total branched, saturated fatty acids (i15:0, a15:0, i16:0, i17:0, and a17:0); and Gram-negative relative abundances by the sum of monoenoic and cyclopropane unsaturated fatty acids (16:1ω7c, 16:1ω7t, cy17:0, 18:1ω7c, and cy19:0). We did not detect any actinobacterial fatty acids.

Soil Microbial CUE and Enzyme Activities

Characterization of microbial CUE at 9 and 15 months was carried out by measuring the amount of a ^{13}C-labeled glutamic acid incorporated into MBC and respired as $^{13}CO_2$–C (Frey et al. 2013; Kallenbach et al. 2015). Fresh soil subsamples were amended with 25 atom% labeled ^{13}C-glutamic acid (50 μg C per g dry soil, $<1\%$ total soil C) and incubated at 25 °C for 22 h. The 22-h incubation time was based on preliminary respiration curves to determine the period when the majority of glutamic acid is utilized but before substrate recycling begins. Following incubation, soils were extracted for ^{13}C-glutamic acid incorporation into MBC by extracting and analyzing MBC as discussed previously and determining its isotopic composition on a 1030 TOC Analyzer (OI Analytical, College Station, TX) interfaced with a PDZ Europa 20–20 isotope ratio mass spectrometer. A 12-ml CO_2 sample was also collected for determining $^{13}CO_2$–C respiration on a GC-isotope ratio mass spectrometer (Thermo Scientific, Bremen, DE).

This approach estimates the amount of C allocated to biomass per unit of C substrate consumed and represents a proxy for microbial CUE. We calculated CUE as $[MB^{13}C/(MB^{13}C + {}^{13}CO_2\text{–}C) \times 100]$, where $MB^{13}C$ and $^{13}CO_2$–C were determined using a standard isotope mixing model equation (Bradford et al. 2013) and represent the amount of substrate incorporated into MBC and the substrate-C respired as CO_2, respectively. We used glutamic acid to estimate microbial community CUE across all substrate-treated soils since it is taken up directly into microbial cells. This approach allows us to more directly compare CUEs across treatments since substrate-C allocation toward enzyme production is minimized and we can better approximate the efficiency of new biomass growth rather than substrate degradation efficiency (Geyer et al. 2016).

The potential activities of microbial extracellular enzymes were measured at 6, 9, 12, and 15 months of incubation. We used a standard fluorometric method to assess hydrolytic enzyme activity by adding 1 g soil dry weight homogenized in a slurry with 50 mM sodium acetate buffer (Saiya-Cork et al. 2002). The buffer solution was adjusted to the average soil pH of 4.9. Soil homogenates were added to black, 96-well microplates with compound-specific fluorescing substrates bound to methylumbelliferone. Oxidative enzyme activity (phenol oxidase and peroxidase), associated with lignin breakdown, was measured spectrophotometrically using clear 96-well microplates.

Statistical and Data Analyses

Data were analyzed using a linear mixed-model two-way analysis of variance (ANOVA) to determine differences in SOM chemistry, SOC concentration, SOC conversion efficiency, SOC stability, and PLFAs and CUE between clay mineralogy and substrate type. Treatment replicate ($n = 5$) was used as a random effect, and clay and substrate were treated as fixed effects. Since the experimental design was unbalanced (there was no syringol treatment within kaolinite soils), analyses for differences in main effects of substrate and mineralogy and their interactions did not include syringol. A one-way ANOVA was used to examine the effect of substrate within a clay treatment, allowing us to include syringol-treated soils in our comparisons. In many analyses the glucose and cellobiose treatments were not statistically different ($P > 0.05$). In these cases, glucose and cellobiose data were averaged and reported as a "sugar" treatment. Pair-wise comparisons between treatments were determined by Tukey's HSD ($P < 0.05$). We used Pearson's correlation analysis to examine relationships between SOC concentrations, microbial CUE, microbial community composition, and SOM chemistry. All ANOVA and correlation analyses were performed in SAS v.9.3 (SAS Institute, 1999) using PROC MIXED and PROC CORR. When necessary, data were normalized using log transformation before analysis of variance analyses. Significance for all analyses was determined at a probability level of $P < 0.05$ unless otherwise stated. Variation around means is reported or shown as standard error.

We also used non-metric multidimensional scaling (NMDS) (PC-ORD; version 4.14) to explore differences in SOM chemistry and community composition. Pyrolysis-gas chromatography/mass spectrometry compound relative abundances and PLFA biomarkers were relativized to maximum and used separately in NMDS matrices. We also used NMDS to determine relevant parameters influencing SOC concentrations for use in our correlation analyses, where SOC concentrations at 9, 12, 15, and 18 months were grouped a priori into low, medium, and high SOC concentrations. The Sorensen (Bray-Curtis) index was used as a distance measure for all ordinations. Final NMDS solutions were considered acceptable if Monte Carlo simulations had stress values <20 and a solution stability of <0.005. Determination of significant groupings was done by the Monte Carlo stress reduction ($P < 0.05$) and multi-response permutation procedure ($P < 0.05$).

Carbon Sequestration and Agroecosystems

Different agricultural management practices affect carbon stocks in soils either positively or negatively. It is essential to adopt agricultural practices that maintain a sufficient level of carbon stocks with the goal of enhancing agricultural production and climate mitigation. The balance between the inputs and outputs of carbon in soil determines the change of soil organic carbon levels (Chan 2008). In agricultural lands, the key factor to maintaining and storing enough organic matter is to increase carbon storage by increasing the level of carbon inputs and decreasing the level of carbon outputs. Many agricultural management practices improve the level of organic matter and increase the carbon input in soil, such as adding crop residues, animal manure, sewage compost and sludge, or reducing soil disturbance and enhancing rotation of crops with high carbon inputs to the soils (Caporali and Marinari 2008). Such management practices maintain soil organic matter and nutrient supplies that promote soil health. Since soil organic matter is the key factor to determine soil productivity and the level of climate pollution, it is critical to maintain its levels by different and promising practices to achieve sustainable agricultural production and climate mitigation.

Carbon sequestration has gained a lot of interest because it is considered a multiple purpose strategy to use in mitigating climate change, protecting the environment, improving diversity, restoring degraded soils, and increasing land productivity (Wang et al. 2010). This process ensures the accumulation of organic carbon in soils by sequestering CO_2 from the atmosphere. Microorganisms have the ability to capture and store carbon in soils through different mechanisms and pathways. Thus, microbial inoculants could be used to increase the level of organic carbon in soils. Moreover, microbial inoculants will provide agricultural lands with the desired manipulated condition to improve yields and productions. Microbial inoculants have the ability to convert low-output agricultural lands to more productive lands. This strategy could enhance soil quality and properties through increasing organic matter, which would lead to increased agricultural yields and climate mitigation.

Conclusion

Soil carbon sequestration is a beneficial strategy that could mitigate climate change, enhance soil quality, increase carbon levels in soil, improve agricultural production, and enhance food security. Soil organic matter contributes in different factors that play major roles in soil health and plant production such as providing nutrients, improving suitable soil ecological conditions, affecting soil quality and physical properties, and regulating important biological processes. The depletion of the soil organic matter could cause serious consequences. Thus, it is critical to maintain enough levels of organic matter in order to maximize the optimal conditions for sustainable agriculture.

Microbial inoculant is one of the major agricultural practices that has been used to acquire desirable characteristics in the soil. Soil ecological conditions for crop production could be optimized by maintaining suitable levels of organic carbon in the soil. This might be achieved by introducing environmentally friendly microbial formulations with carbon sequestering ability to the soil. Different types and species of soil microorganisms could affect carbon sequestration and storage differently in soil. These mechanisms range from different pathways and metabolic activities to the recalcitrant nature of their vegetative tissues and products.

Some bacterial species have been reported to contribute in carbon sequestration and atmospheric CO_2 mitigating. Growth-promoting bacteria have been shown to have potential for carbon sequestration, and they could be studied intensively for use as microbial inoculants for increasing carbon levels in soil. Bacteria with the ability to capture atmospheric CO_2 and/or the ability to sediment carbonates are also promising option as microbial inoculants for increasing the level of soil carbon and reducing atmospheric CO_2 levels.

Fungi also have been proved to sequester more carbon into the soil even than bacteria due to their ability to form soil aggregates, having recalcitrant biomass compounds and by-products. Fungi have three major factors that have been proved to play important roles in the ability of fungi to sequester more carbon in the soil, which are fungal mycelia (living or nonliving), production of glomalin, and the ability to form soil aggregates. Due to these characteristics, fungi participate more in carbon sequestration, and organic matter formation and stabilization. Thus, fungal microbes are good candidates to be used as microbial inoculants to increase organic carbon in soils.

This review highlights the potential of soil microorganisms to be used as microbial inoculants for carbon sequestration and the importance to select and formulate the proper microbial inoculants that are able to sequester carbon in the soil. By achieving that, agricultural production would be raised by the increase of soil organic carbon in soils, which will lead to improved soil quality as well as food security. Carbon sequestration has gained a lot of attention in the past few years due to its potential to mitigate atmospheric CO_2 and fight against climate changes. However, there are not enough studies focused on using microbial inoculants as a carbon sequestering technique to increase soil fertility and mitigate climate change. More efforts should be undertaken to use microbial inoculants as carbon sequestration strategy to mitigate climate change and enhance soil quality and fertility in order to achieve long-term food security and agricultural sustainability.

References

Amundson R (2001) The carbon budget in soils. Annu Rev Earth Planet Sci 29(1):535–562
Averill C et al (2014) Mycorrhiza-mediated competition between plants and decomposers drives soil carbon storage. Nature 505(7484):543–545
Babalola OO (2010) Beneficial bacteria of agricultural importance. Biotechnol Lett 32 (11):1559–1570
Babalola OO, Glick BR (2012) The use of microbial inoculants in African agriculture: current practice and future prospects. J Food Agric Environ 10(3&4):540–549
Baldock J et al (1989) Incorporation of uniformly labeled 13C glucose carbon into the organic fraction of a soil-carbon balance and CP MAS 13C NMR measurements. Soil Res 27 (4):725–746
Bashan Y et al (2014) Advances in plant growth-promoting bacterial inoculant technology: formulations and practical perspectives (1998–2013). Plant Soil 378(1–2):1–33
Berg B, McClaugherty C (2003) Plant litter: decomposition, humus formation, carbon sequestration. Springer, Berlin
Blagodatskaya E, Kuzyakov Y (2013) Active microorganisms in soil: critical review of estimation criteria and approaches. Soil Biol Biochem 67:192–211
Bradford MA et al (2013) Empirical evidence that soil carbon formation from plant inputs is positively related to microbial growth. Biogeochemistry 113(1–3):271–281
Chan Y (2008) Increasing soil organic carbon of agricultural land. NSW Dep Prim Ind 735:1–5
Chen PH et al (2012) Enhancing CO_2 bio-mitigation by genetic engineering of cyanobacteria. Energy Environ Sci 5(8):8318–8327
Drigo B et al (2012) Rapid incorporation of carbon from ectomycorrhizal mycelial necromass into soil fungal communities. Soil Biol Biochem 49:4–10
Ekblad A et al (2013) The production and turnover of extramatrical mycelium of ectomycorrhizal fungi in forest soils: role in carbon cycling. Plant Soil 366(1–2):1–27
Eni D et al (2011) Flood and it impact on farmlands in Itigidi, Abi local government area, Cross River state. Niger J Int Human Soc Sci 1(9):98–104
Fang H et al (2014) Changes in soil heterotrophic respiration, carbon availability, and microbial function in seven forests along a climate gradient. Ecol Res 29(6):1077–1086
Fernandez CW, Koide RT (2011) The role of chitin in the decomposition of ectomycorrhizal fungal litter. Ecology 93(1):24–28
Fernandez CW, Koide RT (2014) Initial melanin and nitrogen concentrations control the decomposition of ectomycorrhizal fungal litter. Soil Biol Biochem 77:150–157
Fischer H et al (2010) Microbial uptake of low-molecular-weight organic substances out-competes sorption in soil. Eur J Soil Sci 61(4):504–513
Frey SD et al (2013) The temperature response of soil microbial efficiency and its feedback to climate. Nat Clim Chang 3(4):395
Geyer KM et al (2016) Microbial carbon use efficiency: accounting for population, community, and ecosystem-scale controls over the fate of metabolized organic matter. Biogeochemistry 127 (2–3):173–188
Ghotbi-Ravandi AA et al (2014) Effects of mild and severe drought stress on photosynthetic efficiency in tolerant and susceptible barley (*Hordeum vulgare* L.) genotypes. J Agron Crop Sci 200(6):403–415
Grandy AS et al (2009) The influence of microbial communities, management, and soil texture on soil organic matter chemistry. Geoderma 150(3–4):278–286
Caporali F, Marinari S (2008) Soil carbon sequestration under organic farming in the Mediterranean environment. Transworld Research Network, pp 39–51
Guckert JB et al (1985) Phospholipid, ester-linked fatty acid profiles as reproducible assays for changes in prokaryotic community structure of estuarine sediments. FEMS Microbiol Lett 31(3):147–158
Han J et al (2013) Induction of calcium carbonate by *Bacillus cereus*. Geomicrobiol J 30(8):682–689

Hasanuzzaman M et al (2013) Physiological, biochemical, and molecular mechanisms of heat stress tolerance in plants. Int J Mol Sci 14(5):9643–9684
Hoorman JJ et al (2009) Using cover crops to convert to MD-till. Crops Soil 42:9–13
Ineson P et al (1995) Quantification of soil carbon inputs under elevated CO 2: C 3 plants in a C 4 soil. Plant Soil 187(2):345–350
Ipcc W (2007) Climate change 2007: the physical science basis contribution of working group I. fourth assessment report of the intergovernmental panel on climate change. Cambridge University Press, Cambridge
Jansen K et al (2014) Douglas-fir seedlings exhibit metabolic responses to increased temperature and atmospheric drought. PLoS One 9(12):1–21
Janzen HH (2004) Carbon cycling in earth systems – a soil science perspective. Agric Ecosyst Environ 104(3):399–417
Kadlec V et al (2012) Soil organic carbon dynamics and its influence on the soil erodibility factor. Soil Water Res 7(3):97–108
Kallenbach C et al (2015) Microbial physiology and necromass regulate agricultural soil carbon accumulation. Soil Biol Biochem 91:279–290
Koide RT et al (2014) Determining place and process: functional traits of ectomycorrhizal fungi that affect both community structure and ecosystem function. New Phytol 201(2):433–439
Komala T, Khun T (2014) Biological carbon dioxide sequestration potential of *Bacillus pumilus*. Sains Malaysiana 43(8):1149–1156
Lal R (2004) Soil carbon sequestration to mitigate climate change. Geoderma 123(1–2):1–22
Lehmann J, Kleber M (2015) The contentious nature of soil organic matter. Nature 528 (7580):60
Li N et al (2015) Fungi contribute more than bacteria to soil organic matter through necromass accumulation under different agricultural practices during the early pedogenesis of a Mollisol. Eur J Soil Biol 67:51–58
Luck J et al (2011) Climate change and diseases of food crops. Plant Pathol 60(1):113–121
Macreadie PI et al (2014) Quantifying and modelling the carbon sequestration capacity of seagrass meadows – a critical assessment. Mar Pollut Bull 83(2):430–439
Mardukhi B et al (2011) Arbuscular mycorrhizas enhance nutrient uptake in different wheat genotypes at high salinity levels under field and greenhouse conditions. C R Biol 334(7):564–571
Miransari M (2010) Contribution of arbuscular mycorrhizal symbiosis to plant growth under different types of soil stress. Plant Biol 12(4):563–569
Nie M et al (2015) Increased plant productivity and decreased microbial respiratory C loss by plant growth-promoting rhizobacteria under elevated CO_2. Sci Rep 5:1–6
Olson KR (2013) Soil organic carbon sequestration, storage, retention and loss in U.S. croplands: issues paper for protocol development. Geoderma 195–196:201–206
Ontl TA, Schulte LA (2012) Soil carbon storage. Nat Educ Knowl 3(10):35
Ortiz N et al (2015) Contribution of arbuscular mycorrhizal fungi and/or bacteria to enhancing plant drought tolerance under natural soil conditions: effectiveness of autochthonous or allochthonous strains. J Plant Physiol 174:87–96
Pal A, Pandey S (2014) Role of glomalin in improving soil fertility: a review. Int J Plant Soil Sci 3 (9):1112–1129
Paul E et al (2001) Dynamics of resistant soil carbon of Midwestern agricultural soils measured by naturally occurring 14C abundance. Geoderma 104(3–4):239–256
Peng Y et al (2010) Progress in microbial and enzyme immobilization of carbon dioxide. Chem Bioeng 27(7):10–13
Peng S et al (2013) The effects of arbuscular mycorrhizal hyphal networks on soil aggregations of purple soil in Southwest China. Soil Biol Biochem 57:411–417
Plante AF et al (2011) Biological, chemical and thermal indices of soil organic matter stability in four grassland soils. Soil Biol Biochem 43(5):1051–1058
Powlson DS et al (2011) Soil carbon sequestration to mitigate climate change: a critical re-examination to identify the true and the false. Eur J Soil Sci 62(1):42–55

Qafoku NP (2015) Climate change effects on soils: accelerated weathering, soil carbon, and elemental cycling. Academic, Advan Agron

Randers J (2012) 2052: a global forecast for the next forty years. Chelsea Green Publishing, Burlington

Rawlins BG et al (2013) Application of a novel method for soil aggregate stability measurement by laser granulometry with sonication. Eur J Soil Sci 64(1):92–103

Saiya-Cork K et al (2002) The effects of long term nitrogen deposition on extracellular enzyme activity in an Acer saccharum forest soil. Soil Biol Biochem 34(9):1309–1315

Sangakkara U et al (2014) Soil quality and crop yields as affected by microbial inoculants in nature farming. Build Org Bridges 3:987–990

Selvakumar G et al (2012) Response of biofertilizers on growth, yield attributes and associated protein profiling changes of blackgram (*Vigna mungo* L. Hepper). World Appl Sci J 16 (10):1368–1374

Sessitsch A et al (2013) The role of plant-associated bacteria in the mobilization and phytoextraction of trace elements in contaminated soils. Soil Biol Biochem 60:182–194

Shindell D et al (2012) Simultaneously mitigating near-term climate change and improving human health and food security. Science 335(6065):183–189

Singh PK (2012) Role of glomalin related soil protein produced by arbuscular mycorrhizal fungi: a review. Agric Sci Res J 2(3):119–125

Singh P et al (2013) Glomalin: an arbuscular mycorrhizal fungal soil protein. Protoplasma 250 (3):663–669

Six J et al (2006) Bacterial and fungal contributions to carbon sequestration in agroecosystems. Soil Sci Soc Am J 70(2):555–569

Smith AP et al (2014) Microbial community structure varies across soil organic matter aggregate pools during tropical land cover change. Soil Biol Biochem 77:292–303

Stockmann U et al (2013) The knowns, known unknowns and unknowns of sequestration of soil organic carbon. Agric Ecosyst Environ 164:80–99

Stocks-Fischer S et al (1999) Microbiological precipitation of $CaCO_3$. Soil Biol Biochem 31 (11):1563–1571

Tan WF et al (2014) Soil inorganic carbon stock under different soil types and land uses on the loess plateau region of China. Catena 121:22–30

Tavasolee A et al (2011) Interactive effects of arbuscular mycorrhizal fungi and rhizobial strains on chickpea growth and nutrient content in plant. Afr J Biotechnol 10(39):7585–7591

Teskey R et al (2014) Responses of tree species to heat waves and extreme heat events. Plant Cell Environ. https://doi.org/10.1111/pce.12417

Trenberth KE (2011) Changes in precipitation with climate change. Clim Res 47(1):123–138

Upadhyay SK et al (2011) Exopolysaccharide-producing plant growth-promoting rhizobacteria under salinity condition. Pedosphere 21(2):214–222

Waksman SA (1936) Humus: Origin, chemical composition, and importance in nature. 1st edn, Baillliere, Tindall & Cox, Covent Garden, London

Wallander H et al (2013) Evaluation of methods to estimate production, biomass and turnover of ectomycorrhizal mycelium in forests soils – a review. Soil Biol Biochem 57:1034–1047

Wang Q et al (2010) Cropping systems to improve carbon sequestration for mitigation of climate change. J Environ Prot 1(3):207–215

Wani P, Khan M (2013) Nickel detoxification and plant growth promotion by multi metal resistant plant growth promoting rhizobium species RL9. Bull Environ Contam Toxicol 91(1):117–124

West JS et al (2015) Impact of climate change on diseases in sustainable arable crop systems: CLIMDIS. Home-Grown Cereals Authority, London, pp 1–67

Wright SF, Upadhyaya A (1999) Quantification of arbuscular mycorrhizal fungi activity by the glomalin concentration on hyphal traps. Mycorrhiza 8(5):283–285

Wu QS et al (2014a) Direct and indirect effects of glomalin, mycorrhizal hyphae, and roots on aggregate stability in rhizosphere of trifoliate orange. Sci Rep 4(5823):1–8

Wu QS et al (2014b) Contribution of arbuscular mycorrhizas to glomalin-related soil protein, soil organic carbon and aggregate stability in citrus rhizosphere. Int J Agric Biol 16:207–2012

Xiao L et al (2015) Effect of carbonic anhydrase on silicate weathering and carbonate formation at present day CO_2 concentrations compared to primordial valves. Scientific reports 5:7733

Zhang Z et al (2011) *Bacillus mucilaginosus* can capture atmospheric CO_2 by carbonic anhydrase. Afr J Microbiol Res 5:106–112

Area Exclosure as a Strategy for Climate Change Mitigation: Case Study from Tigray Region, Northern Ethiopia

19

Samson Shimelse, Tamrat Bekele, and Sileshi Nemomissa

Contents

Abstract

Forests provide multiple benefits for communities. One of these is global climate change combating with carbon sequestration. Exclosures are a particularly important class of restoration and vegetation generating these benefits. Institutional arrangements to govern these commons are believed to substantially influence carbon storage and livelihood contributions, especially when they incorporate local knowledge and participatory decision-making. However, hypothesized relationships between institutional factors and benefits have never been tested on data

S. Shimelse (✉)
Department of Plant Biology and Biodiversity Management, College of Natural and Computational Sciences, Addis Ababa University, Addis Ababa, Ethiopia

College of Dryland Agriculture and Natural Resources, Mekelle University, Mekelle, Ethiopia
e-mail: samshimelse@yahoo.com

T. Bekele · S. Nemomissa
Department of Plant Biology and Biodiversity Management, College of Natural and Computational Sciences, Addis Ababa University, Addis Ababa, Ethiopia
e-mail: tambek07@gmail.com; snemomissa@gmail.com

W. Leal Filho (ed.), *Handbook of Climate Change Resilience*,
https://doi.org/10.1007/978-3-319-93336-8_123

from exclosures found in Tigray. A total of 120 quadrants were sampled using a stratified preferential sampling design technique with flexible systematic model and 180 household respondents were surveyed with multistage stratified random sampling procedure. The differences in carbon stocks and vegetation composition between an exclosure age and free grazing land were assessed using a paired t-test. All exclosures displayed higher ECS (ecosystem carbon stock), and plant species richness, diversity, and aboveground standing biomass than the free grazing lands. Differences in ECS between exclosures and free grazing lands varied between 32.96 and 61.0 t ha^{-1} and increased with exclosure age. Over a period of 20 years, the carbon dioxide sequestered in the investigated exclosures was 223.88 t ha^{-1}, total soil nitrogen increased by 2.93 t ha^{-1} and additionally available phosphorus stocks amounted to 40 Kg ha^{-1}. Carbon is especially sensitive to population pressure, with an increase in population per hectare of 100 people associated with reduced carbon of 96% and 89% carbon per hectare and carbon per household, respectively. More remote areas also tend to have more carbon, with an increase in distance to the town of 1 kilometer associated with a 6–7% increase in carbon. Similarly, the findings on the social and cultural influences will affect exclosure management for other ecological values, such as habitat conservation and water quality, as well as their likelihood to respond to policies that promote these values. Thus, the findings are relevant for a broader range of natural resource issues than restricting only on carbon- and climate change-related issues.

Keywords
Exclosure · Carbon sequestration · Social variables · Tigray (Ethiopia)

Introduction

For exclosures to continue playing their environmental conservation role, socioeconomic needs of local people are very important. In Ethiopia, Biyo Kelala and Tiya exclosures, land productivity is increased due to exclosures (Tefera et al. 2005).

Natural resources, such as forests, are among the primary sources of livelihoods of poor people by providing food, fodder, and fuelwood. In the northern highlands of Ethiopia, forest resources are major sources of livelihoods and comprise 27% of the total household income (Bedru et al. 2009). Similarly, 34% of the household per capita income in Bale mountains, southern Ethiopia (Yemiru et al. 2011), and 39% of the average household income in Dendi district, Ethiopia (Mamo et al. 2007), is generated from forest resources. However, the forest resources in Ethiopia and many developing countries have been deteriorating over time (Sunderlin et al. 2005; Bedru et al. 2009). Deforestation and overexploitation of forests in Ethiopia resulted in reductions in forest and food products and aggravated poverty and malnutrition (Stocking and Murnaghan 2001; Mulugeta et al. 2005). For example, Reusing (1998) estimated the rate of deforestation of 163,600 ha yr^{-1} in the highlands of Ethiopia. Moreover, owing to overgrazing, the natural vegetation in the northern highlands of Ethiopia has virtually disappeared, leaving degraded free grazing lands devoid of vegetation (Betru et al. 2005). For that reason, efforts to rehabilitate degraded natural resources could become a potential strategy for livelihood improvement (Taddese 2001; Kebrom 2001; Shylendra 2002).

In response to the degradation of land resources, "*Baito*" (a system of political leadership and administration at the "*Tabia*" or *Wereda* level and development agents of the Bureau of Agriculture and Rural Development (BoARD) in collaboration with the communities in Tigray, Northern Ethiopia) established exclosures on degraded free grazing lands to combat forest degradation (Wolde et al. 2010).

The negative experiences in managing exclosures using top-down approaches in the "*Derg*" regime and the encouragement of TPLF towards empowering local governance contributed to the more participatory approaches in managing exclosures with locally devised rules. However, Segers et al. (2008) pointed out that the mobilization of farmers in central Tigray tends towards convincing farmers to implement the development programs of the government in central Tigray and was not participatory.

There is an ongoing debate in the literature of communal resources management regarding the more relevant institutional arrangements for improving rural livelihoods using the available communal resources (Ostrom 1990). The debate encompasses views that romanticize the effectiveness of village rules and considering village rules as solutions to most problems in managing communal resources at low transaction costs. On the contrary, there are views that undermine the effectiveness of village rules in facilitating users to have common goals towards managing communal resources in a sustainable manner (Campbell et al. 2001; Makepe 2006). Neither of these extreme views could contribute to solving the challenges in the management of exclosures on its own under the rural context of Tigray, Ethiopia. There is a general consensus among researchers on the need of effective village rules such as taboos to manage forest resources and to enhance socio-ecological benefits (Chisholm 1998; Betru et al. 2005). This need becomes so critical because of the built-up weakness of village rules in Tigray owing to a long history of war, political uncertainty, and government interventions (Yohannes and Waters-Bayer 2007).

Exclosures were also implemented as part of this large-scale land rehabilitation works and contribute to the improved forest resources. We suppose that such positive outcomes could result from the village rules that are used in managing the exclosures. However, studies are lacking on the relevant institutional arrangements that ensure the sustainable management of exclosures. This lack of evidence exists despite the fact that more than three decades have passed since the establishment of most of the exclosures in Tigray. Consequently, in-depth analysis of the effectiveness of village rules in managing exclosures becomes crucial for reinforcing sustainability outcomes such as preventing forest degradation and achieving equal benefit sharing among users. We considered forest degradation, conflicts among users over natural resource use, and meeting high expectations of users to realize economic benefits from exclosures as important challenges in the management of exclosures and evaluated the effectiveness of village rules by examining their roles in addressing those challenges. The following research attempts to analyze the effects of the conversion of free grazing lands to exclosures on ecosystem carbon sequestration, to investigate on exclosures age effect on ecosystem carbon stocks and maximal ecosystem services, to identify which factor controls ecosystem carbon stock restoration, to investigate how exclosure age affects restoration of soil nutrient stocks and properties, to assess the determinants that affect the sustainability of exclosure establishment and management practices at the household level, and to study the collective actions relation with carbon stock in exclosures of Tigray.

Analytical Framework, Data Collection, and Analysis

The analytical framework used draws heavily on knowledge and supposes that communities with more social capital are better able to develop exclosure management systems, leading to higher quality natural resources. We do not directly observe social capitals but instead observed better management and can effectively interpret it, because a large literature has analyzed the components of better community management. Therefore, expect better managed and aged exclosures to sequester more carbon.

Reverse causality is always a potential problem and could not be confidential that good instruments for exclosure management but reverse causality viewed as unlikely, however, because communities have little reason to invest in better social organization when vegetation are abundant (Hyde et al. 1996); if there is reverse causality, we therefore expect to find exactly the opposite of our hypothesis that better collective action is associated with more vegetation carbon.

Finally, though we cannot exclude all possibilities, it is unlikely that a common exogenous variable affects both restoration collective action and carbon stocks. For example, more effective community natural resource management and the social capital that we and the literature believe underpins it are particularly difficult to create from the outside and are instead much more likely to emerge from complicated processes (Ostrom 2009; Agrawal 2007) that have very little to do with larger carbon stocks.

Exclosures of 10, 15, and 20 years old were selected, and each exclosure was paired with an adjacent free grazing land. A total of 120 quadrants were sampled using a stratified preferential sampling design technique with flexible systematic model and 180 household respondents were surveyed with multistage stratified random sampling procedure incorporating household's survey questionnaires, focus group discussions, and key informants discussions were used to interpret the whole results in general. Soil, vegetation, and exclosure management-related data were collected from November 2014 to October 2016. The differences in carbon stocks and vegetation composition between an exclosure age and free grazing land were assessed using a paired t-test. Correlation tests and ordination analyses were also conducted to establish predictive relationships between carbon stocks, vegetation composition, and easily measurable biophysical- and management-related variables. Data analysis also included descriptive statistics, inferential statistical processed through one way ANOVA, t-test, and chi-square test, and binary logistic regression model have been used.

The data used for this study were collected in Tigray region form three *woredas* which are Raya Azebo *Woreda* (Genete, Hawelti, Mehoni *Kebeles*), Atsbi Womberta *Woreda* (Hadinet, Kelesha Emene, Haikmeshal *Kebeles*), and Kilte Awelaelo *Woreda* (Kihen, Negash, Abreha We Atsbeha *Kebeles*) located in Southern and Eastern parts of Tigray region. The *Kebeles* demonstrate degraded vegetation and soils and differ in elevation, rainfall, agro-ecology, access to the main road, distance to regional capital, population distribution and density, and lithology.

Three types of data are used in this study. The first data are from carbon stock estimation study by overall carbon stocks (carbon density) of a given forest obtained from adding all the carbon stocks in different carbon pools together were extrapolated into a hectare. The dry biomass obtained by using allometric equations can be converted into carbon except for soil, which usually measures carbon directly. Second, community characteristics were collected from secondary data sources and *kebele* officials. Third, a total of 180 households were systematically selected from the list of local community members. A structured questionnaire was provided to these households, ten in-depth interviews with key informants and nine focus group discussions were carried out.

In this study, we are interested mainly in differences across *kebeles* rather than carbon levels of the areas outside exclosures. The sample areas come from different altitudes and sizes of exclosures; they represent a diversity of species, size, social, and ecological systems. Therefore, normalize the carbon estimates for population (carbon per household) and vegetation area (exclosure) (carbon per hectare) and, as already noted, take explicit account of species diversity in our biomass calculations.

We use data on exclosure management collected from household heads rather than village administrators. We take this approach because, in developing countries, on-the-ground realities often correspond poorly with policies. This could be for a number of reasons, including leader mis-assessments, attempts to portray local institutions in positive lights, or simple difficulties characterizing social coordination details. For example, we know from the literature on community forests design elements that fairness, clarity of access rules, and appropriate social sanctions are key aspects of successful social coordination (Ostrom 2009; Agrawal et al. 2008). These are subjective and restrained aspects that households are likely to perceive much more accurately than leaders. These indices are continuous, because forest sector coordination in Ethiopia and much of the lower-income developing world is continuous and evolved locally in response to circumstance (Agrawal et al. 2008; Jodha 2008). We take the approach of Glaeser et al. (2002) and treat social capital as an asset that is held at the individual rather than the community level, as suggested by Bowles and Gintis (2002).

Many CF management variables are highly collinear, making it impossible to use all 15 variables in models. We therefore aggregate the information into a higher level index using factor analysis. Based on the results of the factor loadings, we derive one factor with an eigenvalue greater than one. All constituent variables therefore load on a single factor. This single factor explains 85.2% of the variation, indicating that households appear to treat exclosure management as a package with several components.

Social coordination is at the household level. So is distance from households to nearest roads, a critical measure of transport costs and market access and a potential confounder. These features imply that our analysis must also be at the household level. Households within *kebeles* are assumed to have equal access to carbon per household and carbon per hectare in *kebele* forests.

We believe this approach is reasonable, because exclosures are held in common and, as discussed in literature dating back to Gordon (1954), in such situations, average stocks are the correct metric. Households extract forest products from these exclosures subject to management bylaws.

Carbon stored in exclosures is likely affected not only by institutional elements that we hypothesize are important but also community and ecological variables (Chaiyo et al. 2011). We therefore follow the lead of Chhatre and Agrawal (2009) and include several other variables. Physical isolation, because of increased transport costs and reduced market opportunities, is expected to be associated with less extraction of biomass by villagers and therefore more carbon sequestration.

This is consistent with the von Thunen model of deforestation (Hyde et al. 1996). It is also possible that more isolated *kebeles* have systematically different levels of social coordination than more proximate ones, making isolation a potential confounding factor. Because distance to roads can vary widely within *kebeles*, the average value is not terribly meaningful. As already noted, this variable is therefore measured at the household level.

Forest carbon is affected by forest altitude and the type of ecosystem because of variations in temperature, rainfall, and resulting densities and species composition. We therefore include average *kebele* altitude.

Population pressures because of the need to access direct use values important for subsistence (e.g., fuelwood and grass collection) rather than other forest values should be associated with lower carbon stocks. It is also a potential confounding factor if, for example, more competition for natural resources makes social coordination more difficult. More people per hectare of exclosure, which is the variable we measure, implies more population pressure on exclosures, and therefore expect smaller carbon stocks. We also include total area of exclosures, because, all else equal, larger exclosures have more carbon.

Of course, unobserved factors will also influence carbon sequestration in *kebele* vegetation. A key such factor is average *kebele* RF. Such data are not available, because outside some *woreda* towns or some cities, there are very few meteorological stations. To capture this and other unobservables, considered as fixed effects but these dummy variables also are likely to pick up unobserved social variables but not considered in our analysis.

Results and Discussion

Changes in Aboveground Biomass, Species Richness, and Diversity in Relation to Exclosure Establishment

Exclosures have been effective in restoring plant species composition, biomass, and cover of herbaceous species and the exclosure areas were in a better condition than the free grazing areas (Table 1). Similarly, Van Der Westhuizen et al. (2001) argued

Table 1 Plant species richness, diversity, aboveground standing biomass, and herbaceous, shrub, and tree species in exclosures and free grazing lands

		Richness	H	Evenness	Aboveground biomass (t ha^{-1})	Herb + C. Sp. in no.	S. sp. in no.	T. sp. in no.	Eco. Imp. S + T
	Free grazing	76	2.79	0.87	1.62	51	14	11	16
Exclosure	10 year	87	3.52	0.81	4.73	42	27	18	29
	15 year	91	3.68	0.82	9.72	39	29	23	34
	20 year	103	3.82	0.84	13.93	44	31	28	41
P value		0.001	0.150	0.093	0.129	0.014	0.001	0.001	0.001
Sig. codes		***		.		*	***	***	***

The *p*-values are values obtained after paired t-test @ 95% CI or 0.05 significance level
H diversity, *S* shrub, *T* tree, *C* climber, *Eco.Imp* economically important
Sig. codes: 0 '***' 0.001 '**' 0.01 '*' 0.05 '.' 0.1 ' ' 1

that in arid and semiarid rangelands, these parameters are greatly influenced by the effects of gazing pressure, drought, and rainfall variability. Repeated grazing and prolonged drought might lead to a reduction in herbaceous species composition and diversity, which might accelerate decline in rangeland condition. As reported by Angassa (2014), heavy grazing pressure may reduce plant species composition and basal cover. On the other hand, the highest scores for biomass were recorded in aged exclosed sites reflecting the benefits of reduced disturbance such as the effects of heavy grazing, trampling, and inappropriate management interventions (Ahmed 2006). The biomass production in the exclosure areas was better than the free grazing areas, and this might be due to better management practices in the exclosure areas, but the free grazing areas have been deteriorated through continuous overgrazing and mismanagement system of the community (Ahmed 2006; Ibrahim 2016).

Herbaceous species and woody species diversity was higher in the exclosure than the adjacent free grazing area. The higher diversity measured in the exclosure might be explained by increased litter accumulation, improved soil organic matter, and other nutrients inside the exclosures that eventually lead to increased species richness. Exclosures improved vegetation status with respect to species richness, cover, and height, and community structure.

Similarly, Amsalu (2000) suggested that heavy grazing might cause reduction of plant species composition and diversity over time. Angassa and Oba (2010) show that herbaceous biomass and diversity index and evenness of herbaceous species was significantly greater in exclosures than in the free grazed areas.

Woody species appearance in exclosures indicates a long period of protection, allowing regeneration of shrubs and trees (Kebrom 2001; Tefera et al. 2005). The vegetation composed by the woody species and the ground cover of herbaceous species was denser in the exclosure than in the open area. The difference in density was significantly greater for the herbaceous than for the woody species. This indicates that the disturbance in the free grazing area was mainly due to the high grazing intensity throughout the year, and similar result was also reported by Emiru (Emiru et al. 2007). The density of woody vegetation was significantly affected by grazing types, i.e., a higher density and canopy cover of woody species was observed in exclosure site than in the free grazing sites (Augustine and McNaughton 2004; Teshome et al. 2009). Protection influenced the height class distribution of the three woody vegetation attributes, i.e., stem height, canopy height, and canopy cover. Average stem height was higher in the exclosures than in the open grazed areas (Butler and Kielland 2008; Yayneshet 2011).

Ecosystem Carbon Stock Change in Relation to Exclosure Establishment

All exclosures displayed higher ECS, and plant species richness, diversity, and aboveground standing biomass than the free grazing lands. Differences in ECS between exclosures and free grazing lands varied between 32.96 and 61.0 t $ha^{-1}$and increased with exclosure age. Over a period of 20 years, the carbon dioxide sequestered in the investigated exclosures was 223.88 t ha^{-1}. Differences in plant species richness and aboveground standing biomass between exclosures and free grazing lands also increased with exclosure age. The results of this chapter confirm that establishment of exclosures on degraded free grazing lands in the Northern Highlands of Ethiopian is a viable option to restore ECS. The positive correlation of soil C with vegetation canopy cover suggests that restoring the vegetation canopy cover is vital to restoring soil C stocks. An increase in vegetation canopy cover would not only increase litter input into the soil but would also reduce the erosive impact of raindrops, resulting in less soil erosion and consequently less soil C loss (Wolde et al. 2009). A similar mechanism was reported from a study of four different land uses in two ecological regions of Ethiopia, showing that soil organic matter concentration was influenced by the amount and type of vegetation and grasses as well as the intensity of grazing (Yayneshet 2011).

Restoration of badly degraded rangelands soils and ecosystems that were formerly under cultivation, mining, depletion of woody vegetation and restoring soil stability with permanent vegetation is a high potential to sequestrating soil carbon (McDermot and Elavarthi 2014). Exclosures showed higher soil C concentrations, soil C stocks, and aboveground C stocks than the adjacent grazing lands. Exclosures showed higher soil carbon concentration, soil carbon stocks, and aboveground carbon stocks than the adjacent free grazing lands (Table 2). The difference between exclosures and free grazing lands in soil carbon stock, at each exclosure age, ranged from 30.6 to 51.5 t C ha^{-1} whereas in aboveground carbon stock varied between 2.4 and 9.5 t C ha^{-1}. These values translated to soil carbon stock increase of 22.1 to 40.6% and to aboveground carbon stock increase of 40.7 to 74.7% following conversion of degraded grazing lands to exclosures within 10–20 years. Soil carbon stock contributed most of the increase in ECS compare to aboveground carbon in exclosures (Table 2). Woody biomass carbon consistently increased with exclosure age while grass biomass carbon slightly decreased (Table 2). The influence of exclosure age on ECS was linear with time as the increase in exclosure age and it was significant between free grazing and exclosures as well within each age variable exclosures. The minimum and maximum carbon stock was 32.96 and 61.0 t ha^{-1}, in free grazing and 20 years exclosure, respectively (Table 2).

Similarly, Soil C stock increases from 41% to 60% and the aboveground C stock increases from 83% to 87% following conversion of degraded grazing lands to exclosures. Soil C stocks contributed most of the increase in ECS (range 83–90%) compared with aboveground C in exclosures (Wolde et al. 2011). Booker et al. (2013) reported that rangeland management will achieve very little for carbon sequestration. Climate change greatly impacts the soil structure and vegetation on agro-ecosystems, particularly at the arid and semiarid regions where soil water is limited.

The mean carbon stock of the study area was 42.34 ± 9.38 t ha^{-1}. The minimum and maximum equivalent carbon dioxide of the study area was 120.97 and 223.88 t/ha, and the mean is 155.39 ± 34.42 t ha^{-1}. The mean aboveground equivalent carbon dioxide for exclosures was 189.81 t ha^{-1} (Table 2).

Carbon sequestration on rangelands is environmentally friendly, inexpensive, and natural process of mitigating climate change when compared to other methods,

Table 2 Mean biomass and carbon stock of different carbon pools and CO_2 equivalent of free grazing and different ages of exclosures

		Woody biomass			Herbaceous biomass				
	Soil C t ha^{-1}	AGB (t ha^{-1})	BGB (t ha^{-1})	Total t ha^{-1}	(t C ha^{-1})	t ha^{-1}	t C ha^{-1}	Total t C ha^{-1}	CO_2 equivalent t ha^{-1}
Free grazing	30.6	1.62	0.32	1.94	0.97	2.78	1.39		
10 years	39.28	4.73	0.95	5.68	2.84	2.41	1.21		
15 years	43.91	9.72	1.94	11.66	5.83	2.14	1.07	50.81	186.48
20 years	51.51	13.93	2.79	16.72	8.36	2.27	1.14	61.0	223.88
Exclosures average	44.90	9.46	1.89	11.35	5.68	2.27	1.14	51.72	189.81
Mean	37.75	5.54	1.11	6.65	3.33	2.53	1.27	42.34	155.39
Standard error	7.15	3.92	0.79	4.71	2.36	0.26	0.13	9.38	4.42
P	0.003	0.07	0.07	0.07	0.07	0.001	0.001	0.004	0.004
Sig. codes	**	.	.	.	.	***	***	**	**

AGB aboveground biomass, *BGB* belowground biomass, *t C* ton carbon, *t CO_2* ton carbon dioxide
Sig. codes: 0 '***' 0.001 '**' 0.01 '*' 0.05 '.' 0.1 ' ' 1

unlike others, the co-benefits of increasing the SOC pool includes improving biodiversity, enhance agronomic productivity, reduce erosional losses, restore degraded lands, improve the quantity and quality of water resources, and advance global food security (Lal 2011). The literature explains grazing to be one of the management practices that aid in the physical break down, increase rate of residual plant material decomposition and soil incorporation, hence, restoration of badly degraded rangeland soils (Derner and Schuman 2007). Several other studies have shown that grazing intensity and frequency may impact greatly on carbon storage in some areas on rangelands. Schuman et al. (2005) reported that severe drought and heavy grazing can close significant deficits in SOC that was previously stored and also shift the microbial community resulting in further losses in soil C.

In Tunisia, Jeddi and Chaieb (2010) documented that 12-year exclosures enhance the total plant cover, dry matter yield, species richness, and contents of soil organic matter, total nitrogen, and water infiltration rate compared with continually grazed area. Analysis of 115 pastures and other grazing-land studies worldwide by Conant et al. (2001) indicates that soil C levels increase with improved management (i.e., fertilization, grazing management, and conversion from cultivation or native vegetation) and that the greatest C sequestration occurs during the first 40 year of implementation of the management practice. Similarly, Cheng et al. (2011) indicated that 20 year exclusion of livestock grazing significantly increased aboveground and belowground biomass and species richness. Similar trends were also reported from case studies conducted on exclosures in the central and northern highlands of Ethiopia: exclosures had twice the plant species richness and diversity value compared with free grazing lands after 22 years of exclosure establishment (Tefera et al. 2005), an increase in woody species richness of 13 after 8 years of exclosure establishment (Emiru et al. 2007), and an increase in soil organic matter, total N, and available P after 10 years of exclosure establishment (Wolde et al. 2009). Fifteen to 25 years exclosure accumulated relatively more herbaceous biomass than <15 years and >25 years exclosures (Angassa and Oba 2010).

Gebrewahd (2014) also reported that the 5 years enclosed land had significant lower mean of grass species composition compared to the 10 years enclosed land and pointed out that the age had a significant effect on the increment of grass species composition and diversity. The herbaceous species richness declined with increase in age of the exclosures (i.e., <15 years exclosures had more species richness than exclosures of >25 years old (Angassa and Oba 2010).

Consequences for Baseline Studies and Carbon Sequestration in Exclosure Projects

To implement carbon sequestration projects using exclosures, it is necessary to have reliable baseline information which is typically lacking in countries like Ethiopia. In this chapter, we used degraded free grazing lands as the baseline for exclosure projects, and we were able to show that it is possible to make reliable estimates of carbon stocks in degraded free grazing lands using relatively simple methods. This is the first time that this is shown for a semiarid tropical region. Earlier baseline studies that used comparable methods as our present study were restricted to the humid tropics (Powers and Schlesinger 2002; De Koning et al. 2003). We also showed that it is possible to make quantitative predictions of ecosystem carbon increments should exclosures be implemented on degraded free grazing lands. Such predictions are critical to analyzing whether exclosures are also an economically feasible alternative. In the case of Northern Ethiopia, we could make the following prediction: taking an average increment of 28.04 t C ha^{-1} (61.0–32.96 t C ha^{-1}) in exclosures compared to the adjacent grazing lands in 20 years time, the establishment of exclosures could increase ECS by 1.4 t C ha^{-1} $yr.^{-1}$. Using this annual increase in ECS, conversion of only 10% (107,054 ha) of the 1.1 million ha of degraded free grazing lands in the region (Wolde et al. 2010) could sequester approximately 149,876 t C $yr.^{-1}$. If this converted into carbon market, the income is significantly changing the lives of the community. International carbon markets deal in carbon offsets, each offset being equal to one ton of CO_2 sequestered by a forestry project. Carbon offsets are sold to international buyers at a higher price. After deducting from its overheads and commissions payable to brokers, the project pays farmers at the rate of 4.5 USD per t CO_2 in 2008 in the USA (Capoor and Amborsi 2008). For East Africa, after estimation, actual transaction costs 3.64 USD per t CO_2 was recommended by Timothy et al. (2013). Based on this calculation, the price of 149,876 t C $yr.^{-1}$ is 550,045 t CO_2 equivalent, and this will be 2,002,164 USD or 54,058,415 ETB (with the current exchange rate 1 USD ≈ 27 ETB) per annum.

Establishment of exclosures on degraded free grazing lands has the potential to be considered in programs like Reducing Emissions from Degradation and Deforestation (REDD) which aim to generate financial compensation for increasing ECS and enhance sustainable development in developing countries. A crucial condition is, however, that this service is actually rewarded in order to provide sufficient incentives for the establishment of exclosures in the future.

Exclosures in Improving Soil Condition

Free grazing land had a high degree of erosion, due to reduction of vegetation cover by overgrazing and cutting of fuel wood. The rate of soil erosion can be accelerated when grazing intensively exceeds the threshold and plant cover is reduced below critical level in the open grazing land (Gebrewahd 2014). This might be due to the reduced animal and human interference in the exclosures, the revegetating plants reduced raindrop intensity and hindered detachment of soil particles that cause soil erosion. Soil was more eroded and compacted in the free grazing sites than in exclosure area (Teshome et al. 2009). The establishment of exclosures on degraded free grazing is effective in improving soil nutrient content and properties. There was a significant difference ($P < 0.05$) for stoniness cover, grazing, EC, soil organic carbon, and N among the differently aged exclosures and free grazing lands. However, sand, silt clay, slope, CEC, pH, BD, K, P, Mg, and Ca were not significant.

There was a significant difference ($P < 0.05$) for stone cover, EC, soil organic carbon, grazing, and N within groups or with exclosures and free grazing lands. However, there is no any significant difference ($P > 0.05$) within groups or between plots of similarly aged exclosures or free grazing areas (Table 3).

Change in SOC content was closely related to soil texture change under long-term exclosure (Slimani et al. 2010). Zhou et al. (2011) suggested that very long-term exclosure could have a significant impact on soil C, N, and P in exclosures areas. Wolde et al. (2011) pointed out that soil organic matter, total soil nitrogen, available phosphorus, and CEC in exclosures were positively and significantly correlated with woody biomass, vegetation canopy cover, silt content, stone cover, and exclosure age and were inversely and significantly correlated with bulk density and sand content. Mureithi et al. (2014) highlighted that the bulk density of topsoil significantly higher in the open rangeland than in the exclosures. Similar results were reported by Descheemaeker et al. (2006). The average topsoil moisture content was 21–22% (w/w) in exclosure systems and 16% in the open rangeland (Mureithi et al. 2014).

The removal of organic matter from the exclosures by grazing and grass harvesting activities has the greatest influence on soil restoration. Since most of the tropical rangelands are on degraded soils, the potential for sequestering C through rangeland rehabilitation is substantial (FAO 2001). Mureithi et al. (2014) reported that the average N contents were 0.41 mg N/g soil in the open rangeland, increasing to 0.75 mg N/g in the exclosures.

Significant increases in the soil organic carbon, total nitrogen, and microbial biomass contents and stocks were found in the exclosures as compared with the degraded open grazing land. The significant increase of absolute stocks of carbon, nitrogen, and microbial biomass compared to the degraded open rangeland indicates the potential for the restoration of soil quality through range rehabilitation (Mureithi et al. 2014).

Carbon Stock Restoration and Social Attributing Variables

Determinants of farmers' sustainability of exclosure establishment and management practices were assessed. The empirical results from binary logistic regression model showed age, tenure, and off-farm activity were positive and significant predictors of sustainability of exclosure establishment and management practices, while household size, farm size, distance, and number of livestock have a negative effect on exclosure establishment and management practices in the study area and they were not significant except livestock number. These findings strengthen the fact that in order to achieve sustainable exclosure establishment and management practices, institutional and economic factors should be given special attention.

The livelihood implications of exclosures have been a topic for a number of recent case studies across East Africa and indicate negative social impacts in terms of erosion of traditional, collective property rights, emergence of conflicting interests over land and water resources, and a gradual transfer of assets from poor to wealthier households (Angassa and Oba 2010; Beyene 2010; Napier and Desta 2011).

Table 3 Mean values of selected physical and chemical soil properties and site characteristics in exclosures and free grazing lands

	Stone cover (%)	pH (1:5 H_2O)	Bd (gm cm^{-3})	Sand (%)	Silt (%)	Clay (%)	EC	CEC (Cmol kg^{-1} soil)	Exc Na	Exc K (Cmol kg^{-1})	C %	C (t ha^{-1})	N (%)	Total N (t ha^{-1})	P (PPM)	Available P (t ha^{-1})	Mg (Cmol kg^{-1})	Ca (Cmol kg^{-1})
Free G.	60.53	7.33	1.67	41.40	28.73	29.87	0.20	31.61	0.73	4.07	1.04	30.60	0.24	7.38	27.58	0.08	7.81	51.29
10 years	36.87	7.23	1.57	39.13	29.03	31.84	0.24	33.94	0.80	4.56	2.05	39.28	0.25	7.61	28.12	0.10	7.89	54.51
15 years	31.06	7.27	1.52	40.00	29.90	30.10	0.25	37.47	0.96	4.18	2.47	43.91	0.26	9.02	30.24	0.10	7.97	56.91
20 years	28.14	7.12	1.35	41.63	26.00	32.37	0.27	37.52	0.82	3.98	3.56	51.51	0.28	10.31	33.65	0.12	8.61	56.99
Mean	39.15	7.24	1.53	40.29	28.67	31.04	0.23	35.14	0.83	4.20	2.28	41.33	0.26	8.58	29.90	0.10	8.07	54.93
SE	7.35	0.04	0.07	0.64	0.61	0.56	0.02	1.44	0.05	0.13	0.52	4.37	0.01	0.68	1.38	0.01	0.18	1.34
t value	5.32	163.64	22.84	63.07	46.84	55.71	9.65	24.35	17.17	32.90	4.37	9.45	30.16	12.60	21.72	11.66	44.11	40.95
P	0.01	0.07	0.68	0.59	0.86	0.84	0.04	0.06	0.52	0.84	0.01	0.01	0.01	0.01	0.09	0.09	0.15	0.08
Sig.	**						*				**	**	**	**				

pH power of hydrogen, *Bd* bulk density, *EC* electronic conductivity, *CEC* cation exchange capacity, *Exc Na* exchangeable sodium, *K* potassium, *C* carbon, *N* nitrogen, *P* phosphorous, *Mg* magnesium, *Ca* calcium

Exclosures are seen as instrumental in order for households to use land for agribusinesses that includes commercial livestock and agricultural production and in some cases also the buying, selling and/or renting of land (Napier and Desta 2011). Although, Mekuria (2004) supported the idea that farmers benefit from the protective effect of closed areas (closed areas reduce runoff and soil erosion in farmlands located below the closed areas), this is an effect that does not directly affect their income while harvesting of grass and herbs does. The practice of enclosing communal land can be seen as an important management tool for sustainable intensification of drylands within the general framework of a market-driven livestock revolution (Angassa and Oba 2010; Verdoodt et al. 2010; Catley et al. 2013). Local communities view exclosures as an integral part of the afforestation of degraded lands that demonstrates a sacrifice of short-term benefits in favor of long-term visions and benefits. Local communities have also developed a sense of ownership as they gain experience in authority over direct use, participation in decision-making, and establishing their own bylaws (Tefera et al. 2005).

Respondents were asked to assess the questions or statements using Likert scales, with 1 representing complete disagreement and 5 indicating full agreement, and the result was described below in Table 4.

Our main interest is in organizing all variables in to one as SOCIAL PERSPECTIVE. In general, other independent variables are either insignificantly correlated with SOCIAL PERSPECTIVE or have correlations close to zero. From an econometric standpoint, we need not worry about multicollinearity between SOCIAL PERSPECTIVE and any variables examining the whole distribution of PERHHCARBON, PERHACARBON, and SOCIAL PERSPECTIVE but can mask details. We, therefore, break the distribution of SOCIAL PERSPECTIVE down into each of 15 observations each and examine correlations and significance levels (Table 5). The linear regression models of carbon with altitude and exclosure areas size were also given in Table 6 below.

Regarding carbon and social perspectives, all variables in our models are associated with our two measures of carbon (PERHHCARBON and PERHACARBON). SOCIAL PERSPECTIVE, which is the result of factor analysis of our 15 management variables, is positively correlated with carbon and significant at greater than the 1% level for both carbon measures. Other variables have expected correlations. Households living in more remote areas (DISTOWN) and in lower altitude and have higher carbon stocks, as do those with larger forest areas. Households living in *Kebeles* with higher average altitudes have less carbon in their forests as do those living in areas with greater population pressures. As increasing altitude or *Kebeles* with higher altitude and mountainous areas is highly negatively associated with carbon stocks.

We see that in all observations PERHHCARBON and PERHACARBON are negatively correlated with variables related to access, fairness, and benefit but positively correlated with others insignificantly. This pattern shows there is a need for better improvement for a social capital benefit that leads to higher quality collective action and better environmental outcomes. Though these results are

Table 4 Mean and standard deviation (SD) of variables used for constructing collective action (CA) variable

CA attribute	No.	Variable	Mean	SD
Access	1	The rules of access and exclosure use are clear	3.79	0.67
	2	The system for deciding who has access to the exclosure resources is fair	3.79	0.64
Fairness/equity in exclosure products	1	Distribution process is fair and acceptable	3.08	0.79
	2	All other HH have the same allotment of fodder per year	3.01	0.58
Limits/quotas	1	Limitation on how much fuelwood you can collect from the exclosures	2.89	0.98
	2	Limitation on how much fodder collection you can do on the exclosures	2.63	0.98
Monitoring	1	Formally or informally involved in monitoring the exclosures	3.55	0.96
	2	Villagers generally watch who takes forest products from exclosures	3.14	0.8
Participation/ democracy		Responsible bodies (who decide how much each person can take) are democratically chosen	2.54	0.87
Informal sanctions	1	Other residents would be very sad to you if they found that you had taken more than your allotment of fuelwood, fodder or grazing rights?	4.79	0.5
	2	Would you feel uncomfortable or bad if you took more than your allotment of fuelwood, fodder, or grazing rights?	4.18	0.44
Formal sanctions	1	If you took more fuelwood from the exclosures than you were allowed to take, then you would face some sort of punishment?	4.12	0.33
	2	If you took more fodder or did more grazing from exclosures than you were allowed, then you would face some sort of punishment?	4.12	0.33
	3	You could lose some or all of your rights to collect forest products from common lands if you were caught taking more than the amounts you are allowed to take	4.74	0.47
Benefit		Exclosures have more benefits than free grazing lands to the residents around the area	3.02	0.53

Note: Likert scale: 1 = completely disagree with statement, 5 = completely agree

indicative that social coordination leads to better management and more forest carbon, it is possible that effects are spurious.

SOCIAL PERSPECTIVE is positively and significantly correlated with carbon. An approximately one standard deviation change in SOCIAL PERSPECTIVE is estimated to increase or decrease carbon per household and per hectare by 80 to 49% depending on the definition of carbon stock and the model.

We believe there are unobserved social capital variables captured by the variables that explain this correlation. Our empirical model, which focuses on exclosure management, does not capture all aspects of the analytical framework, because it

Table 5 Pearson correlations between social perspective developed from an average of CA, PerHH carbon, PerHA carbon, and other variables ($N = 180$)

		Kebelle	Distance	Altitude	ExclosureArea	PeoplePerHa	SOCIALPERSP	PerHHCarbon	PerHACarbon
Kebelle	Pearson correlation	1	−0.07	0.07	0.91**	0.98**	0.01	−0.90	−0.98
	Sig. (2-tailed)		0.39	0.36	0.001	0.001	0.96	0.08	0.09
Distance	Pearson correlation		1	0.06	−0.07	−0.07	−0.16*	0.06*	0.07*
	Sig. (2-tailed)			0.43	0.36	0.38	0.04	0.03	0.04
Altitude	Pearson correlation			1	−0.25**	0.15*	−0.02	−0.03*	0.04*
	Sig. (2-tailed)				0.001	0.04	0.83	0.05	0.05
ExclosureArea	Pearson correlation				1	0.85**	0.01	0.84**	0.96**
	Sig. (2-tailed)					0.001	0.9	0.001	0.001
PeoplePerHa	Pearson correlation					1	0.01	−0.89**	−0.97**
	Sig. (2-tailed)						1	0.001	0.001
SOCIALPERSP	Pearson correlation						1	0.81*	0.49*
	Sig. (2-tailed)							0.02	0.03
PerHHCarbon	Pearson correlation							1	0.90**
	Sig. (2-tailed)								0.001
PerHACarbon	Pearson correlation								1
	Sig. (2-tailed)								

**Correlation is significant at the 0.01 level (2-tailed)
*Correlation is significant at the 0.05 level (2-tailed)

Table 6 Linear regression models of carbon stock per hectare and per household ($N = 180$)

Variable	Model	Unstandardized coefficients B	Unstandardized coefficients Std. error	Standardized coefficients Beta	t	Sig.	R^2 for the model
PerHACarbon	(constant)	5.05	0.17		29.71	0.00	
	Altitude	0.001	0.00	0.04	0.52	0.61	0.001
	(constant)	9.10	0.09		102.49	0.00	
	Exclosure area	0.001	0.00	−0.96	−44.88	0.00	0.92
PerHHCarbon	(constant)	13.24	0.34		39.10	0.00	
	Altitude	0.001	0.00	−0.03	−0.38	0.71	0.70
	(constant)	19.99	0.34		58.50	0.00	
	Exclosure area	0.001	0.00	−0.83	−20.22	0.00	0.001

does not attempt to analyze the relationship between all social capital variables and carbon. But it proposes also those exogenous social capital variables, such as trust, reciprocity, etc., affect observed exclosure management, but it does not attempt to directly analyze the effect of those variables on carbon.

Conditioning variables, in general, have the expected signs. Increasing average *Kebele* altitude by 100 meters from the average of about 2000 meters is estimated to reduce carbon by 1.4% and 0.7%, and an increase in exclosure area of 100 hectares is estimated to increase CARBONPERHA and CARBONPERHH by about 1.1% and 0.75%, respectively. Carbon is especially sensitive to population pressure, with an increase in population per hectare of 100 people associated with reduced carbon of 96% and 89% CARBONPERHA and CARBONPERHH, respectively. More remote areas also tend to have more carbon, with an increase in distance to the town of 1 kilometer associated with a 6–7% increase in carbon.

The Pearson correlation between social perspectives and carbon was shown in Table 5 and each variable was also seen in detail by Table 7 below.

Conclusion and Implications

Our study showed that the establishment of exclosures on degraded free grazing lands in the northern highlands of Ethiopia is a viable option to restore native vegetation composition, richness, diversity, and aboveground biomass. Consideration of edaphic (e.g., soil total nitrogen, available phosphorus, CEC, exchangeable bases, soil pH, and soil texture) and site (e.g., stoniness, grazing) variables will help to optimize the selection of areas for the establishment of future exclosures. Moreover, our study suggests that with time exclosures may increasingly obtain an important role as refugees and species pool similar to church forests and should be protected and managed in a sustainable manner. However, economic and social

Table 7 Pearson correlations between PerHH carbon, PerHA carbon, and each CA variables ($N = 180$)

		PerHH-Carbo	PerHA-Carbon	Access1	Access2	Fairness1	Fairness2	Limit1	Limit2	Monito-ring1	Monit.2	Particip.	Informal.1	Informal-San2	Formal-san1	Formal-san2	Formal-san3	Benefit
Per HHCarbon	Pearson correlation	1	0.90**	−0.04	−0.01	0.06	−0.15*	0.05	0.03	0.06	−0.02	0.03	0.09	0.03	0.06	0.06	0.09	−0.09
	Sig. (2-tailed)		0.00	0.65	0.88	0.41	0.04	0.55	0.66	0.45	0.78	0.73	0.25	0.69	0.41	0.41	0.26	0.26
PerHA-Carbon	Pearson correlation		1.00	−0.06	−0.04	0.04	−0.17*	0.02	0.03	0.01	−0.04	0.11	0.10	0.02	0.05	0.05	0.10	−0.09
	Sig. (2-tailed)			0.43	0.60	0.56	0.03	0.83	0.74	0.90	0.58	0.16	0.18	0.82	0.49	0.49	0.16	0.25
Access1	Pearson correlation			1.00	0.92**	0.11	0.07	−0.02	0.05	0.09	0.12	0.05	−0.05	0.21**	0.12	0.12	0.06	0.09
	Sig. (2-tailed)				0.00	0.16	0.37	0.83	0.49	0.24	0.11	0.50	0.52	0.01	0.11	0.11	0.45	0.23
Access2	Pearson correlation				1.00	0.14	0.08	−0.01	0.09	0.10	0.14	0.07	−0.05	0.25**	0.17*	0.173*	0.09	0.10
	Sig. (2-tailed)					0.06	0.28	0.85	0.21	0.17	0.06	0.33	0.55	0.00	0.02	0.02	0.21	0.17
Fairness1	Pearson correlation					1.00	0.18*	0.32**	0.08	0.35**	0.36**	0.13	−0.04	0.53**	0.4**	0.4**	0.3**	0.015*
	Sig. (2-tailed)						0.01	0.00	0.31	0.00	0.00	0.09	0.63	0.00	0.00	0.00	0.00	0.05
Farness2	Pearson correlation						1.00	0.23**	−0.09	0.23**	0.36**	0.08	0.19**	0.4**	0.3**	0.3**	0.12	0.9**
	Sig. (2-tailed)							0.00	0.22	0.00	0.00	0.29	0.01	0.00	0.00	0.00	0.10	0.00
Limit1	Pearson correlation							1.00	0.01	0.38**	0.42**	0.05	−0.01	0.53**	0.4**	0.04**	0.32**	0.16*
	Sig. (2-tailed)								0.92	0.00	0.00	0.53	0.89	0.00	0.00	0.00	0.00	0.03
Limit2	Pearson correlation								1.00	0.06	0.14	−0.04	−0.02	0.08	0.09	0.09	0.14	−0.10
	Sig. (2-tailed)									0.40	0.06	0.63	0.78	0.29	0.25	0.25	0.06	0.17
Monitoring1	Pearson correlation									1.00	0.32**	0.02	−0.04	0.52**	0.32**	0.32**	0.19**	0.16*
	Sig. (2-tailed)										0.00	0.78	0.64	0.00	0.00	0.00	0.01	0.04

(*continued*)

Table 7 (continued)

		PerHH-Carbo	PerHA-Carbon	Access1	Access2	Fairness1	Fairness2	Limit1	Limit2	Monito-ring1	Monit.2	Particip.	Informal.1	Informal-San2	Formal-san1	Formal-san2	Formal-san3	Benefit
Monitoring2	Pearson correlation										1.00	−0.03	0.03	0.57^{**}	0.39^{**}	0.39^{**}	0.29^{**}	0.31^{**}
	Sig. (2-tailed)											0.69	0.71	0.00	0.00	0.00	0.00	0.00
Participation	Pearson correlation											1.00	0.01	0.07	0.13	0.13	0.10	0.10
	Sig. (2-tailed)												0.93	0.37	0.09	0.09	0.18	0.20
InformalSan1	Pearson correlation												1.00	0.01	0.06	0.06	0.04	0.23^{**}
	Sig. (2-tailed)													0.89	0.44	0.44	0.61	0.00
InformalSan2	Pearson correlation													1.00	0.49^{**}	0.49^{**}	0.30^{**}	0.36^{**}
	Sig. (2-tailed)														0.00	0.00	0.00	0.00
Formalsan1	Pearson correlation														1.00	1.0^{**}	0.69^{**}	0.27^{**}
	Sig. (2-tailed)															0.00	0.00	0.00
Formalsan2	Pearson correlation															1.00	0.69^{**}	0.27^{**}
	Sig. (2-tailed)																0.00	0.00
Formalsan3	Pearson correlation																1.00	0.10
	Sig. (2-tailed)																	0.21
Benefit	Pearson correlation																	1.00
	Sig. (2-tailed)																	

**Correlation is significant at the 0.01 level (2-tailed)
*Correlation is significant at the 0.05 level (2-tailed)

impacts of exclosures should be included in feasibility studies before establishing exclosures in the future.

The results of this chapter confirm that establishment of exclosures on degraded free grazing lands in the Northern Highlands of Ethiopian is a viable option to restore ECS. Our study showed that ECS in exclosures, in free grazing lands, and the change in ECS following the establishment of exclosures on free grazing lands can be predicted using easily measurable biophysical and management-related indicators. Such information is necessary for the establishment of baseline information for carbon sequestration projects, for evaluation of whether exclosure establishment should be expanded, and for policymakers to take into account the value of exclosures in their management decisions. Although our study showed that exclosures are effective to restore ECS, expansion of exclosures would increase grazing pressure on the remaining free grazing area. Therefore, the decision to establish additional exclosures should also include an economical analysis and an evaluation of the social consequences of such a decision.

Similarly, the results of this chapter confirm that the establishment of exclosures on degraded free grazing lands can be effective in improving soil nutrient content and properties. Our study also confirms that it is possible to generate baseline information and to make a quantitative prediction of the changes in soil nutrient content after the establishment of exclosures using relatively simple field measurements. Such information is critical for evaluating the ecological importance of exclosures and to assist policymakers to consider the value of exclosures in natural resource planning and management. Further studies are needed to investigate the degree to which nutrient enrichment in exclosures is due to soil and nutrient deposition from surrounding areas, as this may affect the sustainability of scaling up the area of exclosures.

There was a significant difference for EC, soil organic carbon and total nitrogen among the land use types. The lowest carbon and nitrogen level of soil was recorded in free grazing lands. Cation exchange capacity, available phosphorus, and exchangeable sodium and potassium were statistically insignificant among exclosure and grazing land.

During fieldwork, it was observed that there are efforts to guard the forest land in the study area from encroachment. However, conversion of vegetated land to cultivation land in steeper slopes is also prevalent. Therefore, unless a solution is devised to strongly protect the hillsides covered with vegetation, the process will increase erosion hazard from the catchment. This, in turn, will have an adverse impact on the long-term ecological balance of the area.

One of the purposes of this study was also to investigate the factors that influence the sustainability of exclosure establishment and management the in the highlands of Tigray. The land users' understanding and perception of exclosure establishment and management practices are important when sustainable natural resources management options are considered. A large-scale mass mobilization undertaken for more than three decades in soil and water conservation practices has minimized flooding and thus soil erosion or degradation both in the farmlands and in exclosures, although it was less practiced in on-farm conservation compared to the exclosures.

Results of the study also indicated that perception of farmers towards benefits of conservation attempts in changing and hence important contribution towards livelihoods has been recorded. However, the benefits are not yet adequate. Therefore, if the people of the study areas are to continue with the community-based soil and water conservation practices, they have to realize tangible net benefits in terms of production and income as well as environmental improvements. The study revealed that the landless people have tackled not only land degradation by applying various hillside conservation methods such as soil/stone bunds, trenches, and tree plantation but also creating alternative income through different activities such as the sales of honey production and growing vegetables.

Thus, in order to achieve sustainable exclosure establishment and management practices, institutional and economic factors should be given special attention.

Although there is no evidence that farmers around the exclosures are particularly concerned about global climate change, communities' social, aesthetic, and ecological values cause them to manage their land to maintain forest cover over time. If carbon sequestration were viewed as a benefit and moral obligation associated with exclosure establishment, farmers might well manage for it with minimal policy intervention.

Farmers in the exclosures have skills and resources to make exclosure establishment and management decisions and some of their practices are compatible with carbon sequestration.

Education and technical assistance will help them to increase their knowledge about how to manage for carbon and avoid practices that reduce carbon sequestration. Like the general public, farmers around exclosures may not be overwhelmingly concerned about climate change.

Indeed, the natures of climate change itself works against risk perception; the causes and consequences of climate change are remote and therefore may seem uncontrollable. People are more likely to take action if they are aware of and view themselves as being vulnerable to climate change, perceive great potential costs of climate change, and believe their individual actions can affect it (i.e., it is a controllable risk).

Community concerns about losing control over their natural resource management decisions and having their privacy violated may limit their enthusiasm for formal management planning, cooperation with additional works and related public agendas, and participation in public programs.

Communities who perceive a risk of being constrained by benefit or fairness will be less likely to respond to climate change mitigation policies than those who anticipate personal benefit or feel that such policy goals are compatible with their own goals.

The lack of overriding economical or social benefits from exclosure establishment and management compounded by concerns about the risks and burdens of entering into additional participation in exclosure or related activities. But, a combination of different types of policy tools will likely have the best outcome for encouraging carbon sequestration on restoring degraded free grazing lands.

This research focused on the carbon sequestration and social variables that are likely to affect exclosure management for carbon sequestration on restorations. However, these same social and cultural influences will affect exclosure establishment and management for other ecological values, such as habitat conservation, air and water quality, as well as their likelihood to respond to policies that promote these values. Thus, the findings are relevant for a broader range of natural resource issues than restricted to carbon and climate change issues.

Acknowledgments The corresponding author acknowledges International Foundation for Sciences (**IFS**) of Sweden under their IFS Grant No: D/5765-1, Department of Plant Biology and Biodiversity Management of AAU, and Mekelle University for their financial and logistical supports.

References

Agrawal A (2007) Common property institutions and sustainable governance of resources. World Dev 29(10):1649–1672

Agrawal A, Chhatre A, Hardin R (2008) Changing governance of the world's forests. Science 320:1460–1462

Ahmed H (2006) Assessment and evaluation of utilization practices of natural pastures and other forages in Basona Worana wereda of North Shoa Zone. School of Graduate Studies of Alemaya University, Dire Dewa, p 174

Amsalu S (2000) Herbaceous species composition, dry matter and condition of the major grazing areas in the middle rift valley, Ethiopia. MSc thesis, Hramaya University, Dire Dewa, 159pp

Angassa A (2014) Effects of grazing intensity and bush encroachment on herbaceous specie and rangeland condition in southern Ethiopia. Land Degrad Dev 25:438–451

Angassa A, Oba G (2010) Effects of grazing pressure, age of enclosures and seasonality on bush cover dynamics and vegetation composition in southern Ethiopia. J Arid Environ 74:111–120

Augustine D, McNaughton S (2004) Regulation of shrub dynamics by native browsing ungulates on east African rangeland. J Appl Ecol 41:47–49

Bedru B, Muys B, Ferdu N, Tollens E, Nyssen J, Deckers J, Mathijs E (2009) The economic contribution of forest resource use to rural livelihoods in Tigray, northern Ethiopia. Forest Policy Econ 11:109–117

Betru N, Jawad A, Ingrid N (2005) Exploring ecological and socio-Economic issues for the improvement of area enclosure management: a case study from Ethiopia. DCG Report No. 38, May 2005, pp 3–30

Beyene F (2010) Locating the adverse effects of rangeland enclosure among herders in eastern Ethiopia. Land Use Policy 27(2):480–488

Booker K, Huntsinger L, Bartolome J, Sayre N, Stewart W (2013) What can ecological science tell us about opportunities for carbon sequestration on arid rangelands in the United States? Glob Environ Chang 23:240–251

Bowles S, Gintis H (2002) Social capital and community governance. The Economic Journal 112 (483):F419–F436

Butler L, Kielland K (2008) Acceleration of vegetation turnover and element cycling by mammalian herbivory in riparian ecosystems. J Ecol 96:136–144

Campbell B, Mandondo A, Nemarundwe N, Sithole B (2001) Challenges to proponents of common property resource systems: despairing voices from the social forests of Zimbabwe. World Dev 29(4):589–600

Capoor K, Amborsi P (2008) State and trends in the carbon market 2008. The World Bank, Washington, DC

Catley A, Lind J, Scoones I (2013) Pathways to sustainability: pastoralism and development in Africa. Dynamic change at the margins. Earth scan from Routledge, Abingdon

Chaiyo U, Garivait S, Wanthongchai K (2011) Carbon storage in aboveground biomass of tropical deciduous forest in Ratchaburi province Thailand. World Acad Sci Eng Technol 58:636–641

Cheng J, Wu G, Zhao L, Li Y, Li W, Cheng J (2011) Cumulative effects of 20-year exclusion of livestock grazing on above- and belowground biomass of typical steppe communities in arid areas of the loess plateau, China. Plant Soil Environ 57:40–44

Chhatre A, Agrawal A (2009) Tradeoffs and synergies between carbon storage and livelihood benefits from forest commons. Proc Natl Acad Sci 106(42):17667–17670

Chisholm N (1998) Community-based natural resource management in Tigray, northern Ethiopia. The World Bank/WBI's CBNRM Initiative, Washington, DC

Conant R, Paustian K, Elliott E (2001) Grazing land management and conversion into grazing land: effects on soil carbon. Ecol Appl 11:343–355

De Koning GH, Veldkamp E, Lopez-Ulloa M (2003) Quantification of carbon sequestration in soils following pasture to forest conversion in northwestern Ecuador. Global Biogeochem Cy 17:1–12

Derner J, Schuman G (2007) Carbon sequestration and rangelands: a synthesis of land management and precipitation effects. J Soil Water Conserv 62:77–85

Descheemaeker K, Nyssen J, Rossi J, Poesen J, Haile M, Moeyersons J, Deckers J (2006) Sediment deposition and pedogenesis in exclosures in the Tigray Highlands, Ethiopia. Geoderma 132:291–314

Emiru B, Demel T, Barklund P (2007) Enclosures to enhance woody species diversity in the drylands of Tigray. East Afr J Sci 1:136–147

FAO (2001) Soil carbon sequestration for improved land management. World Soil Resources Report No. 96, Rome

Gebrewahd A (2014) Herbaceous vegetation restoration potential and soil physical condition in a mountain grazing land of eastern Tigray, Ethiopia. J Agric Environ Int Dev 108(1):81–106

Glaeser E, Laibson D, Sacerdote B (2002) An economic approach to social capital. The Economic Journal 112(483):F437–F458

Gordon HS (1954) The economic theory of a common property resource: the fishery. Journal of Political Economy 62(2):124–142

Hyde W, Amacher G, Magrath W (1996) Deforestation and forest land use: theory, evidence and policy implications. World Bank Res Obs 11(2):223–248

Ibrahim M (2016) Impact of enclosure on plant species composition and biomass production in Ewa Woreda of Afar region state, Ethiopia. J Biodivers Endanger Species 4:157

Jeddi K, Chaieb M (2010) Changes in soil properties and vegetation following livestock grazing exclusion in degraded arid environments of South Tunisia. Flora 205:184–189

Jodha N (2008) Some places again: a restricted revisit to dry regions of India. In: Ghate R, Jodha N, Mukhopadhya P (eds) Promise, trust, and evolution: managing the commons of South Asia. Oxford University Press, Oxford

Kebrom T (2001) Natural regeneration of degraded hill slopes in southern Wello, Ethiopia: a study based on permanent plot. Appl Geogr 21:275–300

Lal R (2011) Sequestering carbon in soils of agro-ecosystems. Food Policy 36:S33–S39

Makepe P (2006) The evolution of institutions and rules governing communal grazing lands in Botswana. East Afr Soc Sci Res Rev 22(1):39–61

Mamo K, Ehrich D, Teberlet P, Sileshi N, Brochmann C (2007) Phylogeography and conservation genetics of a giant *Lobelia* (*Lobelia giberroa*) in Ethiopian and tropical eastern Africa. Mol Ecol 16:1233–1243

McDermot C, Elavarthi S (2014) Rangelands as carbon sinks to mitigate climate change: a review. J Earth Sci Clim Change 5:221

Mekuria W (2004) Impacts of land use changes on soil nutrients and erosion in Tigray, Ethiopia. M.Sc. Thesis, Georg-August University of Gottingen, Germany

Mulugeta L, Karltun E, Olsson M (2005) Soil organic matter dynamics after deforestation along a farm field chronosequence in southern highlands of Ethiopia. Agric Ecosyst Environ 109:9–19
Mureithi SM, Verdoodt A, Gachene CK, Njoka JT, Wasonga VO, De Neve S, Meyerhoff E, Van Ranst E (2014) Impact of enclosure management on soil properties and microbial biomass in a restored semi-arid rangeland, Kenya. J Arid Land 6(5):561–570
Napier A, Desta S (2011) Review of pastoralist rangeland enclosures in Ethiopia. Feinsteain International Centre and Tufts University, November 2011, Addis Abeba
Ostrom E (1990) Governing the commons: the evolution of institutions for collective action. Cambridge University Press, Cambridge
Ostrom E (2009) A general framework for analyzing sustainability of social-ecological systems. Science 325(5939):419–422
Powers J, Schlesinger W (2002) Geographic and vertical patterns of stable carbon isotopes in tropical rain forest soils of Costa Rica. Geoderma 109:141–160
Reusing M (1998) Monitoring forest resources in Ethiopia. Ministry of Agriculture, Addis Ababa
Schuman G, Ingram L, Stahl P, Vance G (2005) Dynamics of long-term carbon sequestration on rangelands in the western USA. In: 20th International Grassland Congress, Dublin, Ireland
Segers K, Dessein J, Nyssen J, Mitiku H, Deckers J (2008) Developers and farmers intertwining interventions: the case of rainwater harvesting and food-for-work in Degua Temben, Tigray, Ethiopia. Int J Agric Sustain 6(3):173–182
Shylendra H (2002) Environmental rehabilitation and livelihood impact: emerging trends from Ethiopia and Gujarat. Econ Polit Wkly 37(31):3286–3292
Slimani H, Aidoud A, Roze F (2010) 30 years of protection and monitoring of a steppic rangeland undergoing desertification. J Arid Environ 74:685–691
Stocking M, Murnaghan N (2001) Handbook for the field assessment of land degradation. Earthscan publications Ltd., London/Sterling, p 169
Sunderlin W, Angelsen A, Belcher B, Burgers P, Nasi R, Santoso L, Wunder S (2005) Livelihoods, forests, and conservation in developing countries: an overview. World Dev 33(9):1383–1402
Taddese G (2001) Land degradation: a challenge to Ethiopia. Environ Manag 27(6):815–824
Tefera M, Demel T, Hulten H, Yonas Y (2005) The role of enclosures in the recovery of woody vegetation in degraded dryland hillsides of central and northern Ethiopia. J Arid Environ 60(2):259–281
Teshome A, Abule E, Lisanework N (2009) Pastoralist's perceptions and rangeland evaluation for livestock production in south eastern Ethiopia. Livest Res Rural Dev 21(7):2009
Timothy R, Pearson H, Brown S, Sohngen B, Henman J, Ohrel S (2013) Transaction costs for carbon sequestration projects in the tropical forest sector. Open access at Springerlink.com. Mitig Adapt Strateg Glob Change. https://doi.org/10.1007/s11027-013-9469-8
Van Der Westhuizen H, Snyman H, Van Rensburg W, Potgieter J (2001) The quantification of grazing capacity from grazing and production values for species in the semi-arid grassland biome of southern Africa. Afr J Range Forage Sci 18:43–52
Verdoodt A, Mureithi SM, Ranst EV (2010) Impacts of management and enclosure age on recovery of the herbaceous rangeland vegetation in semi-arid Kenya. J Arid Environ 74:1066–1073
Wolde M, Veldkamp E, Mitiku H, Kindeya G, Muys B, Nyssen J (2009) Effectiveness of exclosures to control soil erosion and local community perception on soil erosion in Tigray, Ethiopia. Afr J Agric Res 4:365–377
Wolde M, Veldkamp E, Mesfin T, Roland O (2010) Economic valuation of land restoration: the case of exclosures established on communal grazing lands in Tigray, Ethiopia. Land Degrad Dev 22:334–344
Wolde M, Veldkamp E, Corre M (2011) Restoration of ecosystem carbon stocks following enclosure establishment in communal grazing lands in Tigray, Ethiopia. Soil Sci Soc Am J 75(1):246–256
Yayneshet T (2011) Restoration of degraded semi-arid communal grazing land vegetation using the enclosure model. Mekelle University, Mekelle, Ethiopia. Int J Water Res Arid Environ 1(5):382–386
Yemiru T, Roos A, Campbell B, Bohlin F (2011) Livelihood strategies and the role of forest income in participatory-managed forests of Dodola area in the bale highlands, southern Ethiopia. Forest Policy Econ 13(4):258–265

Yohannes G, Waters-Bayer A (2007) Trees are our backbone: integrating environment and local development in Tigray region of Ethiopia. Issue paper no. 145, International Institute for Environment and Development (IIED), London, 38p

Zhou Z, Li F, Chen S, Zhang H, Li G (2011) Dynamics of vegetation and soil carbon and nitrogen accumulation over 26 years under controlled grazing in a desert shrub land. Plant Soil 341:257–268

Balancing Adaptation and Mitigation in the Building Sector of New York State

20

Yasmein Okour, Nicholas B. Rajkovich, and Martha Bohm

Contents

Abstract

Climate change poses a significant threat to commercial and residential buildings. New buildings must be designed, and existing buildings retrofitted, to be resilient to extreme weather and sea level rise. At the same time, designers and building owners need to address energy consumption because the building stock uses nearly 40% of all energy in the United States. Building professionals will shape

Y. Okour (✉)
Department of City Planning and Design, Jordan University of Science and Technology, Irbid, Jordan
e-mail: yfokour3@just.edu.jo

N. B. Rajkovich · M. Bohm
School of Architecture and Planning, State University of New York at Buffalo, Buffalo, NY, USA
e-mail: rajkovic@buffalo.edu; marthabo@buffalo.edu

W. Leal Filho (ed.), *Handbook of Climate Change Resilience*,
https://doi.org/10.1007/978-3-319-93336-8_124

how the building stock responds to climate change. However, despite their critical role, few studies have examined how these professionals balance increasing the resilience of the built environment with greenhouse gas emission reduction efforts. To address this gap, data from 42 semi-structured interviews and an online survey were analyzed. The research focused on the response of New York State building professionals. Results indicate that building professionals are aware of potential climate impacts. However, their understanding of climate impacts are framed by where they live. In addition, while climate change is a concern, these building professionals tend to focus on energy efficiency rather than climate-related resilience. This is potentially due to a lack of information, outdated building codes, a lack of resources, and limited support from their clients.

Keywords
Adaptation · Energy efficiency · Building sector · Climate change · Mitigation

Introduction

In recent years, research on climate change and sectors of the built environment has grown substantially. Buildings are frequently highlighted as key to achieving global greenhouse gas mitigation goals. However, despite the various efforts to reduce greenhouse gas emissions, buildings will continue to experience the impacts of climate change.

The fifth assessment report from the Intergovernmental Panel on Climate Change highlighted that climate change impacts will continue for centuries, even if substantial actions are undertaken to mitigate greenhouse gas emissions (IPCC 2014). As such, adaptation efforts are crucial to withstand the adverse effects of climate change on buildings. Adaptation reflects an understanding of the inevitable risks climate change impacts pose on buildings and the importance of reducing their vulnerability.

Recent efforts toward climate change action within the building sector illustrate the limited attempts to understand how buildings can adapt to climate change. With few exceptions, most guidebooks and initiatives focus on climate mitigation through energy efficiency and renewable energy practices. Moreover, although building professionals are on the frontlines of addressing climate change in the building sector, few studies have explored how building professionals balance climate change adaptation and mitigation efforts.

With the understanding that adaptation measures within the building sector are necessary for an effective climate change action plan, this research addresses the gap in knowledge regarding building professionals' response to climate change. The primary objective was to examine how professionals perceive and respond to climate change impacts on the building stock in New York State.

Literature Review

The long life span of buildings demands us understanding how buildings will respond to shifts in the climate. Generally, climate change poses dual issues for the building sector. On one hand, buildings present a significant potential for effective climate change mitigation, while on the other hand, climate change impacts pose a significant risk to buildings and the built environment.

The IPCC predicted that climate change will have a strong impact on the energy requirements of buildings as their cooling and heating needs are dependent on temperature and weather trends. Global building energy use is expected to double to triple by 2050. Migration patterns, economic development, shifting lifestyles, and appliances all contribute to the increasing energy demand in the building sector (Lucon et al. 2014). In the United States, warmer winters are likely to decrease the net energy demand in the north (space heating), while warmer summers are likely to increase the net energy demand in the south (space cooling) (Scott and Huang 2007). In New York State, the focus of this research, similar patterns are expected to occur. Specifically, a 4% increase in peak demand is projected for New York City with significantly higher energy consumption in the commercial sector in comparison to the residential sector (Hammer et al. 2011).

With an increase in new construction and ongoing warming, the energy demand in buildings will continue to increase. Effective response is challenged by current building performance standards that are based on historical climate data that does not account for future climate trends. Consequently, energy demand will increase, and buildings will continue to contribute to greenhouse gas emissions and climate change.

To date, research and policy development has mainly been driven by the vast interest in reducing energy use and its implications on climate change mitigation. Studies have examined future trends in heating and cooling loads and developed an array of energy conservation measures with the goal of providing better thermal comfort in buildings and reducing greenhouse gas emissions. In contrast to climate change mitigation, literature on climate adaptation and resiliency in buildings is less ubiquitous, and many questions remain unanswered. While adaptation policies have been developed on an international basis, their scope remains limited, and very few have addressed the buildings sector.

According to Roaf et al., "the daunting scale of the necessary changes" has challenged the task of adapting buildings to climate change (Roaf et al. 2009). Research has primarily focused on isolated aspects of adaptation or addressed specific impacts, such as termite control measures (Brennan 2010), disposable infrastructure and relocatable buildings (Gordon 2013), behavioral changes that reduce air exchange rates (Field 2010), nonstructural adaptation (Coley et al. 2012), and community participation to address the quality of indoor environments (IOM 2011).

While such studies are crucial to understanding how buildings can adapt to climate change impacts, none have yet to acknowledge the need for a paradigm shift in the building profession, and that "business as usual" can no longer be accepted (Roaf et al. 2009). This paradigm shift requires changing the conventional practices of how building professionals design, construct, manage, and maintain buildings to maximize their mitigation potential and reduce their vulnerability to climate change impacts. This shift is critical if we are to avoid maladaptation, the selection of strategies that may exacerbate rather than mitigate the impact of climate change.

Adapting to What?

Successful adaptation is contingent on understanding the impacts of climate change on buildings and building-related activities (Hasegawa 2004). Such impacts are characterized by multiple dimensions of diversity, such as the potential risks and hazards, the nature and extent of impacts, and the building components and practices impacted (Hasegawa 2004). As Hasegawa states "climate change impacts on the building sector should not be seen as a single phenomenon but as a complex composite of various kinds of impacts" (Hasegawa 2004). Additionally, the diversity in the geography of impacts should also be considered. For example, Upstate New York is prone to winter weather and ice or hail damage, while in Downstate New York, severe weather events and storms are issues of great concern.

Moser et al. defines this phase as the *understanding stage* of the adaptation process (Moser and Ekstrom 2010). Generally, this stage begins with the awareness or detection of a problem. In the building sector, the stage consists of recognizing climate change and the significant risks it poses on buildings and building operations. Available information and experience with climate change impacts are important drivers of the understanding process.

Within this stage, building professionals are faced with the inherent uncertainty of climate change impacts. While professionals may recognize climate change as an issue, the data required to generate adaptation response may be insufficient. This may lead to "cautious attitude towards adaptation, especially because significant adaptation measures would have to be justified against other priorities" (Hertin et al. 2003). Ultimately, changes in building practices cannot be made if information for sound decision-making is not available.

In addition, before adaptation options can be developed, the problem must cross a threshold of concern. Building professionals must perceive climate change as an important issue that requires immediate action within the sector. There are multiple triggers for crossing the threshold among building professionals, including information, the source of information, personal experience, media, liability concerns, clients, leadership, priorities of firms and organizations, and policy development (Ekstrom et al. 2011; NRC 2009). For example, due to the damage to the building stock caused by Hurricane Sandy, an 11 billion dollar storm in New York, a threshold of concern was crossed, and adaptation efforts were undertaken across New York State.

Adaptation is not merely a technological challenge. A crucial part of the adaptation process is engaging stakeholders. The diversity as explained above highlights the wide scope of stakeholders involved. While at the core of climate change efforts in the building sector are building professionals, stakeholders may consist of non-traditional members such as the insurance industry, building users, community organizations, emergency services, etc. All stakeholders have a role and a responsibility to mitigate and adapt to climate change. Their practices and behaviors shape how the building sector adapts to climate change impacts (Hasegawa 2004).

Generally, adaptation consists of both the outcome and the process of adapting (Smit et al. 2000). While some adaptations can be considered ex post, as in reactive to observed impacts, other adaptations are ex ante, in which actions are taken in response to future projections (Mendelsohn 2000; Smit et al. 2000).

It can be argued that recent adaptation efforts in the building sector in New York State have been of ex post nature. This has been mainly driven by climate disasters attributed to climate change such as Hurricane Sandy (Dubois and Krasny 2016; PlaNYC 2013; Schmeltz et al. 2013). For example, adaptation in response to observed impacts such as flooding has been widely captivated in governmental programs and initiatives, while other projected impacts such as decreasing indoor air quality have not. The prevalence of ex post adaptation represents current limitations in climate change action.

The ability to respond to the challenges of climate change adaptation in the building sector allows action to move forward. Expertise, education, and training background reflect the possible knowledge gaps within professionals attempting to engage in the adaptation process. While both experience and perceived expertise shape the adaptation process, it is crucial to understand the limitations they present. Specifically, experiences and perceived expertise may impact adaptation response by distorting probabilities of events and climate change impacts. Moreover, adaptation efforts shaped by experience may lack comprehensiveness, while efforts based on perceived expertise and skills may lead to ineffective response.

Responses in the Building Sector

Within the building sector, there have been limited efforts to examine the climate change response of building professionals. In one study, Morton et al. explored climate change beliefs within a large engineering firm (Morton et al. 2011). Their results showed that while participants generally perceived climate change as important, seniority impacted the perceived need for adaptation, with higher levels of seniority employees favoring the business-as-usual approach. The study also showed participants emphasized mitigation practices over adaptation.

Another study by Hertin et al. examined the relative role of climate change impacts in driving the adaptation process in UK house builder organizations (Hertin et al. 2003). The study found that limited motivation to engage in climate adaptation work and lacking external pressures impacted adaptation processes (Hertin et al. 2003).

The aim of this research was to explore building professionals' understanding of climate change impacts and engagement in adaptation work. Given the dual implications of buildings in mitigating greenhouse gas emissions and adapting to the impacts of climate change, it is pertinent to examine how stakeholders within the building industry respond to climate change. This chapter examines building professional's perceptions of climate change impacts (adapt to what?), whether they perceive responding to climate change as important, and their engagement in adaptation work (how are they adapting?). Specifically, this chapter addresses the following research questions:

1. *How do building professionals use climate change information?*
2. *Does experience shape how building professionals address climate change?*
3. *How important is responding to climate change for building professionals?*
4. *How are building professionals engaging in climate change adaptation work?*

Methodology

To explore building professionals' understanding of climate change impacts and engagement in adaptation work, the research employed a mixed method approach consisting of two distinct phases: qualitative and quantitative. Generally, a mixed method approach consists of more than one type of research method conducted in either a sequential or concurrent manner (Creswell and Plano Clark 2011; Morse 1991). A sequential mixed method approach was utilized.

Qualitative Method

Over a period of 5 months, semi-structured interviews were conducted with building professionals with extensive experience in climate change adaptation and resiliency in New York State. Professionals residing inside and outside of New York State were recruited. Target participants included architects, engineers, planners, builders, contractors, code officials, energy auditors, university faculty/staff, and building professionals working in governmental agencies, professional organizations, and advocacy groups. Building professionals residing or working in the Upstate and Downstate regions of New York State were recruited allowing for diverse regional representation.

A snowball sampling method (referral sampling) was utilized for recruiting participants. Kuzel indicates that the purpose of snowball sampling is it "identifies cases of interest from people who know people who know what cases are information-rich" (Kuzel 1992). One of the main critiques of snowball sampling techniques is the reliance on a small number of gatekeepers or entry points (Atkinson and Flint 2001; Bloch 2007). To reduce this selection and gatekeeper bias, the researchers increased the number of gatekeepers to professionals in the building sector in New York State (Atkinson and Flint 2001; Bloch 2007).

The interview questions focused on six primary areas: (1) professional and educational background, (2) general perception of climate change, (3) climate change experience, (4) vulnerability of buildings, (5) adaptation strategies, and (6) challenges. Interviews were transcribed using professional transcription service,

Table 1 Characteristics of interview participants by sector and region. (Authors' Interviews 2016)

	Region				
Sector	Out of state	State wide	Upstate	Downstate[a]	Total
Architecture and design	6	8	8	3	**15**
Education and research	5	5	6	2	**12**
Nonprofit and NGO	4	7	7	1	**12**
Construction	2	4	11	1	**10**
Building systems and engineering	1	2	7	1	**8**
Government and policy	2	5	4	1	**6**
Other	1	4	3	2	**8**
Total	**12**	**22**	**24**	**8**	

Some participants identified with more than once sector or region
[a]Downstate includes the following counties: Dutchess, Sullivan, Ulster, Orange, Putnam, Rockland, Westchester, Bronx, New York, Queens, Kings, Richmond, Nassau, and Suffolk

and transcripts were double checked to ensure accuracy. Transcripts and interview memos were analyzed and coded using qualitative data analysis software NVivo (Bazeley 2007).

The analysis consisted of coding the transcripts to reveal emerging themes and patterns. The main goal of qualitative coding is data retention (Richards 2009). Coding as a tool for data retention "allows you to return to the data you want to inspect, interrogate, and interpret" (Richards 2009). The first coding phase included initial coding (open coding) which consists of attribute coding to retain descriptive information on the interviewees and in vivo coding (Richards 2009; Saldaña 2013; Strauss 1987). The second coding includes focused coding (Richards 2009; Saldaña 2013; Strauss 1987), also referred to as pattern coding (Creswell 2014).

In total, we conducted 42 semi-structured phone interviews with 44 building professionals in New York State. The table below illustrates the characteristics of the interview participants. Thirty-five of the interviewees were male professionals (79%), while nine were female professionals (21%). Building professionals were grouped into seven building sectors according to their profession. This included architecture and design, education and research, nonprofit and NGOs, construction, building systems and engineering, government and policymakers, and others (Table 1). Participants were associated with a specific building sector based on their current organization and work. The majority identified with more than one sector.

Quantitative Method

To examine the perception of a larger sample of building professionals in New York State, an online survey was created. The design of the survey instrument utilized the knowledge and input gained from the stakeholder interviews. The structure followed the stakeholder interviews in its objective to illustrate the current perception among building professionals on climate change, climate-related issues, vulnerability of buildings, climate change adaptation strategies, information and resources used, and barriers to climate change adaptation.

The instrument was distributed in June 2016, administered online, and took approximately 15–20 min to complete. The survey instrument mainly consisted of closed-ended questions with follow-up open-ended questions. Validity of the survey was assessed through a survey pretest with three participants. The content of the survey was reviewed by experts in the field who examined the order of questions, the instrument design in relation to research objectives and questions, and the appropriateness of scales used.

The sample for the online survey included building professional working in New York State. Past experience with climate change adaptation and resiliency was not required. The target participants included architects, engineers, builders, contractors, code officials, energy auditors, university faculty, governmental agencies, professional organizations, and advocacy groups.

Non-probability sampling methods were utilized. The survey was sent electronically to four list serves identified by the research team. This included the Society of Building Science Educators (approximately 1000 members), American Institute of Architects Community Resilience Network (102 members), University at Buffalo School of Architecture and Planning Alumni List (2712 members), and a New York State Energy Research and Development Authority (NYSERDA) email list (2199 members). Multiple email reminders were used to increase the response rate in the 2nd and 3rd week after the initial distribution of the survey. Participation in the survey was completely voluntary and responses were anonymized.

Descriptive statistics were used to analyze the data. Summary statistics on the demographics and perceptions of participants were obtained which included identifying frequencies, percentages, and cross tabulations. The survey resulted in a participation rate of 5.2% with a total of 286 building professionals. Table 2 illustrates the characteristics of the respondents based on region and sector. Overall, a total of 75% of respondents were male, 56% of respondents hold a master's degree, while 57% of professionals have over 15 years of professional experience ($n = 151$). Professionals from the architecture and design sector represented that largest group, followed by education and research, nonprofit and NGO, construction, and building systems and engineering.

Results

Climate Change Information

Many interviewees indicated the reliance on government sources, academic journals, news outlets, and online resources as tools to find information on climate change and its impact on the building stock. Similarly, the online survey participants identified several organizations that they rely on for information including professional associations, state agencies such as NYSERDA and New York State

Table 2 Characteristic of online survey participants by sector and region. (Authors' Survey 2016)

	Region			
Sector	Out of state	Upstate	Downstate	Total
Architecture and design	28	41	21	90
Education and research	19	29	8	56
Nonprofit and NGO	7	11	6	24
Construction	5	19	15	39
Building systems and engineering	6	19	9	34
Government and policy	4	8	5	17
Consultation	14	18	18	50
Property owner	2	8	7	17
Property management	1	3	4	8
Planning firm	2	2	1	5

Department of Environmental Conservation, colleges and universities, federal government agencies, professional colleagues, and research institutes. The online survey also highlighted the kinds of resources used by building professionals or their organizations to address climate change in their work (Table 3). Both building codes and standards and climate data are important resources to building professionals.

The interviews provided a more detailed discussion on challenges building professionals face related to climate change information that either fostered or impeded the implementation of climate change-related strategies. One of the main concepts discussed by professionals was the legibility of current available data, including legibility of energy codes and relevant data on energy efficiency.

The complexity of addressing climate change was also a prominent theme. Some building professionals cited the intersection of climate research with hazard mitigation, and risk management may challenge building professionals. In the building sector, variations in the definitions of main concepts, such as resilience, adaptation, and hazards, between different building professionals and research fields is a barrier to effective utilization of various climate change information.

Professionals also acknowledged that the continuously developing science of climate change and its impact on building may be overwhelming to stakeholders. As a result, professionals prefer to stick to industry standards rather than engage in a relatively new and complex science. As one professional stated:

> Where to start? There's a lot of information out there, so how do you get the best access to the information you need.

Uncertainty of climate change data provided new insights into the practices of building professionals. Generally, we found that professionals recognize the uncertainty inherent within climate change data when discussing climate change impacts in their regions. Building professionals highlighted uncertainty as a main reason why the building profession has not effectively responded to climate change demands:

Table 3 Resources used to address climate change impacts on buildings by online survey participants. (Authors' Survey 2016)

Resources used	Total
Building codes and standards	77
Climate data	60
FEMA reports	46
Climate models	40
GIS data	39
Climate projections	38
Meteorological data	23
Meteorological forecasts/projections	22
New York State Hazard Mitigation Plan	20
IPCC reports	19
Typical meteorological year (TMY) data	17
US Global Change Research Program Reports	16
Other hazard mitigation plans (e.g., city/county, region)	15
Insurance data	12
New York Climate Change Science Clearinghouse	11
Northeast Regional Climate Center Data or Reports	11
ClimAID reports	10
HAZUS-MH	7
Great Lakes Integrated Sciences + Assessments Center Data or Reports	6
SHELDUS	4

> I think that [uncertainty] would be quite a good reason why they [building professionals] haven't taken action, because they're confused. Because if they were told… the wind speed will increase by 20%, they could take action, but they're not told that. They're told that maybe the wind speed might increase, it might decrease, we're not sure. That is the situation.

Inaction was also driven by confusion over which climate models to utilize given the uncertainties:

> Right now, if I were to ask a few architects they would say "Which model should we even be using?"

Similarly, most of survey participants identified uncertainty as a major barrier to climate change adaptation and resilience. Approximately 28% of survey participants indicated uncertainty of climate change impacts as an extreme barrier and only 11% as not a barrier. As one survey participants stated:

> Over-adapting or maladapting in fields with very high uncertainty is one of the greater challenges that we face.

While complexity and uncertainty shape response to climate change impacts on the building stock, response can also be limited when considering the nature of information needed by building professionals. Interviewees discussed the misplaced efforts and direction of funding toward data development that does not address the needs of building professionals to respond to the impacts of climate change. Mainly, interviewees discussed the importance of providing data on localized impacts to their work. Professionals indicated the need for scaling down projections to better understand the impacts at the level they operate within.

Such a demand points to a dilemma where building professionals acknowledge challenges with uncertainty yet at the same time call for scaled-down climate change data. One building professional viewed a demand for downscaling projection as unrealistic, given the uncertainties within the projections themselves.

Discussions on uncertainty and downscaling highlight possible confusions within the building profession regarding the process of climate change projections. As a result, building professionals either ignore the challenges presented with uncertainty risking maladaptation or choose to not respond and stick to standard practices. Consequently, some interview participants highlighted the importance of incorporating a robust decision-making process under uncertainty within the building sector. In other words, building professionals should not wait until a set of "required" data is available to respond to climate change impacts.

Experience and Perception of Climate Change Impacts

In the absence of "required" data, professionals utilize experience to fill the data gap. On the one hand, climate adaptation can be made solely on available information, such as climate models and projections, and on the other hand, decisions can be made based on perceived experiences of climate change impacts.

In both the interviews and online survey, we examined the perceived experience with climate change impacts among building professionals. All interview participants agreed that they have experienced climate change impacts. Approximately 80.2% of the online survey respondents agreed that they have experienced climate change in the past, while 84.6% of respondents agreed that they will experience climate change in the future.

Interview participants discussed a range of climate change impacts they have personally experienced. The three most prominent themes included impacts related to temperature (such as warmer temperatures, heat events, precipitation (such as less snow and extreme precipitation), and floods.

While the interview allowed for a general discussion on the different climate change impacts experienced by building professionals, the online survey provided us with a tool to examine regional differences, or the geography of impacts, in experiences with climate change impacts. Table 4 highlights the main climate change impacts perceived by building professionals within Upstate and Downstate New York.

According to Table 4, building professionals in Downstate New York perceive storms and flooding as the most challenging issue to the building stock, while professionals in Upstate New York perceive winter weather and ice or hail damage.

Table 4 Perceived climate-related issues by online survey participants. (Authors' Survey 2016)

Upstate	Downstate
Winter weather, including lake effect snowfall (88.89%)	Storms, including thunderstorms and tornadoes (87.82%)
Ice or hail damage (77.88%)	Flooding or storm surge (87.72%)
Flooding or storm surge (63.79%)	Heatwaves (78.95%)
Storms, including thunderstorms and tornadoes (63.39%)	Hurricanes (78.57%)
Air quality, including air pollution (56.76%)	Air quality, including air pollution (68.42%)
Unsafe or ineffective sewage and/or septic system operation (56.64%)	Sea level rise (67.86%)

Such regional differences are shaped by personal experience with a climate change impact.

Most interviewees discussed the localized nature of climate change impacts. This reflected the understanding that while there will be global impacts because of climate change, there will also be differences in regional and local impacts.

Perceived Importance

Experience with impacts and resources used reflects the possible knowledge gaps within professionals attempting to engage in the adaptation process. However, knowledge gaps cannot be overcome if building professionals do not perceive response to climate change as important. Among the interviewees there was a general agreement that climate change is a significant issue for the building sector (75%). Increasing interest in climate change adaptation post Hurricane Sandy was cited as a significant factor in addition to professional organizations, increasing financial incentives to incorporate energy saving practices, advancement in climate data, increasing insurance focus, and continuous code development.

The priority of engaging in climate change action was also explored. Within different organizations, actions taken toward climate change adaptation and resilience are guided by the identified objectives and mission of the organization. Incorporating the concept of resilience indicates the core role of climate change in the operation of these organizations. Only nine interviewees indicated climate change response as part of their organization's mission.

In the online survey, approximately, 71.1% of the participants reported that it was a priority to their work, while only 56.7% reported that it was a priority to their organization. Examining the geographical differences, approximately 80% of professionals in Downstate New York identified response to climate change impacts as a priority to their work in comparison to 60% in Upstate New York.

Through the interviews, we attempted to gauge why climate change may not be a priority for building professionals. A lack of education, differing priorities, costs, unwillingness to change from known practices, and the politicized nature of climate change were cited as factors. Most interviewees agreed that climate change will become an important issue to the building sector in the future.

While some professionals asserted that not enough is being done within the professional community, some participants reflected on the notion of professional responsibility and the "standard of care" as a potential vital driver of climate change adaptation in the building sector. The standard of care for professional building services should not only be viewed for liability concerns but should also reflect the responsibility of professionals to protect building users. As two interviewees explained it:

> Architects have a responsibility to look at what's expected in the future and design buildings that will provide for changing conditions. That's often called the standard of care and its sort of a key part of the architecture profession. It's usually been focused on safety issues and satisfying building codes and issues like that, but I think we will be seeing an expansion of that standard of care definition to address resilience, was well as to address things like healthy building.
>
> Responsibility for assuming the risk lies with the building owner and the design professional. And that entails building professionals "build beyond the code" and expand the concept of standard of care to address building resilience.

Engaging in Adaptation

With an understanding of the climate change impacts on the building sector and a perceived importance of response, professionals are placed in a position that requires them to engage in climate adaptation. First, we examined the perceived expertise in responding to climate change impacts. The online survey found that approximately 57% participant found that they and the organization they work for have the expertise required to address the potential impacts of climate change on buildings. Regional differences in perceived expertise were also visible. Approximately 68% of professionals in Downstate New York agreed or strongly agreed that they or their organization have the expertise to response to climate change impacts on buildings, in comparison to 50% in Upstate New York.

Despite the perceived expertise, the majority did not engage in adaptation work. Only 48.4% of the online survey responses indicate that they or their organization have worked on a climate adaptation or resilience project. Regional differences in response were also highlighted by the online survey. Approximately 65% of building professionals in Downstate New York indicated that they, or their organization, have worked on a climate adaptation or resilience project related to buildings in comparison to only 35% of professionals in Upstate New York.

The difference between Upstate and Downstate New York might be a reflection of recent hazardous events such as Hurricane Sandy that caused excessive damage to the building stock and infrastructure in New York City and has driven policy development and increased community engagement in climate change adaptation and mitigation efforts in the region. As one interviewee explained:

> I think particularly after Sandy you have a lot of architects and engineers that are working on and have to work on addressing climate change for the immediacy of it.

Another contributing factor for the relatively low response in both regions can be attributed to confusion between mitigation and adaptation. In the online survey, professional indicated building energy efficiency practices when asked to discuss their adaptation and resilience work. The results reflect the findings of a survey of building professionals conducted in the United Kingdom, in which respondents generally discuss mitigation practices rather than adaptation or resilient practices when asked about actions that can be taken by building professionals in response to climate change (Morton et al. 2011).

One interview participant critiqued the building sector's focus on shifting energy cost as a respond to climate change:

> Currently, the way we build, even in respect to building codes that are coming in there.... all they're anticipating is higher energy costs, so they're directly reacting to the energy costs of the house, and not the fact that the climate's going to change.

The interviews also pointed to another issue regarding the nature of adaptation work being implemented. Building professionals discussed resilience and adaptation strategies that mainly responded to the climate change impacts they have experienced. For example, experience with increased pest infestations and termite damage has led some building professionals to incorporate termite barriers in locations that typically are not prone to termite damage. Another building professional stated:

> The design of how we designed roofs on our buildings changed because we began to design them for higher wind-speed because we were noticing, for severe weather, that the wind was creeping up in speed, but also in duration.

Other professionals who have experienced flooding or storms discussed adaptation strategies that responded to these specific challenges. This poses a limitation on the scope of adaptation work that needs to take place. While professionals recognize the variability in the potential impacts, their response has mainly been driven by their direct experiences or expertise in dealing with a specific component of building operations.

Such responses might be attributed to limited data on localized impacts, a need that was reiterated throughout the interviews. Specifically, professionals spoke of the confusion within the design and building community in utilizing best prediction models and projections and the associated interpretation of potential impacts on the context at hand.

Conclusions

The aim was to explore building professionals' understanding of climate change impacts and engagement in adaptation work. Given the dual implications of buildings in mitigating greenhouse gas emissions and adapting to the impacts of climate change, it is pertinent to explore how stakeholders within the building industry respond to climate change. Although this research represents a limited sample of building professionals and adaptation stakeholders in New York State, it provides multiple insights on the adaptation process within the industry.

The findings suggest that building professionals are aware of climate change as a challenge to building practices. While professionals generally utilized a wide variety of resources to access climate change information, the extensive information available may overwhelm professionals. Consequently, professionals might prefer to stick to standard practices rather than engage in a relatively new science. Complexity of climate change data is also a challenge. To better communicate climate change science to building professionals, the delivery mechanisms of data may need to change and a new climate change language appropriate to the buildings industry developed.

With risks of maladaptation and an unstable base for reliable decision-making, the profession is tasked with challenge of addressing uncertainty within the industry. Participants asserted the importance of response to climate change impacts despite the inherent uncertainty. To address this issue, incentives and nonregulatory tools can be adopted as mechanisms to encourage adaptation to move forward.

The concerns with climate change information, particularly in relation to uncertainty, are important for the understanding stage of the adaptation process. Despite the limitation of climate change data, our results show that building professionals rely on perceived experiences of climate change impacts to shape their understanding. While perceived climate change impacts reflected regional differences within the state, firsthand experiences do not represent a comprehensive understanding of all projected impacts. For example, projected climate change impacts that maybe less visible to professionals, such as termite damage, were not discussed.

The need for a finer resolution of data also emerged as a prominent discussion point to address the limitations in perceived impacts. Participants discussed the need for more localized analysis of climate change impacts as a tool to guide building professionals into adopting appropriate and regional specific strategies. As such, comprehensive data that better illustrates the wide scope of climate change impacts on the various building systems is crucial. While a risk of overwhelming professionals may arise as a result, the findings suggest that providing professionals within the industry with adequate information on the climate change impacts would likely facilitate engagement in climate change adaption.

The differences in identified priority of responses illustrate the need to introduce more effective incentives for building professionals and organizations to engage in climate change adaptation. The findings also suggest that while building professionals are aware of the potential impacts of climate change, a better understanding of what adaptation and resiliency entail is required.

The need for a finer resolution of data also emerged as a prominent discussion point to address the limitations in perceived impacts. Participants discussed the need for more localized analysis of climate change impacts as a tool to guide building professionals into adopting appropriate and regional specific strategies. As such, comprehensive data that better illustrates the wide scope of climate change impacts on the various building systems is crucial. While overwhelming professionals may occur, the findings suggest that providing professionals within the industry with adequate information on the climate change impacts would increase climate change adaption in the buildings sector.

Cross-References

- Working Together to Build Climate Resilience in Hudson Riverfront Communities

Acknowledgments This research was supported by the New York State Energy Research and Development Authority.

References

Atkinson R, Flint J (2001) Accessing hidden and hard-to-reach populations: snowball research strategies. Soc Res Updat 33(1):1–4

Bazeley P (2007) Qualitative data analysis with NVivo, 2nd edn. SAGE, London

Bloch A (2007) Methodological challenges for national and multi-sited comparative survey research. J Refug Stud 20(2):230–247. https://doi.org/10.1093/jrs/fem002

Brennan T (2010) Indoor environmental quality and climate change. US Environmental Protection Agency, Washington, DC

Coley D, Kershaw T, Eames M (2012) A comparison of structural and behavioural adaptations to future proofing buildings against higher temperatures. Build Environ 55:159–166. https://doi.org/10.1016/j.buildenv.2011.12.011

Creswell JW (2014) Research design: qualitative, quantitative, and mixed methods approaches, vol 4. SAGE, Thousand Oaks

Creswell JW, Plano Clark VL (2011) Designing and conducting mixed methods research, 2nd edn. SAGE, Los Angeles

Dubois B, Krasny ME (2016) Educating with resilience in mind: Addressing climate change in post-Sandy New York City. J Environ Educ 1–16. https://doi.org/10.1080/00958964.2016.1167004

Ekstrom JA, Moser, SC, Tom M (2011) Barriers to climate change adaption: a diagnostic framework: final project report. California Energy Commission, Sacramento, California

Field RW (2010) Climate change and indoor air quality. Retrieved 15 June 2012

Gordon AD (2013) Disposable infrastructure including relocatable buildings: adapting to climate change. Aust J Water Resour 17(2):152–160. https://doi.org/10.7158/W13-030.2013.17.2

Hammer SA, Parshall L, Leichenko R, Vancura P, Panero M (2011) Energy. In: Rosenzweig C, Solecki W, DeGaetano A, O'Grady M, Hassol S, Grabhorn P (eds) Responding to climate change in New York state: the ClimAID integrated assessment for effective climate change adaptation. New York State Energy Research and Development Authority, Albany

Hasegawa T (2004) Climate change, adaptation and government policy for the building sector. Build Res Inform 32(1):61–64. https://doi.org/10.1080/0961321032000148488

Hertin J, Berkhout F, Gann D, Barlow J (2003) Climate change and the UK house building sector: perceptions, impacts and adaptive capacity. Build Res Inform 31(3–4):278–290. https://doi.org/10.1080/0961321032000097683

IOM (2011) Climate change, the indoor environment, and health. National Academies Press, Washington, DC

IPCC (2014) Climate change 2014: synthesis report. IPCC, Geneva

Kuzel A (1992) Sampling in qualitative inquiry. In: Crabtree BF, Miller WL (eds) Doing qualitative research, vol 3. Sage, Newbury Park

Lucon O, Ürge-Vorsatz D, Ahmed AZ, Akbari H, Bertoldi P, Cabeza LF, ... Vilariño MV (2014) Buildings. In: Edenhofer O, Pichs-Madruga R, Sokona Y, Farahani E, Kadner S, Seyboth K, Adler A, Baum I, Brunner S, Eickemeier P, Kriemann B, Savolainen J, Schlömer S, Stechow C v, Zwickel T, Minx JC (eds) Climate change 2014: mitigation of climate change. Contribution of working group III to the fifth assessment report of the intergovernmental panel on climate change. Cambridge University Press, Cambridge, UK

Mendelsohn R (2000) Efficient adaptation to climate change. Clim Chang 45(3):583–600. https://doi.org/10.1023/A:1005507810350

Morse JM (1991) Approaches to qualitative-quantitative methodological triangulation. Nurs Res 40 (2):120–123

Morton TA, Bretschneider P, Coley D, Kershaw T (2011) Building a better future: an exploration of beliefs about climate change and perceived need for adaptation within the building industry. Build Environ 46(5):1151–1158. https://doi.org/10.1016/j.buildenv.2010.12.007

Moser SC, Ekstrom JA (2010) A framework to diagnose barriers to climate change adaptation. Proc Natl Acad Sci 107(51):22026–22031

NRC (2009) National Research Council: informing decisions in a changing climate. National Academies Press, Washington, DC

PlaNYC (2013) PlaNYC: a stronger, more resilient New York. Retrieved from http://www.nyc.gov/html/planyc/html/about/about.shtml

Richards L (2009) Handling qualitative data: a practical guide, vol 2nd. SAGE, London

Roaf S, Crichton D, Nicol F (2009) Adapting buildings and cities for climate change: a 21st century survival guide, 2nd edn. Architectural Press, Amsterdam

Saldaña J (2013) The coding manual for qualitative researchers, vol 2. SAGE, Los Angeles

Schmeltz MT, González SK, Fuentes L, Kwan A, Ortega-Williams A, Cowan LP (2013) Lessons from hurricane sandy: a community response in Brooklyn, New York. J Urban Health 90 (5):799–809. https://doi.org/10.1007/s11524-013-9832-9

Scott MJ, Huang YJ (2007) Effects of climate change on energy use in the United States. In: T. S. o. G. C. Research (ed) Effects of climate change on energy production and use in the United States. Climate Change Science Program, Washington, DC

Smit B, Burton I, Klein RJT, Wandel J (2000) An anatomy of adaptation to climate change and variability. Clim Chang 45(1):223–251. https://doi.org/10.1023/A:1005661622966

Strauss AL (1987) Qualitative analysis for social scientists. Cambridge University Press, Cambridge, UK/New York

Biodiversity, Ecosystem Degradation, and Climate Change Effects on Livelihoods in the Bitumen Area of Nigeria

21

Mosunmola Lydia Adeleke, O. J. Oluwatosin, O. A. Fagbenro, T. T. Amos, and I. A. Ajibefun

Contents

Abstract

The chapter assessed the biodiversity, ecosystem degradation, and climate change effects on livelihood of fisherfolks in the bitumen area of Ondo State, Nigeria. The chapter was geographically limited to Agbabu in Ondo state, Nigeria, where bitumen was discovered over a hundred years. The chapter examined the socioeconomic characteristics of the respondents, livelihood diversification of the respondents, extent of degradation of the ecosystem, and effects of bitumen exploration on flora, fauna, and cultivated plants in the study area. Both primary and secondary data were used in the chapter. Multi-

M. L. Adeleke (✉) · O. J. Oluwatosin · O. A. Fagbenro
Department of Fisheries and Aquaculture Technology, FUTA, Akure, Nigeria
e-mail: mosunmolalydia@yahoo.com; mladeleke@futa.edu.ng

T. T. Amos
Department of Agriculture and Resource Economics, FUTA, Akure, Nigeria

I. A. Ajibefun
Department of Agriculture and Resource Economics, FUTA, Akure, Nigeria

The Federal University of Technology, (FUTA), Akure, Nigeria

W. Leal Filho (ed.), *Handbook of Climate Change Resilience*,
https://doi.org/10.1007/978-3-319-93336-8_125

stage sampling techniques were used to select the study area and the number of respondents. The result of the socioeconomic characteristics revealed that 53% of the respondents were female. Forty-two percent of the respondents forecasted that the vegetation will be densely populated, while 20% of the respondents forecasted total degradation of habitat and vegetation. All the respondents indicated that the study area is under future threat of climate change. The chapter therefore recommended that government should intervene in the exploitation of bitumen in the area and sensitized the residences which are mainly the fisherfolks on the adverse effect in the nearest future if the misuse of the bitumen continues in the area.

Keywords
Biodiversity · Ecosystem · Degradation · Bitumen

Introduction

Coastal degradation is the deterioration of the coastal environment through depletion of resources such as air, water, and extinction of wildlife. It is any change to the environment perceived to be undesirable. Degradation is also defined as influences that cause a measurable ecological change (Fanos and Naffa 1995). Coastal areas are known for their rich biological diversity, playing major role in the life cycle of many marine organisms, which are referred to as the world's marine production centers (Olawepo 2000).

Environmental degradation is described as grunge changes produced directly or indirectly by human activity which have either long-term, short-term, or immediate effect on the environment. The issue of ecosystem degradation has assumed a global perspective: human activity is the largest contributing factor on ecosystem degradation (Fanos and Naffa 1995). Climate change has significant impacts on marine and freshwater systems that support the world's fisheries. Indeed, many of these impacts have already been observed in scientific studies, including coral bleaching, ocean acidification, and changes in the range of fish stocks in response to warmer ocean temperatures. In addition to these direct ecosystem impacts, fishing fleets and communities will face an increased risk of extreme weather events, rising sea levels, coastal erosion, and other climate-related effects (Cheung et al. 2009).

However, climate change has been said to impact freshwater fisheries through increasing changes in water temperature and nutrient levels and lower dry season water levels in short term while larger changes in river flows as snow melt, thereby reducing their capacity to sustain regular and controlled water flow in the long term. There is also a likely loss of coastal habitats and resources through sea level rise, warming sea temperatures, extremes of nutrient enrichment and invasive species, and migration of other species of fish or redistribution of fish species (Cheung et al. 2008). Coastal fishing communities are faced with exposure to reduced fisheries resources and increased risks of coastal flooding and storm surges.

Allison et al. (2009) showed that most of the 33 countries whose fisheries sectors are most vulnerable to climate change are African countries, because many African countries are semiarid with significant coastal or inland fisheries.

Nigerian Bitumen Company started exploration for bitumen in 1908 along the coastal region stretching from Okitipupa in Ondo State to Lagos Onoh (1983). Bitumen was discovered at Agbabu, Nigeria over a hundred years ago, preceding the discovery of oil by over 50 years. The probable reserve of bitumen and heavy oil in the entire Nigerian belt is about 120 × 4.3 km (Adegoke and Ibe 1982). Bitumen deposits in Ondo State occurred naturally and are found in two forms, namely, bitumen seepages from wells which yield "plain" bitumen and seepages and outcrops of bitumen-impregnated sands generally referred to as "tar sands" or "bitumen sands."

In Nigeria, immense tracts of mangrove forests have been destroyed as a result of petroleum exploitation in the mangroves (Diop 1993; Wikipedia 2006). Exploitation does not only cause degradation to the environment and destroy the traditional livelihood of the region but has affected soil fertility, waterways, habitats for aquatic organisms, and wildlife. This superficial situation continues to attract the interest of environmental observers and calls for regular evaluation of the exploration and exploitation activities in the coastal areas of the country.

Spill during exploration of natural resource has direct effect on the aquatic organisms causing death of great quantity of fish and threat to all possible uses and users of the water. Poor execution or absence of regulations that regulate exploration of natural resources in Nigeria hinders conservation of aquatic resources. Hence, the deposit of bitumen and its exploration in Agbabu (study area) should have effect on the environment, ecosystems, and people directly or indirectly. This chapter therefore looked at the biodiversity, ecosystem degradation, and climate change incidences and effects on livelihoods in Agbabu, Ondo State; in the bitumen area in Nigeria, specifically, the chapter analyzed the effects of bitumen exploration on the socioeconomic activities of the people in the village, the aquatic resources, and the extent of bitumen exploitation in the community in the past years.

This chapter is useful in informing fisherfolks, researchers, extension agents, and the government: Fisherfolks will be able to know the situation of fishing stock available in their community and to understand ways to conserve the resources. This work will serve as a reference material for researchers who want to further the work on biodiversity, ecosystem degradation, and climate change in coastal region of Nigeria with wider implications globally. Extension agents will be able to give solutions to curb effects of ecosystem degradation on the livelihood of the community dwellers. The government, on the other hand, will be able to intervene on the challenges posed on the community by these looming factors and find ways of ameliorating the negative impacts on the society in the bitumen area of the country.

Research Methodology

The study area was Agbabu, in Ondo state, Nigeria. Ondo state covers a land area of 14,793 km^2 (Fig. 1). The population of the state in 2006 census was 3,441,024. The state is located in the southwestern zone of Nigeria. Data was collected through the use of pretested, structured questionnaire administered to 100 randomly selected respondents of Agbabu community. However, additional hands were engaged in the interpretation of the context of the questionnaires to some of the respondents.

Oral interviews and focus group discussion (FGD) were employed to get relevant information from respondents. The analytical tools used include descriptive statistics in analyzing the objectives highlighted and regression analysis for the hypotheses stated in the cause of the study. Descriptive statistics such as mean, mode, frequency table, chart, and percentage were used to describe the socioeconomic variables.

Results and Discussion

Socioeconomic Characteristics of the Respondents

The socioeconomic characteristics of the respondents examined include gender, age, marital status, educational status, level of education, religion, occupation, length of residency, involvement in fish business, period of fishing activities, and household income.

Table 1 represents the distribution of respondents by gender and level of education. The result reveals that the percentage of the male in the community were 47% while that of the female were 53%. This indicates the dominance of female population in the study area and may imply that the activities practiced in the area might not be rigorous and might be woman friendly. The male counterparts might be engaged in other profession to make ends meet. This implies that we have more women in the study area.

The chapter revealed that 38% of the respondents had no formal education, i.e., illiterates/uneducated, while the rest 62% had one form of education, and the other 12% had primary education, 45% had secondary education, 5% had higher formal education, and 2.0%, 1.0%, and 2.0% had NCE, polytechnic, and other equivalent, respectively. The results show that majority of the respondents were educated. The result agrees with other study which had been carried out in the coastal area of Ondo State, Nigeria that showed that majority of the coastal communities are educated because education is the paramount goal of Ondo State (Adeleke 2013). The result might also be due to the geographical location of the study area; there was no accessible road to the community which is a very vital infrastructure that aid development of a country. The livelihoods of the community dwellers will also be adversely affected because the level of education determines the rate of adoption of new technologies and skills for livelihood diversification.

Figure 2 shows the age distribution of the respondents in the study area. The result reveals that the age of the respondents ranges between 20 and above 60 years.

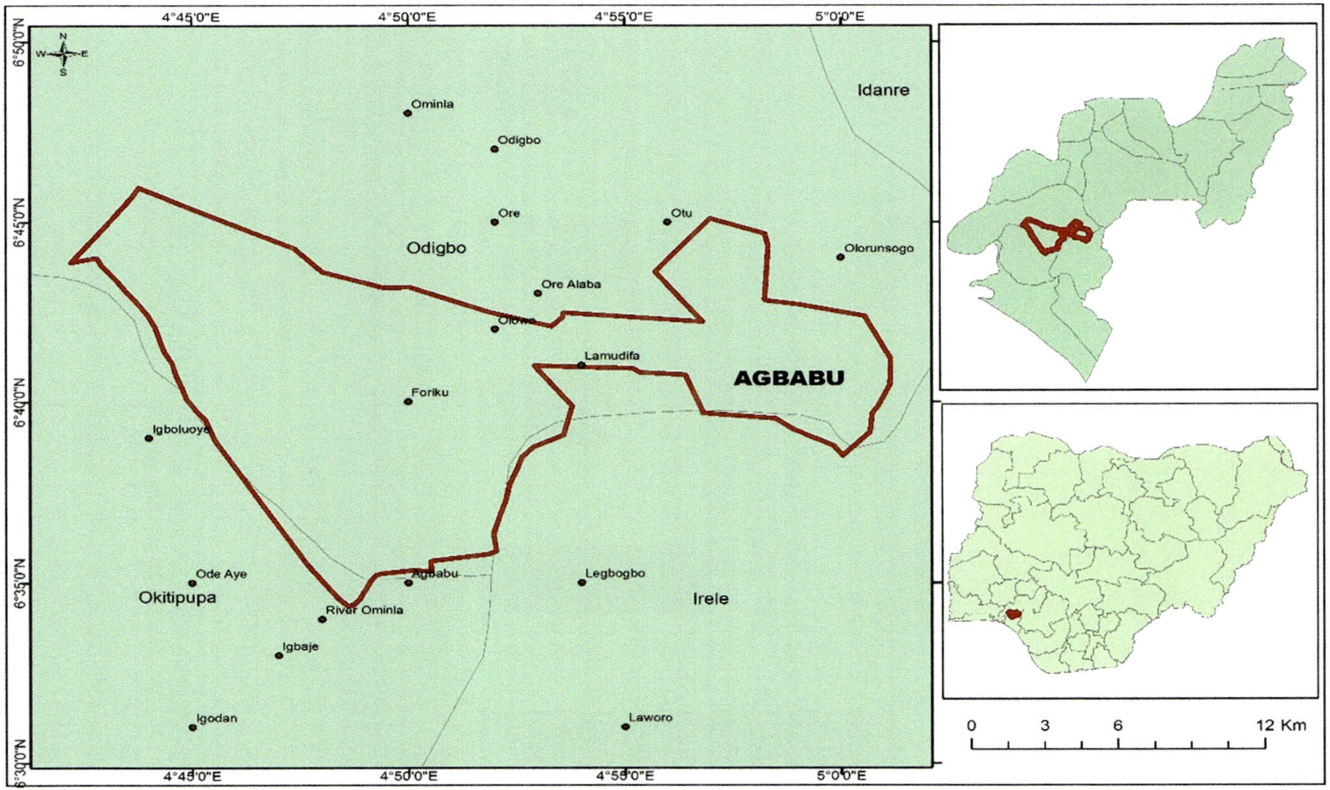

Fig. 1 Map of the study area. (Source: Corresponding Author's Creation: GIS Laboratory, Centre for Space Research and Applications (CESRA), The Federal University of Technology, Akure, Nigeria. (July, 2018))

Table 1 Distribution of respondents by gender and level of education in Agbabu, Ondo State, Nigeria

Parameter		Frequency	Percentage (%)	Cumulative (%)
Gender	Male	47	47.0	47.0
	Female	53	53.0	100.0
Education	Uneducated	38	38.0	38.0
	Primary school	12	12.0	50.0
	Secondary school	45	45.0	95.0
	N.C.E	2	2.0	97.0
	Polytechnic	1	1.0	98.0
	Other	2	2.0	100.0

Source: Authors' Computed Field Data (2015)

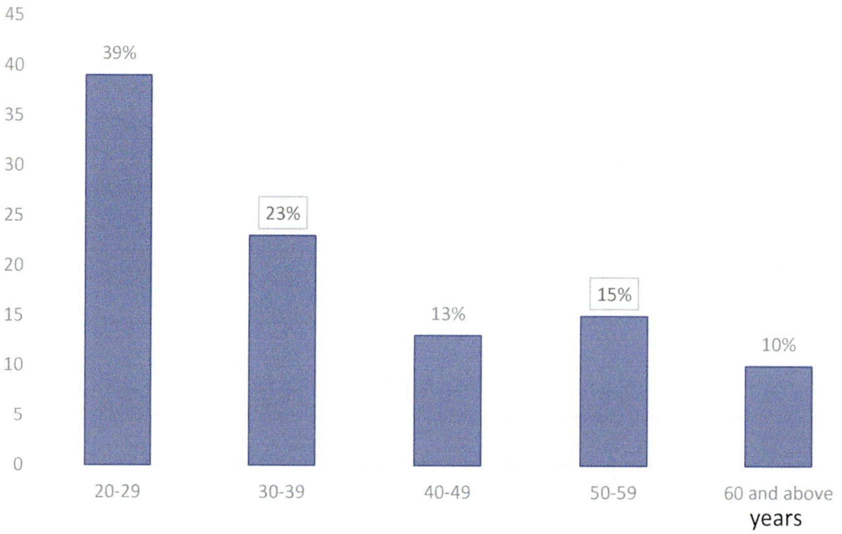

Fig. 2 Age distribution of respondents in Agbabu, Ondo State, Nigeria. (Source: Authors' Computed Field Data 2015)

Thirty-nine percent represent respondents between the age of 20–29 years, 23% (30–39) years, 13% (40–49) years, 15% (50–59) years, while the remaining 10% of the respondents were within the ages of 60 and years and above. The results reveal that 90% of the respondents fall within the economically active group of 20–59 years.

In terms of occupation, 28% of the respondents were traders, 24% were involved in farming (including fishing), and 5% were civil servants. The rest 43% were involve in other jobs majorly as motorcycle riders, few as carpenters and bricklayers. The results of the occupation of the respondents in the study area might be due to the fact that most of the respondents (38%) were uneducated as presented in Table 1.

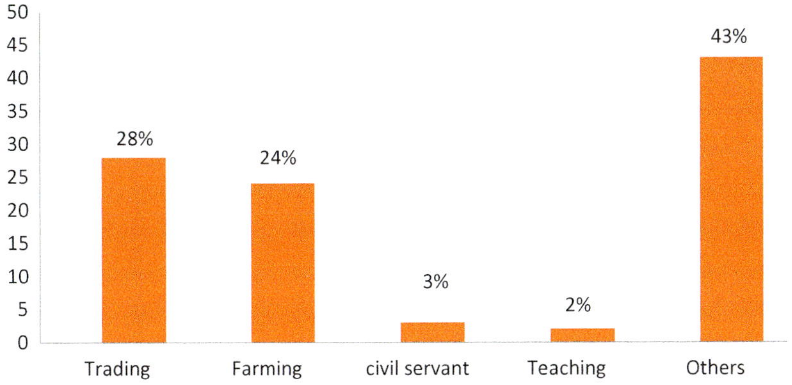

Fig. 3 Distribution of respondents' occupation in Agbabu, Ondo State, Nigeria. (Source: Authors' Computed Field Data 2015)

From the survey and the focus group discussion (FGD), it was gathered from the respondents that most of the youths in the study area were mainly secondary school leavers. This might pose future threat to the environment because the residents might not want to bother about what is happening in their environment since they are uneducated. If there are no urgent interventions by the government of the State, the community might be totally deserted as a result of seeking greener pastures in nearby towns. The livelihoods of the vulnerable groups which include the females and children might also be at future risk.

The distribution of the respondents by marital status as presented in Fig. 4 shows that most of the respondents in the study area (45%) were married, 38% were singles, 3% were divorced, 2% were separated, and 12% were widows. This shows that most of the inhabitants were settled family men and women with responsibilities. The married are actively involved in economic activities than the singles as gathered from the FGD (Fig. 3).

Livelihood Diversification of the Respondents in Agbabu, Ondo State, Nigeria

Table 2 buttresses the various livelihood diversification activities of the respondents in the area as presented in Fig. 3. The results reveal that 41% of the respondents were involved in fisheries and fishing-related ventures/business which comprise of production, processing, distribution, or marketing/trade in fishery products. The rest 59% were engaged in other businesses apart from fishing. Other livelihood sustenance activities identified are traders (28%), farming (24%), civil servants (5%), and motorcycle riders, carpentries, and bricklayers (43%) (Fig. 3). Farming as livelihood activity was done in combination with fisheries. The results from this study are similar to the findings of Anosike and Coughenour (1990) cited by Akanni and

Fig. 4 Marital status of respondents in Agbabu, Ondo State, Nigeria. (Source: Authors' Computed Field Data 2015)

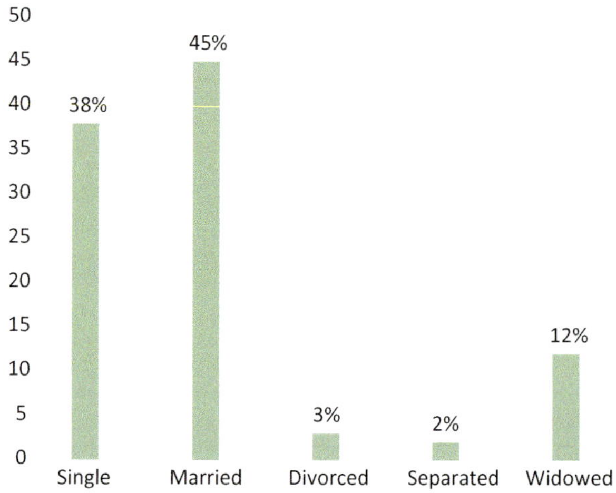

Table 2 Livelihood diversification of the respondents in the study area

Parameter	Percentage (%)	Cumulative percentage (%)
Involvement in fisheries		
No	41.0	41.0
Yes	59.0	100.0
Increased/improved income		
Yes	99.0	99.0
No	1.0	100.0
Effect of bitumen on fishing activities		
Yes	2.0	2.0
No	98.0	100.0
Effect of bitumen on the cost of living		
No	100.0	100.0

Source: Authors' Computed Field Data (2015)

Akinwumi (2007) and Adeleke and Fagbenro (2013) that by diversifying, the fisherfolks may use resources more efficiently and thereby promote sustainability. These results further revealed that the major food production and food security in the area are fisheries based. The chapter revealed that 99% of the respondents' had increased/improved income in the last 10 years as a result of fishing and farming activities. All respondents confirmed affirmatively that bitumen production does not affect the cost of living in the area. This might be a result of the fact that bitumen production has not yet commenced and its effects may have not been felt yet. The effect of bitumen exploration on fishing activities therefore has not been felt yet indicated by 98% of the respondents. The economy of the respondents therefore is likely influenced by the production from fishing and farming only compared to other areas of the country where natural resources like oil is produced and have a profound influence on the economy of adjacent communities. This also confirms the response of some respondents that bitumen is not being tapped and they await government's intervention (Fagbote and Olanipekun 2010). Extraction of bitumen in the area is likely to improve the living standards of the adjacent communities, state, and the country at large.

Extent of Degradation of the Ecosystem

Table 3 presents the response on the state of the vegetation in the past 10 years, the vegetation forecast, and the rate of fishing in the community and prospects of fishing in the next 10 years. It can be deduced from the responses that the vegetation in the past 10 years is denser than it is today as testified by 58% of the respondents. Twenty-two percent of the respondents affirmed that the vegetation of the area is sparsely populated, and 21% of the respondent had no ideas or do not understand what is happening in their environment. Forty-two (42%) of the respondents forecasted that the vegetation of the area will become denser in the next 10 years, and 24% believe that it will be partially degraded, while 20% asserted that the vegetation will be totally degraded, and only 14% were indifferent.

According to the respondents, fishing activity in the community in the last 10 years has been steady as affirmed by 76% of the respondents. The results further revealed that 44% of the respondents feel that fishing output in the area has been relatively steady, while 16% and 4% feel that the fish catch has decreased and moderate decrease, respectively. Monitoring of the ecosystem to determine changes over time is however necessary to be able to give conclusive statement about changes over time. Nevertheless, 99% of respondents signify that all current fish species will be available in the next 10 years because fish stock is not threatened by their fishing activities, while 1% of the respondents did not agree with the future persistence of the fish species. This implies that the respondents might not be aware of climate change presence and its short- and long-term effects on the ecosystems and its ecosystem services. There is, therefore, the need for sensitization on global concepts such as climate change and its effects on ecosystems. Apart from bitumen, other natural mineral resources that are available in the study area include gold and kaolin and occur in large quantities. The respondents' level of awareness designated that 42% were only aware of bitumen, 28% were aware of bitumen and gold, and 5% were aware of bitumen and kaolin, while 25% were aware of all the mineral resources present in the study area.

Effects of Bitumen Exploration on Flora, Fauna, and Cultivated Plant Biodiversity

The response as to the effect of bitumen exploration on flora and fauna biodiversity is shown in Table 4. All respondents had the view that bitumen exploration does not

Table 3 Extent of degradation of the ecosystem in Agbabu, Ondo State, Nigeria

Parameter:	Percentage (%)	Cumulative percentage (%)
State of the vegetation		
Sparsely populated	21.0	21
Densely populated	56.0	77
I do not know	23.0	100.0
Total	100.0	
State of the vegetation in the next 10 years		
Densely populated	42.0	42.0
Total degraded	20.0	62.0
Sparsely populated	24.0	86.0
Indifferent	14.0	100.0
Total	100.0	
Rate of fishing activities		
Frequent	76.0	76.0
Once a while	21.0	97.0
Not often	3.0	100.0
Total	100.0	
Rate of fishing activities in the past 10 years		
Abundant	44.0	44.0
Less abundant	15.0	59.0
Moderate	41.0	100.0
Total	100.0	
Availability of fish species in the next 10 years		
Yes	99.0	99.0
No	1.0	100.0
Total	100.0	
Mineral resources in the study area		
Bitumen	42.0	42.0
Bitumen and gold	28.0	70.0
Bitumen and kaolin	5.0	75.0
All above	25.0	100.0
Total	100.0	

Source: Authors' Computed Field Data (2015)

affect the plant and animal population in the study area. The view of the respondents from the focus group discussion (FGD) shows that bitumen presence or extraction poses no adverse effect on the environment as affirmed by all the respondents.

Given the fact that bitumen extraction is yet to commence in the area, the communities might not be aware of the enormous effects of its exploitation on flora and fauna including cultivated crops. However, as it can be seen from the environmental survey of the research, bitumen that spill over water is sticky in nature, and it reduces the penetration of light and oxygen for both the flora and fauna in the study area. These results conformed to the study of (Fagbote and Olanipekun 2011) that bitumen affects the physicochemical parameters of the soils of farm settlements with bitumen deposit.

Table 4 Effects of bitumen exploration on flora, fauna, and cultivated plant biodiversity in Agbabu, Ondo State, Nigeria

Parameter	Percentage (%)	Cumulative percentage (%)
Effect on flora biodiversity		
No	100.0	100.0
Effect on fauna biodiversity		
No	100.0	100.0
Cultivated plants		
Maize and vegetable	16.0	16.0
Cassava and oil palm	23.0	39.0
Plantain and cocoa	10.0	49.0
All of the above	45.0	94.0
Others	6.0	100.0
Total	100.0	

Source: Authors' Computed Field Data (2015)

Conclusion and Recommendation

The result revealed that Agbabu the area where bitumen is found in Ondo State is not under influence of bitumen exploitation. The research work also reveals that the community lacks a lot of infrastructures and basic amenities. However, if extraction of bitumen should commenced in the community, it will improve the standard of living of the people with better roads, health, and communications facilities.

The future threat on the biodiversity and ecosystem of the community should be patterned in the presence of climate variables and change. The extent of exploitation then should be controlled as any appearance of pollution may lead to damage of the farm settlement, water and aquatic resources, and the general health and well-being of the people of the community. This chapter therefore recommends that the bitumen in the area should be tapped to improve the standard of living of the people. Environmental measures should also be put in place to protect the flora and fauna in Agbabu, Ondo State, Nigeria.

Acknowledgments African-German Network of Excellence in Science: AGNES Grant for Junior Researchers 2015 – supported by AvH and TWAS (The World Academic of Science).

References

Adegoke OS, Ibe EC (1982) The tar sand and heavy crude resources of Nigeria. 2nd International Conference on Heavy crude and Tar-sands, Caracas, Venezuela, chapter 32, pp 280–285, 1982

Adeleke ML (2013) Socioeconomic characteristics of the artisanal fisherfolks in the coastal region of Ondo state, Nigeria. J Econ Sustain Dev. www.iiste.org. ISSN 2222-1700 (Paper) ISSN 2222-2855 (Online) 4(2):2013133

Adeleke ML, Fagbenro OA (2013) Livelihood diversification and operational techniques of the artisanal Fisherfolks in the coastal region of Ondo state, Nigeria. Int J Innov Res Dev. ISSN 2278-0211(online) 2(1):262–273

Akanni KA, Akinwumi JA (2007) Determinants of variations in fish catch levels in artisanal fishing areas. Kainji Lake Basin Annual Report 1999, p 6

Allison E, Perry A, Badjeck MC, Adger WN, Brown K, Conway D, Halls AS, Pilling GM, Reynolds JD, Andrew NL, Dulvy NK (2009) Vulnerability of national economies to the impacts of climate change on fisheries. Fish and Fisheries 10:173–196

Anosike N, Coughenour MC (1990) The socio-economic basis of farm enterprise diversification decisions. J Rural Soc 55:1–24. https://doi.org/10.1111/j.1549-0831.1990.tb00670.x

Cheung WWL, Close C, Lam VWY, Watson R, Pauly D (2008) Application of macro ecological theory to predict effects of climate change on global fisheries potential. Mar Ecol Prog Ser 365:187–197

Cheung WWL, Lam VWY, Sarmiento JL, Kearney K, Watson R, Zeller D, Pauly D (2009) Projecting global marine biodiversity impacts under climate change scenario. Fishand Fish. https://doi.org/10.1111/j.1467-2979.2008.00315.x

Diop ES (1993) Conservation and Sustainable Utilization of Mangrove ecosystem

Fagbote EO, Olanipekun EO (2010) Evaluation of the status of heavy metal pollution of sediment of Agbabu bitumen deposit area, Nigeria. Eur J Sci Res 41(3):373–382

Fagbote EO, Olanipekun EO (2011) Seasonal and depth effects on the physico-chemical parameters of the soils of farm settlements with bitumen deposit. Int Res J Biotechnol 2(9):198–212

Fanos AM, Naffa GN (1995) Long and short term changes of Rosetta, Proceedings on the International conference on Mediterranean coastal environment Tarragona, Spain

Olawepo RA (2000) Environmental Pollution and Management Techniques. In: Jimoh HI, Ifabiyi IP (eds) Contemporary issues in environmental studies. Haytee Press and Publishing Company, Ilorin

Onoh JK (1983) The Nigerian oil economy. St, Martin's Press, New York, pp 12–18, presented at the 2nd Nigerian Economic Summit, Abuja

https://en.wikipedia.org/wiki/2006

Climate Change and Coastal Resilience in Nigeria

22

Mosunmola Lydia Adeleke, Bosede Olufunmilayo Akinwalere, Kehinde Oluwatosin Olajubu, and Gbenga Emmanuel Onibi

Contents

Abstract

Nigerian coast is likely to experience severe effects from flooding due to its geography, climate, vegetation, soils, economic structure, population and settlement, energy demands, and agricultural activities. In this chapter, case study and secondary data were mainly used to assess the effect of flooding and resilience provided as a result of the impacts of climate change in all the eight coastal states in the country in recent years. The result revealed that flooding has effect on the

M. L. Adeleke (✉) · K. O. Olajubu
Department of Fisheries and Aquaculture Technology, FUTA, Akure, Nigeria
e-mail: mladeleke@futa.edu.ng; mosunmolalydia@yahoo.com

B. O. Akinwalere
Department of Agricultural Extension and Communication, FUTA, The Federal University of Technology, Akure, Nigeria

G. E. Onibi
Department of Animal Production and Health, FUTA, The Federal University of Technology, Akure, Nigeria

W. Leal Filho (ed.), *Handbook of Climate Change Resilience*,
https://doi.org/10.1007/978-3-319-93336-8_126

water quality, species distribution, physiology, and population of the aquatic organism in the coastal areas of the country. However, the ecosystem is greatly predisposed and the livelihood of the coastal communities was swayed, hence food insecurity, the fisherfolk in the study area have various inadvertent and intentional adaptation strategies adopted for maximum fish production/catch. It is therefore recommended that Nigeria coastal communities develop local adaptation strategies such as construction of canals and channels, seawalls, and dikes to protect their houses and farms and sustain their livelihoods against climate change and flood disasters.

Keywords
Climate change · Coastal resilience · Flood disaster · Vulnerable and Nigeria

Introduction

Over the years, evidence that climate is changing have been gathered by the Intergovernmental Panel on Climate Change (IPCC) and other sources. The major changes in the coast and marine zones include water temperature, ocean currents, upwelling, and biogeochemistry (IPCC 2007b; Diaz and Rosenberg 2008). A rise in extreme events will have effects on health and lives as well as associated environmental and economic impacts (UNFCCC 2007). Changes in rainfall patterns are likely to lead to severe water shortages and/or flooding. In fact, the world has seen a 5–10% increase in precipitation over the past century. Some computer models predict that the frequency of heavy rainfall events is likely to rise with global warming, further increasing the potential for flooding (NSTA 2007).

Climate is statistically described in terms of means and variability of key weather parameters for a given area over a period of time, usually at least 30 years. Climate change is any change in climate over time, whether due to natural variability or as a result of human activity. Adaptation strategies, therefore, are means that have evolved over time through peoples' long experience in dealing with the known and understood natural variation that they expect in seasons combined with their specific responses to the season as it unfolds.

Consequently, climate change with expected long-term changes in rainfall patterns and shifting temperature zones is expected to have significant negative effects on fisheries, food and water security, and economic growth in Africa. Increased frequency and intensity of drought and flood are expected to negatively affect fishery production (Mano and Nhemachena 2006). Climate change will have greater negative impact on poorer fishing households as they have the lowest capacity to adapt to change in climate conditions. Adaptation has the potential to significantly contribute to reduction in negative impacts from changes in climatic conditions (Allison et al. 2009; Cheung et al. 2010).

In Nigeria, the artisanal fishery occupies a significant position in the economy, by contributing 4% to the total GDP, providing employment for about 6% of the Nigerian population, and supplying 82% of the total domestic fish production

(FDF 2007; FDF&A 2017). Over 200 million of the world populations are full- and/or part-time fisherfolk: fishers, fish processors, traders, and ancillary workers (Teh et al. 2011). Forty-two million people work directly in the fishery sector, with the great majority in developing countries (FAO 2008; Allison et al. 2005; FDF 2007).

Nigeria is vulnerable to climate change impacts. Climate change is expected to impact adversely on wetland functions through change in hydrology, biogeochemistry, and biomass accumulation (Oluowo 2017). Particularly, climate change is expected to cause melting of sea ice, rise in sea levels and loss of habitat, higher storm frequency and intensity, and flooding in the coastal regions, among others. The Nigerian coast is likely to experience severe effects from flooding. The low elevation and topography of the entire Nigerian coastline area make it highly susceptible to flooding, especially at high tides and during the rainy season. Research have shown that Nigeria is already being plagued with diverse ecological problems, which have been directly linked to the ongoing climate change (Idowu et al. 2011).

The fisherfolk, living in areas vulnerable to human-induced climate change, depend for a major part of their livelihood on resources whose distribution and productivity are known to be influenced by climate variability (Allison et al. 2005). Climate change has impacts on fisherfolk already living in poverty (Badjeck et al. 2009). There are some evidence that the inability of fishing households to adapt to environmental change is not only linked to the level of poverty but also to the "specialization trap" where fisherfolk rely mostly on one species or activity (Badjeck et al. 2009; Akanni and Akinwumi 2007). This chapter, therefore, sheds more light on coastal resilience in the presence of climate change in Nigeria. Coastal resilience means building the ability of a community to "bounce back" after hazardous events such as hurricanes, coastal storms, and flooding rather than simply reacting to impacts (National Ocean Service).

Climate Change in Nigeria

Climate change is a global problem which has brought about much anxiety and confusion (Adeleke and Omoboyeje 2016) with Nigeria being highly susceptible to its impacts (IPCC 2007b) especially along the coastline which is prone to sea-level rise, the risk of fierce storms (Apata et al. 2009), floods, and droughts (Sikoki 2013).

Several dedicated studies have been conducted to investigate the impacts of climate change on Nigeria fisheries and aquaculture such as:

The effects of climate change on aquaculture stating that climate change is an additional pressure on top of the many (fishing pressure, loss of habitat, pollution, disturbance, introduced species) which fish stocks already experience and that it is clear that impacts will be adverse with low levels of temperature change especially in the coastal areas (Yazdi and Shakouri 2010)

Effect of climate change and coping strategies on socioeconomic status of fisherfolk in Ilaje-Ese Odo Local Government Areas of Ondo State, Nigeria, stating that there is an increase in rising global temperature and a high risk of extreme flooding as a result of climate change resulting in reduction in fishing activities (Agbebi and Omoniyi 2011)

Occurrence of climate change and its effects on the fishing activities in the coastal region of Ondo State, Nigeria, which was opined by Adeleke and Balogun (2013)

The potential impacts of climate change on fisheries and aquaculture in Nigeria (Ipinjolu et al. 2014)

Climate change, effects, and mitigation strategies on aquaculture with both authors stating that climate change increase, especially through temperature and carbon dioxide, will continuously warm the earth with great effects on fisheries and aquaculture (Anyanwu et al. 2015)

Effects of climate change on aquaculture production and management in Akure metropolis, Ondo State, Nigeria, stating that fish production and food security are at risk without proper adaptive measures on climate change (Adeleke and Omoboyeje 2016)

In summary, the studies done revealed that climate change is real, so also its severe impacts and effects and apt attention are needed to curb further damages as its result and to also promote fishery activities by the people which will further promote their standard of living and boost the nation's economy.

Nigeria Coastal State Description

The coastal states and capitals in Nigeria are Akwa Ibom State (Uyo), Bayelsa State (Yenagoa), Cross River State (Calabar), Delta State (Asaba), Lagos State (Ikeja), Ogun State (Abeokuta), Ondo State (Akure), and Rivers State (Port Harcourt) (Fig. 1). These states shared the nautical kilometers of about 320 km (853 km^2) into the Atlantic Ocean. The coastline is shared among the states as specified: Lagos State (130 km), Ogun State (18 km), Ondo State (180 km), Delta State (126 km), Bayelsa/River States (390 km), and Akwa Ibom/Cross River States (108 km). The coastline lies between latitude 4 10–6 20′N and longitude 2 45–8 35′ E. The terrestrial portion of this zone is about 28,000 km in area, while the surface area of the continental shelf is 46,300 km.

Akwa Ibom State (Uyo) occupies the southeast corner of Nigeria and is bounded in the north by Abia State, in the south by the Atlantic Ocean, in the east by the Cross River State, and in the west by Rivers State. Bayelsa State (Yenagoa) is bounded in the north by Delta State, in the east by Rivers State, and in the south and the west by creeks and rivers spreading through the Atlantic Ocean. Cross River State (Calabar) shares boundaries with Benue State in the north, Enugu and Abia States in the west, the Republic of Cameroon in the east, and Akwa Ibom and the Atlantic Ocean in the south. The coastline is about 105 km and it runs from Imo River to Cross River. Delta State (Asaba) is bordered by Edo State in the north, Ondo in the northwest, Anambra in the east, and Rivers in the southeast. The coastline is about 126 km. Lagos State (Ikeja) is bounded in the

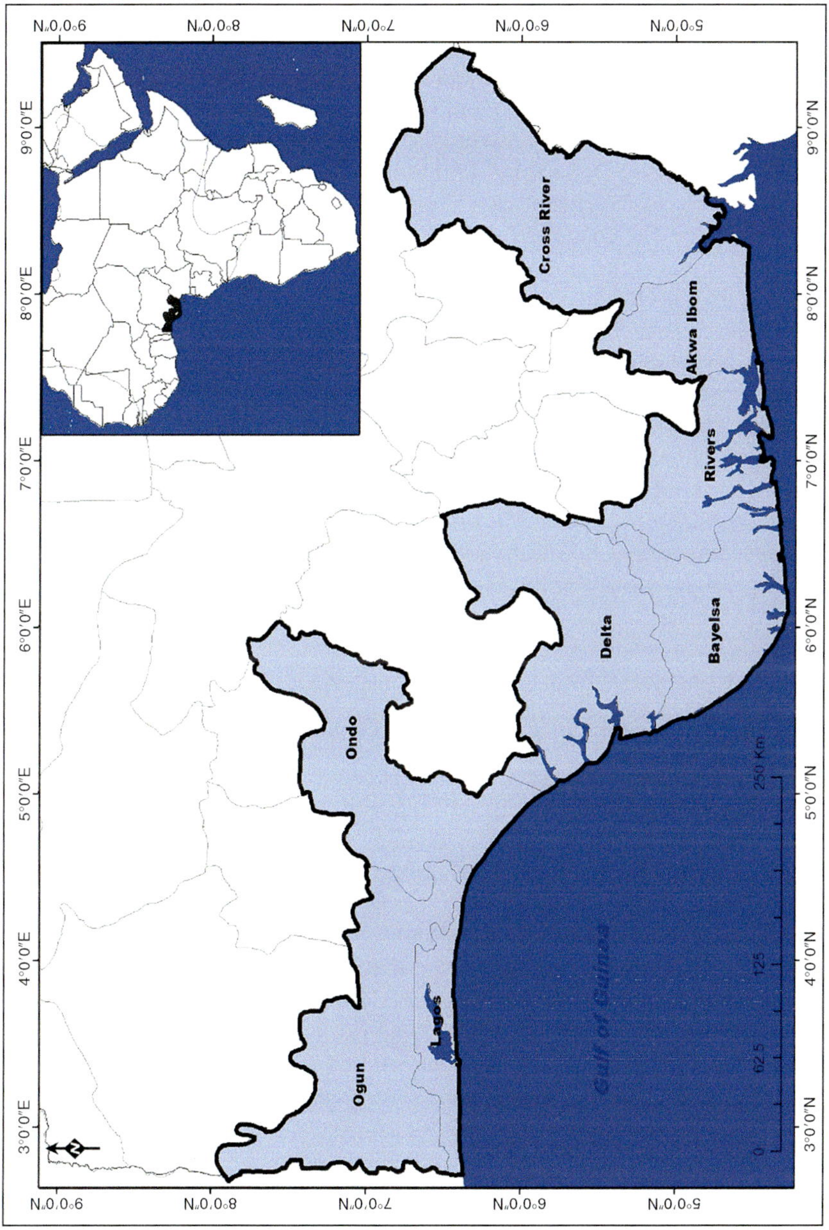

Fig. 1 Map of the coastal states in Nigeria (Source: Corresponding Author's Creation: GIS Laboratory, Centre for Space Research and Applications (CESRA), The Federal University of Technology, Akure, Nigeria (June 2018))

north and east by Ogun State, in the west by Republic of Benin, and in the south by Atlantic Ocean. The coastline is about 130 km long, and it's generally characterized by steep sandy beaches, offshore wave breakers, and littoral drift. Ogun State (Abeokuta) is bounded in the west by the Republic of Benin, in the south by Lagos State and Atlantic Ocean, in the east by Ondo State, and in the north by Oyo State. The Ogun coastline is about 18 km and runs east-west direction. Ondo State (Akure) is bounded by Kogi, Edo, Ogun, Osun, and Ekiti States with a coastline of about 180 km and runs in a northwest to southeast direction. Rivers State (Port Harcourt) is bounded by Delta State, Imo State, Akwa Ibom State, and Bayelsa State. The coastline of Rivers and Bayelsa is about 390 km extending from the Ramos to Imo River. The coastline forms a quadrant of a circle between the Ramos and Brass Rivers, but at the east of Brass River, the coastline runs north of east to the estuary of the Imo River.

The effect of climate change in southern part of Nigeria is enormous. There is a great deal of uncertainty about climate change, but there are some certainties, because the impacts of climate change will not be felt evenly across the world and may not all be negative. Climate change is already happening, and climate variables will continue to fluctuate during the coming years. Fishing communities are equally vulnerable; climate-related dynamics have had serious consequences on evolution of species, fishery variability, and the livelihood of fisherfolk (Adeleke and Balogun 2013). Climate change has drastically reduced fish farming activities in the coastal region of Nigeria, as a result of increased flooding and variation in rainfall pattern and intensity. Climate change is a threat to sustainable development and one of the greatest threats facing mankind today (IFAD 2007; World Bank 2010; Adeloye and Sotomi 2013). It poses significant threats to fisheries on top of many other concurrent pressures such as overfishing, habitat degradation, pollution, introduction of new species, and so on (Brander 2010). The effects of climate change range from the overall reduction in fish production to drastic changes in the quality and quantity of water which is the only fish environment (Ayanwuyi et al. 2012). However, climate change has been said to impact fisheries through the increasing changes in water temperature, nutrient levels, and lower dry season water levels (Adeleke and Omoboyeje 2016).

The ocean and coastal waters are valuable to the survival of our planet and species and not just aquatic species. It is so important that it produces half of the oxygen in the atmosphere and contains 80% of the Earth's living organisms. Some of the oceans' benefits are that it provides seafood and livelihood for millions, medicine to treat diseases, shipping, energy, and tourism and supports coastal economies worldwide. However, human activities are seriously having a negative impact on it and its functions and resources.

Impacts of Climate Change on Coastal Areas in Nigeria

Nigeria is vulnerable to climate change impacts due to its geography, climate, vegetation, soils, economic structure, population and settlement, energy demands, and agricultural activities. The location and size of and the characteristic relief in Nigeria give rise to a variety of climates ranging from tropical maritime climate characterized by the rainforest along the coastal and southern section to the tropical hinterland climate associated with the Sahel in the northeastern section of the country. Impacts on coastal systems are among the most costly and most certain consequences of a warming climate (Nicholls et al. 2007).

The Nigerian coast is likely to experience severe effects from flooding. The low elevation and topography of the entire Nigerian coastline area make it highly susceptible to flooding, especially at high tides and during the rainy season. Research have shown that Nigeria is already being plagued with diverse ecological problems, which have been directly linked to the ongoing climate change (Odjugo and Ikhuoria 2003; NEST 2003).

Climate change is expected to impact adversely on wetland functions through change in hydrology, biogeochemistry, and biomass accumulation. Particularly, climate change is expected to cause melting of sea ice, rise in sea levels, and change in oceanic circulation patterns, species extinction and saltwater intrusion, loss of habitat, higher storm frequency and intensity, and flooding in the coastal regions, among others.

Effects of Climate Change on Fisheries

The links between fisheries and their ecosystems are deeper and more significant than those that exist in mainstream agriculture (FAO 2008b). The productivity of a fishery is tied to the health and functioning of the ecosystems on which it depends for food, habitat, and even seed dispersal (MAB 2009); generally the only control humans can exert over a fishery's productivity is adjustment of fishing effort (Brander 2007). Estuaries, mangroves, coral reefs, and seagrass beds are particularly significant in the provision of ecosystem services, especially as nurseries for young fish (MAB 2009), and are also among the most sensitive and highly exposed to the negative impacts of coastal development, pollution, sedimentation, destructive fishing practices, and climate change. Fish also tend to live near their tolerance limits of a range of factors; as a result, increased temperature and acidity, lower dissolved oxygen, and changes to salinity can have deleterious effects (Roessig et al. 2004). Particular characteristics of the aquatic environment which will be affected by climate change are temperature and primary production.

All marine and aquatic invertebrates (mollusks, crustaceans, worms, etc.) and fish are poikilotherms; their internal temperature varies directly with that of their environment. This makes them very sensitive to changes in the temperature of their surrounding environment. When changes do occur, they move to areas where the external temperature allows them to regain their preferred internal temperature. This "behavioral thermoregulation" (Roessig et al. 2004) results in rapid migrations poleward or into cooler bodies of water (FAO 2008a), corresponding to the poleward shift of climatic zones. As a result, benefits are likely to accrue at higher latitudes, and losses will be experienced in the tropics. Some species will also shift from shallow coastal waters and semi-enclosed areas, where temperatures will increase fastest, into deeper cooler waters (Cheung et al. 2009). Recent predictions suggest this migration alone could reduce maximum catch potential in some areas of the tropics by up to 40% (Cheung et al. 2010), but this may be a conservative estimate as it does not take into account predicted negative effects of climate change on coral reefs or the impact of ocean acidification (Cheung et al. 2009). Recruitment is also strongly affected by climate variability, and some stocks may become vulnerable to overfishing at levels of fishing effort that had previously been sustainable (Easterling et al. 2007).

There are likely to be local extinctions of fish species at the edges of their ranges, especially among freshwater and diadromous species (IPCC 2007a). However, overall extinction rates for marine species are lower than those predicted for terrestrial species (15–37%), in part due to their higher potential for migration (Cheung et al. 2009). Primary productivity is affected by availability of nutrients in the water, which in turn depends on freshwater runoff and ocean mixing as well as levels of light and temperature. In some areas, reduced precipitation could lead to reduced runoff from land, starving wetlands and mangroves of nutrients and damaging local fisheries. In other areas, increased precipitation or increased extreme weather events, including flooding, will lead to excessive nutrient levels in rivers, lakes, and coastal waters as sewage and fertilizer are washed into water bodies causing harmful algal blooms, also known as red tides. With climate change, primary productivity is predicted to decline at lower latitudes (FAO 2008a), where the majority of the world's small-scale fisheries are located, reducing the productivity of the fisheries.

Badjeck et al. (2010) stated that various climate change events due to ocean currents, sea-level rise, rainfall, river flows, storm severity and frequency, salinity, acidification, temperature, lake levels, and thermal structure result in affecting the production ecology; fishing, aquaculture, and associated postharvest operations; communities and livelihoods; and wider society in a short run which then generates long-term effects (impacts) on species composition, production and yield, distribution, seasonality, and diseases; safety and security, efficiency and costs, and infrastructure; loss/damage to livelihood assets, risks to life and health, displacement, and conflict; and adaptation and mitigation costs, market impacts, water allocation, floodplain, and coastal defenses, respectively. Ozor (2009) stated that subtle changes in key environment variables such as temperature, salinity, wind speed and direction, ocean currents, and strength of upwelling due to climate change could sharply alter the abundance, distribution, and availability of fish production in the country. Climate change may directly affect fishery production along many pathways.

The impacts could also be positive. Increased rainfall and flooding will lead to inundation of adjacent land areas, and this will be beneficial to fisheries. It expands the littoral zones and increases the diversity of the ecotones for fish feeding, breeding, and rearing of the juveniles often caused by excessive floods are detrimental to fisheries.

Coastal Resilience

Resilience implies measures taken to decrease the antagonistic effect of an unnatural weather change on human life and nature. These include changing the editing designs; ceasing further improvement on wetlands, surge fields, and near-ocean level; creating crops that are impervious to dry season, warmth, and salt; fortifying general well-being and natural building guard against illnesses; outlining and fabricating new water ventures for surge control and dry season administration; and developing of dikes and tempest surge boundary against ocean-level ascent. Climate adaptation as a wide range of behavioral adjustments that households and institutions make (including practices, processes, legislation, regulations, and incentives) to mandate or facilitate changes in socioeconomic systems aimed at reducing vulnerability to climatic variability and change. Defined adaptation as the decision-making process and the set of actions undertaken to maintain the capacity to deal with current or future predicted change.

Nigerian urban and coastal areas are typical examples of this high level of risk and vulnerability. However, the country needs adaptive and mitigation strategies that will improve the management of the fisheries and aquaculture, protect the integrity of the aquatic ecosystems, respond to the opportunities and threats on livelihoods and food security, and reduce greenhouse gas emissions (FAO 2009). Coastal resilience approaches including assessing risk and vulnerability, identifying solutions, taking action, and measuring effectiveness are the critical elements to be considered in coastal resilience. For capture fisheries, adaptation involves adjusting fishing pressure to sustainable levels. Setting catch limits based on changes in recruitment, growth, survival, and reproductive success can be done via adaptive management, monitoring, and precautionary principles. This may also require changes in vessel or gear types if new fishery opportunities become available. These may come with high costs (for new vessels and/or gear), but these changes may support the continuation of fishing communities.

There are a number of adaptation activities that can be applied in fisheries. These include:

Reduce external stressors on natural systems.

Identify and protect valuable areas.

Invest in safer harbors and landings and measures to improve safety at sea due to increased storm severity as well as improved early warning and forecasting systems for severe weather events.

Promote disaster risk management in general (including disaster preparedness) and protective infrastructure (e.g., "hard" options such as seawalls and flood reservoirs or "soft" options such as buffer zones via afforestation or reforestation of mangroves).

Fishery and aquaculture sectors should fully be integrated into the national policies, and climate change planning needs to be included. Partnerships between private, public, civil society, and NGO sectors are vital for holistic climate change adaptation planning. Standard practice adoption for improved fisheries and aquaculture management (e.g., the FAO Code of Conduct for Responsible Fisheries, precautionary principles, adaptive and ecosystem management) and integrated coastal management for coastal and nearshore fisheries can improve resilience and increase system adaptability.

Coastal Resilience Measures

In Nigeria's coastal communities, common adaptation techniques employed include the use of sandbags along the shore, river embankments, construction of canals and channels, building of dwellings on stilts/raised platforms to prevent homes from being washed away by flood and rising seawater, and government interventions in the form of construction of engineered walls/shore protection embankments.

Other measures include:

Introduction of aquaculture species adapted to high temperatures and changed salinities

Promotion of polyculture and fish-rice rotation in relevant areas

Reduction of land-based sources of pollution (e.g., agricultural and urban runoff) and destructive fishing practices (e.g., fishing with explosives and poisons)

Improved early warning and forecasting systems for severe weather events

Investments in safer harbors and landings and measures to improve safety at sea due to increased storm severity

Standard practice adoption for improved fishery and aquaculture management (e.g., the FAO Code of Conduct for Responsible Fisheries, adaptive and ecosystem management)

Case Study: Adaptation Strategies Adopted by the Artisanal Fisherfolk in the Coastal Area of Ondo State, Nigeria (2011–2012)

The research carried out as a case study by Adeleke (2013), Adeleke and Balogun (2013), and Adeleke and Matthias (2016) revealed that the fisherfolk in the coastal area of Ondo State have different methods of adaptation as presented in Table 1. The result of the research reveals that between the months of January and February, i.e., Harmattan period, this period is characterized by dry wind as a result of increase in temperature, the fisherfolk observed that the fish do run deep inside the water for hiding and catch is low. The inadvertent or intentional coping strategies employed by the fisherfolk include adjusting fishing period to early morning or night, changing the fishing gears to hook and lines, and setting of traps. These methods of adaptation they believe are affordable and effective. Likewise, decrease in temperature usually occurs between November and March during the period of the study in 2011–2012; at this period, the fisherfolk believe that fishes will "come out for food"; hence, catch is high and normal fishing takes place. Fisherfolk believed it affects behavioral

Table 1 Adaptation strategies adopted by the artisanal fisher folks in the coastal area of Ondo State, Nigeria (2011–2012)

S/N	Variables	Effects on fish	Effects on fisherfolk	Adaptive strategies	Characteristics of the adaptive strategies
1.	Increase in temperature	Makes the fish to run deep inside the water for hiding, and catch is low	Get fishes in the small streams	1. Adjusting fishing period to early morning or night	It is affordable and effective
	(January–February)			2. Changing the fishing gears to hook and lines	
	Harmattan			3. Setting traps	
2.	Increase in rainfall	Highest catch of fishes	Fishermen/women go far to the sea and catch big fishes	1. Setting of various types of traps	1. Affordable
	(March–May)	Fishes would come out for food	They go to fish at times 3 times daily	Igun/awa	2. Effective
	(June–July)			Iyanma	3. Available
				Bamboo	
				2. Change the fishing gears to favor the season	
				3. Adjusting fishing period to morning	
				4. Construction of fish walls	
3.	Decrease in temperature	Allows fishes to come out for food	1. Normal fishing takes place	Changing fishing methods by adjusting the fishing period and fishing gears	Environmentally friendly, affordable, and effective
	November–March		2. Catch is high		
4.	Decrease in rainfall	Fishes are exposed because the water that serves as habitat for them might have been dried off	1. It is seasonal	1. Night fishing operation 2.Changing fishing gears to hook and line	1. Affordable
	November–March		2. Catch is high		2. Effective
			3. Normal fishing takes place		3. Available

Source: Authors' case study Ondo State Coastal Area 2013

phenomenon of the fishes and influences the presence of fish and in the way they are caught. At this period, fisherfolk would have maximum fish production, and normal fishing would take place as a result of the favorable temperature. The implication of this is that more fisherfolk would like to fish at this period because catchability would be high with minimum unit effort, which increases the revenue and, hence, increases the standard of living of the fisherfolk.

Furthermore, increase in rainfall was observed between March and July; during the period of the research, the fisherfolk witness highest catch of fishes because they believe that fishes would come out for food. Fisherfolk in the study area believed that it was the best period to catch fish, and they refer to it as the season of "new moon." At this period, fish production would be at the highest peak, because both the fishermen and women go far to the sea covering about 8–14 nautical miles for bigger fishes. Some of the respondents can go for fishing three or more times daily. Fisherfolk made use of various fishing gears and adaptive measures for maximum fish production. Various adaptive methods which the fisherfolk believe were affordable, effective, and available were employed. These include the setting of various types of local/indigenous traps (igun/awa, iyanma, bamboo), changing the fishing gears to favor the season, adjusting fishing period to morning, and constructing fish walls. Similarly, decrease in rainfall was observed in November and March during the course of the study (2011–2012). The inadvertent and intentional adaptation strategies adopted for maximum fish production/catch at this period by fisherfolk for effective and efficient fish production include the attachment of lamp in the bow area of the stern of the canoe; this "kerosene lamp" attracted the fish to jump as a reaction to the light reflection. This fishing strategy usually catches species of the families Belonidae and Mugilidae (Adeleke and Matthias 2016). They also believe that the water that supposed to serve as habitat for the fish might have been dried off; hence, fishes are exposed for easy catch. The result implies that both increase and decrease in rainfall would lead to increase in fish production because fish production/catch is usually very high at this season with little effort.

Conclusion

Coastal zones are highly vulnerable to climate change and its impact. Settlements are eroded and flooded/inundated, and salinization of aquifers highly impedes livelihoods, quality of life, and community development. The coastal communities have developed local but temporary adaptation strategies which may be intentional or unintentional to protect their settlements/dwellings and sustain their livelihoods against climate change-related disasters such as construction of canals and channels, seawalls, and dykes to protect their houses and farms and sustain their livelihoods against climate change and flood disasters. Also, the fisherfolk in the study area have various inadvertent and intentional adaptation strategies adopted for maximum fish production/catch for effective and efficient fish production. It is therefore recommended that adaptation strategies to reduce the vulnerabilities of these coastal communities to climate change be paramount requiring actions that must be undertaken from the community level to the regional, national, and international levels.

References

Adeleke ML, Balogun AA (2013) Occurrence of climate change and its effects on the fishing activities in the coastal region of Ondo State Nigeria. Int J Dev Econ Sustain 1(1):1–13

Adeleke ML, Matthias W (2016) Chapter 13: Adaptation of the artisanal fisher folks to climate change in the coastal region of Ondo State, Nigeria. In: Leal Filho W (ed) Innovation in climate change adaptation, climate change management. Springer International Publishing Switzerland, Switzerland, pp 177–193. https://doi.org/10.1007/978-3-319-25814-0_13

Adeleke ML, Omoboyeje VO (2016) Effects of climate change on aquaculture production and management in Akure Metropolis, Ondo state, Nigeria. Niger J Fish Aquacult 4(1):50–58

Adeloye KA, Sotomi AO (2013) Youth farmers' coping strategies with climate change in Osun state, Nigeria. J Biol Agric 3(14):61–66

Agbebi FO, Omoniyi AO (2011) Effect of climate change and coping strategies on socio-economic status of fisher folks in Ilaje-Ese Odo local government areas of Ondo state, Nigeria. Int J Sci Nat 2(2):339–343

Akanni KA, Akinwumi JA (2007) Determinants of variations in fish catch levels in artisanal fishing areas. Kainji Lake Basin Annual Report 1999, pp 6

Allison EH, Adger WN, Badjeck M-C, Brown K, Conway, D., Dulvy, N.K. (2005) Effects of climate change on the sustainability of capture and enhancement fisheries important to the poor: analysis of the vulnerability and adaptability of fisherfolks living in poverty. Fisheries Management Science Programme project no. R4778J, MRAG, London

Allison EH, Perry AL, Badjeck MC, Adger WN, Brown K, Conway D, Dulvy NK (2009) Climate change and fisheries: a comparative analysis of the relative vulnerability of 132 countries. Fish and Fisheries 10(2):173–96

Anyanwu CN, Amadi-Eke AS, Nwaka DE, Ezeafulukwe CF, Adaka GS (2015) Climate change, effects and mitigation strategies on aquaculture: a review. Agric For Fish 4(3–1):70–72

Apata TG, Samuel KD, Adeola AO (2009) Analysis of climate change perception and adaptation among arable food crop farmers in South Western Nigeria. Contributed Paper prepared for presentation at the International Association of Agricultural Economists' 2009 Conference, Beijing

Ayanwuyi E, Kuponiyi F, Ogunlade A, Oyetoro J (2012) Farmers perception of impact of climate changes on food crop production in Ogbomosho agricultural zone of Oyo state, Nigeria Global. J Human Soc Sci 10(7):33

Badjeck M-C (2008) Vulnerability of coastal fishing communities to climate variability and change: implications for fisheries livelihoods and management in Peru. University of Bremen, Bremen

Badjeck M-C, Mendo J, Wolff M, Lange H (2009) Climate variability and the Peruvian scallop fishery: the role of formal institutions in resilience building. Clim Chang 94(1–2):211–232

Badjeck M, Allison E, Halls A, Dulvy N (2010) Impacts of climate variability and change on fishery based livelihoods. Mar Policy 34:375–383

Bako WS (2005) The role of women in fish processing: handling and marketing in Kainji Lake Basin. In: 19th annual conference of the fisheries society of Nigeria (FISON), 29 Nov–03 Dec 2004, Ilorin, Nigeria

Brander KM (2007) Global fish production and climate change. Proceedings of the National Academy of Sciences 104(50):19704–19714.

Brander K (2010) Impacts of climate change on fisheries. J Mar Syst 79(34):389–402

Cheung WWL, Lam VWY, Sarmiento JL, Kearney K, Watson R, Zeller D, Pauly D (2009) Large-scale redistribution of maximum fisheries catch potential in the global ocean under climate change. Glob Chang Biol 16:24–35

Cheung WWL, Lam VWY, Sarmiento JL, Kearney K, Watson R, Zeller D, Pauly D (2010) Large-scale redistribution of maximum fisheries catch potential in the global ocean under climate change. Glob Chang Biol 16:24–35

Diaz RJ, Rosenberg R (2008) Spreading dead zones and consequences for marine ecosystems. Science 321:629

Federal Department of Fisheries (FDF) (2007) Fisheries statistics of Nigeria. 4th Edition. 1995, 49 pp

Federal Department of Fisheries (FDF) (2017) Nigeria's Fish Production 2010–2015: National Bureau of Statistics (NBS), 22 pp

Food and Agriculture Organization (FAO) (2008): Report of the FAO Expert Workshop on Climate Change Implications for Fisheries and Aquaculture: Rome, Italy, 7–9 April 2008. FAO Fisheries Report no. 870 (Food and Agriculture Organization of the United Nations, 2009.2008); http://tiny.cc/fisheries7.

Food and Agriculture Organization (FAO) (2008a) Climate change implications for fisheries and aquaculture. In: The state of fisheries and aquaculture 2008. FAO, Rome, pp 87–91

Food and Agriculture Organization (FAO) (2008b) Climate change for fisheries and aquaculture. Technical background document from the expert consultation, FAO, Rome. Paper Presented at Climate Change, Energy and Food, High Level Conference on Food Security: The Challenges of Climate Change and Bioenergy, Rome

Food and Agriculture Organization (FAO) (2009) The state of world fisheries and Aquaculture-2008 (SOFIA). FAO Fisheries and Aquaculture Department, Rome

Food and Agriculture Organization (FAO) (2012) The state of world fisheries and aquaculture, Rome. 209 pp

Idowu AA, Ayoola SO, Opele AI, Ikenweiwe NB (2011) Impact of climate change in Nigeria. Iran J Energy Environ 2(2):145–152

Intergovernmental Panel on Climate Change (IPCC) (2007a) In: Core Writing Team, Pauchauri RK, Reisinger A (eds) Climate change 2007: synthesis report – contribution of working groups I, II, and III to the fourth intergovernmental panel on climate change. IPCC, Geneva, 8 pp

Intergovernmental Panel on Climate Change (IPCC) (2007b) In: Parry ML, Canzian OF, Palutikof JP, van der Linden Paul J, Hanson C (eds) Climate change impacts, adaptations and vulnerability, Contribution of working group II to the 4th Assessment Report of the IPCC. University Press Cambridge, Cambridge

International Fund for Agricultural Development (IFAD) (2007) IFAD strategic framework 2007–2010. IFAD, Rome, 35 pp

Ipinjolu J, Magawata I, Shinkafi BA (2014) Potential impact of climate change on fisheries and aquaculture in Nigeria. J Fish Aquacult Sci 2014 9(5):338–344

Mano R, Nhemachena C (2006) Assessment of the economic impacts of climate change on agriculture in Zimbabwe: a ricardian approach. CEEPA Discussion Paper No. 11. Centre for Environmental Economics and Policy in Africa, University of Pretoria

Multi-Agency Brief (MAB) (2009) Fisheries and aquaculture in a changing climate. FAO, Rome, Italy, 6 pp

National Science Teachers Association (NSTA) (2007) Global climate change: resources for environmental literacy. NSTA press, pp. 13–33

Nicholls RJ, Wong PP, Burkett VR, Codignotto JO, Hay JE, McLean RF, Woodroffe CD (2007) Coastal systems and low-lying areas. In: Canziani OF, Parry ML, Palutikof JP, van der Linden PJ, Hanson CE (eds) Climate change 2007: impacts, adaptation and vulnerability, Contribution of Working Group II to the Fourth Assessment Report of the Intergovernmental Panel on Climate Change. Cambridge University Press, Cambridge, UK, pp 315–356

Nigerian Environmental Study/Action Team (NEST) (2003) Climate change in Nigeria. A communication guide for reporters and educators. NEST, Ibadan, pp 5–16

Odjugo PAO, Ikhuoria AI (2003) The impact of climate change and anthropogenic factors on desertification in the semi-arid region of Nigeria. Glob J Environ Sci 2(2):118–126

Oluowo EF (2017) Impact of climate change on aquaculture and fisheries in Nigeria; a review. Int J Multidiscip Res Dev 4(1):53–59

Ozor N (2009) Implications of climate change for national development: the way forward. Debating policy options for national development; Enugu Forum Policy Paper 10; African Institute for Applied Economics (AIAE); Enugu, pp 25–42

Parker LM, Ross PM, O'Connor WA (2009) The effect of ocean acidification and temperature on the fertilization and embryonic development of the Sydney rock oyster *Saccostrea glomerata* (Gould 1850). Glob Chang Biol 15:2123–2136

Roessig JM, Woodley CM, Cech JJ, Hansen LJ (2004) Effects of global climate change on marine and estuarine fishes and fisheries. Rev Fish Biol Fish 14:251–275

Sikoki FD (2013) Fishes in Nigerian waters: no place to hide. Inaugural Lecture Series (No. 100), University of Port Harcourt, 100 pp

Teh LCL Sumaila UR (2013) Contribution of marine fisheries to worldwide employment Global marine fisheries employment FISH FISH 14, 77–88

Teh LSL, Teh LCL, Sumaila UR (2011) Quantifying the overlooked socio-economic contribution of small-scale fisheries in Sabah, Malaysia. Fisheries Research 110:450–458.

United Nations Development Programme (UNDP) (2007) Human development report 2007/2008. Fighting climate change: human solidarity in a divided world. UNDP, New York

United Nations Framework Convention on Climate Change (UNFCC) (2007) Climate change: impacts, vulnerability and adaptation in developing countries. Bonn, UNFCC secretariat

World Bank (2010) World development report; development and climate change. The World Bank, Washington, DC, 424 pp

World Bank (2014) Economic and spatial study of the vulnerability and adaptation to climate change of coastal areas in Senegal. 123 Pages

Yazdi SK, Shakouri B (2010) The effects of climate change on aquaculture. Int J Environ Sci Dev 1(5):378, 5pp

Drivers of Deforestation and Land-Use Change in Southwest Nigeria

23

Mayowa Fasona, Peter Adegbenga Adeonipekun, Oludare Agboola, Akinlabi Akintuyi, Adedoyin Bello, Oluwatoyin Ogundipe, Alabi Soneye, and Ademola Omojola

Contents

Abstract

Land-based mitigation presents viable mechanism for offsetting carbon deficits from land-use change-driven emissions especially in areas with relatively large lowland forests. This chapter reviews key issues about the drivers of land-use change, deforestation, and forest degradation. It also presented a case assessment of perception on drivers of land-use change and deforestation using rapid

M. Fasona (✉) · A. Akintuyi · A. Soneye · A. Omojola
Department of Geography, University of Lagos, Akoka-Yaba, Lagos, Nigeria
e-mail: mfasona@unilag.edu.ng; akin_akintuyi@yahoo.com; aakintuyi@unilag.edu.ng; asoneye@yahoo.com; demola_omojola@yahoo.com

P. A. Adeonipekun · O. Agboola · O. Ogundipe
Department of Botany, University of Lagos, Akoka-Yaba, Lagos, Nigeria
e-mail: p1adeonipekun@yahoo.com; dipod2001@yahoo.com; toyin60@yahoo.com

A. Bello
Department of Urban and Regional Planning, Crescent University, Abeokuta, Ogun State, Nigeria
e-mail: bello.adedoyin@yahoo.com

W. Leal Filho (ed.), *Handbook of Climate Change Resilience*,
https://doi.org/10.1007/978-3-319-93336-8_139

appraisal data collected from 108 households and 57 forest resource users in 17 purposively selected peri-urban and rural communities in the forest zone of Southwest Nigeria. From the appraisal, lumbering and polewood extraction, fuelwood and charcoal production, crop cultivation, urban growth, animal grazing, and transportation remain important proximate land-use change, deforestation, and forest degradation drivers. Population increase and poverty are considered the most important underlying drivers. Response to economic opportunities with regard to cash (tree) and commercial arable crops and high local and export demands for wood substantially drive land-use change and deforestation. National climate change actions, natural resource policies, land tenure, international multilateral commitments, and carbon credit frameworks have very little impacts with regard to land-use change in Southwest Nigeria. Lack of alternative livelihoods undermines people's resilience and further drives deforestation and forest degradation.

Keywords
Land-use change · Deforestation · Forest degradation · Drivers · Alternative livelihood · Logging · Agriculture · Southwest Nigeria

Introduction: Deforestation, Forest Degradation, and Land-Use Change

In most humid-tropical countries, land-use change is a strong factor of climate change. The distribution of sources and sinks of carbon over the land surface is dominated by land use and land-use change (Houghton 2003; Turner et al. 2007). Land-based mitigation including reducing deforestation, promoting afforestation, and protecting forested and woodland areas present viable mechanism for offsetting the carbon deficit with potentials for substantial international climate investment. The relationship between land-use change and climate change on one hand and human livelihoods, poverty, food security, and disaster vulnerability on the other hand is very strong in sub-Saharan Africa. This is because in this region, climate determines the rhythm of life and a very large population depends on renewable natural resources. Modelling studies have suggested that deforestation and unsustainable land-use change have serious impact on the climate system with a general tendency for rainfall and evaporation to decrease (Wang and Eltahir 2000; Taylor et al. 2002; Abiodun et al. 2008; MacKellar et al. 2009). This in turn produces a positive feedback which further degrades the ecosystem (Fasona et al. 2013) and undermines the production of soil, water, and other ecosystem services.

IPCC (2006) defines forest as a land spanning more than 0.5 hectare with trees higher than 5 meters and a canopy cover of more than 10% or trees able to reach these thresholds in situ. This definition excludes land that is perpetually or predominantly under agriculture (including tree crop cultivation such as cocoa, oil palm, rubber, coffee, and banana which are very significant in the African tropical and subtropical agroecosystems). It also does not reckon with trees which are present in areas that are predominantly under urban land use. Forest is thus determined both by the presence of trees and the absence of other predominant land uses. Areas under reforestation (i.e., tree replanting in an area that is under predominant forest cover) that have not yet reached but are expected to reach a canopy cover of 10% and tree height of 5 meters are included as forests. Temporarily unstocked areas, resulting from human intervention or natural causes, which are expected to regenerate are also considered as forest land (IPCC 2006).

Forests are critical to life on the earth. They provide the ecosystem services that support human provisioning needs (including food, clothing, water, shelter, medicines, livelihoods, fiber for industries, etc.) as well as the regulatory services that support climate, air, water, and pollination services. In addition forests also provide the support services to protect soil, stabilize climate, and enhance nutrient cycling and primary productivity. They also support human spiritual and cultural desires.

According to FAO (2016a, b), forest area change is a process of gain (forest expansion) and loss (deforestation). Deforestation according to the United Nations Framework Convention on Climate Change (UNFCCC) refers to the direct, human-induced conversion of forested land to non-forested land. This includes conversion of forest land to other land uses including annual cropland, pasture or rangelands, perennial plants (oil palm, rubber, cocoa, and shrubs), urban lands, and physical infrastructure. The conversion results in significant alteration of the cover and/or causes a reduction in crown cover from the threshold for forest definition to below this threshold (GOFC-GOLD 2016).

Forest degradation, according to GOFC-GOLD (2016), represents a persistent direct human-induced decrease in forest carbon stocks, with measured canopy cover remaining above the threshold for definition of forest and no change in land use. This means that, although the forest cover is not significantly altered or changed, the forest becomes more disturbed. Causes of forest degradation include selective logging and discontinuous smallholder tree crop farms, with evidences in the form of presence of gaps in forest canopy, log decks, network of roads, tracks and footpaths, clearances, isolated patches, and presence of small human dwellings including hamlets, camps, and small villages. Perturbation in the form of forest fires (where the land remains under forest land use) with decline in forest carbon stocks also qualifies as forest degradation. Degradation by overexploitation for fuelwood or other local uses of wood including charcoal production in Africa in woodlands and drier forest areas is often followed by animal grazing that prevents regeneration.

Deforestation, forest degradation, and other unsustainable land-use practices are the result of processes driven by multiple causes occurring at spatially denominated scales (FAO 2016b). They deepen the environment-poverty linkages and amplify climate change impacts (UNDP-UNEP 2009; Fasona et al. 2014) and weaken the resilience of the population. Natural forest area change is a good indicator of the health of the ecosystems and biodiversity dynamics. It is also a defining indicator between the world's social-economic groups. There is a strong correlation between deforestation and forest degradation and poverty. The largest net loss of forest and the largest net gain in agricultural area between 2000 and 2010 were in the low-income group of countries with net forest loss associated with increasing rural population (FAO 2016b). Deforestation accounts for about 17–18% of global anthropogenic GHG emissions, the biggest contributor after energy (Jonathan and Maliheh 2011). Hence, efforts to improve forest and woodland cover are not only climate mitigation measures but adaptation to improve livelihoods and resilience to adverse weather conditions. Land-based adaptation strongly and directly interconnects with the achievement of sustainable development goals (SDG) on zero poverty (goal 1), zero hunger (goal 2), responsible consumption and production (goal 12), climate action (goal 13) and life on land (goal 15), and indirectly to most of the other goals.

Human land-use decisions drive changes in the land-cover (i.e., the natural or cultural artifact on the land). Human land-use decisions are driven by combination of complex underlying drivers (i.e., fundamental social, political, and economic processes beyond immediate control) and proximate drivers (i.e., actions at local levels which could be the result or end product of the underlying drivers) (Verburg et al. 2002; Geist and Lambin 2002). These can either drive land-use change in the positive or negative direction with regard to climate change and environmental management. According to Houghton (2003) and Obioh et al. (1994), some of the proximate drivers of land use and land-use change include cultivation and pastures, abandonment of agricultural lands, logging and wood harvest, reforestation, afforestation, and charcoal and fuelwood extraction. Others are controlled and uncontrolled burning, forest fires, road construction, irrigation, and other infrastructure construction activities. The underlying drivers have remote and yet powerful influence on land transformation and land-use change and act either to accelerate or decelerate the proximate drivers. They include tenure regimes, technological change, international development assistance and pressures, demographic and economic changes (urbanization, industrialization, and infrastructure development), globalization of markets and economies, agricultural changes, social changes (including family size and consumption patterns), and legal frameworks (including land-use laws and policies), among others (Verburg et al. 2002; Geist and Lambin 2002). Globally, underlying factors affecting forest conversion include population growth and changing food consumption patterns, agricultural developments such as changing markets, technological improvements and active policy interventions, land tenure security, and the governance of land-use change (FAO 2016a).

Land-use change and deforestation are key drivers of global environmental change and significantly connect to critical environmental problems including erosion, flood, and droughts; loss of biodiversity; air, soil, and water degradation; and declining primary productivity and ecosystems losses, among others. They also drive natural resource conflicts and clashes and exacerbate poverty especially in Africa. Globally, agriculture causes over 70% of all deforestation. Other majordrivers include timber extraction andlogging, fuelwood collection and charcoal production, uncontrolled fire, and livestock grazing.

Deforestation and Land-Use Change in Nigeria

Africa has the highest continental deforestation rate of 2.8% and net natural forest decline rate of 3.1% from 2010 to 2015 compared to global average of 0.13% and 0.24%, respectively (FAO 2016a). Agriculture is the main driver of land-use change and deforestation in the developing world. Large-scale commercial agriculture accounts for about 40% of deforestation in the tropics and subtropics. Others include local subsistence agriculture (33%), infrastructure (10%), urban expansion (10%), and mining (7%) (Hosonuma et al. 2012; FAO 2016a). Fuelwood collection and charcoal have also been described as major forest degradation drivers for the African continent at about 48%. This suggests local small-scale activities (fuelwood collection, charcoal production, and livestock grazing in forests) are the most relevant in large parts of Africa (Hosonuma et al. 2012).

According to FAO (2014), Nigerian forests occupy about ten million hectares (about 10% of the 92,376,700 hectares total land area). The Nigerian forest resources which abound in the highland and lowland forests, woodlands, and plantations contribute about 2.5% to the GDP and provide employment for over two million people through supply of fuelwood and poles with additional 80,000 people working in the log processing industries (FAO 2014). Oyebo et al. (2010) submit that less than 10% of Nigeria's original forest cover remains intact. In particular, the forest estates from which wood and other products are obtained have been subjected to severe encroachments, vegetation degradation and de-reservation for agriculture, industrial development, and urbanization (FAO 2014).

Logging activities and commercial wood extraction have also become important forest degradation drivers in the forest zones of Nigeria. Nigeria has the highest rate of deforestation in the world at 5% with 410,000 ha of net forest loss annually from 2010 to 2015. Nigeria also ranks ninth by wood removals with 73,103m^3 and 87% used for woodfuel (FAO 2016b). Lack of data, however, remains an issue with the last national forest inventory and national vegetation and land-use map dating back to 1995 (FAO 2014). Within the period 1976 to 1995, agricultural land use increased by 64.4%, natural forest declined by 16.6%, and forest plantation increased by 0.3% (FAO 2014; Oyebo et al. 2010).

According to Oyebo et al. (2010), agriculture expansion, unsustainable wood extraction, infrastructure construction, and bush burning are generally regarded as the proximate drivers of deforestation and forest degradation in Nigeria. The economic factor of higher quick returns from agriculture over forestry is considered one of the underlying drivers. The poor governance factors including outdated forestry laws, land tenure, inability to mainstream land and forestry into government businesses, inadequate human and technical capacities, lack of forest management planning, and high revenue target from forestry sector were also noted as strong drivers. Others include demographic, technological, and cultural factors.

The possibility of compensating developing countries for reduced emissions from deforestation and forest degradation through the Reduced Emissions from Deforestation and Forest Degradation (REDD+) Programme, enhancing forest carbon stocks, sustainable management, and conservation of forests, is attractive because it explicitly recognizes the value of natural forests, has low associated costs, and does not exclusively displace sustainable eco-based livelihoods (Mbow et al. 2012). Countries are encouraged to identify the drivers of deforestation and forest degradation. For most African countries including Nigeria, quantitative national-level information on drivers of land-use change and activities causing deforestation and forest degradation is widely unknown (Hosonuma et al. 2012; FAO 2016a). This limitation together with the challenge of inventory and measurement of forests in terms of the whole stand carbon is tasking their ability to participate in REDD+ (Mbow et al. 2012). Despite the huge and unique forest and woodland assets in Nigeria with enormous potentials to benefit from the REDD+ Programme, only one REDD-readiness assessment (for three sites in Cross River State) has been carried (Oyebo et al. 2010). The Southwest Nigeria has the largest lowland forest and woodlands in the country, yet we found no record of study on the drivers of deforestation and land-use change in the region. The case study report presented here, therefore, assessed the major drivers of land-use change and deforestation in Southwest Nigeria and highlights the extent to which the drivers and their outcomes interface with human livelihoods and reduce resilience at local levels.

Natural Systems Behavior and Land-Use Change and Deforestation Drivers

The concept of *connectivity* may be used to explain the spatial behavior and dimensions of the drivers of ecological change in general and land-use change, deforestation, and forest degradation in particular. Although connectivity in the spatial sense suggests that locations at certain distance are related to each other (Gregory 1985), economic and social activities in one location, regardless of physical distance, can produce a teleconnection effect in another place, and this portrays the trigger-effect relationship in land-use change and deforestation. Globalization of markets is one of such underlying driver of land-use change that produces powerful teleconnection effects (Lambin et al. 2001). Demand for organic industrial raw materials in the developed world drives plantation agriculture and continuously reorients smallholder agriculture in the developing countries (FAO 2016a). Deforestation and forest degradation in Nigeria are also driven by demand for wood products (including timber and charcoal) abroad which further reinforces illegal logging, charcoal production, and wood export at home.

The *hierarchy* theory suggests that higher-level processes constrain lower-level processes, whereas the higher-level processes might emerge from lower-level dynamics (Verburg et al. 2002). This theory is apt for explanation of the nexus between the proximate drivers and the underlying drivers of land-use change and deforestation. A degraded land that is perpetually exposed to the drivers of land degradation will find it difficult to return to its original state without artificial restoration. For example, a large percent of natural (closed) forest in Nigeria has over the years been degraded by agriculture andlogging. With sustained disturbance, such area will eventually transit into open forests and savanna woodlands.

Woodlands are further transited into open savanna through selective fuelwood and charcoal harvest which further opens the land for grazing and seal its potential to return to its climax.

The concepts of *stability* and *resilient* relate to the performance of both natural and human systems (Verburg et al. 2002). Resilience refers to ability of a system (whether ecosystem or human system) to absorb or withstand perturbation. For natural ecosystem, it is the ability of land or forest to withstand the drivers of land-use change before it becomes degraded. For human system, it is the ability of people to continue to derive the benefits (good and services) from the ecosystem under a condition that will not undermine the capacity of the ecosystem to deliver such benefits. Stability refers to ability of an ecosystem to regain original form after perturbation. Resilience and stability combine to dictate the rate at which change in land use will result into permanent change in landscape structure. In the human system, they dictate the rate in which land-use change becomes unsustainable and unable to support healthy ecosystems and the benefits they produce. Ensuring sustainability by meeting the challenges of livelihoods (which depend on short-term ecosystems provisioning service) and preserving the ability of ecosystems to perform long-term regulation (including climate regulation) and supporting functions is a tough task. Communities have a role to play in maintaining the integrity of ecosystems and ensuring that the climate systems continue to support food and water production and other livelihood activities (Fasona et al. 2014).

Perception on Land-Use Change and Deforestation Drivers in Southwest Nigeria

Study Area

The southwest region of Nigeria is roughly defined by latitudes 6° 00′ N and 9° 15′ N and longitudes 2° 45′ E and 6° 00′ E. It covers about 76,852 km^2 (i.e., 8.5% of the Nigeria landmass) and consists of the six states of Ekiti, Lagos, Ogun, Ondo, Osun, and Oyo (see Fig. 1). The region has a population of 32,566,010 (i.e., 20% of Nigeria's population) as projected for 2011 (NBS 2012) and a density of about 423 ppkm^2.

The climate of Southwest Nigeria is tropical, characterized by a subhumid Koppen's Aw climate with distinct wet and dry seasons (Kottek et al. 2006). Average temperature is about 27°C. Rainfall is bimodal with mean annual range between 900 mm and 3000 mm. Due to complex terrain and land surface heterogeneity, the impact of mesoscale processes predominates and accounts for over 75% of total rainfall received in the region (Omotosho and Abiodun 2007). About 60% of the Southwest region is covered by undifferentiated basement complex geological formation. The coastal areas and river floodplains are covered by recent alluvium and coastal plains. The Ewekoro, Ilaro, and Abeokuta sandstone formations dominate the immediate areas outside the coast (Jones and Hockey 1964). Elevation ranges from about 2 to 5 m along the coast to above 600 m around the Idanre and

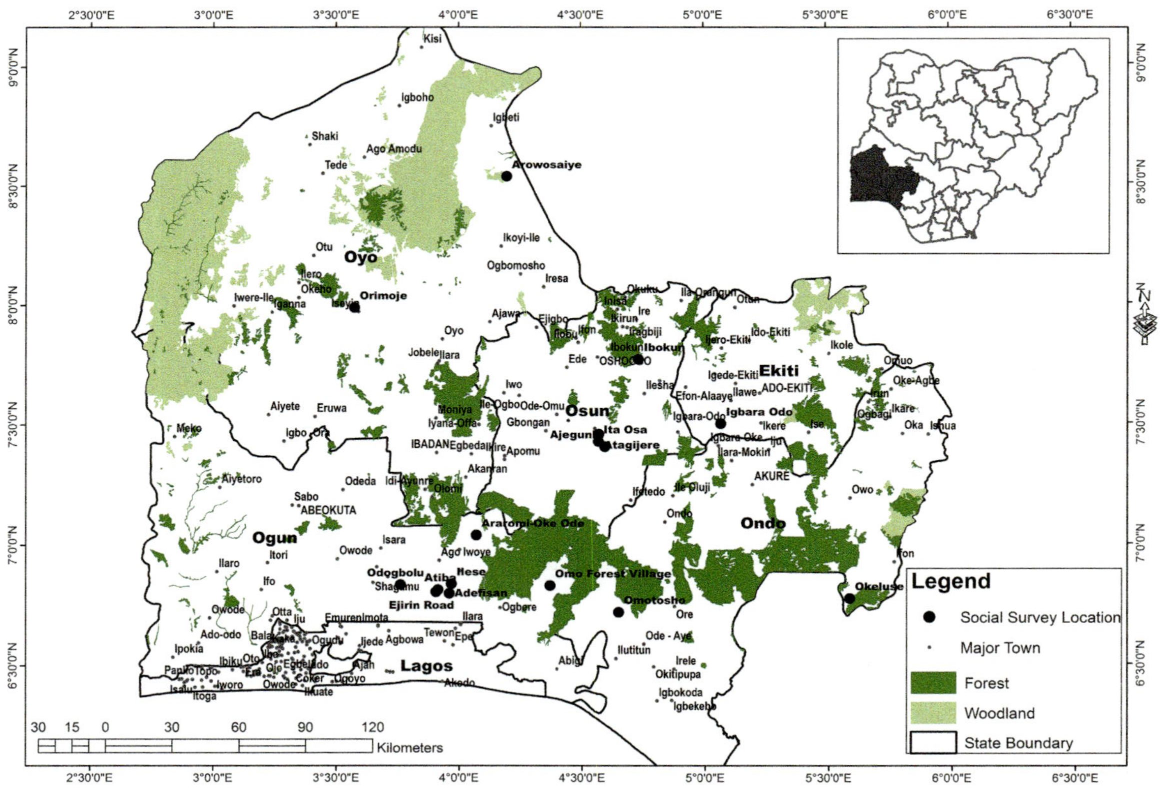

Fig. 1 The Southwest Nigeria showing the major forest and woodland areas and field sample locations. (Source: Adapted from the original data generated in 1995/96 by the Forestry Resources Mapping, Evaluation and Coordinating Unit (FORMECU), Ministry of Environment, Nigeria)

Effon-Ijesha ridge systems. The study area forms the Nigerian hydrological zone 6 with major river systems including the Yewa, Ogun, Oshun, and Osse river basins.

The mangrove forest, rainforest, and savanna woodland ecological zones straddle the Southwest Nigeria from west to east. Thus, the main land-cover types include cropland, mangrove, swamp forest, primary and disturbed forests, forest plantation, savanna woodland and grassland, and urban areas. The region is the most urbanized area of Nigeria and has the largest remaining tracts of lowland forest significantly consisting of the Omo-Shasha-Oluwa forest complex shared by Ogun, Osun, and Ondo States. A computation of land-use statistics was done from the Forestry Resources Mapping, Evaluation and Coordinating Unit (FORMECU) vegetation and land-use data of 1995/1996 which is the only available national land-use and vegetation data in public domain (FAO 2014). It shows that cropland and fallow account for about 67% of the land-cover categories. Others include rainforest (11.4%), savanna woodland (9.7%), swamp forest (4.4%), grassland (2.6%), forest plantation (1.3%), and urban (2.3%).

Field Activity

The activity reported in this chapter is a part of a larger study on land-use change and carbon emissions in Southwest Nigeria. This segment focuses on the drivers of land-use change from the perspective of the people whose livelihoods substantially depend on agriculture and forest and woodland resources. The six states of the Southwest Nigeria was the study space, and communities with high concentration of agriculture and forest-related activities (saw mills, lumbers, wood/plank marketing, log transportation, timber contractor, and charcoal and fire wood sellers) were targeted. The communities were mostly rural and peri-urban (some at the fringes of major urban areas). Field data collection through rapid appraisal was done in 17 communities which were purposively selected across 11 local government areas (LGAs) in 5 of the 6 states. Lagos State was excluded because of its cosmopolitan nature. After the selection of the communities, the choice of the respondents was purely random.

The field survey instruments used were structured questionnaires which were pretested during the reconnaissance survey. Two different questionnaires were designed to capture responses from households and forest resource users. In all 108 households and 57 forest resource users were sampled (Table 1).

The questionnaire for household consists of 3 sections with 33 questions and 282 sub-questions generating over 2000 elements (responses). The forest users' questionnaire consists of 2 sections with 23 questions and 77 sub-questions generating over 600 elements. The coding and processing of the data were done within IBM SPSS Statistics 20 (www.spss.com). The contingency tables were consolidated and analyzed for frequencies and percentage responses. 54.6% of the respondents to the household survey were male and 45.4% female. For the forest users' survey, 75.4% were male and 24.6% female. For both survey, at least 85% of the respondents are married and have mouth to feed, while the rest are either single, widow/widower, or separated. The virile youth within age bracket 31–45 years dominated the respondents (Fig. 2).

Table 1 Sampled communities

SN	Name of community	State	LGA	Households	Forest resource users	Total	Percent
1	Igbara-Odo	Ekiti	Ekiti South	12	5	17	10.3
2	Omo (Area J4)	Ogun	Ijebu East	5	3	8	4.8
3	Araromi-Oke-Ode	Ogun	Ijebu North	4		4	2.4
4	Atiba	Ogun	Ijebu Ode	3	1	4	2.4
5	Ejinrin road	Ogun	Ijebu Ode	7	3	10	6.1
6	Adefisan	Ogun	Ijebu Ode	4		4	2.4
7	Ilese	Ogun	Ijebu Ode	4	7	11	6.7
8	Odogbolu	Ogun	Odogbolu		2	2	1.2
9	Omotosho	Ondo	Okitipupa	12	8	20	12.1
10	Okeluse	Ondo	Ose	16	7	23	13.9
11	Atagijere	Osun	Ife East	1	4	5	3.0
12	Ita-Osa	Osun	Ife East	8	6	14	8.5
13	Ajegunle	Osun	Ife East	3		3	1.8
14	Ibokun	Osun	Obokun	14	3	17	10.3
15	Orimoje	Oyo	Iseyin	3		3	1.8
16	Abaletu Sawmill	Oyo	Iseyin	3	5	8	4.8
17	Arowosaye	Oyo	Ori're	9	3	12	7.3
	Total			108	57	165	100.0

Source: Authors

For both groups, at least 40% have lived in the community for over 20 years, while at least 70% have lived for over 10 years. Seventy-four percent of the forest resource users have been in the business for over 10 years, and 74% also belong to one trade association or another (sawmiller association, wood/plank seller association, farmer association, etc.). Respondents with secondary education accounted for 50% and 42% for household, and forest resource users, respectively, dominate the survey (Table 2).

At least a third of the forest resource users make over N50,000 (about USD140) a month, and 53% make above N40,000 (about USD110), and only 21% of the households make above N50,000, and 35% make above N40,000 monthly.

Livelihoods and Income-Generating Activities

The household type of regular employment is unexpectedly led by trading which accounts for about 44% as shown in Fig. 3. Farming and land-related occupations account for 30.6% and artisans 17%, and only about 6% work in the government sector.

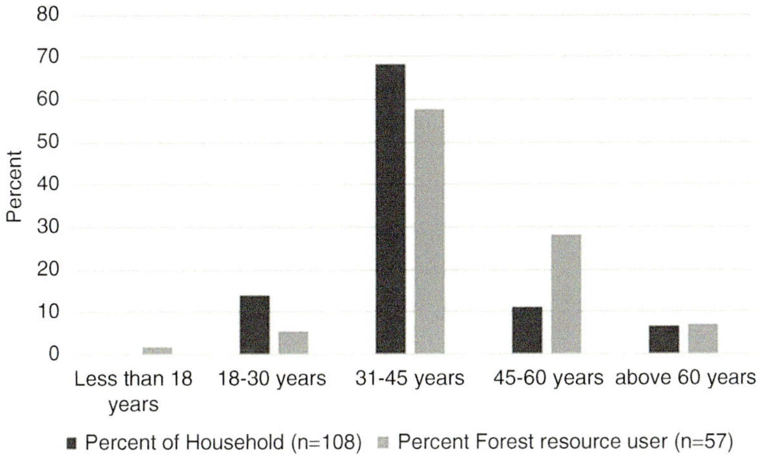

Fig. 2 Age category of respondents

Table 2 Education level of respondents

Education level	Percent household (n = 108)	Percent forest user (n = 57)
No formal education	10.2	14.0
Primary school	16.7	22.8
Secondary school	50.0	42.1
Tertiary education	21.3	21.1
Others	1.9	0.0
Total	100.0	100.0

Source: Authors

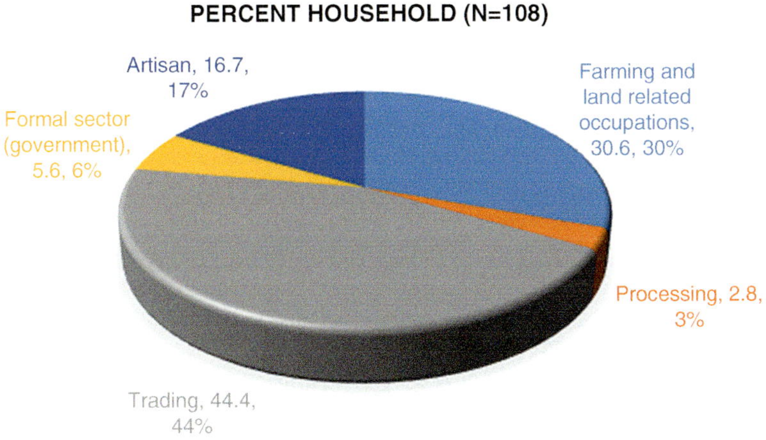

Fig. 3 Household regular employment (n = 108)

Table 3 Regular employment for forest resource users

Regular employment	Percent forest user ($n = 57$)
Sawmiller	26.3
Wood plank seller	35.1
Firewood seller	5.3
Artisan	8.8
Trader	7.0
Timber/wood contractor	7.0
Farmer	10.5
Total	100.0

Source: Authors

Table 4 Major crops grown in the Southwest Nigeria (by percent of household respondents; $n = 108$)

Crops grown	Yes	Moderately important	Extremely important
Cocoa	26.9	3.7	23.1
Plantain	30.6	4.6	25.9
Rubber	2.8	0.9	3.7
Oil Palm	9.3	3.7	8.3
Maize	23.1	4.6	19.4
Cassava	32.4	3.7	28.7
Rice	0.9	0.9	0.9
Beans	3.7	1.9	2.8
Tomatoes	17.6	1.9	14.8
Pepper	15.7	6.5	10.2
Groundnut	3.7	0.9	4.6
Watermelon	3.7	0.0	3.7
Guinea corn	4.6	0.9	4.6
Kola nut	3.7	2.8	2.8
Yam	12.0	1.9	8.3

Source: Authors

Follow-up enquiry suggests that trading which led the regular employment also includes dealings in crops and other farm products as well as selling of wood, plank, and fuelwood. The regular employment for forest resource users is dominated by wood plank seller (35%), sawmiller (26%), and others as shown in Table 3.

Crop cultivation is a very significant livelihood activity as shown in Table 4. At the household level, cassava, plantain, and cocoa are the major crops cultivated. Others include maize, tomatoes, pepper, and yam in that order. The latter crops are normally grown as mixed crops. Southwest region is the heart of Nigeria's cocoa belt where cocoa is the traditional cash crop having been experimented around Lagos (Agege and Otta) in the early 1890s, Ibadan and Abeokuta in the 1890s, Ilesha by 1896, Ife and Ekiti at the eve of the First World War, and Ondo after the war (Fenske n.d.). While cassava is an arable tuber crop cultivated extensively in the savanna woodlands and forest fringes, cocoa and plantain are row crops grown for their commercial values in small private plantations. Where the plantations are in clusters (as can be found in Ondo, Ekiti, Osun, and Ogun States), they can cover several hundreds to a thousand hectares. They are considered to be extremely important cash crops. Plantain and cocoa require very fertile lands, and they are usually cultivated by clearing virgin (primary) forests. Evidence from the field suggests that the cultivation of plantain and cocoa is a major driver of deforestation in the primary forests of Southwest Nigeria.

Cassava is an important staple crop, and its several derivatives provide the major staple foods for Nigerian homes. It is mainly cultivated in the savanna woodlands and in secondary forests. The cultivation of cassava is undoubtedly the major driver of land-use change and clearing of savanna woodland. In recent years, cassava has become an important cash and export crop. It is now extensively cultivated in clusters on both small holder plots and large commercial plots for off-takers who convert them to industrial starch for export to, notably, China. Nigeria has thus become the world's largest producer of cassava with 57, 855, 000 metric tonnes in 2016 (FAO 2016c). This is a connectivity or teleconnection effect where cassava demands abroad drive land-use change to agriculture in Nigeria. The pressure imposed on land-use change and deforestation in Southwest Nigeria by agriculture is, therefore, consistent with FAO (2016a) and Hosonuma et al. (2012). It is also consistent with the submission of Lambin et al. (2001) that peoples' responses to economic opportunities, as mediated by institutional factors, drive land-cover changes. Global forces either amplify or attenuate local factors.

The dearth of viable alternative income that is ecosystem compatible with minimal disturbance or impact on the forest and woodland resources is shown in Table 5. The most significant alternative income activity to farming is trading and marketing (46.3%), including marketing farm products, and lumbering (26%). This is followed at distant by firewood harvesting and sales and processing of agric products (8.3% each). Forest and woodland ecosystem-compatible activities including apiculture; snail farming; fishing and aquaculture; herb, seed, and root collection; collection of other non-timber forest products; crafts making; and working as hired labor are significantly low and unpopular among the respondents. This suggests low awareness about the possibility of generating income from forest and woodland other than crop farming and lumbering and without the usual destructive means.

The situation is the same for the forest user group where 60% have no alternative income sources and 17.5% and 12.3% also engaged in farming and trading, respectively. Thus, intensification of agriculture driven by global connectivity is just as important as the lack of alternative livelihoods in driving land-use change and deforestation in Southwest Nigeria.

Much of the battle to stem deforestation and negative land-use change may have to be fought on the altar of alternative ecosystem-compatible livelihoods. During the fieldwork, some evidence of campaign (and partnership) for alternative livelihoods by some conservation and other nongovernmental organizations were seen in some few isolated villages deep inside the forests (especially in the Omo-Sasha-Oluwa forest complex). However, the greater drivers of land-use change and deforestation are substantially outside and beyond the small villages inside the forest. This

Table 5 Other income-generating activities (by percentage of household respondents; n = 108)

Alternative income activity	Yes	No
Charcoal making	0.9	99.1
Firewood harvesting and selling	8.3	91.7
Lumbering	25.9	74.1
Livestock rearing	4.6	95.4
Fishing/aquaculture	0.9	99.1
Beekeeping	0.0	100
Apiculture (mushroom farming)	0.9	99.1
Snail farming	1.9	98.1
Hunting	2.8	97.2
Herb/seed/root collection	1.9	98.1
Civil service	2.8	97.2
Trading/marketing	46.3	53.7
Collection of non-timber forest product	1.9	98.1
Processing of agricultural produce	8.3	91.7
Hired labor	3.7	96.3
Crafts	5.6	94.4

Source: Authors

suggests that a more robust, widespread, and massive alternative livelihood campaign is required. One way to achieve this will be for the governments to mainstream climate change and alternative livelihoods into the development planning process.

Climate Change and Land-Use and Cover Change Awareness

Figure 4 shows the household perception of vegetation and land-cover change in the last 20–30 years compared to the present. Majority of the respondents are of the opinion that areas covered by urban and agricultural land uses as well as bare surfaces have been increasing. On the other hand, areas covered by forest, forest plantation, woodland, wetland, and grassland are believed to be decreasing by majority of respondents. 47% think that the area covered by water is decreasing, while at least a third also think that there is no change in the area covered by wetland.

Linking the perceived changes in land-use/land-cover categories to climate change, participants were asked if they ever heard of climate change. About 72.2% of households answered in the affirmative, while 27.8% responded in the negative. Thirty-five percent are aware of the relationship between climate change, deforestation, and unsustainable land-use change, while 64.2% are unaware of any connection. The source of their information is varied and diverse including self-observation (19.4%), radio (15.7%), combination of radio and self-observation (11%), television (9.3%), print media (6.5%), from community members (5.6%), and combination of radio and television (5.6%). Some of the connections listed (the question was open and respondents were required to write those connections they know) by those that were aware of

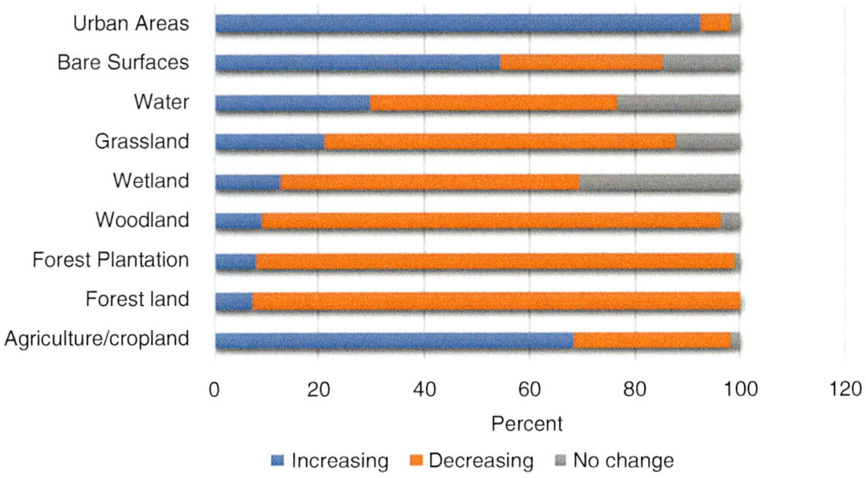

Fig. 4 Household perception on land-use and cover change ($n = 108$)

the possible connection between climate change and unsustainable land-use change include increase in temperature, reduction in rainfall, flooding, and drought.

Again 72% of the forest resource user group has heard about climate change, while 28% has never heard about it. While 31.6% are aware of the relationship between climate change, deforestation, and unsustainable land-use change, 68.4% are unaware of any connection. The sources of their information include radio and self-observation (17.5%), radio (17.5%), self-observation (15.8%), radio and television (8.8%), print media (7%), and television (3.5%). Some of the connections between climate and land-use change known to the forest user group include increasing temperature, decreasing rainfall, flooding, and reduction in land productivity.

In summary, although about 70% of the sampled population has heard of climate change, only about a third is sufficiently aware of the threat of climate change and the connection with their livelihood activities. To improve rural resilience, awareness, and information about climate change, its impacts on rural livelihoods and how some of their livelihood activities invariably act as drivers of climate change and its impacts are important. If people are sufficiently aware of the dangers their short-term activities pose to themselves and their livelihoods on the long run, they may likely be predisposed to making amend.

Land-Use Change and Deforestation Drivers

The complex web-linking social and biophysical processes of land-use change and deforestation are products of the proximate and underlying drivers of change. Several of these drivers often combine to affect land-use change decisions and define the trajectory of change that produces a given outcome for good or bad. In the course of the field survey, household respondents were asked to rank the extent to which these drivers of change affect land-use change decisions and deforestation. Table 6 shows the responses for the proximate drivers. The most important proximate land-use change and deforestation and forest and woodland degradation drivers are lumbering/commerciallogging, fuelwood extraction, polewood extraction, and charcoal production. At least 70% of respondents considered these drivers as having moderate to extreme impact, and at least 40% considered them as having high to extreme impact. In the next set are industrialization, crop cultivation, urban growth, animal grazing, and transportation which at least 60% considered to have moderate to extreme impacts and about a third considered to have high to extreme impact.

The other proximate drivers include rainfall and temperature, construction of market infrastructure, flood, fires, construction of dams and large water impoundments, drought, rural agricultural migration, private constructions, mining, and time of opening and closing of forest reserves for lumbering. For these, at least 40% considered them to have moderate to extreme impacts, and at least 20% considered them to have high to extreme impacts. Afforestation and reforestation efforts are not considered to be significant. This is consistent with the high net deforestation rate for Nigeria (FAO 2016a).

Population increase and poverty are considered the most important underlying drivers of land-use change and deforestation in Southwest Nigeria (Table 7). At least 60% of household respondents rate these two factors to have moderate to extreme impact, and 40% rate them to have high to extreme impact. The next category of factors includes urbanization and industrialization, public attitude, values and beliefs, and technological application in the wood sector. These were ranked moderate to extreme by at least 50% and high to extreme by at least a quarter of the respondents. Other underlying factors that scored at least 40% with moderate to extreme impacts include agricultural commercialization and social and resource conflicts.

Other factors, such as national policies on natural resources, gender, land tenure and property rights, formal policies on forestry, REDD+ and carbon finance framework, government incentives and subsidies, national climate change strategies, activities of development partners, and international multilateral conventions and commitments, among others, have very little impacts with respect to land-use change, deforestation, and woodland degradation in the study area.

These results are consistent with the drivers of land-use change and deforestation established in literature (Lambin et al. 2001) especially as they play out across developing countries and the continent of Africa in particular (FAO 2016a). The low score for land tenure rights underscores the lack of formalized property rights in Nigeria which make people less likely to invest in long-term and long gestation eco-investment such as silviculture. Again, the low score for government policies on natural resources and forestry reinforces the disconnection between government policies and on the field situation where government policies are never translated into actionable plans (Adedayo et al. 2013). Those mechanisms that have the potentials to improve resilience are not popular with rural communities and have no impact on land-use change and deforestation in Southwest Nigeria. These include

Table 6 Perception of proximate land change drivers (by percent of respondents; n = 108)

Proximate land change factors	No impact	Very low impact	Low impact	Moderate impact	High impact	Extreme impact	No response	Sum for moderate to extreme	Sum for high to extreme
Lumbering/commercial logging	0.9	2.8	13.9	19.4	33.3	23.1	6.5	**75.9**	**56.5**
Fuel/firewood extraction	2.8	3.7	13.0	25.0	36.1	13.9	5.6	**75.0**	**50.0**
Polewood extraction	0.9	7.4	11.1	29.6	30.6	12.0	8.3	**72.2**	**42.6**
Charcoal production	2.8	5.6	16.7	27.8	26.9	14.8	5.6	**69.4**	**41.7**
Industrialization	1.9	13.0	12.0	23.1	23.1	22.2	4.6	**68.5**	**45.3**
Crop cultivation	6.5	5.6	21.3	30.6	15.7	17.6	2.8	**63.9**	**33.3**
Urban growth	4.6	9.3	14.8	25.9	17.6	20.4	7.4	**63.9**	**38.0**
Animal grazing	5.6	7.4	13.9	28.7	25.0	10.2	9.3	**63.9**	**35.2**
Transportation	2.8	5.6	21.3	32.4	19.4	9.3	9.3	**61.1**	**28.7**
Rainfall	3.7	13.9	16.7	21.3	26.9	10.2	7.4	**58.3**	**37.0**
Temperature	5.6	13.9	15.7	21.3	19.4	17.6	6.5	**58.3**	**37.0**
Market infrastructure construction	3.7	13.0	17.6	24.1	16.7	12.0	13.0	**52.8**	**28.7**
Flood	8.3	9.3	16.7	25.9	13.9	10.2	15.7	**50.0**	**24.1**
Fires	7.4	9.3	12.0	28.7	11.1	9.3	22.2	**49.1**	**20.4**
Dams and large water impoundments	21.3	10.2	6.5	23.1	13.9	11.1	13.9	**48.1**	**25.0**
Drought	8.3	11.1	13.9	23.1	14.8	10.2	18.5	**48.1**	**25.0**
Rural agricultural migration	10.2	9.3	13.0	23.1	12.0	11.1	21.3	**46.3**	**23.1**
Private constructions	6.5	16.7	18.5	20.4	13.0	11.1	13.9	**44.4**	**24.1**

(*continued*)

Table 6 (continued)

Proximate land change factors	No impact	Very low impact	Low impact	Moderate impact	High impact	Extreme impact	No response	Sum for moderate to extreme	Sum for high to extreme
Mining	10.2	11.1	8.3	24.1	15.7	4.6	25.9	**44.4**	**20.4**
Time of opening and closing of forest for lumbering	22.2	12.0	8.3	12.0	18.5	11.1	15.7	**41.7**	**29.6**
Reforestation efforts	26.9	8.3	10.2	13.0	11.1	15.7	14.8	**39.8**	**26.9**
Terrain	6.5	12.0	16.7	16.7	15.7	5.6	26.9	**38.0**	**21.3**
Afforestation efforts	25.9	13.9	12.0	10.2	4.6	18.5	14.8	**33.3**	**23.1**

Source: Authors

Table 7 Perception of household respondents on the impacts of the underlying drivers (by percent of respondents; n = 108)

Underlying land change factors	No impact	Very low impact	Low impact	Moderate impact	High impact	Extreme impact	No response	Sum for moderate to extreme	Sum for high to extreme
Population increase	5.6	8.3	11.1	22.2	25.0	16.7	11.1	**63.9**	**41.7**
Poverty	6.5	10.2	9.3	21.3	25.9	13.0	13.9	**60.2**	**38.9**
Urbanization and industrialization	7.4	9.3	19.4	17.6	20.4	16.7	9.3	**54.6**	**37.0**
Public attitude, values, and beliefs	8.3	14.8	14.8	23.1	17.6	11.1	10.2	**51.9**	**28.7**
Technological applications in the wood sector (e.g., recycling and use of substitute to wood)	16.7	9.3	8.3	23.1	17.6	11.1	13.9	**51.9**	**28.7**
Agricultural commercialization	4.6	14.8	24.1	23.1	10.2	10.2	13.0	**43.5**	**20.4**
Social and resource conflicts	10.2	11.1	22.2	16.7	13.9	11.1	14.8	**41.7**	**25.0**
Communities livelihood interest	15.7	18.5	11.1	16.7	15.7	7.4	14.8	**39.8**	**23.1**
Economic diversification	8.3	20.4	14.8	12.0	17.6	8.3	18.5	**38.0**	**25.9**
National policy on energy	19.4	17.6	11.1	13.9	13.9	10.2	13.9	**38.0**	**24.1**
Length of fallow period	16.7	18.5	12.0	15.7	18.5	2.8	15.7	**37.0**	**21.3**
National environment and natural resource legislations	21.3	13.9	12.0	15.7	13.9	6.5	16.7	**36.1**	**20.4**
Gender factors (e.g., promoting women farmers)	13.9	20.4	12.0	17.6	12.0	6.5	17.6	**36.1**	**18.5**
Land tenure and property rights	24.1	11.1	9.3	20.4	8.3	7.4	19.4	**36.1**	**15.7**
Private sector economics interests	18.5	14.8	18.5	20.4	10.2	4.6	13.0	**35.2**	**14.8**
Public sector economic interests	27.8	13.9	11.1	13.9	13.9	7.4	12.0	**35.2**	**21.3**
Globalization	21.3	13.9	10.2	14.8	12.0	8.3	19.4	**35.2**	**20.4**
Formal (direct) policies on forestry	19.4	13.0	18.5	13.0	13.9	6.5	15.7	**33.3**	**20.4**
National education policy as relates to the environment	29.6	13.0	8.3	14.8	13.9	4.6	15.7	**33.3**	**18.5**

(continued)

Table 7 (continued)

Underlying land change factors	No impact	Very low impact	Low impact	Moderate impact	High impact	Extreme impact	No response	Sum for moderate to extreme	Sum for high to extreme
Export production and import substitution drive	15.7	21.3	9.3	11.1	13.9	4.6	24.1	**29.6**	**18.5**
REDD and carbon finance framework	24.1	12.0	7.4	9.3	14.8	4.6	27.8	**28.7**	**19.4**
Activities of development partners	24.1	13.9	7.4	9.3	13.9	4.6	26.9	**27.8**	**18.5**
State environment and natural resource legislations	27.8	13.9	9.3	11.1	10.2	6.5	21.3	**27.8**	**16.7**
Informal policies	21.3	12.0	13.9	12.0	9.3	6.5	25.0	**27.8**	**15.7**
National climate change strategy	36.1	11.1	2.8	12.0	9.3	5.6	23.1	**26.9**	**14.8**
Government incentives and subsides	37.0	9.3	7.4	9.3	12.0	4.6	20.4	**25.9**	**16.7**
National economic policy	34.3	11.1	6.5	6.5	13.9	5.6	22.2	**25.9**	**19.4**
Ecological suitabilities	17.6	13.0	12.0	16.7	3.7	3.7	33.3	**24.1**	**7.4**
International multilateral conventions and commitments such as IUCN, RAMSAR, Kyoto protocol, Paris climate agreement, etc.	25.0	15.7	8.3	8.3	8.3	4.6	29.6	**21.3**	**13.0**

Source: Authors

international protocol such as the REDD+ and other carbon finance framework, activities of development partners such as the World Bank to promote climate-compatible livelihoods, national climate strategy documents, and national adaptation plan of actions against climate change, multilateral conventions, and agreements including the Kyoto Protocol and Paris Climate Agreement. This is reinforced by the fact that ecosystem-compatiblealternative livelihoods are hugely unknown to the people.

Conclusion

The Southwest Nigeria is dominated by population whose livelihoods are still substantially tied to the primary natural resource stock through crop production and lumbering which imposes pressure on land-use change. The most important proximate land-use change and deforestation and forest and woodland degradation drivers are lumbering/commercial logging, fuelwood extraction, polewood extraction, and charcoal production. Others include crop cultivation, urban growth, animal grazing, and transportation and industrialization. Population increase and poverty are considered the most important underlying drivers of land-use change. Others are urbanization and industrialization, public attitude, values, and beliefs. The connectivity or teleconnection effect especially with regard to cash crop cultivation (such as cocoa) and arable crops substantially affects land-use change and deforestation. High demand for wood locally and for export also drives deforestation. Thus, responses to economic opportunities as mediated by institutional factors, and amplified by globalization, drive land-use changes in Southwest Nigeria.

Factors such national policies on natural resources, forestry, land tenure and property rights, international multilateral conventions and commitments, REDD+ and carbon finance framework, government incentives and subsidies, national climate change strategies, activities of development partners, etc. have very little impacts with respect to land-use change, deforestation, and woodland degradation. These, combined with low awareness about possibility of alternative ecosystem-compatible livelihoods, undermine and diminish the resilience of people and exacerbate climate change impacts. Expansion of crop cultivation driven by global connectivity is just as important as the lack of alternative livelihoods in driving land-use change and deforestation. Much of the battle to stem deforestation and negative land-use change may have to be fought on the altar of alternative ecosystem-compatible livelihoods. Although about 70% of the sampled population has heard of climate change, only about a third is sufficiently aware of the threat climate change poses to their livelihood activities. To improve rural resilience, awareness, and information about climate change, its impacts on rural livelihoods and how some of the people's livelihood activities invariably act as driver of climate change and its impacts should be emphasized and appropriately publicized. If people are sufficiently aware of the danger which their short-term activities posed to themselves and their livelihood, they may likely be predisposed to making amends.

Land-use change is addressed in international agreements such as the 2030 Agenda on Sustainable Development and the Paris Agreement on climate change. Signatory countries to such agreements are required to ensure that sufficient emphasis is given to land-use change in their national policies and in the implementation of existing policies (FAO 2016a). One way to achieve this will be for the Federal and respective state governments in Nigeria to mainstream climate change into development planning process. This will likely improve the visibility of national and international plans and initiatives on climate actions and elevate the awareness of alternative livelihoods at local levels.

Cross-References

▶ Biodiversity, Ecosystem Degradation, and Climate Change Effects on Livelihoods in the Bitumen Area of Nigeria
▶ Ecosystem-Based Adaptation in Tigray, Northern Ethiopia: A Systematic Review of Interventions, Impacts, and Challenges
▶ Forest Cover Change and Its Impacts on Ecosystem Services in Katimok Forest Reserve, Baringo County, Kenya

References

Abiodun BJ, Pal JS, Afiesimama EA, Gutowski WJ, Adedoyin A (2008) Simulation of West African monsoon using RegCM3 part II: impacts of deforestation and desertification. Theor Appl Climatol 93:245–261. https://doi.org/10.1007/s00704-0222-1

Adedayo V, Fasona M, Olorunfemi F, Elias P, Oloukoi G (2013) An evaluation of policy on natural resource management and its impacts on rural livelihoods in Nigeria. The Nigerian Journal of Business and Social Studies 7(1):49–71

FAO (2014) Global forest resources assessment 2015 country report: Nigeria. FAO, Rome

FAO (2016a) State of the World's forests 2016. Forests and agriculture: land-use challenges and opportunities. FAO, Rome

FAO (2016b) Global forest resources assessment 2015: how are the world's forests changing? 2nd edn. FAO, Rome

FAO (2016c) Food outlook: biannual report on global food markets. FAO, Rome

Fasona M, Oloukoi G, Olorunfemi F, Elias P, Adedayo V (2014) Natural resource management and livelihoods in the Nigerian savanna. Penthouse Publication, Ibadan, 237pages, ISBN: 978-978-52153-2-8

Fasona M, Tadross M, Abiodun B, Omojola A (2013) Some implications of terrestrial ecosystems response to climate change for adaptation in Nigeria's wooded savannah. Environmental Development 5(2013):73–95. https://doi.org/10.1016/j.envdev.2012.11.003

Fenske J (n.d.) The Emergence (or not) of Private Property Rights in Land: Southern Nigeria, 1851 to 1914. Available online: http://economics.yale.edu/sites/default/files/files/Workshops-Seminars/Economic-History/fenske-061129.pdf. Accessed 15 Sept 2016

Geist HJ, Lambin EF (2002) Proximate causes and underlying driving forces of tropical deforestation. Bioscience 52(2):143–150

GOFC-GOLD (2016) A sourcebook of methods and procedures for monitoring and reporting anthropogenic greenhouse gas emissions and removals associated with deforestation, gains and losses of carbon stocks in forests remaining forests, and forestation. GOFC-GOLD Report version COP22–1, (GOFC-GOLD Land Cover Project Office, Wageningen University, The Netherlands)

Gregory KJ (1985) The nature of physical geography. Edward Arnold (Pub) Ltd, London 262p

Hosonuma, N., Herold, M., De Sy, V., De Fries, R., Brockhaus, M., Verchot, L., Angelsen, A., Romijn, E. (2012): An assessment of deforestation and forest degradation drivers in developing countries. Environ Res Lett 7 (2012) 044009.

Houghton RA (2003) Revised estimates of the annual net flux of carbon to the atmosphere from changes in land use and land management 1850–2000. Tellus 55B:378–390

IPCC (2006) IPCC guidelines for national greenhouse Gas Inventories. Prepared by the National Greenhouse Gas Inventories Programme, Eggleston HS, Buendia L, Miwa K, Ngara T, Tanabe K (eds). Published: IGES, Japan

Jonathan MH, Maliheh BF (2011) Forests, agriculture, and climate: Economics and policy issues. A GDAE teaching module on social and environmental issues in economics. Tufts University, Medford, MA, 02155.

Jones HA, Hockey RD (1964) The Geology of Parts of South-western Nigeria: Explanation of 1:250,000 sheets Nos 59 and 68. Geological Survey of Nigeria Bulletin No 31

Kottek M, Grieser J, Beck C, Rudolf B, Ru F (2006) World map of the Köppen-Geiger climate classification updated. Meteorol Z 15(3):259–263

Lambin EF, Turner BL, Geist HJ, Agbola SB, Angelsen A, Bruce JW, Coomes OT, Dirzo R, Fischer G, George PS, Homewood K, Imbernon J, Leemans R, Moran EF, Mortimore M, Ramakrishnan PS, Folke C, Li X, Richards JF, Skanes H, Steffen W, Stone GD, Svedin U, Veldkamp TA, Xu J (2001) The causes of land-use and land-cover change: moving beyond the myths. Glob Environ Chang 11:261–269

MacKellar NC, Tadross MA, Hewitson BC (2009) Effects of vegetation map change in MM5 simulations of southern Africa's summer climate. Int J Climatol 29:885–898. https://doi.org/10.1002/joc.1754

Mbow C, Skole D, Dieng M, Justice C, Kwesha D, Mane L, Gamri M, Vordzogbe V, Virji H (2012) Challenges and prospects of REDD+ in Africa: Desk Review of REDD+ implementation in Africa. GLP Report No 5, GLP-IPO, Copenhagen

Nigerian Bureau of Statistics (NBS) (2012) National Bureau of Statistics Annual Abstracts of Statistics 2012. Available at http://www.nigerianstat.gov.ng/library. Access 16 March 2017

Obioh IB, Oluwole AF, Akeredolu FA, Asubiojo OI (1994) National inventory of air pollutants in Nigeria: emissions for 1988. Environmental Research Laboratory, Ilupeju Press Ltd, Benin City

Omotosho JB, Abiodun J (2007) A numerical study of moisture build-up and rainfall over West Africa. Meteorol Appl 14:209–225. https://doi.org/10.1002/met.11

Oyebo M, Bisong F, Morakinyo T (2010) A preliminary assessment of the context for REDD in Nigeria. Commissioned by the Federal Ministry of Environment, the Cross River State's Forestry Commission and UNDP. Available online at http://www.un-redd.org/AboutUNREDDProgramme/NationalProgrammes/Nigeria/tabid/992/Default.aspx. Last accessed 20 Feb 2018

Taylor CM, Lambin EF, Stephenne N, Harding RJ, Essery RL (2002) The influence of land use change on climate in the Sahel. J Clim 15:3615–3629

Turner BL II, Lambin EF, Reenberg A (2007) The emergence of land change science for global environmental change and sustainability. PNAS 104(52):20666–20671. https://doi.org/10.1073/pnas.0704119104

UNDP-UNEP (2009) Mainstreaming poverty-environment linkages into development planning: a handbook for practitioners. Available online at www.unpei.org. Last accessed 23 July 2012

Verburg PH, Veldkamp A, Espaldon RL, Mastura SS (2002) Modeling the spatial dynamics of regional land use: the CLUE-S model. Environ Manag 30(3):301–405

Wang GL, Eltahir EA (2000) Biosphere–atmosphere interactions over West Africa. II: multiple climate equilibria. Quaternary Journal of the Royal Meteorological Society 126:1261–1280

Tradition Versus Climate Change: Cultural Importance of Indigenous Fruit Tree and Adaptation in Benin, West Africa

24

Laurent Gnonlonfin, Christine N. A. Ouinsavi, Gérard Gouwakinnou, Belarmain A. Fandohan, and Towanou Olivier Houetchegnon

Contents

L. Gnonlonfin (✉) · C. N. A. Ouinsavi · T. O. Houetchegnon
Doctoral School of Agronomic and Water Sciences/Department of Sustainable Management of Natural Resources, Laboratory of Ecology and Forest Research (LERF),
University of Parakou, Parakou, Borgou, Republic of Benin
e-mail: laurentgnonlonfin@gmail.com

G. Gouwakinnou
Laboratory of Ecology, Botanic and Plant Biology (LEB),
University of Parakou, Parakou, Borgou, Republic of Benin

B. A. Fandohan
Université nationale d'agriculture École des sciences et techniques de conservation et de transformation des produits agricoles BP 114, Sakété, Bénin

W. Leal Filho (ed.), *Handbook of Climate Change Resilience*,
https://doi.org/10.1007/978-3-319-93336-8_151

Abstract
While climate variation and its impact on fruit tree species conservation are well documented around the world, few researches tackle the problem in West Africa and particularly in Benin. The present study aimed to evaluate vulnerability of indigenous fruit trees (IFT) to climate change based on the assessment of their diversity (in natural and parklands habitat) and their cultural importance. Objectives were to inventory the diversity of IFT and document knowledge and cultural importance of IFT for a better integration of population's traditions in participatory forests management. Data were collected on fields, market surveys, and interviews. Results showed the importance of indigenous fruit trees' organ trade for different classes of the population, from rural communities around protected areas. Food and magic uses were more important than medicinal uses. This study shows the relative importance of uses and trade of tree organs which could be the threat for species' conservation in Benin. Nevertheless, the place of food tree species in Benin's culture can also be positive for no vulnerability to climate variation.

Keywords
Tradition · Sociocultural importance · Indigenous fruit tree · Trade · Products · Benin

Introduction

Strength growth of the world population is a reality. This strength grows with fast rhythm motion, whereas the resources needed by people for their survival don't increase to the same rhythm. As consequence, one notes the strong pressure on these resources (waters-earths-forests and ores). The excessive emission of greenhouses gas, deforestation, and maladjusted agricultural practices are as many factors that cause climatic changes. These last years, sensitive modifications of the rhythm and rainy regimes observed affected the sub-Saharan Africa considerably. Several reasons explain climatic changes: there are increase temperature and level of the seas. Impacts of climate changes remain the pockets of drought, catastrophic floodings, stormy precipitations, violent winds, and agricultural output. Climate changes can prevent important fringe of human population (especially those in developing countries) and even animal to eat. Benin, like most West African countries, signals an accentuation of impacts of climatic change able to drag the inability of households to assure their food security. Facing this, local communities adopt several strategies allowing them to have a level of acceptable food security.

To date, very few studies on local and indigenous knowledge-based adaptation to climate change have been pursued. Therefore, the assessment of the likely impacts of climate change on forests and forest-dependent people and their vulnerabilities are thus important for enhancing climate change adaptation. Using data from 21 countries across Africa, searcher found a statistically significant positive association between the dietary diversity of children under five and tree cover in their communities. While the communities globally that depend completely on forest foods for their diets are relatively modest in number and size (Colfer 2008), the above African examples illustrate that forest foods often play an important role as nutritious supplements in otherwise monotonous diets. Since the productivity of trees is often more resilient to adverse weather conditions than that of annual crops, forest foods often provide a "safety net" during periods of other food shortages caused by crop failure, as well as making important contributions during seasonal crop production gaps (Shackleton 2004).

Climate change is one of the most serious environmental threats facing mankind worldwide. It affects agriculture in several ways, including its direct impact on food production. Climate change, which is attributable to the natural climate cycle and human activities, has adversely affected agricultural productivity in Africa (Ziervogel et al. 2006). Available evidence shows that climate change is global, likewise its impacts; but the most adverse effects will be felt mainly by developing countries, especially those in Africa, due to their low level of coping capabilities (Nwafor 2007). As the planet warms, rainfall patterns shift, and extreme events such as droughts, floods, and forest fires become more frequent (Zoellick 2009), which results in poor and unpredictable yields, thereby making farmers more vulnerable, particularly in Africa (UNFCCC 2007). Farmers face prospects of tragic crop failures, reduced agricultural productivity, increased hunger, malnutrition, and diseases (Zoellick 2009). It is projected that crop yield in Africa may fall by 10–20% by 2050 or even up to 50% due to climate change (Jones and Thornton 2003), particularly because African agriculture is predominantly rain-fed and hence fundamentally dependent on the vagaries of weather. As the people of Africa strive to overcome poverty and advance economic growth, this phenomenon threatens to deepen vulnerabilities, erode hard-won gains, and seriously undermine prospects for development (Zoellick 2009). There is therefore the need for concerted efforts toward tackling this menace.

Traditional knowledge – the wisdom, knowledge, and practices of indigenous people gained over time through experience and orally passed on from generation to generation – has over the years played a significant part in solving problems, including problems related to climate change and variability (Gyampoh et al. 2015). These traditional knowledge are valuable for informing and guiding indigenous peoples' strategies to prepare for, adapt to, and mitigate climate change.

The strategies that indigenous peoples used historically to identify, understand, and cope with environmental changes are reemerging as important sources of information pertaining to the climate challenges gaining wide public attention today.

Indigenous people have used biodiversity as a buffer against variation, change, and catastrophe; in the face of plague, if one crop fails, another will survive (Salick et al. 2007). In coping with risk due to excessive or low rainfall, drought, and crop failure, some traditional people grow many different crops and varieties with different susceptibility to drought and floods and supplement these by hunting, fishing, and gathering wild food plants (Dossou et al. 2016). The population and diversity of these trees is said to be dwindling in many parts of the continent, and researchers expect that this loss will probably be accelerated in the near future as a result of climate change. Biodiversity loss is expected to harm food and nutrition security, particularly in developing countries and emerging economies. Therefore, there is a need, and an urgency, to gather high-quality data on a wider spectrum of components of the selected indigenous fruits in sub-Saharan Africa.

In Benin, woody perennial parklands are very important for food security, especially during drought periods and food shortage. Indigenous food tree species have important socioeconomic and cultural values for local people (Fandohan et al. 2017). In addition with crops, these species help to enrich physic-chemical soil characteristics which in turn help to increase crop yields. It has been noted that these species population is ageing. Several research studies have been carried out on phenology, reproductive system, vegetative multiplication, natural regeneration, use-values, and distribution of these species' populations in Benin and elsewhere. However, little scientific information is available on climate variation and its impacts on indigenous fruit tree conservation. This is essential to build a local valorization and management scheme for the species. The present chapter was set up to address the following questions: (1) What strategies do local people used historically to cope with environmental changes? (2) What efforts are making people at local level to adjust to climate change they observe?

The main objective of this chapter is to evaluate vulnerability of IFT to climate change based on the assessment of their diversity (in natural and parklands habitat) and their cultural importance (local knowledge on the uses of indigenous fruit trees for developing their value and making management decisions). Specific objectives are to assess (1) the diversity of indigenous fruit trees on the basis of floristic inventory through the study area and (2) document traditional knowledge and cultural importance of IFT found in the Northern Benin.

Study System

Study Species: The Little Understood West African Indigenous Fruit Trees

For one, they are fruit trees native to sub-Saharan Africa, mostly known only in their immediate localities. They are also united by a severe shortage of research data on their nutritional composition. They are monoecious or dioecious species. In order to domesticate for food security, researchers selected 17 indigenous fruit trees from sub-Saharan Africa and set out to gather data on their local knowledge as well as their fruit nutrient composition. In Benin, ten of above indigenous species occur in the Sudanian and Sudano-Guinean zones, where they are found on different types of soil (rocky, hydromorphic, and clayey) and in parkland.

Study Area

The study was conducted in 22 districts (Karimama, Malanville, Kandi, Segbana, Banikoara, Gogounou, Bembereke, Kalalé, Nikki, Perere, N'Dali, Tchaourou, Bassila, Djougou, Waké, Kopargo, Kouandé, Kérou, Natitingou, Tanguiéta, Matéri,

Cobli) of the north of the Republic of Benin (Fig. 1). The climate is characterized by a unimodal rainfall regime (Adomou 2005). It is of the Sudanian and Sudano-Sahelian type with a rainfall between 900 and 1200 mm and two seasons: the rainy season from April to October and the dry season from November 2016 to March 2017. The dry season is marked by the harmattan, a cold-dry and dusty northeast wind which blows from the Sahara desert. The average temperature is 27 °C but can reach 37 °C in March and April. Soils are of the tropical ferruginous type and mostly rocky and shallow throughout the mountain chain. The vegetation is dominated by tree and shrub savannah. The dominant woody species are *Parkia biglobosa*, *Vitellaria paradoxa*, *Adansonia digitata*, *Blighia sapida*, *Tamarindus*

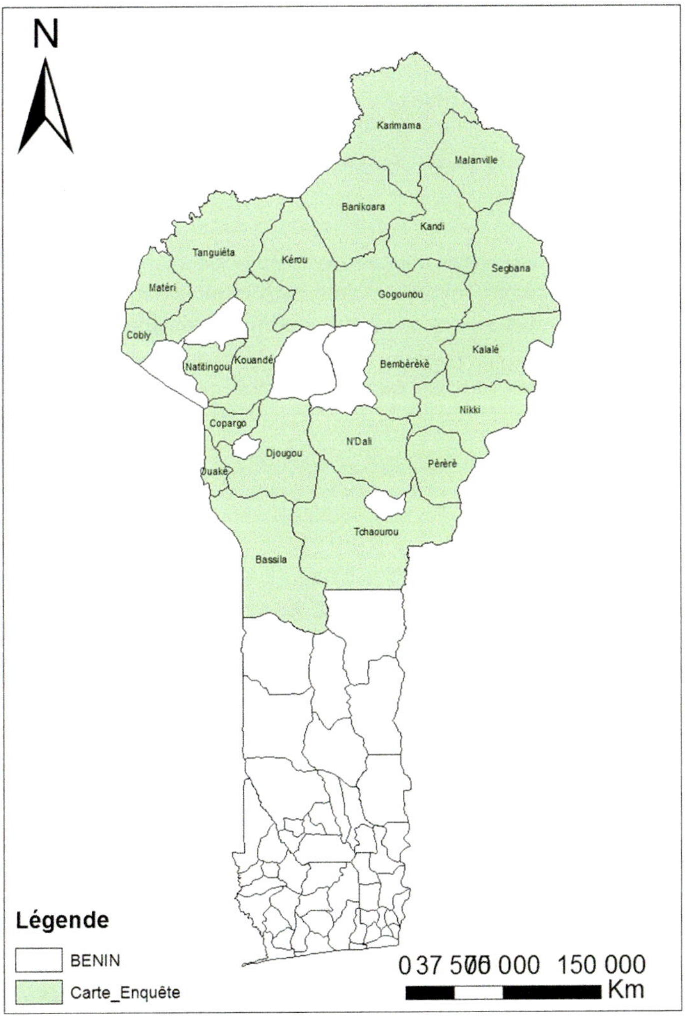

Fig. 1 Study area (Northern Benin)

indica, *Bombax costatum*, and *Khaya senegalensis*. The population of the study area is estimated at 2,941,180 inhabitants, and the farm population was 2,626,430 inhabitants (INSAE 2013). The population in the study area is mostly composed of dominant ethnic groups such as Batonou, Fulani, Dendi, Mokole, Boko, Nagot, Yom, Haoussa, Adja, and Fon. People believe in animist practices and worship divinities represented by physical elements such as clumps of natural stones, large-sized trees, or other specific constructions (Biaou 2006). Livelihood activities carried out by the people of these ethnic groups include agriculture, ranching, fishing, hunting, and processing of agricultural products, trade, and craft. The processing of agricultural products is mainly practiced by women individually or in groups with handicraft equipment. The main processed materials are *Vitellaria paradoxa* nuts (transformed into butter), seeds of *Parkia biglobosa* (transformed with a ready market), the seeds of Sorghum bicolor (transformed into an alcoholic beverage and also used in some traditional ceremonies), and pods of *Arachis hypogea* (transformed into oil) (Biaou 2006).

Study Methods

The study involved field trips and surveys. To assess sociocultural importance of indigenous fruit trees in Northern Benin, two complementary approaches were combined: (1) exploratory survey to identify and select localities where used species are present and (2) ethnobotanical study of species in the selected localities to identify current uses of the species in the study area and incomes from these species.

Exploratory Survey

Informations were obtained through oral and guided questionnaire administered to local user, older household heads, and women. The indigenous fruit plants, which are collected during the field trips, were identified with the aid of Floras of Madame de Souza. Confirmations were carried out using Arbonnier (2002). Field observations and records were made on life specimens. The data on the correct identification including common-English and local names, families, and plant parts used for food are summarized.

Ethnobotanical Study of Species in the Selected Localities to Identify Current Use

Sampling and Data Collection

Ethnobotanical survey was conducted in the dry season (November 2014–April 2015). An exploratory survey was made with 30 informants randomly selected per zone with the aim to determine the number of informants to be interviewed.

The sample size was determined following the normal approximation of binomial distribution of (Dagnelie 2013):

$$n = \frac{U_{1-\alpha/2}^2 \times p(1-p)}{d^2}$$

where n is the total number of surveyed people in the study; $U_{1-\alpha/2}^2$ is the value of the normal random variable for a probability value of $a = 0.05$, $U_{1-\alpha/2}^2 = 1{,}96$; p is the estimated proportion of people in the village which was involved in the production, processing, and/or marketing of indigenous fruit species; and d is the marginal error taken as 5%.

Informants were selected using a random sampling scheme. A semi-structured interview method was used to collect data. The questionnaire was in French and addressed into the interviewee's language with a translator locally engaged. During the survey, in addition to photographs of the whole species, fruits, bark, and leaves were kept and shown to the respondents (Gouwakinnou et al. 2011). Information was gathered through individual interviews. The interview focused on (i) the informant's sociodemographic information, which included the informant's ethnic group, age, sex, and profession and (ii) the informant's knowledge on the uses and cultural importance of the species. Similar uses mentioned were grouped into the same use category (food, medicine, fodder, handcrafts, firewood, packaging, sale, construction, toothpicks, and dyeing) based on the results of previous studies (Marquet et al. 2005; Arbonnier 2002).

Statistical Analysis

A total of 274 informants (188 males and 86 females), with ages ranging between 25 and 100 years, were interviewed. Prior to the analyses, informants were grouped into three generations (young, $25 < \text{age} < 35$; adult, $35 < \text{age} < 65$; and old, ≥ 65 years) following (Koura et al. 2011). The three generations were composed of 40 informants aged less than 35 years, 101 informants aged between 35 and 65 years, and 133 informants aged over 65 years. Seven ethnic groups were surveyed: Batonou, Dendi, Berba, Nagot, Peuhl, Yom-Ditamari, and Haoussa (Table 1). Traditional knowledge on indigenous fruit trees was assessed by the mean of the use-value, which is the mean of the number of total reported uses. The use-value was first calculated globally and afterward per category of uses (Table 2). To this end, the number of uses was calculated per category of uses and aggregated to obtain the total number of reported uses per interviewee. Seven categories of uses were considered and included food, medicine, fodder, firewood, handicraft, sale, and other. Differences among ethnic groups and generations were assessed using generalized linear models with a Poisson error distribution. The full model (main effect with interaction) was built first, and then the effect of each factor (generation, ethnic group, and their interaction) was assessed using the

Table 1 Sample size per ethnic groups, sex, and age

	Age			Sex		
Ethnic groups	Young	Adult	Old	Women	Men	Total
Batonu	11	16	36	13	50	63
Dendi	8	13	18	16	23	39
Berba	3	3	7	6	7	13
Nagot	4	31	19	19	35	54
Peuhl	5	6	16	9	18	27
Yom-Ditamari	3	28	22	11	42	53
Haoussa	6	4	15	12	13	25
Total	**40**	**101**	**133**	**86**	**188**	**274**

Table 2 Index of knowledge and uses calculated for IFT

Index	Calculation	Description
Interviewees diversity value ID = Ux/Ut	ID: Number of uses citations by a given interviewee (Ux) divided by a total number of uses (Ut)	Measure how many interviewees used IFT and how its uses are distributed among the interviewees (Eyog-Matig 1999)
Interviewee equitability value IE = ID/IDmax	IE: Interviewee diversity value (ID) divided by the same index's maximum value (IDmax)	Measure the degree of homogeneity of the interviewees' knowledge (Eyog-Matig 1999)
Consensus value for plant part CPP = Px/Pt	CPP: Number of times a given plant part was cited Px divided by the total number of citations of all parts Pt	Measure the degree of agreement among interviewees concerning the plant part used
Use diversity value UD = Ucx/Uct	UD: Number of indications recorded by category (Ucx) divided by the total number of indications for all categories (Uct)	Measure the importance of the use categories and how they contributed to the total value of used (Eyog-Matig et al. 2002)
Use equitability value UE = UD/UDmax	UE: Use diversity value: UD divided by the index's maximum value (UDmax)	Measures the degree of homogeneity of knowledge about use categories
Consensus value for the form of use CMU = Mx/Mt	CMU: Number of citations for a given form of use (Mx) divided by the total number of citations for all forms (Mt)	Measure the degree of agreement among interviewees concerning the form of use of IFT

Chi-square test with the function *anova* in the R statistical software version 3.3.2. (R Core Team 2017). The cultural importance of the category of uses and plant parts was assessed as the mean of the importance score attributed to each category of uses and plant part. A Kruskal-Wallis test and Mann-Whitney test were used, respectively, to compare the ethnic groups and generations. A table summarizing all the specific uses mentioned for indigenous fruit plants was built. For each specific use, the fidelity level (FL, in %) was computed as the relative frequency of informants who cited that specific use over the sample size. The FL measured the consensus of informants for a specific use.

Results

Diversity, distribution and most parts of Indigenous Fruit Tree in Benin

A total of 26 indigenous fruit trees (15 families) were well-known and used in different ways by local communities in the Republic of Benin (Table 3). The most represented family was Anacardiaceae (five species), followed respectively by Arecaceae and caesalpiniaceae (two species). Twelve families were represented by only one species. Sudanian and Sudano-Guinean zones turned out to be the most diversified with 14 species (10 families) followed by Guinean zone with 8 species (6 families). The common species to the three climatic zones (whole country) are *Adansonia digitata, Annona senegalensis, Blighia sapida, Borassus aethiopium, Diospyros mespiliformis, Parkia biglobosa*, and *Vitex doniana*. The most frequent plant part used for species (cited by at least 31% of participants) were fruit, leaves, bark, and seed.

Differences in Knowledge per Category of Uses Among Ethnic Groups and Generations

Use-value of indigenous fruit trees varied only between the three age groups (p-value <0.05; Table 4). The analysis at the level of categories of uses showed that the use-value for food was similar among ethnic groups and generations (p-value >0.05; Table 4). The use-value for medicinal purposes differed between generations and also among ethnic groups (p-value <0.05; Table 4). Other use-value (magic, firewood, handicraft, sale and other) varied only among ethnic groups (p-value <0.05; Table 4). With regard to the use-value per category of uses of fruit plants among ethnic groups (Fig. 2a) and generations (Fig. 2b), all ethnic groups had a similar food use-value. Dendi used most species for medicinal purposes, whereas Peuhl used them mostly for magic purposes. The "other ethnic groups" valued the less species for food, sale, and handicraft and were followed by Nagot and Yom-Ditamari. The overall use-value was lower for young informants (estimate $= -3.48$; also see Fig. 2b). Similarly, the medicinal use-value of indigenous fruit trees was lower for young informants (estimate $= -4.03$; Fig. 2b).

Concerning the use-value of indigenous fruit tree parts among ethnic groups and generations, fruits were the most used part and were followed by leaves, stem, bark, roots, and seed (Fig. 3a). Seeds and roots are the least used plant parts. Yom-Ditamari, Berba, Dendi, Peuhl, and Batonou were the ethnic groups that valued fruits and leaves the most; while stem was more valued by Nagot followed by Haoussa. Old and Adult informants had a similar use-value for leaves, stem, bark, roots, and seeds. Whereas young informants had a lower use-value for leaves (estimate $= -1.30$), Stem (estimate $= -1.00$), bark (estimate $= -0.81$), and roots (estimate $= -0.02$) than old and adult informants (Fig. 3b).

Table 3 Diversity of indigenous fruit trees in ecological zones of Benin

Fruit tree species	Origin	Family	Occurred zone/ Benin Regions	Most part used
Irvingia gabonensis	Exotic		Whole country	Fruit, nut
Parkia biglobosa	Native/ Indigenous	Mimosaceae	Whole country	Fruit, wood, seed, bark, leaves
Adansonia digitata	Native/ Indigenous	Bombacaceae	Sudanian, Sudano-Guinean	Fruit, seed, leaves, bark
Tamarindus indica	Native/ Indigenous	Caesalpiniaceae	Sudanian, Sudano-Guinean	Pulp, leaves, bark
Vitex doniana	Native/ Indigenous	Verbenaceae	Whole country	Fruit, leaves, bark
Sclerocarya birrea	Native/ Indigenous	Anacardiaceae	Whole country	Fruit, leaves, bark
Pterocarpus santalinoides	Native/ Indigenous	Fabaceae	Whole country	Fruit, leaves, bark
Blighia sapida	Exotic	Sapindaceae	Whole country	Pulp, nut, bark, leaves
Annona senegalensis	Native/ Indigenous	Annonaceae	Whole country	Fruit, root, leaves, bark
Borassus aethiopicus	Native/ Indigenous	Arecaceae	Whole country	Fruit, shoot, wood, root
Gardenia aquala	Native/ Indigenous	Rubiaceae	Whole country	Fruit, leaves
Strychnos innocua	Native/ Indigenous	Loganiaceae	Sudanian, Sudano-Guinean	Fruit, root
Cola spp.	Exotic	Sterculiaceae	Guinean	Fruit, bark, root
Synsepalum dulcificum	Exotic		Sudanian, Sudano-Guinean	Fruit, leaves
Haematostaphis barteri	Native/ Indigenous	Anacardiaceae	Atacora chain	Fruit, leaves, bark, root
Mimusops andongensis	Native/ Indigenous	Sapotaceae	Guinean, South Sudanian	Fruit, leaves, bark, root
Lannea microcarpa	Native/ Indigenous	Anacardiaceae	Sudano-Guinean	Fruit, leaves, root
Mangifera indica	Exotic	Anacardiaceae	Whole country	Pulp, bark, leaves
Diospyros mespiliformis	Native/ Indigenous	Ebenaceae	Whole country	Fruit, wood, bark, root, leaves
Elaeis guineensis	Exotic	Arecaceae	Guinean	Nut, leaves, wood
Vitellaria paradoxa	Indigenous	Sapotaceae	Sudanian, Sudano-Guinean	Nut, leaves, wood, bark
Detarium microcarpum	Indigenous	Caesalpiniaceae	Sudano-Guinean	Pulp, leaves, bark
Cocos nucifera	Exotic		Guinean	Nut, leaves, wood
Balanites aegyptiaca	Indigenous	Balanitaceae	Sudano-Guinean	Fruit, leaves, bark
Anacardium occidentale	Exotic	Anacardiaceae	Guinean, Sudanian	Nut, bark, leaves, root
Carica papaya			Guinean	Fruit, leaves, root

Table 4 Effect of ethnic groups and generations on the overall use-value and category of use: p-value of the likelihood ratio tests

Uses category	Ethnic groups	Generations	Ethnic groups/generations
Food	0.456^{ns}	< 0.001***	0.124^{ns}
Magic	0.002**	< 0.001***	0.204^{ns}
Medicinal	< 0.001***	0.541^{ns}	0.357^{ns}
Firewood	< 0.001***	0.645^{ns}	0.464^{ns}
Handicraft	< 0.001***	0.369^{ns}	0.278^{ns}
Sale	< 0.001***	0.942^{ns}	0.243^{ns}
Other	0.035*	0.034*	0.147^{ns}
Total uses	**0.121^{ns}**	**< 0.001*** **	**0.431^{ns}**
*** = $p < 0.001$;	** = $p < 0.01$;	* = $p < 0.05$;	ns = $p \geq 0.05$

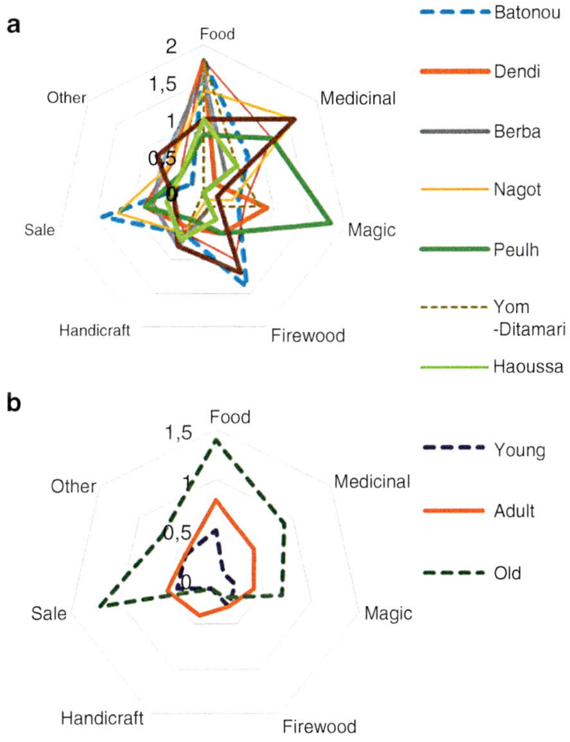

Fig. 2 Differences in the uses of indigenous fruit trees among (**a**) ethnic groups and (**b**) generations

Income from Indigenous Fruit Tree and Local People Adaptation

In most African countries, indigenous fruit trees play a significant role in the livelihood of the population by providing key subsistence products and income. In addition to providing food directly, a multitude of species harvested from natural,

Fig. 3 Differences in the uses of indigenous fruit plant parts among (**a**) ethnic groups and (**b**) generations

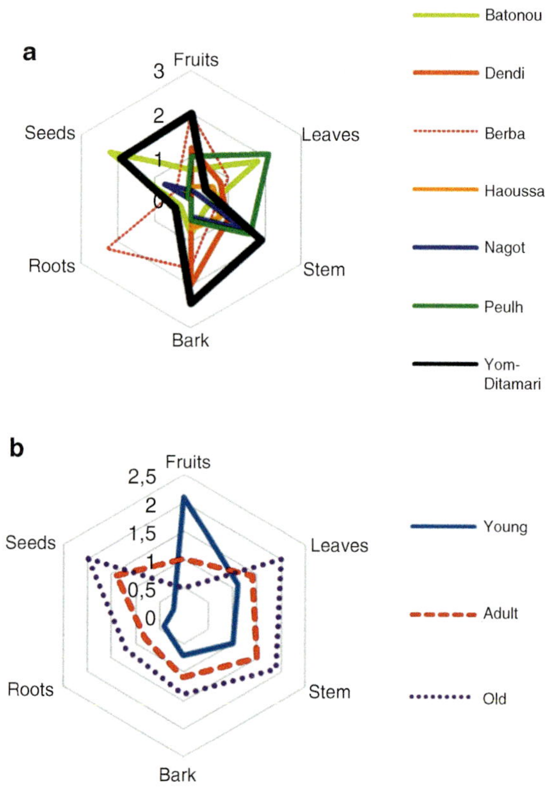

incipiently, and/or semidomesticated forests and woodlands provide a range of resources that are used by harvesters directly for other purposes or are sold for income that can be used to purchase a variety of products, including food.

When there is availability but relatively low indigenous food trees use in areas of dietary need, reasons of adaptation strategy can include the high labor costs involved in collection and processing, low yields, high phenotypic variability (with large proportions of non-preferred produce), and lack of knowledge in the community (Devineau 1999). Regarding the last point, in the Northern Benin, for example, women prepare protein-rich condiments from the seeds of wild *Prosopis* (*Prosopis africana*) trees, respectively, but women in other parts of the country (where the same is found) are not aware of these food values and do not harvest or manage woodlands for it (Koura et al. 2011). Research suggests that knowledge on the use of such species products is often higher among indigenous peoples than among immigrant communities, with knowledge being lost due to social change and "modernization." On the other hand, differences arise between genders and age groups with respect to specialized knowledge and preferences in tree use (Daniggelis 2003). This is illustrated by the different relative use-values assigned to plant products by different-aged respondents in the Batonou and Haoussa communities in the study area, where older people generally had more recall on uses for particular categories of food plants, but both young and old people assigned high use-values to food products (higher than) respondents in their mid-years. When bringing trees from the wild into cultivation, an important aspect is to increase yields: if indigenous trees are perceived as relatively unproductive and can only be produced inefficiently, agricultural landscapes are likely to be dominated by staple crops, with agro-biodiversity (and hence, likely, dietary diversity) reduced (Sunderland 2011). Since many tree species are essentially undomesticated, large increases in yield and quality are often available through selection, supporting cultivation; for example, this is the case for *Allanblackia* (*Allanblackia* spp.), described further in Box 2.2 (Jamnadass et al. 2010). Lack of knowledge on appropriate tree management, however, can be a major limitation (Jamnadass et al. 2011). Increases in efficiency are important for markets, since price to the consumer is a significant factor influencing diet (Glanz et al. 2005). Where limited access to extant forest foods is a major issue, approaches that support access such as the development of community-based forest management plants can be beneficial, but wider efforts are required to include all significant stakeholders and in particular women. Household decision-making regarding food use and practice, mostly made by women, is influenced by levels of knowledge on nutrition (FAO 2001).

Examples from Africa of widely traded tree foods that support farmers' incomes and consumers' choices include the indigenous semidomesticated and widely fruits and exotic mango. New domestic markets for fruit and other proceedings products are developing in Africa as a result of recent investments to source produce locally for juice manufacture and also to meet growing demand from population growth and increased urbanization (Ferris et al. 2014). Worldwide, products supplied from tree-crop systems are fundamental raw materials underpinning the development of small scale to multibillion dollar industries.

Discussion

In case of climate change, local knowledge (perceptions) and resource uses influence the resilience of climate change in many ways. There's a mismatch between policymakers' long-term views (climate amelioration, carbon sequestration, etc.) and farmers' short-term views (fertile forest lands, timber, fruits). Farmers are not naive about benefits of IFT, firewood, timber, nuts, and medicine, resources which drive indigenous people to fell trees (deforestation). There are several factors that influence the choice of indigenous fruit trees in Benin. Aspects such as culture, climate, and socioeconomic status and politics are the main influences on their diversity. In all study area, the coexistence of wild species and weeds has been noted growing together with other farmlands species. This work has demonstrated that indigenous fruit trees (IFT) are multipurpose species in Benin. These species are used in multiple ways. The ethnobotanical index (quantitative measures) allowed for the reaching of consensus among interviewees on the uses of the species. These quantitative measures reflect a consensus among interviewees, based on the understanding that culture represents shared knowledge. The importance and usefulness of collecting local communities' knowledge lie in the fact that such information could help determine the true value of the species, leading to more rational decisions about its sustainable utilization. Knowledge of the local usage of vegetable resources is essential for the elaboration of conservation strategies for future adaptation. Local people are driven by a safety-first strategy: I have to put food and well-being on the table. This means: "forest is great but my food security and health insurance are more urgent."

Overall, local knowledge varied not only according to different ethnic groups but also according to the age and sex of individuals interviewed. The results confirm the significant age and gender differences in the use of medicinal plants shown by other papers (Assogbadjo et al. 2002; Gouwakinnou et al. 2011). All parts of the trees (roots to leaves) are valued and used differently by people. This chapter highlights that in addition to their high nutritional value, these indigenous fruits have another important advantage: they produce even when staple crops fail (e.g., during droughts) and in this way bridge the "hunger gap" during times of food shortage. Indigenous fruit trees can also serve as a source of income for farmers to improve livelihoods. Participatory domestication of IFT enables farmers to grow these trees widely in agroforestry systems and multiplies these gains (coping with negatives effects of climate change in their communities).

Conclusion

One significant manifestation of the marginalization of indigenous peoples from the climate change policy and decision-making is the paucity of references in the global climate change discourse to the existing traditional knowledge on climate change. This chapter emphasized the local knowledge of the indigenous fruit trees dwelling in Benin. Although they are used by several ethnic groups of the study area, they remain undervalued. Our findings established the importance of the species for local people. Further studies should focus on the morphological characterization of the fruits and on their nutrient composition as well as effective propagation methods of species.

Cross-References

- ▶ Adaptation of Sorghum (*Sorghum bicolor* L. Moench) Crop Yield to Climate Change in Eastern Dryland of Sudan
- ▶ Climate Change Adaptation Among Female-Led Micro, Small, and Medium Enterprises in Semiarid Areas: A Case Study from Kenya
- ▶ Climate Change, Vulnerability, and Adaption Under the Small Farming Households of Konso Community, Southern Ethiopia
- ▶ Effect of Climate Variability on Crop Income in Central Highlands of Ethiopia

References

Adomou (2005) Vegetation patterns and environmental gradients in Benin. Implications for biogeography and conservation. PhD Thesis, Wageningen University

Arbonnier (2002) Arbres, arbustes et lianes des zones sèches d'Afrique de l'Ouest. 2ème. [Montpellier]: CIRAD;[Paris]; Museum national d'histoire naturelle 573p.-illus., col. illus. Fr Icones, Anatomy and morphology, Keys

Assogbadjo et al (2002) Connaissances ethnobotaniques et valorisation du baobab (Adansonia digitata) pour la sécurité alimentaire des populations rurales au Bénin. Plant Genet Res Food Secur West Cent Africa 1:66–76.4

Biaou (2006) Land use impact on Vitellaria paradoxa CF. Gaerten. Stand structure and distribution patterns: a comparison of Biosphere Reserve of Pendjari in Atacora district in Benin. Agoforestry Syst 72:205–220

Colfer (2008) Negleted plants of food importance in traditional farming systems of tropical Africa. Londres. RU. Earthscan publications. 52:131–150

Dagnelie (2013) Statistique théorique et appliquée (tome 1) – 2013 ... Principes généraux relatifs aux distributions d'échantillonnage

Daniggelis (2003) Local species highly valued by women and men in Tanzania. In Forests, food security and gender: linckages, disparities and priorities for action. International conference on forests for food security and nutrition. 20 p

Devineau JL (1999) Ecologie des principales espèces ligneuses alimentaires et fourragères dans un système de culture-jachère: Sud-Ouest du Burkina-Faso. Actes du Séminaire International sur la jachère en Afrique tropicale 13–16 avril 1999; Dakar; 1996, 441–450

Dossou EM et al (2016) Analyse de l'impact du changement climatique sur l'aire de distribution actuelle et future de Lannea microcarpa Engl. & K. Krause au Bénin, Afrique de l'Ouest. Afrique Sci 12(1):27–38

Eyog-Matig O (1999) Espèces Ligneuses Médicinales. Compte rendu de la première réunion du Réseau: 15–17 décembre 1999; Station IITA Cotonou, Bénin. In Réseau Institut International des Ressources Phytogénétiques

Eyog-Matig O et al (2002) Espèces Ligneuses Alimentaires. Compte rendu de la première réunion du Réseau: 11–13 décembre 2000; CNSF Ouagadougou, Burkina Faso. In Réseau Institut International des Ressources Phytogénétiques

Fandohan B et al (2017) Usages traditionnels et valeur économique de Synsepalum dulcicum au Sud-Bénin. Bois et Forêts des Tropiques 332(2):2–3

FAO (2001) Evaluation des ressources en produits forestiers non ligneux. Expériences et principes de biométrie au niveau Ecorégional. Rapport Produits forestiers non ligneux 13. ISSN 1020–9727

Ferris et al (2014) Forests and food: addressing hunger and nutrition across sustainable landscapes. Understanding the roles of forests and trees based system in food provision. Incomes from cultivated. OpenBook. IUFRO Publishers

Glanz et al (2005) Value of indigenous fruit trees and its influence on diet in West Africa. In Nutrition and food systems-Global alliance for improved nutrition in low income country. Report 12

Gouwakinnou GN et al (2011) Local knowledge, pattern and diversity of use of Sclerocarya birrea. J Ethnobiol Ethnomed 7(8):1746–4269

Gyampoh BA et al (2015) Using traditional knowledge to cope with climate change in rural Ghana. www.fao.org/docrep/011/i0670e/i0670e14.htm

INSAE (2013) Effectifs de la population des villages et quartiers de ville du Bénin : Institut National de la Statistique et de l'Analyse Economique (INSAE). Cotonou, Bénin. Accessed via-http://www.insae-bj.org/recensement-population.html?file=files/publications/RGPH4/Cahier_Village RGPH4_2013.pdf

Jamnadass et al (2011) The role of wild plants in the native diet in Africa. Agroecosystem 1:45–56

Jones, Thornton (2003) Seasonal traditional fruits and vegetable of Tropical Africa.Challenges of Agricultural Adaptation to Climate Change in Nigeria: a Synthesis from the Literature. Journal of field action. Openedition.org

Koura K et al (2011) Ethnic differences in use values and use patterns of Parkia biglobosa in northern Benin. J Ethnobiol Ethnomed 7(42):1–12
Kuhnlein et al (2009) Edible wild fruit trees and shrubs and their socioeconomic significance in Central Ethiopia. Ethnobot Res Appl 14:183–197
MAEP (2011) Plan Stratégique de Relance du Secteur Agricole (PSRSA). In Plan stratégique opérationnel, Orientations stratégiques de développement du Bénin 2006–2011
Marquet et al (2005) People's needs and the role of indigenous trees in Tanzania. Indigenous Plants: Key Role Players in Community Horticulture
Nwafor (2007) Non wood forest products: The way ahead. Rome, Food and Agriculture Organization, FAO Forestry Paper no. 97. 37p
Odjugo (2010) Proceedings of the twelfth plenary meeting of AETFAT, Symposium VIII, pp 911–928
Ouédraogo A-S (1995) Parkia biglobosa (Leguminosae) en Afrique de l'Ouest: biosystématique et amélioration. Thèse de Doctorat Université Agronomique de Wageningen
R Core Team (2017) R: a language and environment for statistical computing. R Foundation for Statistical Computing, Vienna, Austria. https://www.R-project.org/
Salick et al (2007) Leafing, flowering and fruiting of indigenous fruit trees in savannah parklands in Uganda. Agrofor Syst 60:77–91
Shackleton (2004) Trees for food and fodder in the savanna areas of West Africa. Int Tree Crops J 1:131–141
Sunderland (2011) The contribution of wild plants to human nutrition in the Sahelian zone. J Arid Environ 11(1):61–64
UNFCCC (2007) Convention sur l'utilisation de la diversité biologique. www.un.org/fr/events/biodiversityday/convention.shtml
Vira B et al (2016) Understanding the role of forests and trees-based systems in food provision: Addressing hunger and nutrition across sustainable landscapes (Forests and food). www.worldagroforestry.org/shambashapeup/resources/Forests-Trees-and-Landscapes.pdf
Vodouhê GF et al (2009) Estimating local values of non-timber Forest products to Pendjari biosphere reserve dwellers in Benin. Econ Bot 63(4):397–412
www.fao.org/docrep/011/i0670e/i0670e14.htm
www.worldagroforestry.org/shambashapeup/resources/Forests-Trees-and-Landscapes.pdf
Ziervogel et al (2006) Comparative analysis of diversity and utilization of edible plants in arid and semi-arid areas in Benin. J Ethnobiol Ethnomed 10(80):1–20
Zoellick (2009) Valuation of local use and dynamics of 56 woody species in the Sahel. Biodivers Conserv 13:1961–1990

25 Forest Cover Change and Its Impacts on Ecosystem Services in Katimok Forest Reserve, Baringo County, Kenya

A. Jebiwott, G. M. Ogendi, S. M. Makindi, and M. O. Esilaba

Contents

A. Jebiwott (✉) · S. M. Makindi · M. O. Esilaba
Department of Environmental Science, Egerton University, Egerton, Kenya
e-mail: aliciaphil.chebby@gmail.com; mankindsm@gmail.com; mosesotiali@gmail.com

G. M. Ogendi
Department of Environmental Science, Egerton University, Egerton, Kenya

Dryland Research Training and Ecotourism Centre, Chemeron, Egerton, Kenya
e-mail: gmorara2009@gmail.com

W. Leal Filho (ed.), *Handbook of Climate Change Resilience*,
https://doi.org/10.1007/978-3-319-93336-8_155

Abstract

Forests provide vital ecosystem services to communities living around them, contributing immensely to their livelihoods. Yet, these forests undergo land use changes that threaten the abundance and availability of ecosystem services. This chapter assesses forest cover changes and its impacts on the ecosystem services in Katimok Forest Reserve. Land use changes were quantified using Landsat satellite images for the years 1984, 2001, and 2015. Focus group discussions (FGDs) and household surveys were used to identify and characterize the changes in ecosystem services. Land cover maps were prepared using supervised classification method and post-classification technique was used to detect the changes in forest cover. The results indicate that in the period 1984–2001, there was a decline in dense forest by 4.9%. Open forestland and built-up area had increased by 4.3% and 0.6%, respectively within the same period. These changes led into the reduction and loss of ecosystem services such as loss of certain medicinal plants, decline in mushrooms, and habitat destruction. It is therefore important that the relevant forest resource management agencies formulate sustainable resource utilization options/strategies for the local communities to curb degradation of this life-supporting ecosystem.

Keywords

Ecosystem services · Land use · Forest cover change · Katimok Forest Reserve · Forest ecosystem

Introduction

Forests are among the major world ecosystems whose value lies in the goods and services they provide for the maintenance of both human and environmental welfare. Healthy ecosystems provide goods
and services that are essential for human health and livelihoods, commonly known as Ecosystem Services (Millennium Ecosystem Assessment 2005). Forests sustain major life support systems on earth and are essential for the development of social and economic sectors of many countries. In fact, the expanse of a forest gives the first indication of the relative significance of forest ecosystems in a nation. According to TEEB (2010), forests are estimated to cover about a third of the earth's surface and contains over half of the terrestrial species. In Africa, forests cover approximately 21.4% of the total land area which represent 674 million hectares. The forests of Eastern Africa constitute about 13% of the total land area and Kenya, which is termed as the richest in terms of forest cover in the region, has a forest cover of about 6.07% of the land area (FAO 2008).

Forests and their products play a critical role in the improvement of lives of the local communities. In its Poverty Environment Network (PEN) research, the Center for International Forestry Research (CIFOR 2014) reveals that forests play a major contribution to the livelihoods of communities globally. World Bank

(2013) estimates that approximately 350 million people in forest adjacent communities rely on forests for their livelihoods and 60 million of these are entirely reliant on the forest ecosystems. However, deforestation and forest degradation are threatening the capacity of these ecosystems to support livelihoods globally. This is because the well-being of a forest adjacent community greatly depends upon the services provided by the ecosystem. The continued flow of these services, on the other hand, is dependent upon the human activities which can either increase or decrease the benefits derived from the ecosystem, consequently impacting on their livelihood.

Moreover, global consequences of human-induced land use changes affect a wide array of services that ecosystems provide to answer health, social, cultural, and economic needs. Human activities affect forest cover directly through activities such as harvesting of timber and clearing for settlement and agriculture. According to Millennium Ecosystem Assessment (2005), humans can transform ecosystems in ways that can improve or reduce the benefits derived by the community. When a forest is converted into an agricultural land, crop productivity improves and hence ensuring the supply of food but the functioning of other services such as climate regulation, soil erosion control, pollution control, and other services which are important for human welfare are disrupted. In the long run, the value of services lost through conversion of the ecosystem exceeds the short-term benefits obtained from the economic activity of crop production.

Socioeconomic forces may also have significant indirect influences as they encourage policies and sequences of actions that can trigger degradation and ultimate decline in ecosystem services (Bravo 2008). Despite review of Kenyan forest policies to conserve and manage forests through sustainable utilization, the forests are still undergoing significant negative change. In Kenya, many forests are under pressure owing to increasing human populations, and as a result, these forests have been encroached. Clearing of forested areas for agriculture and settlement are seen in most forests, while in others, the population impinge upon the forests for their products such as timber and charcoal. These human activities threaten forest ecosystems and lead to eventual decline or even loss of services that the ecosystem provides.

Katimok Forest in Baringo County, Kenya, is one of such forests that have undergone degradation despite providing several crucial benefits to the community that include: herbal medicine, fruits, mushrooms, fodder, clean water, employment, recreation, and religious values, as well as the protection of the local people's way of life. Katimok Forest is situated in the semi-arid areas of Baringo County, an area susceptible to environmental and climate change and high rates of food insecurity. Dependence on ecosystem for provision of goods and services in this area is highest owing to the community's limited alternative livelihood options due to aridity of the environment that negatively impacts on household production activities (Baringo County Government 2015; Kahuki and Muniu 2004). Overdependence on the forest and unsustainable utilization of the forest resources translates to considerable deterioration in terms of the ecosystem services and forest cover. Therefore, there is need to embrace sustainable forest resource utilization to safeguard the ecosystem against degradation and thus a reliable source of goods and services for various user-groups This chapter was conceived on the premise that the current resource extraction approaches, as witnessed during the preliminary visit (massive logging and presence of heavy trucks carrying timber to nearby sawmills), are unsustainable and consequently compromising the ability of future generations in accessing the same resources and services. Therefore, this chapter aimed at assessing the effects of forest cover changes on the forest's ability to provide ecosystem services.

Materials and Methods

Description of Katimok Area

Geographical Setting

Katimok Forest Reserve, gazetted 1949, is a forest in Baringo County, Kenya. It lies at an altitude of 2162 m and between Latitude of 0.55000000 and Longitude of 35.75000000. The forest is under the management of Baringo County Government and protected by the Kenya Forest Service. It is the largest block of the current Kabarnet Forest which consists of 13 blocks as shown in Table 1, Fig. 1.

Flora and Fauna

A larger portion of the forest is composed of indigenous trees including *Syzygium guineense* Wall, *Olea africana* Mill., *Prunus africana* Kalkman, *Vitex keniaensis* Turrill, and the endangered *Osyris tenuifolia* L., among other tree species. Exotic plantations, such as *Eucalyptus saligna* Sm., *Cupressus lucitanica* Mill, *Pinus patula* Schiede, and *Grevillea robusta* A.Cunn, also make up the forest, and they were established as early as 1970s on lands which became vacant after the eviction of

Table 1 Kabarnet Forest blocks

Block	Size (ha)
Katimok Forest	1956.59
Saimo Forest	750.9
Tarambas Hill Forest	483.8
Morop Forest	212.6
Kimeto Forest	210.4
Mosegem Forest	202.7
Sokta Hill Forest	163.9
Pemwai Forest	117.7
Chebartigon Forest	103.3
Ketwan Forest	46.6
Cherial Forest	42.5
Kabiok Forest	14.2
Tutwoin Forest	12.1

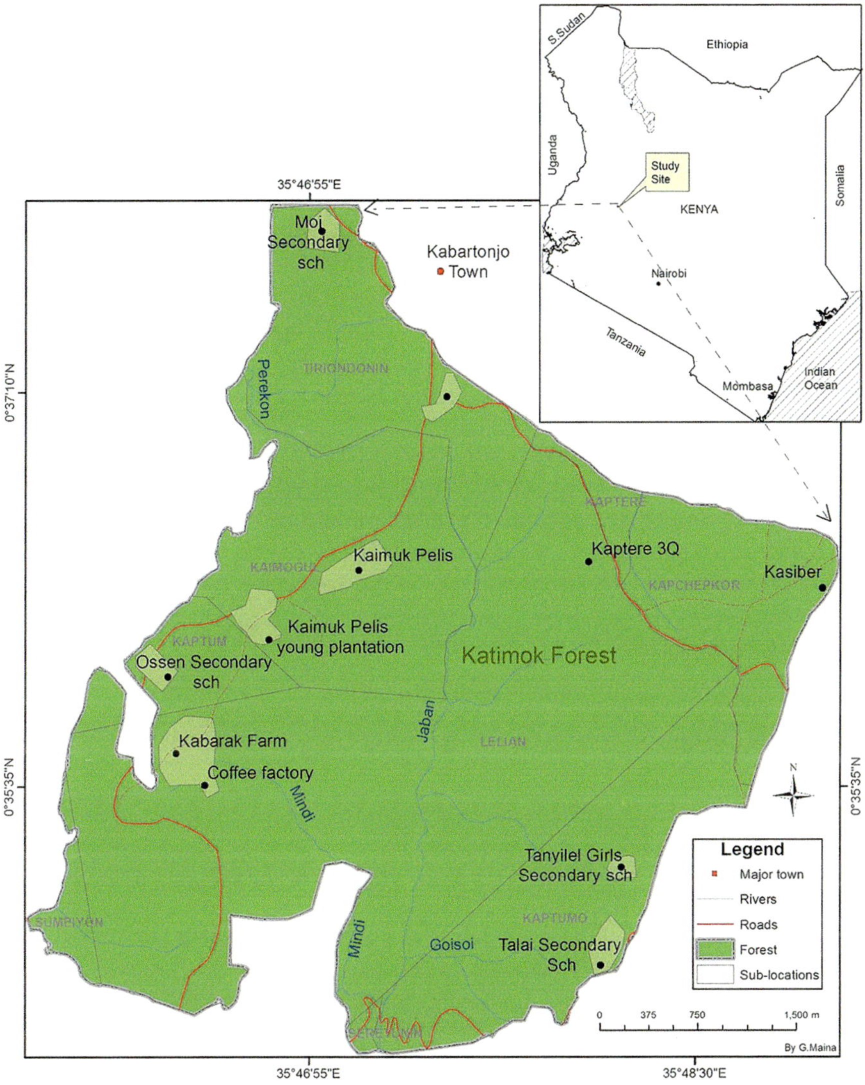

Fig. 1 Map of Katimok Forest Reserve

illegal settlers. The forest is also home to wild animals and birds which include: *Colobus angolensis* P.Sclater, *Papio anubis* (Lesson), *Lepus capensis* Linnaeus, *Coturnix coturnix* Linnaeus, *Numida meleagris* (Linnaeus), and *Madoqua kirkii* Günther, among others.

Climate and Socioeconomic Activities

According to the annual development plan of Baringo County Government (2015), agriculture is the backbone of the forest surrounding community. This is especially because the soils and the climate of the region are conducive for crop and livestock production. The area receives an annual rainfall of about 1000–1500 mm and temperatures ranging from a minimum of 10 °C and a maximum of 30 °C. This coupled with fertile soils make the forest surrounding populations practice production of different varieties of crops such as maize, sorghum, millet, and beans among other crops. In addition, rearing of goats, sheep, cattle, and bee keeping are other common economic activities in the area. Furthermore, the poverty level of the area is high and as such, the communities surrounding the forest are largely dependent on the forest's ecosystem services, such as honey, wild fruits, construction material, fuel wood, agricultural land, water, traditional ceremonies, and fodder among other benefits, for their livelihood. There are four major streams flowing through the forest, and they are *Goisoi, Mindi, Jaban,* and *Perekon* streams.

Population

The forest is surrounded by Ossen, Kapchemungot, Kabarbet, and Kaimugul sublocations which according to the 2009 population census have household populations of 693, 205, 159, and 399, respectively. The population density is 50 persons per km^2 with an annual growth rate of about 2.6%. The age distribution is 0–14 years (48.4%), 15–64 years (48.2%), and 65 and above years (3.3%) (Kahuthu et al. 2005).

Research Design

This chapter employed a social survey design with the aim of gathering information on the ecosystem, its services, and the local community from the selected sample. In addition, information concerning the current status of the ecosystem as well as describing the relationships between and among variables, such as level of awareness and education level, and between forest cover change and ecosystem services, under study was obtained.

Sample Size and Sampling Procedure

The sampling frame was a list of households adjacent to the forest. This chapter used a combination of stratified random sampling and simple random sampling in selecting the required number of household respondents to participate in the study. This was achieved through identification of strata of interest based on their administrative locations and proximity to the forest. The strata were Ossen, Kapchemungot, Kabarbet, and Kaimugul sublocations with household populations of 693, 205, 159, and 399, respectively. Proportionate sampling was then used to draw 48, 14, 11, and 27 households from Ossen, Kapchemungot, Kabarbet, and Kaimugul strata, respectively totaling to 100 households.

Data Collection Methods

Primary Data

Primary data was obtained from the field by employing both qualitative and quantitative data collection techniques. The chapter used the methodological triangulation technique in order to build a valid and reliable data collection plan. The instruments administered for collecting data from the field were:

(a) Household survey
(b) Focus group discussions
(c) Observations
(d) GPS Points

Household Survey: A structured questionnaire was administered to the heads of households that had been selected into the study. The questionnaire mainly served to assess the provisioning, regulatory, and cultural ecosystem services provided by the forest to the community. This was done by assessing the different uses of the different ecosystem services by the households. The questionnaires were administered in conformity with the sample size.

Focus Group Discussions: Focus group discussion (FGD) instrument was developed and administered to the staff from the Kenya Forest Service, members of Community Forest Associations, and local village representatives of the area. A topic guide to aid the discussion was developed beforehand, and brainstorming was used to explore each topic. FGDs were conducted with the purpose of getting additional information or information which may have not been clearly captured through questionnaires.

Observation: To get a clear picture of the status of the forest and its ecosystem services, an observation checklist was used to gather information. The pressures on the forest, e.g., felling of trees, livestock grazing, firewood collection, etc. were recorded. For more detailed information, photographs were taken. Observation gave a clear condition of the forest and hence enabling an understanding of the relationship between the people and the use of the forest's ecosystem services.

Secondary Data

GIS and Remote Sensing Data Acquisition: The datasets used in this chapter were satellite images. All preprocessing and processing activities were done using Idrisi Kilimanjaro and ERDAS IMAGINE image processing software. Satellite images of the study area taken by Landsat Thematic Mapper ™ sensor for the years 1985, and Enhanced Thematic Mapper for 2001 and 2015 were used. The images acquired were for the dry season to minimize errors that might arise from seasonal differences, and near anniversary dates were chosen for consistency within the time periods.

Image Preprocessing

Preprocessing of satellite images is essential so as to remove data acquisition errors. The most common operations applied to an image during the preprocessing stage

Table 2 Land cover description

Land cover type	Description
Dense forest	These are lands that have a compact stock of trees capable of producing timber and other wood products
Open forest land	Composed of lands with scattered patches of trees, farmlands, and lands with small trees, grasses, and shrubs
Built-up area	Areas composed of infrastructure mostly rural villages, schools, hospitals, and roads

include geometric correction, radiometric correction, mosaicking, and subsetting. The images acquired did not need any radiometric and geometric corrections as they were already rectified.

Ground Truth Points

For ground-truthing, each land use type was noted and GPS used to capture the coordinates of the land cover types at each point. These ground control points were used as training samples for supervised classification and also for accuracy assessment.

Land Cover Classification

A supervised classification of the satellite imagery was used to produce land use land cover classes. The following land cover types were generated from the classification as described in Table 2.

Methods of Data Analysis

Data collected from various sources was entered into different software for analysis. Social data was entered into SPSS and Microsoft Excel and GPS ground truth data into ArcGis 10.1, Idrisi Kilimanjaro, and ERDAS IMAGINE. The process of data analysis involved cleaning the questionnaires for errors and coding quantitative data from the household interviews. Descriptive statistics were used for social data, to make cross tabulations, pie charts, bar graphs, and to calculate percentages and means. To determine whether the statistics were significant or not, Chi-square test of independence and Fisher's exact tests were performed.

For digital image processing, false color composites were created using bands 2, 3, and 4 for each of the images. The images were then geo-referenced in UTM projection WGS84 reference ellipsoid. Both supervised and unsupervised classification methods were used to classify the images, but owing to the high accuracy of supervised classification, the change detection base map was prepared using supervised classification.

Post-classification comparison method was used for change detection. In this method, images of different dates were first classified and labeled individually, then the classified images were then compared, and changed areas extracted.

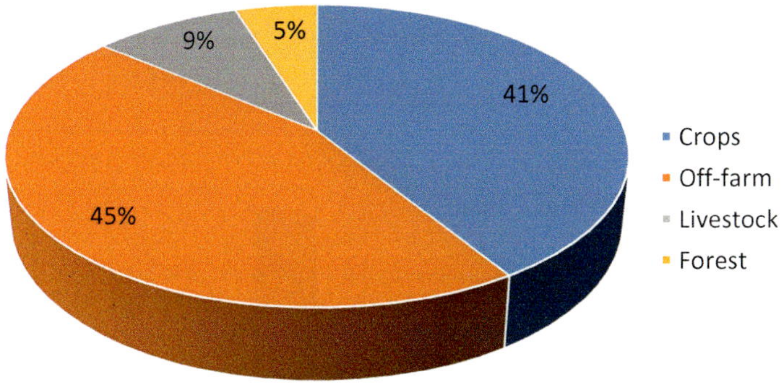

Fig. 2 Sources of income for household

Results and Discussion

General Description of the Respondents

A percentage frequency of the household survey indicated that 62% of the respondents were male and 38% were female. Their education levels varied with 42.7% having up to tertiary level of education, 31.8% secondary education, 14.5% primary education, and 1.8% of them having informal education. The major sources of income for the inhabitants were agriculture, off-farm activities, and products from the forest (Fig. 2).

The major ecosystem goods identified by the respondents included provision of fruits 80%, source of water 90%, fuel wood 94.5%, and timber 83.6%. The results indicate that majority (45.5%) of the respondents relies moderately, that is, they do not entirely depend on forest ecosystem for their livelihood. This implies that they have other sources of income for survival. However, only 16.4% greatly depend on the forest ecosystem for their survival. Although this number may seem small, it may have a great impact that leads to forest degradation possibly due to high extraction of products from the forest. Lastly, 29.1% of the household respondents that least rely on the ecosystem services for their livelihood (Fig. 3) is probably due to diverse sources of income as indicated from the high income from off-farm sources which are all waged and salaried nonfarm work like teaching and carpentry. High percentage of the household income from off-farm activities may be a reflection of high number of members of the community with tertiary education hence engaged either in self-employment activities or salaried jobs. However, a small number of residents entirely rely on the forest ecosystem services as a source of income for their households. These could be the people who may be involved in activities such as logging and sale of timber, building posts, firewood extraction, and production of charcoal.

The respondents' reliance on these ecosystem services varies for different households and also depending on the ecosystem service. The sampled households

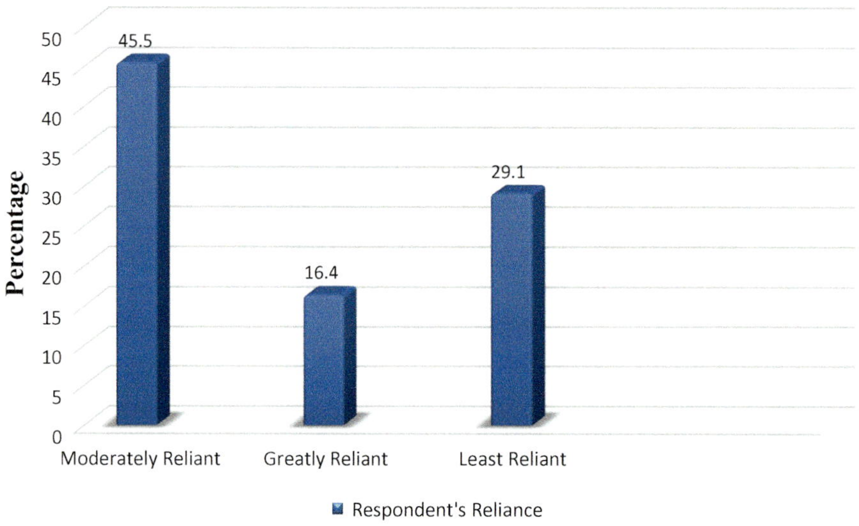

Fig. 3 Extent of reliance on ecosystem services for livelihood

were divided into two categories: those with four or less people were considered small and those with five or more people were considered large households. Smaller households constituted 36% while larger households constituted 64%. The reliance on the ecosystem services was categorized as greatly reliant, moderately reliant, and least reliant. The Fisher's exact test results indicate that there is a significant association ($p < 0.001$, $\alpha = 0.05$) between household size and reliance on the ecosystem services.

Larger household sizes tend to rely more on ecosystem services for income than smaller households who mostly obtain the ecosystem services for domestic uses. These findings are consistent with observations made by Lakerveld et al. (2015). These researchers cite that household size is a demographic factor that influences how a family utilizes the ecosystem services, for example, need for fuel wood and timber for construction. The services that give the most income according to the respondents are timber at 41.7%, charcoal 20.8%, poles 19.4%, and honey 4.2% (Fig. 4). High harvesting of timber and poles (61.1%) for sale may contribute to the degradation of forest ecosystem, hence loss of biodiversity that will affect other forest products such as honey that constitute a source of income for forest adjacent community.

Forest Cover Changes

The summary Table 3 of the land use areas given below shows that in 1984, the dense forest occupied 91.4% of the total land area, and is hence the major land use type, followed by open forest land 8.3% and built-up areas 0.3%. The

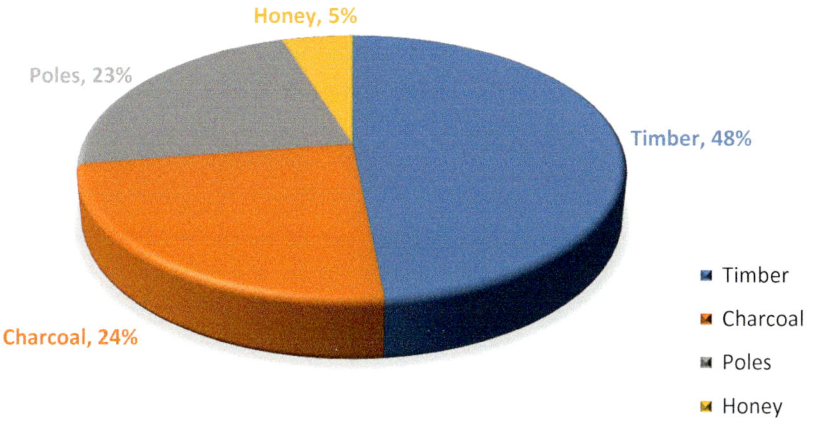

Fig. 4 Income generating from ecosystem services

Table 3 Data analysis summary table

Objective	Variable	Statistical tool
Assess association between forest cover changes and ecosystem services	Forest cover change Ecosystem services	Pearson's chi-square test
Assess relationship between respondent's education level and their awareness on the effects of their derivation of ecosystem services on forest cover	Education level Level of awareness	Pearson's chi-square test
Determine the major sources of income of Katimok residents	Source of income	Pie chart Percentages
Assess the households' extent of reliance on ecosystem services for livelihood and its relationship to household size	Livelihood Household size	Fisher's exact test Bar graph Percentages
Determine the income generating ecosystem services	Ecosystem services	Pie chart Percentages
Determine the forest cover change of Katimok Forest Reserve since 1985 to 2010	Forest cover Land use changes	ERDAS IMAGINE 11 Arc GIS 10.2

summary table also indicates that in 2001, dense forest still remains the major land use type covering 86.5% of the total land area. Open forest land has increased by 4.3% and now covers 12.6% of the total land area. Similarly, built-up area has increased by 0.6%. The results are as shown in Table 4. The comparison of the 1984 and 2001 land use areas indicate a decrease in dense forest land and an increase in open forest land and built-up areas. The dense forest has reduced by 4.9%, and this compensates for an increase in open forest land and built-up area by 4.3% and 0.6%, respectively. This reduction in forest cover can be attributed to

Table 4 Fisher's exact test for reliance on ecosystem services

	Value	df	Asymp. Sig. (2-sided)	Exact Sig. (2-sided)
Pearson's chi-square test	100.000[a]	2	0.000	0.000
Likelihood ratio	130.684	2	0.000	0.000
Fisher's exact test	119.266			0.000
N of valid cases	100			

[a]The minimum expected count - SPSS results

Table 5 Land use land cover areas 1984–2001

Class type	1984 (Area Km^2)	1984 (%)	2001 (Area km^2)	2001 (%)	ΔHa	Change %
Dense forest	*18.81*	*91.4*	*17.80*	86.5	−1.01	−4.9
Open forest land	*1.71*	*8.3*	*2.60*	12.6	0.89	4.3
Built-up areas	*0.05*	*0.3*	*0.17*	0.9	0.12	0.6
TOTAL	**20.57**		**20.57**			

logging, overgrazing, settlement, and conversion of the dense forest into agricultural land as gathered from the focus group discussions. This observation concurs with those of Giliba et al. (2011), who concluded that human activities such as illegal logging, fodder use, agriculture, and honey harvesting are among the causes of forest cover change (Table 5).

The 2001–2015 summary table indicates that the dense forest has now slightly increased up to 90.4% of the total land area, and it still remains the major land use type. This increase in dense forest land has resulted in a decline in open forest land as well as built-up area to 9.1% and 0.5%, respectively. The changes above can be attributed to the eviction of the illegal settlers in the forest in 1988 during the Kenya African National Union regime that led to a decrease in the built-up area as observed in the land use map. Afforestation was carried out in the lands that were left vacant after the eviction, and this led to an increase in the dense forest land and a decline in open forest land. This afforestation was carried out under the Shamba System, now the Plantation Establishment and Livelihood Improvement Scheme (PELIS), which allowed the community to farm on the land as they tend to the tree seedlings until they attain a certain stage. The trees planted on these bare lands were exotic trees which meant that some of the ecosystem services they provided like herbal medicine could not match those of indigenous forest.

Impacts of Forest Cover Change on Ecosystem Services

Based on the focus group discussion, the participants agreed that there have been changes in indigenous forest cover since 1984 in terms of size, density, and species composition. The forest used to be purely indigenous, but the deforested parts have now been replaced with plantation forests. In their view, though the forest cover may have slightly improved owing to the introduction of the PELIS system, the plantation forest which is normally established for commercial rather than restoration purposes, cannot match that of indigenous tree cover. According to their opinion, there is a strong relationship between the changes in indigenous forest cover and ecosystem services, in that, as the indigenous forest cover decreases the ecosystem services also reduce. Therefore, the changes in land use and forest cover of Katimok Forest have had some effects on the ecosystem services that the forest provides. Some of the examples given in the FGDs include:

Loss of Biodiversity

The FGD participants indicated that there was a deteriorating trend of some important plant species and mostly those used for medicinal purposes such as *Piper umbellatum* L. that cures headaches, *Entada abyssinica* Steudel for stomachache, and *Desmodium adscendens* (Sw.). Special flowering plants for bees, e.g., *Combretum schumannii* Engl., have also disappeared, and so the bees go away in search of nectar. The disappearance of the plant as well as the migration of the bees has resulted in a sharp decline in the amount of honey produced. The reduction in the indigenous forest cover has also led to human-wildlife conflicts since their habitat has been fragmented and disturbed. The forest has become open and the animals which used to come out only at night can now be seen even during the day, as they have nowhere to hide. Some animals such as the *Leptailurus serval* (Schreber) have also disappeared and can no longer be found in the forest. There has also been a decrease in the abundance of mushrooms in the forest and the participants indicated that it has become a rare ecosystem service. It should be noted that, though the forest cover may have increased between 2001 and 2015, logging and timber harvesting still continue, and the PELIS program is not helping much in terms of restoring biodiversity since farming leads to forest disturbance. Moreover, this increase in forest cover between 2001 and 2015 is not up to what was there in 1984. There are farmlands and built-up areas like schools and hospitals within the forest as observed. These built-up areas, according to the focus group discussions, were not there before and hence have contributed to the loss in forest cover.

Hindering of Forest Regulation Services

The focus group discussion participants indicated that some of the regulation services of the forest have been hindered such as soil erosion prevention, which is now a common phenomenon during heavy rains. Flooding downstream is now also observed especially in the *Endao* and *Kampi ya Samaki* Rivers. There has also been a reduction in water levels and quantity, and also in water points owing to siltation which eventually lead to the drying up of some rivers. The rivers have also become seasonal. This may be attributed to the change in land use which reduces the forest cover thereby affecting the ability of the forest to act as a catchment area. A study conducted by the Center for Watershed Protection

(2008) and Caja et al. (2018) established that a reduction in forest cover affects the ability of the forest to regulate the quality, quantity, and flow of water. This was measured using water balance models which found that a decrease in vegetation cover affects the water balance of an area. Therefore, it can be argued that, this establishment agrees with the focus group inferences of this study. In general, there has been a reduction in the abundance of the ecosystem services, and in some cases, the services have become extinct.

Increase in Agricultural Produce and Construction Material

However, as they cite a reduction of some of the ecosystem services, the participants also point out that there has been an increase in agricultural production as well as forest products such as timber and other building materials as a result of land use change. Logging increases the demand for timber and also the availability of bare land on which the community are allowed to farm on as they tend to young tree seedlings, a system known as the Plantation Establishment and Livelihood Improvement Scheme (PELIS). This concurs with what Lawler et al. (2014) had established that land use change from forestland to farmland increases the production of food crops and timber among other products but at the expense of a decline in other ecosystem services along with biodiversity. Devisscher (2009) also agrees that changes in land use from forests to agriculture as well as other systems of production affects the abundance of ecosystem services since as the land is modified, the resources undergo exploitation.

Local Community's Perception of the Relationship Between Forest Cover and Ecosystem Services

Most of the respondents are aware of the effects of their derivation of ecosystem services on the forest cover. There is a significant association between this awareness and their education level ($X^2 = 124.690$, df $= 12$, $p < 0.05$). Results of samples of levels of study which are tertiary level, secondary level, primary level, and informal level were compared with those of the multiple categories of awareness on the effects of derivation of ecosystem services on forest cover as shown in Table 6.

Table 6 Land use areas 2001–2015

Class type	2001 (Area km^2)	2001 (%)	2015 (Area km^2)	2015 (%)	ΔHa	Change %
Dense forest	*17.80*	*86.5*	*18.59*	90.4	0.79	3.9
Open forest land	*2.60*	*12.6*	*1.87*	9.1	−0.73	−3.5
Built-up areas	*0.17*	*0.9*	*0.11*	0.5	−0.06	−0.4
Total	**20.57**		**20.57**			

Chi-Square Tests for Levels of Awareness and Education

Those who are educated tend to have knowledge on the effects of derivation of ecosystem services on the forest cover as compared to those who have little or no education. These findings are in agreement with those of Martinez (1998), who cites that social factors such as education level influence the rate of deforestation. Geist and Lambini (2002), in their research, also agree that education level is important in forest ecosystem conservation since knowledge on the ecological functions of a forest enhances sustainable utilization of forest resources. Education shapes the behavior of an individual causing forest destruction.

Based on the focus group discussion, the participants agreed that there is a strong relationship between forest degradation and ecosystem services. As the forest continues to degrade, the ecosystem services also reduce. The Pearson's chi-square test results of the household survey also concur with what the FGD participants had indicated. The tests reveal that there is a strong relationship ($X^2 = 151.581$, df $= 16$, $p < 0.05$) between forest degradation and decrease in provision of ecosystem services overtime (Tables 7 and 8).

Table 7 Education level – level of awareness crosstabulation

			Level of awareness					
			Greatly degrades	Degrades	No effect	Improves	Greatly improves	Total
Education level	Informal	Count	0	1	2	2	1	6
		Expected count	2.6	2.4	0.8	0.1	0.1	6.0
	Primary	Count	1	5	9	0	1	16
		Expected count	6.9	6.4	2.1	0.3	0.3	16.0
	Secondary	Count	5	27	2	0	0	34
		Expected count	14.6	13.6	4.4	0.7	0.7	34.0
	Tertiary	Count	37	7	0	0	0	44
		Expected count	18.9	17.6	5.7	0.9	0.9	44.0
Total		Count	43	40	13	2	2	100
		Expected count	43.0	40.0	13.0	2.0	2.0	100.0

[a]The minimum expected count - SPSS results

Table 8 Chi-square tests

	Value	df	Asymp. Sig. (2-sided)
Pearson's chi-square test	124.690[a]	12	0.000
Likelihood ratio	99.684	12	0.000
N of valid cases	100		

Chi-Square Tests for Forest Degradation and Provision of Ecosystem Services

Household participants were asked to state weather the forest ecosystem services, such as fruits, honey, timber, poles, medicinal plants, firewood, fodder, fiber, charcoal, and water among other goods, had increased or reduced overtime with regard to the degraded forest. The results showed that the ecosystem services have reduced along with forest degradation. According to the research conducted by WWF (2016), forest degradation not only reduces the forest cover but also the ecosystem services that the forest provides and eventually the livelihoods of the people that depend on it (Tables 9 and 10).

Benefits of the Forest to the Community

The forest is the source of livelihood of many of the households in the community living adjacent to the forest as indicated by the data gathered from FGDs

Table 9 Forest cover change – ecosystem services change crosstabulation

			Ecosystem services change					
			Greatly increased	Increased	Remained the same	Reduced	Greatly reduced	Total
Forest cover change	1	Count	2	2	1	0	0	5
		Expected count	0.1	0.3	0.4	4.1	0.3	5.0
	2	Count	0	0	2	3	0	5
		Expected count	0.1	0.3	0.4	4.1	0.3	5.0
	3	Count	0	3	1	1	0	5
		Expected count	0.1	0.3	0.4	4.1	0.3	5.0
	4	Count	0	0	1	75	1	77
		Expected count	1.5	3.9	5.4	62.4	3.9	77.0
	5	Count	0	0	2	2	4	8
		Expected count	0.2	0.4	0.6	6.5	0.4	8.0
Total		Count	2	5	7	81	5	100
		Expected count	2.0	5.0	7.0	81.0	5.0	100.0

Table 10 Chi-square tests

	Value	df	Asymp. Sig. (2-sided)
Pearson's chi-square test	150.005[a]	16	0.000
Likelihood ratio	82.189	16	0.000
N of valid cases	100		

[a]The minimum expected count - SPSS results

and household survey. They obtain socioeconomic as well as cultural benefits from the forest. The forest is a catchment area and the community gets water from the rivers and streams that flow through the forest such as *Sokom, Endao*, and *Kampi Samaki* Rivers. The forest is also important to the community as they rely on it for herbal medicine derived especially from the indigenous plants such as *Garcinia johnstonii*. According to the Focus Group Discussion, the participants also indicated that during the dry season when there is shortage of grass for their livestock, the community relies on the forest for fodder with some of the species like *Trichocladus ellipticus* acting as alternative feed for their livestock (Table 11).

The forest also plays a significant role in fulfilling cultural and spiritual practices of the community. Some cultural benefits of the forest that the participants indicated are that the forest is a ceremonial ground and circumcision is done in the forest. The community also has some designated sacred sites within the forest that are used for prayers: both traditional and modern Christianity. The Governor's Camp is a historical place and a tourist attraction in the forest. Other places of cultural importance are *kebenonin, tuiyobei*, and *kapchumba* where there is a cave. There are also plant species with some cultural significance to the community, and they include *Warburgia ugandensis, Acocanthera schimperi*, and *Euclea divinorum*. These plant species are not available in their farms, and hence the forest is an important source of these species. The forest is also a source of fuel wood which is a readily available form of energy for the community. Moreover, the forest is an ideal place for placing beehives, and as such, it is important for bee farmers.

Other important benefits that the community derive from the forest as gathered from the focus group discussion and as observed include fodder, honey from stingless bee, mushrooms, soil, tree seedlings, farming land, fresh air, habitat for wild animals, recreation, construction material, and wild fruits among other benefits. All these benefits that the community derive from the forest as indicated in the FGDs conducted and field observations agree with the findings of Shackelton et al. (2008). According to these researchers, communities utilize a wide array of ecosystem services such as medicine, food, construction material, and cultural purposes to meet their livelihood needs. Millennium Ecosystem Assessment (2003) findings are also consistent with the current chapter findings in which different communities' value different ecosystems according to the values they attach and obtain from them. The current chapter findings further note that, there are various ecosystem services obtained from the forest and they range from bush meat, water, firewood, wild fruits, as well as spiritual and cultural value.

Challenges Faced in the Management of the Forest

There are many challenges that hinder the sustainable management of the forest as gathered from the focus group discussions. One of the challenges is that the community lacks information and awareness on the significance of trees and the

Table 11 Tree species found in the forest

Local name	Scientific name	Common name	Uses
Tarakwo	*Junipreus procera*	African pencil cedar	Construction material
Auwe	*Polycius kikuyuensis*	Parasol tree	Timber
Arariet	*Ekebergia capensis*	Cape ash	Medicinal for dysentery
Benet	*Podocarpus falcatus*	Yellowwood	Making furniture
Boroa	*Dombeya goetzenii*		Medicinal for indigestion
Kolutwet	*Albizia amara*	Bitter albizia	Fruit and medicinal
Kamilet	*Combretum molle*	Velvet bushwillow	Medicinal for stomachache and headache
Sosionte	*Phoenix reclinata*	Wild date palm	Fruit
	Vitex keniensis	Meru oak	Fruits, medicinal
Bunus	*Eucalyptus saligna*	Sydney blue gum	Building and construction
Yemit	*Olea Africana*	Wild olive	Fruits and oil
Soket	*Warbuginia ugandesis*	Uganda greenheart	Fruit and medicinal for general body pains
Sesia	*Acacia tortilis*	Umbrella thorn acacia	Making furniture
Leketwa	*Carissa edulis*	Simple-spinednum-num	Fruits, medicine for malaria and pain
Kunyukwe	*Prunus Africana*	Red stinkwood	Medicinal for stomachache, malaria, fever
Kuryonte	*Teclea nobilis*	Small-fruited teclea	Fruit tree, timber
Samut	*Cordia Africana*	Sudan teak	Beehives and furniture
Se	*Albizia gummifera*	Peacock flower	Firewood, charcoal
Septa	*Podocarpus latifolius*	Real yellowood	Fruits, making furniture
Ortulet	*Croton megalocarpus*	Croton	Construction material, fuel wood, charcoal
Nerkwo	*Garcinia livingstonei*	African mangosteen	Fruit tree
Lomoiwe	*Syzygium guineense*	Water berry	Fruits, construction material
Koloswo	*Trichocladus ellipticus*	White witch-haze	Fodder, medicine for upset stomach
Tegat	*Bambusa vulgaris*	Bamboo	Construction
Sinendet	*Ficus sycomorus*	Sycamore	Fruits, ceremonial
Kures	*Euphorbia candelabrum*	Candelabra tree	Medicinal for leukemia, tumor, HIV infections

value of conservation, so they take part in forest encroachment. Furthermore, there is no restriction in entering the forest, unlike in the past where people had to seek permission to enter the forest. This encourages illegal practices such as logging and poaching within the forest. Carandang et al. (2013) in their research found out that lack of knowledge and awareness on the benefits of protecting the forest is a sociocultural factor that indirectly causes forest degradation. They also add that laxity on enforcement of the laws further fuels the rise in illegal forest activities.

Focus group discussion respondents indicated that grazing of livestock in the forest has not been controlled as before, where the community used to sit and agree on certain water points and certain boundaries. Fodder would be cut and taken to the animals in the field, and goats were not allowed to browse in the forest. This ensures that the animals could not overgraze in the forest and hence curbing forest degradation. Open grazing in forests adversely affects the growing stock as well as the capacity of the forest to regenerate leading to degradation (Nayak et al. 2008).

Another concern was that the residents are threatened and victimized whenever they report illegal poaching. There is no action taken on the offenders and so people do not care anymore about reporting, so they end up poaching as well. What is more, there are only 8 officers managing the entire 13 blocks of Kabarnet Forests and few staff to monitor a vast expanse of forest. Poor remuneration of KFS officers has also contributed to less effort in management of the forest, and thus they also poach and/or receive bribes from illegal loggers. These findings are in agreement with those of Contreras-Hermosilla (2000) who found out that some forest management authorities often abuse the public powers bestowed on them to unlawfully make themselves rich. He further states that this conduct is especially true for those officials who are poorly paid as they start making prejudiced decisions against those activities that do not attract bribes and hence continued illegal logging.

The participants also mention that commercialization of tree products has made the community look at trees in terms of commercial values, and this has led to illegal logging. There is ready market for timber, poles, and other building materials, and hence they practice illegal logging to earn a living. Goll et al. (2014) in their research also indicated that market demand for forest products is one of the root causes of forest degradation.

Moreover, the FGD participants stated that there is corruption among the forest authorities and those trading in forest products, mostly timber. The authorities are offered handsome bribes to give licenses to some few known individuals. Besides, those licensed persons do not even hail from the community. Therefore, the community feels sidelined as they are not given licenses so that they can also obtain the forest products legally. This makes them practice illegal logging and other forms of forest degradation. This is consistent with the findings of Carandang et al. (2013) who established that corruption is one of the rampant institutional weaknesses that has hindered effective enforcement of forestry laws in developing countries. In addition, they argue that this corruption allows illegal logging to continue as the forest authorities are bribed to issue permits.

Another concern raised in the FGD was that the last batch of the illegal settlers was forcefully evicted in 1988. During this eviction, houses were burnt and property destroyed. For that reason, there is still resentment as evicted persons were not compensated, so they feel neglected and they take part in illegal logging to make up for their lost property. According to FAO when illegal forest settlers are forcefully evicted, they retaliate in form of participating in unlawful forest activities such as illegal logging, and this result in rapid decline of forest resources. Therefore, FAO recommends the resettlement of illegal settlers so as to economically empower the local communities as well as promote forest conservation (FAO n.d.).

Lastly, the FGD results indicates that the community forest associations (CFAs) lack the capacity to make and enforce rules as well as financial resources to carry out their activities. This is consistent with the findings of Chomba et al. (2015) who indicated that the CFAs had no power to even make the basic rules concerning forest management, for example, making decisions regarding the fees to charge the forest users. They also cannot enforce these rules as this mandate is entirely vested on the KFS officers. On top of that, Chomba et al. (2015) adds that the CFAs has no external source of funding but majorly depend on their membership contributions to finance their activities.

Forest Management Interventions

The Kenya Forest Service sows tree seedlings in their tree nurseries and plants them in degraded areas. They also sell the trees to the local people to go and plant them in their farms. This reduces the reliance on the forest in terms of fuel wood and timber since the community has the trees available in the farms. Some of the trees raised in the nursery are listed in Table 12.

The community forest associations (CFAs) also engage in the protection of the forest by offering surveillance and reporting to KFS any illegal activity as well as wildfires that might occur in the forest.

Table 12 Trees grown in Katimok Forest nursery

Names	Description
Eucalyptus saligna Sm.	Tall forest tree with height 30–55 m
Cupressus lusitanica Mill.	Evergreen conifer tree; ovoid crown; up to 40 m tall
Fraxinus pennsylvanica Marshall.	Deciduous tree, medium in size. Grows up to 12–25 m
Podocarpus falcatus Thunb.	Evergreen conifer; grows up to 45 m in height
Prunus africana Hook.f.	Evergreen tree native to Sub-Saharan Africa
Grevillea robusta A.Cunn	Fast growing evergreen tree; 18–35 m in height
Olea hochstetteri Baker.	Evergreen tree; varies from 2 to 15 m in height
Croton megalocarpus Hutch.	Drought resistant tree; upto 36 m high
Warburgia ugandensis Sprague.	Evergreen insect resistant tree
Casuarina equisetifolia L.	Slender evergreen tree with gray-green twigs; 6–35 m tall
Pinus patula Schiede.	30 m tall tree that is moderately drought resistant
Vitex keniensis Turrill.	Straight trunk tree that is endemic to Kenya

Some of the areas have also been fenced off under the PELIS system to keep off any animals that might enter the forest. The forest is manned by KFS forest rangers that can arrest any illegal loggers and grazers in the forest. The rangers are allowed to detain any livestock found illegally grazing in the forest.

Conclusions

The results from this chapter have indicated that there had been a decline in forest cover from 1985 to 2001, then a slight increase between 2001 and 2015, but even this increase did not match the original forest cover. On the overall, comparing the forest cover of 1984 to that of 2015, it is evident that there has been a decline. These changes affected both the availability and the abundance of ecosystem services in the forest since even as the forest cover slightly increased, some of the ecosystem services provided by the original indigenous forest cover could not be recovered. Three major land use types that were identified were dense forestland, open forestland, and built-up area. In general, the forest cover has decreased since 1984, and areas that were once covered with dense forest lands in the early 1980s have been turned into farmlands and build-up areas. Katimok Forest cover change has had some negative profound impacts on the ecosystem services, and this has affected the forest adjacent community who largely depend on the forest for their livelihoods. For instance, there has been loss of certain medicinal plants, reduction in the water levels and water points, decrease in mushroom and honey, and even disappearance of some wild animal and bird species. All these impacts arise from the community's unsustainable utilization of the forest and forest resources through land use. Therefore, this kind of information is useful for the forest management authorities in order for them to protect the forest resources from exploitation.

Cross-References

- ▶ Biodiversity, Ecosystem Degradation, and Climate Change Effects on Livelihoods in the Bitumen Area of Nigeria
- ▶ Drivers of Deforestation and Land-Use Change in Southwest Nigeria
- ▶ Ecosystem-Based Adaptation in Tigray, Northern Ethiopia: A Systematic Review of Interventions, Impacts, and Challenges

References

BCG (2015) Annual development plan 2016/2017. Baringo County Government, Kabarnet

Bravo F (2008) Managing forest ecosystems: the challenge of climate change. Springer Science & Business Media, New York

Caja CC, Ibune NL, Paril JA, Reyes AR, Nazareno JP, Monjardin CE, Uy FA (2018) Effects of land cover changes to the quantity of water supply and hydrologic cycle using water balance models. MATEC Web of Conferences 150:1–4

Carandang A, Bugayong L, Dolom P, Garcia L, Villanueva M, Espiritu N (2013) Analysis of key drivers of deforestation and forest degradation in the Philippines. GIZ, Manila
Center for Watershed Protection (2008) Forests and drinking water. Retrieved March 10, 2016, from http://forestsforwatersheds.org/forests-and-drinking-water/
Chomba S, Iben N, Minang P, Sinclair F (2015) Ilusions of empowerment? Questioning policy and practice of community forestry in Kenya. Ecol Soc 20(3):2
CIFOR (2014) Ecosystem services local benefits, global impacts. Retrieved March 9, 2016, from http://www.cifor.org/library/101/ecosystem-services-%C2%96-local-benefits-global-impacts/
Contreras-Hermosilla A (2000) The underlying causes of forest decline. Occasional paper no. 30. CIFOR, Jakarta
Devisscher T (2009) Review of the economic impacts of climate change in Kenya, Rwanda and Burundi
FAO (2008) FOSA Country report- Kenya. Forestry Department, Nairobi
FAO (n.d.) Socio-economic issues: status of knowledge and experience. Retrieved March 9, 2016, from http://www.fao.org/docrep/006/J0628e/J0628E30.htm
Geist H, Lambini E (2002) Proximate causes and underlying driving forces of tropical deforestation. Bioscience 52:143–150
Giliba R, Mafuru C, Paul M, Kayombo C, Kashindye A, Chirenje L, Musamba E (2011) Human activities influencing deforestation on Meru Catchment Forest Reserve, Tanzania. J Hum Ecol 33:17–20
Goll N, Li J, McKay J (2014) Analysis on the causes of deforestation and forest degradation in Liberia: application of the DPSIR framework. Res J Agric For Sci 2:20–30
Kahuki C, Muniu J (2004) Non-wood forest products in Kenya. Forestry Department, Nairobi
Kahuthu R, Muchoki T, Nyaga C (2005) Baringo district strategic plan. Government of Kenya (Ministry of Planning and National Development), Nairobi
Lakerveld R, Lele S, Crane T, Fortuin J, Baginski S (2015) The social distribution of provisioning forest ecosystem services: evidence and insights from Odisha, India. Ecosyst Serv 14:56–66
Lawler J, Lewis D, Nelson E, Plantinga A, Polasky S, Withey J,. . . Radeloff V (2014) Projected land-use change impacts on ecosystem services in the United States. PNAS 111(20):7492–7497
Martinez K (1998) Social determinants of deforestation in developing countries: a cross national study. Soc Forces 77:567–586
Millenium Ecosystem Assessment (2003) Ecosystems and human well-being: a framework for assessment. (ed: Joseph A, Bennett E & trans: Bennett EM). Island Press, Washington, DC
Millenium Ecosytem Assessment (2005) Ecosystems and human well-being: our human planet: summary for decision makers. Island Press, Washington, DC
Nayak BP, Kohli P, Sharma J (2008) Livelihoods of local communities and forest degradation in India: issues for REDD+
Shackleton C, Shackleton S, Gambiza J, Nel E, Rowntree K, Urquhart P (2008) Links between ecosystem services and poverty alleviation. Ecosystem Services and Poverty Reduction Research Program, London
TEEB (2010) The economics of ecosystems and biodiversity: mainstreaming the economics of nature: a synthesis of the approach, conclusions and reccommendations of TEEB. UNEP, Nagoya, Japan
World Bank (2013, August 28) Forests and poverty reduction. Retrieved March 11, 2016, from http://www.worldbank.org/en/topic/forests/brief/forests-poverty-reduction
WWF (2016) Forests; deforestation causes. Retrieved March 9, 2016, from http://wwf.panda.org/about_our_earth/deforestation/

Ecotourism as an Adaptation Strategy for Mitigating Climate Change Impacts on Local Communities Around Protected Areas in Ghana

26

Yaw Boakye Agyeman

Contents

Abstract

Ecotourism is used as a strategy around protected areas. Its prospects for supporting conservation and livelihoods are well supported by studies. With climate change emerging as a threat to local livelihoods based on natural resources, including ecotourism because of its partial dependence on forest-based attractions, communities have developed strategies to survive. Studies on climate change tend to

Y. B. Agyeman (✉)
Department of Ecotourism, Recreation and Hospitality, University of Energy and Natural Resources, UENR, Sunyani, Brong Ahafo Region, Ghana
e-mail: yaw.agyeman@uenr.edu.gh

W. Leal Filho (ed.), *Handbook of Climate Change Resilience*,
https://doi.org/10.1007/978-3-319-93336-8_159

ignore local capacity to adapt to the climate change. Using Mognori as a case, farming households located in the Savannah Region of Ghana were purposively selected to understand how ecotourism is incorporated into the overall livelihood spectrum and its implications for households' assets and sustenance, particularly where there is observed livelihood uncertainty because of climate change. In-depth interviews revealed how households construct their livelihoods to ensure livelihood security. For the households, ecotourism presents opportunities for livelihood diversification and sustenance. They tend to focus on short-term coping strategies at the household level by using ecotourism as livelihood diversification because it complements livelihood assets and activities. This chapter contributes to mainstreaming climate change in ecotourism development based on protected area attractions and how local communities sustain livelihoods in the context of climate change vulnerability.

Keywords
Ecotourism · Livelihood · Vulnerability · Adaptation · Local communities

Introduction

Ecotourism's prospects in improving recreational opportunities, conservation, and livelihoods are well documented (Regmi and Walter 2017; Reimer and Walter 2013; Snyman 2012). Particularly, it benefits local communities around protected areas in terms of employment, revenue generation, entrepreneurial opportunities, biodiversity conservation, and sociocultural revitalization (Anup et al. 2015; Chirenje 2017; Hunt et al. 2015). Even though others studies indicate that its benefits are limited (Coria and Calfucura 2012; Mbaiwa and Stronza 2010), it continues to serve as an important livelihood strategy for communities living in marginal situations defined by changing climate and other factors. Particularly, climate change has emerged an issue of concern because of the associated impacts on local communities, making them more vulnerable.

In the tourism-climate change literature, it is suggested that climate change may alter the character of the tourism resource base, thereby negatively affecting local destinations' perceived attractiveness with adverse implications for tourism demand and local people's livelihoods (Becken and Hay 2007; Kaján and Saarinen 2013; Scott and Jones 2006). The partial dependence of ecotourism on protected areas and associated biodiversity makes it vulnerable to climate change (Scott 2006). Such vulnerability also creates opportunities for adaptation. The focus on climate change adaptation in tourism is the pursuit of on-going sustainability (Becken and Hay 2007; Farrell and Twining-Ward 2004; Scott 2011). Knowledge about tourism-climate change interaction from an adaptation perspective has increased over the years (see Becken and Hay 2007; Kaján and Saarinen 2013; Scott and Becken 2010). While the contribution of local knowledge in climate change studies is recognized (see Ford and Smit 2004; Ford et al. 2006), knowledge on climate change adaptation at the community level is limited. In a review of the tourism-climate change adaptation discussions, Kaján and Saarinen (2013) concluded that, while perceptions of other stakeholders are captured in the adaptation studies, there is limited knowledge on community perceptions. They emphasized that "a focus on community dimension research in relation to tourism and climate change allows highly contextual adaptation challenges to be met in a more sustainable way" (p. 167). This perception has advantages relating to practicality and implementability of adaptation strategies (Smit and Wandel 2006; van Aalst et al. 2008). An argument is made that scientific knowledge of climate change and its impacts is useful but barely sufficient to serve as a basis for effective local adaptation. As a result, local knowledge, which is specific to the conditions of a place, is also needed (Brace and Geoghegan 2010; Hulme 2008). This place-specific knowledge makes it possible to take into account the knowledge of specific biophysical and social processes in the design of local processes (Berkes 2009). Local knowledge and experience can help build adaptive capacity and resilience (IPCC 2007). Also, Scott and Becken (2010) identify limitation in the tourism-climate change studies in terms of geographic scope, since they focused largely on Western countries. They noted that studies should address the tourism-climate change relationship, particularly in Africa because of the relative importance of tourism to their economies. This knowledge limitation may compromise the prospects of using tourism in improving biodiversity conservation and livelihoods at the local level, particularly in situations where households are experiencing livelihood dynamics because of climate change. Therefore, this chapter aimed at understanding how local communities construct their livelihoods in vulnerability context and adapt to climate change. This knowledge provides a process for integrating local knowledge into livelihood adaptation as communities become more vulnerable to future climate change which is adversely affecting their traditional patterns of livelihoods.

Literature Review

Conceptualizing Ecotourism as Sustainable Livelihood

Ecotourism emerged as an alternative and sustainable form of tourism in the 1980s, driven by the need for sustainable development (Buckley 2009; Fennell 2008). According to Wearing and Neil (2009), it revealed increasing environmental concern, poor linkages, cultural degradation, and disruption of biodiversity conservation associated with conventional tourism and tourists' demand for nature-based and cultural experiences. Its use around protected areas is based on the assumption that it serves as an incentive to reduce unsustainable practices of local communities such as forest degradation, expanding agricultural frontiers, illegal hunting, and firewood collection that undermine conservation and provide alternative livelihoods to sustain the needs of local people, thereby creating a symbiotic relationship between them and protected areas (Coria and Calfucura 2012; Wearing and Neil 2009; Stronza and Gordillo 2008). Using ecotourism to contribute to these needs means shifting focus from industry sustainability to community livelihoods (Tao and Wall 2009a). The emphasis on livelihood is based on the premise that within local communities, engaging in multiple livelihood activities is a norm to ensure livelihood security (Paavola 2008; Scoones 2009; Tao and Wall 2009b).

Sustainable Livelihoods Perspective

The concept of livelihood was popularized by Chambers and Conway (1992). Their definition of livelihood, with minor modifications, has been used by researchers and organizations. This led to the development of different frameworks and the pentagram-based framework of DFID (2004) which captures certain essential elements (Baumgartner and Högger 2004). DFID (2004) defined livelihood as “the capabilities, assets and activities required to make a living. A livelihood is sustainable when it can cope with and recover from stresses and shocks and maintain or enhance its capabilities and assets both now and in the future, while not undermining the natural resource base” (p. 26). However, Ellis (2000) contends that the meaning of the term “capabilities” in the above definition overlaps greatly with assets and activities, and the use of the term “capabilities” can bring confusion. He defined it as “the means of using assets and engaging in activities to make a living” (Ellis 2000). The asset represents a portfolio of five different types: financial, social, physical, human, and natural assets (see Fig. 1). Also, it involves the use of a livelihood strategy, which is an overarching term used to describe a combination of activities and choices that people make in order to achieve their livelihood outcomes (DFID 2004). Central to a livelihood perspective is the notion of adaptive strategy. Helmore and Singh (2001) defined it as “the changes and adjustments people make in their livelihood systems in order to cope under difficult circumstances” (p. 3). The understanding of the adaptive strategy of the local communities is becoming important in the context of climate change (Becken and Hay 2007; Smit and Wandel 2006). The livelihood is also based on local knowledge, combining with appropriate elements from external knowledge, assimilated into the community over time (Tao and Wall 2009a).

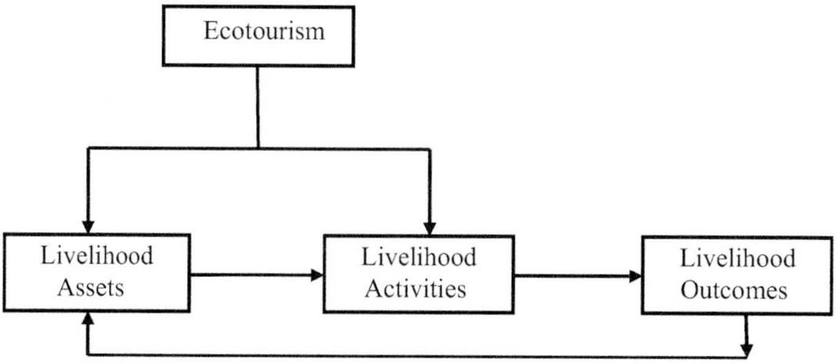

Fig. 1 Sustainable livelihood framework

In the ecotourism literature, livelihood strategy is viewed as a form of strategy in which ecotourism becomes a form of livelihood diversification within the local community (Tao and Wall 2009b). As such, ecotourism has impacts on the local assets and activities, resulting in different outcomes which become investment assets as indicated by the feedback loop.

Climate Change Impacts and Adaptation in Tourism

Climate change is viewed as a major challenge facing tourism. This view is emphasized in the Davos Declaration on Climate Change and Tourism, which declared that climate change "must be considered the greatest challenge to the sustainability of tourism in the 21st century" (Scott and Becken 2010). Its impacts on the different tourism sectors and systems such as skiing, coastal and marine environment, and parks are well discussed and studied. These discourses and studies highlight two broad perspectives, encompassing how tourists will respond to climate change and how tourism destination attractiveness will change in response to climate change (see Becken and Hay 2007; Hall and Higham 2005; Hamilton and Tol 2007). Focusing on the tourists, it is discussed that climate change affects activity choice and travel patterns both domestically and internationally with considerable implications for the tourism enterprises and community livelihoods (Hall and Higham 2005; Hamilton and Tol 2007). Becken and Hay (2007) added that the various components of tourism such as appeal of a destination, transport infrastructure and operations, the resource base, tourist satisfaction and safety, and the viability or sustainability of tourist facilities are affected, changing the attractiveness of destinations and influencing the tourism flow. Also, studies agree that climate change impacts on nature-based tourism are profound in terms of biodiversity loss and landscape appeal. Increase in intense storms or other weather extremes and their consequences such as flooding, landslides, and forest fires adversely affect tourism destinations, altering the character of the tourism resource base, thereby negatively affecting destinations' perceived attractiveness (Becken and Hay 2007; Holden 2008; IPCC 2007; Kaján and Saarinen 2013; Scott and Jones 2006). Particularly, the partial dependence of ecotourism on protected areas and associated biodiversity makes it vulnerable to climate change (Scott 2006). Such vulnerability also creates opportunities for adaptation (Becken and Hay 2007; Scott 2011).

Climate Change Adaptation: Taking Vulnerability Perspective

Climate change adaptation is recognized as important in understanding and reducing the vulnerability of tourism to climate change (Becken and Hay 2007; Scott 2011). Becken and Hay (2007) suggested that adaptation are "those actions or activities that people undertake, individually or collectively, to accommodate, cope with or benefit from, the effects of climate change, including changes in climate variability" (p. 225). It aims at reducing vulnerability to the impacts of climate change (Mercer

2010). The concept of vulnerability is thus central to climate change adaptation. According to Schipper (2007), while there is a significant push for adaptation in development, there is a missing step if vulnerability reduction is not considered central to this. By identifying the points of vulnerability within any given system, the applicable and appropriate adaptation measures can be taken (Smit and Wandel 2006). Thus, understanding vulnerability forms the basis for developing adaptation strategies to facilitate sustainability (Becken and Hay 2007; Farrell and Twining-Ward 2004).

Füssel and Klein (2006) describe four dimensions of vulnerability as a system, the attributes of concern, the hazard, and a temporal reference. The system can refer to a wide selection of coupled human-environmental systems or socio-ecological system (Folke 2006; Turner et al. 2003). The hazard is viewed as having a damaging impact on the attributes of concern, which refer to any valued issues within the local community. The time aspect refers to the temporal reference. Studies suggest a period of 20 years for anticipating a change in climate (Weaver 2011).

In literature, vulnerability is viewed as an outcome or context (O'Brien et al. 2005). Brooks (2003) explained that vulnerability as an outcome emerged from studies that neglect people's adaptation to climate change but focus on human exposure to hazard events. He noted that the definition of IPCC is a case in point. IPCC (2007) defines vulnerability as "the degree to which a system is susceptible to and unable to adapt to adverse effects of climate change" (p. 883). Brooks (2003) labeled this view of IPCC as a biophysical vulnerability. It is argued that adaptation strategies aimed at reducing physical risk fail to reduce exposure and damage and may increase vulnerability instead (Vincent 2007). Also, the view of vulnerability as a context, as indicated by Brooks (2003), emerged from studies of the structural factors that make the human system vulnerable to climate change. Here, vulnerability is an inherent property of a human system arising from its internal characteristics. This view is characterized as social vulnerability, which is influenced by "generic" factors such as poverty, inequality, social networks, health, marginalization, food entitlements, access to resources, and housing quality (Brooks 2003). It is a result of social processes and structures that constrain access to resources undermining people's ability to adapt to climate change. A third view in the climate change discourse (see Cutter 2003; Turner et al. 2003) integrates both the biophysical and social vulnerability, which is consistent with the view of Ford et al. (2006). Ford et al. (2006) view vulnerability as a function of the exposure and sensitivity of a community to climate change effects and its adaptive capacity to deal with that exposure. There is a positive functional relationship between vulnerability and exposure and a negative one between vulnerability and adaptive capacity (Smit and Pilifosova 2001).

Kelly and Adger (2000) state that it would be foolish to suggest that any particular understanding of vulnerability is more or less appropriate in the context of climate change studies. The purpose of the analysis must guide the selection of the most effective definition. The third view is appropriate and guides this chapter as it recognizes the integration of both biophysical and social vulnerability. It is also consistent with the livelihood perspective and sensitive to the local perspective.

According to Turner et al. (2003), both the social and biophysical aspects are essential parts of a comprehensive framework for understanding the vulnerability of coupled human-environment systems.

Theoretical Framework for Vulnerability Assessment

This chapter is guided by ideas about sustainable livelihood and climate change adaptation perspectives, which lead to a framework for the examination of households' adaptation (see Fig. 2). In these two perspectives, the notion of vulnerability is central. It is viewed as the context in which households construct their livelihoods. It is recognized that the changing climatic and social factors at the local scale influence the vulnerability context. These factors describe the biophysical and social conditions, the two aspects of vulnerability, which determine the exposure-sensitivity of households. Exposure-sensitivity is also influenced by the location of the community and livelihoods. The vulnerability context frames the environment in which households construct their livelihoods. It affects their livelihood choices in terms of livelihood strategies, which are determined by the livelihood assets and activities that households engage in, resulting in certain livelihood outcomes. There is an

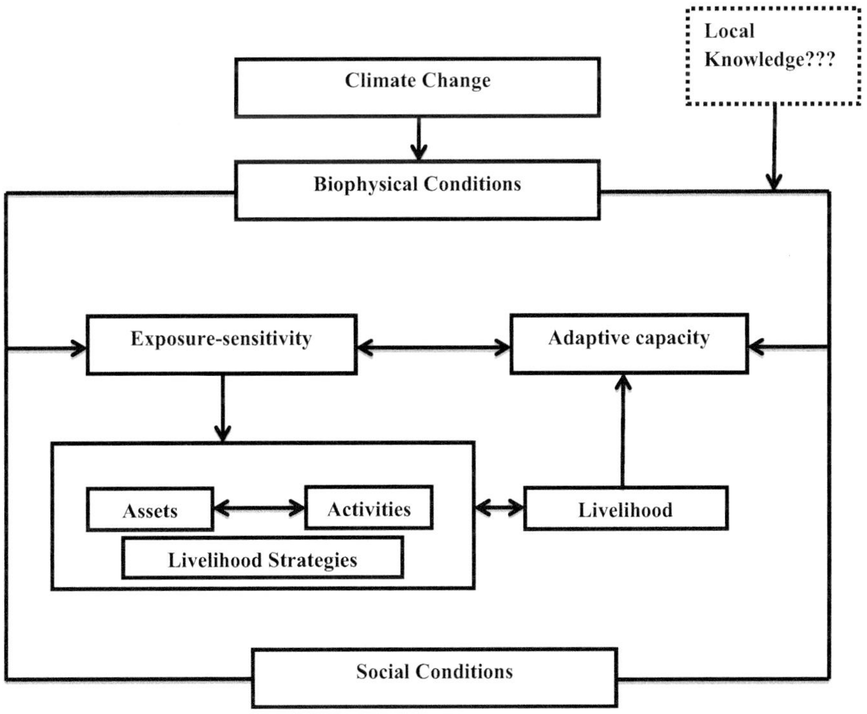

Fig. 2 Framework for vulnerability assessment

interrelationship between livelihood strategies and outcomes in the sense that the outcomes are reinvested into activities generating more assets. The livelihood outcomes affect adaptive capacity of households, which is reflected in the adaptations people employ to deal with biophysical and social conditions. The choices of adaptation need to be both short and long term in order to deal with changing environment or exposure sensitivity as a result of the dynamics of the vulnerable context. Exposure-sensitivity and adaptive capacity are interrelated, as it is explained that, experience with exposure creates learning and knowledge, improving adaptive capacity, while repeated exposure creates low adaptive capacity through assets decline. How the processes are experienced is unknown at the community level.

The assessing vulnerability starts with an assessment of the vulnerability of the community, in terms of who and what is vulnerable, to what stresses, in what way, and what capacity exists to adapt to changing climate. This approach recognizes the importance of involving the local community in assessing vulnerability to current climate change as well as the current adaptations as well as opportunities to reduce future vulnerabilities.

Materials and Methods

Description of Study Area

The study was undertaken at Mognori community, a fringe community of Mole National Park, lying at the southeastern boundary of the Mole National Park. It is a small village of 400 people, who are mainly subsistent agrarian community. It is located in the West Gonja District (see Fig. 3) in the Northern Region of Ghana (Ghana Statistical Service 2012). The district lies on longitude 1° 5′ and 2° 58′ West and Latitude 8° 32′ and 10° 2′ North. It lies in the Guinea Savannah ecosystem, which is characteristically dry due to its proximity to the Sahel and the Sahara. The dryness is influenced by the Harmattan wind, which blows from the Sahara between December and the beginning of February each year (Armah et al. 2010).

The district is characterized by high temperatures that vary between 14 °C (59 °F) at night and 40 °C (104 °F) during the day. The mean monthly temperature is 27 °C, with the maximum in the dry season, between March and April, and the lowest between December and January. The rainfall pattern is erratic. It is bimodal with the average annual precipitation being 1144 mm. The peak of the rainfall is in June–July with a prolonged dry spell in August. The rains can be stormy and torrential, up to 300 mm per hour, causing flooding (Ghana Statistical Service 2012).

Case Study Method

Qualitative case study method was used to explore climate change-livelihood interaction and local adaptation. It's a method widely used in tourism (Stronza and Pegas

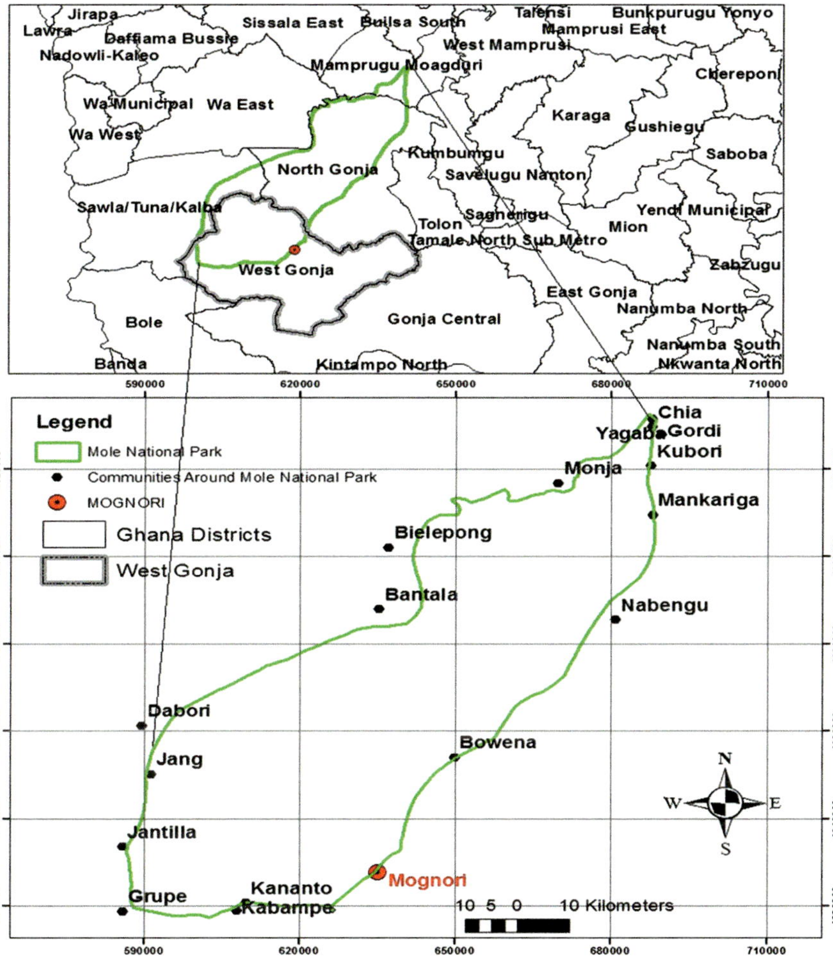

Fig. 3 Map showing the location of Mognori and Mole National Park

2008). Yin (2011) defines case study as "an empirical enquiry that investigates a contemporary phenomenon within its real-life context, when the boundaries between phenomenon and context are not clearly evident, and in which multiple sources of evidence are used." (p. 13). It involves the study of a case, as an example of the phenomenon with the aim to seek understanding of the phenomenon (Veal 2006). It seeks to uncover meaning by analyzing rich, non-numerical information in the context of a particular case. It is particularly appropriate for exploring under-researched phenomenon (Yin 2009). With this in mind, Mognori community was selected as a case to understand households' experiences with the changing climate in the process of constructing their livelihoods and adaptation strategies.

Rationale for Choosing Mognori Community as a Case

Mognori was chosen for the following reasons: (1) the involvement of households in subsistence farming livelihood, (2) changing climatic trend, and (3) livelihood vulnerability and poverty. The district in which Mognori is located is predominantly rural, and its economy is built on the natural resource base with rain-fed agriculture being the predominant livelihood strategy for most people. Agriculture is influenced by low and erratic rainfall, low soil fertility, and frequent wildfires, resulting in post-harvest losses (Armah et al. 2010). Also, while declining poverty in other regions of the country is observed, the district in which Mognori is located is still at a 52–88% poverty level. This is viewed as troubling for a district with an average household size of about eight people, larger than the national average of five per household. According to Sabates-Wheeler et al. (2008), in the absence of climate change mitigation and adaptation measures, poverty could worsen in the district through reduced agricultural yields, and more frequent natural disasters would be experienced. The district is characterized by low adaptive capacity due to low socioeconomic development as a result of low education and high unemployment (Antwi-Agyei et al. 2011). As a result, there is high emigration to the South (Sabates-Wheeler et al. 2008). According to a review of opportunities by Shepherd et al. (2005), in spite of the constraints, agriculture is seen as the mainstay of households, and there are limited opportunities outside of the agricultural sector. However, tourism at Mole National Park, with its rich wildlife, is increasingly providing opportunities for some households in the district (Stanturf et al. 2011). Mognori is one of the communities, close to Mole National Park. Its proximity to park is instrumental in attracting visitors to the community seeking both nature and cultural experiences such as bird-watching, village tour, traditional dancing entertainment, wildlife viewing, and a canoe safari on the Mole River, which connects the community to Mole National Park.

The historical climatic trend in the country indicates an increase in average temperature. Since 1960, the mean annual temperature has risen by 1.0 °C. Generally, the rate of increase is more rapid in the northern rather than southern regions (Würtenberger et al. 2011). Also, the country has experienced decreasing rains since the 1960s, with low levels recorded in the late 1970s and early 1980s. This caused an overall country-wide decreasing trend in the period 1960–2006 to an average of 2.3 mm/month (Stanturf et al. 2011). The rainfall is increasingly erratic. The assessment of future climate change in Ghana based on global circulation models (GCMs) indicates a general rise in mean temperatures. Based on future scenarios for climate change, the models indicate an anticipated increase in mean daily temperatures by 2.5–3.2 °C by 2100. Also, precipitation will decrease by 9–27% by the year 2100, with the range representing spatial variations across the country (Stanturf et al. 2011). It is projected that an estimated temperature increase of 1.7–2.04 °C by 2030 will occur in the Northern Savannah regions, with average temperatures rising as high as 41 °C (Environmental Protection Agency 2011).

Community Entry Protocol

Conducting a study in a community setting requires the understanding of community protocols on access and code of conduct to avoid conflict (Wilson 2001). In this regard, a visit was made to the community to seek permission from the traditional authority. Permission was granted to conduct the study and live in the community.

Recruitment of Participants and Data Collection

As a qualitative approach, recruiting 30 participants was deemed adequate using in-depth interviews (Veal 2006). Because the purpose of the study was to elicit first-hand data which would be representative of the perceptions of the community, 22 households, represented by heads (either a man or woman), were recruited using purposive sampling. The rest could not participate because of engagement outside the community. This sampling technique is based on selecting the interviewees whose insights would be relevant to the phenomenon being studied. Out of the total, 22.7% were females and 77.3% were males, between the ages of 27 and 55 years. They are married and heads of households. They have no formal education except for a few (36.4%) with high school education. The household size ranges from five to eight members. They are primarily agrarian, supporting the subsistent farming with forest and ecotourism-based activities.

Each participant was visited in the evening after returning from the farm. The interviews were conducted in their houses with the permission of the household heads to ensure convenience and comfort. Mason (1993) who conducted his study around Mole National Park advised that the use of formal interviews is insensitive to the culture of the people as they respond inappropriately because of the limited education. So considering the culture of Mognori, which is based on oral tradition, verbal discourse such as conversational interviews allowed them to explain and illustrate their thoughts and feelings. The purpose of the study was explained to each participant. They were made aware that participation in the study was purely voluntary, and they were given letters of consent to sign as a way of agreeing to freely participate in the interviews. But instead of signing, they gave their consent verbally, as they found this comfortable because of their familiarity with the oral tradition in contracts and arrangements. Veal (2006) advised that consent is sought from each interviewee before recording commences. In the process of conducting the interviews, each household head was prompted that the interview was going to be recorded using a tape recorder. Confidentiality and anonymity were ensured by the use of pseudonyms for each participant. The interviews were conducted in the local (i.e., Twi) dialect, familiar with both the researcher and participants. Lastly, participants were informed that the information from them would be used for publications and enhance their adaptation to climate change, to which they agreed verbally. Also, the chief and elders and park management were interviewed as key informants to provided background on livelihoods, changing climate and its impacts on the livelihoods and community adaptation. Observations were made on farms and environment to understanding the impacts of changing climatic conditions. Reports of the park were reviewed, providing good contextual understanding and informing the questions posed to the participants. On the average, each interview lasted for 40 min. The study took 4 months, from December 2011 to March 2012, to complete.

Data Analysis and Interpretation

The interactions and conversations with the household heads were captured with an audio tape and transcribed verbatim into word documents, using pseudonyms to represent each household heads. In most of the conversations, local dialect (i.e., *Twi*), which is well understood and spoken by the household heads, was used. Each interview was captured and transcribed into the English language using Microsoft Word.

In the process of trying to understand households' experiences with the changing climate in the process of constructing their livelihoods and adaptation strategies, I used Nvivo 9 as the data management and interpretation tool. It is a qualitative data analysis (QDA) computer software package produced by QSR International. As a tool for the interpretation of textual material (emerging from interviews), Nvivo served the functions of supporting the storing, retrieval, and manipulation of texts or documents as well as supporting the creation and manipulation of nodes (Gibbs 2002). The software allows users to classify, sort, and arrange information and examine relationships in the data.

Initially, the transcriptions were read thoroughly to correct any grammatical mistakes. After that, they were uploaded into the Nvivo program. Each was read and reread to gain a general framework for understanding each text. Long (2007) is of the view that this process is a way of familiarizing oneself with the contents of the material. This stage was followed by the second stage, involving reading single or sections of the texts to capture the hidden meaning for understanding. Using Nvivo 9, Yin's (2011) idea of disassembling or breaking down the texts into smaller fragments by assigning labels or codes to the fragments in a process called coding was applied. These codes were then reassembled into themes based on the emerging patterns in the texts. This was guided by the integrated theoretical framework (see Fig. 2). This process took about 3 weeks.

Results and Discussion

The findings of the local knowledge on the experiential reality of households constructing livelihoods in the context of vulnerability are discussed highlighting the biophysical and social conditions and the implications on households' livelihoods and the adaptation strategies.

Vulnerability Context and Impacts on Livelihoods

Changes in social and biophysical conditions within the locality create vulnerability conditions within which households construct their livelihoods. The existing social conditions that increase vulnerability to climate change were expressed in terms of crop-raiding, low education, high unemployment, and unsustainable local practices.

Crop raiding is a recurring event in the locality because of the proximity of farming to the boundary of Mole National Park, attracting elephants which cause extensive damage to food crops. Damaged farms receive no compensation from the park, stimulating negative attitude toward wildlife conservation. Farmers either trap or injure wildlife species upon encounter. Also, there is a widespread use of unsustainable practices such as slash and burn, harvesting of fuelwood, illegal logging, charcoal production, and careless use of fire during hunting and land preparation for farming in the dry season. These local practices pose threats to farms and forest patches, in terms of causing extensive clearing and burning of forest and farmlands. The locality also experiences high unemployment, accounting for the youth migration to other towns for employment opportunities, reducing human asset in the locality.

Changes in the climatic elements were observed over time as local knowledge reflects historical climatic changes. These changes were expressed in terms of severity of temperature, speed of the wind, and changes in the magnitude of precipitation related to seasons. Particular emphasis was placed on the increasing temperatures over the years, manifesting in long drought seasons and flooding within the locality during the dry and raining seasons. The high temperature creates drought conditions, with a corresponding increase in the occurrence of wildfire events and general drying of the environment, including drying of water bodies. The dry condition affects crop farming in terms of poor growth and, in addition to recurring bushfire, results in poor yield or destruction of farms. Fire events are not only naturally induced but caused by the careless use of fire during hunting and land preparation. Also, the rainfall pattern is increasingly erratic in the locality causing poor harvests as well. There were also cases of heavy rains during the rainy seasons, with accompanying high winds and flooding, devastating building and other properties, including farms. This event creates unnecessary costs as households need to reconstruct these buildings.

The observation of the social and biophysical conditions and the adverse implications for the environment and livelihoods are congruent with other studies (Akudugu et al. 2012; Laube et al. 2012; Maddison 2007; Stanturf et al. 2011). For instance, a review of local experience in the Northern Ghana suggests a trend of increasing temperature, decreasing rainfall, unreliable amount of precipitation and delays in the onset of the rainy season, windstorms, heavy rains late in the rainy season, increased fires, drying of rivers, and increased flooding events, causing crop and property damage as well as food insecurity (see Akudugu et al. 2012; Stanturf et al. 2011). In effect, the climatic changes and associated social conditions reflect local knowledge about climate change and other factors that place stress on households living on the edge of the habitable zone in Ghana, where climate change impacts are making it difficult to engage in normal livelihood activities. Table 1 below presents a summary of the social and biophysical condition and impacts on farming activities.

Table 1 Social and biophysical conditions and impacts

Vulnerability context	Indicator events	Impacts on assets	Livelihood outcomes
Social conditions	Low education Crop raiding High unemployment Poor local practices	Destruction of *Farms* *Homes* *Forest patches* Youth migration	Low food stocks Low income High cost of rebuilding Fragmentation of habitats Low labor
Biophysical conditions	High temperature Drought Increase fire events erratic precipitation Episodes of heavy rains Flooding	Destruction of; *Farms* *Homes* *Forest patches* Drying of rivers	Low food stocks Low income High cost of rebuilding Fragmentation of habitats and landscape

Livelihood Construction and Adaptation Strategies

Complexities in the construction of livelihoods within the constraints of changing climate were observed. These complexities reflect a portfolio of livelihood choices regarding livelihood strategies such as intensification/extensification, migration, and livelihood diversification, indicating the notion of diversity in survival strategies. Households use a combination of extensification and intensification strategies within the agriculture context, where different crops are planted on a piece of land and some farming activities are relocated elsewhere to avoid crop raiding. Migration, involving the youth, is used to expand livelihood opportunities, even though it potentially undermines local livelihoods in terms of reducing available labor that could support the local economy. However, such migration from farming localities to cities is a phenomenon found all over the world, including Ghana (Sabates-Wheeler et al. 2008). All these strategies are used as attempts to optimize farming outcomes and also ensure survival within a vulnerable context. However, the persisting social and biophysical conditions constrain and frustrate the agrarian lifestyle of households, shifting more attention to alternative livelihood such as ecotourism, as a complementing livelihood option in a vulnerable context. The integration of ecotourism into existing farming livelihood reflects flexibility in households' livelihood choices and decision-making in this vulnerable context. Households in such marginal situations are flexible in incorporating ecotourism as a form of livelihood diversification to improve asset distribution.

The observation of diverse livelihood strategies is found in other studies where households make a living within a constraining context. These studies suggest that the use of diverse strategies is an attempt to achieve households' livelihood resilience, which is consistent with the observation made (see Mbaiwa and Stronza 2010; Shen et al. 2008; Tao and Wall 2009a). However, these studies place emphasis on livelihood diversification, where ecotourism is separated from non-ecotourism activities, ignoring the complex relationships between ecotourism and other activities such as farming, in the process of building livelihood resilience. As observed, this intricate relationship is a two-way process. For instance, the application of extensification and intensification strategies within the agriculture context reflects not only the process of ensuring food and income security from farming but also an attempt to sustain food services to support tourists' sustenance. The second aspect of the relationship is reflected in how ecotourism complements farming. Its complementarities occur in terms of improvement in the financial, natural, human, social, and physical assets as well as creating linkages in the local economy. For instance, its support to households' financial assets evolves from its ability to create diverse employment opportunities, resulting in different streams of income. This creates opportunities for reinvestment into farming by supporting labor hiring and purchasing of equipment and chemicals for weed control. Also, its support to the natural assets is realized through sustainable practices to support conservation and livelihoods. Such practices are observed in local efforts: in conserving indigenous trees and wildlife, seasonal patrolling of wildlife to prevent crop raiding, removing deposits in a river, greenbelt creation to reduce fire outbreak and to protect assets and activities, and early burning to reduce fuel load early in the dry season. It is noteworthy that such change in local behavior is an example of ecotourism practices leading to improved environmental management outside the park, which reduces community susceptibility to fire damage.

Ecotourism has also created opportunities for improvement in households' human assets through a capacity building that support its management. Training such as tour guiding and interpretation, cooking, hygiene, facility maintenance, and dancing and drumming were gained to improve the delivery of services to tourists. They were essential to improving the skills, and knowledge of the local people, with low education and tourism exposure to meet the high demand of the industry (Snyman 2012). Particularly, the training in drumming and dancing was instrumental in revitalizing the culture of the community and drawing tourists from Mole National Park interested in local culture. Ecotourism has also stimulated different social assets such as trust, relationships, and shared decision-making that result in a more cooperative spirit and better community cohesion. The observation indicates that the community's cooperative spirit and shared decision-making stimulated by ecotourism attracted philanthropic support from tourists and external organizations in the form of solar light and borehole construction, sponsorships for children, and revamping of school. It was found that the local bridge and road were revamped to increase tourist access to the community from the park. This also reduced flood risks and increased local access to urban areas. All these impacts indicate ecotourism's ability to complement households' livelihoods and farming, helping to diversify local opportunities within a vulnerable context and related shocks.

These adaptations exist in different combinations, complementing each other to reduce risks, optimize benefits, and ensure households livelihood resilience. The notion of flexibility is induced by seasonality and the nature of the vulnerability, which create opportunities for employing combinations of suitable adaptation strategies. For instance, when and what adaptation is used is influenced by the seasons. The effectiveness of these adaptations depends on how they reduce risks and maximize benefits to households. For instance, observed use of innovation such as greenbelts is claimed to reduce the drastic spread of fire and farm damage during the dry season.

Conclusion

Local knowledge of vulnerability experienced occurs in the form of awareness of the climatic and social factors and changes in livelihood assets and activities. Awareness of these changes stimulates local adaptations to ensure livelihood resilience. Also, local knowledge on adaptation to vulnerability is reflected in three strategies used: migration, intensification/extensification, and livelihood diversification. Particularly, ecotourism emerges as a suitable livelihood diversification, an adaptation strategy with emergent complementarities to farming livelihoods and household survival within a vulnerable context. Its induced adaptations exist in combinations that complement livelihoods in such a way to ensure livelihood resilience, suggesting that local knowledge is reactive, need-based, and short-term. These characteristics indicate the inadequacy of local knowledge and adaptive capacity in terms of future climate change, suggesting a need to move across the local scale to building knowledge and capacity to sustain livelihoods, particularly ecotourism in the face of changing climate because of its partial dependence on nature. It raises the issue of how much adaptation can occur as these conditions continue to change in the future. Therefore, there is the need to continuously build local adaptive capacity, by supporting assets and activities that shape choices on livelihood strategies. Building adequate and relevant capacity with the support of other stakeholders can help local communities explore the opportunities in using ecotourism as an adaptation strategy.

Acknowledgment I would like to express my deepest gratitude to the International Development and Research Centre (IDRC) for providing the needed funding, without which this research project would not have come to successful completion.

References

Akudugu MA, Dittoh S, Mahama ES (2012) The implications of climate change on food security and rural livelihoods: experiences from Northern Ghana. J Environ Earth Sci 2(3):21–30

Antwi-Agyei P, Fraser EDG, Dougill AJ, Stringer LC, Simelton E (2011) Mapping the vulnerability of crop production to drought in Ghana using rainfall, yield and socioeconomic data. Working no. 25. Sustainability Research Institute, Leeds

Anup KC, Rijal K, Sapkot RP (2015) Role of ecotourism in environmental conservation and socio-economic development in Annapurna conservation area, Nepal. Int J Sust Dev World 22(3): 251–258

Armah FA, Yawson DO, Yengoh GT, Odoi JO, Afrifa EKA (2010) Impact of floods on livelihoods and vulnerability of natural resource dependent communities in northern Ghana. Water 2(2):120–139

Baumgartner, R., Högger, R. (2004). In search of sustainable livelihood systems: managing resources and change. New Delhi: Sage

Becken S, Hay JE (2007) Tourism and climate change: risk and opportunities. Channel View Publication, Clevedon

Berkes F (2009) Indigenous ways of knowing and the study of environmental change. J R Soc N Z 39(4):151–156

Brace C, Geoghegan H (2010) Human geographies of climate change: landscape, temporality, and lay knowledge. Prog Hum Geogr 35(3):284–302

Brooks N (2003) Vulnerability, risk and adaptation: a conceptual framework. Centre for Social and Economic Research on the Global Environment, Norwich

Buckley R (2009) Ecotourism: principles and practices. CAB International, Oxford, UK

Chambers R, Conway GR (1992) Sustainable rural livelihoods: practical concepts for the 21st century. Discussion paper no. 296. IDS, Brighton

Chirenje LI (2017) Contribution of ecotourism to poverty alleviation in Nyanga, Zimbabwe. Chin J Popul Resour Environ 15:87–92

Coria J, Calfucura E (2012) Ecotourism and the development of indigenous communities: the good, the bad, and the ugly. Ecol Econ 73:47–55

Cutter SL (2003) The vulnerability of science and the science of vulnerability. Ann Assoc Am Geogr 93(1):1–12

Department for International Development (2004) Sustainable livelihoods guidance sheets. DFID, London

Ellis F (2000) Rural livelihoods and diversity in developing countries. Oxford University Press, New York

Environmental Protection Agency (2011) Ghana's second national communication to the UNFCCC. EPA, Accra

Farrell BH, Twining-Ward L (2004) Reconceptualizing tourism. Ann Tour Res 31(2):274–295

Fennell D (2008) Ecotourism: an introduction, 3rd edn. Routledge, New York

Folke C (2006) Resilience: the emergence of a perspective for social-ecological systems analyses. Glob Environ Chang 16(3):253–267

Ford JD, Smit B (2004) A framework for assessing the vulnerability of communities in the Canadian Arctic to risks associated with climate change. Arctic 57(4):389–400

Ford JD, Smit B, Wandel J (2006) Vulnerability to climate change in the Arctic: a case study from arctic bay, Canada. Glob Environ Chang 16(2):145–160

Füssel H, Klein RJT (2006) Climate change vulnerability assessments: an evolution of conceptual thinking. Clim Chang 75(3):301–329

Ghana Statistical Service (2012) 2010 population and housing census: final results. Ghana Statistical Service, Accra

Gibbs G (2002) Qualitative data analysis: explorations with Nvivo. Open University Press, Philadelphia

Hall CM, Higham J (2005) Tourism, recreation and climate change: international perspectives. Channel View Publication, Clevedon

Hamilton JM, Tol RSJ (2007) The impacts of climate change on tourism and recreation. In: Schlensinger ME, Kheshgi H, Smith JB, de la Chesnaye FC, Reilly JM, Wilson T, Kolstad C (eds) Human-induced climate change: an interdisciplinary assessment. Cambridge University Press, Cambridge, UK, pp 147–155

Helmore K, Singh N (2001) Sustainable livelihoods: building on the wealth of the poor. Kumarian Press, Bloomfield

Holden A (2008) Environment and tourism, 2nd edn. Routledge, New York

Hulme M (2008) Geographical work at the boundaries of climate change. Trans Inst Br Geogr 33(1):5–11

Hunt CA, Durhamb WH, Driscoll L, Honey M (2015) Can ecotourism deliver real economic, social, and environmental benefits? A study of the Osa Peninsula, Costa Rica. J Sustain Tour 23(3):339–357
IPCC (2007) Climate change 2007: impacts, adaptation and vulnerability. Contribution of working group II to the Fourth Assessment Report of the IPCC. Cambridge University Press, Cambridge, UK
Kaján E, Saarinen J (2013) Tourism, climate change and adaptation: a review. Curr Issue Tour 16(2):167–195
Kelly PM, Adger WN (2000) Theory and practice in assessing vulnerability to climate change and facilitating adaptation. Clim Chang 47(4):325–352
Laube W, Schraven B, Awo M (2012) Smallholder adaptation to climate change: dynamics and limits in Northern Ghana. Clim Chang 111:753–774
Long J (2007) Researching leisure, sport and tourism: the essential guide. Sage, London
Maddison D (2007) The perception of and adaptation to climate change in Africa. World Bank Policy Research Working Paper 4308
Mason J (1993) Chaliabuso: indigenous people, the state and alienation: the case of conservation management at mole national park, northern region, Ghana. Unpublished MSc, University of Guelph, Canada
Mbaiwa JE, Stronza AL (2010) The effects of tourism development on rural livelihoods in the Okavango Delta, Botswana. J Sustain Tour 18(5):635–656
Mercer J (2010) Disaster risk reduction or climate change adaptation: are we reinventing the wheel? J Int Dev 22(2):247–264
O'Brien KL, Eriksen S, Schjolden A, Lygaard L (2005) What's in a word? Conflicting interpretations of vulnerability in climate change research. Clim Pol 4:1–16
Paavola J (2008) Livelihoods, vulnerability and adaptation to climate change in Morogoro, Tanzania. Environ Sci Pol 11(7):642–654
Regmi KD, Walter P (2017) Modernization theory, ecotourism policy, and sustainable development for poor countries of the global South: perspectives from Nepal. Int J Sustain Dev World Ecol 24(1):1–14
Reimer JK, Walter P (2013) How do you know it when you see it? Community-based ecotourism in the Cardamom Mountains of southwestern Cambodia. Tour Manag 34:122–132
Sabates-Wheeler R, Sabates R, Castaldo A (2008) Tackling poverty-migration linkages: Evidence from Ghana and Egypt. Social Indicators Research 87(2):307–328
Schipper EL (2007) Climate change adaptation and development: exploring the linkages. Tyndall Centre Clim Chang Res Work Pap 107:13
Scoones I (2009) Livelihood perspectives and rural development. J Peasant Stud 36(4):171–196
Scott D (2006) Climate change and sustainable tourism in the 21st century. In: Cukier J (ed) Tourism research. Department of Geography, University of Waterloo, Canada, pp 175–247
Scott D (2011) Why sustainable tourism must address climate change. J Sustain Tour 19(1):17–34
Scott D, Becken S (2010) Adapting to climate change and climate policy: progress, problems and potentials. J Sustain Tour 18(3):283–295
Scott D, Jones B (2006) Climate change and nature-based tourism: implications for park visitation in Canada. University of Waterloo, Department of Geography, Waterloo
Shen F, Hughey K, Simmons DG (2008) Connecting the sustainable livelihoods approach and tourism: a review of the literature toward integrative thinking. In: CAUTHE 2008 conference, pp 1–14
Shepherd TG, Plumb RA, S.C. Wofsy SC (2005) Preface to JAS specialissue on the Antarctic stratospheric sudden warming and split ozone hole of 2002. J Atmos Sci 62:565–566
Smit B, Pilifosova O (2001) Adaptation to climate change in the context of sustainable development and equity. In: McCarthy et al. (eds) Climate change 2001: impacts, adaptation and vulnerability. Cambridge University Press, Cambridge, pp 877–912
Smit B, Wandel J (2006) Adaptation, adaptive capacity and vulnerability. Glob Environ Chang 16:282–292
Snyman SL (2012) The role of tourism employment in poverty reduction and community perceptions of conservation and tourism in Southern Africa. J Sustain Tour 20(3):395–416

Stanturf JA, Warren ML, Charnley S, Polasky SC, Goodrick SL, Frederick Armah F, Nyako YA (2011) Ghana climate change vulnerability and adaptation assessment. USAID, Washington, DC
Stronza A, Gordillo J (2008) Community views of ecotourism. Ann Tour Res 35(2):448–468
Stronza A, Pegas F (2008) Ecotourism and conservation: Two cases from Brazil and Peru. Human Dimensions of Wildlife 13(4):263–279
Tao TCH, Wall G (2009a) A livelihood approach to sustainability. Asia Pac J Tour Res 14(2):137–152
Tao TCH, Wall G (2009b) Tourism as a sustainable livelihood strategy. Tour Manag 30(1):90–98
Turner BL, Kasperson RE, Matsone PA, McCarthy JJ, Corell RW, Christensene L, Schiller A (2003) A framework for vulnerability analysis in sustainability science. Proc Natl Acad Sci U S A 100(14):8074–8079
van Aalst MK, Cannon T, Burton I (2008) Community level adaptation to climate change: the potential role of participatory community risk assessment. Glob Environ Chang 18(1):165–179
Veal AJ (2006) Research methods for leisure and tourism: a practical guide. Pearson Education Limited, London
Vincent K (2007) Uncertainty in adaptive capacity and the importance of scale, Global Environ Change 17:12–24
Wearing S, Neil J (2009) Ecotourism: impacts, potential and possibilities? 2nd edn. Butterworth-Heinemann, Oxford, UK
Weaver D (2011) Can sustainable tourism survive climate change? J Sustain Tour 19(1):5–15
Wilson S (2001) What is an indigenous research methodology? Can J Nativ Educ 25(2):175–179
Würtenberger L, Bunzeck IG, van Tilburg X (2011) Initiatives related to climate change in Ghana towards coordinating efforts. Retrieved from http://cdkn.org/wp-content/uploads/2012/04/Ghana-initiatives-mapping-climate-change-May2011.pdf
Yin R (2009) Case study research: design and methods, 3rd edn. Sage, Thousand Oaks
Yin RK (2011) Qualitative research from start to finish. The Guilford Press, New York

Influence of REDD+ Pilot Project on Livelihoods of Mangrove Forest Adjacent Communities in Zanzibar, Tanzania

27

Rukia A. Kitula

Contents

Abstract

Climate change is affecting livelihoods of people and environment people live in. The policy to Reduce Emissions from Deforestation and Degradation plus role of conservation, sustainable forest management, and forest carbon stock enhancement (REDD+) has been promoted in developing countries to address climate change impacts and to provide incentives against deforestation. Zanzibar has been involved in piloting REDD+ project between 2010 and 2014. This study was designed to assess influence of REDD+ project on livelihoods in Zanzibar. Questionnaire survey and focus group discussions were used for data collection. Content analysis and descriptive statistical analyses were used in data analysis.

R. A. Kitula (✉)
Institute of Marine Sciences, University of Dar es Salaam, Zanzibar, Tanzania
e-mail: rukiakitula@gmail.com; rkitula@yahoo.com

W. Leal Filho (ed.), *Handbook of Climate Change Resilience*,
https://doi.org/10.1007/978-3-319-93336-8_167

The study revealed that all livelihood capitals were significantly improved and this was associated with implementation of REDD+ project. Financial incentives that were given to each village by the project helped to stimulate economic development and conservation of mangrove forests for carbon credit. The project had promoted participatory process in forest management. It was recommended that long-term projects intending to provide financial support to local communities to initiate alternative income-generating activities have to do so at the beginning of project implementation time. This will enable the project proponent to have enough time to monitor implementation of the project activities.

Keywords
REDD+ pilot project · Community livelihoods · Mangrove forests · Carbon credit · Zanzibar, Tanzania

Introduction

Climate change is affecting livelihoods of people and environment people live in. Reduce Emissions from Deforestation and Degradation plus role of conservation, sustainable forest management, and forest carbon stock enhancement (REDD+) is a policy strategy which is being promoted by international community to address climate change impacts and provide incentives against deforestation in developing countries (UNFCCC 2009). Tanzania is one among the developing countries that has ratified this policy initiative. The incentive under REDD+ scheme is based on performance, that is, to pay forest owners and users to reduce emissions and increase removals. Such payments for environmental services (PES) provides incentives directly to forest owners and users to manage forests better and clear less forests (Angelsen et al. 2011). The carbon-related schemes such as REDD+ have the potential to provide livelihood opportunities for mangrove forest adjacent communities. The schemes can contribute to people's livelihoods in a variety of ways. Financial flows emerging from mangrove REDD+ projects and mangrove forest conservation may alleviate poverty and stimulate economic development within the project area (Chhatre et al. 2012). The distribution of financial rewards for avoiding deforestation can contribute to improve community assets such as school buildings and investment in village infrastructure such as electrification and improved sanitation (King 2012).

Incorporating the mangrove forests into REDD+ project requires allowing communities close to the forest to use the mangroves sustainably. Microfinance and/or alternative livelihood schemes to compensate households for loss of income through the integration of customary use into the management plan to allow communities to use the mangroves for firewood extraction; and the introduction of fuel efficient stoves to reduce the community's charcoal needs. Experience shows that households that own little or no land have few alternative livelihood strategies and that there is a real risk of extraction of mangrove wood for firewood and timber following poor crop harvest (Kitula 2012).

REDD+ actions may spark political change toward better governance, less corruption, and more respect for the rights of vulnerable groups (CARE 2010). On the other hand, weak community governance would make it harder for a REDD+ project to reduce deforestation and degradation to target levels, undermining the chances of achieving the expected financial benefits (Nussbaum et al. 2009).

There is growing interest in Tanzania to participate in the emerging carbon market. CARE International has been involved in piloting mangrove REDD+ project through Community Forest Management in Zanzibar between 2010 and 2014. The project was known in Kiswahili, the national language of Tanzania, as "Hifadhi ya Misitu ya Asili" (which means conservation of natural forests) abbreviated as HIMA (CARE 2010). However, until recently it was not known the influence of HIMA project on livelihoods in Zanzibar. Therefore, this study assessed the influence of REDD+ project on livelihoods in Unguja Island, Tanzania. The study provides viable information to gage performance of the HIMA project in addressing the global REDD+ policy, Tanzania REDD+ strategy, and Zanzibar Forest Policy for improving community livelihoods through mangrove carbon credit.

Methodology

The Study Area

The study was carried out in Unguja Island, Tanzania (Fig. 1). Two villages, namely, Kitogani and Muyuni B, were involved in piloting REDD+ project in mangrove forest between 2010 and 2014. Generally speaking, information on influence of REDD+ pilot project to livelihoods was lacking. So, the area permitted collection of sufficient data important for the study. Kitogani and Muyuni B villages are under the Menai Bay Conservation Unit.

Unguja is part of the United Republic of Tanzania and is located off the eastern coastline of Tanzania. The economy of the Unguja Island is based on agriculture, fishing, animal husbandry, seaweed farming, and tourism. Other economic activities include beekeeping, charcoal making, firewood cutting, tailoring, government employment, building construction, and small-scale business (MACEMP 2009).

Climate

There are four major seasons: the long rains from the end of March to mid-May, cool season with intermittent showers from June to August, the short rains from the end of September to December, and the dry spell which start soon after short rains to the onset of long rains. The average annual rainfall is about 1,700 mm. The temperature is also typically equatorial with very little variation between the seasons. The mean annual maximum temperature is 30.3 °C and the minimum is 23.5 °C.

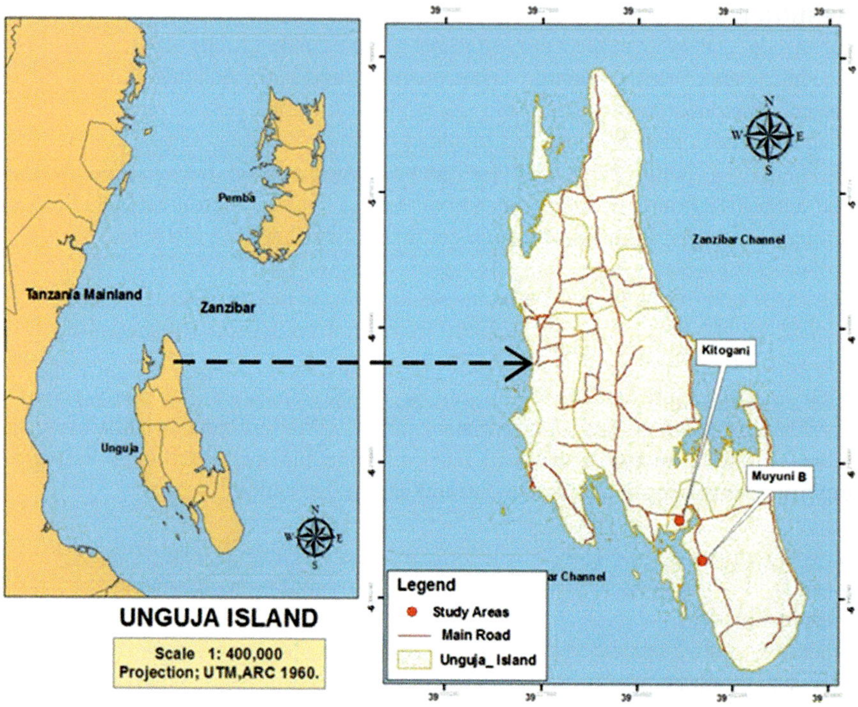

Fig. 1 Map of Ungula Island showing study villages

Research Design, Sampling Design, and Data Collection

Cross-sectional design was adopted in this study. Random sampling was used to select two villages (Kitogani and Muyuni B) that were involved in implementing REDD+ pilot project. Sampling unit in this study was a household. Thirty households were randomly selected and interviewed in each study village to gather information on influence of REDD+ pilot project on livelihoods. A questionnaire was the main research instrument for collecting primary data from households, and it included both open- and close-ended questions. The questionnaires were administered by researcher and assistants. Focus group discussions (FGDs) were conducted with village conservation committees (VCCs) to gather information on REDD+ activities for livelihood improvement and challenges they are facing.

Content analysis was used to analyze qualitative information collected through FGDs and documentary reviews. Descriptive statistical analysis was carried out for quantitative data. Significant differences between categories and villages were judged based on chi-square values at 5% level of significance. A Statistical Package for Social Sciences (SPSS) computer software version 16 was used in analyzing the quantitative data.

Results and Discussions

The influence of HIMA project to livelihoods of the rural mangrove forest adjacent communities was discussed in seven capital assets as follows.

Influence of HIMA Project on Social Capital

Income-generating groups that were identified during the survey were widows tailoring, women pottery, beekeeping, and vegetable farming. These groups were formed by HIMA project to help communities reduce dependence on harvesting mangrove forest for household income. Both men and women who were mostly dependent on mangrove forest-based activities were involved in identifying potential income-generating groups. Gender issue was one of the most important factors that were considered in the formation of groups. During the time of this study, these groups were found to be active. This indicates that facilitating local communities to initiate alternative income-generating activities (IGAs) helps them address a wider range of challenges associated with development and environmental conservation. Also, facilitating local communities to initiate IGAs has potential for compensating their income loss from harvesting mangrove forest and this in turn reduces deforestation and forest degradation, a view that is also shared by Kitula (2012).

Focus group discussions revealed that HIMA project has facilitated villagers in Kitogani and Muyuni B villages to form VCCs to be used as entry point for HIMA project and any other project that is interested on issues pertaining to environmental conservation. Major roles of VCCs are to supervise the implementation of village bylaw on community forest management, to educate other community members on community forest management, to conduct forest patrol, to issue permits regarding the use of natural resources found in the concerned area, and to prepare and present monthly report to village chairpersons. This implies that local communities have been seen as appropriate body to manage forest resources collectively to reduce deforestation and forest degradation, a view that is also shared by Ostrom (1994). Participation of local communities in the management of mangrove forest is very crucial for successful implementation of REDD+ project (King 2012). According to Kajembe et al. (2004), involvement of local people in the management of forest resources facilitates improvement of forest condition, thereby increasing carbon sinks.

Communities' perceptions regarding influence of HIMA project on facilitating communities to form VCCs differed significantly ($p = 0.003$) across the study area (Table 1). More than a half (53.3%) of respondents in Muyuni B village indicated that ability of HIMA project to facilitate communities to form VCCs was rather poor. The main reason given for this was that HIMA project mobilized communities to form a new conservation committee with new faces of committee members as its entry point instead of using the existing one. This approach was complained by local communities as it has resulted into animosity between the former and the new VCC in the area. Focus group discussion revealed that the former conservation committee

Table 1 Respondents' perceptions on influence of HIMA project on formation of VCCs

	Influence of HIMA project in formation of VCCs			
Village names	Poor	Satisfactory	Good	Total
Kitogani (n = 30)	4 (13.3)	6 (20)	20 (66.7)	30 (100)
Muyuni B (n = 30)	16 (53.3)	5 (16.7)	9 (30)	30 (100)
Total (N = 60)	20 (33.3)	11 (18.3)	29 (48.3)	60 (100)
Statistical test	$p = 0.003$	$\chi^2 = 11.463$		

Figures in brackets indicate percentages

Table 2 Perception of Respondents on Influence of HIMA Project on Physical Capital

	Influence of HIMA project on physical capital		
Village names	Satisfactory	Good	Total
Kitogani (n = 30)	6 (20)	24 (80)	30 (100)
Muyuni B (n = 30)	9 (30)	21 (70)	30 (100)
Total (N = 60)	15 (25)	45 (75)	60 (100)
Statistical test	$p = 0.371$	$\chi^2 = 0.800$	

Figures in brackets indicate percentages

members are now providing little support to the new VCC in forest protection for carbon market since they are leading in illegal activities in mangrove forest. Generally, weak collective action in forest protection tends to result into increased deforestation and forest degradation, a view that is also shared by Ostrom (1994).

Influence of HIMA Project on Physical Capital

Perceptions of communities on influence of HIMA project on physical capital did not differ significantly across the study villages with $p = 0.371$, $\chi^2 = 0.800$ (Table 2). About 75% of respondents indicated that influence of HIMA project on physical capital was rather good because the project distributed equally 60 modern beehives and equipment for honey harvesting to 20 beekeepers (10 beekeepers from each village). These people were chosen in order to act as role model in practicing modern beekeeping in the villages. Focus group discussions revealed that before implementation of HIMA project, beekeeping activities in the study villages were done using log hives, and smokers were used to chase away bees. This activity kills many bees and produces honey of poor quality.

Focus group discussions revealed that beekeepers were hanging hives in the mangrove forest. HIMA project has facilitated beekeepers in the study villages to construct apiaries outside the mangrove forest for protecting beehives and honey from thieves. Beekeepers in Kitogani village had started to increase space for keeping hives by constructing one more apiary by using their own resources. These results suggest that communities have already started to realize the benefits accrued from practicing modern beekeeping and would like to expand the enterprise for livelihood improvements and biodiversity conservation.

Respondents were asked to indicate influence of HIMA project in the construction of public facilities. The majority (83.3%) of respondents perceived that HIMA project was very satisfactory in constructing public facilities. This is because the project facilitated local communities to purchase and install milling machine in Muyuni B village. Installing milling machine helps communities to reduce human workload which, in turn, increases time for participating in other developmental activities. The project also facilitated communities in Kitogani and Muyuni B to purchase and install equipment for water supply for irrigating vegetable farms and domestic use. The project also facilitated the groups involved in farming vegetables to have important equipment like jerry cans to support irrigation of vegetables. Type of vegetables grown by farmers included onions, amaranths, tomatoes, sweet peppers, hot peppers, eggplant, and okra. These vegetables were chosen because they were considered as high-value crops. The project also facilitated communities to repair worn-out water pipes in order to make water supply easily accessible by the majority. Facilitating farmers to have reliable source of water for irrigating vegetables reduces dependency on rain water and also minimizes risks and uncertainties in crop production. In addition, promoting non-forest activities has the potential for reducing dependency on forest resources for household income and thereby increasing carbon sequestration.

Influence of HIMA Project on Human Capital

Table 3 gives summary results of perceptions of communities regarding influence of HIMA project on human capital. The results showed that influence of HIMA project on human capital did not differ significantly across the study villages with $\chi^2 = 5.793$, $p = 0.055$. More than a half (60%) of respondents indicated that influence of HIMA project on human capital was rather good, while 25% indicated satisfactory and very few (15%) said to be poor.

The study revealed that VCCs in Kitogani and Muyuni B villages were trained by HIMA project on issues associated with forest governance, carbon market, drivers of deforestation and forest degradation, leadership skills, financial management, gender issues, conflict management, and tackling illegal forest activities at community level. Such training was offered to VCCs in order to strengthen among other things their

Table 3 Perception of respondents on influence of HIMA project on human capital

	Influence of HIMA project on human capital			
Village names	Poor	Satisfactory	Good	Total
Kitogani (n = 30)	4 (13.3)	7 (23.3)	19 (63.3)	30 (100)
Muyuni B (n = 30)	5 (16.7)	8 (26.7)	17 (56.7)	30 (100)
Total (N = 60)	9 (15)	15 (25)	36 (60)	60 (100)
Statistical test	$p = 0.055$	$\chi^2 = 5.793$		

Figures in brackets indicate percentages

legal status and skills on forest management as well as to improve their ability to manage carbon income.

Focus group discussions revealed that income-generating groups that were formed by HIMA project were trained on issues associated with business management. More than a half (58%) of respondents interviewed indicated that they had opportunity to attend the training organized by HIMA project on business management. Training local communities on business management normally strengthens their entrepreneurship skills, and this in turn enhances human capital.

About 75% of respondents perceived that ability of HIMA project to help communities to relieve human workload was rather good, 15% satisfactory, and 10% poor. The main reason given for this was that the project facilitated local communities in the study areas to install milling machine as well as to supply safe water for irrigating vegetable farms and domestic activities. Facilitating local communities to have milling machine and easy access to safe water for various uses tends to reduce human workload for better health, thereby contributing to human capital.

Most (40%) respondents had attained Form II level of education (Fig. 2). It is compulsory for all children in Zanzibar to have basic education (up to Form II level). Both primary and secondary education were offered free of charge in Zanzibar (RGZ 2006). Despite this government initiative, some children from poor families failed to attend schools due to lack of school uniforms, school materials, and examination fees. HIMA project was found to be very helpful in promoting primary and secondary education in the study areas. The project provided school uniforms to children from poorer families to enable them attend schools. This means that HIMA project was very important in addressing the government initiative in promoting formal education of children in Zanzibar. Promoting formal education creates awareness and positive attitudes of communities in sustainable use and management of natural resources (Crawford et al. 2004).

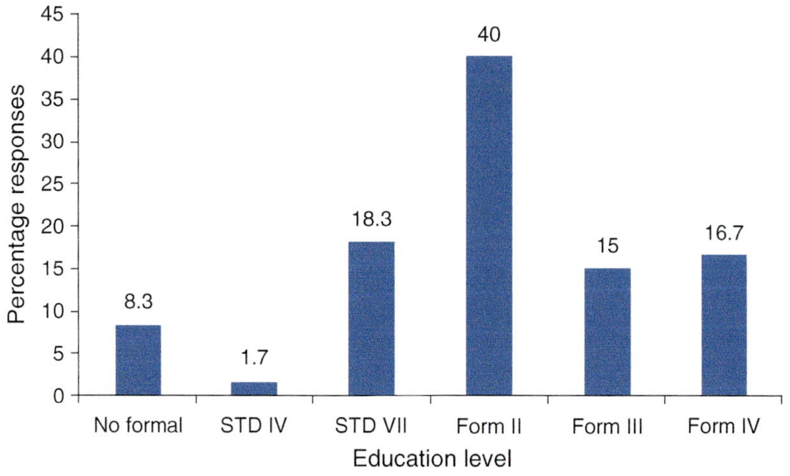

Fig. 2 Level of education of respondents in the study area

Influence of HIMA Project on Financial Capital

The study revealed that Kitogani and Muyuni B villages have received grants from HIMA project as incentive test for REDD readiness in Zanzibar. This grant was distributed to the villages during the fourth year of project implementation to promote forest conservation for carbon market. About TZS 8.3 million were given to each village by HIMA project through JUMIJAZA to stimulate economic development and conservation of mangrove forests for carbon credit. The term JUMIJAZA that means in Kiswahili language *Jumuiya ya Uhifadhi Misitu ya Jamii Zanzibar* is an aggregation entity established by the communities to support HIMA project wide governance and implementation of the project and manage the finances of emission reductions.

Focus group discussions revealed that each VCC established its own modality for the appropriate use of their respective carbon incentive funds. Generally speaking, the funds benefited communities in various ways. About 60% of the total fund was used to initiate alternative income-generating activities; 20% was used for activities associated with improving social services at community level like water supply and installation of milling machine; 10% was used for assisting special groups like women, youth, old people, and disabled, while 10% of the fund was allocated for activities associated with forest protection. This means that the implementation of HIMA project in the study villages has provided financial benefits that delivered wider socioeconomic and ecological benefits, a view that is shared by King (2012).

More than 80% of respondents acknowledged that HIMA project was good in influencing financial capital (Fig. 3). This was due to a fact that the project facilitated communities to initiate small enterprises like pottery, tailoring, beekeeping, and vegetable farming. Nearly TZS 600,000/= per person per year could be accrued from sale of pottery products like cooking lids, cooking pots, and cooking stoves. Regarding beekeeping, about TZS 450,000/= per season per hive could be earned from selling bee products. Focus group discussions revealed that there were potential markets for the vegetables grown in the study area. Vegetables were highly demanded within and outside the village. Estimates show that about TZS 300,000/= per month could be accrued from selling amaranths grown in area of about 0.5 ha. This means that the project has contributed to provide employment and raise income of the communities, particularly through selling non-forest products. These results concur with findings by Balint et al. (2002) who reported that provision of financial incentives to communities facilitates improvement of communities' livelihoods and hence reduces CO_2 emissions.

Influence of HIMA Project on Natural Capital

Perceptions of communities regarding influence of HIMA project on natural capital did not differ significantly ($p = 0.869$) across study villages (Table 4). More than

Fig. 3 Perceptions of communities on HIMA project and financial capital

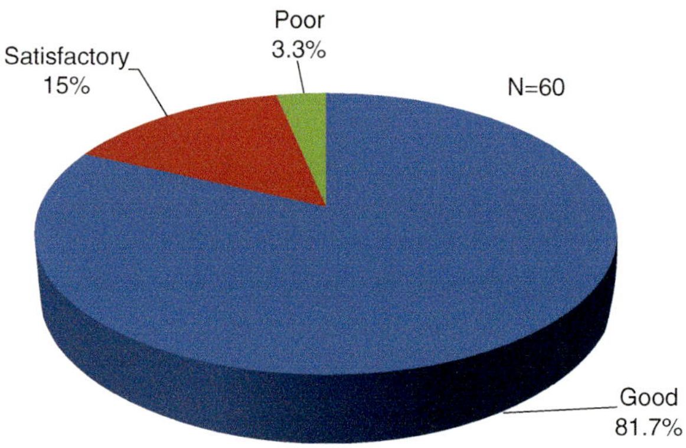

Table 4 Respondents' perceptions on status of mangrove forest in Kitogani and Muyuni B, Zanzibar

Status of mangrove forest	Village names		Total	Statistical test
	Kitogani (n = 30)	Muyuni B (n = 30)	(N = 60)	
Increasing	20 (66,7)	17 (56.7)	37 (61.7)	
Decreasing	5 (16.7)	7 (23.3)	12 (20)	$p = 0.869$
No change	2 (6.7)	2 (6.7)	4 (6.7)	$\chi^2 = 0.719$
Don't know	3 (10)	4 (13.3)	7 (11.7)	
Total	30 (100)	30 (100)	60 (100)	

Figures in brackets indicate percentages

60% of respondents perceived that implementation of HIMA project has helped to improve the status of mangrove forest in the study villages. Key informant interview revealed that HIMA project has facilitated communities in the study area to plant mangrove seedlings in about 10 ha of deforested and degraded mangrove areas. The most important species that were planted by the project included *Avicennia marina*, *Rhizophora mucronata*, and *Ceriops tagal.*

The project also provided local communities free seedlings of upland forest including *Eucalyptus*, *Cupressus lusitanica*, and *Grevillea robusta* to enable them to establish their own woodlots and reduce pressure from harvesting mangrove trees for poles and firewood. These upland tree species were chosen because they are fast growing and drought tolerant. Tree planting activity specifically targeted women who most commonly collect fuelwood.

More than 60% of respondents interviewed indicated that they are using improved cooking stoves for cooking food. The use of this kind of cooking stove tends to reduce amount of wood used for cooking per day. This in turn reduces reliance of fuelwood for cooking and hence increases carbon sequestration. Literature shows that collection of fuelwood is one among the major factors of deforestation in Tanzania (URT 2013).

Table 5 Perception of respondents on influence of HIMA project on political capital

Village names	Influence of HIMA project on political capital			Total
	Poor	Satisfactory	Good	
Kitogani (n = 30)	1 (3.3)	8 (26.7)	21 (70)	30 (100)
Muyuni B (n = 30)	2 (6.7)	9 (30)	19 (63.3)	30 (100)
Total (N = 60)	3 (5)	17 (28.3)	40 (66.7)	60 (100)
Statistical test	$p = 0.782$	$\chi^2 = 0.492$		

Figures in brackets indicate percentages.

Influence of HIMA Project on Political Capital

Perceptions of communities regarding influence of HIMA project on political capital did not differ significantly ($p = 0.782$) across the study villages (Table 5). The majority (66.7%) of respondents indicated that HIMA project was rather good in influencing changes on forest resource management. Focus group discussions revealed that HIMA project has facilitated communities in Kitogani village to prepare and get approval the Community Forest Management Agreement (COFMA). Communities in Muyuni B village were assisted by HIMA project to review the existing COFMA to include issues associated with collaborative forest resource management. This implies that autonomy of the indigenous people in the management of forest resources was recognized and respected.

The study revealed that it was illegal to harvest mangrove forest without a permit. VCCs were responsible in issuing permits to harvest mangrove poles for subsistence use only in case of emergency in a village like house fire. Also, VCC members were found to be responsible in forest patrol. For example, VCC in Kitogani village organized mangrove forest patrol once a week, and it supervised forest patrol on regular basis. On the other hand, VCC in Muyuni B village organized surprising mangrove forest patrol. This indicates that each village has established a patrolled schedule for protecting mangrove forest resources from illegal activities.

Discussions with VCCs in Kitogani and Muyuni B villages revealed that committee members were good in articulating their rights and interests regarding forest resource management. This was probably due to series of training on political leadership and exposure visits that were organized to VCCs by HIMA project. According to Kajembe et al. (2004), involvement of local people in the management of forest resources facilitates improvement of accountability which in turn enhances forest condition.

Influence of HIMA Project on Cultural Capital

Table 6 gives summary results on perceptions of communities regarding influence of HIMA project in improving cultural capital. Perceptions of communities regarding

Table 6 Perception of respondents on influence of HIMA project on cultural capital

	Influence of HIMA project on cultural capital			
Village names	Poor	Satisfactory	Good	Total
Kitogani (n = 30)	2 (6.7)	7 (23.3)	21 (70)	30 (100)
Muyuni B (n = 30)	1 (3.3)	9 (30)	20 (66.7)	30 (100)
Total (N = 60)	3 (5)	16 (26.7)	41 (68.3)	60 (100)
Statistical test	$p = 0.738$	$\chi^2 = 0.608$		

Figures in brackets indicate percentages

influence of HIMA project on cultural capital did not differ significantly with $\chi^2 = 0.608, p = 0.738$ (Table 6). More than a half (68.3%) of respondents perceived that HIMA project was rather good in improving cultural capital in the management of mangrove forests.

Focus group discussions revealed that historically it was uncommon for women to speak out and challenge men especially in the meeting. During the time of this study, this attitude was found to have changed. Both men and women were found to be active during focus group discussions and questionnaire survey. The main reason given for this was due to series of training on gender issues and exposure visits that were organized by the HIMA project to VCCs.

Conclusions

There has been a significant improvement in all livelihood capitals in the study areas that were attributed by the implementation of HIMA project since 2010. The study revealed that HIMA project was successful in promoting IGAs that were geared to substitute activities associated with deforestation of mangrove forests. Promoting alternative livelihood strategies was of particular importance in creating social safeguards. The project enabled communities in the study areas to secure rights over the use and management of forest land and other natural resources that are found in areas of their jurisdictions. This has been incentives for long-term investment in sustainable management of forest resources. Also, it lays the foundation for future expansion of REDD+ model. The study established that involving local communities in forest resource management has significantly increased accountability in forest protection which in turn enhances forest condition. Implementation of HIMA project provided financial benefits that delivered wider socioeconomic and ecological benefits in the study sites. The project has contributed to raise income of the communities, particularly through selling non-forest products. However, the grant was disbursed to communities to promote IGAs and forest protection toward the end of project implementation. There is a need for long-term projects intending to provide financial support to local communities to initiate IGAs at the beginning of project implementation time. This will enable the project proponent to have enough time to monitor project implementation.

References

Angelsen A, Boucher D, Brown S, Merckx V, Streck C, Zarin D (2011) Guidelines for REDD+ reference levels: principles and recommendations. Meridian Institute, Washington, DC, p 16

Balint BE, Forkutsa I, de Freitas ACR (2002) Indicators for the measurement of institutional performance on water management: application Uzbekistan and Ghana. Center for Development Research, Bonn, 37 pp

CARE (2010) Carbon development opportunities from REDD in Zanzibar through community Forest management. In: Carbon credit feasibility study of the *Hifadhi ya Misitu ya Asilia* (HIMA). CARE International in Tanzania, Zanzibar, 58 pp

Chhatre A, Lakhanpa S, Larson AM, Nelson F, Ojha H, Rao J (2012) Social safeguards and co-benefits in REDD+: a review of the adjacent possible. Curr Opin Environ Sustain 4:654–660

Crawford BR, Siahainenia A, Rotinsulu C, Sukmara A (2004) Compliance and enforcement of community-based coastal resource management regulations in North Sulawesi, Indonesia. J Coast Manag 32:39–50

Kajembe GC, Namubiru EL, Shemwetta DTK, Luoga EJ, Mwaipopo C (2004) The impact of rules in forest conservation in Tanzania: case of Kwizu Forest Reserve, Same District, Kilimanjaro. In: Shemwetta DTK, Luoga EJ, Kajembe GC, Madoffe SS (eds) Proceedings of the IFRI East African regional conference, Moshi, pp 92–107, 12–13 January 2004

King L (2012) Including mangrove forests in REDD+. Climate and Development Knowledge Network, Nairobi, 6 pp

Kitula RA (2012) Stakeholders analysis and institutional performance in the use and management of mangrove resources in the Rufiji Delta, Tanzania. Thesis for Award of PhD Degree at Sokoine University of Agriculture, Morogoro, 318 pp

MACEMP (2009) The status of Zanzibar coastal resources: towards the development of integrated coastal management strategies and action plan. Department of Environment, Zanzibar

Nussbaum R, Hoare A, McDermott C, Saunders J, Costa PM (2009) Accelerating transfers of interim finance for REDD+: building absorptive capacity: A report for UK DFID, 44 pp

Ostrom E (1994) Neither market nor state: governance of common-pool resources in the twenty-first century. IFPRI Lecture Series, Washington, DC, 10 pp

RGZ (2006). Zanzibar Education Policy, Ministry of education and Vocational Training. Revolutionary Government of Zanzibar, 152 pp

UNFCCC (2009) Article for the REDD+ mechanism. United Nations Framework Convention on Climate Change, Bonn

URT (United Republic of Tanzania) (2013) National strategy for reduced emissions from deforestation and forest degradation (REDD+). Vice President's Office, Dar es salaam, 67 pp

Impacts of Climate Change on Health: Evidences from Multi-stakeholders in the Western Region of Cameroon

28

Hubert Fudjumdjum

Contents

Abstract

The literature has shown evidence of human's vulnerability to climate change with the significant and emerging threat. The balance of effects on health is negative, and populations in low-income countries are more vulnerable to the adverse effects which raise the need for more attention and effective interventions. This chapter aims to assess the potential health problems induced by climate variabilities such as heat stress, air pollution, waterborne disease, and vector-borne disease on the health of inhabitants in the locality of Bafoussam in the western region of Cameroon. It also highlights the multiple health problems as well as challenges and opportunities in the face of increasing climate variability in that region. For this purpose, the author undertakes a stakeholder's interview ($n = 235$) coupled with the author's observation. The author interviewed farmers ($n = 50$), households ($n = 60$), office workers ($n = 60$), and street sellers ($n = 65$) with questions focused on identifying different health problems caused by the

H. Fudjumdjum (✉)
Research and Transfer Centre Sustainable Development and Climate Change Management, Hamburg University of Applied Sciences, Hamburg, Germany
e-mail: fudjumh@yahoo.fr

W. Leal Filho (ed.), *Handbook of Climate Change Resilience*,
https://doi.org/10.1007/978-3-319-93336-8_174

changing climate. The interviews also sought to explore interviewees' interpretation of climate change impact and awareness. Data were analyzed using SurveyMonkey and Excel data sheet and then presented analytically. Results reveal that climate change-related health problems are increasing in western Cameroon. All participants (100%) have agreed to have perceived a change in their health which is related to climate change. More than 90% (96%) reported that extreme heat and air pollution-related pulmonary diseases had increased mortality and morbidity in the region. Moreover, this paper presents that climate change has intensified harmful interactions between vector-borne and waterborne diseases. To manage that threat, implementation of city's risk management plans for climate change will potentially provide health benefits to inhabitants as well as the adoption of specific proactive measure through the establishment of local early warning system for health effect of climate change.

Keywords
Climate change · Health · Rural community · Water- and vector-borne diseases

Introduction

Anthropogenic activities such as fossil fuel burning have released enough carbon dioxide and other greenhouse gases to trap more heat in the lower atmosphere and affect the global climate. Scientific research shows that rising global temperatures are increasing the possibility of ice melting and the rise of sea levels. If current trends continue, increase in temperature and in the frequency of extreme weather events may lead to severe consequences to the environment as a whole and human health in particular (Leal Filho et al. 2016). Vora (2008) stated that human-driven ecological changes might be contributing to the spread of vector-borne diseases. Following the line, Patz (2000) argue that deforestation may create ecological niches favoring the proliferation of vectors and parasites. Also, Uneke (2008) stated that deforestation had been shown to increase the risk of malaria transmission in sub-Sahara. Harrus and Baneth (2005) also stated that water puddles in deforested areas tend to have lower salinity and acidity than puddles in forests. Such deforested water puddles may be more conducive for the larval development of certain *Anopheles* mosquito species (the vectors of malaria), particularly the most effective vector of malaria in the Amazon.

The resulting flooding of low-lying areas may cause a severe shortage of irrigation water and clean drinking water as well as an increased likelihood of cholera, schistosomiasis, typhoid, hepatitis, malaria, and other waterborne illnesses in the short term (Tosam and Mbih 2015). According to scientific evidence, the increasing extreme weathers such as heat waves, floods, and droughts are becoming more frequent and intense. This increasing of intensity and the variability of weather conditions is having an impact on human's health more especially in rural communities who have insufficient capacity to cope with its impacts (Tosam and Mbih 2015). There is a range of factors characterizing the vulnerability of rural communities to the health impacts of climate change. Among them are age profile, preexisting health conditions, awareness of the risks, socioeconomic status, housing and the surrounding infrastructure as well as exposure of livelihoods, and access to health services. Human health is more affected in two ways. First, it changes the severity or frequency of health problems that are already affected by climate or weather factors; and, second, it creates unusual or unanticipated health problems or health threats in places where they have not previously occurred. Rural communities in Cameroon are facing a range of problems like lack of clean water for household activities, poor hygiene and sanitation, and lack of proper health services which can easily be severed by the changing climate. Due to the fact that those poor communities do not have the necessary tools or knowledge to cope with the changing climate, they are vulnerable to any kind of negative effects or consequences.

Human health is essential before everything else, and this counts a lot for farmers in rural communities in Cameroon because their agricultural practice is traditional which needs more human power. That means the functional health status of the farmers plays a significant role in the agricultural rentability. The economic status of farmers in rural areas in Cameroon is relatively weak. Agriculture is their primary source of income. They sell more of their products at the local market for economic need, and the family consumes the rest, and other parts of it are saved to be planted again in the next season. Deterioration of agricultural products will then affect the daily activities and the health status of the farmer's communities. The lack of prediction of rainfall as in the past tricks farmers' communities into planting at the onset of early rains especially during the short rainy season period, which unfortunately may last only for days succeeded by otherwise more extended periods of dryness resulting in the death of the already germinating seeds. Crop yield fails and food shortages thereby place farmers in an upset situation of having to feed the family barely with anything they can have (Gambo et al. 2010). That exposes them sometimes to some form of emotional torture and a sense of rejection, all as a result of climate generated effects on farm productivity which is also in other side related to the weather condition (Gambo et al. 2010). According to Tosam and Mbih (2015), the resulting flooding of low-lying areas is causing a severe shortage of clean drinking water as well as an increased likelihood of cholera, typhoid, hepatitis, malaria, and other waterborne illnesses in the short term.

The issue of climate change in Cameroon shows numerous publications during the last few years. However few publications emphasize the impact on health. A study conducted by Gambo et al. (2010) to assess the impact of climate change on an indigenous woman in north and southwest of Cameroon shows with high evidence that they are directly exposed to the heat wave, shortage of food and clean water, and increase likelihood of the vector-borne diseases. In their study to describe the environmental condition in schools and to evaluate the impact of heat on the schoolchildren's health during school days in Cameroon cities of Yaounde and Douala, Dapi et al. (2010) establish a significant correlation between daily indoor temperature and the percentages of schoolchildren who felt very hot and had fatigue and headaches in Yaounde. Climate change impact assessment on health in the west and the center region of Cameroon by the Cameroon UNDP Country Study in 2016 reveals that the burden of infectious diseases like cholera, yellow fever, and malaria is high in those regions. Because of the lack of personnel, the early warning system for cholera and meningitis epidemics was acknowledged as not fully operational.

Significant constraints in the health sector were identified as insufficient amounts of human, material, and financial resources (Kovats et al. 2004). Climate change impact on health in Cameroon has presented evidence in the health sector. Malaria, considered in Cameroon as the primary cause of sickness and death, is reported to be increasing with the rising temperatures (MINSANTE 2015). According to the review of the literature, few studies with regard to the assessment of health problems are induced by climate change variabilities in the western region of the country. There is growing consensus that systematic assessment of health effects of climate change in that region is needed to inform the development policies and to include health in the climate change executive agenda of the state government.

Methodology

Tourist literature describes Cameroon as "Africa in miniature" because it exhibits all major climates and vegetation of the continent: coast, desert, mountains, rainforest, and savanna. The territory is roughly triangular with a base of about 800 km and a height of about 1400 km. It covers 475,440 km^2, with a population of about 24.3 million and a growth rate of 2.58% (World Fact Book 2016). The western region of Cameroon is in the central-western portion of the Republic of Cameroon (see Fig. 1) and lies between latitudes 5° 30′ 0″ N and 10° 30′ 0″ E and longitudes 5° 5″ and 10° 5″ E (World Fact Book 2016). It covers about 14,000 km^2 of territory with 1,975,400 inhabitants (2016). It is the smallest of Cameroon's ten regions in the area but has the highest population density: 134/km^2 (347/sq mi) and receives 600–900 cm a year (World Fact Book 2016).

The mean temperature ranges from 22 °C to 29 °C (72°–84 °F) along the coast (Nations Encyclopaedia Africa Cameroon, NEAC 2016b). The region is divided into eight departments: Bamboutos, Haut-Nkam, Hauts-Plateaux, Koung-Khi, Menoua, Mifi, Ndé, and Noun (see Fig. 2).

Review of Climate in the Highlands of Western Region

A decrease of precipitation since 1950 characterizes the agroecological zone of the highlands of western Cameroon. The evolution of annual average precipitation is regressive and weakly significant (Fig. 3) with an overall decrease of 15.33% recorded in 58 years and reduction in mean rainfall of 2.58% per decade during these six last decades.

Evolution of Past Temperatures in the Highlands of the West

In the agroecological zone of the western region, the mean temperature oscillated by 19.56 °C in 1971 to 21.13 °C in 2000, with a difference of 1.57° over 40 years (PNACC 2015). Despite this small variation in temperature in that region, the overall average is 20.38°. The evolution curve shows significant peaks in 2000 and 2004, representing the warmest years (Fig. 4). Rising temperatures of 0.4 °C per decade led to an increase in temperatures of 1.96% per decade in the region.

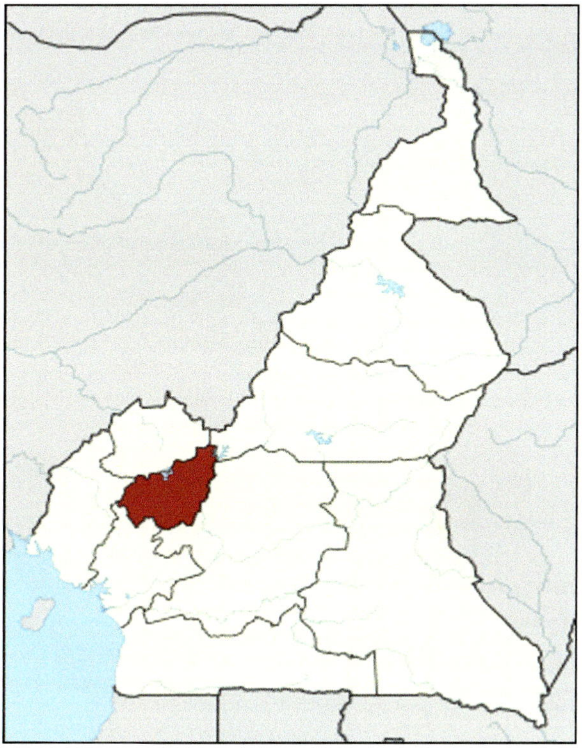

Fig. 1 Location of west region within Cameroon. (Source: Cameroon/media/ File: Provinces of Cameroon EN.svg)

The researcher conducted a study from October till December 2017. The main goal was to understand how inhabitants in the western region of Cameroon perceive climate change, as well as their level of vulnerability, and how they respond to external shocks and any existing health threat. The empirical method was the design used for the study which focused on the literature research, the collection of qualitative data with structured interview following an author observation.

The western region was chosen based on the fact that more than half of the inhabitants in the western region of Cameroon practice agriculture as the primary activity. The western region of Cameroon has a complete representation of the agricultural landscape of Cameroon, and about 60% of food consumed by Cameroonian come from the west region. About 40% of crops harvested in the west region are for exportation.

The respondents participated in the research voluntarily. They have the rights to withdraw from the study at any stage if they wish to do so.

Participant participated in the fundamentally informed consent. Researcher ensured the protection of the privacy and anonymity of participants. Researcher ensured the adequate level of confidentiality of the research data.

The researcher used the following electronic database for literature review: Google Scholar, Biomed, PubMed, Earth and Environment Science, and Web sites. The related Medical Subject Heading keywords were climate change, Cameroon, western region,

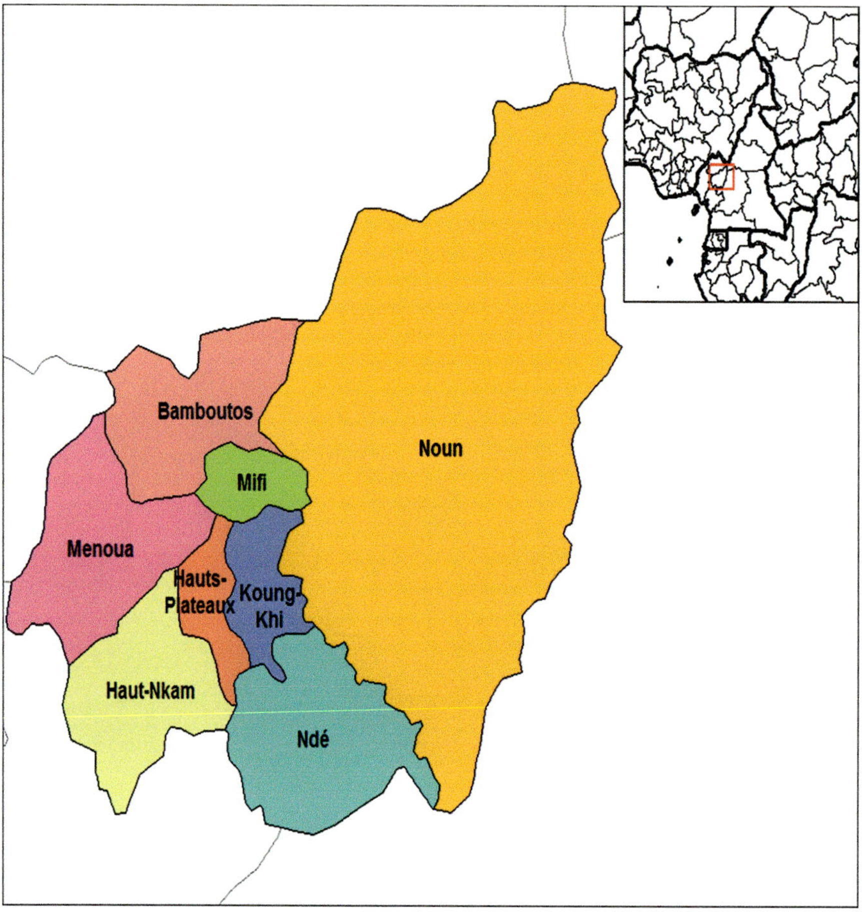

Fig. 2 Division of eight departments of the western of Cameroon. (Source: Cameroon/media/File: Regions of Cameroon EN.svg)

health, and foodborne diseases. The following combinations of keywords were applied to search articles: Cameroon and climate change and health, western region of Cameroon and climate change, and western region and climate change and health. The search article were Journals published on the related topic with no year limit. The search article was only in English. That resulted in 55 articles and abstracts. The researcher filtered articles from topics and abstracts. The researcher went also to Bafoussam in west Cameroon for literature research.

The researcher collected data on temperature and precipitation in the western region at the regional delegation of rural development. With regard to the content, the author structured the questionnaire as coherent as possible. The researcher designed the questionnaires on one page and used a block of two sections. Section 1 was to collect personal data and information about knowledge, attitude,

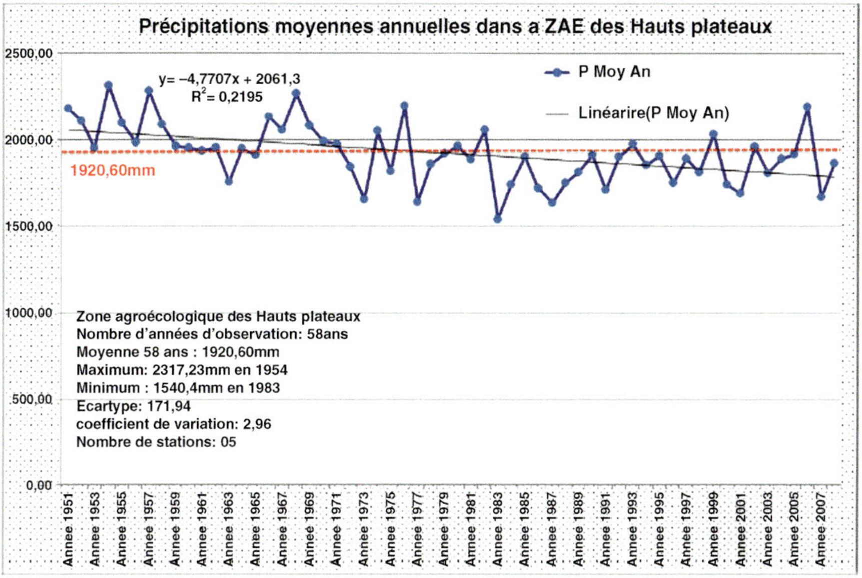

Fig. 3 Evolution of mean annual rainfall of highland of west region 1951–2008. (Source: Tchindjang 2015 in the National Climate Change Adaptation Plan (PNACC) Cameroon 2015)

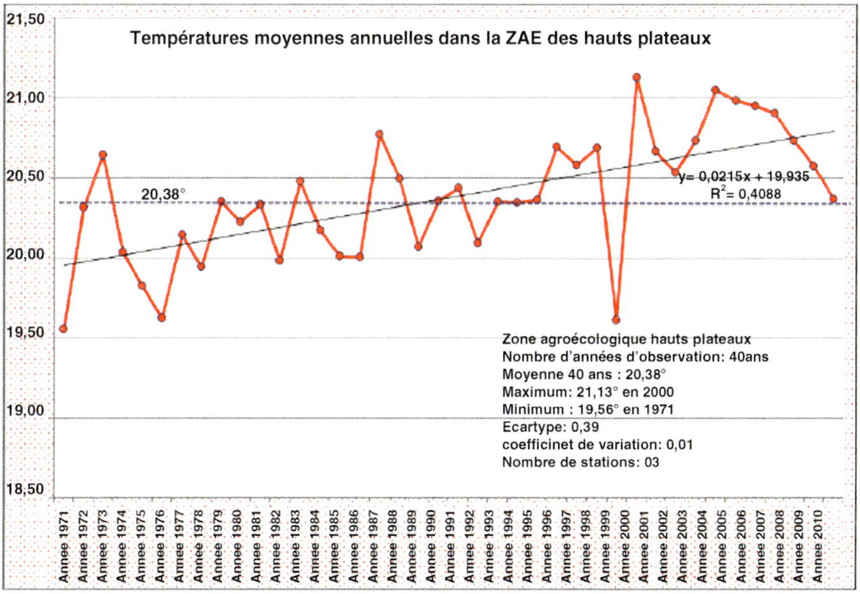

Fig. 4 Evolution of the past temperature in the highlands of the west region. (Source: Tchindjang 2015 in the National Climate Change Adaptation Plan (PNACC) Cameroon)

and the degree of awareness on the issue of climate change and its implication on health. Section 2 was to access information on opinions and views on climate change as well as current effects occurred.

Interview Procedure

The interview was very well-structured and standardized open-ended interview and involves setting precise questions asked to all interviewees. Questions were more focused and precised as they have been designed carefully according to the information needed from interviewees. This approach facilitates faster interviews which are quickly analyzed and compared.

Data Analysis Procedure

To analyze the data, the author uses SurveyMonkey software. First, the researcher creates a survey on the home page of SurveyMonkey, and then he designed a survey by using the questions on the questionnaire sheet. After designing the survey, the author then enters manually each response collected during the study. After entering all responses, the author exported the results of the survey in Excel format for interpretation and comment. Some data were presented using tables, figures, and percentages.

Results and Discussion

This section will report the findings of the research based on the methodology applied to gather information. It will merely state the findings of the research arranged in a logical sequence without bias on interpretation.

Results of the Survey

This section of the chapter presents data of all respondents of the study.

From Table 1, it shows that most respondents representing sixty-one point eight percent (61.8%) were females against thirty-eight point two percent (38.2%) of males.

That also shows that from the category of respondents, women are more active in farming, selling on the street, as well as doing household work.

Knowledge of Participants on Climate Change

This section of the analysis focuses on establishing the responses of respondents if they have heard about climate change.

Table 1 Respondents who participated in the research

Respondents	Gender		Total
	Male	Female	
Street sellers	30	35	65
Office workers	40	20	60
Households	10	50	60
Farmers	10	40	50
Total	90	145	235
% Total	38.2%	61.8%	100%

Source: This table is made based on data from the research

From the Fig. 5, it appears that all participants are aware of climate change. By taking into consideration its impact on health, this proves that a significant number of inhabitants in the western region of Cameroon are aware of climate change variabilities.

Source of Information on Climate Change

As observed in the Fig. 6, most respondents rely on the information from the radio (92.4%) and television (85.7%). It was evident that the easiest way for respondents to get informed is more on the radio than television.

The church (45.7%) also appears to be a useful tool to communicate as about half of the participants have heard about climate change in the church. The low usage of the newspaper (36.8%) could explain the lower incomes or education level of inhabitants. Street theater and drama (0%) appear not to be an essential tool for communication although it might be the best one due to the direct contact with peoples.

The Causes of Climate Change

The Fig. 7 above shows that most participants think that deforestation (98.5%) is the primary cause of climate change followed by the burning of fossil (82.2%). The greenhouse gas (77.5%) is perceived by the participants as the leading cause of climate change.

Effect of Climate Change Already Heard

From the Fig. 8 above, more heat stress and pollution (98.9%) followed by the increase of water- and vector-borne diseases (98.7%) as well as less rain (97.8%) coupled to hotter temperature (95.8%) appear to be observed the most from inhabitants as the main effects of climate change that can occur. From the Fig. 8, it summed that most inhabitants have observed climate change variabilities in the western region.

Common Diseases Observed During the Last Decade

From the Table 2 above, it appears the prevalence of diseases in the table is approximately the same. That shows that the region is impacted by many

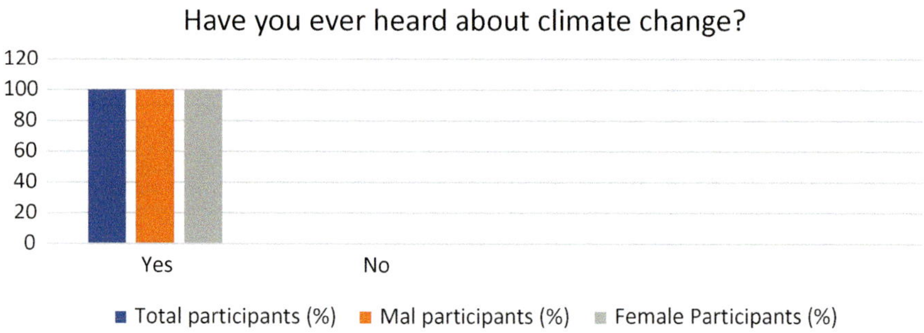

Fig. 5 Knowledge of participants on climate change. (Source: Field of work October to December 2017)

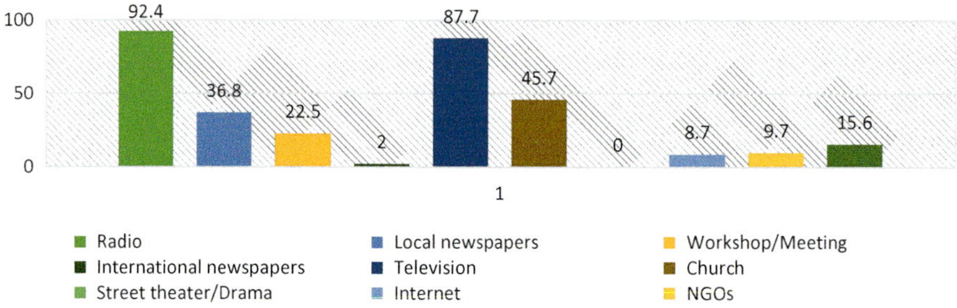

Fig. 6 Source of information on climate change. (Source: Field of work October to December 2017)

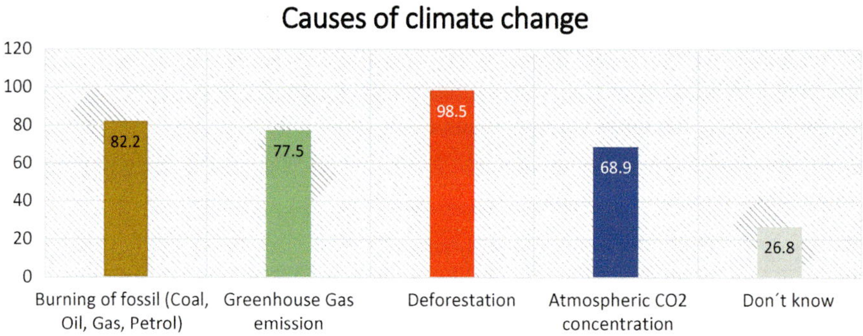

Fig. 7 Causes of climate change. (Source: Field of work October to December 2017)

health problems which are without any doubt driven by extremes of long periods of sunshine followed by long and episodic rainfall in the region (Fig. 9).

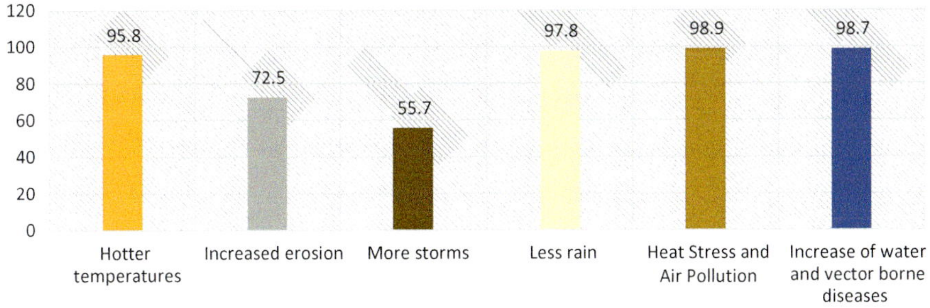

Fig. 8 Effects of climate change already observed. (Source: Field of work October to December 2017)

Table 2 Effects already observed during the last decade

Respondents	Diseases					
	Cholera	Typhoid	Malaria	Heat stress	Pulmonary diseases	Total
Street sellers	12	13	11	14	15	65
Office workers	11	13	13	12	11	60
Households	13	11	14	12	10	60
Farmers	09	10	10	10	11	50
Total	45	47	48	48	47	235
Total (%)	19.2%	20%	20.4%	20.4%	20%	100%

Source: Field of work October to December 2017

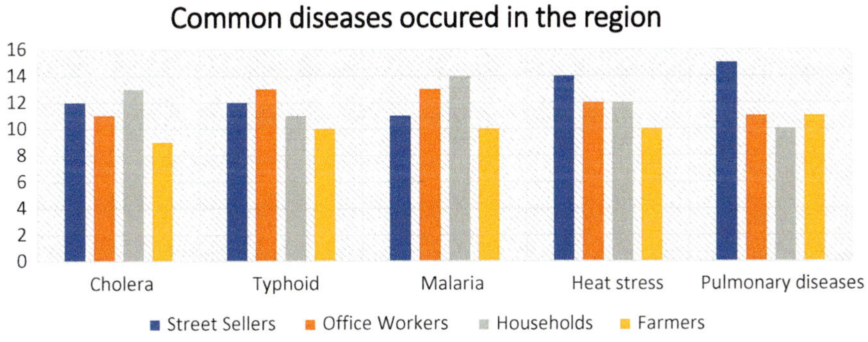

Fig. 9 Effects already observed during the last decade. (Source: Field of work October to December 2017)

From the question which common diseases have already observed in the region during the last 10 years, it appears from the table above that the observation differed from a group of respondents. Pulmonary diseases and heat stress are more observed than other diseases by street sellers compared to other participants.

Discussion

From the answers given by the inhabitants, we can realize that some extreme events like increased heat stress and air pollution coupled with an increase of vector- and waterborne diseases are mainly observed in the region. From the analysis of the results, it is seen that 80% of the population are more exposed to harsh conditions from the environment due to activities which give them their living bread. For instance, we have farmers who work out on open air and are exposed to every kind of extreme weather event. On the other side, we have likewise street sellers who spend all day on the street under terrible weather condition to sell the products. Householders and office workers are almost at the same level of exposure because the living condition at home and the level of comfort and safety in the offices do not require to convert the standard of safe environment against extreme weather condition. A common point which links the respondent is the living condition and the standard of living which is average. Among them, farmer and street seller seemed to play a significant role because the farmer works in the farm under terrible condition and then most of the products harvested are bought by a street seller who is also called "buy and sellam" which mean "buy and sell again." Householders and office workers then buy all that they need for nutrition and daily living.

During the study, many participants have reported that climate change has been observed in the region for more than a decade. This has brought shortened or long rain season, as well as episodic, shortened, extended, and prolonged dry season. Participants like street seller reported an increased incidence of pulmonary diseases which are more probably due to exposure to polluted air. This can be justified in Table 2 where street sellers reported that pulmonary diseases are occurring the most in the region followed by heat stress. Nevertheless, participants mentioned that some diseases like asthma, liver cancer, and lung infections, which in the year 1990 were not manifest in the region, are now very spread.

Conclusion and Policy Recommendations

The result of this study provides essential information on the impact of climate change on the rural community. This study also provides significant measures in collaboration with the local community to find a standard way to stay protected and safe from extreme events occurring in the region. Following this idea, the government should work together with a population of street sellers to design more appropriate zone and safe facilities where they can range and sell their products. Awareness on the indirect and direct health impact of climate change should be intensified because the effective response to extreme event impacts is based on the availability of the quality of information to people and how they easily access it. Community leaders should be mostly involved in the integration of information and new policy.

References

Dapi LN, Rocklöv J, Nguefack-Tsague G, Tetanye E, Kjellstrom K (2010) Heat impact on schoolchildren in Cameroon, Africa: potential health threat from climate change. Glob Health Action 3(1):5610. https://doi.org/10.3402/gha.v3i0.5610

Gambo A, Rabiatou Y, Bamboye G (2010) Indigenous Women and Climate Change in South and North West. Cameroon. Retrieved from http://www.tebtebba.org/index.php/all-resources/category/95-presentations?download=548:Cameroon-308indigenous-women-and-climate-change-in-south-and-north-west-came

Harrus S, Baneth G (2005) Drivers for the emergence and re-emergence of vector-borne protozoal and bacterial diseases. Int J Parasitol 35:1309–1318. https://doi.org/10.1016/j.ijpara.2005.06.005

Indigenous Women and Climate Change in South and North (2017) Retrieved from http://www.tebtebba.org/index.php/all-resources/category/95-presentations?download=548:Cameroon-indigenous-women-and-climate-change-in-south-and-north-west-came

Kovats RS, Edwards SJ, Hajat S, Armstrong BG, Ebi KL, Menne B (2004) The effect of temperature on food poisoning: a time-series analysis of salmonellosis in ten European countries. Epidemiol Infect 132(3):443–453. https://doi.org/10.1017/s0950268804001992

Leal Filho W, Azeiteiro UM, Alves F (2016) Climate change and health- improving resilience and reducing risks. Springer, Cham

MINSANTE (2015) Retrieved 17 Apr 2018 from http://www.minsante.cm/

Nations Encyclopaedia Africa Cameroon, NEAC (2016a) Ministry of Public Health in Cameroon. Retrieved 20 Apr 2018 from https://www.nationsencyclopedia.com/economies/Africa/Cameroon.html

Nations Encyclopaedia Africa Cameroon, NEAC (2016b) Retrieved 20 Apr 2018 from https://www.encyclopedia.com/places/africa/cameroon-political-geography/cameroon

Patz AJ (2000) Effects of environmental change on emerging parasitic diseases. Int J Parasitol 30:1395–1405

Plan National d'Adaptation au Changement Climatique (PNACC) (2015) Retrieved 16 Apr 2018 from http://www.gwp.org/fr/GWP-Afrique-Centrale/PARTENAIRES/Cameroun/PUBLICATIONS/WACDEP-Cameroon/

Tosam MJ, Mbih RA (2015) Climate change, health, and sustainable development in Africa. ResearchGate 17(4):787. https://doi.org/10.1007/s10668-014-9575-0

Uneke C (2008) Deforestation and malaria in sub-Saharan Africa: an overview. Internet J Trop Med 6(1). Retrieved May 2018 from http://ispub.com/IJTM/6/1/11734

United Nations Development Program UNDP Annual Report (2016) Retrieved 15 Apr 2018, from http://www.undp.org/content/undp/en/home/librarypage/corporate/undp_in_action_2008.html

Vora N (2008) Impact of anthropogenic environmental alterations on vector-borne diseases. Medscape J Medicine 10(10):238

World Factbook -Central Intelligence Agency (2016) Retrieved 01 May 2018 from https://www.cia.gov/library/publications/the-world-factbook/geos/us.html

Ecological Limits: Implications for Sustainable Human Capital Development

29

Oluwabunmi Opeyemi Adejumo

Contents

Abstract

The ecosystem is designed to depict the interaction between the natural environment and the human economy. However, an increase in socioeconomic activities has resulted in the recent upheavals of pollution, environmental degradation, and climate change. Therefore, based on the concept of *limits to growth* and following

O. O. Adejumo (✉)
Institute for Entrepreneurship and Development Studies, Obafemi Awolowo University, Ile-Ife, Nigeria
e-mail: jumobum@gmail.com; adejumobum@oauife.edu.ng

W. Leal Filho (ed.), *Handbook of Climate Change Resilience*,
https://doi.org/10.1007/978-3-319-93336-8_177

the ecological theory of growth, this study examines the extent to which the ecosystem can tolerate infiltration beyond which there will be adverse impacts for sustainability in human capital development. Using time series data spanning from 1970 to 2014 in Nigeria, and the breakpoint test for analysis, the threshold value occurs at the point where the environmental polluting rate is 6 kt of CO_2 emissions intensity (kg per kg of energy used). This breakpoint estimate depicts the region beyond which human capital development becomes endangered and probably extinct.

Keywords
Environment · Human capital · Productivity · Threshold · Growth

Introduction

The natural environment remains the ultimate determinant of human survival. It is therefore pertinent to maintain eco-balance between the environment and humans as resources are being utilized for sustenance. According to Groffman et al. (2006), there are environmental limits or threshold that must be taken into consideration in resource utilization, which if overlooked, could result in disastrous conditions for living organisms. Ecological limits or thresholds are the points at which a relative disturbance from external conditions generates a far greater effect in the environment or ecosystem. This disturbance distortion could be as a result of human activities or natural occurrence. Hussen (2000) asserted that if the limits imposed by the ecosystem are passed, the ecosystem may no longer be able to maintain its initial balance. Thus, ecological threshold is a nonlinear concept that shows the responses of ecological or biological arrangements to burdens imposed by human activities or natural processes.

Generally, human economic activities produce both useful commodities (goods and services) and non-useful materials called "waste" or "residuals." This waste usually goes through a reconfiguration in the natural system. Unfortunately, some studies posit that there are limits to the rate at which materials can be recycled through the natural system (Hussen 2000). Furthermore, given the fact that the earth is finite, there exists a theoretical upper limit for resource extraction, harvest, and disposal of waste into the natural environment. This implies that the quality of the environment and the maintenance of life support systems (such as climate regulation and genetic diversity) could be affected adversely in direct proportion to the amount of resource extractions and/or harvesting and the disposal or discharge of waste into the natural environment. Although it has been conceived that income and time-related effects should reduce environmental impacts in countries at all levels of income (Stern et al. 1996; Stern 2004), yet, given the issues of posterity, crucial attention should be attached to the quality, quantity, and rate at which wastes are emitted into the environment.

The basis for the environment-health concept is increasingly becoming prevalent in the field of development thought. Theoretical contributions to development thought emphasize the role of human capital in the process of development (Barro and Sala-i-Martin 1995; Weil 2007, 2009). More so, empirical analyses have revealed that within the field of neoclassical economic thought, the argument is that human capital is a complex asset that consist of some other attributes other than knowledge capital (Knowles and Owen 1995). Health is a form of human capital and has been prominently used in studies conducted in developed economies, (Mushkin 1962; Becker 1964). Some scholars have proved that the omission of health capital from development models has resulted in certain misspecification biases and spurious conclusions on human well-being (Schultz 1961; Mushkin 1962; Knowles and Owen 1995). Besides, several studies on sub-Saharan Africa have concentrated on the educational aspect of human development and have neglected the health component required for comprehensive analysis (Dauda 2010; Adelowokan 2012; Umoru and Omolara 2013; Umoru and Odjegba 2013). This explains why this study is situated within the environment-health context. Therefore, in addition to the environment-health association, the study identifies pollution threshold for the environment, vis-à-vis greenhouse gas emissions, beyond which there will be adverse implications for the human existence.

Literature Review

The Environmental Challenge in a Global Context

This issue of environmental quality sustainability has become a global challenge with the recent consequence of health threats. Also, the clamor on climate change and greenhouse emissions like carbon dioxide, nitrous oxide, and methane from domestic and industrial wastes have aggravated the apprehensions for posterity. The increases in greenhouse gases can be traced to increases in socioeconomic activities, increases in demand, and consumption of human beings, which implies that, as human beings keeps multiplying, there will be an explosive utilization of eco-resources. Incidentally, the ecosystem is by design self-limiting in its availability and supply of natural resources for economic activities; as a result, there is an extent to which the eco-system can cope with resource extraction, utilization, and pollution which emanates from human activities. Despite the limitations of the eco-system, human beings keep multiplying. As of 2015, the world population exceeded 7.4 billion (World Bank 2016), and for the purpose of existence and subsistence, human beings draw their supply from the ecosystem. However, in view of the limitations of the ecosystem, continuous attempts to draw resources without due considerations result in environmental degradation, thus distorting intertemporal balance. Therefore, as withdrawals are made from the ecosystem, plans for replacement, sustainability, and maintaining eco-balance becomes pertinent.

A report by the World Bank in 2005 showed that net additions to fossil fuels consumption, minerals extraction, forest cover decrease, and rising carbon dioxide in the atmosphere can challenge the sustainability of any economy. Besides, it has been posited by report that growth in some countries is not sustainable because of the

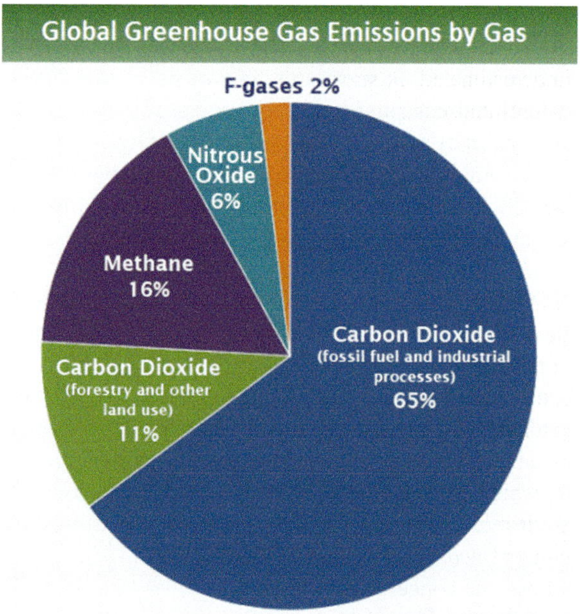

Fig. 1 Global greenhouse gas emissions by gas. (Source: IPCC (2014) based on global emissions from 2010)

depletion of stocks of natural resources and the deterioration in the quality of environmental services (Assessment Millennium Ecosystem 2005). Also, in 2014, the Contribution of Working Group III to the Fifth Assessment Report of the Intergovernmental Panel on Climate Change (IPCC) reported the consumption of fossil fuels as one of the major world sources of greenhouse gas emissions as shown in Fig. 1.

Among the greenhouse gases (GHG), carbon dioxide alone accounts for 76% of total radiative forcing, and the carbon dioxide from fossils and industrial processes alone is 65%, while other gases accounted for the remaining 24%. Although Boden et al. (2011) in a survey observed that Africa's fossil-fuel CO_2 emissions have been low in both absolute and per capita terms, the total emissions for Africa have increased 12-fold since 1950, reaching 311 million metric tons of carbon in 2008. This is however a low figure when compared to the emissions for some single nations like China, the USA, India, Russia, or Japan. Boden et al. (2011) noted that emissions from all fuel sources have grown in the African region over time with liquid and solid fuels now each accounting for approximately 35% and gas fuels accounting for 16.9% of the regional total. Also, only six countries on the continent had annual CO_2 emissions in excess of ten million metric tons of carbon. Among the nations responsible for the regional emissions via fossil fuels and cement production are South Africa, contributing 38%, and the combination of Egypt, Algeria, Nigeria, Libya, and Morocco, contributing 46%.

Meanwhile, studies have noted the importance of economic growth to almost all countries of the world (Ranis et al. 2000; Hsiao and Shen 2003; Allen et al. 2005; Weil 2009). Growth and development are borne out of vibrant economic activities, and most economic activities are somewhat contingent on the environment. However, its negative impact especially in terms of deteriorating environmental quality cannot be underestimated. A situation where overutilization, overexploitation, and pollution of natural and environmental resources begin to challenge human health, productivity, and existence calls for urgent attention. The relationship between the human economy and the natural environment is determined by the laws of entropy and material balances that set constraints on the environment (Sengupta 2010). Incidentally, these constraints are contingent on human development. These potential threats to productivity and human capital development vis-à-vis sustainability requirements have led to the resurgence of the phrase *limits to growth* (Meadows et al. 1972). Basically, the call for growth and eco-limits have become necessary given the global upsurge in economic activities which has led to negative spillovers, thus providing a platform for the ecological theory of growth as an alternative to the classical and neoclassical doctrines (Stokey 1998; Jhingan and Chandar 2008).

Thus, the priority of the ecologist is the issue of saturation, that is the degree to which the eco-system can tolerate continuous disturbances beyond which the environment will become uninhabitable for man. This has brought to the fore the concept *ecological threshold* for sustainable human capital development. In other words, the concern is pursuing an optimum between the eco-system and the human economy. This requires determining the threshold that will not be counter-productive to the environment and living organisms, which are critical inputs to the growth process (Hussen 2000).

The Environment and Human-Health Capital Development

Among the several factors that could assure sustainable development in any economy is a formidable population. This formidability can be expressed in the state and quality of human capital development. Factors that determine these features include human health, education, life expectancy or longevity, and status. Studies have shown that one way of assessing environmental performance is through its effects on human health. Human health has been through effects such as respiratory disorders, waterborne diseases, poisoning from carcinogenic and substances. Thus, since socioeconomic activities remain an ongoing process, the plausibility of increasing and even newer effects on human health through ailments and diseases is inevitable. Already studies have revealed the effects of these degradations on other living organisms via their extinction processes (Mooney and Cleland 2001; Ferrer-i-Carbonell and Gowdy 2007; Drake and Griffen 2010; Archana and Pratibha 2016); the question therefore remains is to what extent should the environment be subverted beyond which man becomes endangered or might equally begin to go into extinction.

According to Hancock (1993), three ecological models link human health, environment, economy social, environmental, and economic dimensions of a healthy and sustainable community. Firstly is the *Mandala of Health*. This is a model of the human ecosystem that presents the determinants of health as a set of nested influences. These influences ranges from the biological, personal characteristics to the ecological factors, as well as the social and political. Secondly is *The Health-Environment-Economy Model* that indicates crucial links between health and environmental and economic well-being. The health-environment economy model hinges on two principles of public-health which are equity and sustainability. The third model juxtaposes the other two models and applies it at a communal level. It is believed that an extension of the individual characteristics, equity, and sustainability at a community or national level will bring about health for all in terms of viability, conviviality, and livability.

As investments in human capital development continues, there are some mitigating factors that reduce or crowd out the outcomes expected from investment in human development. Specifically, if environmental degradation reduces human capital investments, socioeconomic development is slower given the visible and invisible impacts via human health. For instance, the connection between environmental pollution and human capital can be established through an impact on labor productivity (Carson et al. 2010; Hanna and Oliva 2011; Zivin and Neidell 2012). It has been noted that pollution have gradual effects, and on the long run, these effects may transform into large welfare losses when observed at an aggregated level. These aggregation can be evident through welfare impacts like illness cost, labor productivity, hospital attendance, death rate, life expectancy, and length of stay in the hospital. Therefore, the overriding economic policy challenge for the developing economies is how to promote the positive effects of socioeconomic development while limiting its negative impact on human-health capital. This has necessitated a multidimensional approach to environmental-health issues. It involves disciplines such as public health, economics, geographers, ecology, and a host other discipline to critically assessing the eco-threshold for health capital sustainability, so that the growth which creates the wealth of nations will not become the creator of the poverty and death of nations.

Furthermore, environmental degradations and unhealthy conditions hinder the transformation of human skills to human capital (Weil 2007). Also, the poor environmental quality and the environmental deterioration can frustrate human capital investments, thereby challenging the plausibility for sustainable development. This is because environmental degradation through pollution reduce the efficiency of human capital through direct and indirect impacts on health which may be acute or chronic (Viscusi et al. 1991; Steinnes 1992; Alberini and Krupnick 1998; Newell 1998; Patz et al. 2005; Epstein 2005; Furgal and Seguin 2006; McMichael et al. 2007; Costello et al. 2009). Direct acute effects affect today's productivity; direct negative effects are associated with exposure to hazard and illness such as respiratory ailments, neurological diseases, reproductive challenges, and cancerous diseases. Substances such as lead, nitrogen oxides (NOx), sulfur dioxide (SO_2), and other chemicals are examples of the direct effects of environmental degradation on current human capital investments. Indirect long-term effects are generally associated with nutrition quality, pesticide use, and water and air quality (Ozkaynak and Thurston 1987; Brucato et al. 1990; Viscusi et al. 1991; Steinnes 1992; Alberini and Krupnick 1998; Newell 1998; Egan et al. 2009), and they are again related with health problems. Hence, a reiteration of policy pursuit is for societies to invest in the protection of humans to achieve an optimal growth rate.

Empirical Review

Public health scientists in recent times have targeted policy decisions that will address issues on environmental degradation and reduce the risk of illness and mortality. Among major pollutants, the concern for degradation is typical of air and water pollution is highest. This is because they account for almost three million premature deaths globally (WHO 2009).

Yang et al. (2010) proposed that more investment in human capital is the key to solving the dilemma between the fast growing economies and worsening environmental quality. Jost et al. (2006) in an analysis of a society's economic decisions on birth rates, investment into human and physical capital, and polluting emissions employed an optimal control model of the coupled demographic-economic-environmental system. Within this model, it was seen that a society's long-run steady state growth will be optimal if it is characterized by a stable pollutant stocks. In addition, its population and economic growth rates must be stable. Although, this will depend on the possibilities of emission abatement and technical progress that is driven through human capital. Besides, they opined that whether population grows, declines, or is constant in the long-run optimum, depends on the quality of the growing population, production technology, and in particular on clean environment. Jost et al. (2006) added that long-run per consumption growth is only possible by means of continued accumulation of human capital, which is a "clean" substitute for polluting inputs.

In all, in an application of theoretical prepositions to real-life situations, it is summarized that if sustainability is to be assured, output elasticity of human capital must exceed the output elasticity of emissions in order to have long-run growth in per capita consumption. This implies that the rate of human capital development per capita should exceed the rate at which environmental resources are degraded.

An Overview of the Nigerian Environmental Peculiarity

Interestingly, the production of fossil fuels remains one of the mainstay of the national income in Nigeria. Also, the exploration and exportation of crude oil account for 90% of foreign income in Nigeria (Central Bank of Nigeria 2017). Globally, the challenge from fossils is causing a gradual-radical shift to alternative energy sources like solar power, wind power, and biofuels. Thus, the challenge for economic sustainability then begins to arise if the mainstay of the economy is gradually being challenged globally. Prior to before the 1970s, when oil was discovered in large quantity in Nigeria, agriculture used to be the mainstay of the Nigerian economy but had grossly been neglected over the euphoria of crude oil wealth continuously explored by multinational corporations (MNCs). Apart from the dangers attached to oil export such as gluts in the world market, attendant environmental problems are associated with exploration such as gas flaring; as well as the problem connected with oil spillage both on-shore and off-shore.

The rate of oil spill has been on the increase in Nigeria ranging from a spill of 150 barrels in 1970–1581 barrels in 1981, over 10,000 barrels in 1990, and 50,000 barrels in the mid-2000s, (NEST 1991; CBN report, 2009). The immediate consequences of these spillages have been pollution of water bodies, abandonment of agricultural practices, emigration of wildlife, population, resettlement via forced migration, agitations, to mention a few. These negative spillovers have been felt at both the micro- and macrolevels in the form of increased disease burdens and impairment of human health. This is evident in the Environmental Performance Index (EPI) of 2012 which ranked Nigeria 119th out of 132 countries leaving Nigeria only better than Eritrea and South Africa in sub-Saharan region of Africa. Also, the International Human Development Index (HDI) recorded carbon dioxide emission per capita as 0.6 (tonnes); population living on degraded land as 11.5% and change in forest area 1990/2010 as −47.5% (WDI 2014).

A major source of air pollution in Nigeria is the emission of carbonmonoxide (CO) and photochemical oxidants into the atmosphere by motor vehicles, many of which are outdated and exceeded their useful years. In addition to emissions, poor waste disposal in Nigeria has accounted for increases in air pollution in Nigeria, and the spillover of these pollutants usually spread from city centers to suburbs. Emissions in the form of carbon monoxide have its greatest effect on population while oxidants affects people in suburbs more (Anyanwu et al 2003). Also, carbon dioxide (CO_2) emissions are equally potential sources of pollutants that can challenge the existence of human; although carbon dioxide is a nontoxic and nonflammable gas. Meanwhile, undue exposures can engender hazards to life (Keller et al. 2005). CO_2 are basically from the burning of fossil fuels and the manufacture of cement. They include carbon dioxide produced during consumption of solid, liquid, and gas fuels and gas flaring.

Figures 2 and 3 depict a cursory examination of the per capita CO_2 emissions (metric tons per capita) and CO_2 emission intensity of the energy sector (kg per kg of oil used) in Nigeria. The graph showed that CO_2 emissions metric tons per capita in the mid-1970s rose to 1 kt and later fell to 0.9 kt and 0.8 kt in the mid-1980s and mid-1990s, respectively. The reasons for the level of emissions are not far fetched as it was the period of the discovery of oil in commercial quantity in Nigeria. And ever since, Nigeria has been a major exporter of oil products with associated gas flaring practices (see Adejumo and Adejumo 2018). In addition to the period of the oil boom in 1970, this period was when small and medium industries sprang up. These industries have caused pollution from the production of items like steel, tanning, petroleum refining, textiles, pharmaceuticals, and gas flaring (Babayemi et al. 2016). From Fig. 3, carbon emissions as a result of energy used in Nigeria reveals a similar trend with the CO_2 metric ton per capita. In 1970, the emission rate stood at 0.8 kt, which increased to 1.3 kt in the year 1980, but a declining trend was observed in the 1990s to 0.5 kt, which built up again in the early 2000s to an emission rate of about 0.9 kt, and fell again to 0.6 kt in 2013. Although, since the year 2000, there has been a decline in CO_2 emissions metric ton per capita to an average of 0.7 kt; however in Nigeria, major constituents of in

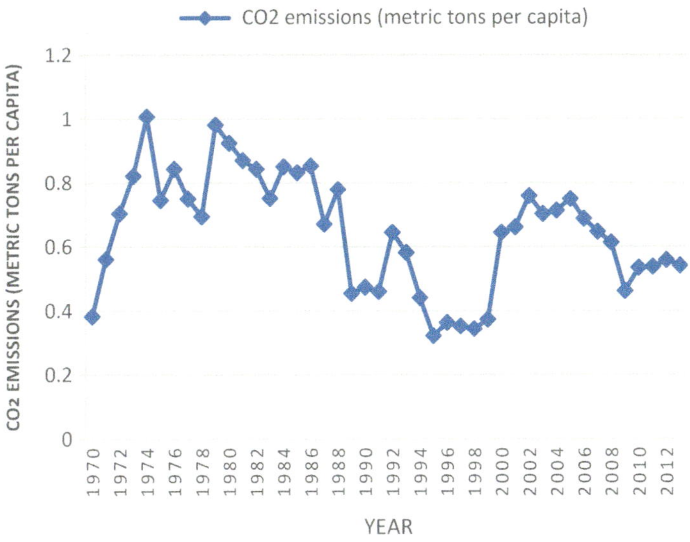

Fig. 2 CO_2(kt) emission (metric ton per capita) in Nigeria (1970–2013). (Source: WDI 2014)

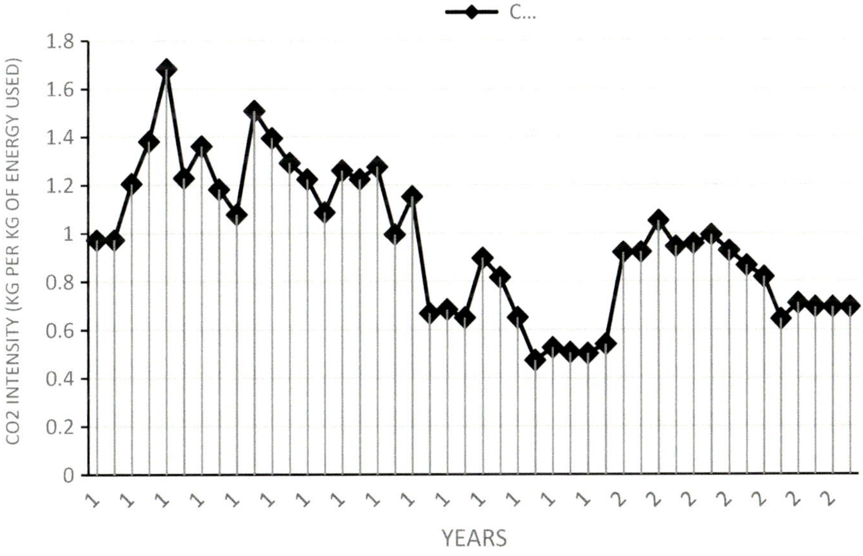

Fig. 3 CO_2(kt) emissions intensity (kg per kg of energy used) in Nigeria (1970–2013). (Source: WDI 2014)

carbon dioxide levels has been basically associated with the burning of fossil fuels. And its threat to environmental quality has been noted by WHO as causing respiratory irritation, impaired pulmonary function, increased susceptibility to diseases, etc. (WHO 1995). As part of the response to global efforts in ensuring clean environments, there has been significant reduction in gas flaring, as these gases are now annexed for cooking and industrial fuels (Oyeshola 2008); nonetheless, the micro-effects of pollution from these emissions are still prevalent in Nigeria.

Atubi et al. (2015) investigated the effects of environmental degradation on human health in nine selected oil communities in Delta State, Nigeria, using structured questionnaires. Specific oil spillage and gas flaring data within the selected communities were also used. Result showed that 95.2% of the people had experienced environmental degradation from the oil-producing communities, and some agreed that most of the effects are still prevalent. The major causes of environmental degradation in the oil-producing communities includes negligence of duty by government agencies charged with oversight duties of monitoring and compliance, neglect of environmental regulations/compliance, corruption, lack of environmental impact assessment (EIA) reports from companies at the beginning of the project and nonsubmission of periodic assessments. These gap in environmental management has however given rise to high degree of sabotage that causes oil spillage in the area. Atubi et al. (2015) therefore stated the necessity for intervention as more than 77.5% of the residents in oil-producing communities were vulnerable to environmental degradation. In addition, high figure of diseases were linked to environmental degradation of air, water, and land, which is rampant in the area. For instance, from Atubi et al. (2015)'s study, a total of 235 diarrhea cases were recorded, 187 asthma cases, 511 cases of eye infection, 90 cases of bronchitis, and 157 cases of skin infection were reported at the hospitals in the area. This therefore places a burden and action on the Federal Government through the Federal Ministry of Environmental Protection Agency (FEPA), Niger Delta Affairs Ministry, and all Policy stakeholders in environment and oil and gas sector. Atubi et al. (2015) called for a revisit and review existing environmental and oil drilling laws in Nigeria may be necessary. This is to ensure that the extant laws keep pace with international and environmental friendly standards.

Beyond this, the pride of Nigeria as the most populous country in Africa and the ninth most populous country in the world is a potential development armament. However, the challenge for inclusion in the league of developed economies has continuously raised concerns among policy and decision-makers in Nigeria. With an estimated population of 160 million, one in every five Africans is a Nigerian; moreover, the country has been undergoing explosive population growth and has one of the highest growth and fertility rates in the world. By UN estimates, Nigeria will be one of the countries responsible for most of the world's total population increase by 2050, (UNICEF 2007). Thus, the sustainability of human capital development in the light of the growing population is fundamental to ensuring economic and sustainable development remains critical for analysis.

Methodology

Theoretical Concept: The Assimilation Model

The assimilation model gives an idea of the threshold of pollution into the environment. Despite the insatiable desire for a clean environment by humans, as long as economic activities remain an ongoing process, the creation of residuals cannot be overruled (Hussen 2000). This residual of economic activities which is commonly referred to as *residuals* can be said to be a by-product of economic activities. These residuals could be in the form of either air waste, water waste, or land waste.

Taking a cue from the first law of thermodynamics, there is no gainsaying that residuals must go somewhere – which is usually the natural environment. Residuals should not necessarily produce any harm to the environment or human health if it is done in moderation; moreover, it has been identified that natural environment has a decomposer population which when given adequate time will transform the waste into harmless materials and could return same as nutrients into the atmosphere. Thus, this self-degrading ability of the natural environment is commonly called *assimilative capacity* (Hussen 2000). As a matter of fact, environmental management scientists are of the opinion that the quality of a particular environmental medium is determined by the extent of its capacity to assimilate waste. Anything beyond the capacity of the environment will indicate exceeding the threshold of the environment which could result in disasters.

According to Pearce (1978), there are some salient issues that must be noted about the environment which include that:

(i) The assimilative capacity of the environment is limited, (and should not be treated as a bottomless sink) thus making it a scarce resource.
(ii) The assimilative capacity of the natural environment depends on the flexibility of the ecosystem and the nature of the waste deposited in it; thus, it will not degrade all waste with equal efficiency. This implies that the way the environment will treat degradable pollutants, such as paper, sewage, etc., is different from the way it will treat persistent or stock pollutant such as plastics, glass, chemicals, radioactive substances, etc. As a result of the difference in time factor in degradation process, some of these substances may take a longer time to render stock pollutants harmless to human health and human capital development.
(iii) The rate at which waste is discharged into the environment affects the ability of the environment to degrade residuals. This goes to explain that continuous pollution can lead to cumulative ecological effect; specifically, by reducing the ability of the environment to withstand further pollution, thus leading to negative multiplier effects on human life and productive activities.

From the foregoing, it can be said that apart from the quality of residuals dissipated into the environment, the quantity and the rate of residuals must be carefully considered for the sake of sustainability and posterity.

Model Formulation

In order to obtain the threshold points and effects in the relationship between environmental quality and human health in Nigeria, carbon intensity (*COI*) was used for environmental quality and two variables were used as proxies for human health (*HH*), life expectancy (*LHH*), and death rate (*DTR*). Carbon intensity is opted for because following the analysis of Fig. 1, carbons appear to be the major GHG in the environment. Nonlinear relationships are assumed among the two pairs of variables. This assumption follows the work of Hansen (1999). According to Hansen 1999), threshold variables are used in their logarithmic form. Therefore, this model takes the natural logarithm of the variables and are implicitly specified as follows:

$$lnHH_t = \mu_1\beta_1 ln(G_t) * I(COI_t \leq \gamma) + \beta_2 \ln(G_t) * I(COI_t > \gamma) + \in_t \quad (1)$$

$$\text{Alternatively, } lnHH_t = \begin{cases} \mu\beta_1 \ln(G_t) + \epsilon_t & COI_t \leq \gamma \\ \mu\beta_2 \ln(G_t) + \epsilon_t & COI_t > \gamma \end{cases} \quad (2)$$

HH_t and COI_t are the threshold variables for the country under study in time period t. G_t are the vectors of other covariates, μ represents the level of the country fixed effect, and ϵ_t is the level of time t's fixed effect.

The Equation (2) indicate clearly that the observations are divided into two regimes depending on whether the threshold variable (HH_t) is smaller or larger than the threshold point (γ). The regimes or periods of carbon intensity are distinguished by differing regression slopes: that is β_1 and β_2 for human-health. According to Hansen (1999), for the identification of the parameters β_1 and β_2, it is required that we assume that the element of G_t and the threshold variable are not time invariant and that ϵ_t are assumed to be independently and identically distributed (iid) with mean zero and finite variance σ^2.

Estimation Technique

The study employed quantitative method of analysis on a time series variable. The period under study is between the years 1970–2014.

The study in determining the ecolimits used threshold points and effects as specified by Eqs. (1) and (2). The equations are in the form of a nondynamic fixed effects, single-country model. Given the assumption of the absence of any unmeasured time-invariant properties (Greene 2005), the bootstrap procedure proposed by Hansen (1999) was carried out to test and obtain threshold (breakpoint) point estimates. Specifically, Chow test estimates were used to generate breakpoints, which further lead to the development of least squares estimates results. The regression results, as suggested by Hansen (1999), gave a clue as to the threshold effects between the variables of interest.

Using the statistical package STATA 12.0 for estimation purposes, it is required that all variables employed in the threshold estimates be in logarithmic form. However, data on variables used are have been left in their original forms as they are measured in percentages and rates. This is so because, irrespective of the unit of measurement, transforming a data into its logarithmic form turns it into elasticities which are explained in percentages.

Table 1 Presentation of variable measurement and sources

Variables	Sources of variable	Measurement of variables
Environmental quality	World Development Indicator, 2015	Carbon dioxide (COI)[a]
Human health	World Development Indicator, 2015	Life expectancy Death rate
Control variable	World Development Indicator, 2015	Fossil fuel (FFL)

[a]CO_2 intensity (kg per kg of energy used) is employed to measure environmental quality, because D. Meadows et al. (1972) noted that while some other pollutants may be related to growth and advances in technology, pollution from carbondioxide and carbon monoxide from fossil fuel powered plants that usually cause irreversible damages in the Earth's climate

Variable Sources and Measurement

The table below summarizes the sources and the measurements of variable (Table 1).

Result

Descriptive Statistics of Variables

Prior the presentation of the result, some preliminary checks on the descriptive nature of the variables are carried out. The description of common sample statistics of the variables employed within the study. The descriptive statistics of data series gives information about the mean, median, minimum value, maximum value, and the distribution of the sample measured by skewness, kurtosis, and the Jarque-Bera statistic. From Table 2, most variables show that all the series displayed a high level of consistency as their mean and median values are within the maximum and minimum values of the series. Besides, the deviations of most data in the series are not really different from their mean value.

The skewness and kurtosis statistics provide useful information about the symmetry of the probability distribution of various data series as well as the thickness of the tails of these distributions, respectively. The data shows that the distribution of the data is normal, given the skewness and kurtosis values. With regard to the skewness of the variable, the rule of thumb may be arbitrary, but the general threshold is one. Since all the variables lie within 1.0 and -1.0, the skewness is not substantial, and it can be said that all the distribution of the variables are symmetrical. Also a Gaussian distribution is expected to have kurtosis of 3.0 (Wooldridge 2013); thus, since all the variables lie within the range of 2 and 3, the distribution is said to be normal.

Correlation Matrix Analysis

The degree of association among included variables was verified using Pearson Correlation technique. The correlation coefficients for the variables are as shown in Table 3.

Table 2 Descriptive statistics of the major variables (what units?)

STATISTICS	COI	LHH	DTR	FFL[a]
Mean	0.959903	46.54121	18.03398	17.62089
Median	0.951633	46.21376	18.3405	18.89029
Maximum	1.682166	52.44215	22.607	22.84479
Minimum	0.47382	41.18102	13.154	5.5543
Std. Dev.	0.299285	2.682336	2.354448	4.503806
Skewness	0.242553	0.345184	−0.24316	−1.45342
Kurtosis	2.308075	3.053556	2.752436	4.157835
Jarque-Bera	1.309163*	0.879041≠	0.54596≠	17.94885**
Probability	0.51966	0.644345	0.761108	0.000127

Note: Critical values of χ^2 at 5% and 1% levels are 5.99 and 9.21, respectively
* (**) denotes the acceptance of the null hypothesis that the variables are normally distributed at 5% and (1%) significant level.
[a]FFL represents fossil fuels. Since the utilization of fossils is a major contributor of CO_2 in the atmosphere, it is introduced as a control variable within the regression estimates for determining COI threshold

Table 3 Correlation matrix

	COI	LHH	DTR	FFL
COI	1	−0.48	0.52954	−0.3122
LHH	−0.48	1	−0.9954	0.59514
DTR	0.52954	−0.9954	1	−0.5996
FFL	−0.3122	0.59514	−0.5996	1

The results showed that the rate of growth of emissions from CO_2 intensity (kg per kg of energy used) (COI) is negatively and weakly correlated with life expectancy (LHH) (0.48) but positively and strongly correlated with death rate (0.5294). Meanwhile, the association between FFL with COI also appeared also negative and weak (0.3122).

Threshold Estimates of the Relationship Between the Environment and Human Health (Life Expectancy) in Nigeria

The task of identifying a precise level of environmental quality (via pollutions/degradation) beyond which human health-capital will be affected required estimating the threshold of pollutants, beyond which it begins to correlate with the stock of health of human beings in Nigeria. The optimal threshold is the level of pollutants that shape environmental quality or a point at which there is a breakpoint.

This breakpoint estimate is used to assess the threshold. Specifically, The Chow estimates show that threshold occurs at the point where the environmental polluting variable is 6 kt as shown in Fig. 4. This implies that when CO_2 emissions intensity (kg per kg of oil used) exceed 6 kt, because the breakpoint exists at this level of pollution, there will be ecological crisis (Table 4).

Based on the proxy life expectancy, regressions for the values of the growth of the health capital both before and after the optimal level are shown in Table 3. In the regression before the breakpoint level, it is clear that a percentage change in either carbon emissions and fossil fuels had an insignificant positive effect on human health, such that health increased by 8.3% ($t = 1.77$; $p = 0.34$). Thus, an increase in pollutions through carbon dioxide emissions (COI) and fossil fuels (FFL) will not affect life expectancy/longevity. But beyond the breakpoint, FFL still maintained an insignificant positive effect on life expectancy, while COI is seen to have a significant negative effect on human health, such that a percentage increase in COI causes

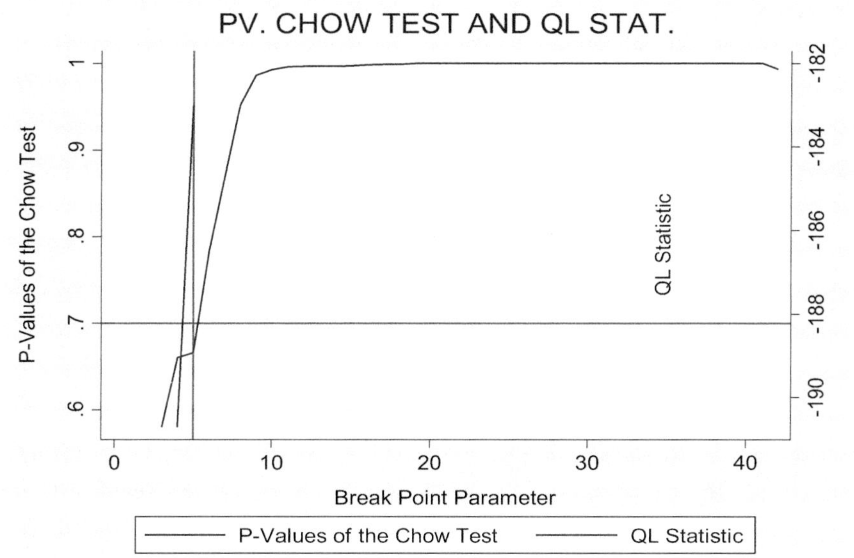

Fig. 4 Breakpoint (optimal point) between environmental quality and health capital (life expectancy) in Nigeria. (Source: Author's computation using Stata 12.0)

Table 4 Estimates of the breakpoint (optimal Level) between environmental quality and human health capital (life expectancy) in Nigeria. (Dependent variable: human health capital – LHH)

Estimates before the breakpoint				
Variable	Coefficient	Standard error	t-statistic	P-value
COI	0.8357479	0.68546	1.22	0.347
FFL	0.2867433	0.1623079	1.77	0.219
Constant	38.94316	0.5066084	76.87	0
Estimates after the breakpoint				
COI	−3.225323	1.115653	−2.89	0.006
FFL	0.1223568	0.1061534	1.15	0.257
Constant	47.77662	2.682374	17.81	0

Source: Author's computation using Stata 12.0

health capital to reduce significantly by 3.2% (t = 2.89; $p = 0.07$). The implication of this result is that fossil deposits in their unused state have little or no effect on human health. However, when fossils are being used up, they can become extremely dangerous to human existence, especially if the rate of utilization remains unchecked.

Threshold Estimates of the Relationship Between the Environment and Human Health (Death Rate) in Nigeria

In a similar vein, a check was performed on the precise level of environmental pollutions/degradation (using CO_2 intensity from energy used) beyond which death rate will be accelerated in Nigeria. Basically, the threshold level of CO_2 intensity was estimated, beyond which it begins to correlate with human-health capital in Nigeria. Based on the threshold estimate of pollutants that shape environmental quality, it was discovered that the threshold occurred at the point where the environmental pollution rate is past 15 kt as shown in Fig. 5. That is, when CO_2 emissions intensity (kg per kg of oil used) exceed 15, death rate will be accelerated in Nigeria.

Regressions for the breakpoint depicts threshold level before and after emissions from pollutants that could accelerate death rate were estimated in Table 5. In the regression, it was discovered that an increase emissions from FFL and COI increases

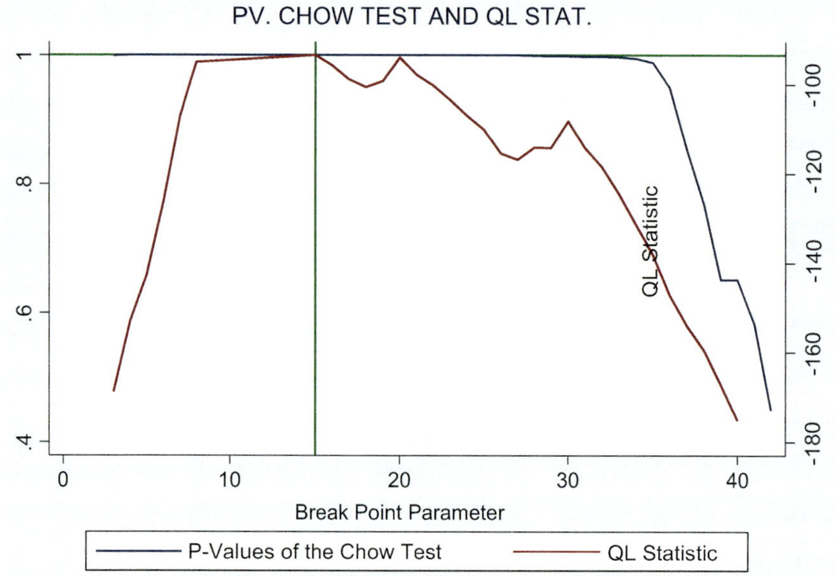

Fig. 5 Breakpoint (optimal point) between environmental quality and health capital (Death rate) in Nigeria. (Source: Author's computation using Stata 12.0)

Table 5 Estimates of the breakpoint (optimal level) between environmental quality and human health capital (death rate-DTR) in Nigeria

Estimates before the breakpoint				
Variable	Coefficient	Standard error	t-statistic	P-value
COI	4.92176	5.643187	0.87	0.4
FFL	0.0265313	0.2691817	0.1	0.923
Constant	10.94845	2.462543	4.45	0.001
Estimates after the breakpoint				
COI	0.9399959	0.277252	3.39	0.002
FFL	−0.2365108	0.0173018	−13.67	0.000
Constant	22.3526	0.5049863	44.26	0.000

Source: Author's computation using Stata 12.0

death rate insignificantly. For instance, a percentage increase in COI caused death rate to increase insignificantly by 4% ($t = 0.8; p = 0.4$) but beyond the breakpoint, an increase in emissions from carbon increases death rate significantly, such that a percentage increase in carbon emissions increases death rate by about 9.3% ($t = 3.3$; $p = 0.002$).

Although an increasingly negative effect is observed from FFL as a pollutant, it only goes to confirm that COI appears to be more harmful to human health in Nigeria. Hence, policies that will reduce carbon emissions should be strengthened in Nigeria, while the policies checking fossil emissions should be upheld and implemented if the principles of sustainability will be imbibed.

Discussion of Results

In determining the ecological limits for human health, it is seen that when the rate of CO_2 intensity from energy used in the atmosphere exceeds 6 kt, there will be adverse reactions on human health; this could be through increased incidence of diseases which could cause a reduction in life span or constitute a significant factor that will contribute to the gross reduction in life expectancy. Similarly, it was discovered that if carbon emissions remain unchecked, and if it rises beyond 15 kt, death rate in Nigeria will be accelerated.

Given the ecological threshold, the question then may be: *What should be the target for pollution or environmental degradation*? In the late 2000s, a discussion on "is zero pollution feasible" was discussed and it was agreed by economists and ecologists alike that zero pollution is neither viable nor desirable – especially given the first law of thermodynamics which states that "matter can neither be created nor destroyed, it can only be transformed." This implies that all processes within the ecosystem work together continuously to maintain a balance for survival (Hussen 2000). Therefore, the concerns for environmental pollution arise when wastes are emitted into the environment beyond its assimilative capacity. Thus it can be affirmed that the socially desirable level of waste discharge or pollutants is that which does not exceed the assimilative capacity of the environment (Soytas and Sari 2009). Also, the economists expanded that the optimal (socially desirable) level of pollution is attained when the marginal social cost (MSC) of waste reduction – in terms of extra output and services sacrificed – is equal to the marginal social benefits (MSB) of waste reduction.

In any ecological succession, there are always natural changes in the species that occur over a period of time, following a pattern from the *pioneer* stage (which is characterized by uncomplicated interrelationships) to *climax* stage (which is characterized by a stable ecosystem that supports interrelations among different organisms) and further to a stage of *diversity* (a stage of eco-resilience to physical environment). With regard to eco-limits, the stability (ability of ecosystem to maintain balance after distortions) and resilience (rate of achieving eco-balance after distortions) of the eco-system cannot be compromised (Ludwig et al. 1997). This is because given the finite nature of the assimilative capacity of the environment and the varying ability of the environment to degradables and stock pollutants, the rate at which pollutants is discharged into the atmosphere must be a topmost policy issue, in order to mitigate cumulative ecological effects.

Moreover, policies that do not check degradations inhibit the ability of an environment to withstand further pollution (Pearce 1978). Therefore, as eco-policies are designed, factors that cause imbalances must be understood, identified, and put to check if balance is to be maintained for healthy living of organisms and humans. Beyond these, an important issue with regard to eco-succession is how it maintains balance or equilibrium. All things being equal, equilibrium is maintained when all elements and processes within the ecosystem are maintained in the long run. For instance, in the atmosphere, it is expected that oxygen is 20%, nitrogen is 79%, argon is 0.9%, and the volume of carbon dioxide is 0.03% (Srivastava 2009); depending on the ecosystem of different regions, distortions from these averages depict the presence of disequilibrium and could be dangerous for human existence.

Thus, energy consumption activities that increases the overall intensity of CO_2 in the atmosphere in Nigeria beyond 6 kt would cause health impairments, threaten human existence, extinction of species, accelerate death rate, and jeopardize human capital development. This is consistent with some of the studies conducted in the USA, where increases in greenhouse emissions, SO_2 and NOx, causes increases in respiratory diseases and pulmonary diseases (Newell 1998; Zivin and Neidell 2012). Despite the fact that waste creation is a part of the ecosystem, the environment by nature is expected to be self-repairing, self-maintaining, and self-regulating, but human biological and industrial wastes could create pollution and distortions in the ecosystem. For instance, increases in population as witnessed in Nigeria could cause an alarming turnout of waste beyond the assimilative capacity of the environment which could lead to a collapse of or an irreversible change of the Nigerian ecosystem. This challenge of irreversibility is evident through improper waste disposal which are common sights of sewage, effluents, and refuse in Nigeria. The production of wastes such as CO_2 emissions, as employed in the model to measure environmental quality, are also increasing steadily in Nigeria. Carbon emissions in

Nigeria are seen to emanate majorly from vehicular and industrial activities especially in urban areas. Continuous and large quantities of carbons however less toxic cannot be handled effectively by a normal ecosystem; thus an overbearing accumulation of such emissions could result in environmental stress that reduces productivity and diversity of natural ecosystem, as well as global warming and climate change.

Conclusion

In all, following the principles of eco-limits, environmental-health threats that emanates from carbon emissions should be prevented, controlled, reduced, and treated where possible. This could be by developing and strengthening capacity to address major diseases like waterborne diseases, chronic respiratory ailments, and cardiovascular diseases. This is because environmental quality, to a great extent, can determine human existence and longevity.

Cross-References

- Climate Change and Population Growth in Pastoral Communities of Ngorongoro District, Tanzania
- Climate Change Implication on Airborne Infections: Roadmap for Nigeria's Health Sector Development
- Drivers of Deforestation and Land-use Change in Southwest Nigeria
- Early Warning System and Ecosystem-Based Adaptation to Prevent Flooding in Ibadan Metropolis, Nigeria
- Impacts of Climate Change on Health: Evidences from Multi-stakeholders in the Western Region of Cameroon

References

Adejumo AV, Adejumo OO (2018) Erratum to: towards a sustainable bioeconomy: principles, challenges and perspectives. In: Leal Filho W, Pociovălişteanu DM, de Brito PRB, de Lima IB (eds) Towards a sustainable bioeconomy: principles, challenges and perspectives. Springer, Cham, pp E1–E1

Adelowokan OA (2012) Growth effect of education and health expenditure in Nigeria (1970–2010). Afr J Sci Res 10(1). ISSN 2220-9433

Alberini A, Krupnick A (1998) Air quality and episodes of acute respiratory illness in Taiwan cities: evidence from survey data. J Urban Econ 44(1):68–92

Allen F, Qian J, Qian M (2005) Law, finance, and economic growth in China. J Financ Econ 77(1):57–116

Anyanwu K, Sheth A, Cardoso J, Miller J, Kochut K (2003) Healthcare enterprise process development and integration. J Res Pract Inf Tech 35(2):83

Archana DC, Tiwari P (2016) Impact of environmental degradation on human health. Int Res J Manag, IT Soc Sci, [S.l.], 3(1):1–8. ISSN 2395-7492

Assessment Millennium Ecosystem (2005) Ecosystems and human well-being (Vol. 5). Island press, Washington, DC
Atubi AO (2015) Effects of oil spillage on human health in producing communities of Delta state, Nigeria. Eur J Bus Soc Sci 4(8):14–30
Atubi AO, Ogbija TE, Ojeh VN (2015) Effects of environmental degradation on human health in selected oil communities in Delta state. J Environ Earth Sci 5(9):72–88
Babayemi JO, Ogundiran MB, Osibanjo O (2016) Overview of environmental hazards and health effects of pollution in developing countries: A case study of Nigeria. Environ Qual Manage 26(1):51–71
Balibey M (2015) Relationships among CO_2 emissions, economic growth and foreign direct investment and the environmental Kuznets hypothesis in Turkey. Int J Energy Econ Policy 5(4):1042–1049
Barro RJ, Sala-i-Martin X (1995) Economic growth, 2nd edn. The MIT Press
Becker GS (1964) Human capital: a theoretical and empirical analysis, with theoretical and special reference to education. National Bureau of Economic Research, New York
Boden TA, Marland G, Andres RJ (2011) Global, regional, and national fossil-fuel CO_2 emissions. Carbon Dioxide Information Analysis Center, Oak Ridge National Laboratory, U.S. Department of Energy, Oak Ridge. https://doi.org/10.3334/CDIAC/00001_V2011
Brucato PF, Murdoch JC, Thayer MA (1990) Urban air quality improvements: a comparison of aggregate health and welfare. J Environ Manag 30:265–279
Burtraw D, Kahn D, Palmer K (2003) CO_2 allowance allocation in the regional greenhouse gas initiative and the effect on electricity investors. Resources for the Future 1616 P NW Washington, DC 20036 Jel: Q2; Q25; Q4; L94
Carson R, Koundouri P, Nauges C (2010) Arsenic mitigation in Bangladesh: a household labor market approach. Am J Agric Econ 93(2):407–414
Central Bank of Nigeria (2009) Statistical bulletin. Central Bank of Nigeria P. M. B. 0187, Garki, Abuja
Central Bank of Nigeria (2017) Statistical bulletin. Central Bank of Nigeria P. M. B. 0187, Garki, Abuja
Costello A et al (2009) Managing the health effects of climate change. Lancet 373(9676): P1693–P1733
Dauda RO (2010) Role of human capital in economic development: an empirical study of Nigerian case. Oxford Business and Economic Conference Programme
Drake JM, Griffen BD (2010) Early warning signals of extinction in deteriorating environments. Nature 467(7314):456
Egan KJ, Herriges JA, Kling CL, Downing JA (2009) Valuing water quality as a function of water quality measures. Am J Agric Econ 91(1):106–123
Epstein PR (2005) Climate change and human health. N Engl J Med 353:14
Ferrer-i-Carbonell A, Gowdy JM (2007) Environmental degradation and happiness. Ecol Econ 60(3):509–516
Furgal C, Seguin J (2006) Climate change, health, and vulnerability in Canadian Northern aboriginal communities. Environ Health Perspect 114(12):1964–1970
Greene W (2005) Reconsidering heterogeneity in panel data estimators of the stochastic frontier model. J Econ 126(2):269–303
Groffman PM, Baron JS, Blett T, Gold AJ, Goodman I, Gunderson LH, Levinson BM, Palmer MA, Paerl HW, Peterson GD, Poff NL (2006) Ecological thresholds: the key to successful environmental management or an important concept with no practical application? Ecosystems 9(1):1–13
Hancock T (1993) Health, human development and the community ecosystem: three ecological models. Health Promot Int 8(1):41–47
Hanna R, Oliva P (2011) The effect of pollution on labor supply: evidence from a natural experiment in Mexico City. National Bureau of Economic Research working paper 17302
Hansen BE (1999) Threshold effects in a non-dynamic panels estimation, testing and inference. J Econ 93:345–368
Hsiao C, Shen Y (2003) Foreign direct investment and economic growth: the importance of institutions and urbanization. Econ Dev Cult Chang 51(4):883–896

Hussen MA (2000) Principles of environmental economics. Routledge, London/NewYork. ISBN:0-415-19570-5(hbk)
IPCC (2014) https://ar5-syr.ipcc.ch/ipcc/ipcc/resources/pdf/IPCC_SynthesisReport.pdf
Jeffery WM (2013) Introductory econometrics, a modern approach. South-Western Cengage Learning ISBN- 13: 978–1-111-53439-4
Jhingan ML, Chandar KS (2008) Environmental economics: theory, management and policy. Vrinda Publications, Delhi. ISBN:978-81-8281-1140
Jöst F, Quaas M, Schiller J (2006) Environmental problems and economic development in an endogenous fertility model (No. 428). Discussion paper series, University of Heidelberg, Department of Economics
Keller K, Hall M, Kim SR, Bradford DF, Oppenheimer M (2005) Avoiding dangerous anthropogenic interference with the climate system. Clim Chang 73(3):227–238
Knowles S, Owen PD (1995) Health capital and cross-country variation in income per capita in the Mankiw-Romer Weil model. Econ Lett 48:99–106
Ludwig D, Walker B, Holling CS (1997) Sustainability, stability, and resilience. Conserv Ecol 1(1). Available online https://www.jstor.org/stable/pdf/26271645.pdf?acceptTC=true&coverpage=false
McMichael AJ, Powles JW, Butter CD, Uauy R (2007) Food, livestock production, energy, climate change and health. Lancet 370(9594):1253–1263
Meadows DH, Meadows DH, Randers J, Behrens III WW (1972) The limits to growth: a report to the club of Rome. Google Scholar
Mooney HA, Cleland EE (2001) The evolutionary impact of invasive species. Proc Natl Acad Sci 98(10):5446–5451
Mushkin SJ (1962) Health as an investment. J Polit Econ 70:S129–S157
NEST (1991) Nigeria's threatened environment: a national profile. Nigeria Environmental Study/ Action Team Publication. ISBN: 978-31203-1
Newell SA (1998) Strategic evaluation and environmental metrics: making use of life-cycle inventories. Ph.D Dissertation, Technology, management and policy, Massachusetts Institute of Technology
Oyeshola DOP (2008) Sustainable development – issues and challenges for Nigeria. Daily Graphics
Özkaynak H, Thurston GD (1987) Associations between 1980 U.S. mortality rates and alternative measures of airborne particle concentration. Risk Anal 7(4):449–461
Patz JA, Campbell-Lendrum D, Holloway T, Foley JA (2005) Impact of regional climate change on human health. Nature 438:17
Pearce DW (1978) Environmental economics, 3rd edn. Longman, London
Ranis G, Stewart F, Ramirez A (2000) Economic growth and human development. World Dev 28(2):197–219
Schultz TW (1961) Investment in human capital. Am Econ Rev 51(1):1–17
Sengupta R (2010) Environmental issues for developing economies. Fundamental economics–Volume. *II*:90
Soytas U, Sari R (2009) Energy consumption, economic growth and carbon emissions: challenges faced by an EU candidate member. Ecol Econ 68(6):1667–1675
Srivastava NY (2009) Environmental pollution. APH Publishing Corporation, New Delhi. ISBN 978-81-313-0477-8
Steinnes DN (1992) Measuring the economic value of water quality: the case of lakeshore land. Ann Reg Sci 26(2):171–176
Stern D (2004) The rise and fall of the environmental Kuznets curve. World Dev 32(8):1419–1439
Stern DI, Common MS, Barbier EB (1996) Economic growth and environmental degradation: the environmental Kuznets curve and sustainable development. World Dev 24:1151–1160
Stokey NL (1998) Are there limits to growth? Int Econ Rev 39(1):1–31
Umoru D, Odjegba OP (2013) Human capital development and the Nigerian economy: a dynamic specification. Developing Country Studies, www.iiste.org ISSN 2224-607X (Paper) ISSN 2225-0565 (Online) 3(3)
Umoru D, Omolara YJ (2013) Labour productivity and health capital in Nigeria: the empirical evidence. Int J Humanit Soc Sci 3(4) [Special Issue – February 2013]
UNICEF (2007) The state of the world's children 2008: Child survival (Vol. 8). Unicef

Viscusi WK, Wesley AM, Huber J (1991) Pricing environmental health risks: survey assessments of risk-risk and risk-dollar trade-offs for chronic bronchitis. J Environ Econ Manag 21(1):32–51
Weil D (2007) Accounting for the effect of health on economic growth. Q J Econ 122(3):1265, MIT Press
Weil ND (2009) Economic growth. Dorling Kindersley (India) Pvt. Licenses of Pearson education in South Asia. ISBN 978-81-317-2481-1
World Bank Group (2016). https://data.worldbank.org/indicator/SP.POP.TOTL?#
World Health Organization (WHO) (1995) The world health report 1995: bridging the gaps. World Health Organization
World Health Organisation (WHO) (2009) Preventing disease through healthy environments: towards an estimate of the environmental burden of disease. Available online http://www.who.int/quantifying_ehimpacts/preventin gdisease/en/index.html. Accessed 23 June 2009
World Development Indicators (WDI) (2014) The world development report. World Bank
Wooldridge MJ (2013) Introductory econometrics, a modern approach. South-Western Cengage Learning. ISBN- 13: 978-1-111-53439-4
Yang H, Zhou Y, Abbaspour KC (2010) An analysis of economic growth and industrial wastewater pollution relations in China. Consilience J Sustain Dev 4(1):60–79
Zivin JG, Neidell M (2012) The impact of pollution on worker productivity. Am Econ Rev 102(7):3652–3673

Relationship Between Climate Change, Natural Disasters, and Resilience in Rural and Urban Societies

30

Safieh Javadinejad, Saeid Eslamian, Kaveh Ostad-Ali-Askari, Mohsen Nekooei, Neda Azam, Hosein Talebmorad, Ali Hasantabar-Amiri, and Mohammad Mousavi

Contents

S. Javadinejad
Water Resource Management, University of Birmingham, Edgbaston, UK
e-mail: sxj228@bhm.ac.uk

S. Eslamian · M. Nekooei · N. Azam · H. Talebmorad
Department of Water Engineering, College of Agriculture, Isfahan University of Technology, Isfahan, Iran
e-mail: saeid@cc.iut.ac.ir; mohsen.nekoei@gmail.com; neda.azam2015@gmail.com; h.talebmorad@gmail.com

K. Ostad-Ali-Askari (✉)
Department of Civil Engineering, Isfahan (Khorasgan) Branch, Islamic Azad University, Isfahan, Iran
e-mail: kaveh.oaa2000@gmail.com; koa.askari@khuisf.ac.ir

A. Hasantabar-Amiri
Department of Civil Engineering, Lenjan Branch, Islamic Azad University, Lenjan, Iran
e-mail: ali.amiri.nyc@gmail.com

M. Mousavi
Department of Civil Engineering, Najafabad Branch, Islamic Azad University, Najafabad, Iran
e-mail: mhd.musavi@yahoo.com

W. Leal Filho (ed.), *Handbook of Climate Change Resilience*,
https://doi.org/10.1007/978-3-319-93336-8_189

Abstract

The government established several strategies to decline the hazards of climate alteration. The most significant of these strategies which usually neglected is recognizing the variances between several communities with regard to resilience indicators against these impacts and improving proper strategy for each of them. This research aims to examine the alteration between urban and rural communities with regard to resilience in the extreme event of climate change.

The present research is applied in a practical way, and all possible methods (such as numerical/analytical modeling and statistical analysis) which are necessary for estimating the value of resilience or monitoring spatial-temporal characteristics of resilience have been used. The results displayed that there is an important difference between the resilience elements in urban and rural districts. Resilience in urban regions usually is influenced by economic parameters. However, social parameters are the most significant element in resilience in rural regions. Also, indicators which depend on spatial variations are critical in both rural and urban areas under the effect of climate change. Therefore, this research analyzed all the individual factors (in rural and urban areas) which can have an effect on resilience under the climate change conditions, so it can help managers to plan a better strategy for resilience during extreme events of climate change.

Keywords

Climate change · Natural hazards · Resilience · Rural and urban societies · Strategy

Introduction

The relationship between climate and communities and its consequences and the behavior of humans have a long history and have become an active ingredient in human life (Tompkins and Adger 2004). At the moment, the fact that humans change the status of the earth is clear, and this important phenomenon has been at the core of broad activities across countries and the United Nations; human can change the earth, and the changes can aggravate by population's excessive growth. Consequently, the need for technology and, as a result, an increase in the demand of the international community for energy carriers and if the same trend continues to increase in demand and energy consumption, severe climate change will occur in the next few decades. Based on the simulation of climate models, it can be expected that by 2100, the Earth's temperature will rise from 1 °C to 5 °C (Pisello et al. 2017). The numbers will change based on different patterns of development and population as well as greenhouse gas emissions. This change range reflects the fact that climate change over the past century and a century later will be more than the changes over the past thousand years (Cameron et al. 2015). One of the most important current issues in the world is competition for access to water resources. Climate change has geopolitical consequences that could be reached in arid areas (Tanner et al. 2015). One of the challenges of climate change is the human spatial dimension of climate change at international borders, in where climate change creates particular challenges (Wyss et al. 2015). Demographic locations in many regions where the rapid expansion of industrialization and agricultural urbanization is a consequence of the intensification of vulnerability to water scarcity are a major concern for climate change (MacDonald et al. 2015). The dangers that climate change will bring to the regions will be far beyond water resources. Resiliency and adaptation to climate change are one of the priorities that has been considered by urban people and farmers. Also, it seems to be a useful practical solution (Eslamian and Eslamian 2017). Adaptive practices in better meeting the acute conditions of meteorology and climate change will help interact and regulate and manage a place to lose opportunities and compensate for the damage caused by climate change (Petrie et al. 2017). Selection of resilience and adaptation is carried out on a regional and global scale, and its actors could be companies and governments or agriculture (Eslamian et al. 2017). Selection of local levels, such as plant diversity, irrigation time, and market responses, including the type of income sources, resources, and credit programs of enterprises, can help to adapt to climate change (Heller et al. 2015). Government responses like eliminating subsidies and maintaining and improving agricultural markets, a technological development that includes the development and upgrading of new products, various advances in water management, and encouraging and educating people to adapt to these changes can enhance adaptation to climate change (Yihdego et al. 2017; Becker et al. 2015). Vulnerability assessment and resilience mechanisms and adaptation to global climate change are critical (Eslamian et al. 2017). It should be noted that as the infrastructure and economic strength of a region are more and more varied, the vulnerability of the area will be less than likely to be affected by maladaptive changes and its resilience and compatibility (Stevens-Rumann et al. 2018; De Souza et al. 2015). The resilience and adaptability potential implies the potential of a system for responding to climate stimuli (Martin et al. 2018; Childers et al. 2015). Consistency and resilience are also defined as changes in human ecosystems in response to observations or changes in response to climate stimuli. Indeed, to demonstrate the urgent need to overcome the issues of climate change publisher, a new model for development is needed (Takeuchi et al. 2018; Nettier et al. 2017). If governments in developing countries consider global climate change in their decisions, adaptation and resilience will be easier. Therefore, to mitigate the effects of climate change, design and presentation of a program that will be achieved with modern technology and indigenous knowledge and local and national policy will ultimately be adapted (Davari et al. 2017; Lafond et al. 2014). Implementing local strategies for climate change can lead to consistency and resilience. Worldwide, several studies have been done to understand urban perceptions and farmers' perceptions of climate change, which has had an impact on livelihoods and agriculture. Although perception is not necessarily consistent with reality, one must pay attention to the socioeconomic challenges facing societies and the impact that climate change has on them. Perception is defined as the process of producing feelings from real-world experience and emphasizes the ability of an individual to take his experiences of nature and natural variables (Abid et al. 2015). Studies show that individuals experience (both urban perceptions and farmers' perceptions) different shocks of climate variables in different social groups, geographic locations, and seasons. There is also a difference in gender and age and the difference in experiences at different levels in the face of climate change. Therefore, addressing the issue of perception and resilience and adaptation requires the collection of empirical data from urban people as well as farmers, and recognizing adaptive practices in a particular place typically involves a different perception of climate change and coping with it.

Resilience to climate change and disasters in urban and rural areas is not the same. Rural areas are dependent on the state for the maintenance of their own business. It is because of their single-product economy during normal times, and so they cannot supply more resources during and after the disaster (Haynes and Tanner 2015; Xu and Grumbine 2014). At the same time, economic diversification in urban areas is an excellent help for resilience in these areas. Lack of resources can reduce flexibility in urban and rural areas.

On the other hand, knowledge and relationships in rural areas and strong social networks can increase their resilience. The full understanding of urban and rural areas regarding resilience is essential because the fields and capabilities are different for managing different risks. For example, a low population in rural areas affects the casualty of villages. Also, ownership in urban areas due to the density and value of structures is likely to be higher than that of villages. Considering that resilience capacity varies in urban and rural communities, to increase capacity, resiliency should be made for each community, taking into account the local capacity indices of the appropriate strategy. Therefore, recognizing their degree of flexibility and providing a suitable approach for coping with the reduction of effects for local resources can increase the resilience capacity in urban and rural areas.

Many studies have been conducted on a resonance that explores some of these studies in this chapter. Whitman et al. (2013) in a study analyzed the resilience of rural settlements against earthquakes. The results showed that social capital is the most important factor in this area. Peng (2015) studied the relationship between the survival of rural settlements and resilience to risks and concluded that increasing the resilience of rural settlements increases its resilience. Liu et al. (2018) measure earthquake resilience and concluded that some of the regions have the best and some others have the worst situation. Despotaki et al. (2018) ranked earthquake resiliency by different factors, and they used some indicators for measuring resilience. The results of their work showed that being away from the hazardous environment is the most important factor in the resilience. Ceridwen (2017) explored the resilience of rural communities in Scotland and concluded that social capital is the most important factor in resilience. Using the exploratory method, Boon (2014) has studied the resilience of the Australian villages to flood and concluded that population stability and socioeconomic indicators are the most important factors in rural resilience. Bozza et al. (2015) have investigated the resilience of urban systems to natural disasters and concluded that physical and social accesses to information are the most important factors of resilience. Suárez et al. (2016) have investigated the resilience indicators of 50 cities in Spain and concluded that sources of income and business are the most important factor in this regard. Young et al. (2018), using the regression model, investigated the resilience of the Wales urban transport systems to the risks. They concluded that the cost of rebuilding is the most important factor in this regard. Boelee et al. (2013) investigated the resilience of rural area under extreme drought in Africa. The study applied t-test by using SPSS software to understand the distance of resilience factors during an extreme drought from resilience factors during normal condition. Another study used a questionnaire method which is filled by institutional managers, to realize the important factors that can increase resilience during extreme droughts. The research found that agricultural insurance development, drought monitoring, and prediction system and damage assessment and using indigenous knowledge can increase resilience in rural areas. A review of studies suggests that so far there has been no study on the spatial analysis of the differences between urban and rural areas regarding resilience to climate change. Therefore, the present chapter focuses on the assessment of adaptive and resilience strategies by urban people and farmers in making adaptation decisions and the factors that lead to a sustainable adaptation in the region. This chapter emphasizes the awareness and understanding of farmers and urban people about climate change and the factors influencing the selection of adaptation strategies. Hence, understanding the perception of farmers and urban people is important concerning the dangers they face. To ensure that farmers and urban populations are prepared to face climate change events and cooperate for awareness and adaptation to climate patterns, farmers and urban people need to focus on their activities, knowledge, and understanding of these changes. The most common models are global climate change forecasts, and adaptation approaches occur locally. Climate change is a global problem but a local solution. In the direction of the catastrophe, climate change is shared between managers and ordinary people. The roles of ordinary people are their perception of the risk of climate change. It is based on the level of perception that can first provide solutions and, secondly, in their daily routine to adjust, being resilient, and be able to adapt. Therefore, this research studies the difference between urban and rural communities concerning resilience to climate change and distinguishes the most important methods and indicators of resilience in urban and rural areas.

Basis Theories, Definitions, and Concepts

The trend of climate change in recent years has shown that human societies have become more vulnerable and risks have increased. However, vulnerability reduction is often ignored after catastrophic events. There are two strategies to deal with catastrophic events: forecasting strategies and resilience strategies; the first is to deal with known problems, and the second is used to deal with unknown problems. In the meantime, resilience has been damaged by the ability of the community to plan and prepare to deal with the unwanted effects of disaster rights and community rehabilitation (Chapin and Brinkman 2016). The concept of resilience in ecological studies was introduced as a way of understanding the nonlinear smell in ecological systems (Levine et al. 2016; Lowe et al. 2015). It also has the capability of a system for self-healing response after an accident that enables the system to absorb the pressures from an accident and can handle future events. Some researchers believe that resilience is one of the characteristics of a system, while others describe it as capacity building for communities, and some describe both the proposition and resonance as the ability of different systems to cope with changes. Cumming et al. (2015) believe that resilience should occur at a national level on a variety of scales, including in individuals, communities, villages, cities, and constituencies. Resilience requires many factors, including population, income growth, and social, economic, and environmental impacts. Resilience approaches can be summarized into three categories (e.g., resilience as sustainability, resilience as a recovery, and resilience as a transform), the common aspect of which is all ability, resistance, and response to pressure.

The first approach is resilience as sustainability: this approach has expanded from ecological studies that define resilience as the ability to return to the previous state. The resonance here is expressed as the amount of disturbance that a system can tolerate or absorb before being transferred to another state. Some scholars consider the rural community beyond what the community suffered to be unable to return to its state of affairs, because a resilient community has a high base and a lot of pressure to take before it crosses the baseline (Paton and Johnston 2017).

The second approach is rescue as a recovery: the recovery approach is related to the ability of society to return to the past from a change or push factor to its original state. Here, resilience is a measure that is measured as the time taken by a community to recover from a change. The resilient society can quickly return to the previous state, while a society that is less resilient may spend more time retrieving or not recovering.

The third approach is resilience as a transform: this approach is more than the capacity of society to respond to change and be adaptive, which, instead of a simple return to the previous state, can mean a change to a new state in a more sustainable environment. This approach is useful to understand how a society can positively change and accepts that change is inevitable and considers it to be a factor in replacing change as a factor in pressure, in which society needs to be restored to its original state. One of the resilient dimensions in the regions is that it varies in different areas and depends on its type; each knows its evaluation methods and its data (Sinclair et al. 2014). Over the past two decades, research has been conducted to explain vulnerability and urban resilience, but less attention has been paid to rural resilience. Some research on the international scene, especially in developing countries, is about the potential risk of accidents to climate change, but there is little information about the classification of specific resiliency features against accidents. Damage may be lower in urban areas for some of the hazards such as intense heat and more earthquakes and less for others such as thunderstorms and floods. Research shows that damage is much different in urban and rural areas. In urban areas, the organization of vulnerable residents is complicated in the long run, while in rural areas, the improvement of the organization of vulnerable residents in an accident leads to the concentration of vulnerable populations. Based on what has been said, it should be recalled that each of the researchers looked at the resilience of societies in various respects. The common aspect in all studies has been the ability, resistance, and response to pressure or change. To promote the resilience of communities against the risks of resiliency, assessment based on local indicators, and identification of the factors affecting it, it is considered essential steps. Therefore, it can be said that rural settlements, due to exposure to climate change damage, have their environmental conditions that are more at risk than urban areas. But the strategies used to mitigate the effects of these hazards are the same for urban and rural communities. For each one, a pattern and method should be provided to fit local indices.

Resilience in Urban Versus Rural Areas Against Climate Change

Before providing clear explanations about differences of resilience in urban and rural areas against climate change, it is necessary to compare climate conditions, vulnerability, and impacts of extreme events in urban and rural regions.

Usually, because of the geography circumstances in rural areas, they are located in somewhere with high elevation and with more vegetation or forest covers in contrast to urban areas. Therefore, average precipitation in rural areas is higher than in urban areas, and the temperature is lower in rural regions (Zeleňáková et al. 2015). So especially in coastal areas and floodplain areas, sea-level rise and the intense storms are more in contrast to urban areas during extreme events of climate change. However, extreme climate change in rural regions can have an effect on urban areas, because normally urban areas are located downstream of rural areas (Thebo et al. 2017).

The vulnerability of rural and urban areas during extreme events of climate change is entirely different and depends on many factors, and for each event, the vulnerability is different. For example, the investigation of flood events showed that there are greater building, infrastructure, electricity supplies industry, and population in urban areas. However, with strong policy, legislative, and great institutional capacity, controlling and managing the flood events are more manageable in contrast to rural areas (however, definitely it depends on the conditions of case study and sustainable planning and institutional arrangement in an area) (Tacoli and Satterthwaite 2014). Another example is about drought events; if an extreme drought happens in a rural area and if the economics of the region are based on agriculture, the loss of local agriculture production during drought events is significant, and resilience of rural area can be lower in contrast to the urban region which is usually established on economic diversity. Therefore, many people from rural areas will migrate to urban areas, which can have an effect on the resiliency in the urban areas as well. The examples displayed although the resilience in the rural and urban area is different, it can affect each other. Thus, for understanding and analyzing the resilience and for examining the factors that can have an effect on both rural and urban resilience, different techniques are investigated and explained in the following sections.

Materials and Methods

To estimate resilience, both modeling simulations and statistical analysis can be used.

Modeling Simulations

Model simulations can be applied to display how richness and variety can raise the resilience of urban regions or farm organizations in farming districts.

Masikati et al. (2015) simulated four typical livestock systems in Southern Africa and modeled the effect of adaptive organization alterations on fodder availability under a range of climatic circumstances.

Richness and evenness are two adaptive strategies. Masikati et al. (2015) discovered that richness and evenness could raise forage availability, especially if mix with two other adaptive strategies which include forage accumulations with significant variation in the optimal growing circumstances and moving the dates of calving.

Also, Cameron et al. (2015) ran simulations to get the range of climate changeability and medium-term effects on two mixed crop-livestock systems (minor, food insecure against great, food secure). The direct and indirect impacts of fertilization on efficiency are an example of the synergistic dimensions of the mixed crop-livestock systems. The integration between the effects of fertilization and target feeding interferences can raise energy production and gross profits and can make higher energy production and total revenues in both farms.

Lowe et al. (2015) showed the positive impacts of a combination of forage production, income stability, and thus climate resilience. The study also displayed rise in benefit from forage a mixt depends on higher precipitation and access to markets.

Statistical Analysis

Using regression analysis can make evidence for linking integration and climate resilience. Heller et al. (2015) showed that unlike mixed horticulture farms, particular crop farms (mainly among market-oriented great holders) are more sensitive under the climate change effects in San Francisco.

Also, the results of the study by using statistical analysis displayed that there is a high correlation between climate circumstances and income. For example with reduced rainfall and increased temperature, the farm income will decrease significantly.

Jakovac et al. (2015) used statistical analysis to estimate crop variety and production losses in drought years in the Amazon area.

Pérez et al. (2015) applied statistical analysis and microeconomic analysis for comparing incomes from mixed crop-livestock farms against two particular farms (only crop or only livestock) under climate alteration. Therefore the study showed how farmers' behavior and actions can have an effect on their incomes under climate change conditions.

Altieri et al. (2015) used statistical analysis with data from 9000 farms to understand the relationship between farmer's selections and climate alteration. The data included rainfall, temperature, water availability, electricity usage, soil data, methods of cropping, etc. The study examined alterations in the likelihood of selecting each system when the climate was unstable. The results showed that a particular crop farm was much sensitive to climate change. However, warmer and drier climate circumstances can raise the probability of combined farming adoption.

Nobre et al. (2016) also represented that mixed farming has better resilience against climate alteration as contrasted to particular systems.

Some factors which can be used for estimating the resilience of the farmers against climate change are land use decisions, farmer's economic rationality, the spatial and temporal variability of farmlands, and current obstacles to the presentation of livestock in the system, like restricted land availability, that could alter in the future under climate extreme events.

These factors can be used and then the statistical analysis can apply different methods like regression, t-test, Vikor test, and so on.

Results

Climate Indicator

As shown in Table 1, climate indicators which are usually used for estimating resilience include precipitation, temperature, characteristics of extreme events, and agriculturally relevant climate metrics.

Resilience Indicators

Resilience indicators are usually used to simulate the effects of climate change and shocks on many outcome variables. Resilience indicators contain income, yield, and the number of labor, production, and revenue variance across time, rates of crop breakdown in space, dietary variety, and danger.

Sometimes resilience indicators were not used directly in the previous studies; however, a relationship between integration and climate resilience is used. For instance, Nobre et al. (2016) focused on the alteration in farming system in a given area; however, they also applied crop breakdown rates as an element of area alteration among harvesting and livestock.

Table 1 The indicators and proxies which can be used for estimating the relationship between climate and resilience

Parameter	Indicators and proxies
Climate	Alteration in (1) temperature, (2) precipitation, (3) catastrophes, and (4) agriculturally relevant climate metrics
Resilience	Measuring the mean and/or variance in (1) agricultural production, (2) agricultural yield, (3) farm income, and (4) household income

Most of the previous works focused on the revenues made by the sale of goods (as resilience indicators) during climate change conditions. Depending on the purpose of a study, multiple outcomes can be estimated.

As shown in Table 2, some resilience indicators are appropriate for shorter time episode; however, some others are appropriate for longer time episode adjustment planning. Also, the table displayed a different group of studies which did different research on resilience. The table revealed that with an equal unit of analysis, temporal episode, there are several climate shocks and resilience indicators. Also, the table showed resistance or recover or both actions against the given events. Besides, the table represented that in some cases, between climate and resilience, there is a nonlinear relationship.

The Main Factors Which Can Affect on the Values of Resilience Under Climate Change

There are some critical factors which can affect the resilience values against climate change which include:

(a) Unit of examination

The unit of analysis for resilience analysis is vital. For example, to assess how agricultural integration and resilience can have an effect on climate, the farm or household data are the units of examination.

Roesch-McNally et al. (2018) discovered that a variation of landscapes and cropping organizations could improve agroecosystem resilience and implementation for US areas. However, McCluney et al. (2014) revealed that biological variation through ecological, spatial, and temporal scales could show how ecosystem services depend on agriculture activities.

Sometimes the results of resilience analysis from different studies cannot be compared because of various units of analysis or different methods. Sometimes the studies analyzed the impacts of farm-level incorporation on farm-level resilience. In those studies, a significant sample of households and observation data (as units of analysis) are used for analyzing the resilience. However, other studies used large scale for units of analysis and resilience. Therefore, different scales need different unit analysis for estimating resilience. It can have an effect on farmer's participation, motivations, decision, and practices.

Table 2 The examples of previous studies which use a few indicators for estimating resilience against the extreme events of climate change

			Resilience features				
Research	Unit of examination	An extreme event of climate change	Indicators/proxies	Statistics	Time episode	level	Is the relationship between climate and resilience factors nonlinear?
	Farm	Drought/ tropical cyclone	Productive properties (especially livestock)	Mean	<20 year	Resistance recovery	Yes
	Farm	Precipitation and temperature	Dietary diversity	Mean	<1 year	Resistance	No
	Household	Drought	Livestock properties	Mean	3 year	Resistance recovery	Yes
	Household	Drought	Livestock properties	Mean/ variance	1 year	Resistance recovery	Yes
	Household	Precipitation	Food security	Mean	1 year	Resistance recovery	Yes
	Local Region	Drought	Crop yield	Mean/ skewness and variance	1 year	Resistance	Yes
	Local Region	Drought	Crop yield and production	Mean	1 year resistance and 40 years adaptation	Resistance recovery	No
	State	Tropical cyclone	Agricultural GDP	Mean	1 year resistance and 40 years adaptation	Resistance recovery	No

Also with more diversity in landscape, more diversity in habits and flexibility in a system can be discovered, and so potential to cope with the hazards of climate change or hard situation of socioeconomic can be increased.

(b) Manage variables

Control variables that can affect the value of resilience include farmer features, organization practices, economic parameters (such as farmer income or price of equipment and maintenance), regional parameters (such as local institutions, population concentration), health and nutrition parameters, and access to data and information. However, the most popular control variables which can be used for resilience analysis are farm and climate variables. Control variables are significantly related to the unit of study.

The studies which worked on a small scale such as household, and farm levels, rarely focused on health and nutrition parameters (as control variables) for resilience analysis. Nevertheless, farm indicators and organization parameters can be used for both small- and large-scale studies.

For improving resilience, the integration between agriculture management (like dates of planting/harvest), economic (like income and prices), climate parameters (like precipitation and temperature), and farm parameters (like the type of soil, irrigation method, yield, fertilizer, and machinery and farm size) are necessary.

(c) The scale of spatial and temporal

The effect of different spatial scales (large or smallholders) on the value of resilience can highlight when extreme climate events occur, and so it can have an impact on resilience strategies and adaptation.

Also, the temporal scale like 1 year or multiple years can affect the planning of resilience significantly. For example, providing the strategy for resilience for multiyear under climate change is different when using it for a given year or season.

(d) Autonomous against combination

The use of autonomous or integration planning depends on the society, policies that resilience can set up on there. Usually in small society, autonomous programs for establishing and improving resilience are better.

(e) The linearity of the association between integration and resilience

Understanding the linear or nonlinear relationship between resilience and integration (especially in farm systems and farm features) is very important. For example, between socio-ecological variables and resilience values under climate change conditions, there is a nonlinear relationship, because the changing speed of socio-ecological variables is very slow.

(f) The efficiency of farm system against rural livelihoods

There is significant relationship between the effectiveness of farm system and rural livelihoods that can affect potential activities for improving resilience, and if the integration between them would be healthy, the resilience can be developed properly.

Examples of Case Studies

In this section, the results of resilience analysis which applied both statistical analysis and modeling simulation are explained. It should be noted that in the first section of "Statistical Analysis" (statistical analysis for examining resilience), the results came from this current research. However, the results showed in section "Modeling Simulation" came from another study to explain the modeling simulation for examining resilience and give an extensive overview of all methods that can be used for analyzing resilience deeply.

Statistical Analysis

To illustrate how society can increase resilience against threats, the examination of the socioeconomic capacity of a physical and natural human institution is necessary.

If the average resilience of the promising average resilience ratio is lower, then there is a different ideal between resilience and survival. Therefore, to analyze the average of real resilience and favorable average, some statistical analysis such as t-test and Friedman test can be used.

If there is not a significant difference between the mean and the mean intermediate, then the zero assumption (H0), based on the same distribution, is not rejected.

The case study of this chapter is United States, and using this method, the results showed that there is a relationship between farmers action and drought (Tables 3, 4, and 5).

In another previous study, for estimating strategy ranking, the ASI index is used to examine the classification of resilience against climate change (Bonfilgio et al. 2018):

$$\mathrm{ASI} = \mathrm{ASn} + \mathrm{ASl} + \mathrm{ASm} + \mathrm{ASh}$$

where ASI is compatibility strategy index and ASn is the abundance of farmers whose compatibility strategy indicator does not matter to them. AS1 is the abundance of farmers whose compatibility strategy indicator is of little importance to them. ASm is the abundance of farmers whose adaptation strategy indicator has a modest value for them, and Ash is the abundance of farmers whose compatibility strategy is of great importance to them.

Based on this indicator, this chapter showed that changing the pattern of crops, selecting resistant species, reducing the amount of land, and cultivating diversity are the most important patterns of adaptation to climate change (Table 6).

Modeling Simulation

To apply the modeling method to investigate the resilience, according to the long-term history of synoptic and kinematic stations, the frequency of climatic hazards in the GIS environment is zoned. Then, considering the social, economic, and biological components, it is possible to set up a questionnaire to understand the perceptions of the people and the related authorities about climate change and then the consistent approaches from their point of view.

Table 3 The mean and level of significance were lower than the desirable variables of increasing the vibrations of farmers against drought

Numerical utility test = 3							
						Confidence interval (95%)	
Variables	Average	T-test	Free degree	Significant value	Difference between desirable	Lower	Upper
Development of insurance of agricultural products	1.46	−47	19	0	−1.55	−1.62	−1.49
Granting credits and loan to suffers against natural hazards	2.35	−2.7	19	0.015	−0.66	−1.16	−0.143
Coordination of agricultural-related acting government forces	2.37	−3.20	19	0.005	−0.73	−1.20	−0.249
Reduce tax or delay in deadline payment	2.83	−0.80	19	0.038	−0.18	−0.65	−0.29
Increase saving	2.49	−2.20	19	0.042	−0.52	−1.00	−0.025
High area of land	2.27	−3.3	19	0.006	−0.74	−1.22	−0.256
Increase knowledge about drought	2.33	−3.2	19	0.007	−0.68	−1.14	−0.23
Increase the income of nonagricultural and economic diversity	2.54	−2.13	19	0.048	−0.47	−0.91	−0.0076
Development of local organizations in the field of agriculture	2.40	−2.74	19	0.014	−0.61	−1.07	−0.141
Enhance local farmers' participation	2.43	−2.43	19	0.026	−0.58	−1.07	−0.079
Attention to the local and nations' knowledge	2.42	−2.40	19	0.028	−0.585	−1.10	−0.074
Drought prognosis and assessment of damage	1.79	−72.7	19	0	−1.22	−1.26	−1.19

(*continued*)

Table 3 (continued)

Numerical utility test = 3							
						Confidence interval (95%)	
Variables	Average	T-test	Free degree	Significant value	Difference between desirable	Lower	Upper
Improve the methods of irrigation and water management	3.02	0.009	19	0.995	0.0011	−0.318	−0.320
Increase the varieties of crop	3.07	0.293	19	0.775	0.0628	−0.388	0.513
Increase spatial continuity in agriculture lands	2.39	−2.478	19	0.024	−0.612	−1.13	−0.095
Improve control of soil erosion	2.29	−3.147	19	0.005	−0.713	−1.19	−0.329
Improve drought-resistant species	2.39	−2.438	19	0.026	−0.616	−1.14	−0.871
Attention to the suitable time for cultivating	3.08	0.417	19	0.683	0.0703	−0.283	0.424
Attention to the cover of irrigation channel	3.13	0.731	19	0.476	0.1297	−0.243	0.5015
Deep plowing in rainy seasons	3.23	1.37	19	0.190	0.23	−0.119	0.57

Table 4 The mean and level of significance were lower than the desirable variables of increasing the vibrations of farmers against drought

Numerical utility test = 3							
						Confidence interval (95%)	
Variables	Average	T-test	Free degree	Significant value	Difference between desirable	Lower	Upper
Politics and governmental supports	2.23	−4.57	19	0	−0.775	−1.13	−0.420
Capacities of economic-social	2.33	−3.48	19	0.003	−0.671	−1.08	−0.268
Local activities	2.82	−1.139	19	0.28	−0.182	−0.517	0.153

Table 5 The ranking of effective factors in resilience farmers based on the distance from the ideal solution

Factors	S	R	Q	Rank	Factors	S	R	Q	Rank
Development of insurance of agricultural products	1	0.5	1	1	Attention to the local and nations' knowledge	0.59	0.27	0.55	3
Granting credits and loan to suffers against natural hazards	0.24	0.13	0.12	15	Drought prognosis and assessment of damage	0.87	0.37	0.87	2
Coordination of agricultural-related acting government forces	0.5	0.15	0.25	11	Improve the methods of irrigation and water management	0.24	0.10	0.08	17
Reduce tax or delay in deadline payment	0.17	0.09	0.01	19	Increase the varieties of crop	0.24	0.11	0.08	16
Increase saving	0.32	0.16	0.21	12	Increase spatial continuity in agriculture lands	0.43	0.22	0.37	6
High area of land	0.43	0.19	0.33	8	Improve control of soil erosion	0.47	0.19	0.35	7
Increase knowledge about drought	0.47	0.18	0.33	9	Improve drought-resistant species	0.44	0.22	0.37	5
Increase the income of nonagricultural and economic diversity	0.39	0.17	0.28	10	Attention to the suitable time for cultivating	0.31	0.13	0.17	14
Development of local organizations in the field of agriculture	0.36	0.13	0.20	13	Attention to the cover of irrigation channel	0.21	0.11	0.08	18
Enhance local farmers' participation	0.47	0.24	0.43	4	Deep plowing in rainy seasons	0.16	0.09	0	20

In a previous study conducted in the United States which used the modeling simulation, the results showed that the most important consequences of climate change are drought, refractory, flood, and frost, which have spatial differences and are the most critical factors of resilience to these consequences of climate change (Figs. 1 and 2). From selling livestock, taking loans and borrowing from relatives, reducing crops and creating new businesses are less dependent on nature.

Discussion

Today, resilience in many scientific fields and a wide range of ecology to the social sciences, psychology, and economics is an important term and has become functional, but so far there is no consensus on the exact definition of resilience that is accepted by all disciplines.

Resilient societies should, like ecosystems, have the ability to withstand disturbances and adapt to changes that are needed. Referencing as a potential system,

Table 6 Analysis of resilience with using ASI

The importance of compatibility for agriculture						
Adaptation and resilience strategy	High	Mean	Low	Unimportant	ASI	Ranking score
Change of crop pattern	171	55	24	–	647	1
Change and rotate the product with different courses	132	84	22	13	501	12
Crop varieties	125	85	32	8	577	3
Finding new jobs (nonagricultural jobs)	115	77	35	25	455	14
Insurance for crops	115	99	21	15	556	6
Selection of salmon-resistant species	142	83	25	5	612	2
Use of new techniques and soil fertility techniques	123	72	35	20	548	7
Water treatment for drainage and reuse the water for irrigation	114	95	35	4	569	4
Change the time of irrigation	102	81	61	8	529	10
Cultivate the crops with the short-term in-growth	101	73	74	2	523	11
Immigrate and leave the farmland	73	91	71	14	477	13
Integration of agricultural systems	67	59	89	35	408	15
Select the species which are resistance against drought	112	101	24	13	562	5
Select products with low water consumption	118	61	55	16	531	8

community or community at risk, to adapt to changes to achieve an appropriate level of performance and structure is known.

Tanner et al. (2015) showed the main features of resilient societies had expressed the thresholds for change, the reorganization of resilience capacity, and the cushioning or recovery after the shock and tensions which are necessary to learn and adapt to it.

Today, there are significant changes in the perception of risks due to climate change, with the prevailing view in focusing on reducing vulnerability to increasing resilience to these risks. Based on this approach, hazard reduction programs seek to create and strengthen the characteristics of fluctuating communities and to concentrate on planning and to manage risks in the sense of resilience.

For example, Arouri et al. (2015) have researched to understand the significant characteristics for resilience during extreme events of drought under the climate change for a rural area of Vietnam. In their research, the questionnaire and t-test technique are applied. They found that socioeconomic resilience can be reached by strengthening social participation, borrowing, saving and reducing costs, not using expensive equipment in agriculture, and choosing new economic activities such as

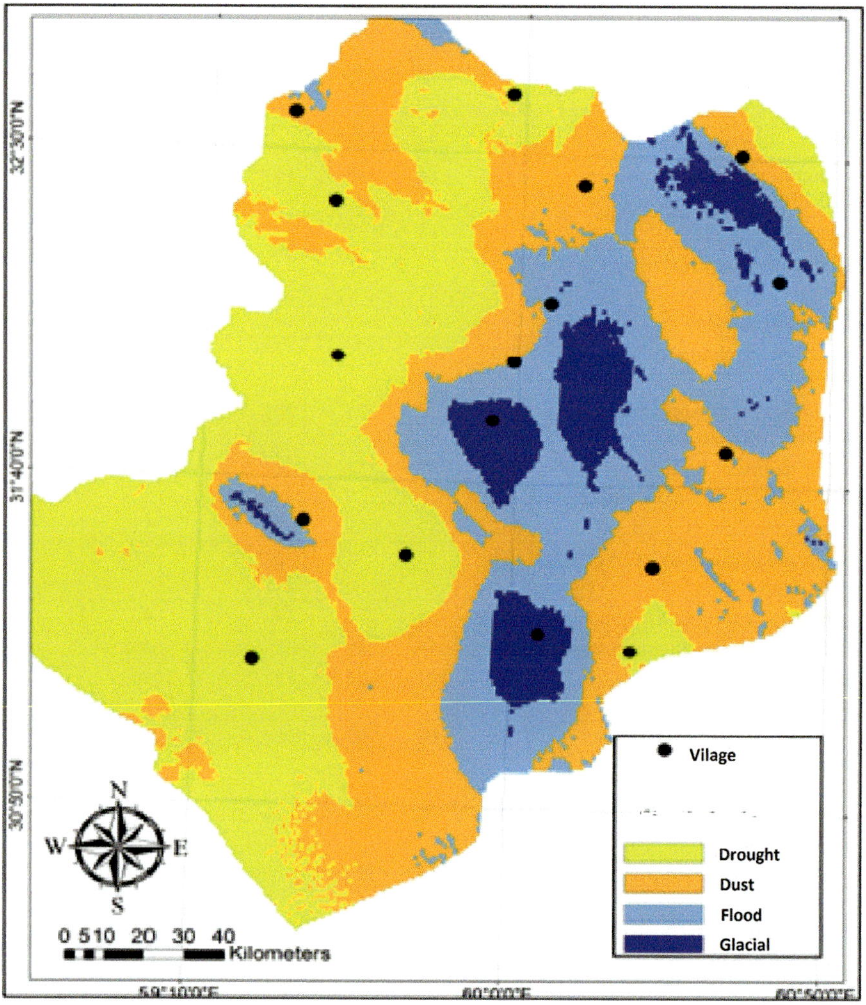

Fig. 1 Distribution of natural hazards due to climate change. (Modified from Borden and Cutter 2008; US government: Disaster data 2018)

camping, selling livestock, collecting medicinal plants for sale, making handicrafts, and keeping turkeys. Also, the study investigated that bio-resilience can be reached by changing crop pattern, selecting of new products which are compatible with dehydration, creating storage pool to save water, and saving food for livestock.

Another study obtained socioeconomic resilience during extreme flood for the rural area of Australia. They applied a questionnaire and ANOVA method (Boon 2014). They found that resilience can be reached through getting a loan and local partnership and help from relief organizations. Also, the study considered that bio-resilience could be reached by creating new dike and building new infrastructure for transforming water to agricultural land.

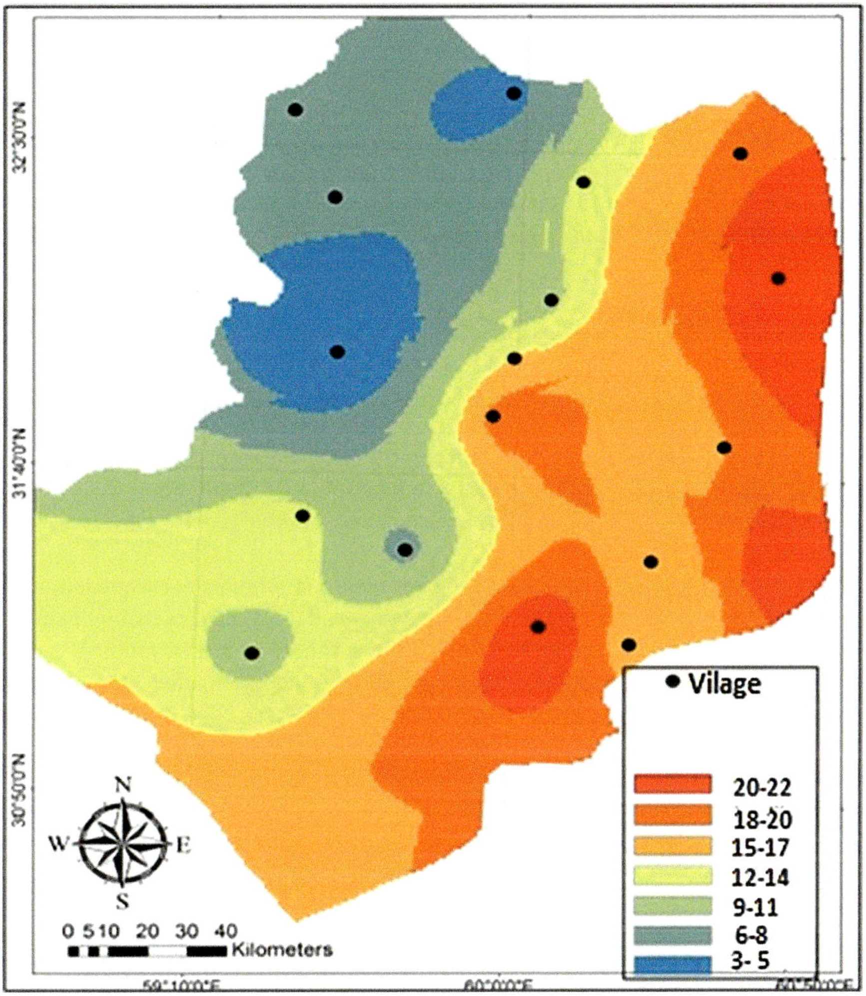

Fig. 2 Frequency of resilience against climate change. (Modified from Disaster Survival Resources 2018)

Barton et al. (2018) focused on increase resilience during glacial for the area of Western Mediterranean. The research explored that the most significant actions which can increase the resilience include the use of resistant crops, using manure, and changing cropping time.

Another example is the research that has been done by Licht et al. (2016). The study obtained the ranking for rising resilience during a dust storm in the area of central Asia. They used the questionnaire and weighting method. They found that the priority for increasing resilience belongs to using large livestock like the camel. The second priority belongs to using windbreaks, and the third priority belongs to planting resistant plants.

The system of resilience is based on three elements: danger, revelation, and susceptibility. So, depending on the condition of climate change, the resilience planning will be different.

Resilience is one of the most critical factors in the realization of sustainability. Due to the dynamism of society's reaction to risks, it is a kind of perspective, and it helps to expand policy choices to confront uncertainty and change. Increasing resilience to hazards can thus improve the adaptability and sustainable livelihoods of the community. Drought, water shortages, and its effects on agricultural production and economic development are one of the major concerns of the world (Haynes and Tanner 2015).

Due to climate change and its effects on surface water and groundwater resources, along with inappropriate management of water resources, communities' vulnerability to these changes has increased and will undoubtedly exacerbate water crises and further increase vulnerability.

Maleksaeidi et al. (2017) described the damage done to rural communities, and the consequences of cyclical droughts indicate that they are not prepared to resist this risk and its implications, as well as the ineffectiveness of crisis management systems.

Hence, the first step in combating drought and modifying its consequences is to recognize and comprehend the results and understand the dimensions of vulnerability and resistance of individuals to promote the threshold of their flexibility.

Of course, often after the occurrence of any drought crisis, there are different reactions from the villagers, especially farmers, which have been observed and the extent of the economic, environmental, and social damage and the resistance of villagers to the dangers of drought in different ways. So understanding the amount of resilience of the population and human groups is involved as an essential step in preventing the spread of this threat.

Resilience depends on geography, time, funding sources, stages of political support, and other elements. There is no standard method for estimating resilience (Nettier et al. 2017).

Beccari (2016) believed that statistical indicators are phenomena that allow comparing and evaluating events at different times and places. Indicators are a tool for assessing and comparing living standards and relative welfare of the community, providing necessary guidance for decision-making in a variety of settings, and helping to measure and promote progress toward sustainable development goals (Woods 2017).

Therefore, the goal of setting indicators is to know the quantitative and exact conditions of a situation at a time and the image of trends and transformations that have been taking place over the years in the community.

The initial conditions work to measure the performance of programs, policies, and workings explicitly designed to improve risk resilience. Assessment and evaluation of the initial conditions that lead to community resilience are not only critical but also measuring the factors contributing to the effects of the work and reducing the capacity of the community at the time of the reaction and the return of society to the same state as the aftermath of the risks.

However, in previous works, comparison the indicators and proxies are missed. So, this chapter collected the critical indicators and different methods to analyze resilience against climate change.

Conclusions

Natural hazards caused by climate change as destructive phenomena are not considered to be disasters in the first stage but are risks that occur repeatedly. Different countries have different management practices to deal with a variety of hazards so that they can implement resilient and risk-reducing programs. The goal of all management practices is to reduce the impact. The occurrence of drought and flood as natural hazards due to lack of proper planning and resilience program can be considered a crisis. Urban and rural households are both exposed to these hazards, but villages are at high risk due to particular environmental conditions more than urban areas. However, the strategies used to mitigate the effects of hazards are the same for both. If the indicators of resilience in urban areas are different, each model should be presented following local indicators. This chapter examines the differences between urban and rural areas regarding resilience to the risks of climate change and extreme events in this chapter. The results of the findings show that there are significant differences between urban and rural areas concerning resiliency indicators. The most significant difference is between individual indicators. For rural areas, social capital and population stability have the most significant impact on rural livelihoods. Employment rates, lack of dependence on primary sectors, property ownership, educational per capita, industry, access to information, and health facilities are prime features of urban resilience. Resilience in urban areas is primarily affected by economic capital, while community capital is the most crucial factor in resilience in rural areas. Investigating the patterns of spatial resonance reflects the difference between city and village concerning resilient indices. Therefore, about the findings of this chapter, the following points are suggested for further research:

1. Considering that the resilience of urban and rural communities varies from risks, they cannot provide the same strategy for them, and each of them needs to be strategically tailored to local indicators and capacities.
2. Due to the low level of institutional and economic infrastructure resilience in rural communities with high employment rates, reducing villagers' dependence on single-industry economics, industrialization of villages, and increasing the per capita of health education facilities can increase the resilience of the town.
3. Considering the factors affecting the resilience capacity in urban and rural communities, providing appropriate services for each location, the degree of adjustment and resilience of families was strengthened.
4. Using the principles of physical planning in urban and rural communities, such as land use planning and rehabilitation, can reduce the vulnerability of the population to a degree of resilience.

5. Improving the mental empowerment of households to address climate change risks and empowering communities is another way of reducing vulnerability.
6. Finally, it is recommended that more studies be done on the resonance concerning local indicators.

References

Abid M, Scheffran J, Schneider UA, Ashfaq M (2015) Farmers' perceptions of and adaptation strategies to climate change and their determinants: the case of Punjab province, Pakistan. Earth Syst Dynam 6(1):225–243

Altieri MA, Nicholls CI, Henao A, Lana MA (2015) Agroecology and the design of climate change-resilient farming systems. Agron Sustain Dev 35(3):869–890

Arouri M, Nguyen C, Youssef AB (2015) Natural disasters, household welfare, and resilience: evidence from rural Vietnam. World Dev 70:59–77

Barton CM, Tortosa JEA, Garcia-Puchol O, Riel-Salvatore JG, Gauthier N, Conesa MV, Bouchard GP (2018) Risk and resilience in the late glacial: a case study from the western Mediterranean. Quat Sci Rev 184:68–84

Beccari B (2016) A comparative analysis of disaster risk, vulnerability and resilience composite indicators. PLoS Curr 8:23–30

Becker AH, Matson P, Fischer M, Mastrandrea MD (2015) Towards seaport resilience for climate change adaptation: stakeholder perceptions of hurricane impacts in Gulfport (MS) and Providence (RI). Prog Plan 99:1–49

Boelee E, Yohannes M, Poda JN, McCartney M, Cecchi P, Kibret S, Hagos F, Laamrani H (2013) Options for water storage and rainwater harvesting to improve health and resilience against climate change in Africa. Reg Environ Chang 13(3):509–519

Bonfilgio NS, Renati R, Pessa E, Penna MP (2018) The influence of resilience factors in the treatment of substances addiction: a multilayer perceptron predictive model. In: 2018 IEEE international symposium on medical measurements and applications (MeMeA). IEEE, pp 1–5

Boon HJ (2014) Disaster resilience in a flood-impacted rural Australian town. Nat Hazards 71(1):683–701

Borden KA, Cutter SL (2008) Spatial patterns of natural hazards mortality in the United States. Int J Health Geogr 7(1):64

Bozza A, Asprone D, Manfredi G (2015) Developing an integrated framework to quantify resilience of urban systems against disasters. Nat Hazards 78(3):1729–1748

Cameron E, Mearns R, McGrath JT (2015) Translating climate change: adaptation, resilience, and climate politics in Nunavut, Canada. Ann Assoc Am Geogr 105(2):274–283

Ceridwen IL (2017) Extreme weather, complex spaces and diverse rural places: an intra-community scale analysis of responses to storm events in rural Scotland, UK. J Rural Stud 54:111

Chapin T, Brinkman TJ (2016) Human adaptation responses to a rapidly changing Arctic: a research context for building system resilience. In: AGU fall meeting abstracts

Childers DL, Cadenasso ML, Grove JM, Marshall V, McGrath B, Pickett ST (2015) An ecology for cities: a transformational nexus of design and ecology to advance climate change resilience and urban sustainability. Sustainability 7(4):3774–3791

Cumming GS, Allen CR, Ban NC, Biggs D, Biggs HC, Cumming DH, De Vos A, Epstein G, Etienne M, Maciejewski K, Mathevet R (2015) Understanding protected area resilience: a multi-scale, social-ecological approach. Ecol Appl 25(2):299–319

Davari A, Eslamian S, Reyhani MN (2017) Sustainable agriculture: building social-ecological resilience. In: Handbook of drought and water scarcity. CRC Press, Boca Raton, pp 201–218

De Souza K, Kituyi E, Harvey B, Leone M, Murali KS, Ford JD (2015) Vulnerability to climate change in three hot spots in Africa and Asia: key issues for policy-relevant adaptation and resilience-building research. Reg Environ Chang 15:747

Despotaki V, Sousa L, Burton CG (2018) Using resilience indicators in the prediction of earthquake recovery. Earthq Spectra 34(1):265–282

Disaster survival resources. Retrieving date on 20 Apr 2018. http://www.disaster-survival-resources.com/us-disaster-statistics.html

Eslamian S, Eslamian FA (eds) (2017) Handbook of drought and water scarcity: environmental impacts and analysis of drought and water scarcity. CRC Press, Boca Raton

Eslamian S, Karami E, Keshavarz M, Maleksaeidi H (2017) Climate change and drought: building resilience for an unpredictable future. In: Handbook of drought and water scarcity. CRC Press, Boca Raton, pp 177–200

Haynes K, Tanner TM (2015) Empowering young people and strengthening resilience: youth-centred participatory video as a tool for climate change adaptation and disaster risk reduction. Children's Geogr 13(3):357–371

Heller NE, Kreitler J, Ackerly DD, Weiss SB, Recinos A, Branciforte R, Flint LE, Flint AL, Micheli E (2015) Targeting climate diversity in conservation planning to build resilience to climate change. Ecosphere 6(4):1–20

Jakovac CC, Peña-Claros M, Kuyper TW, Bongers F (2015) Loss of secondary-forest resilience by land-use intensification in the Amazon. J Ecol 103(1):67–77

Lafond V, Lagarrigues G, Cordonnier T, Courbaud B (2014) Uneven-aged management options to promote forest resilience for climate change adaptation: effects of group selection and harvesting intensity. Ann For Sci 71(2):173–186

Levine NM, Zhang K, Longo M, Baccini A, Phillips OL, Lewis SL, Alvarez-Dávila E, de Andrade ACS, Brienen RJ, Erwin TL, Feldpausch TR (2016) Ecosystem heterogeneity determines the ecological resilience of the Amazon to climate change. Proc Natl Acad Sci 113(3):793–797

Licht A, Dupont-Nivet G, Pullen A, Kapp P, Abels HA, Lai Z, Guo Z, Abell J, Giesler D (2016) Resilience of the Asian atmospheric circulation shown by Paleogene dust provenance. Nat Commun 7:12390

Liu Y, Wei J, Xu J, Ouyang Z (2018) Evaluation of the moderate earthquake resilience of counties in China based on a three-stage DEA model. Nat Hazards 91(2):587–609

Lowe K, Castley JG, Hero JM (2015) Resilience to climate change: complex relationships among wetland hydroperiod, larval amphibians and aquatic predators in temporary wetlands. Mar Freshw Res 66(10):886–899

MacDonald AM, Bonsor HC, Taylor R, Shamsudduha M, Burgess WG, Ahmed KM, Mukherjee A, Zahid A, Lapworth D, Gopal K, Rao MS (2015) Groundwater resources in the Indo-Gangetic Basin: resilience to climate change and abstraction. Nature Geoscience 9:762–766

Maleksaeidi H, Keshavarz M, Karami E, Eslamian S (2017) Climate change and drought: building resilience for an unpredictable future. In: Handbook of drought and water scarcity: environmental impacts and analysis of drought and water scarcity. Taylor and Francis

Martin E, Perine C, Lee V, Ratcliffe J (2018) Decentralized governance and climate change adaptation: working locally to address community resilience priorities. In: Theory and practice of climate adaptation. Springer, Cham, pp 3–22

Masikati P, Tui SH-K, Descheemaeker K, Crespo O, Walker S, Lennard CJ, Claessens L et al (2015) Crop–livestock intensification in the face of climate change: exploring opportunities to reduce risk and increase resilience in southern Africa by using an integrated multi-modeling approach. In: Handbook of climate change and agroecosystems: the agricultural model intercomparison and improvement project integrated crop and economic assessments, part 2. Imperial College Press, London, pp 159–198

McCluney KE, Poff NL, Palmer MA, Thorp JH, Poole GC, Williams BS, Williams MR, Baron JS (2014) Riverine macrosystems ecology: sensitivity, resistance, and resilience of whole river basins with human alterations. Front Ecol Environ 12(1):48–58

Nettier B, Dobremez L, Lavorel S, Brunschwig G (2017) Resilience as a framework for analyzing the adaptation of mountain summer pasture systems to climate change. Ecol Soc 22(4):25–27
Nobre CA, Marengo JA, Seluchi ME, Cuartas LA, Alves LM (2016) Some characteristics and impacts of the drought and water crisis in southeastern Brazil during 2014 and 2015. J Water Resour Prot 8(02):252
Paton D, Johnston D (2017) Disaster resilience: an integrated approach. Charles C Thomas Publisher, Springfield
Peng Y (2015) A comparison of two approaches to develop concentrated rural settlements after the 5.12 Sichuan earthquake in China. Habitat Int 49:230–242
Pérez C, Jones EM, Kristjanson P, Cramer L, Thornton PK, Förch W, Barahona CA (2015) How resilient are farming households and communities to a changing climate in Africa? A gender-based perspective. Glob Environ Chang 34:95–107
Petrie CA, Singh RN, Bates J, Dixit Y, French CA, Hodell DA, Jones PJ, Lancelotti C, Lynam F, Neogi S, Pandey AK (2017) Adaptation to variable environments, resilience to climate change: investigating land, water and settlement in Indus Northwest India. Curr Anthropol 58(1):1–30
Pisello AL, Rosso F, Castaldo VL, Piselli C, Fabiani C, Cotana F (2017) The role of building occupants' education in their resilience to climate-change related events. Energ Buildings 154:217–231
Roesch-McNally GE, Arbuckle JG, Tyndall JC (2018) Barriers to implementing climate resilient agricultural strategies: the case of crop diversification in the US Corn Belt. Glob Environ Chang 48:206–215
Sinclair K, Curtis A, Mendham E, Mitchell M (2014) Can resilience thinking provide useful insights for those examining efforts to transform contemporary agriculture? Agric Hum Values 31(3):371–384
Stevens-Rumann CS, Kemp KB, Higuera PE, Harvey BJ, Rother MT, Donato DC, Morgan P, Veblen TT (2018) Evidence for declining forest resilience to wildfires under climate change. Ecol Lett 21(2):243–252
Suárez M, Gómez-Baggethun E, Benayas J, Tilbury D (2016) Towards an urban resilience index: a case study in 50 Spanish cities. Sustainability 8(8):774
Tacoli C, Satterthwaite D (2014) Seeking an understanding of poverty that recognizes rural-urban differences and rural-urban linkages. In: Urban livelihoods. Routledge, London, pp 75–93
Takeuchi K, Saito O, Matsuda H, Mohan G (2018) Introduction: enhancing resilience against climate and ecosystem changes in rural Asia. In: Resilient Asia. Springer, Tokyo, pp 1–5
Tanner T, Lewis D, Wrathall D, Bronen R, Cradock-Henry N, Huq S, Lawless C, Nawrotzki R, Prasad V, Rahman MA, Alaniz R (2015) Livelihood resilience in the face of climate change. Nat Clim Chang 5(1):23
Thebo AL, Drechsel P, Lambin EF, Nelson KL (2017) A global, spatially-explicit assessment of irrigated croplands influenced by urban wastewater flows. Environ Res Lett 12(7):074008
Tompkins EL, Adger WN (2004) Does adaptive management of natural resources enhance resilience to climate change? Ecol Soc 9(2):12–15
U.S government, Disasters data. Available on 20 Apr 2018. https://catalog.data.gov/dataset?groups=disasters#topic=disasters_navigation
Whitman ZR, Wilson TM, Seville E, Vargo J, Stevenson JR, Kachali H, Cole J (2013) Rural organizational impacts, mitigation strategies, and resilience to the 2010 Darfield earthquake, New Zealand. Nat Hazards 69(3):1849–1875
Woods DD (2017) Essential characteristics of resilience. In: Resilience engineering. CRC Press, Boca Raton, pp 33–46
Wyss R, Luthe T, Abegg B (2015) Building resilience to climate change–the role of cooperation in alpine tourism networks. Local Environ 20(8):908–922
Xu J, Grumbine RE (2014) Building ecosystem resilience for climate change adaptation in the Asian highlands. Wiley Interdiscip Rev Clim Chang 5(6):709–718
Yihdego Y, Webb J, Vaheddoost B (2017) Highlighting the role of groundwater in lake–aquifer interaction to reduce vulnerability and enhance resilience to climate change. Hydrology 4(1):10.

Panteli M, Mancarella P (2015) Influence of extreme weather and climate change on the resilience of power systems: impacts and possible mitigation strategies. Electric Power Syst Res 127:259–270
Young C, Craig JC, Clapham K, Williams S, Williamson A (2018) Stressful life events and resilience among carers of Aboriginal children in urban New South Wales: cross-sectional findings from the Study of Environment on Aboriginal Resilience and Child Health (SEARCH). BMJ Open 8(6):e021687
Zeleňáková M, Purcz P, Hlavatá H, Blišťan P (2015) Climate change in urban versus rural areas. Procedia Eng 119:1171–1180

Printed by Printforce, the Netherlands